PHOTOVOLTAICS

PHOTOVOLTAICS
SYSTEM DESIGN AND PRACTICE

Heinrich Häberlin

Berne University of Applied Sciences, Switzerland

Translated by **Herbert Eppel**

HE Translations, Leicester, UK

A John Wiley & Sons, Ltd., Publication

This edition first published 2012
© 2012, John Wiley & Sons, Ltd

Registered office
John Wiley & Sons Ltd, The Atrium, Southern Gate, Chichester, West Sussex, PO19 8SQ, United Kingdom

Authorised Translation from the German language second edition published by Electrosuisse (2010)

Library of Congress Cataloguing-in-Publication Data

Häberlin, Heinrich.
   Photovoltaics : system design and practice / Heinrich Häberlin; translated by Herbert Eppel.
     p. cm.
   Includes bibliographical references and index.
   ISBN 978-1-119-99285-1 (cloth)
  1.   Photovoltaic power systems–Design and construction. 2.   Photovoltaic power systems–Standards.
3.   Photovoltaic power generation. I.   Title.
   TK1087.H33 2012
   621.31'244–dc23

                                                   2011032983

A catalogue record for this book is available from the British Library.

Print ISBN: 9781119992851

Set in 9/11pt, Times by Thomson Digital, India

To my wife Ruth and my children Andreas and Kathrin, who, while I was writing this book, weren't able to spend as much time with me as they would have liked – and to all those who want to see our society transition to sustainable and responsible electricity generation.

Contents

Foreword

Energy and the concerns it raises for individuals, society at large and the environment, is a more burning issue today than ever before. Evolutions such as climate change, energy security issues, energy market deregulation, energy price fluctuations and the like have made energy the centre of a multi-faceted debate where the need for sustainable energy and improved energy efficiency are taking centre stage as never before. The European Commission's objectives for 2020 in this regard are both courageous and pioneering.

The Stern Report issued by the British government in 2006 noted that global warming and its worldwide economic repercussions constitute the widest-ranging failure of free market mechanisms that the world has ever seen. This report also quantified the economic costs of this evolution. The 'business as usual' attitude that unfortunately still prevails in the business and political communities is going to cost us dearly; and the longer we wait to act, the higher the cost will be. A growing number of politicians, business leaders, and consumers have come to the realization that action must be taken, and that such action will not come cheap.

The timing of this book's publication could not be better. Admittedly photovoltaics is no magic bullet solution for the myriad problems we face but photovoltaics is one of the key technologies that will bring us closest to a sustainable energy supply in the foreseeable and distant future. Many technical and other obstacles remain to be surmounted before photovoltaics can do this, however, the undeniable fact of the matter is that photovoltaics is now the subject of feverish and rapid worldwide development on an industrial scale. The annual growth rates of upwards of 40 percent registered by the photovoltaics industry are all the more remarkable in light of the recent financial crisis, not to mention past economic recessions.

The term photovoltaics is often associated primarily with solar cells and solar modules. As these are the core elements of photovoltaic technology, this mindset makes perfect sense. However, it does not go far enough when it comes to characterizing the energy production of a solar power installation. Only if we regard photovoltaics as an energy system can we begin to make accurate statements concerning its contribution to the energy supply. Moreover, regarding a phenomenon as a system often allows us to connect the dots between theory and practice, paradigm and experience.

As growing numbers of PV installations are put into operation, questions are increasingly being raised concerning their quality and reliability. Only a PV installation that works properly can genuinely contribute to the energy supply. Hence such issues are looming ever larger worldwide. One of the great virtues of the present book is that it places tremendous emphasis on the system related aspects of photovoltaics.

Over the course of his many years of research on the system related aspects of photovoltaics, Prof. Häberlin has accumulated unparalleled experience that is richly detailed in his numerous publications on the subject. He has now made this experience available to a wider public, via this book, which in using its author's experience as a springboard, provides a wealth of insights and highly practical information concerning the design and operation of PV installations. In so doing, this book also addresses an increasingly pressing problem, namely that rapid growth in the photovoltaics industry and other renewable energy sectors will increase the need for qualified individuals in this domain. Thus education and training are a matter of growing importance in this regard.

I would like to on one hand express my gratitude to Prof. Häberlin for having written this book, and on the other congratulate him on his willingness to share his photovoltaics expertise with a wider audience via this book, which I feel will make a significant contribution to the advancement of photovoltaic technology. I am confident that this richly detailed and very complete book will enable numerous photovoltaics engineers, researchers and other professionals to gain greater insight into photovoltaics, and particularly into the practical aspects of this intriguing field.

Stefan Nowak, Ph.D.
Chairman of the International Energy Agency Photovoltaics Power Systems
Programme (IEA PVPS)
St. Ursen, Switzerland

Preface

The PV industry has experienced an exponential growth since the appearance of the first edition of this book in German in 2007 and ever-larger PV installations are being realized in an ever-growing number of countries. The tremendous interest shown in the first edition of this book, as well as the extensive positive feedback it elicited, show that photovoltaics, which is discussed comprehensively and in all its complexity in this book, is a topic of tremendous importance nowadays.

Therefore the second edition published also in German in 2010 goes into even greater depth on a number of matters and also explores some aspects of and insights into photovoltaics that were not contained in the first edition. For this second edition, I have extensively revised, updated and expanded the material from the first edition in such a way that, using the information in this book, designers can make very close sizing and yield estimates for PV installations at any site worldwide between 40 °S and 60 °N. In addition to updating my own extremely extensive PV installation monitoring data, I have also included energy yield figures from other countries, thanks to the generosity of the relevant operators who kindly provided data concerning their installations.

Considering the success of the two German books and the expressed interest also from many English speaking PV engineers, it was decided to translate this book into English to make my extended PV experience available also to them. The present book is an exact translation into English of the extended second German edition published in 2010.

Since the finalisation of the German book and the translation, prices for PV modules have considerably dropped further especially in 2011. For large quantities of crystalline modules prices of 1 € per Wp or even somewhat less were offered in autumn 2011, i.e. even less than the lowest values indicated in the book. Therefore the price of PV electricity is now already very close to competiveness with conventional electricity in many countries.

Heinrich Häberlin
Ersigen
October 2011

About the Author

After earning a Masters degree in electrical engineering from ETH Zurich, Heinrich Häberlin became a staff researcher at the ETH Microwave Lab. During this time he developed hardware, software and numerous learning applications for a computerized teaching and learning system that was used at ETH until 1988. He earned a doctorate in 1978, based on the thesis he wrote concerning this system.

From 1979 to 1980, Dr. Häberlin headed a team at the Zellweger Company that worked on the development of hardware and software for microprocessor-based control of a complex short-wave radio system.

In late 1980 he was appointed a professor to the engineering school in Burgdorf, Switzerland, where from 1980 to 1988 besides electrical engineering he also taught computer science.

Prof. Häberlin has been actively involved in photovoltaics since 1987. In 1988 he established the Bern University of Applied Sciences Photovoltaics Lab, where he and his staff mainly investigate the behaviour of grid-connected systems. Since 1989 he has also been operating his own PV installation. In 1988 he began testing various PV inverters and in 1990 he initiated a series of lab experiments concerning PV installation lightning protection. Prof. Häberlin's PV Lab has, since 1992, also been continuously monitoring more than 70 PV installations, mainly under the auspices of research projects commissioned by Switzerland's Federal Office for Energy. His lab also carries out specialized measurements of PV installation components for the relevant manufacturers and also works on various EU projects.

Prof. Häberlin has been teaching photovoltaics at Bern University of Applied Sciences since 1989. He is a member of Electrosuisse, ETG, the Swiss TK82 panel of experts on PV installations, and the IEC's International Photovoltaics Standards Committee TC82.

Acknowledgements

I would like to thank all of the private and public sector organizations that kindly provided photos, graphics, data and other documentation for this book. Without these elements, I would not have been able to provide such a detailed and in-depth account of the relevant issues.

I would also like to express my heartfelt gratitude to my past and current assistants, who carried out investigations and analyses in connection with numerous research projects and whose work products I have used in this book. I also owe a debt of gratitude to the following organizations that commissioned and financed the aforementioned research projects: the Swiss Ministry of Energy; the Swiss Ministry of Science and Education; the Bern Canton Office of Water and Energy Resource Management; and various power companies (namely, Localnet, Bernische Kraftwerke, Gesellschaft Mont Soleil, Elektra Baselland and Elektrizitätswerk der Stadt Bern).

My former assistants Christoph Geissbühler, Martin Kämpfer and Urs Zwahlen, and my current assistants Luciano Borgna and Daniel Gfeller, read the manuscript for the first edition of this book in German and pointed out errors and elements that were unclear. My colleagues Dr Urs Brugger and Michael Höckel read certain sections of the first-edition manuscript and provided helpful suggestions.

The manuscript for the second book in German was read by my current assistants Daniel Gfeller, David Joss, Monika Münger and Philipp Schärf, who likewise pointed out errors and elements that were unclear.

I would like to express my gratitude to all of these individuals for their assistance.

Heinrich Häberlin

Ersigen
January 2010

Note on the Examples and Costs

Many of the chapters in this book contain examples whose numbers were in many cases calculated using spreadsheet programs that round off the exact numbers that were originally input. On the other hand, for reasons of space many of the numbers in the tables in Appendix A have been rounded off to two decimal places. Hence, when used for actual calculations, these rounded-off numbers may under certain circumstances differ slightly from the counterpart numbers indicated in the examples.

In several sections of this book, costs of PV modules or PV systems or feed-in tariffs are given in euros. In some cases or examples, especially where the situation in Switzerland is discussed, costs in Swiss francs (SFr) are expressed as their equivalent in euros (€). The exchange rate used in this case was the exchange rate during the finalization of the German book, i.e. 1 Swiss franc is equivalent to about 0.67 euros. Due to the problems on the financial markets, there were extreme variations of this exchange rate in 2010 and 2011 (variation between about 1 SFr ≈ 0.67 € and 1 SFr ≈ 1 €). For actual values in Swiss francs the actual exchange rate has to be used.

Note on the Examples and Costs

List of Symbols

Symbol	Name	Metric
a	Depreciation rate	%
a_A	Battery depreciation rate	%
a_E	Depreciation rate for electronic installations	%
A_G	Total surface area of a solar generator field (aggregate module surface area)	
a_G	Solar generator depreciation rate	%
A_L	Space required for a ground-based or rooftop solar generator field	m^2
AM	Air mass number	—
a_{MB}	Relative number of shaded modules per string	—
a_{MM}	Relative number of modules per string that exhibit power loss	—
A_Z	Solar cell surface area	m^2
C	Battery capacity	F
C	Capacitance	F
C_E	Solar generator earthing capacitance	F
CF	Capacity factor	—
c_T	Temperature coefficient for a solar generator's MPP output	K^{-1}
di/dt_{max}	Maximum current curve in a lightning leader stroke	kA/μs
d_V	Relative voltage rise at the grid link point	—
e	Electron charge (scope of an electron or proton charge) ($e = 1.602 \cdot 10^{-19}$ A s)	A s
e	Basis for natural logarithms: e = 2.718 281 828	—
E	Energy (in general)	kWh, MJ
e_A	PV installation surface-related grey energy	kWh/m^2 MJ/m^2
E_{AC}	PV installation AC power output	kWh
E_D	Mean daily DC power used by a stand-alone installation	Wh/d
E_{DC}	PV installation DC power output	kWh
$E_{DC\text{-}S}$	Mean DC power output per day and string for a stand-alone installation with MPT	Wh/d
EF	Yield factor = L/ERZ	—
E_G	Band gap energy (usually expressed in eV; 1 eV = $1.602 \cdot 10^{-19}$ J)	eV
E_H	Mean daily energy yield of a hybrid generator	Wh/d
E_L	Total energy produced by a PV installation during its service life	kWh, MJ
e_P	Peak-power-related grey energy in PV installations	kWh/W, MJ/W
ERZ	Energy payback time (time needed to produce grey energy)	a
f	Frequency	Hz
FF	Filling factor for a solar cell, solar module or solar generator	—
FF_i	Idealized filling factor	—

G	Global irradiance (power/surface), usually indicated for the horizontal plane	W/m^2
G_B	Direct beam irradiance, usually indicated for the horizontal plane	W/m^2
G_D	Diffuse irradiance, usually indicated for the horizontal plane	W/m^2
GE	Grey energy	kWh, MJ
G_{ex}	Extraterrestrial irradiance	W/m^2
G_G	Global irradiance on the solar generator plane	W/m^2
G_o, G_{STC}	Irradiance under STC: $G_o = 1$ kW/m^2	W/m^2
H	Total irradiation (energy/area), usually indicated for the horizontal plane	kWh/m^2 (MJ/m^2)
H_B	Total irradiation; direct beam irradiation (usually indicated for the horizontal plane)	kWh/m^2 (MJ/m^2)
H_D	Total irradiation; diffuse irradiation (usually indicated for the horizontal plane)	kWh/m^2 (MJ/m^2)
H_{ex}	Total extraterrestrial irradiation on a plane parallel to the horizontal plane outside of the Earth's atmosphere	kWh/m^2 (MJ/m^2)
H_G	Total irradiation, irradiation on the solar generator plane (energy/area)	kWh/m^2 (MJ/m^2)
I	Electric current (general)	A
i_A	Lightning current in a down-conductor	A
i_{Amax}	Peak lightning current in a down-conductor	A
I_{Dceff}	Figure for the DC input current of a stand-alone inverter	A
I_F	Passband current	A
I_L	Charging current	A
i_{max}	Peak lightning current voltage	A
I_{MPP}	MPP current	A
I_{PV}	Solar generator current	A
I_R	Solar module reverse current (= passband current in a solar cell diode)	A
I_S	Saturation current of a diode or solar cell	A
I_{SC}	Short-circuit current of a solar cell, solar module or solar generator	A
I_{SC-STC}	Short-circuit current under STC	A
I_{SN}	Nominal string fuse voltage	A
i_{So}	Short-circuit current induced by lightning current in a conductor loop	A
i_{Somax}	Peak induced short-circuit current in a conductor loop	A
i_V	Displacement current induced in a PV installation by a distant lightning strike	A
i_V	Varistor current induced by lightning current	A
$I_{V8/20}$	Requisite nominal varistor current (for an 8/20 μs waveform)	A
J_{max}	Peak current density	A/m^2
J_S	Saturation current density	A/m^2
k	Boltzmann's constant $= 1.38 \cdot 10^{-23}$ J/K	J/K
K_A	Battery costs for stand-alone installations	euros
k_B	Annual operating costs	euros/year
k_B	Shading correction factor (1 for no shading, 0 for full shading)	—
k_C	Proportion of lightning current in a down-conductor	—
K_E	Costs for electronic components such as inverters	euros
k_G	Solar generator correction factor	—
K_G	Costs attributable to a solar generator, site modification, wiring, and so on	euros

k_I	Total harmonic current distortion	—
K_J	Total annual PV installation operating costs	euros/year
k_{MR}	Correction factor for deriving M_{Mi} from M_{MR} for a module frame	—
K_N	Usable battery capacity	Ah
K_S	Cost savings for roof tiles and the like for PV installations that are integrated into buildings	euros
k_T	Temperature correction factor	—
K_x	Battery discharge capacity expressed as x number of hours $(K_x = f(x))$	Ah
L	Inductance (in general)	H
L	Installation lifetime (in years)	a
L_C	Capture losses	h/d
L_{CM}	Miscellaneous capture losses	h/d
l_{CM}	Standardized miscellaneous non-capture losses: $l_{CM} = y_T - y_A$	—
L_{CT}	Thermal capture losses	h/d
l_{CT}	Standardized thermal capture losses: $l_{CT} = y_R - y_T$	—
L_S	Conductor loop inductance	H
l_S	Standardized system capture losses: $l_S = y_A - y_F$	—
L_S, L_{BOS}	Balance of system losses	h/d
M	Mutual inductance (in general)	H
M_i	Effective mutual inductance, based on total lightning current i	H
M_{Mi}	Effective mutual inductance of a module (based on total lightning current i)	H
M_{MR}	Module frame mutual inductance (based on $i_A = k_C \cdot i$)	H
n_{AP}	Number of parallel-connected batteries	—
n_{AS}	Number of series-connected batteries	—
N_D	Mean annual number of direct lightning strikes	—
N_g	Number of lightning strikes per square kilometre and year	—
n_I	Inverter efficiency (energy efficiency)	—
n_{MP}	Number of parallel-connected modules in a solar generator	—
n_{MS}	Number of series-connected modules in a string	—
n_{MSB}	Number of shaded modules per string	—
n_{MSM}	Number of modules per string that exhibit power loss	—
n_{SP}	Number of parallel-connected strings in a solar generator	—
n_{VZ}	Full-cycle service life of a battery	—
n_Z	Number of series-connected cells	—
n_{ZP}	Number of parallel-connected strings in a solar module	—
P	Effective power	W
p	Interest rate that is to be applied to depreciation	%
P_A	Solar generator DC power output	W
P_{Ao}	Effective (measured) peak solar generator output under STC	W
P_{AC}	AC-side output	W
P_{AC1}	Maximum connectable single-phase nominal inverter output	W
P_{AC3}	Maximum connectable triphase nominal inverter output	W
P_{ACn}	AC-side nominal output of an inverter or a PV installation	W
P_{DC}	DC-side output	W
P_{DCn}	DC-side nominal inverter output	W
PF	Packing factor	—
P_{Go}	Nominal solar generator peak output under STC (aggregate P_{Mo})	W
P_{GoT}	Temperature-corrected nominal solar generator peak output	W
P_{max}	Maximum output (equates to P_{MPP} under STC)	W
P_{Mo}	Nominal module output under STC, according to the vendor's data	W

P_{MPP}	MPP of a solar cell, solar module or solar generator	W
P_{use}	PV installation output power	W
PR	Performance ratio $= Y_F/Y_R$	—
pr	Instantaneous performance ratio $= y_F/y_R$	—
PR_a	Annual performance ratio	—
p_{VTZ}	Maximum allowable area-specific solar cell power loss	W/m^2
Q	Watless power (> 0 when inductive)	var
Q_D	Mean daily load consumption for a stand-alone installation	Ah/d
Q_H	Mean daily hybrid solar generator charge for a stand-alone installation	Ah/d
Q_L	Lightning current charge (up to a few hundred milliseconds)	A s
Q_L	Mean daily charge consumption ($> Q_D$) for a stand-alone installation	Ah/d
Q_{PV}	Mean daily charge provided by a solar generator	Ah/d
Q_S	Lightning current charge (surge current of less than 1 ms duration)	As
Q_S	Mean daily string charge for a stand-alone installation	Ah/d
R	Resistance (in general); real component of a complex impedance $\underline{Z}$	Ω
R_{1L}	Real component of complex single-phase impedance in a conductor between a transformer and grid link point (for inverter connection purposes)	Ω
R_{1N}	Real component of complex single-phase grid impedance (ohmic component)	Ω
R_{1S}	Real component of complex single-phase interconnecting line impedance (with impedance $\underline{Z}_S$) between a grid link point and inverter	Ω
R_{3L}	Real component of complex triphase impedance in a conductor between a transformer and grid link point (for inverter connection purposes)	Ω
R_{3N}	Real component of complex triphase grid impedance (ohmic component)	Ω
R_{3S}	Real component of complex triphase interconnecting line impedance (with impedance $\underline{Z}_S$) between a grid link point and inverter	Ω
$R(\beta,\gamma)$	Global radiation factor $= H_G/H$	—
$R_a(\beta,\gamma)$	Annual global radiation factor $= H_{Gd}/H_a$ (ratio of annual irradiance figures)	—
R_B	Direct beam radiation factor $= H_{GB}/H_B$ (as in the tables in Section A4)	—
r_B	Lightning sphere radius	m
R_D	Diffuse radiation factor $= H_{GD}/H_D$	—
R_D	Ground resistance of a grounding installation	Ω
R_i	Inner resistance of a battery or the like	Ω
R_L	Power lead resistance (real component of $\underline{Z}_L$)	Ω
R_M	Shielding resistance; resistance in the cladding of a shielded conductor	Ω
R_N	Inner grid resistance (real component of $\underline{Z}_N$)	Ω
R_P	Parallel resistance	Ω
R_R	Frame reduction factor	—
R_S	Series resistance of a solar cell or conductor loop	Ω
R_T	Medium-voltage transformer resistance (real component of $\underline{Z}_T$)	Ω

R_V	Equivalent (linearized) resistance in a varistor replacement source	Ω
S	Apparent output	VA
S_{1KV}	Single-phase grid short-circuit current at the grid link point	VA
SF	Voltage factor	—
S_{KV}	Triphase grid short-circuit current at the grid link point	VA
s_{min}	Minimum safety gap for hazardous proximities	m
S_{WR}	Apparent triphase output of an inverter	VA
T	Absolute temperature	K
T_C	Cell temperature (variant of T_Z)	°C
T_o, T_{STC}	Reference STC temperature (25 °C)	°C
T_U	Ambient temperature	°C
t_V	AC full-load hours (installation full load P_{ACn})	h
t_{Vb}	AC full-load hours for a PV installation whose power limitation is $P_{AC\text{-}Grenz} < P_{ACmax}$	h
t_{Vm}	AC full load hours for a PV installation based on the installation's peak AC output P_{ACmax} (normally P_{ACmax} differs from P_{ACn})	h
t_{Vo}	PV installation full-load hours, including peak output P_{Go} (under STC)	h
T_Z	Cell temperature	°C
t_Z	Battery depth of discharge	—
T_{ZG}	Irradiance-weighted cell and module temperature	°C
V	Voltage (in general)	V
V_{1N}	Grid phase voltage for a replacement source under open-circuit conditions	V
V_{1V}	Phase voltage at the grid link point	V
V_{1WR}	Phase voltage at the inverter connection point	V
V_{BA}	Bypass diode voltage under avalanche conditions	V
V_G	Battery charge limiting voltage (gassing voltage)	V
V_L	Battery charging voltage; output voltage of an MPT charge controller	V
V_M	Peak voltage induced by lightning current in a module	V
v_{max}	Maximum induced voltage	V
V_{MPP}	MPP voltage	V
$V_{MPPA\text{-}STC}$	PV installation or solar generator MPP voltage under STC	V
V_N	Concatenated grid voltage for a replacement source under open-circuit conditions	V
V_{OC}	Open-circuit voltage of a solar cell, solar module or solar generator	V
$V_{OCA\text{-}STC}$	Open-circuit PV installation voltage under STC	V
V_{Ph}	Theoretical photovoltage $= E_G/e$	V
V_{PV}	Solar generator voltage	V
V_R	Inverse voltage	V
V_{RRM}	Diode inverse voltage	V
V_S	PV installation system voltage	V
V_S	Peak voltage induced by lightning current in a string	V
V_V	Concatenated voltage at the grid link point	V
V_V	Equivalent (linearized) voltage in a varistor replacement source	V
V_V	Peak voltage induced by lightning current in wiring	V
V_{VDC}	Varistor DC operating voltage specified by the vendor	V
V_{WR}	Concatenated voltage at the inverter connection point	V
V_{max}	Peak potential increase relative to remote ground	V

X	Reactance (in general); imaginary component of an impedance $\underline{Z}$	Ω
X_{1L}	Imaginary component of complex single-phase impedance in a conductor between a transformer and grid link point (for inverter connection purposes)	Ω
X_{1N}	Imaginary component of complex single-phase grid impedance (reactance)	Ω
X_{1S}	Imaginary component of complex single-phase interconnecting line impedance (with impedance $\underline{Z}_S$) between a grid link point and inverter	Ω
X_{3L}	Imaginary component of complex triphase impedance in a conductor between a transformer and grid link point (for inverter connection purposes)	Ω
X_{3N}	Imaginary component of complex triphase grid impedance (reactance)	Ω
X_{3S}	Imaginary component of complex triphase interconnecting line impedance (with impedance $\underline{Z}_S$) between a grid link point and inverter	Ω
X_L	Power lead reactance (imaginary component of $\underline{Z}_L$)	Ω
X_N	Grid reactance (imaginary component of $\underline{Z}_N$)	Ω
X_T	Medium-voltage transformer reactance (imaginary component of $\underline{Z}_T$)	Ω
Y_A	Array yield, i.e. full-load P_{Go} hours	h/d
y_A	Standardized solar generator power $= P_A/P_{Go}$	—
Y_F	Final yield, i.e. full-load P_{Go} hours	h/d
y_F	Standardized output power $= P_{use}/P_{Go}$	—
Y_{Fa}	Specific annual energy yield	kWh/kWp and h/a
Y_R	Reference yield, i.e. full-load solar hours	h/d
y_R	Standardized irradiance $= G_G/G_o$	—
Y_T	Temperature-corrected reference yield	h/d
y_T	Temperature-corrected standardized irradiance $= y_R \cdot P_{GoT}/P_{Go}$	—
$\underline{Z}$	Complex impedance $\underline{Z} = R + jX$ (in general)	Ω
Z	Impedance amount (AC resistance)	Ω
$\underline{Z}_{1L}$	Complex single-phase line impedance at the transformer grid link point	Ω
$\underline{Z}_{1N}$	Complex single-phase grid impedance	Ω
$\underline{Z}_{3L}$	Complex triphase line impedance at the transformer grid link point	Ω
$\underline{Z}_{3N}$	Complex triphase grid impedance	Ω
$\underline{Z}_L$	Complex power lead impedance	Ω
$\underline{Z}_N$	Complex grid impedance	Ω
$\underline{Z}_S$	Complex grid impedance in the inverter interconnecting line	Ω
Z_N	Amount of grid impedance	Ω
$\underline{Z}_T$	Complex medium-voltage transformer impedance	Ω
Z_W	DC cable wave impedance	Ω
ΔV_V	Voltage rise at the grid link point	V
ΔV_{WR}	Voltage rise at the inverter connection point	V
α	Lightning protection angle	°
β	Solar generator angle of incidence	°
γ	Solar generator azimuth	°
δ	Solar declination	°
η_{Ah}	Battery ampere-hour efficiency	—
η_E	PV installation energy efficiency	—

η_M	Solar module efficiency	—
η_{MPPT}	MPP tracking efficiency; degree of grid inverter adaptation	—
η_{MPT}	Global efficiency (tracking plus conversion) of an MPT charge controller	—
η_{PV}	Solar cell efficiency	—
η_S	Spectral efficiency	—
η_T	Theoretical efficiency	—
η_{tot}	Global efficiency $= \eta \cdot \eta_{MPPT}$	—
η_{WR}	Mean inverter efficiency for PV installation sizing and yield calculation purposes (recommended: η_{tot}, if available)	—
η, η_{UM}	Conversion efficiency	—
φ	Phase angle between V and I	°
φ	Latitude (used to determine R_B for irradiance calculations using the three-component method)	°
ν	Frequency (variant of f for very high frequencies, e.g. light frequency)	Hz
ρ	Specific resistance of a material (usually metal)	$\Omega \ mm^2/m$
ρ	Reflection factor of a surface for reflection radiation calculation purposes	—
ψ	Grid impedance angle (phase angle) for grid impedance $\underline{Z}_N$	°
ψ_1	Phase angle for single-phase grid impedance $\underline{Z}_{1N}$	°
ψ_3	Phase angle for triphase grid impedance $\underline{Z}_{3N}$	°

1

Introduction

1.1 Photovoltaics – What's It All About?

Photovoltaics is a technology involving the direct conversion of solar radiation (insolation) into electricity using solar cells. Interest in photovoltaics has grown exponentially in many countries over the past decade, with worldwide photovoltaic sector growth since 1997 ranging from 30 to 85%.

A solar cell is essentially a specialized semiconductor diode with a large barrier layer which, when exposed to light, allows for direct conversion into DC electricity of a portion of the energy in the light quanta or photons arriving at the cell (Figure 1.1).

As individual solar cells generate very low voltage, a number of such cells are connected in series and are combined into a so-called solar module. Higher output can be obtained by wiring a number of modules together to create solar generators, which can be of any size.

The first usable solar cell was developed in 1954, and solar cells were first used for technical purposes in connection with space flight. Virtually all satellites that have been put into orbit around the Earth since 1958 are powered by solar cells, which were originally called solar batteries. The high cost of these early solar cells posed no obstacle to their use, since they were extremely reliable, lightweight and efficient. Following the 1973 oil crisis, interest in renewable energy increased, particularly in terms of solar power. Interest in photovoltaics has grown even further since the Chernobyl accident in 1986, which spurred the development of simpler and cheaper solar cells for terrestrial applications. This first generation of solar cells was initially used to supply electricity to remote locations (e.g. telecommunication facilities, holiday homes, irrigation systems, villages in developing countries and so on, as for instance shown in Figures 1.2–1.9). For such applications, photovoltaics has long since been an economically viable energy resource.

The USA was in the vanguard of PV development and use in the 1980s, at which time various multi-megawatt PV power plants that had been built in desert regions were converting solar cell DC power into AC power that was being fed into the public grid. Many of these installations integrated single- or dual-axis solar trackers. The first such installation, which had 1 MW of power, was realized in 1982 in Hesperia, followed by a second, 6.5 MW, installation in Carrizo Plain. Both of these installations have since been dismantled. In the late 1980s, the Chernobyl disaster aroused interest in grid-connected systems in Europe as well. For many years Europe's largest PV installation (3.3 MWp) was the facility in Serre, Italy, which was connected to the grid in 1995. In recent years, PV power plants of 5 MWp and more have become increasingly prevalent in Germany and Spain. At the time this book went to press, the largest such plant was the 60 MWp PV power station in Olmedilla de Alarcon, Spain (around 150 km west of Valencia), which began operating in 2008. Various, even larger PV power stations are under construction or in the planning stages, including a 2000 MWp facility in China.

Photovoltaics: System Design and Practice. Heinrich Häberlin.
© 2012 John Wiley & Sons, Ltd. Published 2012 by John Wiley & Sons, Ltd.

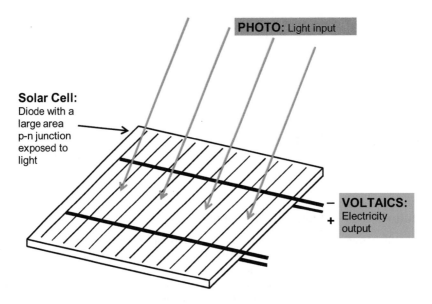

Figure 1.1 Solar cells convert light into electricity. The term photovoltaic is derived from *photo*, the Greek word for light, while 'voltaic' refers to volt, which is the unit for electric voltage

The largest PV power station in Switzerland (1342 kWp) is the BKW facility atop the new Stade de Suisse football stadium in Bern, where the first phase of the installation (855 kWp) began operating in 2005. Two years later the facility was expanded to full nominal capacity, and as at December 2009 was still the world's largest football stadium PV installation. Another large Swiss PV installation is the 555 kWp

Figure 1.2 The 400 Wp solar generators at the Monte Rosa Hostel in the late 1980s. Such hostels today commonly have PV installations with at least 3 kWp of power (Photo: Fabrimex/Willi Maag)

Figure 1.3 Solar home system in India. Even a 50–100 Wp PV installation appreciably raises the standard of living in developing countries (Courtesy of DOE/NREL)

Figure 1.4 Lighthouse PV installation (Photo: Siemens)

Figure 1.5 Solar-powered emergency phone in the California Desert (Courtesy of DOE/NREL)

Mont Soleil power station, which is located at 1270 m above sea level and was connected to the grid in 1992. Information concerning other large-scale PV installations is available at www.pvresources.com. Various grid-connected systems around the world are shown in Figures 1.10–1.16.

The PV efficiency of today's commercially available solar cells for terrestrial applications ranges up to 22.5%. Although this figure is likely to rise in the coming years, the efficiency of PV installations is limited by the laws of physics.

Figure 1.6 Solar-powered village well in Mondi, Senegal. As the well water is easy to store, such simple installations do not need a battery to supply power in the night; this in turn considerably reduces system costs (Photo: Siemens)

Figure 1.7 A 6 kWp stand-alone AC installation with a battery in the Chinese village of Doncun-Wushe (Photo: Shell Solar/SolarWorld)

The insolation used by solar cells constitutes an inexhaustible energy resource, unlike oil, uranium and coal, which are the most widely used energy resources today. But these resources are finite and will run out in a few decades in the case of oil and uranium, and in a few centuries in the case of coal. However, a basic problem with terrestrial photovoltaics is that many solar power applications are subject to fluctuating insolation by virtue of the sunrise and sunset cycle, cloudy and rainy weather, and seasonal changes in insolation. Hence in order for PV installations to provide an uninterrupted supply of electricity, it is necessary to use energy accumulators, which considerably drive up the cost of PV electricity. However, an existing power grid can be used as an accumulator to a certain extent, without the need for additional energy storage facility expenditures.

PV energy is extremely clean and environmentally friendly, and engenders no noise pollution, waste-gas emissions or toxic waste. Moreover, silicon, the main raw material used for solar cells, is one of the most abundant substances on Earth (in the form of sand) and can be readily disposed of when solar cells are retired. The substances used in silicon solar cells are non-toxic and environmentally friendly. Thus electricity from silicon solar cells can in a sense be thought of as 'power from sand and Sun'.

Figure 1.8 PV telecommunication installation at a remote location in Sipirok, Indonesia (Photo: Siemens)

Figure 1.9 PV installation that supplies power to Lime Village, Alaska. This is a 4 kWp stand-alone AC installation with a battery and a backup diesel generator (Courtesy of DOE/NREL)

Figure 1.10 A 3.18 kWp grid-connected system on the roof of a home in Burgdorf, Switzerland; the installation has been in operation since 1992

Figure 1.11 A 60 kWp grid-connected system on the roof of the Electrical Engineering Department at Bern University of Applied Sciences in Burgdorf. This installation has been in operation since 1994. © Simon Oberli, www.bergfoto.ch (details from Neville Hankins) (Photo: Simon Oberli)

Figure 1.12 The grid-connected system (1152 kWp) owned by Bern University of Applied Sciences on the Jungfraujoch Mountain, which is located at 3454 m above sea level and was connected to the grid in 1993. On inception, the Jungfraujoch facility was the world's highest-altitude PV installation

Figure 1.13 The grid-connected 555 kWp system on Mont Soleil, which is located at 1270 m above sea level and was commissioned in 1992

Figure 1.14 The grid-connected system (3.3 MWp) in Serre, Italy. The installation was commissioned in 1995 (Photo: ENEL)

Figure 1.15 The 5 MWp Leipziger Land grid-connected system in Germany that was commissioned in 2004 (Photo: Geosol)

Solar cells are maintenance free and extremely reliable. If they are protected from environmental influences by a fault-free packing, monocrystalline and polycrystalline solar cells have a life expectancy of 25 to 30 years.

In contrast to solar thermal systems, which can only use direct beam radiation, PV installations can also convert into electricity the diffuse portion of insolation, which accounts for a major portion of total irradiated energy in northern climates such as Central Europe. In Switzerland's Mittelland region and large portions of Germany and Austria, diffuse radiation accounts for more than 50% of total radiation. Another important difference between solar thermal systems and PV installations is that the latter are not subject to minimum size restrictions, which means that low-energy-consumption devices such as watches, calculators and the like can be solar powered. Conversely, owing to the modularity of PV installations, they are not subject to an upper size limit either – which means that PV power plants with peak capacity in the gigawatt range can in principle be realized in desert regions [1.1].

But PV technology does have one major drawback, namely its high cost, which means that in most countries PV installations are economically viable solely for consumers who are located at a considerable distance from the nearest grid hook-up. On the other hand, exporting solar power to the public grid is not an economically viable proposition in light of the large difference between the cost of generating solar

Figure 1.16 The grid-connected system in Springerville, Arizona. Currently 4.6 MWp, the capacity of this installation is being expanded incrementally to around 8 MWp (Photo: Tucson Electric Power Company; www. greenwatts.com)

power and the price of grid electricity. However, this did not prevent numerous environmentally conscious idealists and companies from realizing grid-connected systems in the late 1980s and early 1990s, once mass-produced PV inverters in the 0.7 to 5 kW range came onto the market. Ever since the 1986 Chernobyl disaster and the subsequent freeze on building new nuclear power plants in many countries, some European power companies have expressed an interest in the possible realization of PV power plants. This interest has been largely driven not so much by a desire to produce large amounts of electricity, but rather by the wish to gain experience with this relatively new technology. However, only a limited number of individuals and power companies are willing to take on the risk and cost of building such PV installations and thus spur advances in PV technology.

In the 1990s, a handful of countries established PV subsidy programmes to promote the realization of grid-connected systems. But unfortunately the steady increase in demand for solar cells was not nearly high enough to bring about the kind of economies of scale and mass production that would have allowed for lower solar cell prices. Such economies of scale only occur in cases where sustained long-term demand for a product necessitates substantial investments in production equipment that hold out the prospect of a robust return.

By far the most efficient types of subsidies for grid-connected systems are cost-recovery feed-in tariffs, which were first introduced in 1991 by the local power company in Burgdorf, Switzerland, and which have since become known as the Burgdorf model (Figure 1.17). Under this framework, the owners of grid-connected systems that were built between 1991 and 1996 were to receive the equivalent of €0.67 per kWh for electricity fed into the grid for a period of 12 years, whereby this emolument was spread across the electricity prices paid by all customers. In conjunction with the fixed federal and cantonal subsidies amounting to a few euros per watt of peak power that were available back then, this framework constituted an emolument system that allowed for the depreciation of grid-connected systems during their lifetime. This system also resulted in the realization of a great many such systems during the ensuing years. This type of subsidy spurs not only the realization of grid-connected systems, but also the efficient operation of them; in turn, this is the most efficient way to promote advances in PV technology. A modified form of the Burgdorf model was subsequently implemented in several other cities such as Aachen in Germany, and with the passage of Germany's Renewable Energy Act (EEG) in 2000 was elevated to the status of a nationwide framework – which in turn occasioned an exponential growth in

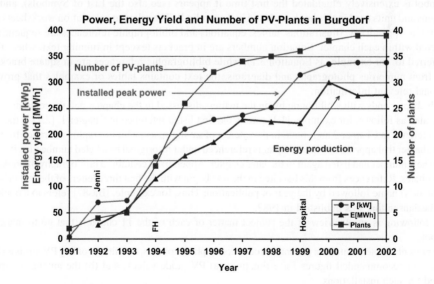

Figure 1.17 Installed peak PV capacity following the introduction of the Burgdorf model (€0.67 per kWh emolument from the local power company). The second spurt of growth in 1999 and 2000 was attributable to installation realized in connection with the 'solar electricity exchange'

Germany's PV sector and a rapid increase in installed PV capacity. A similar Swiss law (known as the KEV law), which was adopted in early 2009, has also prompted robust growth in installed PV capacity. Unlike Germany's EEG, Switzerland's KEV law calls for numerous, very low-cost coverage limits and these have inhibited growth in Switzerland's PV sector.

Since the advent of the first 3 kW inverter (the SI-3000) in 1987, the number of commercially available inverters has shot up and they are now far more reliable than they used to be. In addition, since the mid 1990s, so-called module inverters in the 100 to 400 W range have been available. Such module inverters allow for connection of anywhere from one to a handful of modules that are mounted directly on or in close proximity to the solar module. A large selection of mass-produced triphase inverters in the 20 to 500 kW range are also available.

The price of a PV installation can be reduced somewhat by integrating the system into a newly constructed building. In other words, the installation is mounted not on the outside of the building, but instead forms part of the building's shell or fulfils a key function such as shading. This kind of arrangement also lends itself to a far more aesthetically pleasing and creative realization, but can nonetheless negate the original savings by dint of architect's fees and the cost of the mass-produced modules or laminates. Such building-integrated PV installations were first realized during the 1990s. In countries such as France and Switzerland the higher emoluments that have been recently instituted for PV electricity exported to the grid from building integrated installations have spurred interest in this type of solution.

Product researchers around the world are working feverishly on the development of inexpensive and more efficient solar cells, while at the same time PV system technology is being optimized. Worldwide solar cell production has been rising at a rate of 30 to 85% annually since 1997, and increased production volumes have translated into lower prices. A combination of manufacturer learning curves and a doubling of production volumes could potentially occasion a decrease of around 20% in solar module prices. One thing is certain, however: the use of photovoltaics is set to account for a steadily increasing proportion of overall electricity production in the coming years.

1.2   Overview of this Book

The central portion of this book comprises a comprehensive introduction to PV technologies. Each term or symbol is extensively elucidated the first time it appears (see also the List of Symbols), and all acronyms and units used in this book are listed in Appendix B. The relevant calculation worksheets can be found in Appendix A. Illustrations, tables, equations and bibliographic references are sequentially numbered within each chapter. Equation numbers are in brackets (except in running text, where they are referred to, for example, as Equation 3.1), while bibliographic references are in square brackets. Apart from numerous photographs and diagrams, the text contains tables or examples that provide additional relevant information.

Each chapter ends with a bibliography on the matters discussed in the chapter, and the references are enumerated as follows: for example, [1.1] means the first listed reference in Chapter 1, [2.1] is the first listed reference in Chapter 2 and so on. In the interest of saving space, a bibliographic reference from an earlier chapter that appears in a later chapter is referenced under its original bracketed number, i.e. it is not given a new number and listed again in the later chapter. Appendix B contains a bibliography of books on photovoltaics. References from this list cited in the text begin with the first three letters of the author's (or first author's) name followed by the year of publication. Thus, for example, [Bas87] refers to a book by Uwe Bastiansen that was published in 1987.

The following is an overview of the subject matter of each of the 11 chapters that go to make up this book.

The present chapter contains a general introduction to photovoltaics, definitions of key PV terms and an overview of recommended figures for estimating: (a) PV guide values; and (b) the amount of space required for such installations.

Chapter 2 discusses solar insolation on horizontal and tilted surfaces. Two insolation calculation methods are presented, one very basic and easy to use, the other more robust but also more labour intensive to allow basic shading scenarios to be taken into account. Detailed charts for locations between 60°N and

40°S and that apply to both methods are provided in Appendix A. PV installation energy yield can be determined far more precisely if solar generator surface irradiation arising from local solar insolation is known.

Chapter 3 is concerned with the following: the structure and functional principle of solar cells; the impact of various solar cell materials and technologies on solar cell efficiency; solar cell designs; the solar cell manufacturing process; and possible solar cell development in the future.

Chapter 4 centres around solar modules, laminates and solar generators, and discusses in detail solar cell and solar module wiring options, particularly in scenarios involving partially shaded solar generators or module data scatter. This chapter also discusses solar generator configuration and how solar generators can be integrated into buildings or infrastructure installations.

Chapter 5 contains detailed information on the layout and configuration of PV installations. The chapter contains two main sections, the first discussing stand-alone systems and the second grid-connected systems, and also provides information on the key components of PV systems, particularly batteries and inverters. This chapter also discusses the electromagnetic compatibility of PV installations and the problems that can arise when such installations are connected to the public grid.

Chapter 6 begins with a brief introduction to lightning protection and then goes on to discuss in detail lightning current and surge protection measures for PV installations.

Chapter 7 is concerned with normalized presentation methods for PV installation energy yield and power output, which are extremely useful when it comes to determining whether PV installations are functioning properly. Such presentation methods also simplify and systematize the energy yield calculations discussed in the subsequent chapter.

Chapter 8 discusses PV installation sizing and energy yield calculations, based on the insolation calculations and normalized energy yield presentation methods described in Chapters 2 and 7 respectively. Detailed tables for use with these methods can be found in Appendix A.

Chapter 9 takes up the subject of the economic viability of PV installations from both financial and energy standpoints.

Chapter 10 describes selected grid-connected systems (both building-integrated and ground-based PV installations) that are currently in operation in Germany, Switzerland, Spain, Italy, Australia and the USA. The energy yields of these installations and the consequent learning curves are also discussed.

Chapter 11 briefly reviews the topics discussed in the book as a whole and offers a number of cautious predictions on the future of photovoltaics.

Appendix A begins with detailed weather data charts, as well as insolation calculation tables for tilted surfaces worldwide between 60°N and 40°S. Also included are: (a) insolation maps for the entire world; (b) figures that enable estimated horizon-induced shading to be folded into the sizing calculations for sites in the northern hemisphere; (c) PV installation energy yield calculation charts; and (d) numerous maps showing insolation around the world.

Appendix B contains a list of key web sites, a bibliography of books on photovoltaics, a list of the acronyms used in this book, a list of useful conversion factors, and a number of key physical constants. The List of Symbols and Acknowledgements appear separately.

1.3 A Brief Glossary of Key PV Terms

This section provides a brief glossary of key PV terms [SNV88], all of which are discussed in detail later in this book and illustrated with diagrams and photos.

1.3.1 Relevant Terminology Relating to Meteorology, Astronomy and Geometry

Solar radiation: Radiation originating from the Sun, in the 0.3 to 3 μm wavelength spectrum.
Solar spectrum: Distribution of solar radiation intensity as a function of wavelength or frequency.
Direct solar radiation: Solar radiation arriving at a plane directly from the solar disc.
Diffuse solar radiation: Solar radiation arriving at a plane after scattering by atmospheric particles (e.g. water droplets, clouds) or ambient reflection.

Global radiation: The sum of direct and diffuse solar radiation (i.e. aggregate radiation originating from the Sun) arriving at a level surface.

Global irradiance G:

Power density (power/area) of the global radiation arriving at a plane. Unit: W/m^2.

Irradiation H (radiation energy):

Energy density (energy/area) of the global radiation arriving at a plane within a certain time interval, calculated by integration of irradiance G over this time interval. Common time intervals are one year (a), one month (mt), one day (d) or one hour (h). Units: kWh/m^2 and MJ/m^2 (kWh/m^2 is more expedient for PV applications). Conversion: $1\,\text{kWh} = 3.6\,\text{MJ}$ and $1\,\text{MJ} = 0.278\,\text{kWh}$.

Pyranometer: Instrument for measuring global radiation (global irradiance G) on a level surface over the whole wavelength range between approx. 0.3 and 3 µm. Based on the thermoelectric principle, pyranometers are highly accurate but expensive instruments that are mainly used by weather services.

Reference cell: A calibrated solar cell for measuring global radiation G on a level surface. Reference cells are much cheaper than pyranometers. Like an actual solar cell, a reference cell only utilizes a portion of total incident insolation and is calibrated such that under standard conditions (standard spectrum AM1.5, where $G = 1\,\text{kW/m}^2$) it exhibits the same insolation as a pyranometer. In practice there are discrepancies ranging up to several per cent between the values indicated by pyranometers and reference cells, depending on the weather conditions.

Solar altitude h_S: Angle between the direction of the Sun (centre of the solar disc) and the horizontal plane.

Solar azimuth γ_S: For $\varphi > \delta$, the angle between south and the projection of the direction of the Sun on a horizontal plane. For $\varphi < \delta$, the angle between north and the projection of the direction of the Sun on a horizontal plane. In both cases, $\gamma_S < 0$ for deviations to the east, $\gamma_S > 0$ for deviations to the west (φ = latitude, δ = solar declination; for details see Section 2.1.2).

Solar generator tilt angle β: Angle between the solar cell plane and horizontal.

Solar generator orientation (solar generator azimuth) γ: In the northern hemisphere, the angle (clockwise) between south and the normal projection (vertical) of the solar cell and horizontal. In the southern hemisphere, the angle (anticlockwise) between north and the normal projection of the solar cell and horizontal. In both cases, $\gamma < 0$ for deviations to the east, $\gamma < 0$ to the west.

Relative air mass number (AM): Ratio of (a) the actual atmospheric mass (optical thickness) through which solar radiation travels to (b) the minimum possible atmospheric mass at sea level (applicable when the Sun is at its zenith). The following applies in this regard:

$$\text{Air mass AM} = \frac{1}{\sin(h_S)} \cdot \frac{p}{p_o} \tag{1.1}$$

where p = local air pressure and p_o = air pressure at sea level.

Examples (for locations at sea level):

AM1: $h_S = 90°$
AM1.1: $h_S = 65.4°$
AM1.2: $h_S = 56.4°$
AM1.5: $h_S = 41.8°$ (convenient mean value for Europe)
AM2: $h_S = 30°$
AM3: $h_S = 19.5°$
AM4: $h_S = 14.5°$

According to Equation 1.1, AM values of less than 1 can occur in mountainous regions (e.g. in the Alps during summer).

1.3.2   PV Terminology

Crystalline silicon (c-Si): Silicon that has solidified into atoms arranged in a crystal lattice, i.e. crystals.
Monocrystalline silicon (sc-Si, mono-c-Si): Silicon that has solidified into a single large crystal.
Production of this substance is very energy intensive and pulling of the single crystal is time consuming.
Polycrystalline or multicrystalline silicon (mc-Si): Silicon that has solidified into many small crystals
(crystallites) in any orientation. Energy consumption for production is significantly lower than for
monocrystalline silicon. Polycrystalline and multicrystalline are often used as synonyms, although in
some cases a distinction is made between the raw material (polycrystalline silicon) and the material used
for producing solar cells after the silicon is cast into ingots and sliced into wafers (multicrystalline silicon,
or mc-Si; see Section 3.5.1).
Amorphous silicon (a-Si): Silicon whose atoms are not arranged in a crystal lattice.
Solar cell: Semiconductor diode with a large barrier layer exposed to light, which generates electrical
energy directly when sunlight strikes it.
Solar module: A number of galvanically connected solar cells (usually connected in series) arranged in a
casing to protect against environmental influences.
Solar panel: A unit consisting of several mechanically joined solar modules (often pre-wired) that are
delivered pre-assembled and are used for configuring larger solar generators. In technical jargon, solar
panel is often incorrectly used as a synonym for solar module.
Solar generator (array): A series of solar panels or solar modules that are arranged on a mounting rack
and wired to each other (including the mounting rack).
Solar generator field (array field): An arrangement of several interconnected solar generators, which
together feed a PV system.
Photovoltaic system (PV system, PV installation): Aggregate components used for direct conversion of
the energy contained in solar radiation into electrical energy.
Grid-independent PV system (autonomous system, stand-alone system): A PV electricity generation
system that is not connected to the public grid. Such systems usually require batteries to store energy in the
night and to balance load peaks. This storage system increases electricity costs.
Grid-coupled PV system (grid-connected system): A PV system that is connected to the public grid,
which is used as a storage medium, i.e. any excess energy is fed into the grid while energy is obtained from
the grid at times of insufficient local production.
Current/voltage characteristic ($I = f(V)$): Graphic display of current as a function of the voltage of a
solar cell or a solar module (at a specific irradiance G and cell temperature).
Open-circuit voltage V_{OC}: Output voltage of a solar cell or solar module in open-circuit condition (no
current), at a specific irradiance G and cell temperature.
Short-circuit current I_{SC}: Current in a short-circuited solar cell or solar module, i.e. with output voltage
0 V, at a specific irradiance G and cell temperature.
Standard test conditions (STC): Usual test conditions, as follows, for the purpose of specifying solar cell
and solar module guide values: irradiance $G_o = G_{STC} = 1000$ W/m^2, AM1.5 spectrum, cell temperature 25 °C.
Peak power P_{max}: Maximum output power (product of voltage × current) of a solar cell or solar module at
a specific insolation and solar cell temperature (usually under STC) at the maximum power point (MPP).

Metric: 1 watt $= 1$ W and 1 Wp (watt peak).

Peak power is often expressed in Wp rather than W to indicate that it is a peak power value under
laboratory conditions, which is rarely reached under real operating conditions.
Fill factor FF (of a solar cell or a solar module): Ratio of peak power to the product of multiplying open-
circuit voltage by short-circuit current at a specific insolation and solar cell temperature. In practice FF is
always less than 1. The following equation applies to the fill factor FF:

$$FF = \frac{P_{max}}{V_{OC} \cdot I_{SC}} \qquad (1.2)$$

Packing factor PF of a solar module:
Ratio of total solar cell area to total module area (including the frame). PF is always less than 1.
Photovoltaic efficiency or solar cell efficiency η_{PV}:
Ratio of peak power P_{max} of a solar cell to the radiation power arriving at the solar cell. Solar cell efficiency decreases somewhat at lower insolation and higher temperatures.

The following equation applies to solar cell efficiency (where A_Z = solar cell area):

$$\eta_{PV} = \frac{P_{max}}{G \cdot A_Z} \tag{1.3}$$

Solar module efficiency η_M:
Ratio of peak power of a solar module to the radiation power arriving across the whole module area (including the frame). In practice η_M is always less than η_{PV}.

The following therefore applies to module efficiency:

$$\eta_M = \eta_{PV} \cdot PF \tag{1.4}$$

Energy efficiency (utilization ratio, system efficiency) η_E:
Ratio of the usable electrical energy produced by a PV system to the solar energy incident on the whole solar generator area over a certain period (e.g. a day, month or year).

The following applies: $\eta_E < \eta_M$.

1.4   Recommended Guide Values for Estimating PV System Potential

The figures indicated in this section will enable the reader to estimate guide values for PV installations. More precise calculation methods are described in later chapters.

1.4.1   Solar Cell Efficiency η_{PV}

Solar cell efficiency, which is mainly of interest to solar cell researchers and manufacturers, is nowadays (2009) as follows for commercially available solar cells under STC ($G = G_o = 1$ kW/m^2, AM1.5 spectrum, cell temperature 25 °C):

For monocrystalline Si solar cells (sc-Si): $\eta_{PV} = 13$–22.5%
For polycrystalline (multicrystalline) Si solar cells (mc-Si): $\eta_{PV} = 12$–18%
For amorphous Si solar cells (a-Si): $\eta_{PV} = 3.5$–8.5%
   (the higher figures apply to a-Si triple cells only)
For CdTe solar cells: $\eta_{PV} = 7.5$–12%
For CuInSe$_2$ (CIS) solar cells: $\eta_{PV} = 8$–12.5%

The efficiency of the counterpart lab cells is several percentage points higher.

Solar cell efficiency is inherently limited, however, by: (a) the fact that sunlight is composed of various colours comprising varying and sometimes not fully usable light quantum energy; and (b) the principles of semiconductor physics (see Chapter 3).

1.4.2   Solar Module Efficiency η_M

Solar module efficiency is mainly of interest to PV installation designers and installers. Its calculation should always be based on total module surface area and not just on solar cell surface area. The efficiency of currently available (2009) solar modules under the test conditions referred to in Section 1.4.1 is as follows:

For monocrystalline Si solar modules (sc-Si): $\eta_M = 11\text{--}19.5\%$
For polycrystalline (multicrystalline) Si solar modules (mc-Si): $\eta_M = 10\text{--}16\%$
For amorphous Si solar modules (a-Si): $\eta_M = 3\text{--}7.5\%$
 (the higher figures apply to a-Si triple cells only)
For CdTe solar modules: $\eta_M = 7\text{--}11\%$
For CuInSe$_2$ (CIS) solar modules: $\eta_M = 7.5\text{--}11.5\%$

1.4.3 Energy Efficiency (Utilization Ratio, System Efficiency) η_E

PV installation energy efficiency is always considerably lower than the module efficiency of the installation's solar modules, on account of the additional power losses that occur in the installation's wiring and diodes on the DC side and on account of the following: module data scatter; module soiling and shading; elevated solar cell temperatures; elevated reflection resulting from non-perpendicular incident light; non-usable energy under very low-insolation conditions; DC to AC conversion loss in the inverter (for grid-connected systems) and battery storage loss (stand-alone installations).

The energy efficiency η_E of Central European grid-connected systems with monocrystalline solar cells ranges from around 8 to 14.5% over the course of any given year.

Low though the energy efficiency attainable using today's commercial components may seem now, as in the past, the efficiency of commercially available solar cells will increase steadily, albeit not as quickly as the efficiency of laboratory solar cells; in turn this means that PV installation efficiency as a whole will improve considerably in the long run.

1.4.4 Annual Energy Yield per Installed Kilowatt of Peak Installed Solar Generator Capacity

Thanks to the modularity of PV installations, they can be realized in greatly varying sizes. In view of this fact, when comparing annual energy yields, the specific annual energy yield kWh/kWp/a derived from peak installed solar generator power P_{Go} should be indicated either as: (a) the annual value Y_{Fa} for the so-called final yield Y_F in kWh/kWp/a; or (b) after truncating via kW, the number of full load hours t_{Vo} (in hours/year) for peak installation capacity P_{Go} (at STC):

$$\text{Specific annual energy yield } Y_{Fa} = t_{Vo} = \frac{E_a}{P_{Go}} \qquad (1.5)$$

where E_a = annual energy yield (E_{AC} for grid-connected systems) and P_{Go} = peak solar generator capacity.

By rearranging Equation 1.5 and inserting the Y_{Fa} value (if known; see the recommended values in Table 1.1), the potential annual energy yield for a PV installation of a specific size can be readily estimated.

Apart from the fact that Alpine installations exhibit higher annual energy yield, their winter energy production is far greater than for installations in low-lying areas. For example, a façade-mounted installation in the high Alps can attain winter energy output relative to total annual yield ranging from around 45 to 55%, as opposed to 25 to 30% for typical installations in Central European lowland areas.

The energy yield of selected Swiss grid-connected systems has been monitored since 1992 via a joint BFE–VSE project, whose data mainly stem from annual reports filed by installation owners based on monthly meter readings. These data reveal that from 1995 to 2008 the mean energy yield of the monitored

Table 1.1   Recommended values for determining energy yield for fixed solar panel installations

Installation site	Y_{Fa} (kWh/kWp/a) and t_{Vo} (h/a)
Façade-mounted installation in a low-lying area in Central Europe	450–700
Suboptimal or foggy sites in low-lying regions in Central Europe	600–800
Well-situated or optimally situated site in a low-lying area in Central Europe	800–1000
Sites in inner Alpine valleys, in the Alpine foothills, or just south of the Alps	900–1200
Southern Europe; Alpine regions	1000–1500
Desert regions	1300–2000
Recommended approximate value for flashover calculations	1000

installations was 833 kWh/kWp/a, a figure that should be interpreted in light of the following: (a) most of these installations are located in lowland areas; (b) many of these installations integrate components that are a number of years old (owing to the fact that few new installations have been realized in recent years); and (c) some of the installations are suboptimally oriented or are subject to shading.

The German solar advocacy organization Solarförderverein maintains a Web-based information service (www.pv-ertraege.de) that gathers information in the manner described above. However, the data available on this web site are more up to date as they are entered monthly. Noteworthy in this regard is that, owing to the realization of numerous new PV installations in recent years, Germany's mean energy yield for 1995 to 2008 is 869 kWh/kWp/a, which is higher than the figure for Switzerland despite that country's higher insolation (Figure 1.18).

From the low number of new PV installations realized in Switzerland, up until 2006, the technical improvement entailed by the new installations was for the most part cancelled out by the age of older installations. In contrast, because of the passage, in 2000, of Germany's EEG, it predominantly has new installations which, from their optimized components, use the available insolation more efficiently. The rate of new PV installation realization did not begin to pick up in Switzerland until 2007 (see Sections 1.4.7 and 1.4.9).

A similar monitoring programme realized in Austria in the mid 1990s revealed an annual energy yield amounting to 803 kWh/kWp/a, which means that the situation was apparently much the same as in Switzerland [1.6].

The use of dual-axis solar trackers in lieu of stationary mounting structures allows for more efficient use of direct beam energy and increases PV installation energy yield by 25 to 40%, albeit at the cost of higher

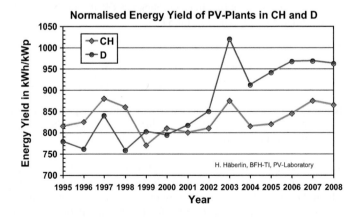

Figure 1.18   Mean specific energy yield for grid-connected systems from 1995 to 2008 in Germany and Switzerland. Early in this period, Switzerland's energy yield outpaced that of Germany because of Switzerland's higher insolation. But as from 2000, when Germany's Renewable Energy Act (EEG) was enacted, the proportion of new and technically superior installations in Germany increased substantially. *Source*: [1.4], www.pv-ertraege.de

capital expenditures for mechanical and control components. These expenditures mainly pay off with large PV power plants in desert regions.

1.4.5 PV Installation Space Requirements

The following equation applies to the nominal peak power P_{Go} of a solar generator field with n_M solar modules, peak capacity P_{Mo} and module surface area A_M ($G_o = 1\,\text{kW/m}^2$):

$$P_{Go} = n_M \cdot P_{Mo} = n_M \cdot A_M \cdot G_o \cdot \eta_M = A_G \cdot G_o \cdot \eta_M \qquad (1.6)$$

Thus the total area needed A_G for the solar generator field is determined as follows:

$$A_G = n_M \cdot A_M = \frac{P_{Go}}{P_{Mo}} A_M = \frac{P_{Go}}{G_o \cdot \eta_M} \qquad (1.7)$$

The ground or roof area needed to mount a solar generator is

$$A_L = \text{LF} \cdot A_G. \qquad (1.8)$$

where:

n_M = number of solar generator solar modules
P_{Go} = nominal peak solar generator power (at STC)
P_{Mo} = peak solar module power at G_o
G_o = global irradiance for which P_{Mo} is defined ($G_o = G_{STC} = 1\,\text{kW/m}^2$)
A_G = total solar generator field area
A_M = surface area of a solar module
η_M = solar module efficiency
A_L = space required for a ground-based or rooftop solar generator field
LF = Land Factor (between around 2 and 6 in Central Europe) for the avoidance of reciprocal shading in cases where a series of solar generators are arranged behind each other in a large installation.

1.4.6 Cost per Installed Kilowatt of Peak Power

Relative to the figures indicated in [Häb07], current prices for large numbers of solar modules have decreased and as of December 2009 were as follows, per watt of peak power:

For crystalline silicon solar modules: €1.6–3.2
For thin-film CdTe solar cells: €1.5–3
For thin-film amorphous Si solar cells: €1.5–3

The manufacturing costs for certain components have also gone down considerably, e.g. for CdTe, around US$0.9 per Wp (€0.65 per Wp), which is extremely good for manufacturers' bottom lines. The tremendous production capacity increase on the part of many manufacturers in recent years is likely to result in further price reductions. Solar cell costs have also dropped considerably, and for crystalline Si cells now range from about €1 to €1.9 per Wp.

Laminates are nearly as expensive as framed modules, with some manufacturers selling them at a price that is a few percentages points lower than for framed modules of the same type.

Table 1.2   German feed-in tariffs, in eurocents per kWh, for PV installations that went into operation in 2010 or later, pursuant to the 2008 version of the EEG. These tariffs (those indicated here were valid as at December 2009) will be paid at a constant level for 20 years and will be subject to a further reduction as from 2011 (for further information visit www.erneuerbare-energien.de)

Installed capacity	Ground-based PV installations (cents per kWh)	Installations mounted on buildings or noise barriers (cents per kWh)
Up to 30 kW	28.43	39.14
30–100 kW	28.43	37.23
100–1000 kW	28.43	35.23
More than 1000 kW	28.43	29.37

New types of thin-film solar cell modules (e.g. CdTe) exhibit high efficiency, sell for far less than crystalline modules Si modules, and since 2006 have come into increasing use in Germany for large-scale ground-based PV installations as the statutory feed-in tariffs are lowest for them, according to Germany's EEG (see Table 1.2). Triple-cell amorphous Si modules are often mounted on roofs as well. But if these modules are unduly inefficient, this considerably ramps up the costs of wires and other system components, as well as the amount of space needed for the installation.

In keeping with the Staebler–Wronski effect, amorphous silicon solar cells are prone to light-induced degradation early on, resulting in an efficiency drop of 10 to 30% during the first few months of operation. Hence the life span of these cells may be shorter than that of monocrystalline and polycrystalline solar cells. Moreover, as amorphous Si cells (particularly the single-film variety) are less efficient, they require a larger installation space. For these reasons, amorphous Si cells are mainly used in small devices such as calculators, watches, radios and the like, and are rarely integrated into large PV installations for energy production purposes.

As at December 2009, the cost of a complete grid-connected system ranged from €2.7 to €7 per Wp. The lower price applies to large-scale ground-based PV installations with thin-film solar cells in Germany, while the higher price applies to small building-integrated monocrystalline cell installations.

1.4.7   Feed-in Tariffs; Subsidies

In the early 1990s, various Swiss power companies had begun subsidizing PV electricity by paying relatively high feed-in tariffs for PV installation electricity exported to the grid. The first power company to opt for this extremely effective subsidy instrument (now referred to as the Burgdorf model) for the technological advancement of photovoltaics was Industriellen Betriebe Burgdorf (IBB; now known as Localnet), which paid the equivalent of €0.77 per kWh for a 12-year period for all PV installations that went into operation before 1997. This exemplary European model was then implemented by certain municipalities in a modified form, and was later instituted in Germany, where it is referred to as the Aachen model.

However, a major breakthrough in the development of PV technology was the nationwide implementation in 2000 of cost-recovery feed-in tariffs in Germany under the EEG, which prompted a rapid rise in installed PV capacity. The tariff limitations called for by the original law were abolished in the amended version of the EEG that was adopted in 2004. The Act was amended again in 2008.

In 2003, Austria imposed a nationwide performance-related limit of 15 MWp on total installed PV capacity and subsequently amended its EEG to include the payment of cost-recovery feed-in tariffs for 13 years at a rate of 60 cents per kWh [1.7]. However, the capacity limit was soon reached and installed PV capacity in Austria rose only slightly (see Figure 1.21).

Switzerland finally instituted a modicum of nationwide PV subsidies via passage in 2008 of the KEV law governing cost-recovery feed-in tariffs (see Table 1.3), which are financed by a surcharge equivalent to 0.4 eurocents per kWh on electricity rates. Hence total Swiss subsidies for all renewable energy technologies are currently limited to the equivalent of around €220 million a year. Moreover, owing to the

Table 1.3 Swiss feed-in tariffs in eurocents per kWh (1 SFr. equivalent to 0.67€) as at February 2010 for PV installations commissioned in 2010 or later. These tariffs, which vary according to installation mounting modality and size, will be paid at a constant level for 25 years as from commissioning and will be reduced by 8% annually for each year of operation thereafter. For further information and actual valid values visit www.swissgrid.ch

Installed capacity	Ground-based PV installations (eurocents per kWh)	Surface mounted (eurocents per kWh)	Integrated (eurocents per kWh)
Up to 10 kW	36	41	49
10–30 kW	30	36	41
30–100 kW	28	34	37
More than 100 kW	27	33	34

very low spending limit imposed for photovoltaics (around 5% of the total amount at present), these subsidies involve an extremely bureaucratic application procedure and the available funds were exhausted two days after they became available. This ceiling will be raised to 10%, then 20% and then 30% when PV costs are equivalently less than €0.33, €0.27 and €0.20 higher than grid electricity, respectively.

Some Swiss power companies sell solar power at cost-recovery tariff prices (e.g. with a surcharge equivalent to €0.54 on the standard rate). This power is derived partly from the power companies' own PV installations and partly from so-called solar power exchanges operated by electricity contractors, which provide electricity at cost-recovery tariff prices that are guaranteed for an extended period. In view of the enormous profits that this higher priced electricity entails for the power companies, there is of course little demand for it; this programme is only a drop in the ocean in terms of bringing about a monumental increase in installed PV capacity and instituting substantial subsidies for PV technology development.

Although Switzerland had far and away the highest per capita installed PV capacity in Europe in the early 1990s (see Figure 1.22), it is now ranked somewhere in the middle and is unfortunately one of the European laggards when it comes to growth in installed PV capacity. The institution of PV electricity feed-in tariffs in 2007 has spurred this growth to some extent (see Table 1.4 and Figure 1.21), although the low multi-level limits on PV subsidies will retard PV sector growth for the foreseeable future. Swiss PV

Table 1.4 Peak installed PV capacity (in MWp) in selected countries, according to IEA data [1.16]

	1997	1998	1999	2000	2001	2002	2003	2004	2005	2006	2007	2008
Australia	18.7	22.5	25.3	29.2	33.6	39.1	45.6	52.3	60.6	70.3	82.5	104.5
Austria	2.2	2.9	3.7	4.9	6.1	10.3	16.8	21.1	24	25.6	27.7	32.4
Canada	3.4	4.5	5.8	7.2	8.8	10	11.8	13.9	16.7	20.5	25.8	32.7
Denmark	0.4	0.5	1.1	1.5	1.5	1.6	1.9	2.3	2.7	2.9	3.1	3.3
England	0.6	0.7	1.1	1.9	2.7	4.1	5.9	8.2	10.9	14.3	18.1	22.5
France	6.1	7.6	9.1	11.3	13.9	17.2	21.1	26	33	43.9	75.2	179.7
Germany	41.8	53.8	69.4	114	195	278	431	1034	1897	2727	3862	5340
Israel	0.3	0.3	0.4	0.4	0.5	0.5	0.5	0.9	1	1.3	1.8	3.03
Italy	16.7	17.7	18.5	19	20	22	26	30.7	37.5	50	120	458.3
Japan	91.3	133	209	330	453	637	860	1132	1422	1709	1919	2144
Korea (S)	2.5	3	3.5	4	4.8	5.4	6	8.5	13.5	34.7	77.6	357.5
Mexico	11	12	12.9	13.9	15	16.2	17.1	18.2	18.7	19.7	20.8	21.75
Netherlands	4	6.5	9.2	12.8	20.5	26.3	45.9	49.5	51.2	52.7	53.3	57.2
Norway	5.2	5.4	5.7	6	6.2	6.4	6.6	6.9	7.3	7.7	8	8.34
Portugal	0.5	0.6	0.9	1.1	1.3	1.7	2.1	2.7	3	3.4	17.9	67.95
Sweden	2.1	2.4	2.6	2.8	3	3.3	3.6	3.9	4.2	4.8	6.2	7.91
Switzerland	9.7	11.5	13.4	15.3	17.6	19.5	21	23.1	27.1	29.7	36.2	47.9
Spain				1	3	7	11	22	45	143	655	3354
USA	88.2	100	117	139	168	212	275	376	479	624	831	1169
TOTAL	305	385	509	715	974	1318	1809	2832	4154	5584	7841	13412

subsidies fall far short of those in many other countries. Some Swiss legislators are currently making efforts to raise or abolish these statutory limits.

Spain, France, Italy and a number of other European countries have in recent years instituted feed-in tariffs that have spurred marked growth in installed PV capacity. An example of how extremely successful such subsidies can be is provided by Spain, whose feed-in tariffs: (a) are not subject to a national installed-capacity limit for PV installations commissioned up to September 2008; (b) are far more economically beneficial than in Germany by virtue of Spain's far higher insolation relative to Germany; and (c) in 2008 resulted in a jump in installed PV capacity to more than 3 GWp. However, a national capacity limit has since been imposed in order to keep the costs of this programme within reasonable bounds.

As the tariffs and regulations in this domain change all the time, a discussion of the applicable laws has been forgone here. Such information is better obtained from professional journals, which are more up to date and for some larger countries are available in a nationwide edition [1.9]. Switzerland's originally planned feed-in tariffs for 2010 were reduced owing to the major price reductions that occurred in 2009 (see Table 1.3). German feed-in tariffs (see Table 1.2) are slated for further reduction as at 1 July 2010, amounting to 11%, 15% and 16% depending on the type of installation. For further information in this regard, see the web sites indicated at Tables 1.2, 1.3 and in Appendix A.

1.4.8   Worldwide Solar Cell Production

Figure 1.19 displays worldwide production of solar cells from 1995 to 2008 according to [1.8] (1995–2000) and [1.9] (2001–2008). These data were published each year during the spring.

It is noteworthy here that since 1997 worldwide solar cell production has been largely unaffected by the overall economic situation in that this sector has exhibited steady annual growth ranging from 30 to 85% – a trend that has prompted a number of solar cell manufacturers to expand their production capacity. Most solar cells are still made of silicon (Si). The proportion of thin-film solar cells made of cadmium telluride (CdTe), which is currently the most inexpensive technology, has increased substantially. Other materials such as copper indium diselenide ($CuInSe_2$, CIS) or copper indium gallium diselenide (Cu(In, Ga)Se, CIGS) are used for only a fraction of the thin-film solar cells made worldwide.

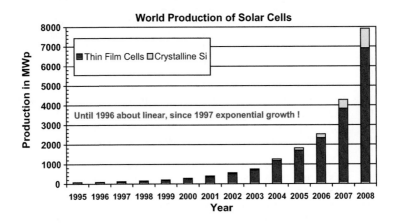

Figure 1.19   Worldwide solar cell production (*Sources*: [1.8] and [1.9]). Of the thin-film solar cells manufactured in 2008, around 51% were made of CdTe, around 41% were made of amorphous or micromorphous silicon and around 8% were made of CIS and CIGS [1.9]

1.4.9   Installed Peak Capacity

It is difficult to estimate the total installed worldwide capacity of all PV installations used for energy production (as opposed to calculators and the like), because reporting such data is not required by law.

It would seem that, as at the end of 2008, some 14.8 GWp of PV capacity was installed worldwide, around 9.6 GWp of it in Europe [1.17]. This figure: (a) was obtained by using the 1987 figure indicated in [1.10], which was then updated using the worldwide 1988 to 2008 production figures in [1.8] and [1.9]; and (b) is predicated on the assumption that 96% of all crystalline solar cells are used to generate electricity. Inasmuch as thin-film solar cells were initially used for the most part in consumer products, it was assumed that 30, 40, 50, 60 and 70% of these cells were used in PV installations as from (respectively) 1991, 2000, 2005, 2007 and 2008. Figure 1.20 displays the peak worldwide PV power thus calculated for 1995 to 2008.

Until the mid 1990s, the lion's share of worldwide installed PV capacity was concentrated in stand-alone installations. Thus, for example, in 1994 grid-connected systems accounted for only around 20% of world output, whereas the figure for stand-alone installation systems was 61% and for consumer products 19% [1.12]. However, since then there has been a steady increase in the proportion of worldwide PV production accounted for by grid-connected systems in industrialized countries such as Germany, Japan and Spain that offer relatively high feed-in tariffs. By the end of 2008, a minimum of 86% of worldwide installed PV capacity was apparently concentrated in grid-connected systems; in IEA states (Table 1) the figure was around 94% (13.4 GWp of installed capacity) [1.16].

Installed capacity as at the end of 2008 was around 47.9 MWp in Switzerland [1.16], where PV installations were at first mainly realized in the form of numerous, small stand-alone installations for remote sites such as holiday homes, telecommunication installations and so on. However, Switzerland has seen the advent of numerous 1–1000 kWp grid-connected systems realized by private citizens, power companies and municipalities. These installations accounted for around 44.1 MWp of Switzerland's total installed PV capacity as at the end of 2008 [1.16].

German peak installed PV capacity as at the end of 2008 amounted to 5340 MWp, 5300 MWp of which was grid connected [1.16].

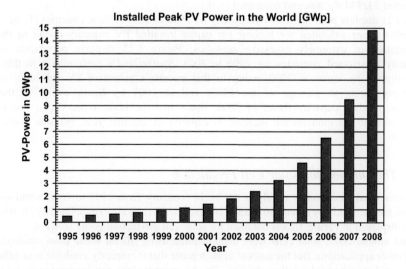

Figure 1.20   Worldwide installed solar generator capacity as at the end of the year, from 1995 to 2008. The new cells made in any given year are factored in only partly, as they must first be integrated into modules and then into PV installations. The 2004 figure is consistent with that indicated in [1.11]

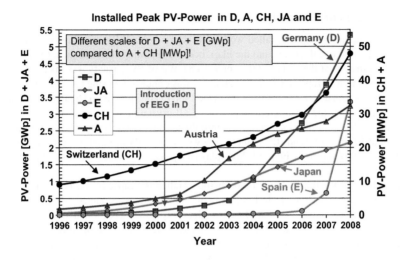

Figure 1.21 Total installed peak PV capacity in selected countries from 1996 to 2008 (*data source*: [1.16]). The 2008 figures are consistent with the estimates in [1.9], [1.16], [1.17], [1.18] and elsewhere

Austrian peak installed PV capacity as at the end of 2008 amounted to 32.4 MWp, 29 MWp of which was grid connected [1.16].

Owing to an extensive subsidy programme introduced in the late 1990s, Japan's total installed PV capacity as at the end of 2008 amounted to 2144 MWp, around 2053 MWp of which was grid connected [1.16].

In Spain, which only recently began subsidizing solar power in earnest (for installations commissioned prior to 1 October 2008), peak installed PV capacity as at the end of 2008 amounted to 3354 MWp, of which around 3323 MWp was grid connected [1.16].

Figure 1.21 displays peak installed PV capacity for 1996 to 2008 in these countries [1.16].

Switzerland long exhibited the highest per capita installed PV capacity owing to the early implementation of extremely extensive subsidies. Figure 1.22 displays per capita installed PV capacity in selected countries for 1996 to 2008. Switzerland's performance in this regard was outstripped by Japan in 2000 owing to that country's extensive PV subsidies, in 2002 by Germany owing to passage of the EEG, and in 2007 by Spain owing to the rapid increase in installed capacity there. In 2008, Spain had the highest per capita installed PV capacity owing to the monumental increase in capacity during that year, but was overtaken by Germany in 2009.

1.4.10   The Outlook for Solar Cell Production

Despite occasional recessionary trends in the world economy, from 1997 to 2008 world solar cell production grew annually by around 30 to 85%. If these growth rates are extrapolated by 30, 40 or 50%, the trend displayed in Figure 1.23 is obtained for solar cell production going forward.

Up until around 2007, solar cells were manufactured using silicon waste from semiconductors made for other applications. But the amount of such waste that is currently available is insufficient to cover the rapid growth in solar cell production. This has prompted the development in recent years of alternative silicon production modalities to meet the demand entailed by the growth shown in Figure 1.23. Solar cells not subject to particularly high efficiency requirements can be made using a

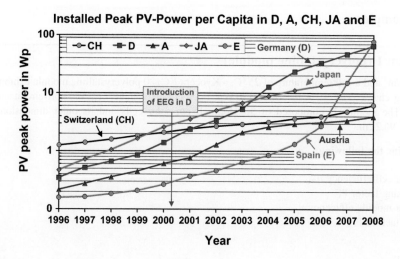

Figure 1.22   Per capita installed peak PV capacity in selected countries from 1996 to 2008, based on the figures in Figure 1.21

somewhat impure material known as solar grade silicon, which is more inexpensive to manufacture. A shortage of silicon in the face of rising demand for this material has spurred efforts on the part of various manufacturers to find innovative methods that will reduce the amount of silicon per Wp needed to make solar cells – for example, by using thinner silicon wafers (see Chapter 3).

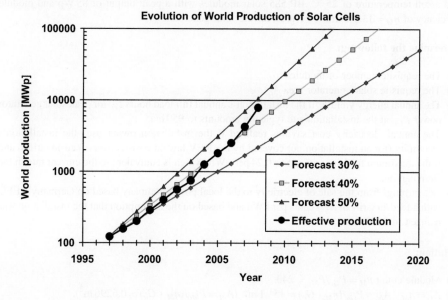

Figure 1.23   Projected worldwide solar cell production rate. These figures were obtained by extrapolating the growth figures for 1997 to 2008, during which period the growth rates were roughly as follows: 30% from 1997 to 1999; 35% from 2000 to 2003; 40% from 2004 to 2006; 70% from 2006 to 2007; and 85% from 2007 to 2008

1.5 Examples

Problem 1 (polycrystalline module)

At $1\,kW/m^2$ and a cell temperature of $25\,°C$, a Kyocera KC120 polycrystalline module exhibits the following characteristics: peak power $120\,Wp$; open-circuit voltage $21.5\,V$; short-circuit current $7.45\,A$. The module is composed of 36 $15\,cm \cdot 15\,cm$ solar cells and its outer dimensions are $96.7\,cm \cdot 96.2\,cm$.

Determine the following:

(a) Solar cell efficiency η_{PV}.
(b) Packing factor PF.
(c) Solar module efficiency η_M.
(d) Fill factor FF.

Solution:

(a) For a solar cell: $P_{max} = P_{Zo} = P_{Mo}/36 = 3.333\,W \Rightarrow \eta_{PV} = P_{max}/(G_o \cdot A_Z) = 14.8\%$.
(b) Packing factor $PF = 36 \cdot A_Z/A_M = 0.81\,m^2/0.9303\,m^2 = 0.8707$.
(c) Module efficiency $\eta_M = \eta_{PV} \cdot PF = 12.9\%$.
(d) Fill factor $FF = P_{Mo}/(V_{OC} \cdot I_{SC}) = 0.749$.

Problem 2 (20 kWp PV installation)

A PV installation should exhibit peak solar generator power amounting to $P_{Go} = 20.4\,kWp$ at $1\,kW/m^2$ and a cell temperature of $25\,°C$. BP 585 solar modules with a peak output of $85\,Wp$ and module efficiency of $\eta_M = 13.5\%$ are to be used for the solar generator.

Determine the following:

(a) The requisite number of modules n_M.
(b) The requisite solar generator area A_G.
(c) The annual energy yield E_a if the installation's annual full load hours t_{Vo} for peak solar generator power P_{Go} at the installation site typically amounts to $950\,h/a$.
(d) The annual electricity cost savings realized if the installation owner uses the installation's electricity (for an installation not covered by the KEV law on cost-recovery feed-in tariffs) and if the maximum feed-in tariff is SFr 0.24 per kWh, which is equivalent to the current rate in the Bern canton.
(e) Net earnings from the sale of electricity to the local power company based on Germany's EEG with a feed-in tariff of 57.4 cents per kWh and based on the assumption that the installation was realized in 2004.

Solution:

(a) Module count $n_M = P_{Go}/P_{Mo} = 240$.
(b) $A_G = n_M \cdot A_M = P_{Go}/(\eta_M \cdot G_o) = 151.1\,m^2$. $(A_M = P_{Mo}/(\eta_M \cdot G_o) = 0.6296\,m^2)$.
(c) $E_a = P_{Go} \cdot t_{Vo} = P_{Go} \cdot Y_{Fa} = 19\,380\,kWh/a$.
(d) $S_{Electricity} = E_a \cdot$ SFr 0.24 per kWh = SFr 4651.20 per year.
(e) $S_{Electricity} = E_a \cdot$ €0.574 per kWh = €11 124 per year.

Problem 3 (replacing energy produced by a nuclear power plant with PV electricity)

A nuclear power plant with 950 MW of installed capacity is operated for 7700 hours per year. This amount of energy is to be provided by PV power plants at three different sites, with fixed solar module mounting structures and 14% module efficiency.

Determine the following:

For all three sites specified below under (a), (b) and (c):

The peak power P_{Go} of the solar generator at STC, the total solar generator area A_G and the amount of land or roof area A_L for
(a) Switzerland's lowland region (Mittelland): annual full load hours $t_{Vo} = 900$ h/a, land factor $LF = 3$.
(b) Alps: annual full load hours $t_{Vo} = 1400$ h/a, land factor $LF = 2$.
(c) Sahara Desert: annual full load hours $t_{Vo} = 1900$ h/a, land factor $LF = 1.6$.

Solution:

$E_a = 950$ MW $\cdot$ 7700 h/a $= 7.315$ TWh/a.

(a) $P_{Go} = E_a/t_{Vo} = 8.128$ GWp, $A_G = P_{Go}/(\eta_M \cdot G_o) = 58.06$ km^2, $A_L = LF \cdot A_G = 174.2$ km^2.
(b) $P_{Go} = E_a/t_{Vo} = 5.225$ GWp, $A_G = P_{Go}/(\eta_M \cdot G_o) = 37.32$ km^2, $A_L = LF \cdot A_G = 74.64$ km^2.
(c) $P_{Go} = E_a/t_{Vo} = 3.850$ GWp, $A_G = P_{Go}/(\eta_M \cdot G_o) = 27.50$ km^2, $A_L = LF \cdot A_G = 44.00$ km^2.

1.6 Bibliography

[1.1] K. Kurokawa: *Energy from the Desert*. James & James, London, 2003. Summary of a study performed for IEA-PVPS-Task 8:VLS-PV.
[1.2] IEA-PVPS Task 1 National Survey Report 2007, Switzerland.
[1.3] F. Jauch, R. Tscharner: 'Markterhebung Sonnenenergie 2007'. Ausgearbeitet durch Swissolar im Auftrag des BFE, June 2008.
[1.4] T. Hostettler: 'Solarstromstatistik 2008 mit massivem Zubau'. *Bulletin SEV/VSE*, 9/09.
[1.5] U. Jahn, W. Nasse: 'Performance von Photovoltaik-Anlagen: Resultate einer internationalen Zusammenarbeit in IEA-PVPS Task 2'. 19. Symposium PV Solarenergie, Staffelstein, 2004, pp. 68ff.
[1.6] H. Wilk: 'Gebäudeintegrierte Photovoltaiksysteme in Österreich'. 11. Symposium Photovoltaische Solarenergie, Staffelstein, 1996, pp. 173ff.
[1.7] A. Gross: 'Deckel für PV–Entwicklung am Beispiel Österreich'. 19. Symposium PV Solarenergie, Staffelstein, 2004, pp. 23ff.
[1.8] Photovoltaic Insider's Report, 1011 W. Colorado Blvd, Dallas, TX.
[1.9] *Photon – Das Solarstrom-Magazin*. ISSN 1430-5348. Solar-Verlag, Aachen.
[1.10] G.H. Bauer: 'Photovoltaische Stromerzeugung (Kap.5)', in [Win89], p. 139.
[1.11] S. Nowak: 'The IEA PVPS Programme – Towards a Sustainable Global Deployment of PV'. 20th EU PV Conference, Barcelona, 2005.
[1.12] W. Roth, A. Steinhüser: 'Marktchancen der Photovoltaik im Bereich industrieller Produkte und Kleinsysteme'. 11. Symposium Photovoltaische Solarenergie, Staffelstein, 1996, pp. 59ff.
[1.13] IEA-PVPS Task 1 National Survey Report 2007, Germany.
[1.14] IEA-PVPS Task 1 National Survey Report 2007, Austria.
[1.15] IEA-PVPS Task 1 National Survey Report 2007, Japan.

[1.16] Report IEA-PVPS T1-18:2009: Trends in Photovoltaic Applications – Survey Report of Selected IEA Countries
        between 1992 and 2008.
[1.17] 'Baromètre photovoltaïque – Photovoltaic barometer'. Studie von EUROBSERV'ER, March 2009, in *Systèmes
        Solaires – le journal du photovoltaïque*, 1–2009. **Note:** In cases where data from other sources were unavailable,
        the estimated 2008 and 2009 figures from this study were used.
[1.18] 'EPIA's Global Market Outlook'. European Photovoltaic Industry Association, Brussels, April 2009.
[1.19] 'Solarwirtschaft – grüne Erholung in Sicht'. Bank Sarasin, Basle, November 2009.

2

Key Properties of Solar Radiation

2.1 Sun and Earth

For billions of years now, the Sun has been producing energy via nuclear fusion, which converts hydrogen nuclei into helium nuclei. This process releases energy, in accordance with the equation $E = m \cdot c^2$, by virtue of the fact that the mass of the helium nuclei that are produced is somewhat lower than that of the hydrogen nuclei from which the helium nuclei stem. The output of this gargantuan nuclear reactor amounts to roughly $3.85 \cdot 10^{26}$ W or $3.85 \cdot 10^{17}$ GW, or roughly 10^{17} more than the thermal output of a 1200 MW nuclear power plant. Hence solar energy is in fact nuclear energy in the truest sense of the term. But luckily the nuclear reactor we call the Sun is a very great and reassuringly safe distance from the Earth.

The Earth rotates around the Sun once per year in a slightly elliptical orbit at a mean distance of 149.6 million km (see Figure 2.1). Because solar radiation is distributed evenly in all directions, irradiance S_{ex} at the outer edge of the Earth's atmosphere (at a plane that is perpendicular to the direction of the Sun) averages $S_o = 1367 \pm 2$ W/m²; this is known as the solar constant. Moreover, due to the slightly elliptical nature of the Earth's orbit around the Sun, S_{ex} varies somewhat over the course of any given year, reaching a maximum of 1414 W/m² in early January (Earth at perihelion $\approx$ 147.1 million km) and a minimum of 1322 W/m² (Earth at aphelion $\approx$ 152.1 million km).

The amount of solar energy that reaches the Earth is well over 1000 times higher than all of the energy we actually use. In view of the fact that the energy consumed by humans is having dire consequences for the Earth's environment, it is an intriguing as well as very important task for us to capture this mighty flow of energy so that we can use it to power human endeavours.

2.1.1 Solar Declination

The Earth rotates around the Sun once a year and at the same time rotates around its own axis once a day. The Earth's axis is at an angle of 23.45° relative to the plane of the Earth's path around the Sun; this is known as the ecliptic. Because of the Earth's rotation, the Earth's axis retains its orientation in space all year round – which means that over the course of any given year the Sun's rays come from varying directions relative to the equatorial plane. During the summer, the Earth's axis is inclined towards the Sun, and in the winter it is inclined away from the Sun. This phenomenon is known as the **solar declination δ**, which is defined as the angle between the direction of the Sun and the equatorial plane, or the angle at which the Earth's axis is inclined towards the Sun (see Figure 2.2). Hence at a northern hemisphere location at latitude φ, the highest elevation of the Sun at noon is not constant throughout the year, but is instead $h_{Smax} = 90° - \varphi + \delta$. In the northern hemisphere, $\delta > 0$ in summer, but $\delta < 0$ during winter. The amount of solar radiation over the course of a given year is primarily determined by latitude φ and solar declination δ.

Photovoltaics: System Design and Practice. Heinrich Häberlin.
© 2012 John Wiley & Sons, Ltd. Published 2012 by John Wiley & Sons, Ltd.

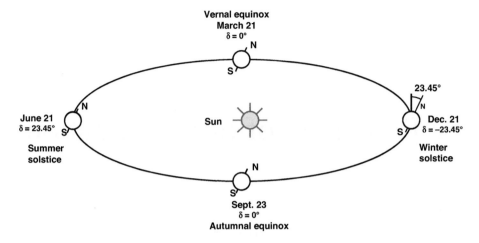

Figure 2.1 The Earth's orbit around the Sun and the position of the Earth's axis over the course of a year

Figure 2.3 shows the solar declination δ over the course of a year, with the days of the year numbered sequentially from 1 to 365, i.e. 1 January is day no. 1 (dy $= 1$) and 31 December is day no. 365 (dy $= 365$).

2.1.2   The Apparent Path of the Sun

To be able to use solar energy, we need to know the apparent path of the Sun over the course of a day, although in most cases we do not need to determine the exact position of the Sun at any given time. This in turn reduces the amount of calculation needed to determine the Sun's path, and the longitude of the location in question can be disregarded.

To compute the Sun's path, we need to know the hour angle ω_S, which is defined as the angle between the local longitude coordinate of the location in question and the latitude coordinate at which the Sun is currently passing through the meridian, i.e. the highest point reached by the Sun's daily path. The fact that the Earth rotates around its axis every 24 hours (1 rotation $= 360°$) results in the following equation for the hour angle ω_S, where ST $=$ true solar time (or local time) in hours (0 to 24 hours), which holds true at the location in question:

$$\text{Hour angle } \omega_S = (\text{ST} - 12) \cdot 15° \tag{2.1}$$

The position of the Sun as seen from the Earth is described via the angles h_S and γ_S (see Figure 2.4). The following equation applies to latitude φ, solar declination δ and hour angle ω_S:

$$\sin h_S = \sin \varphi \sin \delta + \cos \varphi \cos \delta \cos \omega_S \tag{2.2}$$

$$\sin \gamma_S = \frac{\cos \delta \sin \omega_S}{\cos h_S} \tag{2.3}$$

These equations allow for characterization of the Sun's path at any location and time. Figure 2.5 shows the Sun's path on selected days of the year at locations with latitude $\varphi = 47°$. The mean solar path for

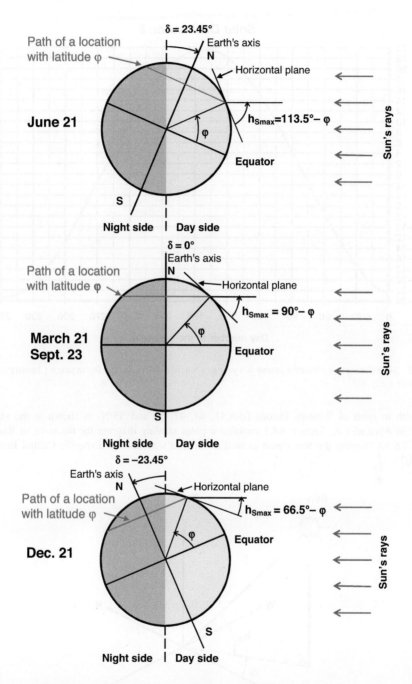

Figure 2.2 Sunlight and the Sun's path at a location at latitude φ in the northern hemisphere at the time of the summer solstice, the equinoxes and the winter solstice. In the southern hemisphere, the latitude is $\varphi < 0°$ and the seasons are reversed, i.e. the summer solstice occurs on 21 December and the winter solstice on 21 June

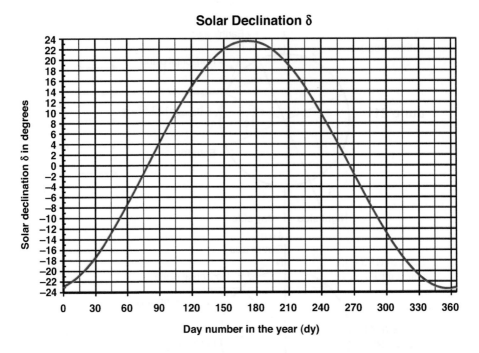

Figure 2.3   Solar declination δ over the course of a year as a function of day number (dy in year): 1 January = day 1;
31 December = day 365

each month in most of Western Europe ($\varphi = 41$, 44, 47, 50 and 53°N) is shown in the shading
diagrams in Appendix A. Section A8.1 contains a polar shading diagram for the town of Burgdorf
(47.1°N, 7.6°E) showing the Sun's path as well as the Sun's position at a specific Central European
Time (CET).

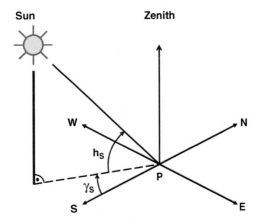

Figure 2.4   Description of the Sun's position (as seen from point P on the Earth) via the angles h_S and γ_S. **Solar
altitude h_S:** Elevation of the Sun above the horizontal plane. **Solar azimuth γ_S:** Angle between the connecting line to
the Sun projected onto the horizontal plane from the south (for $\varphi > \delta$, always applies for $\varphi > 23.5°$) respectively from
the north (for $\varphi < \delta$, always applies for $\varphi < -23.5°$). In this context, γ_S for western values is greater than 0 and for
eastern values is less than 0. (φ = latitude; δ = solar declination.)

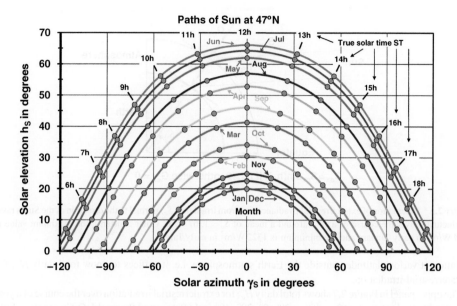

Figure 2.5 Solar paths for the mean day of each month at latitude $\varphi = 47°N$

At sunrise and sunset, $h_S = 0$. Rearranging Equation 2.2 yields the following for sunrise hour angle ω_{SR} and sunset hour angle ω_{SS}:

$$\cos \omega_{SS} = \cos \omega_{SR} = -\tan\varphi \tan\delta \qquad (2.4)$$

Using Equation 2.1, the time of sunrise and sunset, as well as the astronomical duration of sunshine, can be determined from ω_{SS} and ω_{SR}.

2.2 Extraterrestrial Radiation

Using the relationships sketched out in Section 2.1 (see [2.1] and [2.2] for further details), we can determine the **extraterrestrial irradiance** G_{ex} for any location on Earth, i.e. the irradiance at the outer edge of the Earth's atmosphere at a horizontal plane that is parallel to the Earth's surface. G_{ex}, which takes into account the daily fluctuations (resulting from the Earth's rotation) and the annual fluctuations (resulting from the inclination of the Earth's axis and the rotation of the Earth around the Sun) for irradiance, lends itself to mathematical calculation in that only the celestial relationships are taken into account without regard for weather effects.

As Figure 2.6 shows, the following simple relationship exists between G_{ex} and S_{ex}:

$$G_{ex} = S_{ex} \cdot \sin h_S, \text{ where } h_S > 0; \text{ otherwise } G_{ex} = 0 \qquad (2.5)$$

Owing to the fact that, on passing through the Earth's atmosphere, solar radiation is attenuated to varying degrees by reflection, scatter and absorption, the solar radiation that reaches the Earth is normally less than G_{ex}, except when enhanced by clouds. Hence in practice G_{ex} is also the upper limit for possible irradiance G on the horizontal plane of the Earth's surface.

In sizing PV installations or calculating their energy yield, we normally do not need to know the exact curve for irradiance G as a function of time. It often suffices to know total irradiation H (irradiated energy) per unit area for a certain period of time such as a day or month. This also holds true for determining solar

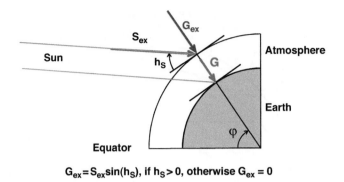

$$G_{ex} = S_{ex}\sin(h_S), \text{ if } h_S > 0, \text{ otherwise } G_{ex} = 0$$

Figure 2.6   Calculation of extraterrestrial irradiance G_{ex} from irradiance S_{ex} normal to the direction of the Sun's rays. S_{ex} fluctuates over the course of a year around a mean of $S_o = 1367\,\text{W/m}^2$ (solar constant); the maximum value is $1414\,\text{W/m}^2$ in early January and the minimum is $1322\,\text{W/m}^2$ in early July

radiation at various latitudes outside the Earth's atmosphere, i.e. it suffices to know total daily H_{ex} for extraterrestrial irradiation.

The upper panel in Figure 2.7 shows total daily H_{ex} for extraterrestrial irradiation over the course of a year at various latitudes ($\varphi = -60°, -40°, -20°, 0°, 20°, 40°$ and $60°$) around the world. Owing to the a little shorter distance between the Earth and Sun, the summer maxima in the southern hemisphere are somewhat higher than in the northern hemisphere. The lower panel in Figure 2.7 contains details concerning the situation in the northern hemisphere in the $\varphi = 38°$ to $56°$ area, where (as the diagram shows) during the summer the H_{ex} maxima are virtually the same, i.e. the longer days at the higher latitudes almost completely compensate for the lower solar altitude. Of course, in winter H_{ex} decreases considerably with increasing latitude φ, because at higher latitudes the solar altitude is lower and the days are shorter.

2.3   Radiation on the Horizontal Plane of the Earth's Surface

As noted in Section 2.2, when solar radiation passes through the Earth's atmosphere it is attenuated by reflection, absorption and scatter. Hence global irradiance G_H on the horizontal plane of the Earth's surface is weaker than extraterrestrial irradiance G_{ex}, and accordingly irradiated energy H_H on the horizontal plane of the Earth's surface is weaker than total extraterrestrial irradiation H_{ex}. In the interest of keeping the number of indices used in this book within reasonable bounds, subscript H (for horizontal) is generally not indicated for either irradiance or total irradiation on the horizontal plane, i.e. $G_H = G$ and $H_H = H$. However, for these parameters on other planes, such as the solar generator plane, a subscript is always indicated in order to avoid confusion.

Global irradiance G on the horizontal plane of the Earth's surface for sunlight that passes through the Earth's atmosphere at 90° (AM1, see Section 1.3.1) with clear skies is still roughly 1 kW/m^2 and in moderate climates such as Central Europe, a mean of AM1.5 is the norm (in Switzerland the AM value at noon varies between about 1 in the summer and 3 in the winter). Global irradiance for solar radiation after traversing AM1.5 is still 835 W/m^2 [Sta87]. On very clear summer days without any atmospheric haze, global irradiance is likewise around 1 kW/m^2 at noon on surfaces that are perpendicular to solar radiation. On a bright day with scattered clouds G can go as high as 1.3 kW/m^2 for a few seconds (a phenomenon known as cloud enhancement).

Sunlight scatter in the Earth's atmosphere substantially attenuates direct sunlight, resulting in diffuse light that comes out of the sky from all directions. This diffuse radiation accounts for roughly half of the total annual solar radiation in the flat areas of Central Europe. Solar radiation at such locations during the winter is diffuse for the most part.

Inasmuch as most PV systems do not concentrate solar radiation, they use not only direct solar radiation but the whole global radiation, which is the aggregate of direct and diffuse solar radiation. The output of such PV systems is roughly proportional to global radiation on the solar generator area.

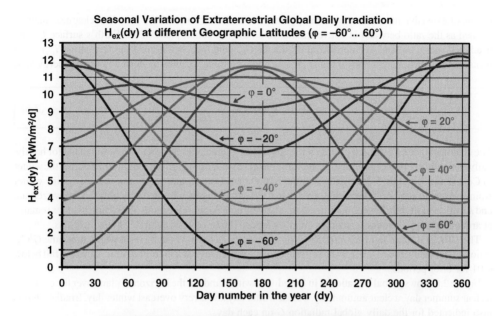

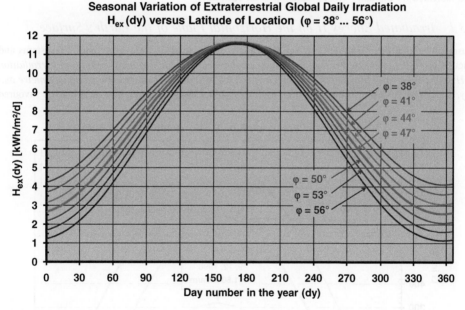

Figure 2.7 Daily extraterrestrial radiation worldwide over the course of a year H_{ex} (dy) at various latitudes as a function of day number (dy), where 1 January = day 1 and 31 December = day 365. Upper panel: world overview graph for $\varphi = -60°, -40°, -20°, 0°, 20°, 40°$ and $60°$. Lower panel: detail graph for the northern hemisphere for $\varphi = 38°$, $41°, 44°, 47°, 50°, 53°, 56°$

Hence global radiation incident on the horizontal plane of the Earth's surface fluctuates daily on account of the Earth's rotation and annually owing to the 23.5° tilt of the Earth's axis relative to the Earth's orbit. However, these two fluctuations, which lend themselves to mathematical calculation and can be used to calculate extraterrestrial radiation G_{ex} and H_{ex} respectively (see Section 2.2), are superimposed by weather-induced fluctuations in irradiance. At most locations, these fluctuations can only be quantified by means of extensive measurements using pyranometers. In this context, the so-called clearness index – defined as the ratio between effective irradiance on the horizontal plane of the Earth's surface G (and irradiated energy H) and extraterrestrial irradiance G_{ex} (and total extraterrestrial irradiation H_{ex}) – can measure the atmosphere's optical transmittance and determine weather-induced power and energy loss:

$$\text{Clearness index } K_H = \frac{H}{H_{ex}} \tag{2.6}$$

The clearness index K_H, which is often referred to in the literature as K_t or Kt, can be indicated for hourly, daily or monthly periods. In Europe, the monthly clearness index ranges from roughly 0.25 during the winter in Central Europe to 0.75 during the summer in Southern Europe, and varies considerably in the winter from one year to another. The clearness index in flat areas during the summer is roughly 0.5 in Switzerland and Austria and slightly lower in Germany. At relatively high Alpine elevations, the clearness index remains at around 0.5 all year round, ranging as high as around 0.6 in very favourable Alpine locations.

The H/H_{ex} ratio can fall below 0.05–0.1 on extremely overcast winter days, at which times the G/G_{ex} ratio can fall below roughly 0.1. In heavy rain or in the presence of storm fronts, G/G_{ex} can even fall below 0.01 for brief periods.

Figure 2.8 shows global radiation G in Burgdorf, Switzerland, on the horizontal plane over the course of a clear summer day, a clear autumn day, a clear winter day and a very overcast winter day. Irradiation H is also indicated for the daily global radiation G on each day.

2.3.1 Irradiated Energy H on the Horizontal Plane of the Earth's Surface

Inasmuch as total daily irradiation H (irradiated energy) normally suffices for sizing PV systems and calculating H, the tables in this book only indicate the *mean monthly values for total daily irradiation of H on the horizontal plane of the Earth's surface*, expressed as *kilowatt hours per square metre and day*, or *kWh/m²/d*. Using kilowatt hours for PV systems in lieu of megajoules, which is favoured

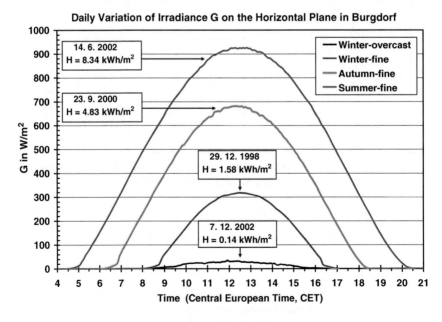

Figure 2.8 Global radiation G on the horizontal plane in Burgdorf, Switzerland (47°N), on a clear summer day (14 June 2002), a clear autumn day (23 September 2000), a clear winter day (29 December 1998) and a very overcast winter day (7 December 2002)

by physicists, is very practical for PV systems as it substantially reduces the number of conversions that are needed.

The figures thus obtained are also highly useful in that they indicate the number of full-load solar hours, i.e. the number of hours during which the Sun must have shone ($G_o = 1 \, kW/m^2$) to radiate this level of energy onto the surface of the solar receiver. The number of full-load solar hours is also referred to as the reference or irradiance yield Y_R (see Chapter 7).

A day should be used as the reference period for calculations involving stand-alone systems, as this allows for sizing of the system without regard for the number of days in a month (n_d). To obtain the monthly amount, multiply the indicated mean daily irradiation for the month in question by the number of days in the month. To obtain the annual amount, multiply the annual mean value (as expressed in $kWh/m^2/d$) by 365.

Table 2.1 lists the mean monthly values for total daily H on the horizontal plane of the Earth's surface at a number of locations in Europe and elsewhere. Section A2 contains more extensive tables in this regard for the following: 35 cities and 16 mountain locations in Switzerland; 27 German cities; 7 Austrian cities; 62 cities elsewhere in Europe; 25 cities in Africa; 27 cities in Asia; 27 cities in North America; 16 cities in Latin America; and 9 cities in Australia and New Zealand. It should be noted that in these tables, in addition to the H values (which are bolded), total daily H_D for diffuse radiation is indicated, as this value is needed to calculate radiation for three-component models. All of these values were computed using [2.4].

It is noteworthy that in Austria and some southern German cities, global radiation values vary considerably for cities that are relatively near each other in the Alps and in flat areas. This phenomenon is mainly attributable to either additional direct beam radiation in the presence of high-pressure systems at higher altitudes, or the frequent absence of fog in inner Alpine valleys. Cities that are located in inner Alpine valleys surrounded by high mountains (e.g. Wallis) also exhibit pronounced global radiation spikes during the summer owing to the relatively clear skies in such areas.

The mean monthly irradiation values indicated fluctuate substantially, particularly during the winter (see Figure 2.9), when more than a 100% difference between the lowest and highest values can occur, although the mean annual values fluctuate less than this. Fluctuations are particularly high in December, which tends to be a low-sunshine month. Figure 2.9 shows the mean monthly global irradiation in Burgdorf from 1992 to 2008 on the horizontal plane of the Earth's surface, expressed as $kWh/m^2/d$. These measurements were realized by the PV Lab at Bern Technical University using a Kipp&Zonen CM 11 pyranometer.

Figures 2.10–2.14 show the mean daily global horizontal radiation in Central Europe for the entire year (Figure 2.10), as well as the months of March (Figure 2.11), June (Figure 2.12), September (Figure 2.13) and December (Figure 2.14). Figures 2.15–2.19 show the mean daily global horizontal radiation for Europe and its surroundings for the entire year (Figure 2.15), as well as the months of March (Figure 2.16),

Table 2.1 Mean monthly values for total daily H on the horizontal plane of the Earth's surface (in $kWh/m^2/d$) for various locations in Europe and elsewhere

Location	Average values for daily irradiation of the horizonal plane in $kWh/m^2/d$												
	Jan	Feb	Mar	April	May	June	July	Aug	Sept	Oct	Nov	Dec	Year
Aswan	4.99	6.00	6.96	7.85	8.25	8.81	8.40	8.04	7.37	6.24	5.32	4.78	6.90
Berlin	0.60	1.20	2.28	3.57	4.85	5.25	5.04	4.32	2.95	1.63	0.72	0.43	2.74
Bern	1.06	1.76	2.79	3.72	4.68	5.20	5.69	4.82	3.56	2.06	1.13	0.84	3.12
Davos	1.68	2.66	4.02	5.01	5.58	5.70	5.88	5.01	3.95	2.78	1.72	1.34	3.77
Jungfraujoch	1.65	2.65	3.97	5.53	6.15	6.42	6.31	5.47	4.46	3.19	2.16	1.57	4.12
Cairo	3.42	4.41	5.56	6.59	7.46	7.96	7.81	7.23	6.28	5.06	3.78	3.10	5.72
Kloten ($\approx$Zurich)	0.91	1.66	2.69	3.77	4.78	5.20	5.59	4.73	3.38	1.92	0.94	0.70	3.02
Locarno	1.42	2.00	3.06	3.57	4.39	5.50	5.95	5.17	3.72	2.17	1.44	1.13	3.29
Marseilles	1.80	2.45	3.89	5.14	6.19	6.96	7.05	6.09	4.63	3.00	1.92	1.49	4.21
Munich	1.03	1.80	2.88	4.01	5.04	5.43	5.40	4.61	3.53	2.13	1.13	0.79	3.14
Seville	2.52	3.26	4.70	5.35	6.62	7.20	7.58	6.51	5.38	3.86	2.50	2.16	4.80
Vienna	0.86	1.54	2.71	3.81	5.12	5.38	5.45	4.67	3.22	2.09	0.96	0.65	3.03

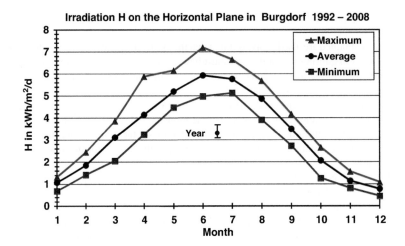

Figure 2.9 Mean monthly values for global daily irradiation H in Burgdorf from 1992 to 2008 on the horizontal plane of the Earth's surface, expressed in kWh/m²/d. The lowest, mean and highest readings are indicated for each month during the period. The elevated peak for April resulted from an exceptionally sunny April 2007, as was also the case for the April reading in Figure 2.26

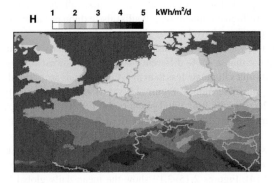

Figure 2.10 Mean annual global irradiation H on the horizontal plane of the Earth's surface in Central Europe (in kWh/m²/d)

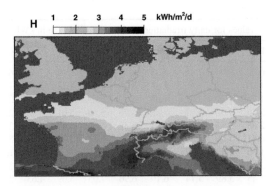

Figure 2.11 Mean global irradiation H on the horizontal plane of the Earth's surface in the month of March in Central Europe (in kWh/m²/d)

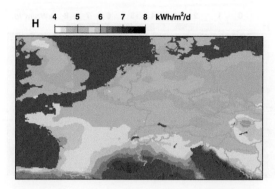

Figure 2.12 Mean global irradiation H on the horizontal plane of the Earth's surface in the month of June in Central Europe (in kWh/m²/d)

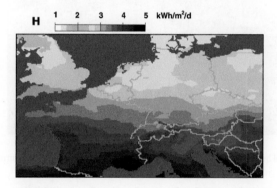

Figure 2.13 Mean global irradiation H on the horizontal plane of the Earth's surface in the month of September in Central Europe (in kWh/m²/d)

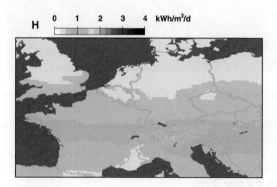

Figure 2.14 Mean global irradiation H on the horizontal plane of the Earth's surface in the month of December in Central Europe (in kWh/m²/d)

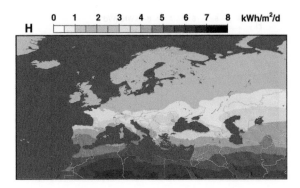

Figure 2.15 Mean annual global irradiation H on the horizontal plane of the Earth's surface in Europe and its surroundings (in kWh/m²/d)

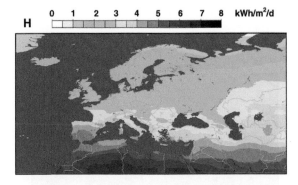

Figure 2.16 Mean global horizontal irradiation in the month of March in Europe and its surroundings (in kWh/m²/d)

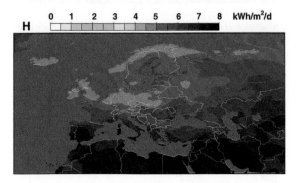

Figure 2.17 Mean global horizontal irradiation in the month of June in Europe and its surroundings (in kWh/m²/d)

June (Figure 2.17), September (Figure 2.18) and December (Figure 2.19). These maps were created using the *European Solar Radiation Atlas (ESRA)* [2.6].

In reading these maps (in colour), it should be borne in mind that the same colour was not used in all cases for a specific radiation level, so as to allow for the representation of minute gradations. The scale for each map is indicated on the map itself.

And for the counterpart maps for Europe and elsewhere, see Figures 2.15–2.19.

Section A8 contains additional maps showing annual global horizontal irradiation in kWh/m²/d. Figures A8.2–A8.9 show annual global radiation respectively as follows: the entire world; the Alpine

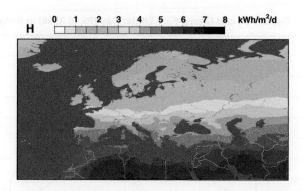

Figure 2.18 Mean global horizontal irradiation in the month of September in Europe and its surroundings (in kWh/m²/d)

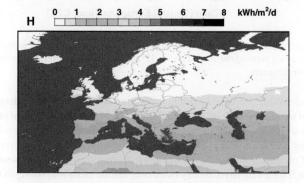

Figure 2.19 Mean global horizontal irradiation in the month of December in Europe and its surroundings (in kWh/m²/d)

countries; Germany; Africa; Asia; the Oceanic region; North America; Latin America. Figure A8.10 shows the annual energy yield for an optimally oriented PV system in Europe [2.7].

2.4   Simple Method for Calculating Solar Radiation on Inclined Surfaces

We have so far discussed irradiance on horizontal surfaces. Irradiance on a solar generator and thus its energy yield can be increased in higher latitudes by tilting the generator towards the Sun, i.e. at an angle β relative to the horizontal plane (see Figure 2.20). This allows for optimized use of direct beam radiation.

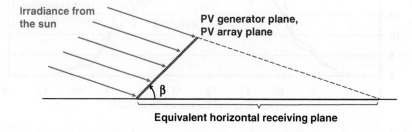

Figure 2.20   Tilting a solar generator at an angle β relative to the horizontal plane can increase irradiance and therefore also irradiation at the solar generator plane

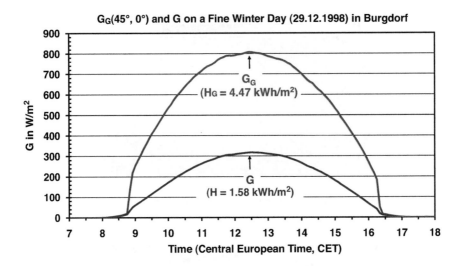

Figure 2.21   Irradiance G measured in Burgdorf over the course of a very clear winter day on the horizontal plane and irradiance G_G on a tilted surface pointing due south with $\beta = 45°$ and $\gamma = 0°$. The measurements were obtained using two Kipp&Zonen CM 11 pyranometers

The potentially achievable increase in irradiance varies considerably depending on the angle of incidence, season and weather conditions. Figure 2.21 shows the irradiance measured in Burgdorf (a) over the course of a clear winter day on the horizontal plane on a surface pointing due south at an angle of incidence $\beta = 45°$ and (in Figure 2.22) (b) under the same conditions on a very overcast day.

H_G far exceeds H on a clear winter day owing to he optimized use of direct beam radiation, but is somewhat lower than H on a very overcast day (with only diffuse radiation) owing to the lack of diffuse radiation on the tilted surface (also see Section 2.5).

This method greatly increases energy yield, notably in low-fog areas (see Figure 2.23), and in all settings somewhat reduces diffuse irradiance on the solar generator, which under certain circumstances

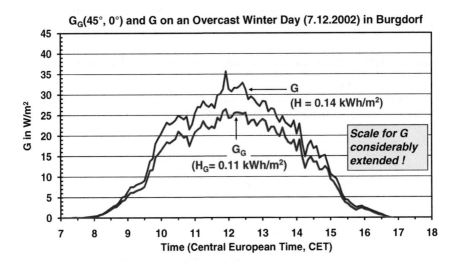

Figure 2.22   Irradiance G measured in Burgdorf over the course of a very overcast winter day on the horizontal plane and irradiance G_G on a tilted surface pointing due south with $\beta = 45°$ and $\gamma = 0°$. The measurements were obtained using two Kipp&Zonen CM 11 pyranometers

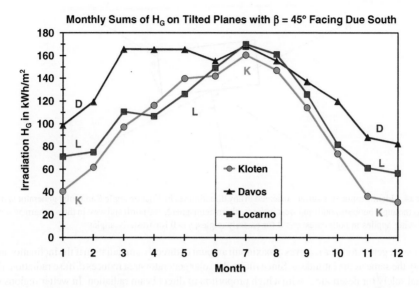

Figure 2.23   Monthly global radiation H_G on a solar generator area that is pointing due south with inclination angle $\beta = 45°$ relative to the horizontal plane. These measurements were taken in the following locations: Kloten, which is a typical city in the eastern portion of Switzerland's Mittelland region; Davos, which is a typical Alpine city; and Locarno, which is a typical city on the southern slopes of the Alps

can be partially offset by high levels of diffuse ground reflection radiation resulting from the presence of snow, light-coloured concrete and the like.

The winter energy yields indicated for an inclined surface in Figure 2.23 are considerably higher than the calculable total monthly H obtained by multiplying the data in Table 2.1 by n_d (number of days in the month). In other words, tilting the solar generator area greatly increases monthly energy yield during the winter.

Direct beam radiation on solar generator areas peaks when the Sun is positioned vertically above the generator surface. Because solar altitude and azimuth vary over the course of any given day, a complex biaxial solar tracking mechanism is needed to maintain this solar-radiation-maximized state, which can increase energy yield by roughly 25 to 40% (see Section A3, Figure A8.11 and [2.11]). This strategy increases the possible number of full-load hours by a comparable amount, i.e. more solar energy can be fed into the grid before problems with peak load in the grid arise (also see Section 5.2.7).

However, this increase can only be achieved if the various devices do not shadow each other when the Sun is low, which means that the devices must exhibit sufficient reciprocal clearance; but for this, a relatively large property is needed. Solar tracking systems are mainly suitable for use in open spaces, and work less well on buildings. Moreover, in view of the considerably higher installation and maintenance costs entailed by solar tracking systems, the lower reliability of more complex systems and the fact that solar cell prices are decreasing steadily, before installing a solar tracking system it is advisable to determine if perhaps the money invested in such a system might not be better spent on additional solar modules.

Therefore solar generators are normally installed at a fixed inclination angle β and in the northern hemisphere are oriented southwards as far as possible (20° to 30° deviation from due south reduces the energy yield only to a minor extent). Seasonal adjustment of the inclination angle (e.g. in Switzerland, 55° in winter and 25° in summer) of a solar generator area that is pointing due south increases the energy yield by a few percentage points, but entails a somewhat more complex mounting structure.

For installations without seasonal variation of inclination with a fixed angle β, the question of course arises as to the optimal value for β. In locations with ideal conditions (i.e. clear, sunny weather all year

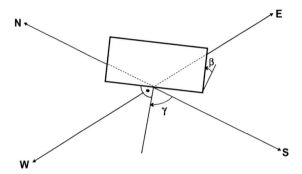

Figure 2.24   Identification of a surface oriented in any direction via inclination angle β and solar generator azimuth γ, where γ is the angle between south and west in the northern hemisphere, and north and west in the southern hemisphere. The following applies in both cases: $\gamma < 0$ for easterly angles; $\gamma > 0$ for westerly angles

round), a solar generator area receives a maximum amount of direct beam radiation if the inclination angle β is roughly the same as the latitude φ. Since tilting the solar generator area reduces diffuse radiation, $\beta = |\varphi|$ is optimal solely for desert areas with a high proportion of direct beam radiation. In wetter regions with a high proportion of diffuse radiation (i.e. flat areas in Central Europe), a lesser incline such as $\beta = 30°$ is preferable in cases where the solar generator area is rarely covered with snow owing to the mild local climate. On the other hand, a steeper angle of incidence such as $\beta = 60°$ increases winter energy yield and makes it easier for snow to slide off the solar generator area in harsher winter climates. A steeper inclination angle should definitely be used in mountainous regions owing to the frequent snowfall there.

For solar generator areas at any angle or in any orientation (inclination angle β, azimuth γ, see Figure 2.24), monthly irradiated energy on the solar generator plane can be calculated using the *global irradiation factor* $R(\beta,\gamma)$, based on irradiated energy H on the horizontal plane [2.1], [2.2], [Bur83], [Häb91], [Häb07], [Lad86]:

$$\text{Irradiated energy incident on the solar generator plane}: \ H_G = R(\beta, \gamma) \cdot H \qquad (2.7)$$

This equation can of course also be used to determine the monthly mean daily global irradiation and for monthly global irradiation.

The global irradiation factor $R(\beta,\gamma)$ is determined by β, γ and the ratio of diffuse to global radiation, i.e. in principle by local climate, and can of course not be determined for any location. Defined in [2.1] are three reference stations for Switzerland (Kloten, Davos and Locarno) that allow for approximate characterization of the various types of climate in Switzerland and their $R(\beta,\gamma)$ values in their respective designated zones for the calculation of irradiation on tilted surfaces. This method can be applied to Germany as well by simply adding the relevant location in southern Germany (Munich), central and western Germany (Giessen) and northeast Germany (Potsdam). The method can be applied to Austria, which is at approximately the same latitude as Switzerland and is also an Alpine country, through use of the $R(\beta,\gamma)$ values from Switzerland and southern Germany. Also provided in are the relevant values for Marseilles, Seville and Cairo that can be used for rough calculations for Southern Europe, North Africa and the Middle East.

For reasons of space, no reference station information could be provided in this book for the tropics or the southern hemisphere, where the three-component model described in Section 2.5 should be used. In view of the six-month time difference in the advent of the seasons, values from reference stations at similar latitudes can also be used. To be on the safe side, the Kloten $R(\beta,\gamma)$ values should be used for the Mitteland region of Switzerland, as well as for southern Germany and Austria, since Kloten has a relatively foggy climate and thus an above-average proportion of diffuse radiation. In most flat regions (apart from those in close proximity to bodies of water), the proportion of direct beam radiation and thus the $R(\beta,\gamma)$ values are somewhat higher in winter.

Table 2.2 Regions typified by the reference stations listed in Section A3

Reference station	Region typified by the reference station
Davos	Regions at an elevation of 1200 m or higher in Switzerland, southern Germany, Austria, France and Italy
Giessen	Locations in northwest Germany, northern France and Benelux countries
Cairo	Locations in northeast Africa, the Middle East and the Sahara
Kloten	Flat regions in Switzerland, southern Germany, Austria and central France
Locarno	Locations on the southern slopes of the Alps and in inner Alpine valleys (e.g. Wallis)
Marseilles	Locations in southern France, northern Spain, and northern and central Italy
Munich	Locations in southern Germany and in Austria's inner Alpine valleys
Potsdam	Locations in northern and eastern Germany and in Poland
Seville	Locations in southern Spain, southern Italy, the North African coast (W) and Greece

Table 2.2 lists the selected global irradiation factors $R(\beta,\gamma)$ for the aforementioned nine reference stations for solar generator surface that are pointing due south. Deviations from due south amounting to $20°$ to $30°$ have little effect on the R values.

Section A3 contains a more detailed table of $R(\beta,\gamma)$ values for the nine reference stations, as well as for other orientations, including those that deviate from due south.

The R values in Tables 2.3 and 2.4 were calculated using [2.4] and may differ slightly from the values based on older data in [Häb91] or [Lad86].

Table 2.3 Global irradiation factors $R(\beta,\gamma)$ for southerly oriented solar generator surfaces for each month of the year for the nine reference stations

		\multicolumn{13}{c}{Global irradiation factors $R(\beta,\gamma)$ for tilted planes facing due south ($\gamma = 0°$)}

Location	β	Jan	Feb	Mar	April	May	June	July	Aug	Sept	Oct	Nov	Dec	Year
	30°	1.67	1.46	1.27	1.11	1.01	0.98	0.99	1.05	1.16	1.32	1.53	1.73	1.16
Davos	45°	1.90	1.59	1.33	1.10	0.95	0.91	0.92	1.00	1.15	1.39	1.69	1.98	1.17
	60°	2.01	1.66	1.32	1.04	0.85	0.80	0.82	0.91	1.09	1.39	1.78	2.13	1.12
	30°	1.33	1.31	1.16	1.09	1.02	0.99	1.00	1.05	1.13	1.22	1.32	1.35	1.09
Giessen	45°	1.44	1.39	1.16	1.06	0.96	0.92	0.94	1.01	1.13	1.26	1.38	1.45	1.06
	60°	1.48	1.41	1.12	0.99	0.87	0.82	0.84	0.92	1.07	1.24	1.41	1.50	0.99
	30°	1.33	1.23	1.12	1.01	0.93	0.90	0.91	0.97	1.07	1.20	1.31	1.37	1.06
Cairo	45°	1.40	1.26	1.08	0.94	0.83	0.78	0.80	0.88	1.03	1.20	1.36	1.44	1.01
	60°	1.39	1.21	1.00	0.82	0.68	0.62	0.65	0.75	0.93	1.15	1.34	1.44	0.91
	30°	1.34	1.28	1.16	1.06	1.01	0.98	1.00	1.05	1.13	1.21	1.26	1.34	1.08
Kloten	45°	1.42	1.33	1.17	1.03	0.94	0.91	0.93	1.00	1.13	1.24	1.31	1.45	1.05
	60°	1.47	1.33	1.12	0.95	0.84	0.80	0.82	0.91	1.06	1.21	1.31	1.48	0.98
	30°	1.47	1.27	1.16	1.03	0.99	0.98	0.99	1.06	1.14	1.20	1.33	1.49	1.10
Locarno	45°	1.61	1.33	1.16	0.99	0.93	0.90	0.92	1.00	1.13	1.22	1.42	1.62	1.08
	60°	1.68	1.32	1.11	0.91	0.83	0.79	0.81	0.91	1.06	1.19	1.43	1.68	1.01
	30°	1.57	1.35	1.22	1.09	1.01	0.97	0.99	1.06	1.18	1.31	1.49	1.61	1.14
Marseilles	45°	1.73	1.43	1.24	1.06	0.94	0.89	0.91	1.01	1.18	1.37	1.61	1.79	1.12
	60°	1.80	1.43	1.20	0.97	0.82	0.76	0.79	0.91	1.11	1.35	1.65	1.89	1.05
	30°	1.47	1.35	1.19	1.08	1.01	0.98	1.00	1.05	1.16	1.27	1.38	1.48	1.11
Munich	45°	1.60	1.44	1.21	1.05	0.95	0.92	0.93	1.01	1.15	1.31	1.47	1.64	1.09
	60°	1.70	1.47	1.18	0.97	0.85	0.81	0.83	0.92	1.10	1.30	1.51	1.73	1.02
	30°	1.40	1.30	1.19	1.09	1.03	1.00	1.01	1.07	1.16	1.26	1.33	1.44	1.10
Potsdam	45°	1.56	1.38	1.21	1.07	0.98	0.94	0.95	1.03	1.16	1.32	1.43	1.56	1.08
	60°	1.60	1.40	1.18	0.99	0.89	0.84	0.85	0.95	1.11	1.31	1.47	1.61	1.01
	30°	1.43	1.28	1.17	1.04	0.97	0.94	0.95	1.01	1.13	1.25	1.35	1.46	1.10
Seville	45°	1.53	1.32	1.16	0.99	0.89	0.84	0.86	0.95	1.10	1.28	1.42	1.58	1.07
	60°	1.56	1.30	1.10	0.89	0.76	0.71	0.72	0.83	1.02	1.24	1.42	1.61	0.98

Table 2.4 Blank table for H_G calculation using the simple method described in the text

	Jan	Feb	Mar	April	May	June	July	Aug	Sept	Oct	Nov	Dec	Year	Unit
H														kWh/m²
$R(\beta,\gamma)$														
H_G														kWh/m²

When using Equation 2.7, it is essential that the correct H values are used, namely *those of the installation location* and not of the reference station. *Only* the $R(\beta,\gamma)$ values from the reference station should be used.

For calculating H_G values it is best to use Table 2.4 or Table A1.1, which contain the monthly H values and the $R(\beta,\gamma)$ factors. H_G can then be readily computed as the product of the two aforementioned factors.

Total annual H_G is determined by tallying the monthly totals. Ideally, the calculation of mean annual H_{Ga} of total daily H_G should be based on total annual irradiation (see Section 2.5.6), although what is actually obtained is simply the mean of the monthly mean of total daily irradiation, with a very minor error.

Like irradiation on the horizontal plane H, the global irradiation factor $R(\beta,\gamma)$ and thus irradiation H_G are subject to considerable statistical fluctuation, particularly in the winter. Figure 2.25 shows, for 1992–2008, the global irradiation factor $R(45°,0°)$, while Figure 2.26 shows the corresponding irradiation H_G on a plane where $\beta = 45°$ and $\gamma = 0°$ (measured in Burgdorf, which is at 47°N). Inasmuch as low H values translate into low $R(\beta,\gamma)$ values, the H_G fluctuations are usually greater than for H.

2.4.1 Annual Global Irradiation Factors

As is the case with the mean monthly values and monthly totals, an *annual global irradiation factor* $R_a(\beta,\gamma)$ can be determined for the *mean annual values or annual totals H_{Ga}*. To this end, the mean annual value or total annual H_{Ga} for global irradiation on the solar generator plane can be determined directly from the mean annual or total annual H_a for global irradiation on the horizontal plane:

$$\text{Annual incident irradiation on the solar generator plane:} \quad H_{Ga} = R_a(\beta, \gamma) \cdot H_a \qquad (2.8)$$

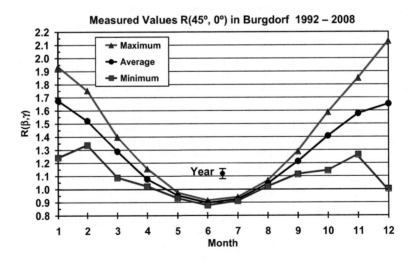

Figure 2.25 Mean monthly global irradiation factors $R(\beta,\gamma)$ for 1992–2008 as measured by the Bern Technical University PV Lab, for $\beta = 45°$ and $\gamma = 0°$ (readings from Burgdorf). The lowest, mean and highest values are indicated for each month. Measurement scatter tends to be relatively high, notably in the winter, whereas the mean annual value is very constant. *Note*: Measurement readings obtained with Kipp&Zonen CM 11 pyranometers

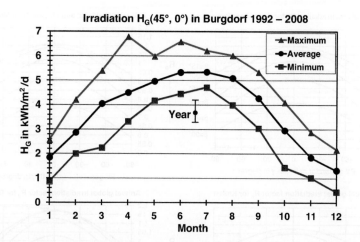

Figure 2.26 Mean monthly global irradiation for 1992–2008, as measured at the Bern Technical University PV Lab, in a plane $\beta = 45°$ and $\gamma = 0°$ (readings from Burgdorf). The lowest, mean and highest values are indicated for each month. Measurement scatter was extremely pronounced throughout the winter, where the ratio between the highest and the lowest value varied between about 2 and 5. The relative fluctuation was far lower during the summer (the elevated April value is attributable to an unusually sunny April 2007). *Note:* Measurement readings obtained with Kipp&Zonen CM 11 pyranometers

Note: An annual global radiation factor R_a should not be based on the mean of the monthly R factor, but should instead comprise the quotient of the correctly computed annual H_G and H totals for the relevant reference station.

Use of the annual global irradiation factor R_a of a reference station for rapid H_{Ga} determination is very quick and efficient as no monthly values need be computed. However, since in most cases the monthly H values at the station in question differ somewhat from that of the reference station, the H_{Ga} value thus determined is somewhat less exact than the result obtained (albeit at the cost of a far more extensive calculation) by determining the annual total from the monthly H_G totals. Hence, if the monthly H_G totals have already been calculated, it is best to determine the annual total from this monthly total.

Figures 2.27 and 2.28 show this for the annual global radiation factors $R_a(\beta,\gamma)$ calculated using [2.4], for the reference stations north and south of the Alps.

As these graphs show, the maximum annual yield in flat Central European regions is roughly $\beta \approx 30°$ and $\gamma \approx 0°$, whereas in the Alpine location of Davos, $\beta \approx 40°$ is optimal for obtaining maximum energy yield. Despite its southern location, in the city of Marseilles a somewhat steeper angle $\beta \approx 33°$ is optimal, while $\beta \approx 28°$ and 24° are optimal in Cairo and Seville respectively. The maxima are relatively shallow, i.e. up to 15° deviation from the optimum with β and up to 30° deviation from due south with γ, and have only a negligible effect on $R_a(\beta,\gamma)$.

In flat areas, where the lion's share of annual energy yield is generated during the summer, a slight eastward deviation ($\gamma = -5°$ to $-10°$) may even be optimal. In contrast to the older model from [2.1], the calculation model that was used in [2.4] does not take account of the fact that irradiation on summer afternoons is often somewhat lower than in the morning due to thunderstorms.

The smallest inclination angle that should be used for PV systems is $\beta \approx 20°$ so as to allow rainfall to clean the solar generator area and so that winter snow can slide off the solar generator.

Inasmuch as solar module efficiency decreases somewhat at higher temperatures, the energy efficiency of PV systems is somewhat lower in the summer than in the winter, i.e. irradiated energy

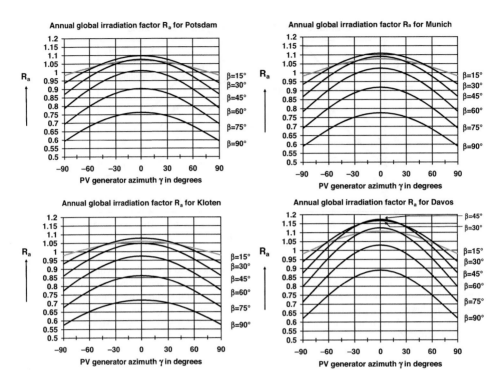

Figure 2.27 Annual global irradiation factors for four reference stations north of the Alps. Note that the Giessen graph (not shown) and Kloten graph are virtually the same. The values shown here were computed using [2.4]

on the solar generator will be used less efficiently in the summer than in the winter. Moreover, in the presence of higher β values winter snow slides off the solar generator sooner, dirt striations at the lower edges of the modules are narrower, and rainfall cleans the solar generators more efficiently.

As a result of many years of observing numerous PV systems, I have reached the conclusion that, in Central Europe, in the long term somewhat steeper angles of incidence are often more efficient than the optimal roughly 30° angle for maximum total annual irradiation on the generator surface. As Figures 2.27 and 2.28 show, in Europe the difference in the annual global radiation factor $R_a(\beta,\gamma)$ for $\beta = 30°$ and 45° is only 2 to 3%. Many PV systems acquire a permanent layer of dirt over the years despite rain- and snow-induced self-cleaning, which is far more efficient at an inclination angle around 45° than with small inclination angles. Hence, all other conditions being equal, such PV systems tend to be less dirty over the long term. This permanent layer of solar generator dirt, which develops within a few years especially at the lower edges, can provoke a power loss of up to 10%, which is far greater than the minor loss caused by a slightly steeper angle of incidence.

Hence the cause of long-term optimal PV system energy yield (in kWh/kWp) is better served through the use of a somewhat steeper angle of incidence than the angles obtained via irradiance optimization. In short, 35°–40° angles should be used in flat areas and 45°–50° should be used in low-altitude Alpine locations. For higher-altitude Alpine locations, steeper angles of incidence ranging from 60° to 90° are best, owing to the frequency of winter snow covering in such areas. From an economic efficiency standpoint (optimal power yield during the winter in lieu of maximum annual power yield) the ideal inclination angle in Central Europe is probably in the region of 45°–60°.

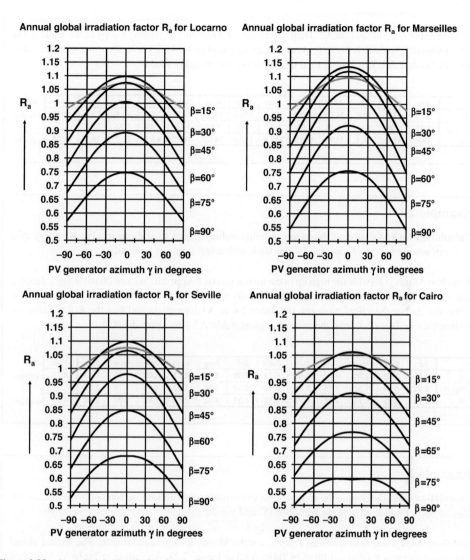

Figure 2.28 Annual global irradiation factors for four reference stations south of the Alps. The values shown here were computed using [2.4]

2.4.2 Elementary Radiation Calculation Examples for Inclined Surfaces

Example 1

Calculation task: Determine the total monthly and annual irradiated energy H_G for a 30° inclination angle, at a site in Berlin that faces south.

Solution: Use the Potsdam reference station, since the site in question is located in a flat northern German area. First determine total monthly H by multiplying n_d (number of days per month) by the relevant H value for irradiance on the horizontal plane, which can be found in Table 2.1 or in the more detailed Table A2.3 in Section A2. Then determine total monthly H_G (using a copy of the Table 2.4 or A1.1 template), for which the $R(\beta,\gamma)$ values can be found in Table 2.3 or in the more detailed Table A3.2 in Section A3. The annual total is the sum of the monthly totals.

	Jan	Feb	Mar	April	May	June	July	Aug	Sept	Oct	Nov	Dec	Year	Unit
H	18.6	33.6	70.7	107.1	150.4	157.5	156.2	133.9	88.5	50.5	21.6	13.3	1000	kWh/m²
$R(\beta,\gamma)$	1.40	1.30	1.19	1.09	1.03	1.00	1.01	1.07	1.16	1.26	1.33	1.44		
H_G	26.0	43.7	84.1	116.7	154.9	157.5	157.8	143.3	102.7	63.7	28.7	19.2	1098	kWh/m²

Example 2

Calculation task: Determine the mean monthly values H_G of total daily irradiated energy at a southern oriented site in Montana, Switzerland, with a 60° angle of incidence.

Solution: As this is a Swiss site high up in the mountains (see Appendix A), use Davos as the reference station. Find the mean monthly values for the daily total of H for Swiss locations in Table A2.2 in Appendix A. For the calculation, use the Table 2.4 or A1.1 template and find the $R(\beta,\gamma)$ value for Switzerland in Table 2.3 or in the more complete Table A3.1 in Appendix A.

	Jan	Feb	Mar	April	May	June	July	Aug	Sept	Oct	Nov	Dec	Year	Unit
H	1.61	2.53	3.88	4.77	5.61	5.74	6.08	5.28	4.13	2.80	1.68	1.35	3.78	kWh/m²
$R(\beta,\gamma)$	2.01	1.66	1.32	1.04	0.85	0.80	0.82	0.91	1.09	1.39	1.78	2.13		
H_G	3.24	4.20	5.12	4.96	4.77	4.59	4.99	4.80	4.50	3.89	2.99	2.88	4.24	kWh/m²

Example 3

Calculation task: Determine the annual mean of the daily totals and annual total of irradiated energy H_{Ga} in Nice, France, on a plane with $\beta = 45°$ and $\gamma = 30°$.

Solution: As this site is in southern France, use the Marseilles reference station. The annual global irradiation factor is $R_a(\beta,\gamma) = 1.08$ as in Table A3.3 or Figure 2.28. According to Table A2.5, in Nice $H_a = 4.03\,\text{kWh/m}^2/\text{d}$ or $1471\,\text{kWh/m}^2/\text{a}$.
Hence $H_{Ga} = R_a(\beta,\gamma) \cdot H_a = 4.35\,\text{kWh/m}^2/\text{d}$ or $1589\,\text{kWh/m}^2/\text{a}$.

Example 4

Calculation task: Determine annual H_{Ga} in Cairo at a southerly oriented site with a 30° angle of incidence.

Solution: Use Cairo as a reference, since this site is in the Middle East. The annual global irradiation factor is $R_a(\beta,\gamma) = 1.06$ as in Table A3.3 or Figure 2.28. According to Table 2.1 or the more complete Table A2.7 in Appendix A, in Cairo $H_a = 5.72\,\text{kWh/m}^2/\text{d}$ or $2088\,\text{kWh/m}^2/\text{a}$.
Hence $H_{Ga} = R_a(\beta,\gamma) \cdot H_a = 6.06\,\text{kWh/m}^2/\text{d}$ or $2213\,\text{kWh/m}^2/\text{a}$.

2.5 Radiation Calculation on Inclined Planes with Three-Component Model

Although the irradiance calculation method for inclined surfaces described in Section 2.4 is simple, it cannot take precise account of local climate differences, since the number of reference stations is limited. Inasmuch as reference station R factors vary greatly, particularly in the winter, intermediate levels are needed. Another shortcoming of the method discussed in the previous section is that horizon shading cannot be taken into account with a reasonable amount of effort.

The three-component model, which breaks down inclined-surface irradiance into direct, diffuse and reflected radiation, allows for considerably more precise calculation of irradiance on inclined surfaces [2.1], [2.2], [Bur83], [Eic01], [Luq03], [Mar94], [Qua03], [Win91].The model considerably improves accuracy, notably for mean and total monthly irradiation, and obviates the need to use reference stations. Moreover, as the horizon or nearby buildings mainly affect direct beam radiation, the three-component model also takes approximate account of shading at any angle of inclination β. Using this model to calculate monthly winter figures is also a safer bet, as the model tends to avoid overestimation of winter energy yield. Although use of the three-component model is more labour intensive than the simple model discussed in the previous section, this workload can be greatly reduced if both global irradiation H and diffuse irradiation H_D in the horizontal plane are known for the site in question. To this end, Tables A2.1–A2.10 indicate both of these irradiation values for various locations. These tables were prepared using [2.4]. Table 2.5 contains these figures for a number of locations, while more complete tables (Tables A2.1–A2.10) in this regard can be found in Appendix A. In these tables, the H values are bolded and the H_D values are in regular typeface. The present book contains all the figures necessary for use of the three-component model at any site between 60°N and 40°S.

2.5.1 Components of Global Radiation on the Horizontal Plane

Not all of the global irradiance G arriving at the horizontal plane of the Earth's surface is direct beam irradiance G_B. To varying degrees, extraterrestrial solar irradiance G_{ex} is reflected, absorbed or scattered as it traverses the Earth's atmosphere, and some of this solar radiation that is reflected by the clouds or scattered reaches the Earth's surface as diffuse irradiance G_D. As can be readily observed on a very overcast day, the intensity of diffuse radiation is independent of direction, which means that in a low-horizon location, diffuse radiation comes from all directions with equal intensity.

Table 2.5 Mean values for the daily total H of global irradiation and H_D for diffuse irradiation on the horizontal plane for various locations (for further locations see Section A2); all figures are in kWh/m²/d

Location		Jan	Feb	Mar	April	May	June	July	Aug	Sept	Oct	Nov	Dec	Year
Berlin	**H**	**0.60**	**1.20**	**2.28**	**3.57**	**4.85**	**5.25**	**5.04**	**4.32**	**2.95**	**1.63**	**0.72**	**0.43**	**2.74**
52.5°N, 13.3°E, 33 m	H_D	0.45	0.82	1.42	2.06	2.57	2.80	2.69	2.28	1.69	1.05	0.54	0.34	1.56
Bern (-Liebefeld)	**H**	**1.06**	**1.76**	**2.79**	**3.72**	**4.68**	**5.20**	**5.69**	**4.82**	**3.56**	**2.06**	**1.13**	**0.84**	**3.12**
46.9°N, 7.4°E, 565 m	H_D	0.71	1.09	1.63	2.19	2.63	2.81	2.69	2.36	1.86	1.27	0.78	0.58	1.72
Davos	**H**	**1.68**	**2.66**	**4.02**	**5.01**	**5.58**	**5.70**	**5.88**	**5.01**	**3.95**	**2.78**	**1.72**	**1.34**	**3.77**
46.8°N, 9.8°E, 1560 m	H_D	0.78	1.12	1.62	2.23	2.68	2.87	2.78	2.43	1.91	1.34	0.88	0.66	1.77
Cairo	**H**	**3.42**	**4.41**	**5.56**	**6.59**	**7.46**	**7.96**	**7.81**	**7.23**	**6.28**	**5.06**	**3.78**	**3.10**	**5.72**
30.1°N, 31.2°E, 16 m	H_D	1.26	1.47	1.76	1.99	2.05	2.01	1.99	1.89	1.73	1.50	1.30	1.18	1.68
Marseilles	**H**	**1.80**	**2.45**	**3.89**	**5.14**	**6.19**	**6.96**	**7.05**	**6.09**	**4.63**	**3.00**	**1.92**	**1.49**	**4.21**
43.3°N, 5.4°E, 5 m	H_D	0.79	1.11	1.49	1.90	2.16	2.18	2.02	1.85	1.58	1.24	0.87	0.70	1.49
Seville	**H**	**2.52**	**3.26**	**4.70**	**5.35**	**6.62**	**7.20**	**7.58**	**6.51**	**5.38**	**3.86**	**2.50**	**2.16**	**4.80**
37.4°N, 6.0°W, 30 m	H_D	1.08	1.41	1.75	2.22	2.37	2.40	2.15	2.11	1.82	1.51	1.19	0.99	1.75
Vienna	**H**	**0.86**	**1.54**	**2.71**	**3.81**	**5.12**	**5.38**	**5.45**	**4.67**	**3.22**	**2.09**	**0.96**	**0.65**	**3.03**
48.2°N, 16.4°E, 170 m	H_D	0.62	1.01	1.59	2.17	2.61	2.81	2.71	2.35	1.81	1.24	0.69	0.48	1.67

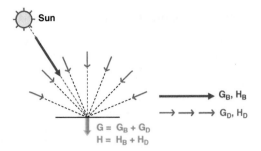

Figure 2.29 Components of global radiation G and H on the horizontal plane, as the sum of direct beam radiation (G_B and H_B) and diffuse radiation (G_D and H_D)

Hence global irradiance G on the horizontal plane comprises the total of direct beam and diffuse irradiance (see Figure 2.29):

$$\text{Global irradiance on the horizontal plane}: \; G = G_B + G_D \qquad (2.9)$$

Of course this same equation applies to incident energy arriving at the horizontal plane:

$$\text{Irradiated energy arriving at the horizontal plane}: \; H = H_B + H_D \qquad (2.10)$$

2.5.2   Radiation Reflected off the Ground

Some of the solar radiation that arrives at the Earth's surface is reflected by it, for the most part diffusely. This usually has no bearing on the calculation of radiation on a horizontal plane, except if it is surrounded by very high snow-covered mountains. However, as an additional component this reflected radiation is relevant for calculating irradiance on inclined surfaces.

The reflected irradiance G_R from a non-reflecting surface is diffuse and at the same time proportional to irradiance G onto this surface. This reflected irradiance distributes itself in the half space just above the surface (see Figure 2.30). In conjunction with reflection factor ρ, the following applies to the irradiance G_R that is reflected by the aforementioned surface:

$$\text{Irradiance reflected by the ground}: \; G_R = \rho \cdot G \qquad (2.11)$$

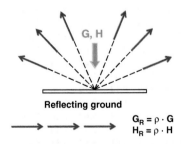

Figure 2.30   Formation of reflected radiation G_R and H_R from the global radiation G and H reflected off the ground

Table 2.6   Guide values for the reflection factor ρ

Type of surface	Reflection factor ρ (albedo)
Asphalt	0.1–0.15
Green forest	0.1–0.2
Wet ground	0.1–0.2
Dry ground	0.15–0.3
Grass-covered ground	0.2–0.3
Concrete	0.2–0.35
Desert sand	0.3–0.4
Old snow (depending on how dirty it is)	0.5–0.75
Newly fallen snow	0.75–0.9

Sources: [Bur83], [Mar03], [Qua03], [Win91]

A similar equation also applies to the energy reflected by the ground:

$$\text{Energy reflected by the ground}: H_R = \rho \cdot H \qquad (2.12)$$

The reflection factor ρ, which is determined by the type of surface involved, must range from 0 to 1. This factor is often determined by whether the ground is wet or dry. Table 2.6 contains approximate guide values for the reflection factor ρ. Inasmuch as calculations of PV system energy yield often involve the use of a mean monthly figure, for ρ a mean reflection factor often needs to be used for the month in question.

2.5.3   The Three Components of Radiation on Inclined Surfaces

As Figure 2.31 shows, irradiance G_G and irradiation H_G arriving onto an surface with inclination angle β referred to the horizontal plane comprises the following three components:

- Direct beam radiation G_{GB} and H_{GB}.
- Diffuse radiation G_{GD} and H_{GD} (from diffuse solar radiation).
- Radiation reflected by the ground G_{GR} and H_{GR} (likewise diffuse).

Hence the following applies:

$$\text{Irradiance } G_G \text{ on the solar generator plane}: G_G = G_{GB} + G_{GD} + G_{GR} \qquad (2.13)$$

$$\text{Energy arriving at the solar generator plane}: H_G = H_{GB} + H_{GD} + H_{GR} \qquad (2.14)$$

These values of course differ from their counterpart values in the horizontal plane. The sections that follow discuss in greater detail the various components of irradiated energy (total radiation) that arrives at inclined surfaces.

2.5.3.1   Direct Beam Radiation Arriving at an Inclined Surface

Using a number of simplifying presuppositions [2.1], [2.2], the following equation can be formulated with the direct beam irradiation factor R_B for direct beam irradiation H_{GB} arriving at an inclined surface:

$$H_{GB} = R_B \cdot H_B = R_B \cdot (H - H_D) \qquad (2.15)$$

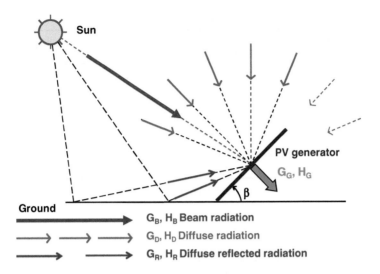

Figure 2.31 Irradiance G_G and irradiation H_G arriving at an angle of incidence relative to a horizontal plane is composed of direct beam radiation, diffuse solar radiation and diffuse reflected radiation

The direct beam irradiation factor R_B is determined by latitude φ, solar generator inclination angle β and solar generator azimuth γ, i.e. $R_B = f(\beta, \gamma, \varphi)$, and can be calculated geometrically. As this calculation is extremely complex, the mean monthly R_B values for 42 possible solar generator orientations and for 49 latitudes between 60°N and 40°S were computed using the theory in [2.2]. These values can be found in Table A4 in Appendix A. The mean monthly R_B values were determined using the average day of the month concerning radiation at 47°N, i.e. the day where daily extraterrestrial irradiation H_{ex} was closest to the monthly mean. In the interest of making it easier to find the correct R_B value, the latitude of each of the various locations is indicated in the radiation data tables for these locations (Table 2.5 or the tables in Section A2). If necessary, the tabular R_B values can be interpolated. Table 2.7 contains selected R_B values for southerly oriented surfaces.

Table 2.7 Selected mean monthly R_B values for southerly oriented surfaces (additional values can be found in Section A4)

| | | \multicolumn{12}{c}{Beam irradiation factors $R_B(\beta, \gamma)$ for tilted planes facing due south ($\gamma = 0°$)} | | | | | | | | | | | |
Latitude	β	Jan	Feb	Mar	April	May	June	July	Aug	Sept	Oct	Nov	Dec
$\varphi = 47°$N	30°	2.25	1.80	1.44	1.20	1.06	1.00	1.03	1.13	1.33	1.66	2.09	2.46
	45°	2.66	2.02	1.53	1.18	0.98	0.90	0.94	1.09	1.37	1.83	2.44	2.96
	60°	2.89	2.11	1.50	1.08	0.84	0.75	0.79	0.97	1.31	1.87	2.62	3.26
	90°	2.76	1.86	1.16	0.68	0.41	0.32	0.36	0.55	0.93	1.59	2.45	3.19
$\varphi = 50°$N	30°	2.49	1.92	1.51	1.24	1.08	1.02	1.05	1.17	1.38	1.76	2.29	2.77
	45°	3.00	2.20	1.62	1.24	1.02	0.94	0.97	1.13	1.44	1.97	2.71	3.40
	60°	3.31	2.33	1.61	1.15	0.89	0.79	0.83	1.02	1.40	2.04	2.96	3.80
	90°	3.24	2.11	1.29	0.76	0.47	0.37	0.41	0.61	1.04	1.78	2.84	3.81
$\varphi = 53°$N	30°	2.82	2.08	1.58	1.28	1.11	1.05	1.08	1.20	1.44	1.88	2.54	3.22
	45°	3.47	2.42	1.72	1.30	1.06	0.97	1.01	1.18	1.52	2.14	3.08	4.04
	60°	3.88	2.60	1.74	1.22	0.94	0.83	0.88	1.08	1.50	2.25	3.40	4.58
	90°	3.90	2.42	1.44	0.84	0.53	0.42	0.46	0.68	1.15	2.02	3.35	4.71

2.5.3.2 Diffuse Radiation Arriving at an Inclined Surface

A surface inclined at an angle β is exposed to a smaller portion of the sky from which diffuse radiation stems. If the horizon is very deep, diffuse radiation is determined for an inclined surface (G_{GD} and H_{GD}) as follows, based on the diffuse radiation at the horizontal plane (G_D and H_D):

> Irradiance G_{GD} of diffuse radiation arriving at an inclined surface:
>
> $$G_{GD} = (\tfrac{1}{2} + \tfrac{1}{2}\cos\beta) \cdot G_D = R_D \cdot G_D$$
>
> (2.16)

Accordingly, the following equation applies to irradiation H_{GD} that stems from diffuse radiation and that arrives at an inclined surface:

> $$H_{GD} = (\tfrac{1}{2} + \tfrac{1}{2}\cos\beta) \cdot H_D = R_D \cdot H_D$$
>
> (2.17)

In line with the direct beam irradiation factor discussed earlier, the R_D factor can also be referred to as the diffuse radiation factor. This also applies to G_{GD} and H_{GD}. The R_D factor is much easier to calculate, as it is determined only by the inclination angle β, but independent from by latitude and season; R_D is always between 0 and 1:

> Diffuse radiation factor $R_D = (\tfrac{1}{2} + \tfrac{1}{2}\cos\beta)$
>
> (2.18)

As the inclination angle begins to increase from 0, diffuse radiation decreases only slightly at first since the inclination angle of radiation from areas of the sky that are no longer present is initially very shallow. The larger the inclination angle, the greater the radiation loss. A vertical surface such as a building façade receives only half of the available diffuse radiation.

2.5.3.3 Radiation Reflected by the Ground on an Inclined Surface

A surface inclined by an angle β is also exposed to a portion of the Earth's surface and thus receives a portion of the radiation reflected by it. Reflected radiation arriving at an inclined surface is calculated using the reflection factor ρ, based on global radiation on the horizontal plane (G and H).

The following applies to irradiance G_{GR} that is reflected by the ground onto an inclined surface:

> $$G_{GR} = (\tfrac{1}{2} - \tfrac{1}{2}\cos\beta) \cdot G_R = (\tfrac{1}{2} - \tfrac{1}{2}\cos\beta) \cdot \rho \cdot G = R_R \cdot \rho \cdot G$$
>
> (2.19)

The following applies to irradiation H_{GR} from ground reflection radiation:

> $$H_{GR} = (\tfrac{1}{2} - \tfrac{1}{2}\cos\beta) \cdot H_R = (\tfrac{1}{2} - \tfrac{1}{2}\cos\beta) \cdot \rho \cdot H = R_R \cdot \rho \cdot H$$
>
> (2.20)

The R_R factor can also be referred to as the effective portion of reflective radiation. R_R, which also applies to G_{GR} and H_{GR}, is very easy to calculate, and is determined solely by inclination angle, but not by latitude or season; R_R is always between 0 and 1:

> Effective portion of reflective radiation $R_R = (\tfrac{1}{2} - \tfrac{1}{2}\cos\beta)$
>
> (2.21)

As the inclination angle begins to increase from 0, the ground reflection radiation arriving at an inclined surface increases only slightly at first since the angle of incidence is still very shallow. The greater the inclination angle, the greater the reflected radiation arriving at the surface. A vertical surface such as a building façade receives 50% of the available ground reflection radiation, which is actually appreciable for a snow-covered surface.

2.5.3.4 Total Irradiated Energy Arriving at an Inclined Surface in the Presence of a Low Horizon

The sum total of direct beam radiation, diffuse radiation and ground reflection radiation equals irradiation H_G for the entirety of an inclined surface with a low horizon:

$$H_G = H_{GB} + H_{GD} + H_{GR} = R_B \cdot (H - H_D) + R_D \cdot H_D + R_R \cdot \rho \cdot H \qquad (2.22)$$

where:

H = irradiation from global radiation on the horizontal plane
H_D = irradiation from diffuse radiation on the horizontal plane
R_B = direct beam irradiation factor as in Table 2.7 or Section A4
R_D = diffuse radiation factor, $R_D = \frac{1}{2} + \frac{1}{2} \cos \beta$
R_R = effective portion of reflective radiation, $R_R = \frac{1}{2} - \frac{1}{2} \cos \beta$
β = inclination angle of the inclined surface relative to the horizontal plane
ρ = reflection factor (albedo) of the ground in front of the solar generator

Equation 2.22 allows for the calculation of irradiated energy for a *single-row array* of solar generators installed out in the open or on a roof, in the presence of a low horizon.

2.5.4   Approximate Allowance for Shading by the Horizon

How irradiated energy on the horizontal plane stemming from direct beam radiation is distributed across the individual hours of a day is described in [2.1]. This information can be entered in the mean Sun's path diagram for each month, which provides shading diagram for the relevant latitude. Figure 2.32 shows an example of this type of shading diagram for 47°N.

Each point along the monthly mean Sun's path corresponds to a specific portion of daily direct beam radiation in accordance with the relevant assessment diagram (see Figure 2.33).

To use a shading diagram, first enter in it the solar generator orientation γ as a vertical line. Then enter in the diagram the horizon towards which the solar generator is oriented. Then, using the assessment diagram (see Figure 2.33), determine the total weighting $\sum$GPS for all points along the Sun's path for the month of interest (often only a few winter months). Depending on the deviation $\Delta\gamma$ from the solar generator orientation, the weighting of a point will be 3% ($0° < |\Delta\gamma| \le 30°$), 2% ($30° < |\Delta\gamma| \le 60°$) or 1% ($60° < |\Delta\gamma| \le 80°$–$110°$ according to the value of β). In determining shading-induced energy loss, each point below the horizon line is also weighted 3%, 2% or 1% in accordance with the assessment diagram (see Figure 2.33). The sum total of the weighting percentages for the shaded points below the horizon equals the total weighting $\sum$GPB for the shaded points. Now, using $\sum$GPB and $\sum$GPS, the shading correction factor k_B can be determined for the month of interest, whereby k_B, which is between 0 (completely shaded) and 1 (not shaded), is calculated as follows:

$$\text{Shading correction factor } k_B = 1 - \sum \text{GPB} / \sum \text{GPS} \qquad (2.23)$$

where:

$\sum$GPS = total weighting for shaded points along the Sun's path for the month of interest
$\sum$GPB = total weighting for all points along the Sun's path for the month of interest

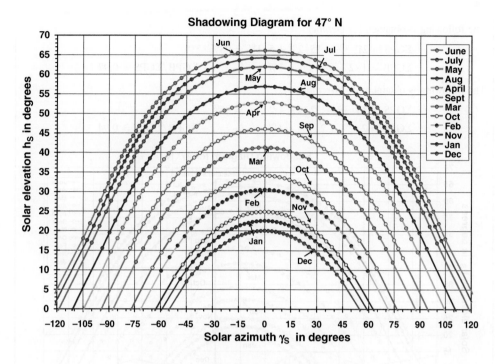

Figure 2.32   Shading diagram for locations at 47°N

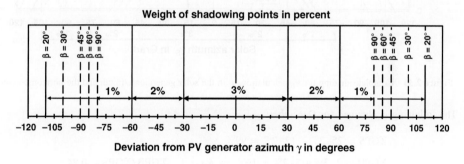

Figure 2.33   Assessment diagram for a shading diagram as in Figure 2.32

During the winter, ΣGPS is almost always 100% for relatively low γ (solar generators that are oriented almost due south). In such a case, Equation 2.23 can be simplified as follows: $k_B \approx 100\% - \Sigma GPB$.

Example of Calculation of the Shading Correction Factor k_B

Using Figure 2.34, the calculation procedure for shading correction factor k_B for a solar generator at 47°N with $\beta = 45°$ and $\gamma = 15°$ will now be described. After the solar generator azimuth $\gamma = 15°$ and the horizon have been entered in the shading diagram for 47°N, it emerges that shading occurs only from October to February. Hence k_B needs to be determined for this period.

The following is obtained for *October*:

$$\Sigma GPS = 21 \cdot 3\% + 15 \cdot 2\% + 5 \cdot 1\% = 98\%$$

$$\Sigma GPB = 1 \cdot 2\% + 1 \cdot 1\% = 3\% \Rightarrow k_B = 1 - \Sigma GPB / \Sigma GPS = 0.97$$

The following is obtained for *November*:

$$\Sigma GPS = 23 \cdot 3\% + 12 \cdot 2\% + 2 \cdot 1\% = 95\%$$

$$\Sigma GPB = 2 \cdot 2\% + 2 \cdot 1\% = 6\% \Rightarrow k_B = 1 - \Sigma GPB / \Sigma GPS = 0.94$$

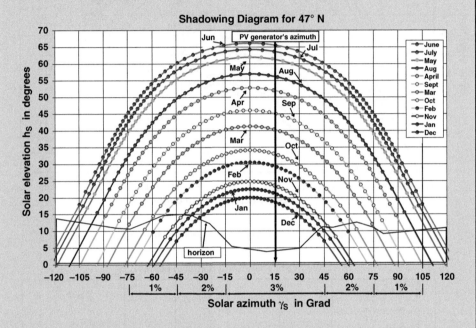

Figure 2.34 Complete shading for k_B calculation, with the solar generator azimuth and the horizon entered

The following is obtained for *December*:

$$\Sigma GPS = 25 \cdot 3\% + 12 \cdot 2\% = 99\%$$

$$\Sigma GPB = 2 \cdot 3\% + 5 \cdot 2\% = 16\% \Rightarrow k_B = 1 - \Sigma GPB / \Sigma GPS = 0.84$$

The following is obtained for *January*:

$$\Sigma GPS = 24 \cdot 3\% + 12 \cdot 2\% + 1 \cdot 1\% = 97\%$$

$$\Sigma GPB = 1 \cdot 3\% + 4 \cdot 2\% + 1 \cdot 1\% = 12\%$$

$$\Rightarrow k_B = 1 - \Sigma GPB / \Sigma GPS = 0.88$$

The following is obtained for *February*:

$$\Sigma GPS = 21 \cdot 3\% + 14 \cdot 2\% + 4 \cdot 1\% = 95\%$$

$$\Sigma GPB = 1 \cdot 2\% + 2 \cdot 1\% = 4\% \Rightarrow k_B = 1 - \Sigma GPB / \Sigma GPS = 0.96$$

With $k_B = 1$ for the remaining months.

Section A5 contains shading diagrams for 41, 44, 47, 50 and 53°N that were realized using the same method, as well as an assessment diagram with the same scale, which can be used as a master copy. These documents allow for sufficiently accurate calculation of the shading correction factor k_B for Central Europe and much of Southern Europe.

The shading correction factor k_B can in turn be used to calculate irradiated direct beam energy in a manner that takes account of shading. This calculation is performed in the guise of a correction of Equation 2.15 by multiplying by k_B:

Direct beam energy on an inclined surface: $H_{GB} = k_B \cdot R_B \cdot H_B = k_B \cdot R_B \cdot (H - H_D)$ (2.24)

2.5.4.1 Shading Induced by Neighbouring Buildings and Other Wall-Like Structures

It often happens that a solar generator is shaded by a wall-like structure such as a neighbouring building or stacked rows of solar generators (see Figure 2.37). It is a relatively simple matter to integrate into a shading diagram the horizon that is masked by such elements so as to allow for calculation of the shading factor k_B.

Such a situation is depicted in Figure 2.35, for which the procedure is as follows. First determine the maximum elevation α_{max} below which the edges that cast the shadows are visible from the solar generator site (point P). The direction (or more precisely the azimuth) where this maximum elevation $\alpha_{max} = \arctan(h/d)$ occurs is referred to as γ_B. In cases of stacked rows of solar generators (see Figure 2.37), γ_B of course equals the solar generator azimuth γ. The elevation α under which the edges that cast these shadows are visible from P can be determined as follows using the azimuth difference $\Delta\gamma$ between the direction of view and γ_B:

Elevation of a wall-like element (height h, distance d from P):

$$\alpha = \arctan(\cos \Delta\gamma \cdot \tan \alpha_{max}) = \arctan\left(\cos \Delta\gamma \cdot \frac{h}{d}\right) \qquad (2.25)$$

Figure 2.36 shows α as a function of $\Delta\gamma$ with α_{max} as a parameter. This type of curve can readily be integrated into a shading diagram as shown in Figure 2.32. This method, which was used for Example 2 in Section 2.5.8, can also be used in cases where the shadow-casting edges do not reach α_{max} and γ_B. In such a case, α_{max} is determined based on the estimated extension of the relevant edges. Then the

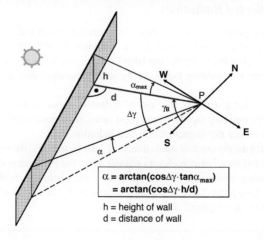

Figure 2.35 Situation arising from shading by a wall-like element

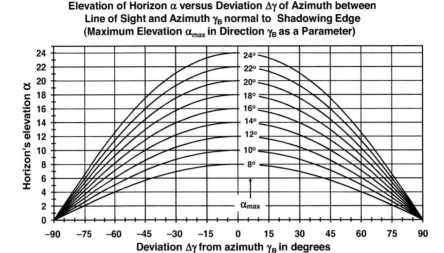

Figure 2.36   Horizon elevation α as a function of difference $\Delta\gamma$ between the azimuth of the direction of view and the azimuth γ_B normal to the edge that casts a shadow

curves are entered as in Figure 2.36, but only up to γ in the shading diagram for which the structure ends as viewed from P. This method also allows two or more neighbouring buildings or the like to be taken into account. Taking account of shading by nearby elements is complicated by the fact that solar generators are not point-like in nature. In some cases an approximate shading calculation can be realized using the calculation for a mean radiation point such as in the centre of the solar generator. That said, it should be borne in mind that solar generators are extremely sensitive to shading (see Chapters 3 and 4).

In the interest of avoiding excessive shading loss in winter, α_{max} should always be $>1°$ lower for smaller γ_B values (see Figure 2.35) than the maximum solar altitude $h_{Smax} = 66.5°-\varphi$ (see Figure 2.2, $\varphi = $ latitude).

2.5.5   *Effect of Horizon and Façade/Roof Edge Elevation on Diffuse Radiation (Sky and Reflected Radiation)*

The horizon facing a solar generator is often somewhat elevated, which reduces not only direct beam radiation in the winter, but also to some extent diffuse radiation all year round. When PV systems are installed on buildings (an increasingly popular modality), diffuse radiation arriving at solar generators can be additionally reduced by the building façade or by solar generators that are installed above other generators in a stack.

In cases of single-row solar generators, this loss of diffuse radiation is partially offset by additional reflected radiation from the building façade. But in most cases, solar generators are installed in multiple rows or in stacks, which reduce the impact of reflected radiation.

The distant horizon or (in the presence of stacked rows of solar generators) the near horizon can often be approximately characterized by a horizon elevation angle α_1 in the γ direction (see Figure 2.37).

In the interest of avoiding unduly elevated shading in winter, here too α_1 should be smaller than $h_{Smax} = 66.5°-\varphi$. Where, for a given d, α_1 contains a specific value, the clearance a between the rows of the various arrays is as follows (see Figure 2.37):

$$a = d\cos\beta + h\cot\alpha_1 = d(\cos\beta + \sin\beta\cot\alpha_1) \qquad (2.26)$$

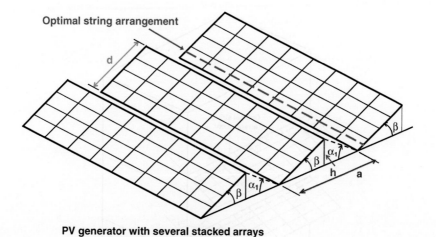

Optimal string arrangement

PV generator with several stacked arrays

Figure 2.37 Solar generator consisting of multiple horizontal arrays arranged in a series of rows. Angle α_1 should be somewhat smaller than the minimum solar altitude in winter $h_{S\max} = 66.5° - \varphi$ ($\varphi = $ latitude)

Building-integrated solar generators (see Figure 2.38) that are also stacked (see Figure 2.39) can also be characterized by the effect of façade/roof edge height, in respect to the generator plane, resulting from a second elevation angle α_2.

In such a case, it is advisable (using Figure 2.40) to understand such effects and quantify them approximately. Inasmuch as the effects of nearby structures are not homogeneous across the entire solar generator field, it is difficult to quantify such effects precisely. The goal here, however, is to gain insight into the main factors that come into play in the presence of various diffuse radiation horizons.

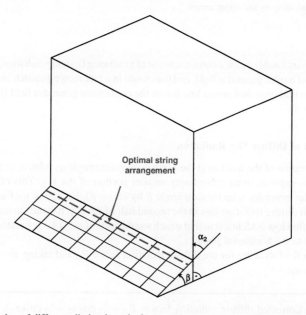

Optimal string arrangement

Figure 2.38 Reduction of diffuse radiation in a single-row solar generator installed on a building façade

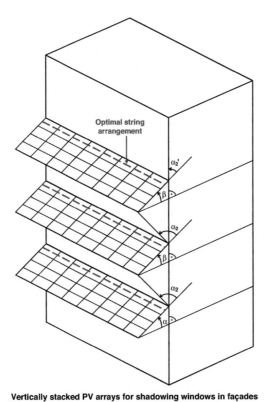

Vertically stacked PV arrays for shadowing windows in façades

Figure 2.39 Solar generator composed of stacked arrays for façade window shading. The façade and the arrays above it reduce diffuse radiation on the lower arrays, which in the morning and evening also lose some direct beam radiation owing to transient shading by the upper arrays

These differences inevitably cause a certain amount of irradiance G_G and irradiation H_G inhomogeneity in various portions of a solar generator field, and thus result in a radiation mismatch among the various solar generators; this in turn provokes power loss across the entire solar generator field (for more on mismatch loss, see Chapter 4).

2.5.5.1   Impact of Diffuse Sky Radiation

Owing to the elevation of the horizon at the horizon elevation angle α_1 relative to the horizon plane, the solar generator is exposed to an accordingly smaller portion of the sky. This effect can be offset by adjusting the solar generator's inclination angle β by $(\alpha_1 + \beta)$. Façade or roof edge elevation α_2 also provokes a partial diffuse radiation loss in the second half of the sky. It stands to reason that the extent of this diffuse radiation loss is identical to that which would occur if the solar generator were set to this same angle α_2 relative to the horizontal plane.

This results in the following for the diffuse radiation factor R_D and taking account of the aforementioned observations:

$$\text{Corrected diffuse radiation factor: } R_D = \tfrac{1}{2}\cos\alpha_2 + \tfrac{1}{2}\cos(\alpha_1 + \beta) \qquad (2.27)$$

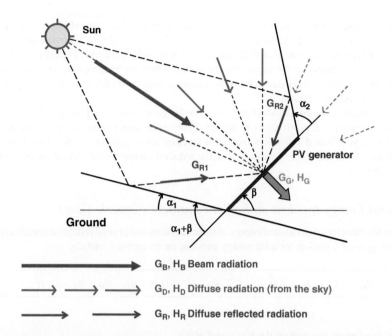

Figure 2.40 Calculation method for radiation on a solar generator with inclination angle β for a generic scenario (with horizon elevation α_1 and façade/roof edge elevation α_2 relative to the solar generator plane)

Using this corrected diffuse radiation factor in Equations 2.16 and 2.17, the corrected irradiance G_{GD} and corrected diffuse irradiation H_{GD} can be determined.

2.5.5.2 Impact of Radiation Reflected by the Ground or a Façade

In the case of a single-row solar generator that is remote from elements such as a building façade, global radiation is somewhat reduced in the plane that is inclined by α_1, relative to horizontal global radiation, whereas the solar generator is exposed to the inclined plane at angle $(\alpha_1 + \beta)$. In the case of relatively low α_1 values, it is safe to assume that the global radiation arriving at the plane with an angle of inclination α_1 relative to the horizontal plane is roughly the same as the global radiation on the horizontal plane. This allows for calculation of a corrected effective portion of reflected radiation:

$$\text{Corrected portion of reflected radiation: } R_R = \tfrac{1}{2} - \tfrac{1}{2}\cos(\alpha_1 + \beta) \qquad (2.28)$$

Inasmuch as the surfaces that form an elevated horizon are not as evenly illuminated as presupposed above (particularly when it comes to the elements near the horizon that have the greatest impact on reflected radiation), irradiance on inclined surfaces is usually overestimated. Hence it is more in keeping with real-world conditions, and also simpler, to calculate the effective portion of reflected radiation R_R using Equation 2.21, despite the horizon elevation α_1.

In single-row solar generators near a building façade (see Figure 2.38), often $\alpha_2 = 90° - \beta$, i.e. in addition to ground reflection radiation G_{R1} and H_{R1}, façade reflection radiation G_{R2} and H_{R2} also need to be taken into account (see Figure 2.40). But in order to compute G_{R2} and H_{R2} accurately, the global radiation G_F and H_F incident on the façade must first be determined. If α_2 is relatively low, this latter calculation is hardly worth the effort. Owing to the relatively low value for the effective portion of reflected radiation R_{R2}, in most instances calculation of the global radiation incident on the façade can be dispensed with, except for very bright façades such as metal ones with a relatively high α_2. Forgoing this calculation to some extent compensates for the slight reduction in ground reflection radiation provoked by the reduced global radiation G and H on the plane in front of the solar generator. For as Figure 2.38 shows, diffuse radiation is reduced by the façade on the surfaces that are immediately in front of the solar generator, as in Equation 2.27. Hence in most such cases it suffices to take account of ground reflection radiation as in Equations 2.19 and 2.20.

Ground reflection radiation for stacked rows of solar generators (see Figure 2.37) is lower than for a single-row generator, particularly in winter. This can be approximately taken into account by using a reduced year-round reflection factor. In stacked solar generator fields (see Figure 2.39), façade radiation from the upper arrays is usually reduced owing to the virtual absence of façade reflection radiation.

2.5.6   Total Energy Incident on Inclined Surfaces (Generic Case)

Adding together direct beam irradiated energy, diffuse radiation and ground reflection radiation results in the following general equation for total energy incident on an inclined surface:

$$H_G = H_{GB} + H_{GD} + H_{GR} = k_B \cdot R_B \cdot (H - H_D) + R_D \cdot H_D + R_R \cdot \rho \cdot H \qquad (2.29)$$

where:

H = global irradiation incident on the horizontal plane
H_D = diffuse irradiation incident on the horizontal plane
k_B = shading correction factor as described in Section 2.5.4 (non-shaded: $k_B = 1$)
R_B = direct beam irradiation factor as in Table 2.7 or the tables in Section A4
R_D = corrected diffuse radiation factor, $R_D = \frac{1}{2} \cos \alpha_2 + \frac{1}{2} \cos(\alpha_1 + \beta)$
R_R = effective portion of reflective radiation, $R_R = \frac{1}{2} - \frac{1}{2} \cos \beta$
α_1 = horizon elevation in the γ direction
α_2 = façade/roof edge elevation relative to the solar generator plane
β = inclination angle of the inclined surface relative to the horizontal plane
ρ = reflection factor (albedo) of the ground in front of the solar generator as in Table 2.6

Calculation of H_G values using the three-component model is best done using a table in which the monthly H and H_D values have been entered, as well as any other values needed for calculations entailing various monthly values. Then, after determining the monthly values H_{GB}, H_{GD} and H_{GR}, add these values together to obtain H_G. Table 2.8 or Table A1.2 can be used as a template for the aforementioned table.

Table 2.8   Tabular template for determining H_G using the three-component model

$R_D = \frac{1}{2} \cos\alpha_2 + \frac{1}{2} \cos(\alpha_1 + \beta) =$							$R_R = \frac{1}{2} - \frac{1}{2} \cos \beta =$						
	Jan	Feb	Mar	Apr	May	June	July	Aug	Sept	Oct	Nov	Dec	Year	Unit
H														kWh/m²
H_D														kWh/m²
R_B														
k_B														
$H_{GB} = k_B \cdot R_B \cdot (H - H_D)$														kWh/m²
$H_{GD} = R_D \cdot H_D$														kWh/m²
ρ														
$H_{GR} = R_R \cdot \rho \cdot H$														kWh/m²
$H_G = H_{GB} + H_{GD} + H_{GR}$														kWh/m²

Before filling the table, the diffuse radiation factor R_D and the effective portion of reflected radiation R_R (these values are the same for all months) should be determined.

To determine the exact mean annual value H_{Ga} of total daily irradiation values H_G, proceed as follows: (a) determine the monthly total by multiplying the mean monthly total by the number of days n_d in each month (28, 29, 30 or 31); and then (b) using the result of (a), determine the annual total. Determine total daily irradiation H_G by dividing the annual total by the number of days in the year (365 or 366). As noted in Section 2.4, H_{Ga} is also obtainable as the mean of the monthly mean of total daily irradiation, albeit in most cases with a negligible error. To differentiate between the monthly or annual mean of total daily irradiation and monthly or annual total irradiation, it is best to use a clear unit of time as the denominator, where kWh/m²/d is the mean of total daily irradiance, kWh/m²/mt is a monthly total and kWh/m²/a is an annual total. The values obtained from this calculation all represent the same parameter – namely, the yield during the period entered.

Although the calculations in Table 2.8 can be readily carried out using a calculator, they are much quicker when performed on a computer using a spreadsheet program such as Excel.

Meteonorm 4, 5 or 6.1 [2.4, 2.5, 2.12], which can be run on Windows computers and sells for a few €100, are programs that readily allow for the calculation of irradiation on both the horizontal plane and an inclined surface for many locations in Europe and worldwide.These programs use a somewhat optimized model (the Perez model) for radiation calculations. For further information concerning this model, see [2.2], [Eic01] or [Qua03]. In this model, energy H incident on the horizontal plane is referred to as H_Gh, while the energy H_G incident on the solar generator plane is referred to as H_Gk. The model also allows the effect of horizon shading to be taken into account. The program is worth purchasing for individuals who often need to perform these kinds of calculations. Some radiation calculation freeware is also available online (see Section 8.4).

2.5.7 Retrospective Calculation of Irradiation Incident on Inclined Solar Generators, Using Global Radiation Readings on the Horizontal Plane

In any given year, mean monthly irradiation readings can deviate considerably from the long-term mean values indicated in this book, particularly during the winter. However, such discrepancies mainly relate to direct beam irradiation H_B. When the H values that differ considerably from the mean values are converted to H_G values on the solar generator plane using the $R(\beta,\gamma)$ factor (see Section 2.4), a major error arises because R factors are determined by the ratio of H_B to H_D. In such cases as well, using the three-component model (see above) in conjunction with the following simple method allows for a far realistic determination of the H_G values that actually occur: 80% of the H value exceedances relative to mean monthly values in the tables is allocated to direct beam irradiation H_B, while only 20% is allocated to diffuse irradiation H_D. The H values that fall short of the tabular values are evenly divided between direct beam irradiation H_B and diffuse irradiation H_D. In cases where the H values fall *far* short of the tabular values, H_B can of course never be negative, and in extreme cases $H_D = H$ would occur.

Examples:

1. The December reading in Kloten was $H = 1.10 \, \text{kWh/m}^2$ in lieu of $H = 0.70 \, \text{kWh/m}^2$. Hence, to calculate H_G retrospectively for December, $H = 1.1 \, \text{kWh/m}^2$, $H_D = 0.60 \, \text{kWh/m}^2$ and $H_B = 0.50 \, \text{kWh/m}^2$ were used.
2. If, on the other hand, the December Kloten reading was only $H = 0.50 \, \text{kWh/m}^2$ in lieu of $0.70 \, \text{kWh/m}^2$, the calculation would be performed using $H = 0.50 \, \text{kWh/m}^2$, $H_D = 0.42 \, \text{kWh/m}^2$ and $H_B = 0.08 \, \text{kWh/m}^2$.

2.5.8 Examples of Radiation Calculations with the Three-Component Method

---

Example 1

PV system in Bremen with a single-row solar generator with $\beta = 30°$, $\gamma = 0°$. Mean ground reflection factor ρ: December, 0.3; January, 0.4; February, 0.35; rest of the year, 0.2.

Calculation task:

(a) Monthly and annual mean of total daily irradiation, as well as annual total H_G incident on a solar generator that is out in the open.
(b) Monthly and annual mean of total daily irradiation, as well as total annual H_G incident on the same solar generator, if the latter is mounted on a dark-coloured southern façade of a very tall building (see Figure 2.38).

Solution:

(a) $\alpha_1 = 0°$, $\alpha_2 = 0°$. H_G calculation using Table 2.8:

Energy yield calculation for grid-connected PV plants (with three-component model)

$R_D = \frac{1}{2}\cos\alpha_2 + \frac{1}{2}\cos(\alpha_1 + \beta) =$					0.933			$R_R = \frac{1}{2} - \frac{1}{2}\cos\beta =$			0.067			
	Jan	Feb	Mar	Apr	May	June	July	Aug	Sept	Oct	Nov	Dec	Year	Unit
H	0.60	1.30	2.09	3.57	4.78	4.54	4.61	3.96	2.67	1.56	0.77	0.41	2.57	kWh/m²
H_D	0.44	0.84	1.36	2.05	2.56	2.74	2.67	2.26	1.64	1.01	0.55	0.32	1.54	kWh/m²
R_B	2.82	2.08	1.58	1.28	1.11	1.05	1.08	1.20	1.44	1.88	2.54	3.22		
k_B	1	1	1	1	1	1	1	1	1	1	1	1		
$H_{GB} = k_B \cdot R_B \cdot (H - H_D)$	0.45	0.96	1.15	1.95	2.46	1.89	2.10	2.04	1.48	1.03	0.56	0.29	1.37	kWh/m²
$H_{GD} = R_D \cdot H_D$	0.41	0.78	1.27	1.91	2.39	2.56	2.49	2.11	1.53	0.94	0.51	0.30	1.44	kWh/m²
ρ	0.4	0.35	0.2	0.2	0.2	0.2	0.2	0.2	0.2	0.2	0.2	0.3		
$H_{GR} = R_R \cdot \rho \cdot H$	0.02	0.03	0.03	0.05	0.06	0.06	0.06	0.05	0.04	0.02	0.01	0.01	0.04	kWh/m²
$H_G = H_{GB} + H_{GD} + H_{GR}$	0.88	1.77	2.45	3.91	4.91	4.51	4.65	4.20	3.05	1.99	1.08	0.60	2.84	kWh/m²

Total annual H_G: $H_{Ga} = 365 \, \text{d/a} \cdot 2.84 \, \text{kWh/m}^2/\text{d} = 1037 \, \text{kWh/m}^2/\text{a}$.

(b) $\alpha_1 = 0°$, $\alpha_2 = 60°$. H_G calculation using Table 2.8:

Energy yield calculation for grid-connected PV plants (with three-component model)

$R_D = \frac{1}{2}\cos\alpha_2 + \frac{1}{2}\cos(\alpha_1 + \beta) =$					0.683			$R_R = \frac{1}{2} - \frac{1}{2}\cos\beta =$			0.067			
	Jan	Feb	Mar	Apr	May	June	July	Aug	Sept	Oct	Nov	Dec	Year	Unit
H	0.60	1.30	2.09	3.57	4.78	4.54	4.61	3.96	2.67	1.56	0.77	0.41	2.57	kWh/m²
H_D	0.44	0.84	1.36	2.05	2.56	2.74	2.67	2.26	1.64	1.01	0.55	0.32	1.54	kWh/m²
R_B	2.82	2.08	1.58	1.28	1.11	1.05	1.08	1.20	1.44	1.88	2.54	3.22		
k_B	1	1	1	1	1	1	1	1	1	1	1	1		
$H_{GB} = k_B \cdot R_B \cdot (H - H_D)$	0.45	0.96	1.15	1.95	2.46	1.89	2.10	2.04	1.48	1.03	0.56	0.29	1.37	kWh/m²
$H_{GD} = R_D \cdot H_D$	0.30	0.57	0.93	1.40	1.75	1.87	1.82	1.54	1.12	0.69	0.38	0.22	1.05	kWh/m²
ρ	0.4	0.35	0.2	0.2	0.2	0.2	0.2	0.2	0.2	0.2	0.2	0.3		
$H_{GR} = R_R \cdot \rho \cdot H$	0.02	0.03	0.03	0.05	0.06	0.06	0.06	0.05	0.04	0.02	0.01	0.01	0.04	kWh/m²
$H_G = H_{GB} + H_{GD} + H_{GR}$	0.77	1.56	2.11	3.40	4.27	3.82	3.98	3.63	2.64	1.74	0.95	0.52	2.45	kWh/m²

Total annual H_G: $H_{Ga} = 365 \, \text{d/a} \cdot 2.45 \, \text{kWh/m}^2/\text{d} = 894 \, \text{kWh/m}^2/\text{a}$.

Example 2

PV system with a horizontally mounted solar generator field (see Figure 2.37) located on Mont Soleil (1270 m) in Switzerland, with $\beta = 45°$, $\gamma = -30°$. The horizon elevation in the direction of the solar generator azimuth is $\alpha_1 = 12°$. The reduced ground reflection radiation resulting from the horizontal arrangement is taken into account via the somewhat reduced reflection factor $\rho = 0.2$ (May–October), $\rho = 0.3$ (November and April) and $\rho = 0.4$ (December–March).

Calculation task:

(a) Monthly and annual mean of total daily irradiation and annual H_G total on the upper edge of the topmost solar module ($H_{G\text{-}top}$) of an array.
(b) Monthly and annual mean of total daily irradiation and annual H_G total on the lower edge of the bottom solar module ($H_{G\text{-}bottom}$) of an array.
(c) Irradiance $G_{G\text{-}top}$ at the upper edge of the topmost PV module and $G_{G\text{-}bottom}$ at the lower edge of the bottom PV module of an array, where $G = G_D = 200\,\text{W/m}^2$ and $\rho = 0.2$.

Solution

(a) $H_{G\text{-}top}$: $\alpha_1 = 0°$, $\alpha_2 = 0°$ · H_G calculation using Table 2.8 (the front module does not shade the upper module edges, thus $k_B = 1$ all year round):

Energy yield calculation for grid-connected PV plants (with three-component model)

$R_D = \frac{1}{2}\cos\alpha_2 + \frac{1}{2}\cos(\alpha_1 + \beta) =$				0.854			$R_R = \frac{1}{2} - \frac{1}{2}\cos\beta =$			0.146			
	Jan	Feb	Mar	Apr	May	June	July	Aug	Sept	Oct	Nov	Dec	Year	Unit
H	1.36	2.10	3.09	3.96	4.54	4.98	5.57	4.79	3.63	2.42	1.44	1.14	3.25	kWh/m²
H_D	0.65	0.98	1.45	1.91	2.23	2.42	2.46	2.14	1.66	1.14	0.71	0.55	1.53	kWh/m²
R_B	2.40	1.85	1.42	1.13	0.96	0.89	0.92	1.05	1.29	1.68	2.21	2.66		
k_B	1	1	1	1	1	1	1	1	1	1	1	1		
$H_{GB} = k_B \cdot R_B \cdot (H - H_D)$	1.70	2.07	2.33	2.32	2.22	2.28	2.86	2.78	2.54	2.15	1.61	1.57	2.20	kWh/m²
$H_{GD} = R_D \cdot H_D$	0.56	0.84	1.24	1.63	1.90	2.07	2.10	1.83	1.42	0.97	0.61	0.47	1.31	kWh/m²
ρ	0.4	0.4	0.4	0.3	0.2	0.2	0.2	0.2	0.2	0.2	0.3	0.4		
$H_{GR} = R_R \cdot \rho \cdot H$	0.08	0.12	0.18	0.17	0.13	0.15	0.16	0.14	0.11	0.07	0.06	0.07	0.12	kWh/m²
$H_G = H_{GB} + H_{GD} + H_{GR}$	2.34	3.03	3.75	4.12	4.25	4.50	5.12	4.75	4.07	3.19	2.28	2.11	3.63	kWh/m²

Total annual H_G: $H_{Ga} = 365\,\text{d/a} \cdot 3.63\,\text{kWh/m}^2\text{/d} = 1325\,\text{kWh/m}^2\text{/a}$.

(b) $H_{G\text{-}bottom}$: $\alpha_1 = 12°$, $\alpha_2 = 0°$ · H_G calculation using Table 2.8. First determine the shading correction factor k_B using Figures 2.32 and 2.33. To do this, make a copy, cut it out, place it below the shading diagram and read off the shading point weightings (Figure 2.41).

Calculation of the shading correction factor k_B as in Section 2.5.4:

For December:
$$\Sigma\text{GPS} = 19 \cdot 3\% + 12 \cdot 2\% + 6 \cdot 1\% = 87\%$$
$$\Sigma\text{GPB} = 3.5 \cdot 3\% = 10.5\% \Rightarrow k_B = 1 - \Sigma\text{GPB}/\Sigma\text{GPS} = 0.88$$

For January:
$$\Sigma\text{GPS} = 19 \cdot 3\% + 12 \cdot 2\% + 6 \cdot 1\% = 87\%$$
$$\Sigma\text{GPB} = 2 \cdot 3\% = 6\% \Rightarrow k_B = 1 - \Sigma\text{GPB}/\Sigma\text{GPS} = 0.93$$

For November:
$$\Sigma\text{GPS} = 19 \cdot 3\% + 12 \cdot 2\% + 6 \cdot 1\% = 87\%$$
$$\Sigma\text{GPB} = 1.5 \cdot 3\% = 4.5\% \Rightarrow k_B = 1 - \Sigma\text{GPB}/\Sigma\text{GPS} = 0.95$$

For February:
$$\Sigma\text{GPS} = 20 \cdot 3\% + 11 \cdot 2\% + 8 \cdot 1\% = 90\%$$
$$\Sigma\text{GPB} = 1 \cdot 3\% = 3\% \Rightarrow k_B = 1 - \Sigma\text{GPB}/\Sigma\text{GPS} = 0.97$$

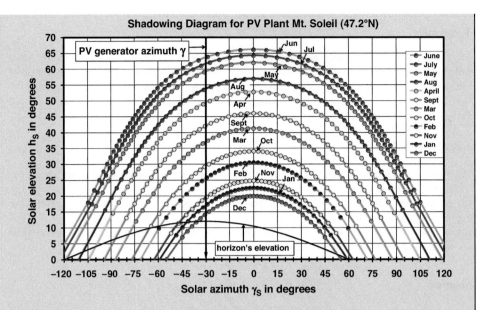

Figure 2.41 Shading diagram for the Mont Soleil installation as in Example 2, with the solar generator azimuth entered and the horizon seen from the lowest module edge

Now determine H_G using Table 2.8:

Energy yield calculation for grid-connected PV plants (with three-component model)

			$R_D = \frac{1}{2}\cos\alpha_s + \frac{1}{2}\cos(\alpha_s + \beta) =$		0.772			$R_R = \frac{1}{2} - \frac{1}{2}\cos\beta =$		0.146				
	Jan	Feb	Mar	Apr	May	June	July	Aug	Sept	Oct	Nov	Dec	Year	Unit
H	1.36	2.10	3.09	3.96	4.54	4.98	5.57	4.79	3.63	2.42	1.44	1.14	3.25	kWh/m²
H_D	0.65	0.98	1.45	1.91	2.23	2.42	2.46	2.14	1.66	1.14	0.71	0.55	1.53	kWh/m²
R_B	2.40	1.85	1.42	1.13	0.96	0.89	0.92	1.05	1.29	1.68	2.21	2.66		
k_B	0.93	0.97	1	1	1	1	1	1	1	1	0.95	0.88		
$H_{GB} = k_B \cdot R_B \cdot (H - H_D)$	1.58	2.01	2.33	2.32	2.22	2.28	2.86	2.78	2.54	2.15	1.53	1.38	2.17	kWh/m²
$H_{GD} = R_D \cdot H_D$	0.50	0.76	1.12	1.47	1.72	1.87	1.90	1.65	1.28	0.88	0.55	0.42	1.18	kWh/m²
ρ	0.4	0.4	0.4	0.3	0.2	0.2	0.2	0.2	0.2	0.2	0.3	0.4		
$H_{GR} = R_R \cdot \rho \cdot H$	0.08	0.12	0.18	0.17	0.13	0.15	0.16	0.14	0.11	0.07	0.06	0.07	0.12	kWh/m²
$H_G = H_{GB} + H_{GD} + H_{GR}$	2.16	2.89	3.63	3.96	4.07	4.30	4.92	4.57	3.93	3.10	2.14	1.87	3.47	kWh/m²

Total annual H_G: $H_{Ga} = 365\,\text{d/a} \cdot 3.47\,\text{kWh/m}^2\text{/d} = 1267\,\text{kWh/m}^2\text{/a}$.

(c) This scenario entails diffuse radiation only: $G = G_D = 200\,\text{W/m}^2$, $G_B = 0$

$$\Rightarrow G_{G-top} = R_{D-top} \cdot G_D + R_R \cdot \rho \cdot G = 177\,\text{W/m}^2.$$

$$\Rightarrow G_{G-bottom} = R_{D-bottom} \cdot G_D + R_R \cdot \rho \cdot G = 160\,\text{W/m}^2.$$

In this scenario (presence of diffuse irradiance only), the radiation incident on the bottom edge of an array is only about 90% of that incident on the top edge. Although this effect is less pronounced in the presence of direct beam irradiance (e.g. 600 W/m² incident on the solar generator), there is still a 2% discrepancy between the two edges. This results in a radiation mismatch that is unavoidable for this particular installation.

Note: The actual Mont Soleil PV system has an inclination angle $\beta = 50°$ and a mean solar generator orientation of about $\gamma = -26°$. Inasmuch as a table showing the R_B values for this parameter could not be included here for reasons of space, the next closest tabular values were used for the example. The resulting discrepancy is relatively minor.

Example 3

PV system in Kampala with a single-row solar generator with $\beta = 20°$, $\gamma = 0°$. From April to September the installation is inclined northwards and from October to March southwards, with a view to increasing the power yield and minimizing the accumulation of dirt. Ground reflection factor: $\rho = 0.2$.

Calculation task: Monthly and annual mean of total daily irradiation, as well as annual total H_G incident on a solar generator that is out in the open.

Solution: $\alpha_1 = 0°$, $\alpha_2 = 0°$. H_G calculation using Table 2.8:

Energy yield calculation for grid-connected PV plants (with three-component model)

$R_D = \frac{1}{2}\cos\alpha_s + \frac{1}{2}\cos(\alpha_s + \beta) =$		0.970					$R_R = \frac{1}{2} - \frac{1}{2}\cos\beta =$			0.030			
	Jan	Feb	Mar	Apr	May	June	July	Aug	Sept	Oct	Nov	Dec	Year	Unit
H	4.87	4.99	5.16	4.80	4.61	4.51	4.42	4.68	5.00	4.93	4.82	4.69	4.78	kWh/m²
H_D	2.20	2.26	2.30	2.20	2.10	2.04	2.03	2.14	2.25	2.24	2.19	2.14	2.17	kWh/m²
R_B	1.10	1.04	0.95	1.01	1.09	1.13	1.11	1.05	0.96	1.01	1.08	1.12		
k_B	1	1	1	1	1	1	1	1	1	1	1	1		
$H_{GB} = k_B \cdot R_B \cdot (H - H_D)$	2.94	2.84	2.72	2.63	2.74	2.79	2.65	2.67	2.64	2.72	2.84	2.86	2.75	kWh/m²
$H_{GD} = R_D \cdot H_D$	2.13	2.19	2.23	2.13	2.04	1.98	1.97	2.08	2.18	2.17	2.12	2.08	2.11	kWh/m²
ρ	0.2	0.2	0.2	0.2	0.2	0.2	0.2	0.2	0.2	0.2	0.2	0.2		
$H_{GR} = R_R \cdot \rho \cdot H$	0.03	0.03	0.03	0.03	0.03	0.03	0.03	0.03	0.03	0.03	0.03	0.03	0.03	kWh/m²
$H_G = H_{GB} + H_{GD} + H_{GR}$	5.10	5.06	4.98	4.79	4.81	4.80	4.65	4.78	4.85	4.92	4.99	4.97	4.89	kWh/m²

Blue: Winter in northern hemisphere, inclination towards south   Red: Summer in northern hemisphere, inclination towards north

Total annual H_G: $H_{Ga} = 365\,\text{d/a} \cdot 4.89\,\text{kWh/m}^2/\text{d} = 1785\,\text{kWh/m}^2/\text{a}$.

Owing to the seasonal orientation adjustment at this equatorial site, less dirt accumulates on the solar generator and the monthly yields are more consistent; surprisingly, the annual yield is 2.3% higher than the irradiation on the horizontal plane.

Example 4

PV system in Perth with a single-row solar generator with $\beta = 35°$, $\gamma = 0°$. Mean year-round ground reflection factor: $\rho = 0.3$.

Calculation task: Monthly and annual mean of total daily irradiation, as well as annual total H_G incident on a solar generator that is out in the open.

Solution: Using Table 2.8:

Energy yield calculation for grid-connected PV plants (with three-component model)

$R_D = \frac{1}{2}\cos\alpha_s + \frac{1}{2}\cos(\alpha_s + \beta) =$				0.910			$R_R = \frac{1}{2} - \frac{1}{2}\cos\beta =$				0.090			
	Jan	Feb	Mar	Apr	May	June	July	Aug	Sept	Oct	Nov	Dec	Year	Unit
H	7.75	6.92	5.73	4.18	3.14	2.74	2.86	3.74	5.03	6.31	7.50	7.92	5.31	kWh/m²
H_D	2.35	2.20	1.90	1.58	1.25	1.07	1.15	1.41	1.77	2.11	2.30	2.43	1.79	kWh/m²
R_B	0.87	0.98	1.15	1.37	1.62	1.76	1.70	1.47	1.23	1.03	0.90	0.84		
k_B	1	1	1	1	1	1	1	1	1	1	1	1		
$H_{GB} = k_B \cdot R_B \cdot (H - H_D)$	4.70	4.63	4.40	3.56	3.06	2.94	2.91	3.43	4.01	4.33	4.68	4.61	3.93	kWh/m²
$H_{GD} = R_D \cdot H_D$	2.14	2.00	1.73	1.44	1.14	0.97	1.05	1.28	1.61	1.92	2.09	2.21	1.63	kWh/m²
ρ	0.3	0.3	0.3	0.3	0.3	0.3	0.3	0.3	0.3	0.3	0.3	0.3		
$H_{GR} = R_R \cdot \rho \cdot H$	0.21	0.19	0.16	0.11	0.09	0.07	0.08	0.10	0.14	0.17	0.20	0.21	0.14	kWh/m²
$H_G = H_{GB} + H_{GD} + H_{GR}$	7.05	6.82	6.29	5.11	4.29	3.98	4.04	4.81	5.76	6.42	6.97	7.03	5.71	kWh/m²

Total annual H_G: $H_{Ga} = 365\,\text{d/a} \cdot 5.71\,\text{kWh/m}^2\text{/d} = 2084\,\text{kWh/m}^2\text{/a}$.

2.6 Approximate Annual Energy Yield for Grid-Connected PV Systems

Calculating site-specific irradiation on the solar generator plane as described in Section 2.4 or 2.5 provides a far more accurate estimate of annual energy yield for a grid-coupled PV system than with the approximate guide values in Section 1.4.4. However, inasmuch as most stand-alone systems are unable to use all of the available PV installation energy, the calculation for these installations is more complex and thus will be discussed in Chapter 8.

Multiplying total annual H_{Ga} of H_G by the mean annual sum performance ratio PR_a allows for an approximate determination of specific annual energy yield Y_{Fa} (for more on specific annual energy yield, see Section 1.4.4 and Chapter 7; for more on precise yield calculations, see Chapter 8):

$$\text{Approximate specific annual energy yield}: Y_{Fa} = \frac{E_a}{P_{Go}} = PR_a \frac{H_{Ga}}{1\,\text{kW/m}^2} \qquad (2.30)$$

Dividing by $1\,\text{kW/m}^2$ entails no calculation effort, but merely provides the correct unit. If the annual mean value of total daily irradiation of H_G is available in lieu of total annual sum of H_G, to obtain the requisite total annual H_{Ga}, multiply the annual daily mean by the number of days in the year, i.e. 365. The performance ratio PR_a for most grid-connected PV systems ranges from around 65 to 85%. Reasonably sized installations normally achieve a PR_a exceeding 70%, and good systems reach values exceeding 75%. Excellent systems at sites with relatively low summer temperatures (e.g. in the mountains) can even achieve PR_a values of up to around 80%. The rule-of-thumb PR_a value for good PV systems in Central Europe is around 75%.

The annual performance ratio PR_a can be expressed as the product of three variables (see Chapter 7):

$$\text{Annual performance ratio } PR_a = k_{Ta} \cdot k_{Ga} \cdot \eta_I = k_{Ta} \cdot k_{Ga} \cdot \eta_{WR} \qquad (2.31)$$

where:

k_{Ta} = mean annual temperature correction factor. If no exact data are available, the following approximate estimate applies in the presence of T_{Ua} (mean annual ambient temperature in °C) for PV systems with crystalline solar modules:

$$k_{Ta} \approx 1 - 0.0045(T_{Ua} + 2) \qquad (2.32)$$

k_{Ga} = mean annual generator correction factor. Depending on tolerance, this value normally ranges from around 75% (for systems with numerous modules with unduly low P_{Mo}, small β, sporadic winter snow cover, problems with shading, dirt, maximum power tracking by inverters, mismatch and so on) to 90% (for new and very good systems with modules that actually deliver the nominal rated output P_{Mo}). k_G values exceeding 90% are achievable if effective module power output exceeds P_{Mo} or for PV systems with solar generators that integrate uniaxial or biaxial solar tracking mechanisms.

η_{WR} = mean (total) inverter efficiency (European efficiency, average efficiency for radiation conditions in Europe); η_{WR} for selected inverters is indicated in Tables 5.11 and 5.12. Modern inverters achieve η_{WR} values ranging from around 90 to 97%, depending on size.

2.6.1 Examples of Approximate Energy Yield Calculations

The radiation calculations used here are from the examples in Section 2.4.2.

Example 1: Installation in Berlin with $\beta = 30°$, $\gamma = 0°$

$H_{Ga} = 1098\,\text{kWh/m}^2/\text{a}$ (and $3.01\,\text{kWh/m}^2/\text{d}$), $k_{Ga} = 86\%$, $\eta_{WR} = 96\%$, $T_{Ua} = 9.0\,°\text{C}$

$\Rightarrow$ With Equation 2.32, $k_{Ta} = 0.951 = 95.1\%$, while with Equation 2.31, $PR_a = 78.5\% = 0.785$

$\Rightarrow Y_{Fa} = PR_a \cdot H_{Ga}/(1\,\text{kW/m}^2) = 862\,\text{kWh/kWp/a}$ (and $2.36\,\text{kWh/kWp/d}$).

Example 2: Installation in Montana, Switzerland, with $\beta = 60°$, $\gamma = 0°$

$H_{Ga} = 4.24\,\text{kWh/m}^2/\text{d}$ and $1548\,\text{kWh/m}^2/\text{a}$, $k_{Ga} = 88\%$, $\eta_{WR} = 94\%$, $T_{Ua} = 5.7\,°\text{C}$

$\Rightarrow$ With Equation 2.32, $k_{Ta} = 0.965 = 96.5\%$, while with Equation 2.31, $PR_a = 79.8\% = 0.798$

$\Rightarrow Y_{Fa} = PR_a \cdot H_{Ga}/(1\,\text{kW/m}^2) = 3.39\,\text{kWh/kWp/d}$ and $1236\,\text{kWh/kWp/a}$.

Example 3: Installation in Nice, France, with $\beta = 45°$, $\gamma = 30°$

$H_{Ga} = 4.35\,\text{kWh/m}^2/\text{d}$ and $1589\,\text{kWh/m}^2/\text{a}$, $k_{Ga} = 85\%$, $\eta_{WR} = 93\%$, $T_{Ua} = 15.3\,°\text{C}$

$\Rightarrow$ With Equation 2.32, $k_{Ta} = 0.922 = 92.2\%$, while with Equation 2.31, $PR_a = 72.9\% = 0.729$

$\Rightarrow Y_{Fa} = PR_a \cdot H_{Ga}/(1\,\text{kW/m}^2) = 3.17\,\text{kWh/kWp/d}$ and $1158\,\text{kWh/kWp/a}$.

Example 4: Installation in Cairo with $\beta = 30°$, $\gamma = 0°$

$H_{Ga} = 6.06\,\text{kWh/m}^2/\text{d}$ and $2213\,\text{kWh/m}^2/\text{a}$, $k_{Ga} = 87\%$, $\eta_{WR} = 95\%$, $T_{Ua} = 21.4\,°\text{C}$

$\Rightarrow$ With Equation 2.32, $k_{Ta} = 0.895 = 89.5\%$, while with Equation 2.31, $PR_a = 73.9\% = 0.739$

$\Rightarrow Y_{Fa} = PR_a \cdot H_{Ga}/(1\,\text{kW/m}^2) = 4.48\,\text{kWh/kWp/d}$ and $1636\,\text{kWh/kWp/a}$.

2.7 Composition of Solar Radiation

Solar radiation is a mixture of light of varying wavelengths, some of which are visible to the naked eye and some of which are not. If solar radiation intensity is represented in a graph as a function of wavelength, the solar radiation spectrum is obtained (see Figure 2.42). The radiation for this spectrum at the edge of the Earth's atmosphere (AM0) differs somewhat from the mean radiation incident on the Earth's surface in

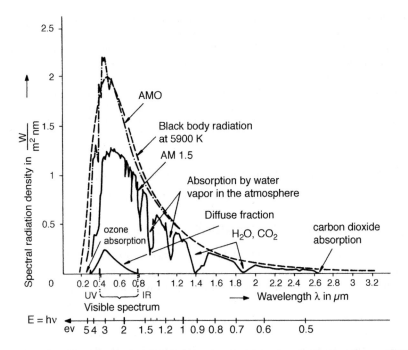

Figure 2.42   Spectrum of solar radiation: (i) Intensity as a function of wavelength and photon energy, (ii) Ultraviolet (UV) spectrum: $100\,\text{nm} < \lambda < 380$ nm, (iii) Visible spectrum: $380\,\text{nm} < \lambda < 780\,\text{nm}$, (iv) Infrared (IR) spectrum: $780\,\text{nm} < \lambda < 1\,\text{mm}$, (v) AM0: extraterrestrial radiation spectrum, (vi) AM1.5: spectrum of radiation on the Earth's surface following penetration of 1.5 times the density of the atmosphere

Central Europe (AM1.5), by virtue of the fact that a portion of this radiation is lost through atmospheric reflection, absorption and scatter.

As is well known, light consists of both waves and particles. In order to understand how solar cells work, it is essential to be aware of the fact that light is composed of a large number of individual light quanta (light particles) or photons, each of which exhibits a very specific energy E, which in turn has a very specific relationship with wavelength and frequency. The following equation applies to photon energy:

$$E = h \cdot v = h\frac{c}{\lambda} \tag{2.33}$$

where:

E = photon energy (often indicated as electron volts (eV) rather than joules, where
    $1\,\text{eV} = 1602 \cdot 10^{-19}\,\text{J}$)
v = frequency (Hz)
λ = wavelength (m)
h = Planck's constant $= 6.626 \cdot 10^{-34}\,\text{W s}^2$
c = speed of light $= 299\,800\,\text{km/s} = 2.998 \cdot 10^8\,\text{m/s}$

A side-by-side comparison of the AM1.5 and AM0 spectra in the solar spectrum (Figure 2.42) clearly shows that certain wavelength ranges are absorbed in whole or in part by specific elements of the atmosphere. According to [Sta87], the AM1.5 spectrum (see Figure 2.42) equates to an irradiance of $835\,\text{W/m}^2$, which is the irradiance that occurs at sea level after solar radiation has passed through 1.5 times the density of the atmosphere. On the other hand, solar cell output P_{max} and efficiency η_{PV} are usually determined using the AM1.5 spectrum, increased by the factor $1.198 = 1/0.835$, which equates to an irradiance of $1\,kW/m^2$ (see the bolded black curve in Figure 3.19 and the tabular values in [Gre95]).

Hence the spectral intensity distribution of solar radiation is a key factor for PV systems since a photon can only release electrical energy in a solar cell if the photon carries at least one specific energy (band gap energy E_G) whose nature varies according to the material involved. Lower-energy photons do not promote energy conversion. With higher-energy photons, only a portion of the energy – namely, the band gap energy – can be used (see Chapter 3). The band gap energy of monocrystalline silicon is around 1.1 eV. On overcast days, chiefly diffuse radiation is available in the visible spectrum, i.e. the percentage of visible light in the spectrum is somewhat greater (see Figure 2.42). Hence, under such conditions reference cells register somewhat higher irradiance relative to pyranometer readings (see Figure 2.47).

2.8   Solar Radiation Measurement

Solar radiation can be accurately measured solely on the horizontal plane, though obtaining such measurements on the solar generator plane is no easy matter. However, in view of the fact that accurate radiation readings are essential for determining whether PV systems are working properly, the key sensors that are used for such measurements and the problems associated with these devices will now be briefly discussed.

2.8.1   Pyranometers

In the field of meteorology, virtually all irradiance and irradiated energy are measured worldwide using pyranometers,which are composed of a thermopile comprising a large number of series-connected thermocouples. The black receiver at the tip of a thermocouple absorbs a broad band of total irradiated energy in the $300\,nm < \lambda < 3\,\mu m$ wavelength spectrum. A double hemispheric glass dome prevents additional reflection on the glass at small angles of incidence and helps avert condensation on the glass under unfavourable temperature to moisture ratios. A pyranometer's output voltage is (with a certain amount of dwell) exactly proportional to irradiance, but is relatively weak since the voltage generated by thermocouples is extremely low. With $G = 1\,kW/m^2$, the output voltage of a typical commercial pyranometer such as the Kipp&Zonen CM 11 is around 5 mV.

This relatively low output voltage cannot be processed directly, and must therefore be amplified. Design engineering a precise amplifier that is able properly and accurately to amplify signals ranging from around $10\,\mu V$ to around 10 mV necessitates a certain amount of electrical engineering acumen. Substantial measurement errors can occur with poorly conceived measuring set-ups (e.g. improper shielding, substandard measurement amplifiers and so on).

Inasmuch as pyranometer sensitivity gradually decreases (of the order of tenths of a per cent per year), these devices should be calibrated every two years in settings where extreme accuracy is needed. Reasonably priced pyranometer calibration services are available from specialized organizations such as the World Radiation Centre in Davos. These calibrations can be readily replicated. Good-quality and accurate pyranometers cost anywhere from €1300 to €2000 and their drying agent needs to be replaced annually. Figure 2.43 shows an example of a pyranometer that is used to measure global radiation incident on the horizontal plane.

2.8.2   Reference Cells

Reference cells are precisely calibrated silicon solar cells that are used to measure radiation and are well packed to protect them from the environment. At prices ranging from €200 to €400, they are considerably less expensive than pyranometers. Short-circuit current in a solar cell can safely be assumed to be proportional to the irradiance G that occurs at the reference cell (see Chapter 3). Many reference cells integrate a precise measurement shunt that converts the short-circuit current into a voltage (e.g. about 30 millivolts with $G = 1\,kW/m^2$) and an additional sensor that allows for precise cell

Figure 2.43   Kipp&Zonen CM 11 pyranometer for measurement of global radiation incident on the horizontal plane

temperature measurement (e.g. a PT-100 sensor, thermocouple, or the like). Figure 2.44 shows two such reference cells.

Unlike pyranometers, solar cells only use a portion of the solar spectrum, i.e. photons whose energy exceeds the band gap energy E_G (see Section 2.7 and Chapter 3). Although solar cell calibration technicians try to take account of this particularity as much as possible, as the solar spectrum is not always fully homogeneous (including the presence of exactly the same G and the same AM number), solar cell calibration discrepancies can arise. Many vendors or calibration providers calibrate these reference cells using artificial light sources, which are also used for solar module measurements. Although such light sources are similar to natural sunlight, they do not have exactly the same spectrum as natural sunlight. [Ima92] proposes a reference cell calibration method for use in natural sunlight. Hence, precise and replicable reference cell calibration poses a problem for PV systems.

Figure 2.44   Two reference cells with integrated cell temperature measurement sensors; the ESTI cell is shown above, and the Siemens M1R is below. The M1R is the centre measurement cell, around which eight non-contacted cells are arrayed, thus allowing for highly accurate readings of the thermal conditions in a module

Reference cells should be mounted at a point on the solar generator plane where the solar radiation reflects the mean solar radiation for the site as a whole. Wherever possible, the same cell technology should be used for reference cells that is used in the PV system being measured. If the reference cells are the same types as those used in the solar module, and if they possess a flat glass surface like the cells shown in Figure 2.44 (and preferably are made out of the same type of glass as the module glass), the reference cells will be affected by various key factors that affect the reference cells in exactly the same way the solar modules are affected. Hence variations in the solar spectrum, as well as additional reflection from direct beam radiation incident at small angles of incidence, have the exact same impact on reference cells as on solar modules, thus enabling integrated temperature sensors to provide a point of reference for mean module temperatures that are measured in the solar generator. Reference cells mounted in this fashion are a useful tool for monitoring the correct functionality of a PV system (see Chapter 7). Reference cells with glass domes are less suitable for this purpose since they reflect radiation at a low angle of incidence to a lesser extent and differently relative to solar modules.

The ESTI reference cell shown in Figure 2.44 was widely used in the 1990s in particular. These cells are for the most part well calibrated and exhibit only minor discrepancies between the readings for the various cells. Unfortunately, they also have an endemic defect: that is, radiation at a low angle of incidence can penetrate the lateral edge of the cell, whereupon this radiation is bounced onto the measurement cell via total reflection from the glass surface – particularly in the presence of the kind of white backsheet that is used on many of these sensors. This defect can cause the cell readings to be several per cent too high. ESTI reference cells with a black backing sheet (as in Figure 2.44) are less prone to this error and thus should be used in order to obtain accurate measurements.

2.8.3 Pyranometer Versus Reference Cell Measurements

Synchronous measurement of irradiance on the solar generator plane using a pyranometer, together with a reference cell calibrated by the module vendor (see Figure 2.45), yield revealing results. Most of the H_G readings obtained using reference cells are several percentage points lower than those obtained with a highly accurate CM 11 or MC 21 pyranometer.

Figures 2.46–2.49 show this type of comparison of readings obtained using Siemens M1R reference cells (with 2% accuracy) versus the readings obtained with pyranometers at a PV installation in Liestal

Figure 2.45 Synchronous measurement of irradiance on the module plane using a Siemens M1R mono-c-SI reference cell and a heated and ventilated Kipp&Zonen CM 11 pyranometer

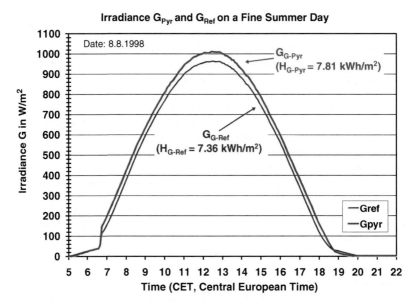

Figure 2.46 Synchronous measurement of irradiance on the module plane using a Siemens M1R mono-c-SI reference cell and a heated and ventilated Kipp&Zonen CM 11 pyranometer on a sunny summer day at a PV system with a 30° inclination angle in Liestal (elevation 340 m). Interestingly, the pyranometer readings are several percentage points higher

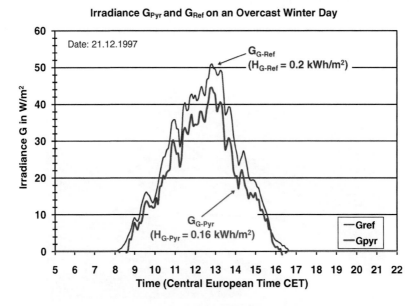

Figure 2.47 Synchronous measurement of irradiance on the module plane using a Siemens M1R mono-c-SI reference cell and a heated and ventilated Kipp&Zonen CM 11 pyranometer on an overcast winter day at a PV system with a 30° inclination angle in Liestal (elevation 40 m). Interestingly, the pyranometer readings are considerably lower

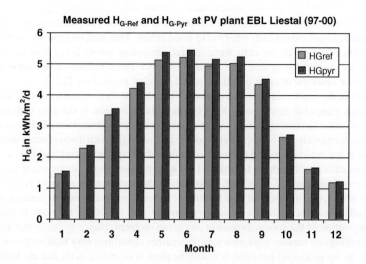

Figure 2.48 Synchronous measurement of the monthly mean of total daily irradiation H_G on the module plane using a Siemens M1R mono-c-SI reference cell and a heated and ventilated Kipp&Zonen CM 11 pyranometer on a sunny summer day at a PV system with a 30° inclination angle in Liestal (elevation 340 m) from 1997 to 2000. Interestingly, the multi-year mean pyranometer readings are 4.2% higher

(inclination angle 30°, elevation 340 m) and one on the Jungfraujoch (inclination angle 90°, elevation 3454 m). A more precise analysis reveals that on sunny days in particular (see Figure 2.46) the reference cell readings are generally considerably lower than the pyranometer readings, whereas on days with little direct beam radiation the crystalline reference cells often register somewhat more radiation (see Figure 2.47).

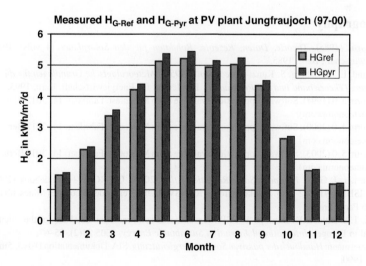

Figure 2.49 Synchronous measurement of the monthly mean of total daily irradiation H_G on the module plane using a Siemens M1R mono-c-SI reference cell and a heated and ventilated Kipp&Zonen CM 21 pyranometer on a sunny summer day at a PV system with a 90° inclination angle on the Jungfraujoch (elevation 3454 m) from 1997 to 2000. Interestingly, the multi-year mean pyranometer readings are 4.2% higher

Inasmuch as there are clear days and overcast days during any given year, these effects on the monthly and annual mean readings cancel each other out to some extent. That said, as a rule the multi-year mean irradiation readings from reference cells were several percentage points lower than the pyranometer readings at the same fixed PV system (see Figures 2.48 and 2.49). Similar and in some cases larger discrepancies were observed at other installations and by other researchers [2.10].

The spectrum of the light source used by vendors to calibrate their reference cells ($G = 1\,kW/m^2$) probably contains somewhat more usable light for the solar cells than is the case with natural light. However, in view of the fact that (a) vendors use the same devices to calibrate their modules and reference cells, and (b) vendor module power outputs are specified accordingly, such discrepancies should be taken into account in order to obtain more precise calculations of PV system energy yield.

Such discrepancies could perhaps be reduced for measurement purposes by using the calibration procedure proposed by [Ima92]. However, inasmuch as the spectrum of sunlight is determined not only by the air mass (AM) number but also by current atmospheric water vapour content, calibrations of reference cells in natural sunlight against pyranometers will inevitably be prone to random errors due to the composition of the solar spectrum at any given moment. Hence pyranometer and reference cell G and H readings on various days with similar weather conditions may well vary – in the present case, as noted, by up to several per cent. It would be ideal if reference cells that are highly weather resistant were available from an internationally recognized test lab for each cell technology and if such reference cells were calibrated extremely accurately relative to stable primary standards of the same type.

This spectral mismatch between natural sunlight and the light sources used for calibration is the source of some of the discrepancies in the solar generator correction factor k_G and the anticipated value for the ideal case, namely 1 (see Section 2.6 and further details in Chapters 7 and 8). Consequently, calculation simulation software for PV system energy yield, which (as is normally the case) meteorologists use in tandem with pyranometer radiation readings and that do not consider spectral mismatches, mostly provide energy yield figures that are too high by several percentage points, even if other key factors such as reduced radiation attributable to low angles of incidence, horizons, shading, power loss and so on are duly taken into account and the requisite additional work is performed.

2.9   Bibliography

[2.1]   *Meteonorm (1985). Theorie, Daten, Rezepte, Rohdaten für den Solarplaner* (4 vols). Bundesamt für Energiewirtschaft, Bern, 1985.

[2.2]   J. Remund, E. Salvisberg, S. Kunz: *Meteonorm (1995) – Meteorologische Grundlagen für die Sonnenenergienutzung (Theorieband und PC-Programm).* Bundesamt für Energiewirtschaft, Bern, 1995.

[2.3]   Meteonorm 3. 0 (1997). Software application available from Meteotest, Fabrikstr. 14, 3012 Bern, Switzerland. (www.meteonorm.com).

[2.4]   Meteonorm 4. 0 (2000). Software application available from Meteotest, Fabrikstr. 14, 3012 Bern, Switzerland. (www.meteonorm.com).

[2.5]   Meteonorm 5. 0 (2003). Software application available from Meteotest, Fabrikstr. 14, 3012 Bern, Switzerland. (www.meteonorm.com).

[2.6]   K. Scharmer, J. Greif: *European Solar Radiation Atlas* (2000). CD-ROM + Guidebook (290 pages), A4 format, ISBN 2-911762-22-3. Available from TRANSVALOR, Presses de l'Ecole des Mines, 60 Bd St Michel, F-75006 Paris, France.

[2.7]   M. Šúri, T.A. Huld, E.D. Dunlop: 'PV-GIS: A web-based solar radiation database for the calculation of PV potential in Europe'. *International Journal of Sustainable Energy*, 2005 24(2), 55–67.

[2.8]   M. Zimmermann: *Handbuch der passiven Sonnenenergienutzung.* SIA-Dokumentation D 010, 5th edition.SIA, Zurich, 1990.

[2.9]   P. Valko: 'Solardaten für die Schweiz' Docu-Bulletin *3/1982.* Verlag Schweizer Baudokumentation, Blauen.

[2.10] T. Degner, M. Ries:'Evaluation of Long-Term Performance Measurements of PV Modules with Different Technologies'. 19th EU PV Conference, Paris, 2004.

[2.11] T. Huld, M. Suri, E. Dunlop: 'Comparison of Potential Solar Electricity Output from Fixed-Inclined and Two-Axis Tracking Photovoltaic Modules in Europe'. *Progress in Photovoltaics: Research and Applications*, 2009 16, 47–59 (www.interscience.wiley.com).

[2.12] Meteonorm 6.1 (2009). Software application available from Meteotest, Fabrikstr. 14, 3012 Bern, Switzerland (www.meteonorm.com).

Additional References

[Bur93], [DGS05], [Eic01], [Gre95], [Häb91], [Häb07], [Her92], [Hu83], [Ima92], [Lad86], [Luq03], [Mar94], [Mar03], [Qua03], [See93], [Sta87], [Wen95], [Wil94], [Win91].

Yields as secure as the gold in Fort Knox.

Since 1936, the United States government has safely stored its gold at Fort Knox. Gold has been a dependable investment for centuries. Wise investors today put their money in PV. So it's only natural that Fort Knox is equipped with a PV plant. No wonder the security experts have chosen inverters by KACO new energy. But watch out: Some would even steal for maximum PV yields. We say why bother when you can simply buy a KACO inverter. They are the safest investment around. Ask a dealer today!

KACO new energy. We turn passion into power.

www.kaco-newenergy.de

3

Solar Cells: Their Design Engineering and Operating Principles

3.1 The Internal Photoelectric Effect in Semiconductors

Each atom is composed of a positively charged nucleus and a shell comprising negatively charged electrons (charge $e = -1.602 \cdot 10^{-19}$ A s). When a photon collides with an atom, the photon's energy, $E = h \cdot v$, can be transferred to an electron. In this case the photon is absorbed.

In the external photoelectric effect (e.g. in the case of highly ionizable alkali metals such as Li, Cs and so on), the electron can be liberated from the material (e.g. photoemission in a photocathode) if the energy, $E = h \cdot v$, carried by the photon is higher than the energy E_A needed by an electron to leave the material.

On the other hand, in solar cells made of semiconductor materials the internal photoelectric effect comes into play. A photon that carries sufficient energy, $E = h \cdot v$, can liberate an electron from the crystal lattice or lift it out of the valence band into the conduction band.

In semiconductors, electrons are normally bonded to the outermost shell in the crystal lattice; these are known as valence electrons. In order for an electron to be liberated from its lattice, a minimum amount of additional energy is needed; this is known as band gap energy E_G. This situation is illustrated by the semiconductor band model (Figure 3.1).

Strictly speaking, the aforementioned situation applies solely to temperatures close to the absolute zero point. When semiconductor temperature rises, the crystal-lattice atoms begin oscillating around their respective steady positions, with the result that some of the valence bonds are broken and the electrons thus liberated migrate to the conduction band; this phenomenon is known as intrinsic conductivity. The stronger the band gap energy, the fewer electrons migrate to the conduction band, i.e. the lower the electrical conductivity of the material at a given temperature. On the other hand, the higher the temperature of a given semiconductor material, the more electrons migrate to the conduction band and electrical conductivity increases accordingly.

Liberation of an electron from a valence bond creates a hole in the crystal lattice. An electron from an adjoining atom's valence bond can fall into a crystal-lattice hole, in which case the original hole disappears but a new one is created elsewhere. Hence, like a free electron, a hole can move freely within a semiconductor and can also promote conductivity. If a free electron happens to collide with a hole, it will fall into it, i.e. the electron and hole recombine.

In the case of radiation onto semiconductor materials, photons with sufficient energy, $h \cdot v > E_G$, can lift an electron out of the valence band into the conduction band, whereupon the photon is absorbed, a

Photovoltaics: System Design and Practice. Heinrich Häberlin.
© 2012 John Wiley & Sons, Ltd. Published 2012 by John Wiley & Sons, Ltd.

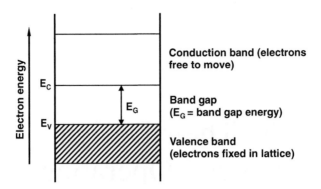

Figure 3.1 Semiconductor band model. The allowable energy levels in a solid are no longer discrete as in atoms, but are instead distributed owing to the proximity of other atoms to the energy bands. The width of the prohibited zone, band gap energy E_G, is determined by the semiconductor material used. The lower threshold of the valence band and the upper threshold of the conduction band are often omitted

hole appears in the valence band and a free electron appears in the conduction band (see Figure 3.2). With directly absorbent semiconductors, micro-thin material (around 1 μm) can fully absorb all photons with sufficient energy, whereas with indirectly absorbent semiconductors such as crystalline silicon, the optical path in the semiconductor material needs to be at least 100 μm in order for all low-energy photons (red light and near-infrared light) to be reliably absorbed. Hence, directly absorbent semiconductors are mainly used to make thin-film solar cells entailing low material use, whereas more materials are needed for indirectly absorbent semiconductors due to the requisite minimum optical path; or, if thinner material is used, the effective optical path is extended using special techniques (see Figure 3.26).

The electrons generated by absorption of a photon and the consequent hole are in very close proximity to each other. Hence in semiconductors with no electric field that separates electrons and holes from each other, the electrons soon fall back into their respective holes, with the result that the photon energy deflagrates to no avail and does only heat up the semiconductor. Photons where $h \cdot v < E_G$

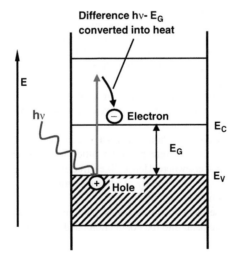

Figure 3.2 Internal photoelectric effect: photons where $h \cdot v > E_G$ lift an electron out of the valence band into the conduction band and are absorbed

do not allow an electron to be lifted out of the valence band into the conduction band and are thus not absorbed.

If an external voltage source in an irradiated semiconductor generates an electric field, this field separates the holes from the electrons generated by the absorbed photons, thus engendering a photo-conductor or photoresistance whose conductivity is proportional to irradiance but which is nonetheless a passive element that unfortunately cannot produce electricity.

However, under certain circumstances (e.g. at a junction between a p-type and n-type semiconductor), strong internal electric fields can also be generated without an external voltage source. Hence these fields are used to separate photon-engendered electrons from their respective holes and thus leverage the energy resulting from these electrons and holes being separated. This is the mechanism that forms the basis for and is realized in solar cells. Hence to understand how a solar cell works, it is necessary briefly to discuss semiconductor doping and the conditions that obtain at the p–n junction.

3.2   A Brief Account of Semiconductor Theory

Semiconductors are materials whose electrical conductivity is lower than that of conductors but higher than that of non-conductors. Silicon (Si), which is the most widely used semiconductor material today, is abundantly available around the globe and ecologically friendly. The other semiconductor materials that can be used for technical applications are germanium (Ge), selenium (Se), gallium arsenide (GaAs), gallium phosphide (GaP), indium phosphide (InP), cadmium sulphide (CdS), cadmium telluride (CdTe) and copper indium diselenide ($CuInSe_2$ or CIS, sometimes with a small amount of gallium added to form copper indium gallium diselenide, $Cu(In,Ga)Se_2$, or CIGS). One of the key parameters for the characteri-zation of semiconductor properties is band gap energy E_G. Table 3.1 shows the band gap energy and absorption mechanisms for some of the most important semiconductors.

Figure 3.3 shows the spatial structure of a silicon semiconductor crystal. Silicon has four valence electrons in its outermost shell. In order to establish a stable electron configuration (rare-gas configuration with eight electrons), each silicon atom along with four adjacent atoms form a covalent bond, where each atom controls an electron at each bond; thus a bond consists of two electrons. In a silicon crystal, eight electrons are arrayed around each silicon atom, which means that the desired electron configuration has been attained.

In view of the relative complexity of the spatial structure shown in Figure 3.3, the schematic diagram of a crystal structure as in Figure 3.4 is often used.

3.2.1   Semiconductor Doping

Although semiconductor conductivity (see Section 3.1) at temperatures above the absolute zero point is considerably higher than the conductivity of insulators, it is nonetheless very low. This conductivity can be substantially increased by adding suitable external atoms in a process known as semi-conductor doping.

Table 3.1   Band gap energy and absorption mechanisms for selected semiconductor materials

Semiconductor	Abbreviation	Band gap energy E_G (eV)	Absorption mechanism
Germanium	Ge	0.66	Indirect
Copper indium diselenide	$CuInSe_2$ (CIS)	1.02	Direct
Crystalline silicon	c-Si	1.12	Indirect
Indium phosphide	InP	1.35	Direct
Gallium arsenide	GaAs	1.42	Direct
Cadmium telluride	CdTe	1.46	Direct
Amorphous silicon	a-Si	≈1.75	Direct
Cadmium sulphide	CdS	2.4	Direct

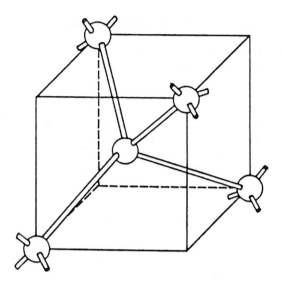

Figure 3.3   Spatial crystal structure of crystalline silicon (c-Si)

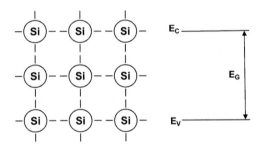

Figure 3.4   Crystal lattice and band model of undoped silicon (intrinsic Si), $E_G = 1.12\,\text{eV}$

If (as in Figure 3.5) a silicon atom is replaced by a phosphorus atom with five valence electrons in the outermost shell, one of these electrons cannot form a bond with one of the four adjacent atoms. As a result, the electron can readily liberate itself from its atomic nucleus, which can then regain its positive charge. Hence a phosphorus atom 'donates' an electron to the crystal lattice and is thus referred to as the donor,

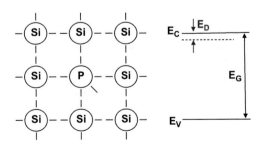

Figure 3.5   Crystal lattice and band model of silicon (n-Si) that is doped with donors (P)

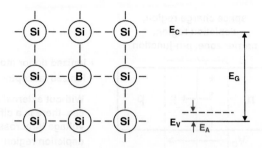

Figure 3.6   Crystal lattice and band model of silicon (p-Si) that is doped with acceptors (P)

while the counterpart electron is referred to as the donor electron. In the band model, the energy of the donor electron is only slightly below (by E_D) the band threshold for the conduction band, which means that only a minute amount of energy needs to be obtained from the temperature-induced motion for the donor electron to migrate to the conduction band. Donor electrons engender conductivity by means of negative charge carriers; thus the semiconductor is n-conductive.

On the other hand if (as shown in Figure 3.6), a silicon atom on the outermost shell is replaced by a boron atom with only three valence electrons, only three of the four boron atoms bonding with the adjacent silicon atoms can be saturated with electrons, i.e. one of the four bonds is missing an electron, thus engendering a hole. An electron from an adjoining atom's valence bond can fall into this hole, in which case the original hole disappears, a new one is created elsewhere and the boron atom becomes negatively charged.

In the band model, the energy available at the space for the electron not provided by the acceptor atom for the fourth adjacent atom is only slightly higher (by E_A) than at the upper edge of the valence band. Hence an electron in the valence band needs only a minute amount of energy from the temperature-induced motion to fill the aforementioned space and in so doing leaves a hole in the valence band.

A hole in a valence band can move as freely as a free electron in the conduction band and is thus also referred to as a defective electron. A boron atom can 'accept' an electron and is thus called an acceptor. Hence acceptors contribute to conductivity through holes in the valence band, i.e. they are positive charge carriers, thus rendering semiconductors p-conductive.

Key to understanding the p–n junction process is that donor atoms that have donated an electron constitute permanently integrated positive charges (ions) in the crystal lattice; conversely, acceptor atoms that have absorbed an electron constitute permanently integrated negative charges (ions) in the crystal lattice.

Hence semiconductor doping is actually a process involving targeted semiconductor contamination that can only be carried out in a controlled manner if the semiconductor is extremely pure – which for silicon means one external atom per 10^{10} silicon atoms. Needless to say, attaining this level of purity is an extremely cost-intensive process, prompting manufacturers of inexpensive solar cells to reduce this cost by reducing the requisite purity gradient of the base material through the use of so-called solar silicon in lieu of purer electron silicon.

In addition to P, other five-valence elements such as As, Sb or Bi can serve as donors. As for acceptors, B, Al, Ga or In can also be used.

3.2.2   The P–N Junction

When it comes to understanding the main operating principle of solar cells, it suffices to investigate what is known as a homogeneous junction, i.e. the junction between p- and n-conducting semiconductors made of the same base material. A homogeneous junction automatically engenders a space charge zone and thus a

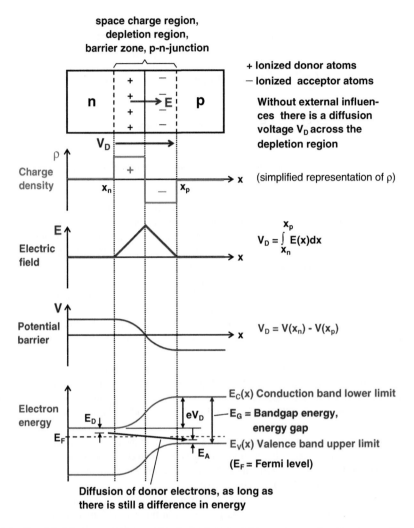

space charge region,
depletion region,
barrier zone, p-n-junction

+ Ionized donor atoms
− Ionized acceptor atoms

Without external influen-
ces there is a diffusion
voltage V_D across the
depletion region

Charge density — (simplified representation of ρ)

$$V_D = \int_{x_n}^{x_p} E(x)dx$$

Electric field

Potential barrier — $V_D = V(x_n) - V(x_p)$

Electron energy

$E_C(x)$ Conduction band lower limit
E_G = Bandgap energy, energy gap
$E_V(x)$ Valence band upper limit
(E_F = Fermi level)

Diffusion of donor electrons, as long as
there is still a difference in energy

Figure 3.7 Simplified depiction of the scenario at a p–n junction without an external voltage source. The electrons diffused from the n-zone to the p-zone fill the holes in the latter. The positively charged donor atoms engender a positive charge in the n-zone, while the now negatively charged acceptor atoms engender a negative charge in the p-zone. These space charges engender an electric field in the boundary layer. Diffusion voltage V_D is created via the space-charge region and thus also a potential difference. As a result, the energy bands in the p-zone are increased by eV_D. Electron diffusion comes to a halt when there is no longer enough energy available for it

strong electric field, which can be used to separate the electron–hole pair resulting from the internal photoelectric effect – as can of course the space charge zones and electric fields associated with the p–n junction between different base materials. In this scenario, known as a heterogeneous junction, the p- and n-doped parts are composed of chemically heterogeneous materials, or a Schottky junction can occur, i.e. a junction between a semiconductor and a metal.

Figure 3.7 illustrates such a p–n junction without an external voltage source. Here, electrons are diffused into the p-zone from the n-zone, where they fill holes. The positively charged donor atoms left behind engender a positive space charge in the n-zone, while the now negatively charged acceptor atoms engender a negative space charge in the p-zone. These space charges create an electric field in the

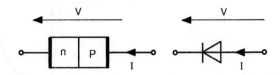

Figure 3.8   Structure and wiring symbols of a semiconductor diode

boundary layer that initially impedes further electron diffusion and ultimately brings it to a halt. The barrier layer thus created at the boundary between the n- and p-material is now devoid of freely moving charge carriers. Diffusion voltage V_D is created via the diffusion zone, thus also engendering a potential difference.

Although what we have said thus far accounts for the creation of a diffusion voltage V_D, we are still in the dark as to its level, which is a key factor for solar cells as it determines the maximum possible open-circuit voltage V_{OC}. In all solar cells, V_{OC} is lower than V_D.

When the effects of electron diffusion from the n-zone to the p-zone in the band model are taken into consideration, it becomes clear that the consequent lower potential V on the p-side induces an increase in the p-side energy bands (the electrons have more energy on account of their negative charge resulting from lower potential). The electrons diffused from the n-zone to the p-zone can gain energy until the lower valence band edge in the p-zone increases to the point where there is no longer a substantial difference between the energy level of the donor electrons on the n-side and that of the acceptor holes on the p-side. Inasmuch as the energy of the donor electrons is E_D lower than the bottom conduction band edge, and the energy of the acceptor holes is E_A higher than the upper valence band edge, eV_D must be somewhat lower than band gap energy E_G.

The following applies to homogeneous p–n junctions at ambient temperature:

$$\text{Diffusion voltage } V_D \approx E_G/e - (0.35 - 0.5)\, \text{V} \tag{3.1}$$

Hence the diffusion voltage is roughly 0.35–0.5 V lower than the so-called theoretical photovoltage V_{Ph} (see Equation 3.16), which is determined by dividing the band gap energy E_G by the electron charge ($e = 1.602 \cdot 10^{-19}$ A s).

If a metal contact is integrated into the n-zone and p-zone, a semiconductor diode results (see Figure 3.8). If such a diode is briefly short-circuited, despite the diffusion voltage at the p–n junction the flow of current is still blocked. In such a case, space and contact charges are immediately created at the contact points between the metal and semiconductor, and this exactly compensates for the diffusion voltage.

3.2.3   Characteristic Curves of Semiconductor Diodes

A semiconductor diode is composed of a p–n junction with metallic electrical terminals (see Figure 3.8). If a positive voltage V is applied in the direction shown (from p to n), holes from the p-side and electrons from the n-side penetrate the barrier layer, whereupon the space charges, diffusion voltage and potential difference between the n-zone and p-zone disappear. The numerous majority carriers comprising electrons on the n-side and holes on the p-side inundate the barrier layer, thus allowing for a substantial flow of current that is conducted by the diode.

But if, on the other hand, a negative voltage V is applied, even more n-side electrons and p-side holes will flow out of and away from the boundary layer, whereupon the space charge zone is enlarged, the voltage passing through the barrier layer will exceed diffusion voltage V_D by the level of the external voltage source, and the potential difference between the n-zone and p-zone will increase. The inverse current is now very low as a result of the thermally generated minority carriers (n-zone holes and p-zone electrons), which can penetrate the potential barrier; thus the diode forms a barrier.

Figure 3.9   Characteristic curve of a semiconductor diode (principle)

The characteristic diode curve $I = f(V)$ (Figure 3.9) is roughly the following;

$$I = I_S \left(e^{eV/nkT} - 1 \right) = I_D \text{ (in Figure 3.12)} \tag{3.2}$$

where:
V = voltage applied at the diode (from p to n)
I = current conducted by the diode ($= I_D$ in the solar cell equivalent circuit shown in Figure 3.12)
I_S = saturation current (idealized inverse current)
e = elementary charge $= 1.602 \cdot 10^{-19}$ A s
n = diode quality factor ($1 < n < 2$; this factor is close to 1 in most cases)
k = Boltzmann's constant $= 1.38 \cdot 10^{-23}$ J/K
T = absolute temperature (in K)

Equation 3.2 is particularly suitable for describing the characteristic line in the forward direction ($V > 0$). In the reverse direction ($V < 0$) the inverse current of diodes rated for higher voltages is not constant, as is meant to be the case according to Equation 3.2, but increases somewhat with increasing voltage. At very high reverse voltages, reverse current may suddenly increase very much creating a breakdown and the diode is destroyed, unless the current is limited to non-hazardous values by external circuitry.

3.3   The Solar Cell: A Specialized Semiconductor Diode with a Large Barrier Layer that is Exposed to Light

3.3.1   Structure of a Crystalline Silicon Solar Cell

As Figure 3.10 shows, a monocrystalline or polycrystalline silicon solar cell (see Figure 3.11) is composed of a large diode with a barrier layer that is exposed to light. In order for as many light quanta as possible to arrive at a point near the barrier layer, the (usually n-Si) semiconductor zone facing the light must be ultrathin (e.g. 0.5 μm). Moreover, the metallic front electrical contacts that are needed to achieve low internal resistance cannot be allowed to shade more than a minute portion of the active solar cell area. An anti-reflection coating is applied to the outside of the solar cell area to minimize reflection.

When light quanta arrive at the solar cell, those light quanta where $h \cdot v > E_G$ can be absorbed by the crystal lattice, thus allowing for creation of an electron–hole pair secondary to the internal photoelectric effect. The strong electric field E in the barrier layer quickly separates this electron–hole pair before it can recombine. On account of the electrons' negative charge, they are subject to a force whose direction is opposite to that of the field direction, and thus accumulate in the n-zone. The holes migrate in the field direction and accumulate in the space charge-free portion of the p-zone.

Since the charges have been separated, the space charge (and thus the electric field strength in the barrier layer) is reduced until it is no longer strong enough to separate electron–hole pairs. It is at this point that the solar cell reaches its open-circuit voltage V_{OC}. In this process, the voltage conducted through the barrier layer becomes far weaker than diffuse voltage V_D, but does not quite reach 0. Hence solar cell open-circuit voltage V_{OC} is always somewhat lower than V_D.

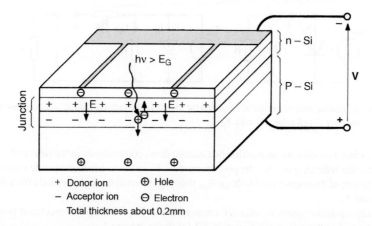

Figure 3.10   Basic structure of a crystalline silicon solar cell (aggregate thickness roughly 150–300 μm)

But if, on the other hand, the front and back contacts are briefly shorted via a conductive connection, the load carriers engendered by the internal photoelectric effect immediately flow out of the relevant zones. Neither the volume charge nor the electric field is attenuated, diffusion voltage V_D continues to flow through the barrier layer, and a maximum current for the defined irradiance is conducted, i.e. short-circuit current I_{SC}, which in a solar cell is proportional to irradiance G.

In most cases the n-layer on the cell area is doped considerably more than the p-layer, a difference that is sometimes indicated by marking the n-layer n^+. Inasmuch as a solar cell must be electrically neutral as a whole, the volume charge zone extends much further into the p-zone than is suggested by the simplified depiction in Figure 3.10.

3.3.2   Equivalent Circuit of a Solar Cell

A non-irradiated solar cell is a standard semiconductor diode that allows forward current to flow from the p-side to the n-side if the voltage is directed from p to n via the diode. When the diode is exposed to light, photocurrent I_{Ph} is also generated that is proportional to irradiance G and flows from the n-side to the p-side. This arrangement can be readily represented using an equivalent circuit from an ideal current source I_{Ph} and a diode – if necessary with an added series resistance R_S and parallel resistance R_P (see Figure 3.12).

Figure 3.11   Left: a polycrystalline or multicrystalline silcon solar cell; right: a monocrystalline solar cell (Photo: AEG)

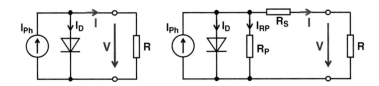

Figure 3.12 Simplified diagram of an equivalent circuit for a loaded solar cell (no-load, $R = \infty$; shorted, $R = 0$). I_{Ph} is proportional to irradiance G and solar cell area A_Z

Under open-circuit conditions, an equalization current flows continuously in the solar cell's equivalent circuit, and thus the following occurs: the photocurrent I_{Ph} from the n-side to the p-side resumes flowing (under the influence of the open-circuit voltage V_{OC} thus engendered) through the diode from the p-side to n-side as current I_D.

The following equation applies to solar cell current I in the simplified equivalent circuit (without R_S or R_P) as a function of voltage V, using Equation 3.2 for diode current I_D:

$$I = I_{Ph} - I_D = I_{Ph} - I_S \left(e^{eV/nkT} - 1 \right) = I_{Ph} - I_S \left(e^{V/V_T} - 1 \right) \tag{3.3}$$

where:
I_{Ph} = photocurrent ($\sim G$) = solar cell short-circuit current I_{SC} (in the simplified circuit)
I_S = saturation current in the reverse direction, which increases roughly exponentially as temperature rises (doubles at approximately 10 K intervals)

The remaining terms for this equation are as in Equation 3.2.
The following abbreviation is also useful:

$$\text{Thermal (diode) voltage } V_T = nkT/e \tag{3.4}$$

where n = the diode quality factor, which ranges from 1 to 2. At 25 °C, $V_T = 25.7$ mV for $n = 1$.
Inverting Equation 3.3 also allows the calculation of solar cell voltage V in the simplified equivalent circuit:

$$V = \frac{nkT}{e} \ln \left(1 + \frac{I_{Ph} - I}{I_S} \right) = V_T \ln \left(1 + \frac{I_{Ph} - I}{I_S} \right) \tag{3.5}$$

$I = 0$ is used for the calculation of open-circuit voltage V_{OC}:

$$V_{OC} = \frac{nkT}{e} \ln \left(1 + \frac{I_{Ph}}{I_S} \right) = V_T \ln \left(1 + \frac{I_{Ph}}{I_S} \right) \approx V_T \ln \left(\frac{I_{Ph}}{I_S} \right) \text{ for } I_{Ph} \gg I_S \tag{3.6}$$

Inasmuch as I_{Ph} is proportional to irradiance G, short-circuit current I_{SC} is also proportional to irradiance in the simplified model, whereas the correlation between open-circuit voltage V_{OC} and irradiance is much weaker. V_{OC} decreases as temperature increases by virtue of the fact that I_S increases exponentially as temperature rises and more than compensates for the increase in kT.

The interrelationships are somewhat more complex for the complete equivalent circuit with R_S and R_P. Although I_{SC} is still approximately I_{Ph} and thus approximately proportional to G, I and V cannot be represented by a closed expression and thus an iteration is necessary. To calculate I as a function of V, a value such as V_i (a value near any desired V) at the inner diode is applied, and then the resulting current I is computed as follows:

$$I = I_{Ph} - I_S \left(e^{eV_i / nkT} - 1 \right) - \frac{V_i}{R_P} \tag{3.7}$$

The effective output voltage at the solar cell terminals is then determined as follows:

$$V = V_i - R_S \cdot I \tag{3.8}$$

Equation 3.8 can also be realized in accordance with V_i, and this term can then be used in Equation 3.7. This yields an equation for I, which, however, can only be realized via iteration:

$$I = I_{Ph} - I_S \left(e^{e(V + R_S \cdot I)/nkT} - 1 \right) - \frac{(V + R_S \cdot I)}{R_P} \tag{3.9}$$

In addition to the single-diode model discussed above, the two-diode model is also used. In this model, in lieu of only one diode (see Figure 3.12) where $1 < n < 2$, two separate diodes where $n_1 = 1$ and $n_2 = 2$ respectively are used [Eic01], [Qua03], [DGS05]. However, this method is more labour intensive on account of the additional parameter. The two-diode model is mainly suitable for highly precise simulations of solar cell characteristics using calculation software. A full basic understanding of the behaviour of solar cells and modules can be achieved via the single-diode model discussed above.

3.3.3 Characteristic Curves of Solar Cells

If a solar cell uses the same metering direction for both voltage and current like with standard diodes (load metering system), the characteristic curves shown in Figure 3.13 for illuminated and non-illuminated solar cells are obtained; this constitutes the idealized case using diodes with good reverse properties. The characteristic curve of an irradiated solar cell exhibits the same form as that of a non-irradiated cell and is simply shifted by I_{SC} in the negative current direction by virtue of the fact that photocurrent and diode

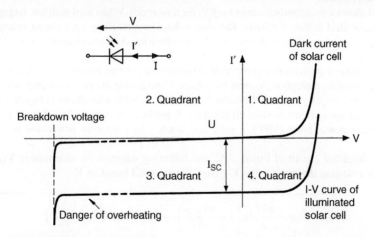

Figure 3.13 Idealized characteristic curve of a solar cell (illuminated and not illuminated) with an idealized good reverse blocking behaviour (low reverse current)

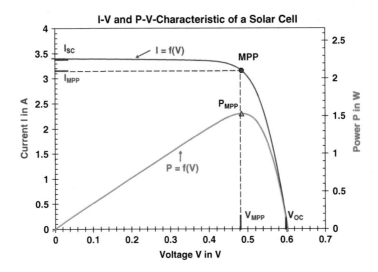

Figure 3.14 Characteristic curves $I = f(V)$ and $P = f(V)$ of a monocrystalline silicon solar cell with a cell area of approximately $102 \, \text{cm}^2$, irradiance G amounting to $1 \, \text{kW/m}^2$ and $25 \, ^\circ \text{C}$ cell temperature

current flow in opposite directions (in comparing Figures 3.13 and 3.12, note that $I' = -I$). The solar cell consumes power in quadrants 1 and 3, but it *generates* power when the cell is operated in quadrant 4.

No problems arise when solar cells are wired to solar modules, so long as measures are taken to ensure that: (a) the load on a shaded solar cell in the forward direction does not exceed more than about I_{SC} (which the cell can tolerate without problems under open-circuit conditions); and (b) in the reverse direction it is not subjected to unduly high voltages, so as to avoid unduly high cell power loss. Unduly high stress can overheat solar cells and thus ruin an entire solar module.

For further details concerning solar cell characteristics under unusual operating conditions in power quadrants 1 and 3, see Chapter 4.

For solar cells, the characteristic curves of power quadrant 4 are of overriding importance as this is where they generate power. Hence in most cases only this characteristic curve is indicated, whereby the metering directions for V and I shown in Figure 3.12 are used (generator metering system) so as to ensure that V and I are positive.

Figure 3.14 shows a characteristic curve $I = f(V)$ for a solar cell. When used with low voltages, a solar cell is a virtually ideal source of current. But once voltages approaching open-circuit voltage V_{OC} are reached, the current drops off fairly sharply (the diode in the equivalent circuit as in Figure 3.12 begins to conduct current).

In addition to current, power is also a key parameter since solar cells are used to produce electrical energy. Power is determined by multiplying current by voltage. Determining the power for each point along the characteristic curve $I = f(V)$ results in the curve $P = f(V)$, which is likewise shown in Figure 3.14. When a solar cell is in an open-circuit or short-circuit state, it produces no power. At a defined point known as the maximum power point (MPP), a solar cell reaches its maximum power and thus the value $P_{max} = P_{MPP}$.

Using the simplified circuit of Figure 3.12, the following equation for determining V_{MPP} can be obtained by rearranging the product $P = V \cdot I$ from Equation 3.3 based on V:

$$V_{MPP} = V_{OC} - \frac{nkT}{e} \ln\left(1 + \frac{eV_{MPP}}{nkT}\right) = V_{OC} - V_T \ln\left(1 + \frac{V_{MPP}}{V_T}\right) \tag{3.10}$$

In the interest of using solar cell power optimally, a connected consumer should be installed in such a way that it operates in as close proximity as possible to the MPP. However, this is easier said than done, since the MPP site is determined by irradiance, temperature, manufacturing tolerance and ageing. A device that ensures that a consumer is always operating at the MPP is known as a maximum power point tracker (MPPT) or maximum power tracker (MPT).

The maximum power, $P_{max} = P_{MPP} = V_{MPP} \cdot I_{MPP}$, that a solar cell can produce at the MPP is always lower than the value obtained by multiplying open-circuit voltage V_{OC} by short-circuit voltage I_{SC}. Inasmuch as a PV system needs to be able to withstand both open-circuit and short-circuit current, the ratio of P_{max} to $V_{OC} \cdot I_{SC}$ is a key solar cell measurement value, along with efficiency. This ratio is known as the fill factor:

$$\text{Fill factor FF} = \frac{P_{max}}{V_{OC} \cdot I_{SC}} \tag{3.11}$$

The fill factor for commercially available solar cells ranges from around 60 to 80%, while this factor for lab cells can go as high as about 85%.

For the simplified equivalent circuit of Figure 3.12, the fill factor can be determined quite accurately using the following equation [Gre95], [Wen95]:

$$\text{Idealized fill factor FF}_i = 1 - \frac{1 + \ln\left(\frac{V_{OC}}{V_T} + 0.72\right)}{1 + \frac{V_{OC}}{V_T}} \tag{3.12}$$

As Equation 3.12 shows, a high V_{OC} and low V_T (low n close to 1, low temperature) promote a high fill factor.

The R_S and R_P loss in the complete equivalent circuit diagram (see Figure 3.12) reduces the fill factor still further. According to [Gre95], approximately the following is obtained with characteristic resistance $R_{CH} = V_{OC}/I_{SC}$:

$$\text{FF} \approx \text{FF}_i\left(1 - \frac{R_S}{R_{CH}}\right) \cdot \left(1 - \frac{\left(\frac{V_{OC}}{V_T} + 0.7\right)\frac{R_{CH}}{R_P}\left(1 - \frac{R_S}{R_{CH}}\right)\text{FF}_i}{\frac{V_{OC}}{V_T}}\right) \tag{3.13}$$

As noted, a solar cell's characteristic curve is determined by irradiance and solar cell temperature. Figure 3.15 shows the characteristic curves for the solar cell in Figure 3.14, with irradiance as parameter. Here, the short-circuit current is proportional to irradiance, whereas open-circuit voltage increases only slightly as irradiance rises.

This also means that solar cell voltage can be quite high even in the presence of very low irradiance – for example, at dusk. This fact should be taken into account when installing and servicing PV systems that exhibit relatively high voltages.

Inasmuch as the ratio between voltage at the MPP V_{MPP} and open-circuit voltage V_{OC} fluctuates only slightly, the MPP voltage is also somewhat lower for lower irradiance. In addition, under extremely low irradiance conditions, the voltage flowing through the parallel resistance R_P weighs more heavily on balance and results in an additional voltage reduction at the solar cell. On account of this reduced V_{MPP} in the presence of low irradiance and the current flowing through R_P, solar cell efficiency is also lower in such situations. Figure 3.16 shows the efficiency of the solar cell in Figure 3.15, as a function of irradiance.

Figure 3.17 shows the characteristic curves for the solar cell in Figure 3.14, with cell temperature as parameter. In Si solar cells, open-circuit voltage decreases as temperature increases, by around 2 to 2.4 mV/K (the temperature coefficient of V_{OC} is -0.3% per K up to -0.4% per K); this decrease is attributable to the fact that diode threshold voltage in the equivalent circuit diagram decreases as for a standard Si diode. In this context, MPP voltage of course decreases as temperature rises. The fill factor FF likewise decreases as temperature rises (according to [Wen95], the FF temperature coefficient is

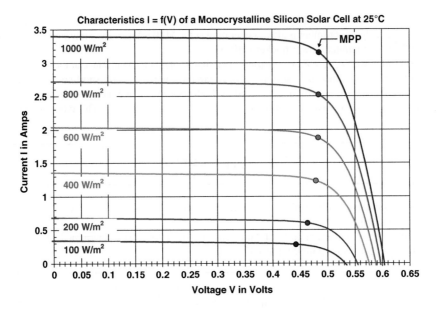

Figure 3.15 Characteristic curves $I = f(V)$ for the solar cell as in Figure 3.14, with cell temperature of 25 °C and irradiance G as a parameter (AM1.5 spectrum)

roughly -0.15% per K). Inasmuch as short-circuit current increases only slightly as temperature rises (the I_{SC} temperature coefficient is only about $+0.04$ to $+0.05\%$ per K), power P_{max} at the MPP also decreases as temperature increases.

For crystalline silicon solar cells, the temperature coefficient c_T for P_{max} ranges from about -0.4 to -0.5% per K. Inasmuch as PV system cell temperature can be 20 to 40 °C above the ambient temperature in the presence of high irradiance, this can provoke a considerable power drop and thus reduced efficiency at high temperatures; these parameters are often underestimated, however (see Figure 3.18).

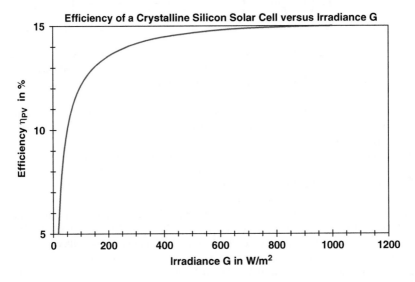

Figure 3.16 Efficiency of the monocrystalline Si solar cell shown in Figure 3.15 at a cell temperature of 25 °C, as a function of irradiance G (AM1.5 spectrum, operated at the MPP)

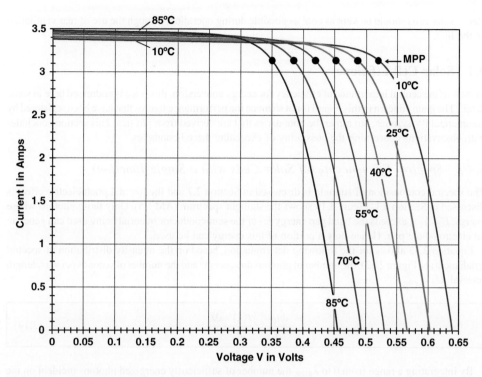

Figure 3.17 Characteristic curves $I = f(V)$ for the solar cell as in Figure 3.14, with irradiance $G = 1\,kW/m^2$ (AM1.5 spectrum) and with cell temperature as parameter

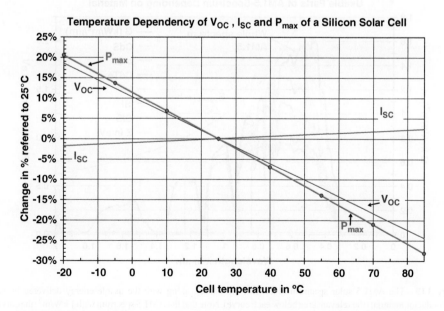

Figure 3.18 V_{OC}, I_{SC} and P_{max} as a function of cell temperature in a crystalline silicon solar cell

Hence solar cells should be kept as cool as possible during operation, through the use of rear ventilation or the like.

3.4 Solar Cell Efficiency

In view of the central importance of efficiency for energy conversion, this issue is addressed here in some detail. The main concern in this regard is not so much the performance figures that have been registered by commercial PV systems, but rather the boundaries that are derived from physics. This section concludes with observations concerning the possibility of exceeding these boundaries.

3.4.1 Spectral Efficiency η_S (of Solar Cells with a Single Junction)

The spectral composition of sunlight is discussed in Section 2.7 and the internal photoelectric effect is discussed in Section 3.1. Figure 3.19 shows the sunlight spectrum (AM 1.5). Only those photons whose energy, $E = h \cdot v$, exceeds the band gap energy E_G of the semiconductor material being used can generate an electron–hole pair. Hence only a portion of this energy can be used.

Equation 2.33 allows for the following determination, based on the intensity distribution of spectral irradiance (see Figure 2.42), the number of photons dn_{Ph} per m^2 and the number of seconds per wavelength interval $d\lambda$:

$$\frac{dn_{Ph}}{A} = \frac{I(\lambda) \cdot d\lambda}{h \cdot \frac{c}{\lambda}} \qquad (3.14)$$

By integrating a range from 0 to λ_{max}, the number of sufficiently energized photons incident on the solar cell per area unit and per second is obtained. Assuming that an ideal solar cell material is available that can separate each of these electron–hole pairs, the maximum possible current density

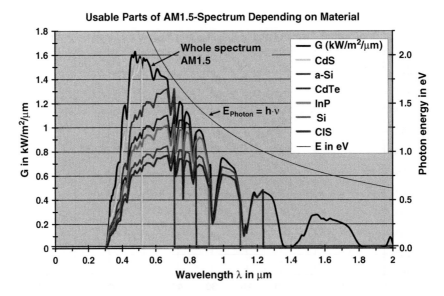

Figure 3.19 The AM1.5 solar spectrum (1 kW/m^2, top curve), along with the usable energy delivered by each semiconductor material (the relevant area below each curve). Note that the AM1.5 spectrum with 1 kW/m^2 (the curve in Figure 2.42, increased by a factor of 1.2) is used for most solar cell measurements in PV systems. *Data source*: [Gre95]

$J_{max} = J_{SCmax}$ can be determined for such an ideal solar cell, based on the number of photons per area unit:

$$\text{Maximum current density } J_{max} = \frac{e}{h \cdot c} \int_0^{\lambda max} I(\lambda) \cdot \lambda \cdot d\lambda \qquad (3.15)$$

where:

$J_{max} = J_{SCmax} = I_{SCmax}/A_Z =$ maximum possible short-circuit current/cell area

$\qquad I(\lambda) =$ spectral intensity distribution as in Figures 2.42 and 3.19 (AM1.5)

$\qquad \lambda_{max} = h \cdot c/E_G$, i.e. the maximum wavelength that is sufficient to separate an electron–hole pair in the presence of a given band gap energy E_G

This method allows for the determination of J_{max} as a function of E_G. [Gre86] contains a graph for J_{max} for AM0 up to 60 mA/cm². [Gre95] contains a table indicating the J_{max} values as a function of E_G for the (increased) AM1.5 spectrum with 1 kW/m². Figure 3.20, which shows J_{max} as a function of E_G, was elaborated using these figures.

Whereas in the AM0 spectrum current density rises relatively steadily as band gap energy E_G decreases, this increase in the AM1.5 spectrum is less regular. As Figures 3.19 and 2.42 show, radiation intensity is extremely low in certain wavelength ranges of the AM1.5 spectrum (according to the photon energy range), as a result of greater light absorption by atmospheric water and carbon dioxide. Consequently, current can increase only slightly.

If we presume that all of the energy E_G obtained from this ideal solar cell can be output to the external electric circuit, then spectral conversion efficiency η_S can be calculated as a function of band gap energy E_G.

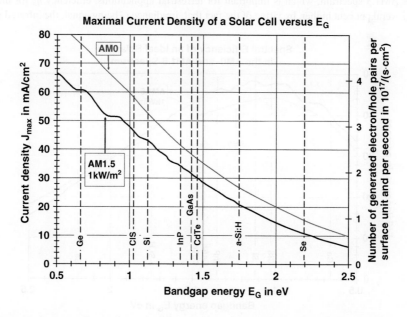

Figure 3.20 Maximum achievable current density J_{max} (number per second and unit of area for generated electron–hole pairs) for an ideal solar cell, as a function of semiconductor material band gap energy E_G

The voltage for this ideal solar cell is determined as follows (e = elementary charge):

$$\text{Theoretical photovoltage } V_{Ph} = \frac{E_G}{e} \tag{3.16}$$

The following applies to area-specific power P/A_Z of this ideal solar cell (area A_Z):

$$\text{Area-specific power of an ideal solar cell} : \frac{P}{A_Z} = V_{Ph} \cdot J_{max} \tag{3.17}$$

The following then applies to spectral efficiency η_S:

$$\text{Spectral efficiency of an ideal solar cell} : \eta_S = \frac{P}{G \cdot A_Z} = \frac{V_{Ph} \cdot J_{max}}{G} \tag{3.18}$$

Here, G is the irradiance on the solar cell.

Figure 3.21 shows this spectral efficiency η_S as a function of E_G for the AM0 and AM1.5 spectra with 1 kW/m^2, which is generally used to determine the efficiency of terrestrial solar cells.

If E_G is low, then virtually all photons will generate electrons and current density J_{max} will be high. However, the energy released per electron E_G will be low, and thus V_{Ph} will likewise be low. It therefore follows that spectral efficiency in the ideal solar cell will be low. In the presence of elevated E_G values, only a minute proportion of the photons will be able to generate an electron, and thus current density J_{max} will decrease to a low value. As E_G is high, also V_{Ph} is high. Therefore spectral efficiency in the presence of high E_G values is likewise low owing to the low current density. The optimum lies somewhere in between.

For the AM1.5 spectrum, which is important for terrestrial applications, efficiency η_S for mean E_G values is several per cent higher, by virtue of the fact that, irradiance G being equal, the infrared portion

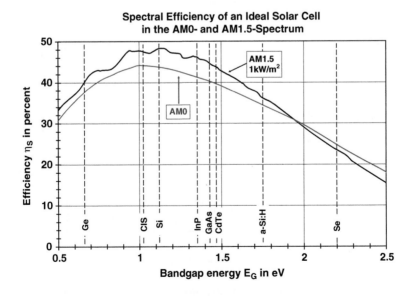

Figure 3.21 Spectral efficiency η_S of an ideal solar cell as a function of semiconductor material band gap energy E_G

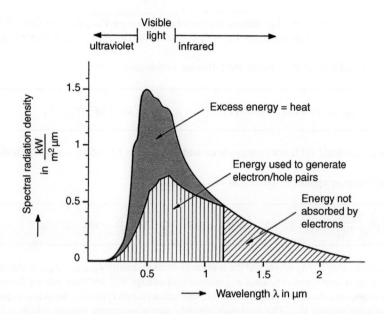

Figure 3.22 Spectral energy absorption in a crystalline silicon solar cell. Photons with unduly low energy, $h \cdot v < E_G$ ($\lambda > 1.11\,\mu\mathrm{m}$), are not absorbed by the semiconductor material, and thus their energy is unusable. In photons where $h \cdot v > E_G$, the difference, $h \cdot v - E_G$, is likewise unusable and is converted into heat in the semiconductor material. In this ideal solar cell, only the energy represented by the vertically hatched area can be converted into electrical energy

of this spectrum contains somewhat less energy that is usable for a solar cell, relative to the AM0 spectrum. For the same reasons as for $J_{\max}$, the η_S curve for the AM1.5 spectrum is more irregular than for the AM0 spectrum.

The curves in Figures 3.20 and 3.21 apply solely to the aforementioned ideal solar cells and not to real-world solar cells. Figure 3.22 shows the relationships that apply in such an ideal silicon solar cell (according to the suppositions made in the present section).

3.4.2   Theoretical Efficiency η_T (of Solar Cells with a Single Junction)

The suppositions laid out in Section 3.4.1 are extremely rough and are based solely on the principle of photon absorption in semiconductors, but not on the other relevant aspects of semiconductor physics that come into play here. Far more realistic results can be obtained if it is *also* assumed that the electric field in the barrier layer of a semiconductor diode is used to separate the photon-generated electron–hole pair. The simplified equivalent circuit of Figure 3.12 is used for this purpose, but without R_S and R_P. This method allows for the calculation of theoretical efficiency η_T in an idealized solar cell as a function of band gap energy E_G, based on spectral efficiency η_S and using Equations 3.6, 3.11 and 3.12. This method duly takes account of the key laws of semiconductor physics, while still leaving intact the following idealized suppositions:

- The solar cell will be operated at the MPP.
- All photons where $h \cdot v > E_G$ will generate an electron–hole pair.
- All electron–hole pairs will be separated.
- Shading attributable to front contacts will be negligible.
- The solar cell will be operated at a cell temperature of 25 °C (room temperature).
- The diode characteristic curve of the non-irradiated solar cell will be consistent with Equation 3.2.
- The effect of R_S and R_P will be negligible, i.e. no power loss will occur.

This method takes account of two key factors: (a) the solar cell current V_{MPP} in terms of its MPP will be far lower than the theoretical photovoltage $V_{Ph} = E_G/e$; and (b) the current I_{MPP} at the MPP will be lower than I_{SC}.

Equation 3.11 realized for P_{max} yields the following in this case:

$$P_{max} = FF_i \cdot V_{OC} \cdot I_{SC} = FF_i \cdot \frac{V_{OC}}{V_{Ph}} V_{Ph} \cdot I_{SC} = FF_i \cdot SF \cdot V_{Ph} \cdot I_{SC} \qquad (3.19)$$

where FF_i is the idealized fill factor in accordance with Equation 3.12 and $SF = V_{OC}/V_{Ph}$ is the voltage factor.

Hence the following applies to theoretical efficiency η_T:

$$\eta_T = \frac{P_{max}}{G \cdot A_Z} = FF_i \cdot SF \cdot V_{Ph} \frac{I_{SC}}{A_Z} \cdot \frac{1}{G} = FF_i \cdot SF \cdot V_{Ph} \frac{J_{SC}}{G} = FF_i \cdot SF \cdot \eta_S \qquad (3.20)$$

It therefore follows that I_{SC}/A_Z equates to the maximum current density $J_{SCmax} = J_{max}$ in Equation 3.18.

Hence, to achieve maximum efficiency η_T, open-circuit voltage V_{OC} and thus voltage factor SF need to be as high as possible. According to Equation 3.6, saturation current I_S needs to be as low as possible for a high open-circuit voltage V_{OC}. The minimum possible saturation current density, which is $J_S = I_S/A_Z$, can be estimated using the following equation [Gre86]:

$$J_S = K_S \cdot e^{-E_G/kT} = K_S \cdot e^{-V_{Ph}/V_T} \qquad (3.21)$$

where:
$V_{Ph} = E_G/e$, i.e. theoretical photovoltage according to Equation 3.16
$V_T = nkT/e$, i.e. thermal diode voltage according to Equation 3.4

According to [Gre86], in order to obtain a reasonably accurate estimate of the minimum possible J_S, it is necessary to apply something along the lines of $K_S = 150\,000$ A/cm^2. For a silicon solar cell where $E_G = 1.12$ eV, a value of around 18 fA/cm^2 for J_S is obtained with this method at 25 °C. More recent findings show that a minimum of $J_S \approx 5$ fA/cm^2 may even be attainable [3.1]. This results in a K_S value of around 40 000 A/cm^2.

Inasmuch as $I_{Ph}/I_S \approx I_{SC}/I_S = J_{SC}/J_S$, the J_S used for Equation 3.21 can be applied to Equation 3.6, thus resulting in a very simple equation for maximum possible open-circuit voltage V_{OC}:

$$V_{OC} = V_{Ph} - V_T \ln \frac{K_S}{J_{SC}} \qquad (3.22)$$

Based on the values used for Equation 3.15 and those indicated for $J_{SCmax} = J_{max}$ in Figure 3.20, using Equation 3.6 or 3.22, the maximum possible open-circuit voltage V_{OC} and the voltage factor SF in a solar cell can now be determined as a function of E_G, for diode quality factor $n = 1$:

$$\text{Voltage factor } SF = \frac{V_{OC}}{V_{Ph}} \qquad (3.23)$$

Figure 3.23 shows the maximum values determined for open-circuit voltage V_{OC} and the voltage factor SF in a solar cell as a function of E_G (for diode quality factor $n = 1$) at STC (1 kW/m^2, AM1.5, 25 °C), based on the above suppositions.

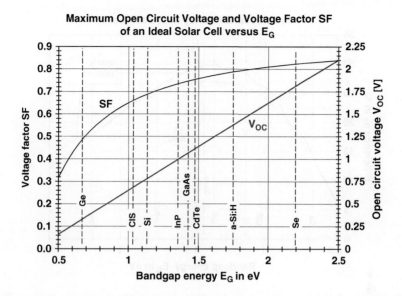

Figure 3.23 Maximum open-circuit voltage V_{OC} and voltage factor SF for an idealized solar cell where $n = 1$, as a function of semiconductor material band gap energy E_G

By the same token, using the maximum V_{OC} value thus determined, the idealized fill factor FF_i can be calculated. Figure 3.24 shows FF_i as a function of E_G for $n = 1$ at STC (1 kW/m², AM1.5, 25 °C).

On account of the logarithmic dependency that comes into play here, with 1 kW/m² irradiance the voltage factor SF and the idealized fill factor FF_i are determined to only a minor extent by spectrum type (AM0 or AM1.5). As can be readily seen in Equations 3.22 and 3.12, both V_{OC} and FF_i decrease as the diode quality factor n increases, which means that optimally low n values close to 1 yield the best results.

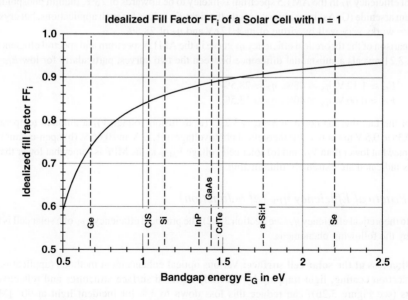

Figure 3.24 Idealized fill factor FF_i as for Figure 3.23, determined using Equation 3.12 for a solar cell where $n = 1$, as a function of semiconductor material band gap energy E_G

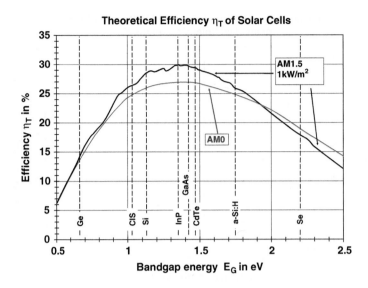

Figure 3.25 Theoretical efficiency η_T of an ideal solar cell as a function of band gap energy E_G for semiconductor material for $n=1$ in the AM1.5 spectrum with $1\,\text{kW/m}^2$ and in the AM0 spectrum

Using Equation 3.20 and multiplying the curves in Figures 3.21, 3.23 and 3.24, the theoretical efficiency η_T of a solar cell for the AM0 and AM1.5 spectra with $1\,\text{kW/m}^2$ can now be obtained. Figure 3.25 shows η_T as a function of band gap energy E_G of solar cell material with $1\,\text{kW/m}^2$ and a cell temperature of 25 °C. As with spectral efficiency η_S, theoretical efficiency η_T for the AM1.5 spectrum is several percentage points higher than for the AM0 spectrum, owing to somewhat lower infrared content.

Hence semiconductor materials whose E_G ranges from around 0.8 to 2.1 eV are particularly well suited for use in solar cells. The roughly 1.1 to 1.6 eV range is particularly suitable by virtue of the fact that theoretical efficiency η_T in the AM1.5 spectrum is likeley to be upwards of 28%. Indium phosphide (InP) and gallium arsenide (GaAs) have almost exactly the right E_G values for such applications, but crystalline silicon also works very well by virtue of its 1.12 eV and η_T of 28.5%.

A comparison of the theoretical efficiency η_T curve in the AM1.5 spectrum and spectral efficiency η_S as in Figure 3.21 reveals a substantial difference between the two curves, particularly for low E_G:

For c-Si: $E_G \approx 1.12\,\text{V}$, $\eta_S \approx 48\%$, $\eta_T \approx 28.5\%$.
For Ge: $E_G \approx 0.66\,\text{V}$, $\eta_S \approx 40\%$, $\eta_T \approx 13.5\%$.

In view of the fact that (as noted in Section 3.2.2) (a) diffusion voltage V_D at ambient temperature is roughly 0.35 to 0.5 V lower than the theoretical photovoltage of 1.12 V and 0.66 V, (b) open-circuit voltage V_{OC} is somewhat lower than V_D, and (c) solar cell voltage V_{MPP} at the MPP is somewhat lower than V_{OC}, the values indicated are extremely illuminating.

3.4.3   Practical Efficiency η_{PV} (at a Junction)

Relative to theoretical efficiency η_T (see Section 3.4.2), the practical efficiency η_{PV} of a solar cell is further reduced by the following phenomena:

- **Reflection loss at the solar cell surface:** Various optical enhancement methods (application of an anti-reflection coating, light trapping by specially textured surface structures and reflective back contacts (see Figure 3.26)) can reduce this loss down to 1% for incident light at 90° [Mar94]). Reflection loss is higher for non-perpendicular incident light, mainly because of glass reflection (see Figure 8.1).

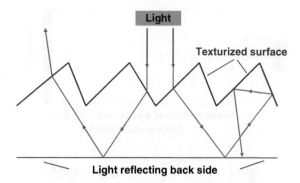

Figure 3.26 Light trapping in a c-Si solar cell: solar cell reflection loss is reduced and solar cell light absorption is improved through the use of suitably designed (i.e. textured) surfaces and reflective back contacts. In this process, incident light is refracted at the solar cell surface in such a way that it traverses the silicon at an angle, is reflected on the back, and is captured by the silicon to the maximum possible extent via total reflection. The multiple light reflections thus achieved in the cell elongate the light path by a factor of upwards of 20 and allow for full absorption of virtually all photons. This in turn means that solar cells can be considerably thinner, i.e. recombination is reduced using the same material quality and material use is also reduced. With these methods, and despite the reduced cell thickness, the light traverses a path that is long enough to allow for full absorption of even low-energy (i.e. red light, near-infrared) photons [Gre95]. This effect can be further enhanced through the use of such methods on the back of the cell, or through integration of an optical lattice. In addition, an anti-reflection coating is applied to the textured area of most solar cells

- **Recombination loss:** In some materials, silicon being one of them, photon penetration depth is determined by wavelength. Hence not all electron–hole pairs separated by the internal photoelectric effect are generated in or near the space charge zone, where they can be separated by the field instantaneously. In c-Si some energized photons are already absorbed on the n-layer area (see Figure 3.10), whereas some low-energy photons with $h \cdot v > E_G$ are not absorbed until they are beyond the space charge zone.

  Some electrons that are generated by low-energy photons (in red light or near-infrared light) at unfavourable locations (far beyond the space charge zone) recombine with the abundant holes in the p-zone before being diffused in the space charge zone, where the electric field can retain them via pre-recombination transfer to the n-site, thus allowing them to be definitively shunted to the outer circuit. This recombination mechanism is particularly prevalent at lattice imperfections and on the semiconductor surface (e.g. at the back contact).

  In today's solar cells, design engineers try to minimize back-contact recombination loss by creating a back-area field (BSF) via substantial additional doping (p$^+$) on the p-field area (see Figure 3.27), thus increasing I_{SC}, V_{OC} and P_{max}.

  Hence the extent to which photons with $h \cdot v > E_G$ are efficiently leveraged in working solar cells is determined by photon wavelength. Figure 3.28 shows the spectral quantum efficiency for a number of solar cell materials, and represents the state of the art in 1986.
- **Self-shading attributable to opaque front electrodes:** Front electrode optimization through the use of buried contact cells (inserting front contacts in grooves made using lasers; see Figure 3.29) can reduce self-shading loss from opaque front electrodes by a considerable amount (normally by up to several percentage points). For cells with back contacts, contact is established for the n- and p-zones from the back of the cell, thus completely obviating self-shading (see Figure 4.4).
- **Ohmic loss in semiconductor materials:** R_S and R_P in the equivalent circuit shown in Figure 3.12 result in solar cell ohmic loss, which in c-Si solar cells can be reduced in the n-zone via relatively high doping (n$^+$).
- **Lower efficiency attributable to temperature:** As with P_{max} (see Figure 3.18) efficiency η_{PV} declines with rising temperature, whereby the characteristic value for the temperature coefficient of c-Si solar cell efficiency is -0.004 to -0.005 per K. In such cases, the somewhat higher open-circuit voltage of

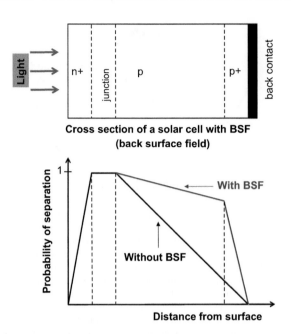

Figure 3.27 By application of a more heavily doped coating (p +) just in front of the back contact, a back-area field (BSF) and thus a potential barrier are created that deflect electrons away from the solar cell area, where they are particularly prone to recombination. This in turn greatly increases the chances that an electron will be liberated by a low-energy photon near the back contact in such a way that recombination will be averted and the electron will be diffused while still in the space charge zone, where the electron can be definitively separated from the ubiquitous holes in the p-field and can thus contribute to the outer current flow. Thus realization of a BSF increases the likelihood p_T that a photon-generated electron can be successfully separated and thus be made to flow through the outer circuit. This in turn increases solar cell I_{SC} an (by virtue of reduced I_S) also V_{OC} [Gre95]

newer c-Si solar cells has a beneficial effect on this coefficient in that the relative change in the presence of higher V_{OC} is somewhat lower.

Solar cell efficiency is an extremely important parameter for PV system operation. Needless to say, both researchers and vendors in the PV sector would like to be able to indicate the highest possible efficiency

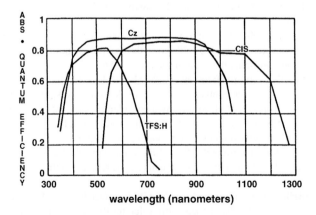

Figure 3.28 Spectral quantum efficiency of the following materials: **Cz**, monocrystalline Si (sc-Si); **TFS:H**, amorphous thin-film Si (a-Si:H); **CIS**, thin-film CuInSe₂. State of the art as of 1986. Arco Solar [3.3]/Willi Maag

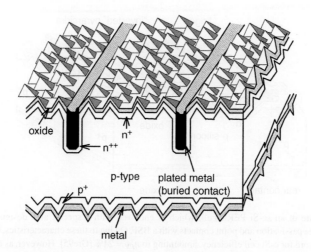

Figure 3.29 Reduced self-shading via a front electrode buried-contact arrangement. The front contact is realized via narrow laser grooves. A specially designed back contact increases light reflection, while higher p^+ doping creates a BSF that reduces recombination. BP Solar makes this type of solar cell. UNSW, Centre for Photovoltaic Engineering

levels for their research outcomes or products (as the case may be); these values for commercially available products are indicated in Section 1.4.1.

Efficiency levels exceeding those of commercially available products by several percentage points have been achieved in laboratory settings. However, a certain amount of fudging has gone on in pursuit of the highest possible solar cell efficiency levels, in that many researchers base efficiency solely on active surfaces (i.e. those not shaded by front contacts) and thus automatically 'achieve' higher values. In some cases, the efficiency values indicated are missing key data such as spectrum AM count, cell area, cell temperature and so on, thus often making it impossible to compare the values in the literature head to head. The solar cell efficiency η_{PV} data in Table 3.2, achieved in *laboratory* settings, can be regarded as reliable.

Figure 3.30 shows the structure of a PERL cell developed by a research team led by M. A. Green from the University of New South Wales, which has long since held the record (as at 2009) for the highest monocrystalline Si solar cell efficiency. The efficiency η_{PV} of this cell, which integrates virtually all of the efficiency optimization technologies that are realizable today, is very close to the theoretical limit as in Figure 3.25.

Table 3.2 The lab efficiency level η_{PV} that has been attained (and confirmed by impartial observers) with small cells (area 1 to 4 cm^2) and only one p–n junction, at STC (AM1.5 spectrum, 1 kW/m^2, cell temperature 25 °C) [3.2]. The differences between the data shown here and the data in [Häb07] are partially attributable to a minor change that was effected in the standard AM1.5 spectrum in 2008 (new IEC 60904–3 : 2008) [3.2]. Reproduced by permission of Wiley-Blackwell

Material	η_{PV}
GaAs	26.1%
GaAs (thin film)	26.1%
sc-Si (monocrystalline)	25.0%
mc-Si (multicrystalline)	20.4%
InP	22.1%
Cu(In,Ga)Se$_2$ (CIGS)	19.4%
CdTe	16.7%
a-Si	9.5%

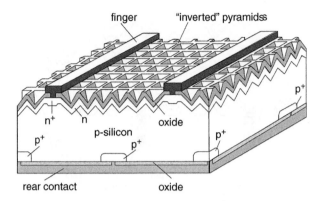

Figure 3.30 Structure of an sc-Si PERL cell, which is endowed with a special area design featuring inverted pyramids, bilateral area passivation and point contacts with a BSF. Owing to these characteristics, the PERL solar cell currently holds the record for c-Si cell efficiency, amounting to $\eta_{PV} = 24\%$ [Gre95]. However, as a result of the many process steps entailed by this cell, producing it is still extremely cost intensive. UNSW, Centre for Photovoltaic Engineering

Figure 3.31 shows the characteristic *V–I* curve for such a solar cell. Noteworthy here is that the open-circuit voltage V_{OC}, short-circuit current density J_{SC} and fill factor FF are considerably higher than with commercially available monocrystalline solar cells.

A key factor for commercially available solar cells is not only high efficiency, but also a reasonable price. It is no easy matter to integrate all available efficiency optimization technologies into an industrial production process at a reasonable cost. However, the efficiency of commercially available solar cells is bound to increase somewhat in the coming years, in view of the above lab values.

3.4.4 Efficiency Optimization Methods

3.4.4.1 Optimized Concentration of Sunlight

According to Equations 3.6 and 3.22, the open-circuit voltage increases logarithmically with short-circuit current I_{SC}, which is directly proportional to irradiance G. Hence according to Equation 3.23, the voltage factor SF increases with rising irradiance G, as does (according to Equation 3.12) the idealized fill factor

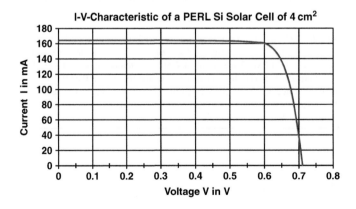

Figure 3.31 *V–I* characteristic curve for a PERL solar cell with a cell area of $A_Z = 4 \, \text{cm}^2$ [Gre95]. At STC, this cell attains a current density J_{SC} of 41 mA/cm^2, V_{OC} of 710 mV, a fill factor FF of 83% and efficiency η_{PV} of 24%, and in so doing comes very close to the theoretical limits as shown in Figures 3.20, 3.23, 3.24 and 3.25

Figure 3.32 Field test of concentrator cells. Here, sunlight is concentrated on relatively small solar cells using lens systems. To attain a reasonable energy yield, concentrator cells should always be tracked biaxially (Courtesy of DOE/NREL)

FF_i; thus, according to Equation 3.20, theoretical efficiency also increases. Hence optical concentration of sunlight can somewhat optimize solar cell efficiency still further. However, to do this it is necessary to increase doping so that the series resistance of such a concentrator cell remains sufficiently low. In addition, inasmuch as the absolute value of non-converted radiant power increases as irradiance rises, it is essential that such solar cells be cooled. Moreover, only direct beam radiation can be used for the optical concentration of sunlight.

Owing to the preponderance of diffuse radiation in Central Europe, the use of concentrator solar cells (see e.g. Figures 3.32 and 3.33) is not particularly worthwhile. According to [3.11], magnitude 232 sunlight concentration and 28.8% efficiency were attained for a small lab concentrator cell (0.05 m²) made of GaAs. The record for small silicon concentrator cells ($A_Z = 1 \, m^2$) is 27.6% efficiency with magnitude 92 concentration.

3.4.4.2 Tandem and Triple Solar Cells

Thin-layer solar cells (i.e. cells that are only a few micrometres thick) made of semiconductor materials with differing band gap energy E_G can be stacked (optical series connection, see Figure 3.34).

Figure 3.33 Close-up of a concentrator cell installation where sunlight is concentrated using mirrors that focus their light onto the solar cell at the point. The relatively large (black) cooling elements for each of these solar cells are clearly visible (Courtesy of DOE/NREL)

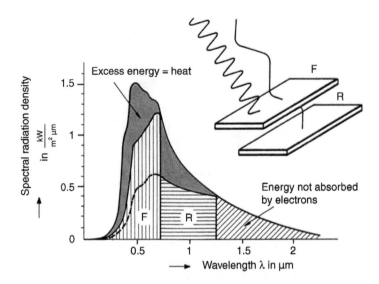

Figure 3.34 Structure and spectral energy absorption of an a-Si and CIS tandem cell, which allows for optimized use of the solar spectrum. The front cell F uses the energy represented by the vertically hatched area (blue curve in Figure 3.19), while the back cell R uses the energy represented by the horizontally hatched area, which is the energy portion with higher λ that was not yet used by the front cell (total potential energy production of the back cell alone shown by dark red curve in Figure 3.19).

If E_G for each junction is defined in such a way that the front solar cell (i.e. the cell facing the Sun) has the highest E_G and the back cell the lowest, the energy from the photons not absorbed by the front layers can be used by the layers that are further back, since photons with $h \cdot v < E_G$ simply pass through semiconductor material. This of course increases the spectral efficiency of the entire arrangement, as a comparison of Figures 3.34 and 3.22 clearly shows.

The relationships shown in Figure 3.34 are obtained with an arrangement such as a two-layer tandem solar cell made of a-Si (thin-film silicon, TFS), where $E_G \approx 1.75$ eV, and CIS, $E_G \approx 1$ eV. As can be seen in Figure 3.28, the spectral quantum efficiencies of both materials complement each other very well. As early as 1986, Arco Solar made such experimental tandem solar cells whose efficiency η_{PV} was 14% with a 65 m^2 area [3.3].

The concept of series connecting both tandem cell layers not only optically but also electrically has great appeal, as it greatly simplifies the manufacturing process (see Figure 3.36). However, with this concept it is difficult to achieve that the bottom cell always produces the exact same current as the top cell via the residual light that filters down, since the current must always be the same for series connections. If this is not the case, the weaker cell determines overall current, thus immediately reducing efficiency. This problem can be solved by electrically insulating the two optically series-connecting solar cells using light-coupling film, which allows for series connection of solar cell groups comprising different types of solar cells with about the same voltage. These groups can then be parallel connected without any difficulty using a four-terminal arrangement. This concept was used in the Arco Solar a-Si/CIS tandem solar cells mentioned above (see Figure 3.35).

Transparent contact materials are used on both the front and back of tandem solar cells to allow sunlight to reach the rear cell. For this purpose in most cases tin and zinc oxide layers are used. Only for the rear contact of the back cell a metal contact can be used.

Optical and electrical series connection has now also been successfully realized. Figure 3.36 shows the cross-section of such a tandem cell with optical and electrical series connection.

Uni-Solar sells a triple-cell module (three cells with varying levels of band gap energy and arranged one behind the other) made of amorphous silicon (with slightly varying additives in the amorphous silicon

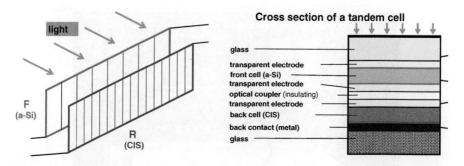

Figure 3.35 Structure of an experimental Arco Solar four-terminal tandem solar cell made of a-Si and CuInSe$_2$ (CIS) (after [3.3]). In this arrangement, the two cells are electrically insulated from each other, thus allowing for the use of any connection method desired

cells), which, apart from being somewhat more efficient than individual amorphous cells, is above all far more stable. Figure 3.51 in Section 3.5 shows a cross-section of such a cell.

The procedure described in Section 3.4.2 also allows for the determination of the theoretical efficiency η_T of tandem solar cells with two p–n junctions, i.e. electrically stand-alone (four-terminal) cells. At STC, η_T for such a cell reaches a maximum of around 40% with $E_G \approx 1.75\,\text{eV}$ and $E_G \approx 1\,\text{eV}$ in the front and back cell respectively (see Example 4 in Section 3.7).

The highest efficiency achieved by a lab tandem cell to date (GaInP and GaAs) in the AM1.5 spectrum with 1 kW/m^2 at 25 °C is 30.3%, while the highest efficiency achieved to date for a lab triple cell (GaInP, GaAs and Ge) under the same conditions is 32%. The area of each of these cells was around 4 cm^2 [3.2].

Particularly high efficiency levels can be reached by combining the two efficiency optimization techniques, i.e. by building tandem or triple concentrator cells. The currently confirmed record efficiency for a tandem concentrator cell (GaAs and GaSb, $A_Z = 0.05\,\text{cm}^2$) with four terminals (as in Figure 3.35) with 100-fold sunlight ($G = 100\,\text{kW/m}^2$) is $\eta_{PV} = 32.6\%$ [3.2]. In 2005, some vendors were hoping that by the end of 2006 triple concentrator cells would be available that could provide 40% efficiency and that by 2010, 45% efficiency could even be achieved [3.4]. In July 2008, under 140-fold sunlight a triple concentrator cell made of GaInP, GaAs and GaInAs with $A_Z \approx 0.1\,\text{cm}^2$ attained $\eta_{PV} = 40.8\%$ [3.2], while in January 2009 under 454-fold sunlight a triple cell made of GaInP, GaInAs and Ge with $A_Z \approx 0.05\,\text{cm}^2$ reached $\eta_{PV} = 41.1\%$ [3.11]. By splitting the light into three different spectral ranges and using cells specially designed for them, $\eta_{PV} = 42.7\%$ and 43% were reached in 2007 and 2009 respectively without unduly elevated solar concentration [3.12].

Although research on tandem and triple solar cell techinques is still in its infancy, it is safe to assume that intensive R&D will ultimately allow for considerably higher solar cell efficiency. However, it will be a while longer until such advances can be applied to industrially produced solar cells.

3.4.4.3 Efficiency Limits – Onwards and Upwards?

The efficiency limits discussed in this section are based on a concept that has been widely accepted for decades, to the effect that each photon generates only a single electron–hole pair. However, more than a decade ago the idea was developed that this unshakeable law may not apply in certain cases [Gre95], which means that (a) efficiency η_T could conceivably be closer to spectral efficiency η_S and (b) efficiency in the neighbourhood of 40% may be within reach. The operant concepts here are the impact ionization effect, quantum well structure cell and subband absorption cell. That said, it should also be noted that, to date, using these various principles, no one has made an actual functioning solar cell with an efficiency exceeding the theoretical efficiency for single-junction solar cells (see Figure 3.25). Hence such concepts should be regarded with a certain amount of scepticism.

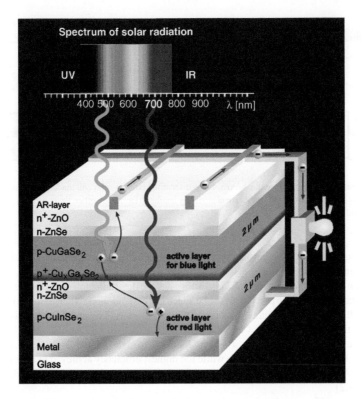

Figure 3.36 Cross-section of a two-terminal tandem cell with optical and series connection. The front cell's band gap energy E_G is similar to that of amorphous silicon (roughly 1.75 eV), while the back cell is made of CIS with E_G about 1 eV. At the contact area between the front and back cell, the holes generated in the front cell recombine with the electrons generated in the back cell. Helmholtz-Zentrum Berlin

3.5 The Most Important Types of Solar Cells and the Attendant Manufacturing Methods

This section describes some of the most important types of solar cells in use today, as well as the most widely used manufacturing processes for silicon solar cells.

3.5.1 Crystalline Silicon Solar Cells

As the structure and function of crystalline silicon (c-Si) solar cells were discussed in Section 3.1 (Figure 3.10), the present section mainly describes the manufacturing process for such cells.

In order to achieve full absorption of all photons where $h \cdot v > E_G$, the crystalline silicon solar cells must allow for at least a 100 μm path of light through the silicon. For reasons of mechanical stability (a key factor during the manufacturing process in particular), material thicknesses ranging from 150 to 300 μm are often used, so as to ensure that the condition for full radiation absorption is met.

The base material for silicon manufacturing is silicon dioxide (SiO_2), which occurs abundantly in nature in the form of quartz sand or as large quartz crystals, and is reduced in electric furnaces using charcoal, resulting in metallurgical silicon with approximately 98% purity. After being ground very finely, this raw silicon is reacted with hydrochloric acid (HCl), resulting in trichlorosilane ($SiHCl_3$), whose boiling point in liquid form is 31 °C and which is ultra-purified via repeated distillation. In a final step,

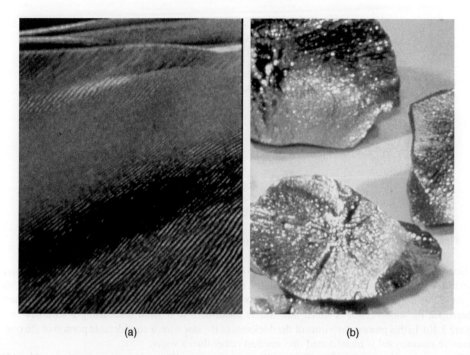

(a)   (b)

Figure 3.37 Left: base material for silicon manufacturing, quartz sand, which is abundantly available in many deserts around the world. Right: polycrystalline silicon (Photo: Fabrimex/Arco Solar/Willi Maag)

ultrapure polycrystalline silicon is obtained via gaseous $SiHCl_3$ and hydrogen (H_2) using a reactor that is heated electrically to 1000–1200 °C. Figure 3.37 shows the base material (sand) and the polycrystalline silicon that is made from it.

Monocrystalline solar cells are made by melting polycrystalline silicon in a crucible in the presence of inert gas. After a crystal nucleus attached to a tension rod has been immersed in the molten silicon, a monocrystal is drawn out of the fused silicon via continuous rotation of the rod (Czochralski method; see Figure 3.38). The maximum drawing speed is around 30 cm/h.

Figure 3.38 Pulling a silicon monocrystal (sc-Si) using the Czochralski (CZ) method for monocrystalline solar cell manufacturing (Photo: Fabrimex/Arco Solar/Willi Maag)

Figure 3.39 Finished round silicon monocrystal (sc-Si) during a quality test (Courtesy of DOE/NREL)

Hence the monocrystal manufacturing process is slow, energy intensive and expensive. To obtain a high packing factor (PF) in today's solar cell modules, the originally round monocrystal thus produced (see Figure 3.39) is processed in such a way that a virtually square rather than round form is obtained, whereupon the monocrystal is sliced into wafers about 0.15 to 0.3 mm thick using a wire saw (see Figure 3.40). In this process, by virtue of the thickness of the saw wire, a considerable portion of the cost-intensive monocrystal is transformed into sawdust rather than a wafer.

The manufacturing process for polycrystalline or multicrystalline solar cell wafers is considerably simpler and not nearly as energy intensive. The ultrapure polycrystalline silicon is simply cast into square blocks. The polycrystalline silicon rods thus obtained already have the desired square shape and can be sliced into polycrystalline wafers using a wire saw (see Figure 3.41). Because more lattice imperfections occur at crystallite grain boundaries that promote recombination in the charge carriers generated in the solar cells, polycrystalline solar cells are somewhat less efficient.

Using these monocrystalline or polycrystalline wafers, solar cells are then manufactured via numerous processes, an overview of whose number and complexity is shown in Figure 3.42.

Owing to the 100 µm minimum cell thickness that is required for crystalline solar cells and the material loss resulting from sawing the silicon rods, manufacturing these cells is extremely energy intensive.

Figure 3.40 Completed monocrystal (now almost square) after a few wafers have been sliced from it using a wire saw. The almost completely severed wafers are visible on the right edge of the monocrystal (Photo: Fabrimex/Arco Solar/ Willi Maag)

Figure 3.41   Polycrystalline or multicrystalline block (mc-Si) after being sawn into rectangular segments, from which 40 wafers are then sliced as shown in Figure 3.40 (Photo: DOE/NREL)

The polycrystalline solar cell manufacturing process is simpler (no monocrystal drawing is required) and thus less energy intensive. For more on the energy used for solar cell manufacturing, see Chapter 9.

In view of the relatively low voltage of individual solar cells, in order to achieve reasonable voltage values a large number of cells need to be series connected (see Chapter 4) via a process that for crystalline solar cells is relatively labour intensive and has yet to be fully automated.

In the interest of avoiding material loss occasioned by silicon rod sawing, processes have been developed that allow polycrystalline bands to be drawn directly from fused silicon. One such process is known as the EFG process (see Figure 3.43). Figure 3.44 shows an EFG installation at ASE, an American company that makes commercial solar modules using wafers realized via the EFG process. Here, thin cylindrical octagonal tubes are pulled rather than massive monocrystals, whereupon wafers are produced from the eight even lateral surfaces.

3.5.1.1   Spherical Solar Cells

Spherical solar cells were developed in the early 1990s by Texas Instruments using a large number of silicon globules, 1 mm in diameter, with a concentric p–n junction (see Figure 3.45). These globules are placed near each other and pressed together in double-layer aluminium foil with insulation between the layers. The globules are ground in such a way that one layer of the foil abuts the n-layer and the other the p-layer. With the aluminium foil being placed between the globules, the aluminium generates additional diffuse light, which likewise ramps up current production.

The advantage of these solar cells lies in their mechanical flexibility and the fact that, due to a relatively simple manufacturing process that lends itself to mass production (i.e. no monocrystals, casting or sawing), they can be made using relatively impure silicon. Texas Instruments repeatedly announced over a lengthy period that the product was slated to go into pilot production, but no effective product appeared on the market. Spheral Solar Power (SSP) recently brought out initial products of this type from pilot production (module with 200 Wp and η_M up to 9.5%).

3.5.2   Gallium Arsenide Solar Cells

Compared with other semiconductor materials, gallium arsenide (GaAs) has a band gap energy ($E_G = 1.42 \, \text{eV}$) that is virtually ideal for both the AM1.5 and AM0 spectra. The theoretical efficiency η_T in the AM1.5 spectrum at ambient temperature is around 30%, and is nearly 27% for the AM0 spectrum (see Figure 3.25). As GaAs is a directly absorbent semiconductor, full light absorption is attainable with ultrathin material, which means that GaAs is suitable for both crystalline and thin-film solar cells.

Hence GaAs solar cells for the AM1.5 spectrum and 1 kW/m^2 provide the highest absolute efficiency among all types of cells with only a single p–n junction (values achieved using small cells where $A_Z \approx 1 \, \text{cm}^2$: $\eta_{PV} = 26.1\%$ for both crystalline and thin-film GaAs solar cells, according to [3.2]). Moreover, owing to their higher band gap energy, GaAs solar cells are able to withstand far higher

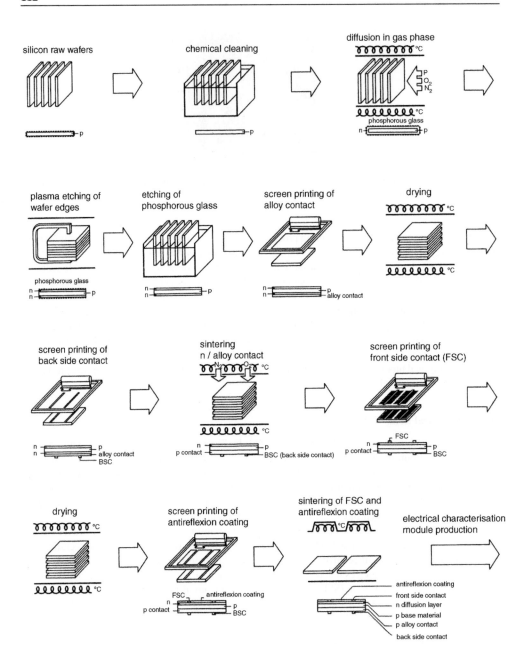

Figure 3.42 The workflow from wafer to finished solar cell (Based on AEG documents)

temperatures than silicon solar cells and have lower temperature coefficients for V_{OC} and P_{max} than silicon solar cells (around -0.2% per K for GaAs according to [Hu83]). Hence such cells make very good concentrator cells, i.e. solar cells that use highly concentrated sunlight (up to *well over $100\,kW/m^2$*).

According to [3.11], in early 2009 such a GaAs concentrator cell with $A_Z \approx 0.05\,cm^2$ and 232-fold sunlight concentration attained 28.8% efficiency. Even higher efficiency is attainable with GaAs containing tandem and triple concentrator cells (see Section 3.4.4.2).

Principle of EFG Cell Manufacturing Process

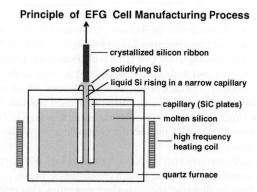

Figure 3.43 EFG band pulling process for wafer manufacturing

Figure 3.44 Industrial production of EFG wafers at ASE (Photo: DOE/NREL)

Basic Layout of Spherical Solar Cells

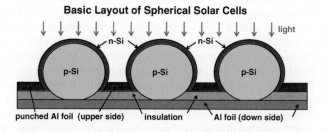

Figure 3.45 Structure of spherical solar cells. Using p-doped silicon globules, a p–n junction is created via area n-doping. These globules are then pressed together in aluminium foil. After the n-layer has been etched away, the globules are abutted against a second layer of foil

That said, it should also be noted that GaAs solar cells have two major drawbacks. First, unlike abundantly available silicon, the base material for GaAs solar cells is relatively scarce, and thus obtaining the requisite ultrapure quality is very cost intensive (according to [3.5], 27% of the mass of the Earth's crust is composed of silicon, whereas only 0.000 15% is made of Ga and only 0.000 05% of As). The GaAs manufacturing process is also very cost and energy intensive on account of the extremely high purity standards that must be met. Another drawback is that As is highly toxic, which means that disposing of GaAs solar cells at the end of their service life poses a major ecological problem. Moreover, if such cells catch fire they release highly toxic and carcinogenic substances such as As_2O_3 [Hu83]. Thus in view of the environmental friendliness of solar cells, the use of such problematic substances should be avoided despite their excellent efficiency.

Inasmuch as GaAs solar cells are very cost intensive for the reasons described above, they are used almost exclusively as concentrator cells with very high concentration factor, or for space exploration applications on account of their excellent radiation resistance properties. Owing to the preponderance of diffuse radiation in Central Europe, the use of GaAs solar cells fortunately is not particularly worthwhile. Hence they will not be discussed further in this book. For more on these cells see [Hu83], [Joh80], [Gre86].

3.5.3   Thin-Film Solar Cells

In many of the semiconductors that are suitable for solar cell manufacturing, only a film 0.5 to 5 µm thick is needed for full absorption of photons where $h \cdot v > E_G$. This massively reduces material and thus energy costs, if the semiconductor properties needed for solar cell functionality can be realized in such thin films. Against this backdrop, the developmental efforts of many researchers and vendors centre around thin-film solar cells, by virtue of the cost reduction potential they offer [3.3], [3.6], [3.7].

Among the semiconductors that are suitable for the manufacture of thin-film solar cells are amorphous silicon (a-Si), CdTe, $CuInSe_2$ (CIS), $Cu(In,Ga)Se_2$ (CIGS), $CuInS_2$, CdS/Cu_2S and GaAs. Using high-efficiency light-trapping techniques and suitable back-contact designs, researchers and vendors also seek to create on suitable substrates polycrystalline thin-film solar cells that likewise use far less material (maximum 10 µm). For thin-film solar cells, high base-material costs are less of a factor and the semiconductor material quality standards are lower. Moreover, the efficiency of such cells under low-irradiance conditions often falls off less rapidly than is the case with crystalline silicon solar cells, and in some cases is even somewhat higher than under elevated irradiance conditions.

In thin-film solar cells, ultrathin layers of suitable semiconductor material are vapour deposited onto inexpensive backings made of glass, metal, ceramic, plastic or the like. Such backings are referred to as substrates when they face away from the light and transparent superstrates when they face the light. The various layers are deposited successively from the gas phase at relatively low temperatures (only a few 100 °C), such that considerably less energy is expended than for the manufacture of crystalline solar cells. If the manufacturing workflow is deftly structured, the requisite series connections can also be realized for the various solar cells while deposition is being carried out. Since deposition is an automatic and monolithic process, as for the production of integrated circuits, the thin-film solar cell manufacturing process lends itself to substantial cost savings and energy efficiency measures far more readily than is the case with crystalline technology. Thin-film solar cells can also be realized as tandem cells or triple cells (see Section 3.4.4.2), which in principle allow for optimized use of solar energy (see Figure 3.34). However, considerable progress will need to be made in order for thin-film solar cells to become commercially available and provide stable efficiency amounting to, say, 15 to 20%, as is the case with today's monocrystalline solar cells.

In this section, we will first discuss in somewhat greater detail amorphous silicon thin-film solar cells, as they are composed of environmentally-friendly materials and are already available from various vendors. This will be followed by a very brief discussion of other materials.

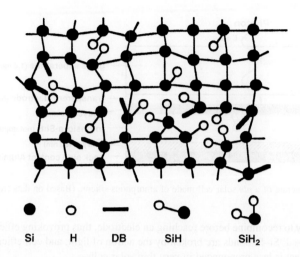

Si   H   DB   SiH   SiH₂

Figure 3.46   Amorphous silicon containing mainly hydrogenated broken bonds (DB stands for dangling bond, i.e. a non-hydrogenated bond) [3.6]. Reproduced by permission of Arvind Shah

3.5.3.1   Amorphous Silicon (a-Si) Solar Cells

As amorphous silicon exhibits a more random structure than crystalline silicon (see Figure 3.46), not every silicon atom is able to bond with four adjacent atoms, as is the case with crystalline silicon. Free electrons thus form dangling bonds that trap charge carriers that move freely in the semiconductor material. Amorphous silicon (a-Si) becomes a viable semiconductor material only if most of these dangling bonds are saturated with hydrogen atoms, in which case the material is often referred to as Si:H. However, in practice it is not possible for all dangling bonds to be saturated with hydrogen [3.6].

One of the main problems with amorphous silicon solar cells is the stability of their electrical properties in that, relative to baseline, the efficiency of such cells decreases by 10 to 30% during the first few months of operation, a phenomenon known as the Staebler–Wronski effect. Incident photons are able to break weak bonds between silicon atoms (e.g. because of an unduly large gap between the atoms), causing other bonds to break (see Figure 3.47). As a result, the charge carriers generated by the internal photoelectric

light

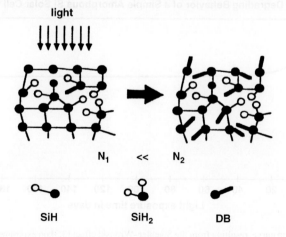

N_1 $\ll$ N_2

SiH   SiH₂   DB

Figure 3.47   Principle of the Staebler–Wronski effect [3.6]: The effects of light cause weak bonds to be broken, resulting in additional dangling bonds. Reproduced by permission of Arvind Shah

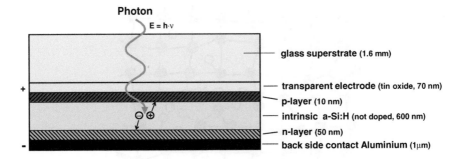

Figure 3.48 Structure of a *pin* solar cell made of amorphous silicon. (Based on data from Arco Solar)

effect are more likely to recombine before reaching an electrode, thus provoking efficiency loss. After a time, all 'compromised' Si–Si bonds are broken by the action of light, and cell efficiency plateaus at a lower level. This effect is less pronounced in very thin solar cells.

The band gap energy E_G values of amorphous silicon solar cells can be decreased by adding Ge, can be increased by adding C, and can be modified either way via partial saturation of broken bonds with substances other than hydrogen (e.g. F). This in turn allows for the generation of tandem or even triple solar cells with three active junctions.

Figure 3.48 shows the structure of an amorphous silicon solar cell, whereintegration of an intermediate layer consisting of non-doped (intrinsic) amorphous silicon allows for the creation of an electric field throughout virtually the entire cell. This facilitates the transport of photogenerated charge carriers and partially makes up for the fact that the semiconductor properties are of lesser quality.

Figure 3.49 shows the degradation of thin-film solar cells made of amorphous silicon. As can be seen here, most of the efficiency loss is provoked by the Staebler–Wronski effect during the first three weeks of exposure to the Sun.

Thin-film solar cells normally have a lower fill factor FF than crystalline silicon solar cells owing to the higher series resistance in the thin films and defects in the semiconductor material. Such cells also exhibit higher open-circuit voltage relative to crystalline silicon solar cells on account of the higher band gap energy of amorphous silicon (see Figure 3.50).

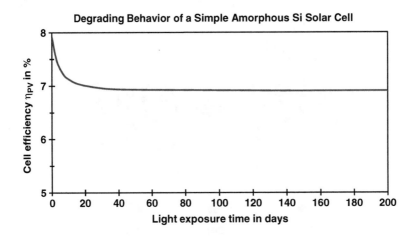

Figure 3.49 Degradation curve, resulting from the Staebler–Wronski effect [3.3] on exposure to light, of amorphous silicon thin-film solar cells (a-Si:H) (*data source*: Arco Solar). Solar cell efficiency plateaus at a substantially lower level after three weeks

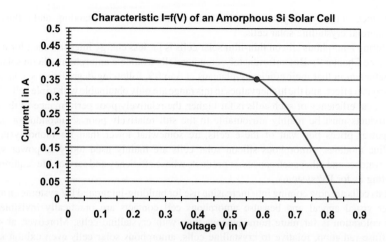

Figure 3.50   Characteristic curve $I = f(V)$ of an amorphous silicon (a-Si:H) thin-film solar cell with $1 \, kW/m^2$, AM1.5 spectrum and cell temperature of $25\,°C$ according to Arco Solar data (module G100 cell, $\eta_{PV} = 5.5\%$). Relative to a crystalline silicon solar cell, open-circuit voltage V_{OC} is considerably higher and the fill factor FF (57%) is somewhat lower

Adding a specific proportion of germanium (Ge) to the silicon reduces band gap energy E_G in the amorphous material, thus in turn allowing for the realization of amorphous tandem and triple solar cells that are far less prone to degradation secondary to the Staebler–Wronski effect. Such cells exhibit higher and far more stable efficiency and lower initial degradation than is the case with conventional single-film amorphous cells.

Figure 3.51 shows a cross-section of a Uni-Solar triple amorphous solar cell composed of three optically and electrically series-connected ultrathin cells with varying germanium content and whose front, middle and rear cells process blue, green and red light respectively. However, the cell efficiency of

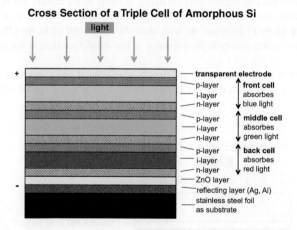

Figure 3.51   Structure of a Uni-Solar triple solar cell composed of three optically and series-connected ultrathin amorphous silicon cells. The manufacturing process used for these cells (successive deposition of thin silicon film on a stainless-steel foil substrate) is highly suitable for mass production. According to Uni-Solar, small-scale lab cells of this type exhibit up to 13% efficiency [Mar03], while 12.1% efficiency has been confirmed by an impartial investigator [3.2]

solar cell products of this type is just under 9%, the triple structure notwithstanding, and is thus only half that of commercial crystalline solar cells.

Manufacturing amorphous silicon thin-film solar cells is far less energy intensive than for monocrystalline cells (see Chapter 9). But unfortunately, commercially available amorphous silicon solar cells are still very inefficient in that their efficiency plateaus at 3 to 9% following degradation secondary to the Staebler–Wronski effect, and the higher values in this range are only obtainable with triple cells. Inasmuch as the theoretical efficiency of such cells is far higher, the relatively poor performance of the currently available products must be mainly attributable to the still relatively poor semiconductor properties. The peak power prices per watt of these cells are somewhat lower than for monocrystalline and polycrystalline modules. Amorphous silicon solar cells are mainly used today as power sources in calculators, watches and other such consumer products whose efficiency and service life requirements are not as exacting as for PV systems.

Amorphous cells are now coming into increasing use for building-integrated PV systems on account of their greater visual appeal and because amorphous cell contacts are practically invisible and the consequent coloration is far more homogeneous than with crystalline cells. Moreover, at sites with elevated diffuse radiation, relative to crystalline cells, amorphous solar cells even exhibit somewhat higher specific energy yield (kWh produced per Wp of solar generator power) as they are more compatible with the diffuse radiation spectrum (see Figure 2.42). Amorphous triple cells (see Figure 3.51) are particularly well suited for building-integrated PV systems and for the manufacture of specialized hybrid products for this purpose (electricity-producing roof elements).

3.5.3.2   Cadmium Telluride (CdTe) Solar Cells

The 1.46 eV band gap energy of cadmium telluride is very close to maximum theoretical efficiency (see Figure 3.25), and this material also lends itself very well to the manufacture of thin-film solar cells. Unlike amorphous silicon, cadmium telluride exhibits excellent stability and no degradation secondary to the effects of light. Figure 3.52 shows a cross-section of a cadmium telluride thin-film solar cell, whose active films are only a few micrometres thick. Such cells have attained up to 16.7% efficiency in the lab [3.2] and are relatively easy to manufacture. US-based First Solar makes a large number of commercial cadmium telluride modules measuring 120 cm · 60 cm with power outputs of 60, 62.5, 65, 67.5, 70, 72.5, 75 and 77.5 Wp, efficiency ranging from $\eta_M = 8.3$ to 10.8%, and fill factors FF ranging from 62 to 68%. Hence the consequent cell efficiency is likely to be around 1% higher for each such module. The production costs for these solar cells, as of December 2009, was just under $0.9 per Wp (€0.6 per Wp) and are likely to decrease. As with other thin-film solar cells, the efficiency of cadmium telluride cells decreases more slowly in the presence of low irradiance than is the case with crystalline silicon solar cells.

However, cadmium (Cd) is ecologically unfriendly, and even if only very thin cadmium films are needed, some cadmium solar cells may not lend themselves to proper disposal as hazardous waste at

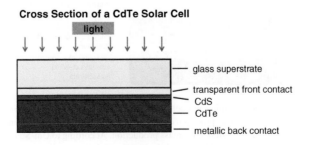

Figure 3.52   Cross-section of a cadmium telluride solar cell (After [Mar03])

Cross Section of a CIS Solar Cell

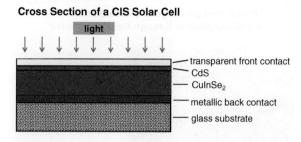

Figure 3.53 Cross-section of a CIS solar cell. (After [Mar03])

the end of their service lives; this in turn will result in some of the highly toxic cadmium ending up in the environment without having been disposed of properly. That said, according to [Mar03], cadmium telluride is highly stable. Moreover, as cadmium telluride only breaks down at temperatures exceeding 1000 °C, it is not released into the environment in case of fire [DGS05].

3.5.3.3 Copper Indium Diselinide (CuInSe₂) and Cu(In,Ga)Se₂ Solar Cells

Ultrathin (0.5 μm) films of copper indium diselenide (CIS), whose band gap energy is around 1 eV, absorb virtually all $h \cdot v > E_G$ photons; CIS also exhibits high stability under the effects of light. But unfortunately indium is a very rare element. Figure 3.53 shows the cross-section of a thin-film CIS solar cell.

The band gap energy of CIS can be substantially altered by replacing a portion of the indium (In) with gallium (Ga) and a portion of the selenium (Se) with sulphur (S). Replacing 10 to 20% of the indium with gallium yields CuInGaSe₂ (CIGS), whose band gap energy is around 1.1 eV (roughly the same as c-Si). The open-circuit voltage of CIGS (around 500 to 720 mV) is also considerably higher than with pure CIS cells (400 to 550 mV). CIGS cells also exhibit a better fill factor FF. According to [3.11], small-scale CIGS cells ($A_Z = 1$ cm²) attain $\eta_{PV} = 19.4\%$ efficiency. Commercial CIS modules from vendors such as Würth Solar (and, previously, Shell Solar) exhibit cell efficiency η_{PV} ranging from 10.5 to 12.5%. The interconnections and an electron microscope cross-section for a CIGS solar cell (light incident from above) are shown in Figures 3.54 and 3.55 respectively.

Many vendors make CIS and CIGS solar cells using a thin intermediate cadmium sulphide (CdS) film, which, owing to the presence of toxic cadmium, is also environmentally problematic. However, the search for cadmium substitutes is ongoing. With their relatively low band gap energy, CIS and CIGS are highly suitable for use as back-side solar cells behind a high-band-gap-energy front cell in tandem arrays (see Section 3.4.4.2, Figure 3.35 and Figure 3.36).

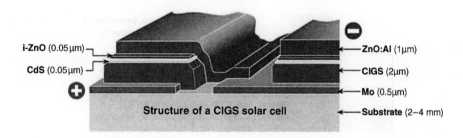

Figure 3.54 Cross-section and interconnections for a CIGS solar cell, for which glass, plastic and other low-cost substrates are used. *Source*: Würth Solar

Cross-section image of CIGS solar cell
Light illumination is through the ZnO:Al layer

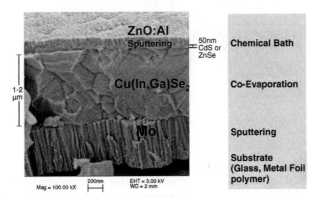

Figure 3.55 Electron microscope cross-section of a CIGS solar cell, whose manufacturing process begins with the substrate, as follows: (1) deposition of the molybdenum back contact; (2) deposition of the CIGS film; (3) deposition of a thin CdS film or an intermediate ZnSe film; (4) realization of a highly conductive ZnO:Al front electrode. Light incidence is from above (*Source*: Dr A. Tiwari, ETH Zurich)

3.5.3.4   Microcrystalline and Amorphous Silicon Tandem Solar Cells

For about the last 10 years, researchers have been attempting to develop ultrathin microcrystalline solar cells using far smaller crystals than those used in today's polycrystalline and multicrystalline solar cells. However, to achieve this, light must be trapped and retained in the cell very efficiently so as to attain sufficient current density and photon absorption. Crystal miniaturization also substantially reduces open-circuit voltage V_{OC} relative to that of crystalline solar cells. Moreover, the miniaturized lattice that is obtained with microcrystalline solar cells obviates the Staebler–Wronski effect. Efficiency amounting to $\eta_{PV} = 10.1\%$ has already been achieved for small-scale lab cells $2\,\mu m$ thick ($A_Z \approx 1.2\,cm^2$) [Mar03], [3.2].

Microcrystalline solar cells are very suitable for use as back cells in optically and electrically series-connected tandem arrays with an amorphous silicon front cell. This arrangement is known as a micromorph solar cell. Inasmuch as only the front cell is amorphous, the Staebler–Wronski effect is far less pronounced in the cell as a whole than is the case with pure amorphous cells. Initial efficiency amounting to $\eta_{PV} = 14.5\%$ has already been achieved for small-scale lab cells ($A_Z \approx 1\,cm^2$) [Mar03]. Figure 3.56 shows a cross-section of this type of microcrystalline and amorphous silicon tandem cell.

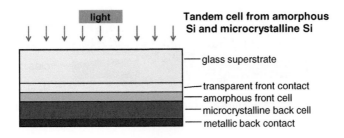

Figure 3.56   Cross-section of a micromorph tandem cell with an a-Si front cell and a microcrystalline (μc) Si back cell (After [Mar03])

3.5.4 Dye Sensitized Solar Cell (DSSC; Photoelectrochemical Solar Cells, Grätzel Solar Cells)

Two decades ago a research team at EPFL in Lausanne led by Dr M. Grätzel developed a new type of solar cell known as a Dye Sensitized Solar Cell (DSSC), which is based on a completely different principle than conventional solar cells. DSSCs integrate not merely a semiconductor, but rather a series of elements: a titanium dioxide (TiO_2) semiconductor, a monomolecular dye film deposited on the semiconductor, and an electrolyte solution. Figure 3.57 shows the structure of a DSSC.

In a DSSC, when a sufficiently energized photon strikes a dye molecule, the photon is absorbed and the molecule enters an excited state, i.e. the energy level of one of the molecule's valence electrons increases. This excited molecule then injects an electron into the titanium dioxide's conduction band and is thus ionized.

In the interest of preventing this electron from recombining before long, an electrolytic 'redox mediator' (referred to here for reasons of simplicity as I/I^-, but in practice referred to as I_3^-/I^- or the like) almost instantaneously (in about 10 ns) injects an electron into the ionized dye molecule, thus shifting the molecule to a neutral state and obviating recombination.

The electrons e^- generated by photon absorption are shunted, via the semiconductor, to the bottom electrode and from there, releasing some energy, to the opposite (upper) electrode via the outer electric circuit. This in turn closes off the circuit by virtue of the fact that the electrolytic redox mediator oxidized to I by the dye is diffused to the opposite electrode, where it is reduced to I^- on incorporation of an electron (see [3.8] for a more complete and detailed function description with numerous illustrations).

According to [3.2], a small-scale (0.22 cm²) DSSC achieves $\eta_{PV} = 11.2\%$ efficiency at STC. According to [3.8], the efficiency of such cells in the presence of weak irradiance, $G = 100 \text{ W/m}^2$, is considerably higher than with $G = 1000 \text{ W/m}^2$. Moreover, the somewhat simpler production process for DSSCs would presumably allow for inexpensive mass production.

Although there have been reports about DSSCs time and time again over the past two decades in the media and at conferences on photovoltaics, it needs to be borne in mind that DSSCs are still a lab development, and that at the date this book went to press, these modules had yet to be produced commercially and no datasheets were available for them. This technology may well be promising, but its long-term stability and practical usefulness remain to be proven. To the best of my knowledge, DSSCs have yet to be used for long-term energy production. Moreover, experience has shown that it can take a very long time before such new types of solar cells can be successfully produced in a factory. Only time will tell whether the great expectations concerning DSSCs will come to fruition.

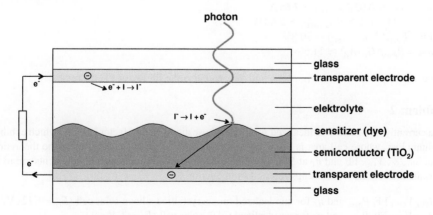

Figure 3.57 Cross-section of a DSSC (After 3.8])

3.6   Bifacial Solar Cells

Some of the types of solar cells that have been discussed so far can trap in principle incident light bilaterally, which opens up promising new applications in that bifacial modules need not necessarily be tilted towards the Sun but can also be oriented vertically (e.g. in a north–south direction). On the other hand, normally oriented solar generators with bifacial modules whose rear side also captures light reflected off the ground yield 10 to 15% more energy. The company Solar Wind Europe sells pilot-production bifacial modules in the 40 to 120 W power range, although the two sides of these modules do not make equally good use of incident light, i.e. the light incident on one side is used more efficiently than that incident on the other side (e.g. only about half of the light incident on the rear side is used). In addition, at Expo 2005 in Japan, Hitachi unveiled a pilot-production bifocal module. Only time will tell the extent to which such modules will be able to gain a reasonable market share and whether larger vendors will begin making such products.

3.7   Examples

Problem 1

A silicon solar cell whose cell area is $A_Z = 225\,cm^2$ exhibits V_{OC} of 600 mV, I_{SC} of 7.5 A and P_{max} of 3.4 W at STC (1 kW/m², AM1.5, 25 °C).

This solar cell is now being operated in the presence of the same irradiance but at a cell temperature of $T_Z = 75\,°C$. It can be presumed that the temperature dependence of this module will be as in Figure 3.18.

Calculation task:

(a) V_{OC} (with $T_Z = 75\,°C$)
(b) I_{SC} (with $T_Z = 75\,°C$)
(c) P_{max} (with $T_Z = 75\,°C$)
(d) FF (with $T_Z = 75\,°C$)
(e) η_{PV} (with $T_Z = 75\,°C$)

Solution:

(a) $V_{OC} \approx (1 - 0.2)V_{OC(STC)} \approx 480\,mV$
(b) $I_{SC} \approx (1 + 0.02)I_{SC(STC)} \approx 7.65\,A$
(c) $P_{max} \approx (1 - 0.235)P_{max(STC)} \approx 2.60\,W$
(d) $FF = P_{max}/(V_{OC} \cdot I_{SC}) \approx 70.8\%$
(e) $\eta_{PV} = P_{max}/(G_o \cdot A_Z) \approx 11.6\%$

Hence, as temperature rises, V_{OC}, P_{max} and η_{PV}, and also the fill factor FF, decrease.

Problem 2

For a currently non-realizable idealized solar cell whose diode quality factor is $n = 1$, which exhibits no ohmic loss, whose cell area is $A_Z = 100\,cm^2$, and which in practice also reaches the theoretical limits for J_{SC}, V_{OC}, FF and η_T at STC (1 kW/m², AM1.5, 25 °C), the following elements are to be determined using the relationships referred to in Section 3.4:

(a) I_{SC}, V_{OC}, FF, P_{max} and η_T for an idealized monocrystalline silicon solar cell ($E_G = 1.12\,eV$).
(b) I_{SC}, V_{OC}, FF, P_{max} and η_T for an idealized CdTe solar cell ($E_G = 1.46\,eV$).

Solution:

(a) According to Equation 3.4, $V_T = 25.68$ mV for $n = 1$ at 25 °C.
According to Figure 3.20, $J_{SC} \approx 43.1$ mA/cm² $\Rightarrow I_{SC} \approx A_Z \cdot J_{SCmax} = 4.31$ A.
Using Equation 3.22 or an approximation based on Figure 3.23, the following is obtained:
$V_{OC} = 767$ mV ($K_S = 40\,000$ A/cm²).
Using Equation 3.12 or an approximation based on Figure 3.24, $FF_i = 0.857$ is obtained.
Using Equation 3.19, the following holds true: $P_{max} = V_{OC} \cdot I_{SC} \cdot FF_i = 2.83$ W.
Using Equation 3.20, or directly based on Figure 3.25, $\eta_T = P_{max}/(G_o \cdot A_Z) = 28.3\%$.

(b) According to Equation 3.4, $V_T = 25.68$ mV for $n = 1$ at 25 °C.
According to Figure 3.20, $J_{SC} \approx 30.1$ mA/cm² $\approx I_{SC} \approx A_Z \approx J_{SCmax} = 3.01$ A.
Using Equation 3.22 or an approximation based on Figure 3.23, the following is obtained:
$V_{OC} = 1.098$ V ($K_S = 40\,000$ A/cm²).
Using Equation 3.12 or an approximation based on Figure 3.24, $FF_i = 0.891$ is obtained.
Using Equation 3.19, the following holds true: $P_{max} = V_{OC} \cdot I_{SC} \cdot FF_i = 2.94$ W.
Using Equation 3.20, or directly based on Figure 3.25, $\eta_T = P_{max}/(G_o \cdot A_Z) = 29.4\%$.

Problem 3

For the currently non-realizable idealized solar cell as in Example 2(a) whose diode quality factor is $n = 1$, whose characteristics have been provisionally defined, which exhibits no ohmic loss with a cell area of $A_Z = 100$ cm², and that in practice likewise reaches the theoretical limits for $J_{SC}, V_{OC},$ FF and η_T, using the relationships referred to in Section 3.4 calculate the theoretical efficiency η_T in monochromatic red light ($\lambda = 0.77$ μm) with $G_o = 1$ kW/m² and cell temperature $T_Z = 25$ °C.

Solution:

According to Equation 2.33, the energy of such a red-light photon is $E_{Ph} = 2.58 \cdot 10^{-19}$ J $= 1.61$ eV.
The number of photons per second and unit area arriving at the solar cell is as follows:
$n_{Ph}/(A \cdot t) = G_o/E_{Ph} = 3.88 \cdot 10^{17}$ photons/cm² s $\Rightarrow$ current density $J_{SC} = n_{Ph} \cdot e/(A \cdot t) = 62.1$ mA/cm².
Using Equation 3.22, it follows that maximum possible open-circuit voltage is $V_{OC} = 777$ mV ($K_S = 40\,000$ A/cm²).
Using Equation 3.12, the idealized fill factor $FF_i = 0.858$.
Using Equation 3.19, $P_{max} = FF_i \cdot V_{OC} \cdot J_{SC} \cdot A_Z = 4.14$ W is obtained.
Using Equation 3.20, it follows that $\eta_T = (P_{max}/A_Z)/G_o = 41.4\%$.
Hence solar cell efficiency in monochromatic light can be considerably *greater* than in the AM1.5 spectrum.

Problem 4

A currently non-realizable idealized tandem solar cell whose diode quality factor is $n = 1$, which exhibits no ohmic loss (as in Example 2) and whose cell area is $A_Z = 100$ cm², is composed of two electrically isolated solar cells that are optically connected behind each other (four-terminal tandem cell). The band gap energy E_G of the front and back cell is 1.75 and 1 eV respectively. Inasmuch as this idealized example presupposes that the front cell will absorb all sufficiently energized photons and that all insufficiently energized protons will be allowed through unimpeded, the possible current density (based on Figure 3.20) of the rear cell will be reduced by exactly the current density of the front cell. Based on these suppositions, the following elements are to be calculated using the relationships referred to in Section 3.4, at STC (1 kW/m², AM1.5, 25 °C):

(a) $I_{SC}, V_{OC},$ FF, P_{max} and η_T for an idealized front solar cell ($E_G = 1.75$ eV).
(b) $I_{SC}, V_{OC},$ FF, P_{max} and η_T for a rear solar cell ($E_G = 1$ eV).

(c) P_{max} and η_T for the cell as a whole, assuming that both cells of the tandem array are operated at MPP.

Solution:

(a) **Front solar cell ($E_G = 1.75$ eV):**
According to Equation 3.4, $V_T = 25.68$ mV for $n = 1$ at 25 °C.
According to Figure 3.20, $J_{SC} = J_{SC\text{-}F} \approx 20.7$ mA/cm² $\Rightarrow I_{SC\text{-}F} \approx A_Z \cdot J_{SCmax} \approx 2.07$ A.
Using Equation 3.22 or an approximation based on Figure 3.23, the following is obtained:
$V_{OC} = 1.378$ V ($K_S = 40\,000$ A/cm²).
Using Equation 3.12 or an approximation based on Figure 3.24, $FF_i = 0.909$.
Using Equation 3.19, the following holds true: $P_{max} = P_{max\text{-}F} = V_{OC} \cdot I_{SC} \cdot FF_i \approx 2.59$ W.
Using Equation 3.20, or directly from Figure 3.25, $\eta_T = P_{max}/(G_o \cdot A_Z) = 25.9\%$.

(b) **Rear solar cell ($E_G = 1$ eV):**
According to Equation 3.4, $V_T = 25.68$ mV for $n = 1$ at 25 °C.
According to Figure 3.20, with $E_G = 1$ eV, $J_{SC} \approx 47.7$ mA/cm².
However, for a tandem cell, $J_{SC\text{-}F} = 20.7$ mA/cm² must be subtracted from the front cell: $\Rightarrow$
$J_{SC\text{-}R} = J_{SC} - J_{SC\text{-}F} = 27.0$ mA/cm² $\Rightarrow I_{SC\text{-}R} \approx A_Z \cdot J_{SC\text{-}R} \approx 2.70$ A.
Using Equation 3.22 or an approximation based on Figure 3.23, the following is obtained:
$V_{OC} = 635$ mV ($K_S = 40\,000$ A/cm²).
Using Equation 3.12 or an approximation based on Figure 3.24, $FF_i = 0.835$ is obtained.
Using Equation 3.19, it then follows that: $P_{max} = P_{max\text{-}R} = V_{OC} \cdot I_{SC\text{-}R} \cdot FF_i \approx 1.43$.
Using Equation 3.20 (but *not* directly based on Figure 3.25), $\eta_T = P_{max}/(G_o \cdot A_Z) \approx 14.3\%$ is obtained.

Hence the back cell converts into electrical energy an additional 14.3% of total energy arriving at the solar cell. However, as the higher-energy photons have already been processed by the front cell, referred to the residual light that penetrates to the back cell, the efficiency of the back cell is considerably higher.

(c) **Total for the tandem array:**
Total $P_{max} = P_{max\text{-}F} + P_{max\text{-}R} = 4.02$ W. Using Equation 3.20, the following is obtained:
Theoretical efficiency of the entire tandem array: $\eta_T = P_{max}/(G_o \cdot A_Z) \approx 40.2\%$.

3.8 Bibliography

[3.1] R.M. Swanson:'How Close to the 29% Limit Efficiency Can Commercial Silicon Solar Cells Become?'. 20th EU Photovoltaic Solar Energy Conference, Barcelona, 2005.

[3.2] M.A. Green et al.: 'Solar Cell Efficiency Tables (Version 33)'. *Progress in Photovoltaics: Research and Applications*, 2009, **17**, 85–94 (www.interscience.wiley.com).

[3.3] W. Maag: 'Photovoltaische Dünnfilmzellen – Der grosse Aufschwung'. *Bulletin SEV*, 6/1987, S. 323ff.

[3.4] G. Herning, A. Kreutzmann: 'Konzentration bitte – Die Renaissance der konzentrierenden Solarmodule'. *Photon*, 7/2005, S. 62ff.

[3.5] A. Shah, R. Tscharner: 'Technologien für Solarzellen und Solarmodule'. *Bulletin SEV*, 23/1991, S. 11ff.

[3.6] A. Shah, H. Curtins et al.: 'Die Abscheidung von amorphem Silizium im VHF-GD-Prozess'. IMT, Universität Neuenburg, 1988.

[3.7] H. Curtins, A. Shah: 'Photovoltaik: Strom aus Sonnenlicht'. *Technische Rundschau*, 25/1988, S. 70ff.

[3.8] P. Bonhôte, A. Kay, M. Grätzel: 'Photozellen mit Energieumwandlung nach Pflanzenart'. *Bulletin SEV*, 7/1996, S. 11ff.

[3.9] Y. Tawada, H. Yamagishi, K. Yamamoto: 'Mass Productions of Thin Film Silicon PV Thin Film Modules'. *Solar Energy Materials & Solar Cells*, 2003, **78**, 647.

[3.10] Presseinformation 01/09: 'Weltrekord: 41.1% Wirkungsgrad für Mehrfachsolarzellen am Fraunhofer ISE', 14 January 2009.

[3.11] M.A. Green *et al.*: 'Solar Cell Efficiency Tables (Version 34)'. *Progress in Photovoltaics: Research and Applications*, 2009, **17**, 320–326 (www.interscience.wiley.com).

[3.12] M.A. Green, A. Ho-Baillie: '43% Composite Split-Spectrum Concentrator Solar Cell Efficiency'. *Progress in Photovoltaics: Research and Applications*, 2010, **18**, 42–47 (www.interscience.wiley.com).

Additional References

[Bur83], [Eic01], [Gre86], [Gre95], [Häb91], [Häb07], [Her92], [Hu83], [Ima92], [Lad86], [Luq03], [Mar94], [Mar03], [Qua03], [See93], [Sta87], [Wen95], [Wil94], [Win91].

MaXimise your yield!

A properly matured Swiss product.

SolarMax inverters are made to deliver full capacity and a clever cooling system helps them keep their cool at all times. And that is good for you. Because maximum efficiency and the very best reliability give you both the best-possible yields and a carefree life as well.

And that is no surprise because every SolarMax is a genuinely Swiss product with typical Swiss virtues: Highest quality materials, no compromises in their fabrication, and a generous warranty whose benefits you will probably never need anyway. And if you do, our After Sales Service will respond quickly and reliably.

Whether you are planning a PV system for a single-family home or a large solar power plant: SolarMax has the right product for you. Nothing cheesy about that.

20
20 years Swiss Quality and Experience

 Easy installation

 High, constant efficiency

 Swiss Quality

 Maximum reliability

 Competent After Sales Service

 Maximum revenues

4

Solar Modules and Solar Generators

4.1 Solar Modules

Commercial crystalline silicon solar cells have open-circuit voltage V_{OC} ranging from around 0.55 to 0.72 V at a cell temperature of 25 °C. With a cell area of $A_Z \approx 100\,\text{cm}^2$ (4 inch wafer) such cells have short-circuit current I_{SC} ranging from 3 to 3.8 A; with $A_Z \approx 155\,\text{cm}^2$ (5 inch wafer) the figure ranges from around 4.6 to 6 A; with $A_Z \approx 225\,\text{cm}^2$ (6 inch wafer) the figure ranges from around 6.8 to 8.5 A; and with $400\,\text{cm}^2$ (8 inch wafer) the figure ranges from around 13 to 15 A. For optimal power yield (MPP; see Section 3.3.3), the voltage is around $V_{MPP} \approx 0.45$ to 0.58 V. Hardly any appliance can be operated at such low voltages. Hence voltages that are usable for PV system operation can only be generated using multiple solar cells wired in series.

If higher voltages are needed for a specific application, a series of solar cells must be wired in parallel. Series and parallel connection of solar cells allow for interconnection of an unlimited number of such cells to create massive solar generator fields with many megawatts of power.

In some cases it is useful to wire in series anywhere from around 32 to 72 solar cells and to house them in a single enclosure to protect them against the environment. The entities thus created are referred to as solar cell modules, solar modules or simply modules.

Solar modules with 36 cells, operating voltages from 15 to 20 V and output ranging from 50 to 200 Wp are very widely used, since a viable 12 V power supply is unobtainable with only one module and battery. On the other hand, such modules comprising 72 cells for operating voltages ranging from 30 to 40 V are suitable for stand-alone 24 V system voltage installations. Modules with output up to around 200 Wp that measure $1.5\,\text{m}^2$ and weigh around 18 kg can be readily handled by one person.

The output of the largest mass-produced polycrystalline module currently on the market (sold by Schott) is 300 Wp (80 cells wired in series). The largest monocrystalline module with an output of around 315 Wp (96 cells wired in series) is made by Sunpower (see Table 4.1 for details concerning these modules). Larger modules for integration into buildings can be custom ordered from specialized vendors.

The life span of such modules is largely determined by how well they are protected against the ambient environment. Some vendors indicate a 30-year life span and grant 2- to 5-year full warranties and in some cases limited performance guarantees for 10 to 26 years. The fronts of most such products are well protected against hail via specially tempered low-iron glass that is highly transparent. In addition, solar cells are hermetically packed in a transparent plastic material such as ethyl vinyl acetate (EVA). The rear protection elements are made of plastic or glass, depending on the manufacturer. A classic solar module integrates relatively thin (e.g. 3–4 mm) glass and has a robust metal frame (usually made of aluminium) that provides the requisite mechanical stability and good edge protection. Modules with thin-film silicon solar cells or the like also have plastic frames.

Photovoltaics: System Design and Practice. Heinrich Häberlin.
© 2012 John Wiley & Sons, Ltd. Published 2012 by John Wiley & Sons, Ltd.

Table 4.1 Key technical data for selected mass-produced silicon solar cell modules from a number of large manufacturers. This information was provided by the vendors and makes no claim to completeness or accuracy. The efficiency of HIT cells (Heterojunction with Intrinsic Thin-layer cells) is optimized through deposition of an amorphous silicon film on a crystalline cell, which also somewhat increases the voltage per cell. Laminate versions of many modules are available (indicated in the 'Frame' column with 'L'), and slight deviations from the dimensions and weights shown can be specially ordered. All electrical values refer to STC. Per-cell voltages are higher for triple-cell and highly efficient silicon cell modules, but are somewhat lower for CIS modules relative to conventional crystalline silicon cells

Company	Type	Cells	Frame	Cell	P_{max} [W]	V_{OC} [V]	I_{SC} [A]	Mass [kg]	Length [mm]	Width [mm]	A_M [m^2]	η_{IM} [%]
Aleo	S-18-I230	60	Alu	poly	230	36.6	8.44	22	1660	990	1.64	14.0
BP Solar	BP 790	36	Alu	mono	90	22.4	5.4	7.7	1209	537	0.65	13.8
BP Solar	BP 7195	72	Alu	mono	195	44.9	5.6	15.4	1593	790	1.26	15.5
BP Solar	BP 380	36	Alu	poly	80	22.1	4.8	7.7	1209	537	0.65	12.3
BP Solar	BP 3165	72	Alu	poly	165	44.2	5.1	15.4	1593	790	1.26	13.1
First Solar	FS-275	116	L	CdTe	75	92	1.2	12	1200	600	0.72	10.4
Isofoton	I-75S/12	36	Alu/L	mono	75	21.6	4.67	9	1224	545	0.67	11.2
Isofoton	IS150/12	72	Alu/L	mono	150	21.6	9.3	14.4	1590	790	1.26	11.9
Isofoton	IS150/24	72	Alu/L	mono	150	43.2	4.7	14.4	1590	790	1.26	11.9
Kyocera	KC130GHT-2	36	Alu	poly	130	21.9	8.02	12.2	1425	652	0.93	14.0
Kyocera	KC175GHT-2	48	Alu	poly	175	29.2	8.09	16	1290	990	1.28	13.7
Kyocera	KC200GHT-2	54	Alu	poly	200	32.9	8.21	18.5	1425	990	1.41	14.2
Schott	ASE260DG-FT	120	Alu	EFG	268	71.4	5	41	1605	1336	2.14	12.5
Schott	Poly 300	80	Alu	poly	300	48.9	8.24	41.5	1685	1313	2.21	13.6
Sanyo	HIP-215 NHE	72	Alu	HIT	215	51.6	5.61	15	1570	798	1.15	17.2
Sharp	ND162E1F	48	Alu	poly	162	28.4	7.92	16	1318	994	1.31	12.4
Sharp	NT175E1	72	Alu	mono	175	44.4	5.4	17	1575	826	1.3	13.5
Sharp	NU180E1	48	Alu	mono	180	30	8.37	16	1318	994	1.31	13.7
Solarworld	SW225 poly	60	Alu	poly	225	36.6	8.17	22	1675	1001	1.68	13.4
Solarworld	SW185 mono	72	Alu	mono	185	44.8	5.5	15	1610	810	1.3	14.2
Shell	ST40	42	Alu	CIS	40	23.3	2.7	7	1219	328	0.42	9.4
Siemens	*M55**	*36*	*Alu/L*	*mono*	*55*	*21.7*	*3.4*	*5.7*	*1293*	*330*	*0.43*	*12.9*
Sunpower	SPR 95**	32	Alu	mono	95	21.2	5.85	7.4	1038	527	0.55	17.4
Sunpower	SPR 220**	72	Alu	mono	220	48.3	5.95	15	1559	798	1.24	17.7
Sunpower	SPR 315**	96	Alu	mono	315	64.6	6.14	24	1559	1046	1.63	19.3
Uni-solar	US-32	11	Alu	3·a-Si	32	23.8	2.4	4.8	1366	383	0.52	6.1
Uni-solar	US-64	11	Alu	3·a-Si	64	23.8	4.8	9.2	1366	741	1.01	6.3
Würth Solar	WS 31046	ca.33	Alu	CIS	55	22	3.56	9.7	905	605	0.55	10
Würth Solar	WS 11007/80	ca.69	Alu	CIS	80	45.5	2.5	12.7	1205	605	0.73	11

Examples partially from 2007, possibly some modules no longer available now.

*No longer available on the market.

**Ground + pole of PV array by a high resistance for full power.

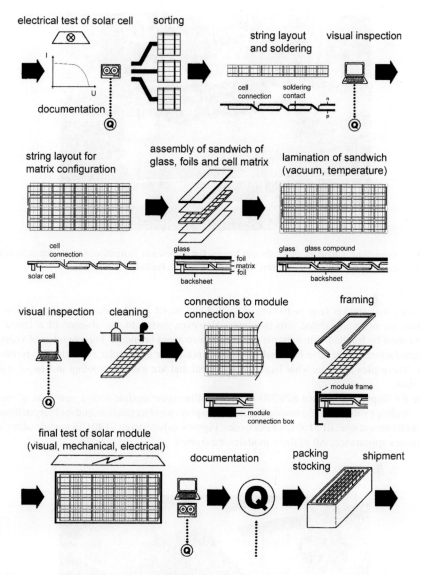

Figure 4.1   The manufacturing process for crystalline silicon solar cells (Partially based on documents from AEG)

The module manufacturing process for crystalline solar cells is relatively labour intensive. Figure 4.1 displays the workflow for the production of such solar cells. Figures 4.2 and 4.3 display some of the intermediate stages in the production process.

The above packing method is extremely cost intensive in that the requisite materials – particularly for modules with aluminium frames – consume considerable amounts of grey energy, i.e. energy that is used to manufacture these elements. That said, anyone who has seen a solar module that has been exposed to the relatively damp Central European climate for a lengthy period will instantly understand why such modules need extremely robust protection against the elements. By the same token, it is no easy matter to manufacture building windows that will remain perfectly watertight and transparent for two to three decades without maintenance.

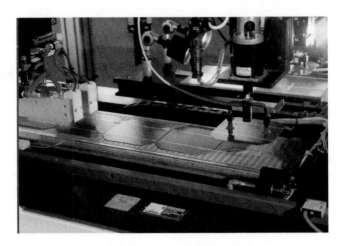

Figure 4.2 Assembling individual solar cells into strings. For purposes of illustration, a somewhat older but more readily understandable apparatus is shown here. © Solar-Fabrik AG, Freiburg

Laminates with thicker (e.g. 6–10 mm) glass that can be set into façades or roofs as is done with plate glass are often integrated into buildings. However, owing to the absence of a frame, such laminates must be handled with extreme care in order to avoid breakage. For a number of years now, some manufacturers have also been selling special plastic or glass solar roof tiles for rooftop PV systems. These tiles are somewhat larger than normal and are used for roofing in lieu of standard roofing tiles.

Figure 4.4 displays a finished SPR220 monocrystalline solar module with a new kind of rear that obviates shading by the front electrodes. Figure 4.5 displays a polycrystalline and multicrystalline solar module with conventional BP3160 front electrodes. Figure 4.6 displays an ST40 CIS module with a highly homogeneous appearance. All of these modules are framed.

Figure 4.3 In the laminator, the sandwich comprising the cover glass, films and the intermediate, pre-wired strings are baked together in a vacuum. For purposes of illustration, a somewhat older but more readily understandable apparatus is shown here. © Solar-Fabrik AG, Freiburg

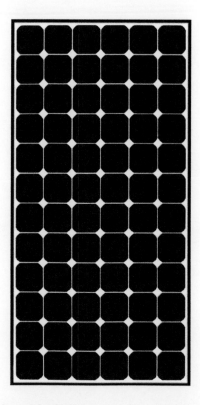

Figure 4.4 Sunpower SPR220 solar module with 72 monocrystalline silicon solar cells and a new type of rear contact (220 Wp, $\eta_M = 17.7\%$) (Photo: Sunpower Corporation)

Figure 4.7 displays a laminate with polycrystalline cells. Figure 4.8 displays a Uni-Solar US64 amorphous silicon triple-cell solar module (solar cell structure as in Figure 3.51). Figure 4.9 displays the Newtec SDZ36, which was one of the first solar roof tiles to come on the market, integrates 24 monocrystalline solar cells, and can be mounted on a roof in lieu of standard tiles; such tiles can also be walked on.

Framed modules are superior to their laminate counterparts in terms of handling, mechanical stability and lightning protection. However, in framed modules – particularly those mounted flush with the roof – over time a permanent layer of grit forms between the outermost cells and the frames that reduces energy yield. Hence the frames on the outside of such modules (i.e. the side facing the Sun) should be as low as possible and approximately 5–15 mm clearance should be left on all sides between the cells and module frames.

The solar cell module wiring diagrams in this chapter use the symbol shown in Figure 4.10. The triangle at the positive end is somewhat similar to the diode connection symbol in Figure 3.8.

In cases where higher voltages are needed, a series of modules must be wired in series into a string. For stronger current, a series of modules or strings is wired in parallel.

In the past, in order to display characteristic waveforms for a 36-cell solar module in a product and vendor neutral manner, the then widely used (but since discontinued) Siemens M55 module with 36 cells of 103 mm × 103 mm wired in series and 55 Wp of rated STC power output was employed.

But as in recent years it has become increasingly difficult to obtain detailed characteristic curves from vendors for their modules under various operating conditions, in the interest of obtaining data that are as consistent as possible with the scant data disclosed by vendors and with my own measurements for selected modules, the characteristic curves in Figures 4.11–4.15 were computer generated using the full equivalent

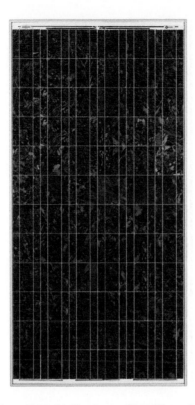

Figure 4.5 BP Solar BP3160 solar module with 72 polycrystalline silicon solar cells and a conventional contact system (175 Wp, $\eta_M = 12.7\%$) (Photo: BP Solar)

circuit as in Figure 3.12. Figure 4.11 provides an overview of the I–V characteristics of the M55 module at cell temperatures of 25 and 55 °C, and with three different insolation levels (100, 400 and 1 kW/m²).

Figure 4.12 displays the I–V characteristic curves for an M55 module at a constant cell temperature of 25 °C and various levels of insolation. The short-circuit current is exactly proportional to insolation, whereas at low insolation the open-circuit voltage is almost as high as for 1 kW/m². Maximum output P_{MPP} at the MPP exhibits a somewhat larger increase relative to insolation, i.e. at lower insolation efficiency is also somewhat poorer than with 1 kW/m².

Figure 4.13 displays the I–V characteristic curves for an M55 module at various cell temperatures T_Z and constant 1 kW/m² insolation. Short-circuit current increases very slightly as temperature rises, but this induces a reduction in the open-circuit voltage, filling factor and maximum output at the MPP; as a result efficiency declines sharply as temperature rises.

The I–V characteristic curves in Figure 4.12 for various insolation levels and at a constant cell temperature display module characteristics under laboratory conditions, but provide insufficient information concerning practical applications, since insolation will of course make a solar module quite hot. Depending on mounting modality, module design and wind conditions, cell temperature T_Z at 1 kW/m² insolation normally ranges from around 20 to 40 °C above ambient temperature T_U. If the module's available electrical power is drawn off by the outer circuit, T_Z will be somewhat lower than at open or short circuit, where insolation is converted into heat. This phenomenon, which is attributable to the law of conservation of energy, can be used for purposes such as thermographic searches for inactive modules in large solar generator fields.

The cell temperature increase relative to ambient temperature T_U attributable to module design can be determined using nominal operating cell temperature (NOCT), which is defined as the fixed temperature

Figure 4.6   Shell Solar ST40 CIS solar module (40 Wp). Photo: Shell Solar/SolarWorld

the module would exhibit in the AM1.5 spectrum at open circuit at an ambient temperature of 20 °C, 1 m/s wind speed and $G_{NOCT} = 800$ W/m^2 irradiance. Assuming that the temperature increase relative to T_U is proportional to module irradiance G_M and solar generator irradiance G_G, the following is obtained for T_Z:

$$\text{Cell temperature } T_Z = T_U + (\text{NOCT} - 20\,°\text{C}) \cdot \frac{G_M}{G_{NOCT}} \qquad (4.1)$$

Figure 4.7   Solarfabrik SF125 laminate with a 36-string series array ($P_{max} = 125$ Wp). © Solar-Fabrik AG, Freiburg

Figure 4.8   Uni-Solar US64 amorphous solar module, $P_{max} = 64$ Wp, with triple cells as in Figure 3.51. As the cells are relatively large and, according to the vendor, also bypassed via a bypass diode, these modules are far more shading tolerant than standard crystalline cell modules (Photo: Uni-Solar)

Figure 4.9   Newtec walkable plastic solar roof tile with 24 monocrystalline solar cells, $P_{max} = 36$ Wp. This product, which was one of the first reasonably sized solar roof tiles to come on the market, could be series interconnected to form strings without screws using a simple insertion system and took the place of four standard roof tiles

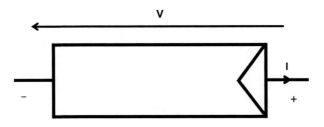

Figure 4.10   Solar module connection symbol

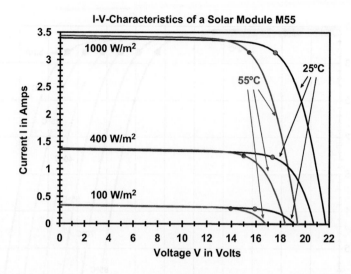

Figure 4.11   Overview of the *I–V* characteristics of the Siemens M55 module (55 Wp, 36 cells connected in series) for three different insolation levels (100, 400, 1 kW/m^2) and cell temperatures of 25 and 55 °C. The maximum power points (MPPs) are also indicated on each curve

where: T_U is the ambient temperature in °C; NOCT is the nominal (normal) operating cell temperature in °C according to the datasheet; $G_M = G_G$, i.e. irradiance at the solar module and solar generator; $G_{NOCT} = 800$ W/m^2 is the irradiance at which NOCT is defined; and NOCT in conventional modules ranges from around 44 to 50 °C.

Hence it is more in keeping with operational PV systems to plot a solar module's *I–V* characteristic curves for various insolation levels, *but* at constant ambient temperatures. The characteristic curves for an M55 module are illustrated in Figure 4.14 for 25 °C ambient temperature (e.g. for an average summer day) and in Figure 4.15 for 5 °C ambient temperature (e.g. for an average winter day).

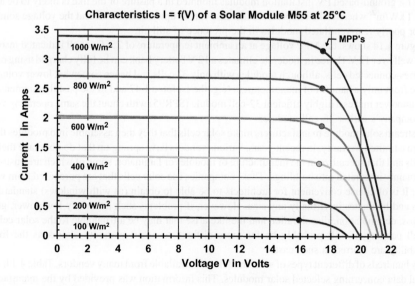

Figure 4.12   Characteristic $I = f(V)$ curves for the M55 monocrystalline solar module, at various insolation levels and a cell temperature of 25 °C

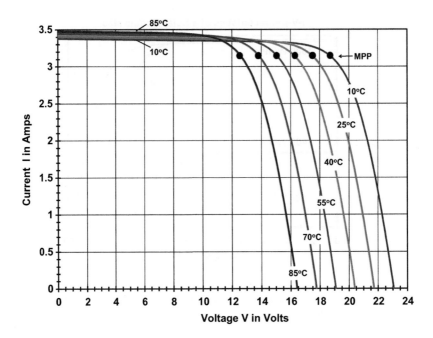

Figure 4.13 Characteristic $I = f(V)$ curves for the M55 monocrystalline solar module, at various cell temperatures and 1 kW/m^2 insolation

These characteristic curves are predicated on a 30 °C temperature increase relative to ambient temperature at 1 kW/m^2 insolation. These figures would be reached, for example, with still air and average mounting conditions, i.e. attached to a building but with good rear ventilation. For ground-based PV installation modules, which are continuously wind cooled, the temperature increase is appreciably lower and the voltage somewhat higher than shown in Figures 4.14 and 4.15. For example, the temperature increase for ground-based PV installation modules mounted in a pasture or the like is likely to be around 22 °C at 1 kW/m^2, whereas the temperature increase can be upwards of 40 °C and the voltage somewhat lower for poorly rear-ventilated modules in the presence of still air.

As Figure 4.14 shows, the MPP voltage at an ambient temperature of 25 °C for all indicated insolation levels is well over 12 V. Hence in moderate climates a 12 V battery bank can be fully charged using only 32 or 33 series-connected cells, although modules with only 33 cells and hence somewhat lower voltage are available from only a handful of manufacturers (e.g. the Isofoton I-47 and in the past the Siemens Solar M50). Sunpower makes a highly efficient 32-cell module (SPR95) with about the same operating voltage as the company's earlier 36-cell module.

Mainstream solar module manufacturers make solar cells that they then assemble into modules that are subject to a limited warranty. In recent years, various vendors have sprung up that do not make their own solar cells and that specialize in the manufacture of modules or laminates, some of which are customized and are mainly integrated into buildings. These companies use solar cells that are purchased from various vendors. It is of course convenient for architects to be able to obtain (as with windows) standard solar modules and laminates in virtually any size. However, if problems such as unduly low power, gradual power loss, delamination, discoloration and the like arise that may be attributable to the solar cells and solar cell packing, it will probably be far more difficult to assert warranty claims as the lines of responsibility are blurred in such cases.

Today hundreds of different types of solar modules are available from many vendors. Table 4.1 lists key technical data concerning selected solar modules. This information was provided by the manufacturers themselves and makes no claim to completeness. Each year in its February issue, *Photon* magazine publishes more complete technical data concerning several hundred modules.

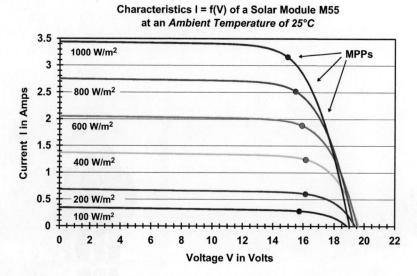

Figure 4.14 Characteristic $I = f(V)$ curves for the M55 monocrystalline solar module, at various insolation levels and 25 °C ambient temperature

In addition to solar modules whose cells are all wired in series, hybrid solutions that combine series and parallel wiring are also available, particularly for larger modules.

Various problems can arise when solar cells and solar modules are wired to solar generators, and it is necessary to take account of these problems to avoid damage under unusual operating conditions (see Section 4.2).

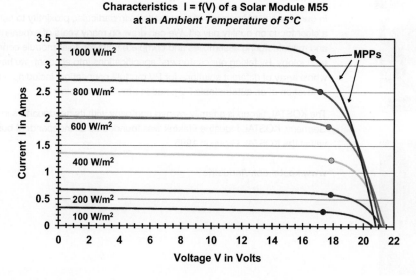

Figure 4.15 Characteristic $I = f(V)$ curves for the M55 monocrystalline solar module, at various insolation levels and 5 °C ambient temperature

Smart connections – our philosophy

Smart connections.

With our network – both internally and externally – we generate the conditions to open up new paths: paths that we tread together with our partners and customers.

In our core line of business of photovoltaics in particular, proximity to our customers and a clear focus on quality pay off. We can draw on many years of experience in developing and manufacturing assemblies and complete units for solar module connection technology. By taking our customers' specifications into account, we have built up a whole array of different solutions for PV module connection, including our very latest innovation of an automatable PV junction box, for example.

The KOSTAL Group is a family-owned company with its headquarters in Lüdenscheid, Germany. KOSTAL Industrie Elektrik was founded as an independent business division within the KOSTAL Group in 1995.

www.kostal.com/industrie

INDUSTRIAL ELECTRONICS

4.2   Potential Solar Cell Wiring Problems

4.2.1   Characteristic Curves of Solar Cells in all Quadrants

Solar cells need to be wired to solar modules or solar generators in such a way as to safeguard the individual cells against damage from unusual operating conditions resulting from overload and/or overheating. In practical terms, this means that operation of any given cell in the first and third quadrants of the characteristic diode curve (see Figure 3.13), where the cell absorbs power rather than outputting it, should be avoided.

If, in the event of a malfunction (e.g. during shading), a solar cell can nonetheless be operated in the first and third power quadrants of the characteristic diode curve, measures must be taken to prevent the current passing through and output by the cell from becoming unduly strong. In order to see exactly how such scenarios should be handled, we need to take a closer look at the complete characteristic curves of actual solar cells.

As solar cell vendors do not provide technical data concerning solar cell operation in the first and third power quadrants, the Bern University of Applied Sciences PV Lab decided to investigate a number of commercial silicon solar cells and modules.

Figure 4.16 displays the characteristic curves that the PV Lab obtained for a monocrystalline solar cell with $A_Z \approx 102 \text{ cm}^2$ surface area. The metering devices used for these investigations were the same as those used for voltage and current in a standard diode (appliance counting arrow system), i.e. the current indicated by these devices corresponds to I' in Figure 3.13. The first characteristic curve quadrant represents the non-conducting direction of the diode, the third quadrant represents the passband direction of the diode and the fourth quadrant represents the active solar cell areas to which power is output. On the other hand, the solar cell absorbs power in quadrants 1 and 3. In most cases, the diode reaches its avalanche point at a reverse voltage exceeding 15–25 V [4.4].

4.2.1.1   Reverse Current Characteristics in the Diode Passband

If solar cell voltage exceeds open-circuit voltage V_{OC} owing to an external source, a current $I' > 0$ flows through the diodes in the solar cell affected, thus causing the diode to begin operating in the passband.

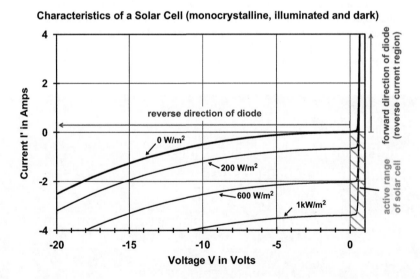

Figure 4.16   Characteristic curves of a Siemens monocrystalline solar cell with $A_Z \approx 102 \text{ cm}^2$, 25 °C cell temperature in all quadrants, and with no sunlight (appliance counting arrow system). In most cases, the diode reaches its thermal avalanche point at a reverse voltage ranging from 20 to 30 V, although this can occur earlier with some types of cells [4.14]

This phenomenon can also be referred to as reverse current in connection with current direction I under normal solar cell operating conditions.

A 1995 diploma thesis at the PV Lab investigated the quadrant 1 operating characteristics of various monocrystalline solar cells [4.1]. Analogue measurements of a Würth CIS module were also realized in 2006, at which time the solar cells, which were in a dark state, were subjected to successively higher passband currents I' for 15 minutes at a time (see Figure 3.13). The characteristic curves of these solar cells were then transferred to a solar simulator, where changes in the characteristic curves were investigated. The filling factor FF was particularly sensitive to solar cell damage. In the solar simulator investigations, all cells tolerated an $I_R = 3 \cdot I_{SC\text{-}STC}$ reverse current for 15 min periods without any measurable characteristic curve change ($I_{SC\text{-}STC}$ = short-circuit current at STC power output). On being subjected to $I_R = 3 \cdot I_{SC\text{-}STC}$, at which time 800 to 900 W/m^2 of area-specific power loss occurred in the solar cells, the temperature in the solar cells rose to 25 °C above ambient temperature. Some of the cells only began showing characteristic curve changes on being exposed to diode passband current and reverse current ranging from $4.5 \cdot I_{SC\text{-}STC}$ to $6 \cdot I_{SC\text{-}STC}$, although other cells exhibited no measurable characteristic curve change despite having been exposed to these currents. In addition, some modules were exposed to $3 \cdot I_{SC\text{-}STC}$ for up to 30 min without exhibiting any characteristic curve change.

These measurements were realized with the solar cells in the dark at an ambient temperature ranging from 20 to 25 °C. But of course when solar cells at 1 kW/m^2 insolation are subjected to such reverse currents, cumulative temperature increases occur. Insolation results in a cell temperature increase of 20–40 °C. The area-specific loss secondary to the reverse current induced further temperature increases With $3 \cdot I_{SC\text{-}STC}$, owing to the larger temperature differences resulting from slightly higher insolation, the temperature increased by around 20 °C more. Hence, normally the temperature of a solar cell subjected to such operating conditions will increase by some 50 °C above ambient temperature. At a maximum allowable cell operating temperature of 90–100 °C (as for the temperatures indicated in some manufacturers' module datasheets), an ambient temperature of 40–50 °C is allowable for this scenario.

In similar Fraunhofer Institut investigations where the modules were heated to more than 150 °C, no damage was observed in modules that were briefly exposed to up to $7 \cdot I_{SC\text{-}STC}$ short-circuit current [4.15]. But in modules that were exposed to such temperatures for longer periods, damage was observed in the embedding material (EVA film) as there are no vendor-specified modules for such temperatures. Moreover, there is no guarantee that internal module connections can reliably conduct such currents over a lengthy period as the modules are not rated for these operating conditions.

Hence it is safe to assume that in the first power quadrant all commercial modules can withstand short-circuit current $I_{SC\text{-}STC}$ at STC power output, including as passband current (and as reverse current I_R based on the normal current direction for solar cell operation), without being damaged. As the measurements demonstrated, solar cells can also withstand $I_R = 2 \cdot I_{SC\text{-}STC}$ to $3 \cdot I_{SC\text{-}STC}$ without any difficulty. That said, the PV system design process would be much simpler if exact figures concerning the maximum allowable module reverse current I_R (and possibly at various temperatures as well) were included in module datasheets. For example, $I_R = 3.67 \cdot I_{SC\text{-}STC}$ reverse current is indicated for the Shell Ultra 85-P module. In some cases, I_R is indirectly indicated for this module via indication of a maximum value I_{max} for series-connected fuses, in which case $I_R \approx 1.1 \cdot I_{max}$ can be presupposed. Such information is essential for determining how many module strings can be wired to each other in parallel without using string fuses.

4.2.1.2 Solar Cell Characteristics under Voltage Reversal Conditions in the Diode Cut-Off Region

If the current $I = -I'$ generated in a solar cell's active area exceeds the cell's short-circuit current I_{SC} owing to an external factor, it is necessary for the voltage in the cell affected to be negative, whereupon the cell begins operating in the third power quadrant or in the diode cut-off region. To obtain the characteristic curve for the third quadrant, the dark I–V curve (0 W/m^2) was measured for a number of cells and a characteristic curve was selected from it. The cut-off region characteristics of a go-live solar cell are far poorer than those of standard silicon diodes (see Figure 4.16), a phenomenon that is exacerbated by large manufacturing tolerances. The dark I–V curve is also influenced by temperature. A certain amount of power loss, which increases cell temperature, always occurs in a solar cell in the presence of more than 1 V. This in turn alters the operating point and thus the I–V curve to some extent.

4.2.1.3 Approximate Values for Allowable Total Area-Specific Thermal Loss

When a solar cell is operated in the first or third quadrants, electrical power loss and the heat resulting from solar radiation cause the cell to heat up. Assuming a maximum possible ambient temperature T_U of 40–50 °C and (as in the solar module datasheet) an allowable maximum cell operating temperature of 90–100 °C, under average cooling conditions a roughly 50 °C increase relative to ambient temperature is allowable. This allowable increase (around 20 W per cell for a 100 cm^2 cell) will be reached if the entire module is exposed to the same thermal conditions in the presence of total area-specific loss of around 2 kW/m^2.

Maximum area-specific thermal loss p_{VTZ} in the solar cell is given by

$$p_{VTZ} = G_Z + \frac{P_{VEZ}}{A_Z} \approx 2\,\text{kW/m}^2 \approx 20\,\text{W/dm}^2 \approx 200\,\text{mW/cm}^2 \tag{4.2}$$

where: G_Z is the solar cell irradiance (in practice, $G_Z = G_G$); A_Z is the solar cell area; and P_{VEZ} is the electrical loss ($V \cdot I$) in the solar cell during operation in the first or third power quadrant as shown in Figure 4.16.

In view of the relatively conservative nature of the suppositions that form the basis for Equation 4.2, in most cases applying it entails no risk. During operation in the first quadrant, i.e. the penetration area, the entire module is exposed to the same load. And in view of the relatively low voltage in this quadrant ($V_F \approx 0.8$ V), and as noted in Section 4.2.1.1, reverse current $I_{R\text{-}Mod}$ rather than power loss can be the limiting factor in such cases.

The worst-case scenario for operation in quadrant 3 arises if, while a module is partially or completely shaded, only some of its cells are operating in the non-conducting direction and if the remaining, fully insolated cells are being heated by the module's aggregate output. However, inasmuch as the immediately adjacent cells help to cool the module in this scenario, most vendors make somewhat less conservative suppositions for it and allow p_{VTZ} values such as the following for individual partly shaded cells:

$$\text{Allowable value range for individual third-quadrant cells}: p_{VTZ} \approx 2.5 \text{ to } 4\,\text{kW/m}^2 \tag{4.3}$$

In principle, this approximate p_{VTZ} value can be used to determine the number of bypass diodes needed per module. The higher the p_{VTZ} limit value, the fewer bypass diodes per cell are needed (see Section 4.2.2.2).

4.2.2 Wiring Solar Cells in Series

The voltage in solar cells wired in series is cumulative, which means that the cumulative voltage of all cells wired in series n_Z is n_Z times as high as the voltage in one cell. Owing to the power source characteristics of solar cells, the level of current in such series-wired configurations is determined by the weakest cell.

This scenario will now be discussed in terms of an example involving two solar cells wired in series (see Figure 4.17). The short-circuit current of cell B is only half that of cell A owing to partial shading via a leaf or bird droppings or the like. The I–V characteristics of the circuit arrangement as a whole, which arise from adding together the voltages from the two cells in the presence of equal current, approximately equate to the I–V characteristic curve of the weaker cell B extended by a factor of 2 along the voltage axis. The maximum power P_{max}, at the MPP, of the overall I–V characteristic curve does not equate to the cumulative maximum power of cells A and B, but is instead only slightly more than twice the maximum power of the weaker cell.

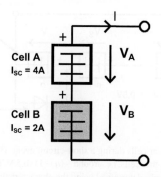

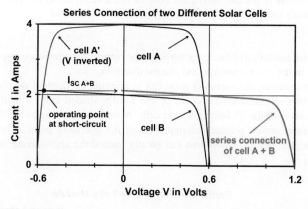

Figure 4.17 Individual characteristic curve and overall *I–V* characteristics of two series-connected solar cells, where one of the cells is partly shaded

If this series connection is short-circuited, $V_B = -V_A$, and thus becomes negative, the power produced by the stronger cell A is converted to heat in cell B by virtue of the fact that the cell B current and voltage are both flowing in the same direction (operation in the third quadrant as shown in Figure 4.16). If only two cells are wired in series, cell B can withstand this scenario without any difficulty.

The scenario illustrated in Figure 4.17 is far less common for identically insolated series-connected solar cells since, owing to manufacturing tolerances, the characteristic curves of the two cells cannot be absolutely identical. Hence the maximum power of series-connected solar cells is always somewhat lower than the cumulative maximum power of the individual cells. Vendors try to minimize these so-called mismatch losses by incorporating into a module only cells whose I_{MPP} current at the MPP is as similar as possible. To this end, each cell is measured after being manufactured and is assigned to the appropriate I_{MPP} group (see cell characterization at the beginning of Figure 4.1).

4.2.2.1 Hot Spots

Most PV installations contain more than two series-connected cells. In such cases a short circuit in a partially/completely shaded cell and the resulting far greater voltage load on the cell in question can provoke a hot spot, as the cell is then subjected to all of the voltage produced by the other cells (see Figure 4.18).

Figure 4.19 displays the characteristic curves for the scenario illustrated in Figure 4.18 for a 36-cell module with a cell area of around $102 \, \text{cm}^2$ (without bypass diodes); 35 of the cells have $1 \, \text{kW/m}^2$ insolation and the aggregate characteristic curve is designated as K35. For locally shaded characteristic curves (whose mean insolation is indicated in the figure), voltage *V* in the direction illustrated is negative.

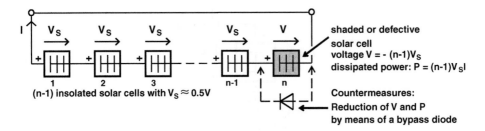

Figure 4.18   Series connection of *n* solar cells during a short-circuit event. Here, a shaded or defective solar cell is subjected to the aggregate voltage of all other cells, i.e. around $-(n-1) \cdot 0.5$ V. The consequent operating point and the *V* and *P* of the shaded cell can be precisely determined using the characteristic curves in Figure 4.19. This overload voltage can be substantially reduced through the use of bypass diodes (dashed line)

In order to allow for determination of the operating point, the barrier characteristic curves as in Figure 4.16 are also shown here in the reverse voltage and current direction.

For a short-circuit scenario, the operating point can in principle be deemed simply to comprise the intersection of the barrier characteristic curve of the locally shaded cell (with the corresponding insolation) and the curve for the 35 fully insolated cells. Nonetheless, owing to the thermal instability of barrier characteristic curves, precise determination of the operating point is no easy matter. In many cases, however, the defined power loss can greatly exceed the approximate value indicated by

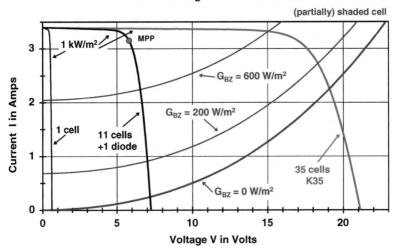

Figure 4.19   Characteristic curves for a short-circuited 36-cell module (cell area $A_Z \approx 102\,\mathrm{cm}^2$, cell temperature $T_Z = 25\,°\mathrm{C}$) with one cell partly shaded. The insolation for 35 of the cells is AM5.1 and 1 kW/m^2, while the **irradiance** G_{BZ} for the partly shaded cell is limited to that shown in the relevant curves. The operating point comprises the intersection of (a) the barrier characteristic curves for locally shaded cells and (b) the curve for the 35 fully insolated cells. In the interest of illustrating the conditions in a scenario involving 12 cells and a bypass diode (see Figure 4.23), the resulting characteristic curve of 11 fully insolated cells (increased by a diode flow voltage) is also shown here. The operating point for a conductive bypass diode is the intersection of the above characteristic curve and the barrier characteristic curves of the locally shaded cells

In the presence of a bypass diode, a partly shaded cell will be subjected to a lesser load and its area-specific thermal loss will be lower than the approximate value from Equation 4.3

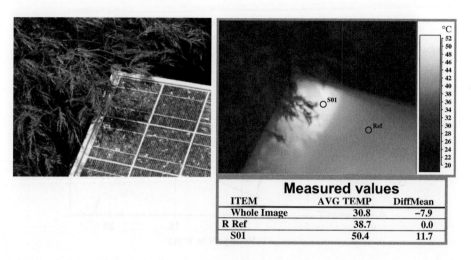

Measured values		
ITEM	AVG TEMP	DiffMean
Whole Image	30.8	−7.9
R Ref	38.7	0.0
S01	50.4	11.7

Figure 4.20   Formation of a hot spot in a solar cell that is in normal operating mode and partly shaded by a shrub (the solar generator is operating at the MPP). The thermographic image shows that the reference module temperature ('Ref') is 38.7 °C but that the temperature of the insolated portion of the shaded cell has increased to 50.4 °C ('S01')

Equation 4.3, resulting in damage not only to partly or fully shaded solar cells, but also to the solar module as a whole.

Use of a bypass diode for each cell is of course very expensive. Hence one diode is normally used for each group of solar cells, e.g. for 12–24 cells (see Figure 4.23). By way of illustration of the conditions that obtain within such a group in this situation, Figure 4.19 displays the relevant characteristic curve for the scenario involving one bypass diode for 12 cells.

Noteworthy here is the fact that the highest power loss in the shaded cell is induced not by full shading, but rather by a partial shading scenario where the characteristic curve of the shaded cell traverses an element such as the MPP or $I–V$ curve of the cell group (in the example, $n = 12$) that is bypassed by a bypass diode. The voltage in conductive bypass diodes is roughly the same as in an insolated cell.

Although a scenario where a solar module is operated at short-circuit current constitutes an unusual load, such a scenario is well within the realm of possibility. But partly shaded solar cells also heat up during normal operation, albeit to a lesser degree (see Figure 4.20).

Figure 4.21 displays the characteristic curves for the module illustrated in Figure 4.19 (35 fully insolated cells, 1 partly shaded cell), while a 12 V battery bank is being charged. In order to allow for determination of the operating point for the partly shaded cell, in addition to characteristic curve K35 for the 35 fully insolated cells the residual voltage (curve K35/12 V) for the partly shaded portion of the cell is also shown. Here, too, the operating point for the partly shaded cell comprises the intersection of the barrier characteristic curve for this cell (with the relevant insolation) and curve K35/12 V. Although the power loss attributable to the partly shaded cell is appreciably lower than at short-circuit current, it is still in the range of the approximate value from Equation 4.3, i.e. the partly shaded cell can become very hot, but without causing any damage.

Partial shading of a solar module solar cell also greatly alters the module's characteristic curves and drastically reduces maximum power P_{max} at the MPP. Figure 4.22 displays the resulting characteristic curves for the 36-cell solar module in Figure 4.19 (35 cells at 1 kW/m^2 and 1 partly shaded cell with the indicated mean irradiance G_{BZ}).

4.2.2.2   Bypass Diodes in Solar Modules

Parallel connection of a standard bypass diode to each solar cell obviates hot-spot formation. If a cell is shaded or defective, the bypass diode allows the current to bypass the remaining solar cells, in which case

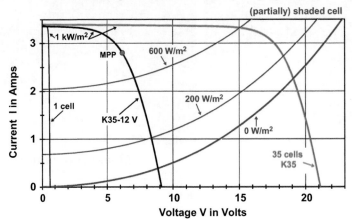

Figure 4.21 Characteristic curves for a 36-cell module with $A_Z \approx 102\,\text{cm}^2$, cell temperature $T_Z = 25\,°\text{C}$ and a partly shaded cell, while a 12 V battery bank is being charged; 35 of the cells have $1\,\text{kW/m}^2$ insolation (curve K35), while the partly shaded curve (in red) only has the insolation indicated. The partly shaded cell is subjected to the voltage (reduced by 12 V) of the 35 fully insolated cells (curve 35/12 V). The operating point is the intersection of the characteristic curve and the barrier characteristic curves of the locally shaded cells

the negative voltage in the jeopardized cell is only around 0.6 to 0.9 V, and with Schottky diodes 0.3 to 0.5 V (bypass diode forward voltage; see Figure 4.18).

Use of a bypass diode for each cell is the optimal solution, but it is also very expensive and far from indispensable. Inasmuch as a solar cell operated in the third quadrant (i.e. in the non-conducting direction)

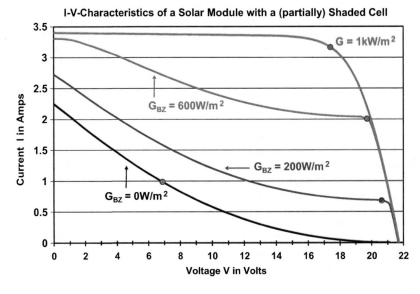

Figure 4.22 *I–V* characteristic curves (as in Figure 4.19), without bypass diodes, with an $A_Z \approx 102\,\text{cm}^2$ cell area, $T_Z = 25\,°\text{C}$ cell temperature and one cell partly shaded. Although only 1 of the 36 cells is partly shaded (resulting in irradiance G_{BZ}), the characteristic curves are drastically altered and power at the MPP decreases sharply. If bypass diodes are integrated into the module, the characteristic curves change somewhat, depending on circuit configuration

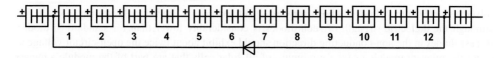

Figure 4.23 Bypass diodes protect a group of series-connected solar cells against hot-spot formation. One bypass diode should be used for about 12–24 solar cells

can withstand up to a few volts without any difficulty, before the approximate value (from Equation 4.3) is reached for the maximum area-specific thermal power loss, depending on the vendor's suppositions in this regard, one bypass diode for a group of 12–24 series-connected solar cells suffices (see Figure 4.23). In the case of a locally shaded cell in a group that is protected by a bypass diode, the current produced by the solar cells outside such a group is able to bypass the group via the bypass diode. The situation within such a bypass diode group is roughly the same as what would occur if the entire module were operated at short-circuit current, even if this is not actually the case. Bypass diodes need to be sized for somewhat higher than normal currents owing to possible short-term overload caused by cloud enhancements ([4.11] for $\geq 1.25 \cdot I_{SC-STC}$).

With conductive bypass diodes, the voltage in the shaded cells exceeds the voltage in the unshaded cells by the amount of the voltage drop at the diodes, i.e. it is about the same as the voltage generated by the bypass diode group under normal operating conditions and thus does not pose a problem. Figure 4.19 also displays a curve (for 11 cells plus one diode) representing the voltage at the partly shaded solar cell, for a scenario where one bypass diode is integrated for each group of 12 cells as shown in Figure 4.23. In this configuration, the maximum possible thermal power loss p_{VTZ} at a partly shaded cell is around 2.5 kW/m^2, which means that no hot-spot damage will occur.

Proper use of bypass diodes for series-connected solar cells can thus reliably obviate hot-spot formation in individual shaded cells and the attendant damage. Under normal operating conditions, bypass diodes induce no power loss, and for reasons of safety should be integrated into all PV installations whose system voltage exceeds 12 V, except in cases where the solar cells exhibit controlled avalanche point characteristics or have pre-installed bypass diodes.

Many commercial solar cell modules integrate the requisite bypass diodes or such diodes can be integrated into the module junction boxes. In most cases one bypass diode for 10–24 solar cells suffices. The smaller a bypass diode group, the less sensitive the module is to partial shading and the higher the cost. It would be helpful if vendors' module datasheets indicated the size of a bypass diode group for the module in question, or at least the number of bypass diodes per module.

If, as is the case with most modules, bypass diodes are integrated for groups of series-connected solar cells rather than for each individual cell, solar module power decreases disproportionately when any individual cell is shaded. Bypass diodes do not ameliorate this situation very much unless a large number of modules are series connected in a string, in which case the bypass diodes do help to prevent an unduly large power drop in the series string in the case of local shading of individual cells or an entire module.

Bypass diode sizing for maximum module operating temperature [4.11] is

$$\text{Forward current } I_F \geq 1.25 \cdot I_{SC-STC} \qquad (4.4)$$

$$\text{Reverse voltage } V_R \geq 2 \cdot V_{OC} \qquad (4.5)$$

Bypass diodes integrated into a module also need to be adequately cooled. The larger the solar cell area and the higher the solar cell voltage, the greater the need for cooling. Whereas for a module whose I_{SC} is around 3.5 A, inexpensive non-cooled, plug-in 6 A diodes are sufficient, 12 A diodes, which need to be cooled, must be used for larger cells (e.g. 15 cm · 15 cm). This problem can be partly solved using Schottky diodes, whose forward-direction voltage drop is only around 0.3 to 0.5 V, although the dielectric strength of such diodes is considerably lower than for standard Si diodes. Bypass diodes in the non-conducting direction should exhibit about twice the open-circuit voltage of either the relevant module or (at a minimum) of all solar cells that are part of the module's bypass diode group. That said, for reasons of lightning protection the highest possible inverse current should be used for bypass diodes (see Section 6.7.7).

4.2.2.3 Scenarios Where Bypass Diodes Are Unnecessary

For up to 12 V PV plants, where a module short circuit cannot occur, modules without bypass diodes can normally be used. However, modules for higher-voltage installations should always be protected by bypass diodes, or the cells used need to exhibit highly specific cut-off region characteristics. Bypass diodes in larger-scale solar modules and solar cells are prohibitively expensive owing to the cooling capacity needed for these devices. If cell barrier capacity is strategically decreased, controlled avalanche point characteristics can be achieved.

It would be preferable here for each solar cell to integrate a bypass diode, but of course in such a way that the other characteristics of the cells are not degraded. Figure 4.24 displays the complete characteristic curves in all power quadrants for this putative solar cell, which would not need bypass diodes.

Figure 4.25 displays the characteristic curves at short-circuit current for the 36-cell solar module in Figure 4.19. Inasmuch as the voltage in partly shaded cells remains low at all insolation levels, power loss is likewise low, which means that the absence of bypass diodes would pose no risk for the cells. Various vendors have tried to make such solar cells, but none have been commercialized as yet as their guide values go downhill under normal operation.

Bypass diodes should only be dispensed with if the module vendor guarantees that the solar cells used in the modules integrate bypass diodes or exhibit controlled shutdown behaviour. However, dispensing with diodes means that the key tasks performed by these devices in larger systems, such as obviating excessive power drops in partly shaded cell series strings, will not be performed as efficiently as would otherwise be the case. Hence a configuration of one bypass diode per module should be adopted for systems that are prone to considerable shading at certain times of the day or year.

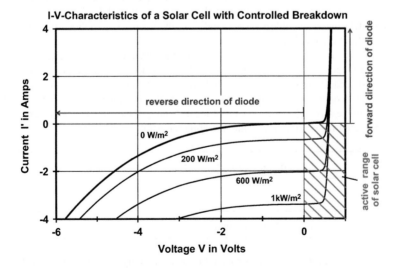

Figure 4.24 *I–V* characteristic curves for a (putative and not commercially available) monocrystalline solar cell ($A_Z \approx 102 \, cm^2$) with controlled avalanche point characteristics in all quadrants, with and without insolation, and with a cell temperature of 25 °C (usage metering system)

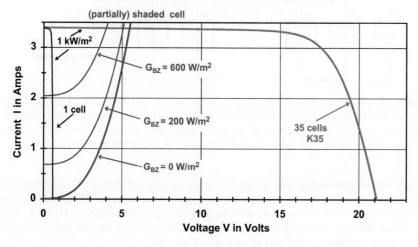

Figure 4.25 Characteristic curves of a short-circuited 36-cell module whose cell temperature is $T_Z = 25\,°C$, one of whose cells is partly shaded, and whose avalanche point characteristics are shown in Figure 4.24; 5 of the cells have $1\,kW/m^2$ insolation (curve K35), while the partly shaded curve (in red) only has the irradiance G_{BZ} indicated. The operating point is the intersection of curve K35 and the barrier characteristic curves of the locally shaded cells. Such a module would not require bypass diodes

4.2.3 Parallel-Connected Solar Cells

Only solar cells that use the same technology, come from the same manufacturer and are of the same type should be wired in parallel to each other. The current of parallel-connected solar cells is cumulative, but the voltage remains the same as for only one cell. Inasmuch as manufacturing tolerances can induce discrepancies between the characteristic curves of individual solar cells at the same insolation, the maximum power of parallel-connected solar cells is always somewhat lower than the aggregate maximum power of the individual cells. This so-called mismatch loss in parallel-connected cells can be minimized by using solar cells whose MPP voltage V_{MPP} is as similar as possible.

Like series-connected solar cells, parallel-connected solar cells are subject to critical operating states that need to be managed. To this end, it is necessary first to determine the effect of *one* parallel-connected cell being shaded. Under certain circumstances, in such cases the shaded solar cell can be operated in the first power quadrant (see Figures 3.13 and 4.16), i.e. in the forward direction, and can thus have the effect of an appliance.

This state poses the greatest hazard for shaded cells in cases where the entire module is at open-circuit voltage, such that the voltage generated by the non-insolated cells is nearly the same as the open-circuit voltage. In such a case, the shaded cell receives power from all insolated cells adjacent to it (see Figure 4.26). The results of an investigation of this scenario with crystalline silicon solar cells are illustrated in Figure 4.27, which displays the characteristic curves of the $1\,kW/m^2$ insolation cells and the dark characteristic curve for a fully shaded cell. The operating point for the shaded cell in such a case is the intersection of the shaded cell and insolated cell curve.

Parallel connection of an insolated and non-insolated solar cell, each of which exhibits a cell temperature of $25\,°C$, results in operating point A_1, which is a completely benign situation since the resulting reverse current in the shaded cell is far below $I_{SC\text{-}STC}$. However, it is safe to assume that in an actual PV system a solar cell at $1\,kW/m^2$ insolation will be around $30\,°C$ higher than a shaded cell. Based on this assumption, the reverse current for operating point A_2 will be lower still. But for parallel-connected

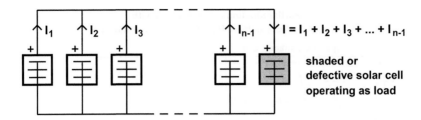

Figure 4.26 When *n* solar cells are parallel connected, a shaded or failed solar cell can absorb the output of the *n* − 1 insolated cells and act as an appliance. Parallel connection of an unlimited number of crystalline silicon solar cells of the same type is not critical for the possible reverse current in a partial-shading scenario. However, if cell failures are reliably managed in such a situation, then no more than three or four cells should be parallel connected

shaded cells with an unlimited number of 1 kW/m^2 insolation solar cells with 25 °C cell temperature, the resulting operating point A$_3$ will initially present reverse current that is around half of short-circuit current I_{SC} at STC power output. This is clearly shown in the solar cell equivalent circuit (see Figure 3.12), since a reverse current induces a voltage drop at the series resistance R_S, such that the diode voltage in the equivalent circuit diagram is lower than at open-circuit current.

In the interest of allowing for manufacturing tolerances between solar cells, a second curve (blue) was inserted for the shaded cell, which under the same current exhibits only 95% of the voltage of a normal cell. This results in the attendant operating points B$_1$, B$_2$ and B$_3$ under otherwise identical conditions, whereby the reverse current that arises in such cases is likewise lower than at I_{SC-STC}. Hence an unlimited number of solar cells of the same *type* can be parallel connected without difficulty, in terms of reverse current and under a shading condition.

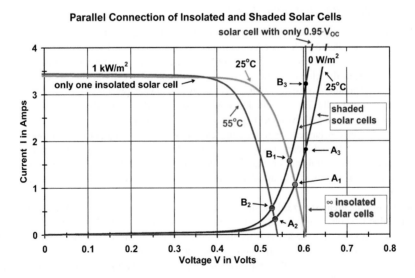

Figure 4.27 Characteristic curves, at 1 kW/m^2 insolation, of crystalline silicon solar cells with a cell temperature of 25 °C and 55 °C (as indicated), and the dark characteristic curve of a normal solar cell (dark blue line), with reduced threshold voltage attributable to manufacturing tolerances (blue line). The characteristic curve for insolated cells with an unlimited number of parallel-connected solar cells at V_{OC} is the vertical line. The operating point is the intersection of the characteristic curve of the insolated cell and the dark characteristic curve. Even in the worst-case scenario, reverse current I_R is lower than I_{SC-STC} at STC power output. Hence an unlimited number of crystalline silicon solar cells of the *same type* can be parallel connected without generating excessive reverse current

If the relevant module is under load, the voltage via the parallel-connected cells is lower, and the current in the shaded cell is far lower. The module current decreases by approximately the total amount of the current previously input by the now shaded cell, thus in turn reducing module power, albeit to a far lesser degree than if all cells are series connected.

The worst-case scenario here is an imperfect short circuit in a cell, where the current from all remaining intact cells is fed into the short circuit, thus heating up the residual resistance in the failed cell. The best way to safeguard against this rare, but by no means impossible, malfunction scenario is to parallel-connect a maximum of around three or four cells, thus obviating the potentially catastrophic effect of limited power loss in the failed cell secondary to low voltage. In the worst-case scenario, cell calefaction will provoke a saturated short circuit (contact sweating) or open circuit (wire meltdown), both of which events will permanently reduce power but will probably not knock out the module affected.

Series connection of groups of n_{ZP} parallel-connected modules in a matrix arrangement engenders higher voltages that are less sensitive to local shading than is the case with series connection of all solar cells. However, in order to avoid shading-induced damage, it is still necessary to realize the measures described in Section 4.2.2, such as using cooled bypass diodes that are rated for $1.25 \cdot n_{ZP} \cdot I_{SC}$ current in a group of 12–24 solar cells.

4.3   Interconnection of Solar Modules and Solar Generators

Interconnecting solar generators and solar modules is prone to essentially the same problems as interconnecting solar cells. Modules are basically super solar cells with higher voltages and stronger currents. However, owing to the higher voltages and stronger currents exhibited by solar generators, more robust preventive measures are needed in view of the far higher outputs entailed by a malfunction. Thus the modules used should be rated for total system voltage at a minimum, i.e. total available direct current in PV installation go-live mode.

A large solar generator should be subdivided into groups that can be switched individually in the event of a failure and that can also be maintained separately without the need to shut down the entire generator. To this end, one should make sure that all solar generator fuses, circuit breakers and switches are expressly rated for DC operation at the relevant voltage ($\geq 1.2 \cdot V_{OCA\text{-}STC}$) and current ($V_{OCA}$ is the system open-circuit voltage). Even under low voltages (above 24 V), shutting down direct current is far more difficult than alternating current owing to the absence of a zero crossing in direct current. In most cases, AC materials can only be used for DC systems at very low voltages (e.g. in lieu of 230 V AC, for 24–48 V DC only). Owing to possible cloud enhancements (see Section 2.3), cables, diodes, bypass diodes and fuses need to be sized in such a way as to ensure that at maximum temperature 1.25 (or, better, 1.4) times the short-circuit current $I_{SC\text{-}STC}$ of the relevant string can be conducted for an unlimited period [4.10]. Moreover, cooling elements should be integrated into the system diodes if necessary.

4.3.1   Series Connection of Solar Modules to a String

In wiring solar modules of the same type in series with a view to ramping up system voltage, it is necessary to bypass each module using a bypass diode, if such devices are not already integrated into the system. For most high-voltage modules, two to six diodes should be used for each module (see Section 4.2.2.2) and the vendor's data and recommendations in this regard *must be followed to the letter*.

Series connection of multiple (n_{MS}) solar modules is referred to as a string (see Figure 4.28). In cases where a series (n_{SP}) of such strings is parallel connected, as a rule an element that safeguards against reverse current and overload in the string cables, such as an additional diode (string diode, reverse current diode, blocking diode (red) and/or a fuse or circuit breaker (green)), should be wired in series in each string. For test purposes, the strings should be arranged on either side of the other solar

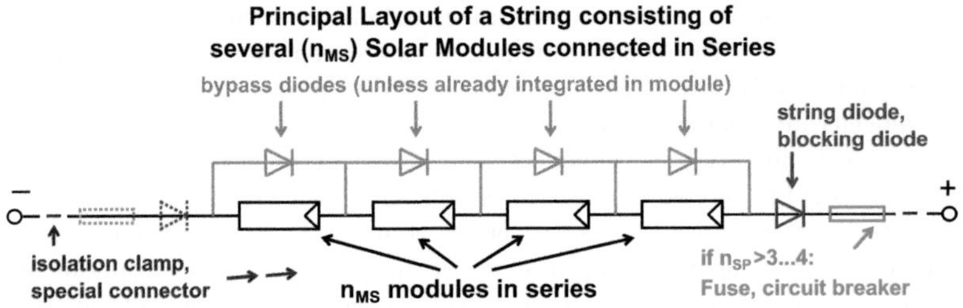

**Principal Layout of a String consisting of
several (n_{MS}) Solar Modules connected in Series**

bypass diodes (unless already integrated in module)

string diode,
blocking diode

isolation clamp,
special connector

n_{MS} modules in series

if $n_{SP} > 3...4$:
Fuse, circuit breaker

Figure 4.28 In the string that results from series connection of solar modules of the same type, bypass diodes need to be integrated into each module if this was not done at the factory. In such cases, if possible the string should be installed in two segments on either side of the remaining solar generators so that these two string segments can be serviced and measured separately. For protection class II modules and wires/cables (i.e. with dual or reinforced insulation, which is standard nowadays), in most cases fuses and any string diodes that are used need only be installed on one side. For optimally reliable grounding and short-circuit protection at any given point along the string, fuses or circuit breakers and any string diodes that are being used should be installed on either side of the string. Nowadays special PV plugs are used in lieu of disconnect terminals

generators, i.e. at least one fuse, disconnect terminal or a special PV plug (blue) should be installed on either side. Such special disconnect-enabled plugs are also frequently used to interconnect modules, appreciably reduce string wiring work and allow a string to be divided into various segments where $V_{OC} < 120$ V (low voltage). This is useful for troubleshooting in the event of a failure. Wherever possible, protection class II modules should be used for high-voltage systems.

4.3.1.1 String Diodes

A string diode prevents a string module from being operated under reverse current (forward direction in solar cells) in the event of a malfunction attributable to shading, failure or the like. A string diode should present an inverse voltage that is appreciably higher than the maximum operating open-circuit voltage of the string (e.g. $2 \cdot V_{OCA\text{-}STC}$) and that can indefinitely conduct at least 1.25 times the module short-circuit current $I_{SC\text{-}STC}$ of the string – although it should be noted that the factor for Alpine PV plants is around 1.5 rather than 1.25. If one side of a solar generator string is grounded, the diode should always be installed on the ungrounded side. A functioning string diode provides completely reliable protection against reverse current, including in cases of wire failure or a severe malfunction in a string module. However, as string diodes are prone to atmospheric power surges, relatively strong currents such as those exceeding 4–6 A need to be cooled, and always provoke a voltage drop and thus power loss – although this is of minor importance in the presence of relatively high system voltages. Nowadays, string diodes are mainly used in relatively large PV installations with numerous parallel strings and operating voltages exceeding 100 V. A proven and low-cost solution for string diodes is to integrate into them rectifier bridges which are well insulated and are available for up 35 A and 1400 V applications. A single-phase bridge of this type is sufficient for two strings, and a triphase bridge is sufficient for three strings.

Since, as noted in Section 4.2.1.1, (a) full protection against reverse current is unnecessary and (b) string diode cooling elements for large-scale module currents are prohibitively expensive, the tendency for some time now has been to dispense with these diodes if possible, except in special cases such as test installations that are frequently rewired. Dispensing with string diodes simplifies the system, enhances system reliability and also marginally improves efficiency, particularly for low system voltages. In cases where a stand-alone system is needed in order to obviate nocturnal battery discharge by the PV generator of a reverse current diode, this can also be achieved by installing only one such diode near the charge controller or through the use of a charge controller that performs this function.

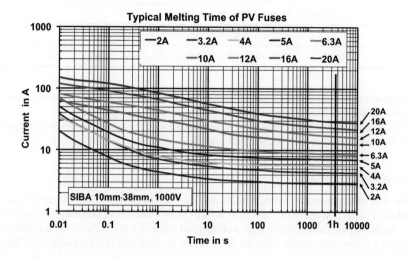

Figure 4.29 Correlation between current and melting time for specialized PV fuses – here 1000 V DC Siba fuses (10 mm in diameter, 38 mm long). For these fuses, $k_f \approx 1.45$ and thus guaranteed one-hour fusing current is $I_f \approx 1.45 \cdot I_{SN}$

4.3.1.2 String Fuses

It is also easier to safeguard a PV system against unduly high reverse current that could damage the modules and the string cables, by installing a DC-compatible fuse on one side of the string and a disconnect terminal on the other. Under certain weather conditions, brief cloud enhancements exceeding $1\,kW/m^2$ can occur that subject the system fuses to additional stress. Hence, in order to ensure that the fuses last a reasonably long time, their rated current I_{SN} should be a factor k_{SN} higher than the string's STC short-circuit current $I_{SC\text{-}STC}$, whereby k_{SN} contains a hybrid safety factor for an elevated ambient fuse temperature (45 °C), for cloud enhancements up to about $1.3\,kW/m^2$ and for reverse load. If one side of the PV generator string is grounded, the fuse should be installed on the opposite side. String fuses should be rated for $> 1.2 \cdot V_{OC\text{-}STC}$ DC and for the *entire* current range, which means that gPV or possibly gR fuses must be used. *Fuses rated solely for short-circuit protection (type aR) are unsuitable and can cause a smouldering fire.* Thus

> Nominal current I_{SN} of string fuses : $I_{SN} = k_{SN} \cdot I_{SC\text{-}STC}$, where $1.4 < k_{SN} < 2$ (4.6)

The higher the ratio between actual system current and nominal current, the more likely a fuse is to cut off a circuit. According to IEC standards, fuses need to be able to conduct nominal current for an unlimited period, but once guaranteed fusing current $I_f = k_f \cdot I_{SN} > I_{SN}$ is reached, they must shut down the circuit within one hour (typical value, $I_f = 1.45 \cdot I_{SN}$; Figure 4.29). Thus

> Guaranteed fusing current I_f within one hour : $I_f = k_f \cdot I_{SN}$ (typical value for $k_f \approx 1.45$)   (4.7)

Inasmuch as more finely graded fuses are available today (see Table 4.2), the recommended upper limits for k_{SN} relative to the values in [Häb07] can be defined somewhat more narrowly.

In desert regions with elevated fuse ambient temperatures such as 55 °C, k_{SN} should be at least 1.5, and for Alpine systems at least 1.6 owing to possible snow reflection, although in such cases the upper limit

Table 4.2 Examples of commercial PV fuses (here, Siba fuses)

Examples of commercially available PV fuses (gPV):										
Size 10 · 38 mm, $V_{DC} \leq 1000$ V:	2A	3A	4A	5A	6A	8A	10A	12A	16A	20A

for k_{SN} may need to be slightly above 2 (e.g. 2.1), in order for commercial fuses to be usable. Needless to say, all string wiring should be rated for currents $\geq I_f = k_f \cdot I_{SN}$.

In addition, a string can be segmented for servicing purposes using fuses and disconnect terminals, once the system has been shut down via the main DC switch.

4.3.1.3   String Circuit Breakers

A more practical, albeit more expensive, solution is to use DC circuit breakers (preferably two-or four-pole models in $+$ and $-$) in lieu of fuses, as these devices allow a string to be shut down while under load. Like fuses, the higher the ratio between system current and nominal current, the sooner a circuit breaker will be tripped, i.e. there are characteristic shutdown curves for each type of circuit breaker, as illustrated in Figure 4.9 ($k_f \approx 1.45$). Circuit breakers rated for extremely high currents often integrate magnetic instantaneous tripping mechanisms rated for 5 to 15 times I_{SN} or the like. Such circuit breakers should be used, at a minimum, for arrays in relatively large solar generators. If only a unipolar circuit breaker is used, at least one disconnect terminal or a suitable PV plug should be mounted on the other side of the string.

4.3.2   Parallel-Connected Solar Modules

Only solar modules that use the same technology (e.g. crystalline silicon solar cells) and that present equivalent V_{OC} and V_{MPP} values can be parallel connected; if possible, the same type of module should of course be used. The number of parallel-connected modules is designated as n_{MP}. Inasmuch as solar modules are actually super solar cells with elevated voltages and currents, the observations in Section 4.2.3 also apply to modules that are wired to each other in parallel. Even in cases where numerous modules of the same type are directly parallel connected, shading of individual intact modules does not engender any hazardous reverse current, even if the solar generator is at open circuit.

In the interest of reliable handling of the very rare instances of short circuits in a module, no more than three or four modules should be parallel connected to each other, since the power and thus the hazard entailed by parallel-connected modules are considerably greater than for parallel-connected cells. However, in order to do this the wire size must be large enough and precautions must be taken to prevent reverse current from arising from connected appliances such as battery banks and inverters. For a relatively large number of parallel-connected modules ($n_{MP} > 4$), a fuse (nominal current around 1.4 to 2 times module short-circuit current I_{SC-STC}) should be series connected for each module in order to allow for reliable shutdown of a module short circuit (see Figure 4.30). However, in view of the elevated cost of such arrangements, they are a rarity in actual PV systems.

If these parallel-connected modules are series connected to other parallel-connected modules (see Section 4.3.4), a single bypass diode should be integrated for the entire parallel connection that can conduct at least 1.25 times the aggregate short-circuit current of all modules, i.e. $1.25 \cdot n_{MP} \cdot I_{SC-STC}$. This is necessary because, if a series of smaller bypass diodes is used, internal current distribution will be imperfectly harmonized, and this can provoke module damage secondary to overheating and failure of the small diodes that are integrated into the modules.

4.3.3   Solar Generators with Parallel-Connected Series Strings

In this frequently used arrangement for larger solar generators, the modules that are used to generate the requisite system voltage are first wired to series strings comprising n_{MS} modules (see Figure 4.28). To achieve higher aggregate current, a number (n_{SP}) of such strings are wired in series. Figure 4.31

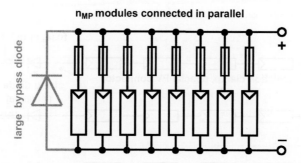

Figure 4.30 n_{MP} parallel-connected modules with module fuses. Parallel-connected modules of the same type generate only low reverse current under local shading. In order to allow for reliable handling of a PV generator malfunction, however, only three or four modules should be directly parallel connected. Module fuses should be used for $n_{MP} > 4$ scenarios, and if such groups are series connected the parallel-connected module groups should be bypassed via a large bypass diode that can conduct 1.25 times the aggregate short-circuit current of all modules

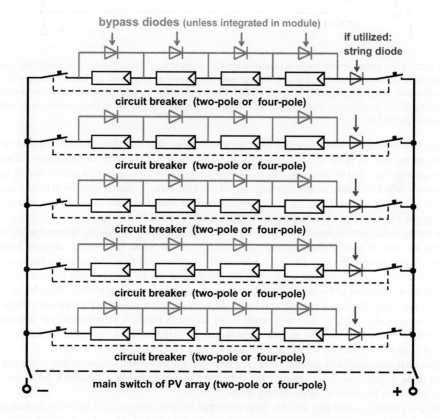

Figure 4.31 Solar generators with parallel-connected series strings that provide optimal protection against all possible anomalies. The various strings are decoupled from each other via string diodes so as to obviate reverse current in functioning string diodes. In the event of string diode failure (often induced by a short circuit), the reverse current is limited still further by the circuit breakers; each string can be decoupled from its circuit breaker by the generator busses and measured separately

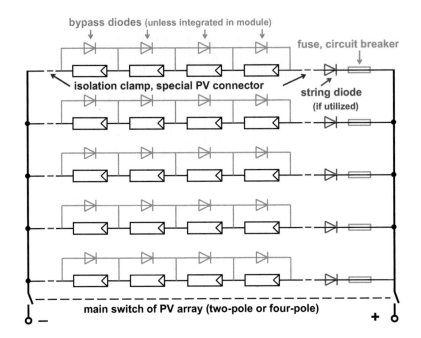

Figure 4.32   Solar generator with low-cost parallel-connected series strings (fuses, disconnect terminals and special PV plugs in lieu of circuit breakers). Local shading of individual module strings does not induce excessive reverse current. However, string fuses should be integrated for more than three to four parallel-connected strings, so as to allow for handling of catastrophic solar generator failure induced by module or partial-string short circuits, or grounding faults (see Figure 4.33)

displays an arrangement where optimal solar generator protection is realized to take account of all possible anomalies. As noted in Section 4.3.1.1, the string diodes shown here can be dispensed with.

Two separate fuses can be substituted for bipolar circuit breakers, or unipolar circuit breakers or fuses can be used – which of course should only be removed when the entire system is shut down. In such a case, however, as is illustrated in Figure 4.32, a disconnect terminal should be installed on the opposite side of the string, or a special disconnect-enabled PV plug should be used (as indicated).

The wiring configuration shown in Figure 4.31 also makes it easier to monitor system operation by allowing for current measurement in individual strings. In the event of a string anomaly such as excessively low current, the failed string can be simply disconnected from the generator busses using the circuit breakers, whereupon the anomaly in the failed string can be localized and corrected without interfering with operation of the intact array. Hence this arrangement (or the simpler version in Figure 4.32 involving fuses and disconnect terminals) is by far the most prevalent modality for medium to large PV plants. However, the simpler connection arrangement shown in Figure 4.32 provides only suboptimal protection for awkwardly located ground connections. To obtain optimal protection in such a case, *bilateral* fuses should be used for the positive and negative conductors in each string.

Before solar generators with parallel-connected series strings are hooked up, one should verify that the various string voltages are approximately the same, so as to avoid needless mismatch power loss, or reverse current in the event of major wiring errors.

Cast bridge rectifiers are highly suitable for systems that integrate string diodes, whereby the following elements should be used: two or three low-cost insulated string diodes with a shared cathode in a well-insulated casing that readily cool down by virtue of being mounted on a metal plate.

When modules wired in series are shaded, their power drops instantaneously. However, the greater the number of modules in a series string, the lower the shading-induced power in the individual modules. Owing to the higher system voltage that comes into play here, a sizeable portion of the current in the non-shaded modules will flow into the shaded module via the bypass diodes. Such mismatch losses can be mitigated by measuring the individual modules and only series-connecting modules in a string that present similar I_{MPP} current levels at the MPP (for more on mismatch loss see Section 4.4).

4.3.3.1 The Role of Reverse and Short-Circuit Current in Solar Generator Malfunctions

Figure 4.33 illustrates the scenario involving a severe module string malfunction induced by module short circuits, mechanical damage resulting from an accident, or a catastrophic string wiring error. Such events occur now and again (a) in building-integrated PV plants or infrastructure installations such as highway or railway line sound barriers, or (b) in the event of an earth fault in a system whose poles are grounded.

In such a case, in the absence of string diodes or fuses (i.e. n_{SP} strings wired to each other in parallel), the following reverse current I_R flows into the intact solar generators:

$$\text{Maximum reverse current } I_R \text{ for directly parallel-connected strings}: \ I_R \approx (n_{SP} - 1) \cdot I_{SC} \quad (4.8)$$

where: I_{SC} is the short-circuit current in a solar module or string; and n_{SP} is the number of parallel-connected strings.

The actual current in a short circuit is I_{SC} higher, i.e. the following holds true for short-circuit current I_K for strings that are wired to each other in parallel:

$$\text{Short-circuit current } I_K = I_{SC} + I_R \approx I_{SC} + (n_{SP} - 1) \cdot I_{SC} \approx n_{SP} \cdot I_{SC} \quad (4.9)$$

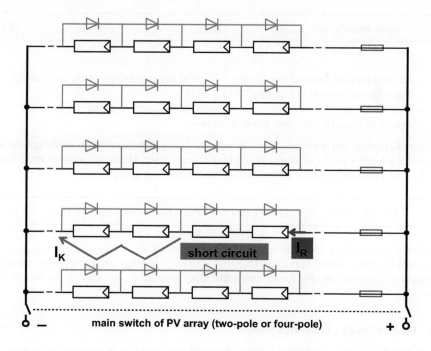

Figure 4.33 Reverse current I_R in a diode-free string resulting from an event such as a string short circuit involving a number of modules, caused by a severe malfunction such as the mechanical effects of an accident, or a bypass diode short circuit induced by lightning. This same scenario can also occur if too few modules are available owing to a major string wiring error

The following scenarios can arise, depending on solar generator configuration:

- In the presence of functioning string diodes, the short-circuit current I_K that arises is limited to the short-circuit current I_{SC} of a string, i.e. $I_K = I_{SC}$ and $I_R = 0$.
- In the absence of string diodes or if they are short-circuited, but in the presence of string fuses with k_{SN} as in Equation 4.6 and k_f as in Equation 4.7, I_K initially equates to aggregate solar generator short-circuit current $n_{SP} \cdot I_{SC}$. Thus, depending on the scope of n_{SP}, the transient reverse current I_R induces a greater or lesser temporary overload in the module through which I_R flows. In the longer term, however, the maximum reverse current from the intact array is around $I_R \approx k_f \cdot k_{SN} \cdot I_{SC\text{-}STC}$, since otherwise the fuse would be tripped, i.e. $I_K \leq I_{SC} + k_f \cdot k_{SN} \cdot I_{SC\text{-}STC}$ applies. $I_R = 0$ for decoupled fuses; the sole remaining continuous current is the short-circuit current I_{SC} of a string, i.e. $I_K = I_{SC}$.
- In the case of directly parallel-connected strings, the short-circuit current remains at around the same level as the short-circuit current $I_K = n_{SP} \cdot I_{SC}$ of the solar generator as a whole, such that for higher n_{SP} values string wiring can incur severe damage, as can the string modules through which reverse current $I_R \approx (n_{SP} - 1) \cdot I_{SC}$ flows.

The situation is somewhat more complex for extremely large-scale solar generators comprising n_{TG} parallel-connected arrays (TG) with $n_{SP\text{-}TG}$ parallel strings each (i.e. for which $n_{SP} = n_{TG} \cdot n_{SP\text{-}TG}$) (see Figure 4.34).

The cable array (TGK) used for large-scale solar generators should be protected against reverse current from the solar generator as a whole, via fuses, or preferably circuit breakers. $I > I_{SN}$ may also occur briefly until these devices are activated.

As the array wiring current is higher than string current, in accordance with Equation 4.6 the following approximate guide value applies to nominal current $I_{SN\text{-}TG}$ for such array fuses:

$$\text{Array fuses } I_{SN\text{-}TG}: \quad I_{SN\text{-}TG} = k_{SN\text{-}TG} \cdot I_{SC\text{-}STC\text{-}TG} = k_{SN\text{-}TG} \cdot n_{SP\text{-}TG} \cdot I_{SC\text{-}STC} \qquad (4.10)$$

where:

$k_{SN\text{-}TG}$ = the ratio between nominal fuse current $I_{SN\text{-}TG}$ and nominal current in the array wiring (approximate value as in Equation 4.6: 1.4 to 2)

$I_{SC\text{-}STC}$ = short-circuit current in a solar module or string at STC power output

$n_{SP\text{-}TG}$ = number of parallel-connected strings in an array

In the event of a malfunction in a large solar generator, a $k_f \cdot I_{SN\text{-}TG}$ current can flow back through the array conductor for a lengthy period in the presence of intact array fuses. Hence the possible string reverse current I_R in a large solar generator comprising a series of arrays is as follows:

$$\text{Possible return current } I_R \text{ for arrays}: \quad I_R \approx k_f \cdot I_{SN\text{-}TG} + (n_{SP\text{-}TG} - 1) \cdot I_{SC} \qquad (4.11)$$

Even with low $n_{SP\text{-}TG}$ values, appreciable reverse current can occur, e.g. with $n_{SP\text{-}TG} = 3$, $k_{SN\text{-}TG} = 2$ and $k_f = 1.45$ at STC power output I_R reaches $10.7 \cdot I_{SC\text{-}STC}$. Hence in large multi-array solar generators, fuses should be installed not only for the individual array conductors, but also for the strings of the various arrays.

4.3.3.2 Directly Parallel-Connected Strings

Inasmuch as solar modules are actually super solar cells that exhibit high voltages and currents, the observations in Sections 4.2.3 and 4.3.2 also apply to directly parallel-connected strings of the same type. Even in cases where numerous strings of the same type are wired to each other in parallel, shading of

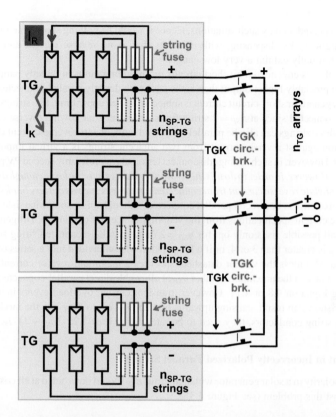

Figure 4.34   Reverse current I_R induced by a string malfunction in a very large solar generator comprising n_{TG} arrays with $n_{SP\text{-}TG}$ parallel-connected strings each. Inasmuch as n_{TG} is usually very high and much greater than 1, the cable array (TGK) needs to be protected, via fuses or preferably circuit breakers, against reverse current from the solar generator as a whole. Since (a) $n_{SP} \gg 1$, reverse current in a string can be extremely high in the event of a malfunction; and (b) in such cases the reverse current comes not only from the $(n_{SP\text{-}TG} - 1)$ intact strings of the native arrays, but also from the $(n_{TG} - 1)$ intact arrays. Hence in such cases as well, string fuses should be used, including for low $n_{SP\text{-}TG}$ values. Integrating cooled diodes into each array that is sized for $1.25 \cdot n_{SP\text{-}TG} \cdot I_{SC\text{-}STC}$ is also a viable solution

individual intact strings does not engender hazardous reverse current, even if the solar generator is at open circuit. That said, prior to direct parallel connection of strings, it is essential to verify that the strings are wired correctly and present identical open-circuit voltage. Hence it is indispensable to mount disconnect terminals or special PV plugs for each string.

It may be advisable to dispense completely with anti-reverse current elements if particularly well-insulated protection class II solar modules and dual-insulation cables that are resistant to grounding faults and short circuits (as recommended in [4.2]) are used. In addition, [4.19] reports that aR fuses (which do not protect an entire area) have caused fires. However, just because use of the wrong type of fuse has caused fires does not mean that fuses should be eschewed – providing of course that the right ones (namely, type gPV or gR) are used. But as the solar module and laminate cover glass as well as the relatively thin back foils used in these devices are extremely fragile and easily damaged by impacts, PV plants comprising numerous parallel-connected strings are prone to damage that can induce grounding faults, or short circuits in individual modules or module groups (see Section 4.3.3.1). Wiring errors can also occur from time to time. Electrical engineering students are always told (in accordance with what is undeniably the state of the art for electricity distribution) to design systems that can withstand the worst-case scenario, as well as the damage induced by such situations. Hence it would be nothing short of negligent to reduce system costs marginally by dispensing with elements that avoid reverse current and protect string conductors, which usually exhibit a very low gauge.

The elevated voltages engendered by short circuits or grounding faults in directly parallel-connected strings result in appreciably higher output than is the case with directly parallel-connected solar cells or modules. As solar generator short-circuit current is subject to a fixed upper limit, two strings can always be directly parallel connected since at $n_{SP} = 2$ string fuses can never shut down a reverse current.

In most cases, three strings can also be parallel connected without anti-reverse-current elements, since effective protection against reverse current from two adjacent strings is a virtual impossibility using commercial fuses. However, in such settings disconnect terminals or preferably special PV plugs should be used on both sides. *However, in order to do this the string wire gauges must be large enough and precautions must be taken to obviate reverse current from connected appliances such as battery banks and inverters.*

Four strings can often also be directly parallel connected, although this is somewhat risky in the absence of explicit reverse current specifications from the vendor, in the event of damage. Diligent planners that wish to rule out all possible risks only opt for $n_{SP} = 2$ direct parallel-connected string solutions.

However, if n_{SP} is greater than 3 or 4, or if a PV generator is subdivided into a series of arrays, string fuses should be used, unless the vendor datasheet specifies a maximum reverse current $I_{R\text{-}mod}$ for the relevant module, such that the maximum number n_{SP} of allowable direct parallel-connected strings can be determined using Equation 4.8 or 4.11. However, transient current overload events induced by cloud enhancements or snow can result, causing upwards of $1.25 \cdot I_{SC\text{-}STC}$ to flow into the modules. Hence to avoid damage to string conductors, they need to be rated for at least $1.25 \cdot (n_{SP} - 1) \cdot I_{SC\text{-}STC}$.

4.3.3.3 Current in Incorrectly Polarized Parallel Strings

Incorrect string polarity in a solar generator with numerous n_{SP} should be avoided at all costs, as fuses can do nothing against this problem (see Figure 4.35).

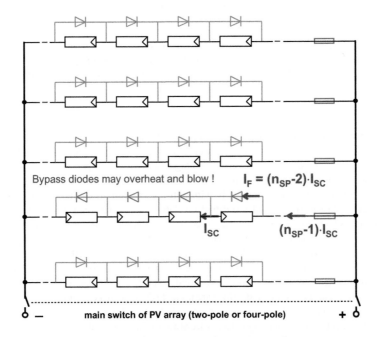

Figure 4.35 Solar generator with parallel-connected series strings, one of which is incorrectly polarized. Bypass diodes can be damaged beyond repair in such a case in a system with a high n_{SP} value

If a solar generator with n_{SP} parallel-connected strings is incorrectly polarized, the generator will in effect be short-circuited via the bypass diodes of the incorrectly polarized string, and (as shown in Figure 4.35) a current of around $(n_{SP} - 1) \cdot I_{SC}$ will flow through the string. As the modules of such a string are usually as insolated as the other modules, an I_{SC} current will flow into them as well, and an $(n_{SP} - 2) \cdot I_{SC}$ current will flow through the bypass diodes. In such a case, inasmuch as bypass diode capacity is very low, at $n_{SP} \geq 4$ all such diodes in the incorrectly polarized string could easily be damaged beyond repair before the string fuses are tripped.

4.3.4 Solar Generators with Solar Module Matrixing

In this infrequently used wiring configuration, groups comprising n_{MP} parallel-connected solar modules and a bypass diode are defined (see Section 4.3.2). The n_{GS} of such groups are then series connected with a view to ramping up system voltage (see Figure 4.36).

The intermeshing of serial and parallel connections partly compensates for internal asymmetries. Such solar generators are not as adversely affected by the partial shading of individual modules or slight differences between module characteristic curves. If the shaded modules (which represent 25% of the total) in Figure 4.36 were fully shaded, the generator would incur only a 25% power loss. Absent cross-connections, this would comprise a solar generator with eight parallel strings for four modules wired in series. If the installation fed by this solar generator continued in operation using only the original MPP voltage V_{MPP} of the unshaded generator, the output power of such a solar generator would decrease to practically zero.

However, one of the main drawbacks of matrixing is that the attendant intermeshing greatly complicates the task of localizing malfunctions (e.g. module failure), which mainly manifest themselves as a voltage loss in the group with the defective module, and to a lesser extent as reduced current across the entire solar generator.

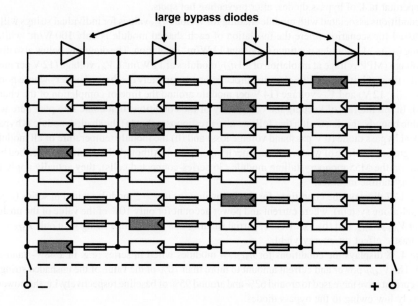

Figure 4.36 Solar generator matrixing realized by series connecting various module groups of n_{MP} parallel-connected solar modules. Each parallel-connected module group is bypassed by a large (and cooled) bypass diode that can conduct at least 1.25 times the aggregate current of all modules. On shading of the shaded modules in the diagram, matrixing limits power to only 25%. However, installation and maintenance of such solar generators are far more cost intensive

Moreover, it is virtually impossible to monitor internal voltages continuously as this task is extremely labour intensive. To localize the defective module in the event of a malfunction, the entire generator must be shut down – which is whymatrixing is rarely used. However, it might well be possible to integrate small matrixing arrays into very large PV plants in such a way that the arrays form a string for the solar generator and are parallel connected as described in Section 4.3.3.

4.4   Solar Generator Power Loss Resulting from Partial Shading and Mismatch Loss

As noted in Section 4.2.2, power in a solar module and the relevant string decreases dramatically if individual solar cells or entire modules are partly shaded.

Power loss also occurs, albeit to a lesser degree, if module characteristic curves are mismatched as the result of unavoidable manufacturing tolerance discrepancies, namely unequal P_{max}, V_{MPP} and I_{MPP}. Such scenarios are referred to as solar generator mismatches, which can also be induced by unshaded modules with identical solar module characteristic curves if a string module does not exhibit exactly the same insolation as the result of a factor such as unequal diffuse radiation conditions. Hence, effective solar generator power P_{Ao} is always lower than rated solar generator power $P_{Go} = n_M \cdot P_{Mo}$ as in Equation 1.6.

4.4.1   Power Loss Induced by Module Shading

Partial or complete shading can provoke genuinely drastic power loss in a string comprising only a few modules (see Figure 4.22). In a string with n_{MS} modules wired in series that integrate bypass diodes, the power loss induced by partial shading of individual modules increases with higher n_{MS}. In such a case, the unshaded modules shift their operating point towards an open-circuit state, the voltage in these modules rises, and current flows via the bypass diodes to the partly or fully shadowed module. This is the second most important task of bypass diodes, after preventing hot spots.

The conditions associated with module shading for various n_{MS} values for individual strings will now be discussed for scenarios where the insolation of each shaded module is only 100 W/m^2 while the remaining $(n_{MS} - 1)$ modules continue to present 1 kW/m^2 insolation. The diagrams below also display V_{MPP} voltage (MPP voltage at insolation of all n_{MS} modules at 1 kW/m^2), V_{A1} voltage (12 V per module around the time of initiation of the charging process for a completely discharged battery whose nominal power is $n_{MS} \cdot 12$ V) and V_{A2} voltage (14 V per module around the time of completion of the charging process). In the interest of keeping the complexity of the conditions that come into play here within reasonable bounds, the discussion that follows also presupposes that (a) the shaded module is bypassed by an ideal bypass diode (0 V passband voltage loss) and (b) the module temperature of all modules is 40 °C (aggregate average of winter and summer temperatures). Although the examples described below are based on the M55 monocrystalline module (36 series-connected cells), they actually apply to all types of crystalline modules.

Figure 4.37 displays the conditions for the following: (a) $n_{MS} = 2$ modules wired in series (e.g. in a 24 V stand-alone system), where current and power account for only 10% of the value of the unshaded string for V_{MPP}, V_{A2}; and (b) V_{A1}. Hence, partial shading of a module in a string with only $n_{MS} = 2$ has a very drastic effect on energy yield.

Figure 4.38 displays the conditions for $n_{MS} = 4$ modules wired in series (e.g. in a 48 V stand-alone system). The V_{MPP} power and current amount to more than 70% of the value of the unshaded string. V_{A2} and V_{A1} current have increased to around 62% and around 95% of baseline respectively, i.e. the power loss is relatively low owing to the bypass diodes.

Figure 4.39 displays the conditions for $n_{MS} = 9$ modules wired in series (e.g. in a 110 V stand-alone system). Owing to the bypass diodes, the V_{MPP} power and current amount to more than 70% of the value of the unshaded string. V_{A2} and V_{A1} current have returned to around 97% and nearly 100% of baseline respectively.

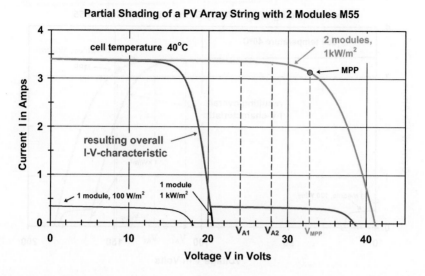

Figure 4.37 Solar generator string comprising two M55 modules (e.g. for a 24 V stand-alone system) with shading of a module that is bypassed by an ideal bypass diode and the following conditions obtain: cell temperature $T_Z = 40\,°C$; one module at $1\,kW/m^2$ insolation; $V_{MPP} = MPP$ voltage of the fully insolated string; $V_{A2} = 14\,V$; $V_{A1} = 12\,V$ per module

A more general investigation can be realized for string power in n_{MS} modules, n_{MSB} of which are shaded (i.e. irradiation is still $100\,W/m^2$), whereas the remaining modules $(n_{MS} - n_{MSB})$ present full $1\,kW/m^2$ insolation. Figure 4.40 displays an example for M55 modules where $T_Z = 40\,°C$ cell temperature and relative string power is a function of the relative number of shaded modules $a_{MB} = n_{MSB}/n_{MS}$. Although the

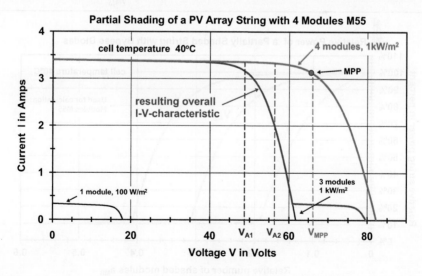

Figure 4.38 Solar generator string comprising four M55 modules (e.g. for a 48 V stand-alone system) with shading of a module that is bypassed by an ideal bypass diode and the following conditions obtain: cell temperature $T_Z = 40\,°C$; one module with $1\,kW/m^2$ insolation; $V_{MPP} = MPP$ voltage of the fully insolated string; $V_{A2} = 14\,V$; $V_{A1} = 12\,V$ per module

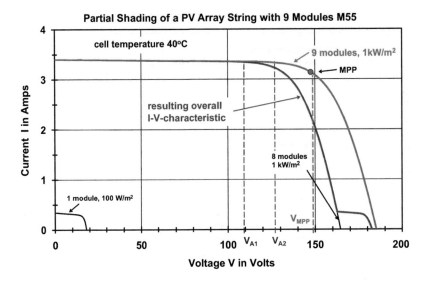

Figure 4.39 Solar generator string comprising nine M55 modules (e.g. for a 110 V stand-alone system) with shading of a module that is bypassed by an ideal bypass diode and where the following conditions obtain: cell temperature $T_Z = 40\,°\mathrm{C}$; one module with $1\,\mathrm{kW/m^2}$ insolation; $V_{MPP} = $ MPP voltage of the fully insolated string; $V_{A2} = 14\,\mathrm{V}$; $V_{A1} = 12\,\mathrm{V}$ per module

calculation here was based on the characteristic curves of the M55 module, the resulting curves apply to all crystalline modules. Thus

$$\text{Relative number of shaded modules } a_{MB} = \frac{n_{MSB}}{n_{MS}} \qquad (4.12)$$

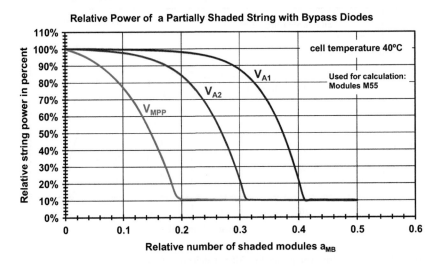

Figure 4.40 Relative output of a crystalline module solar generator, with shading of individual modules (bypassed via an ideal bypass diode), as a function of the relative number of shaded modules a_{MB} according to Equation 4.12. Suppositions: cell temperature $T_Z = 40\,°\mathrm{C}$; modules at $1\,\mathrm{kW/m^2}$ insolation; shaded module with $100\,\mathrm{W/m^2}$ insolation. $V_{MPP} = $ MPP voltage of the fully insolated string; $V_{A2} = 14\,\mathrm{V}$ and $V_{A1} = 12\,\mathrm{V}$ per 36-cell module

where n_{MS} = number of modules per string (total number) and n_{MSB} = number of shaded modules per string.

As Figure 4.40 shows, when the fully insolated string is operated at MPP voltage V_{MPP}, a power loss already occurs with low a_{MB} values, whereby this power loss is small at first and then increases. However, if the string uses a lower voltage V_{A2} (14 V per module) or V_{A1} (12 V per module), the power loss is virtually nil with low a_{MB} values and only begins climbing as these values increase. Indeed, owing to the bypass diodes, the power loss induced by shading of individual modules is gradual in strings with large numbers of modules, whereby power only begins dropping precipitously when a substantial number of modules is shaded.

4.4.2   Mismatch Loss Attributable to Manufacturing Tolerances

In cases of discrepancies between the characteristic curves of the modules in a specific string (i.e. one or more string modules exhibit power loss), string power loss is greater than would be expected based on the reduced aggregate output of all string modules. This additional power loss is attributable to a phenomenon known as mismatch loss, which will now be discussed in connection with various n_{MS} string values.

All modules Exhibit 1 kW/m^2 insolation and a cell temperature of $T_Z = 25\,°C$. Moreover, it is assumed that, owing to manufacturing tolerances, one module will exhibit the same characteristic curve as a module with only 900 W/m^2 insolation. *However*, the remaining ($n_{MS} - 1$) modules will exhibit *normal* characteristic curves at 1 kW/m^2 insolation. The diagrams below also display V_{MPP} voltage (MPP voltage under insolation of all n_{MS} modules at 1 kW/m^2), V_{A1} voltage (12 V per module around the time of initiation of the charging process for a completely discharged battery bank whose nominal power is $n_{MS} \cdot 12$ V) and V_{A2} voltage (14 V per module around the time of completion of the charging process). For reasons of simplification, it is assumed that all modules are bypassed via an ideal bypass diode (passband voltage loss 0 V) and that the module temperature of all modules is 25 °C. These examples are likewise based on the M55 module.

Figure 4.41 displays a series string comprising $n_{MS} = 4$ M55 modules, where output and voltage are 5.2% and 10% lower for V_{MPP} and V_{A2} respectively than for an intact string comprising the exact same type of module. These power losses are appreciably higher than the purely arithmetic 2.5% power loss in the string (three modules with 100% nominal module output P_{Mo} and one for which the figure is 90%). Even if the string is operated in the resulting new MPP of the string, its output will still be around 4.7% lower than for an intact string. Under V_{A1} conditions, however (intact string operating at far below MPP), the power loss is only around 2.5% owing to the available voltage buffer and is thus close to the arithmetic string power loss.

Figure 4.42 displays a series string comprising $n_{MS} = 9$ M55 modules. where output and voltage are 4.2% and 1.5% lower for V_{MPP} and V_{A2} respectively than for an intact string comprising the exact same type of module. These power losses are appreciably higher than the purely arithmetic 1.1% power loss in the string (eight modules with 100% nominal module output P_{Mo} and one for which the figure is 90%). Even if the string is operated at the resulting new MPP of the string, the output is still around 3.1% lower than for an intact string. Only with V_{A1} operation (intact string operating at far below MPP) is the power loss virtually nil, owing to the large available voltage buffer, and is thus lower than the arithmetic string power loss.

A more general investigation can be realized for mismatch loss in a string with n_{MS} modules, n_{MSB} of which exhibit power loss whereas the remaining modules ($n_{MS} - n_{MSB}$) exhibit normal characteristic curves. Figure 4.43 displays the resulting mismatch loss (at 1 kW/m^2 insolation) compared with nominal string output ($n_{MS} \cdot P_{Mo}$) as a function of the relative number of modules with power loss $a_{MM} = n_{MSM}/n_{MS}$. The resulting power loss under V_{MPP} operation of the intact string (V_{MPP} curve), under the resulting MPP operation of the string (MPP curve), and, for purposes of comparison, the purely arithmetic loss (ΣP_M curve), are indicated for 5% and 10% module power loss relative to P_{Mo}. The V_{MPP} curve is of interest here for a solar generator with numerous parallel-connected strings, for determining the effect of mismatch loss on an individual string. On the other hand, the MPP curve is the determining factor for a solar generator

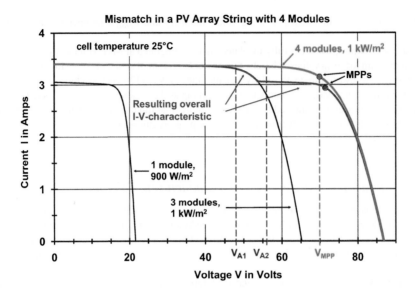

Figure 4.41 Characteristic curves of a solar generator string comprising four M55 modules (cell temperature $T_Z = 25\,°C$) where $G = 1\,kW/m^2$ (e.g. for a 48 V stand-alone system). One of these modules exhibits a 10% power loss (same characteristic curves as for a normal module at $900\,W/m^2$). All modules are bypassed by ideal bypass diodes. V_{MPP} = MPP voltage of the fully insolated string; $V_{A2} = 14\,V$ and $V_{A1} = 12\,V$ per module

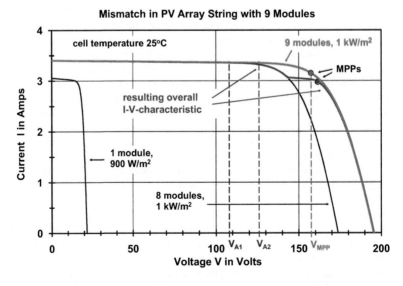

Figure 4.42 Characteristic curves of a solar generator string comprising nine M55 modules (cell temperature $T_Z = 25\,°C$) where $G = 1\,kW/m^2$ (e.g. for a 110 V stand-alone system). One of these modules exhibits a 10% power loss (same characteristic curves as for a normal module at $900\,W/m^2$). All modules are bypassed by ideal bypass diodes. V_{MPP} = MPP voltage of the fully insolated string; $V_{A2} = 14\,V$ and $V_{A1} = 12\,V$ per module

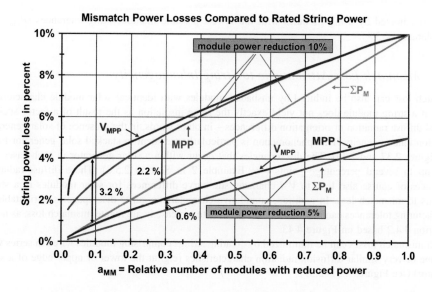

Figure 4.43 Mismatch loss in a solar generator string comprising Siemens M55 modules, as a function of the relative number of modules with power loss a_{MM} according to Equation 4.13. Suppositions: cell temperature $T_Z = 25\,°C$; $1\,kW/m^2$ insolation in all modules; module loss 5% and 10%. All modules are bypassed by ideal bypass diodes. V_{MPP} curve: operation of an intact string under MPP voltage V_{MPP}; MPP curve: operation under the new MPP of the resulting V–I characteristic curve; ΣP_M curve: purely arithmetic power loss induced by module power loss

comprising a single string, or in cases where all n_{SP} parallel-connected strings of a larger solar generator have the exact same number of modules with power loss. The following is defined for Figure 4.43:

$$\text{Relative number of modules with power loss: } a_{MM} = \frac{n_{MSM}}{n_{MS}} \qquad (4.13)$$

where n_{MS} = number of modules per string (total number) and n_{MSM} = number of modules per string with power loss (5% and 10%).

The curves in Figure 4.43 presuppose that at equal insolation the current in the weaker modules will be 10% and 5% lower than for a normal module. In other words, it was assumed that, in the solar cell equivalent circuit shown in Figure 3.12, the power source current will be proportionally weaker and that the other circuit elements will remain unchanged.

As Figure 4.43 shows, even with a relatively small number of modules with power loss (i.e. low a_{MM} values), as well as with MPP voltage V_{MPP} operation of an intact string (V_{MPP} curve) and under operation with the newly resulting string MPP (MPP curve), appreciable mismatch loss occurs that far exceeds the expected purely arithmetic loss (ΣP_M curve).

A comparison of these curves with the 5% and 10% module power loss shows that mismatch loss rises disproportionately as power loss increases. Under MPP voltage V_{MPP} operation of an intact string, the power loss is greater than under operation of the newly resulting string MPP. This effect is particularly pronounced for low a_{MM} values and 10% module power loss, where deviations translate into purely arithmetic output of more than 2% (operation under the new MPP) and 3% (V_{MPP} operation of the intact string using modules with no power loss). With 5% module power loss, the curves for operation under the new and old MPP are virtually the same and the maximum deviation from pure arithmetic power loss is 0.6%. It is noteworthy here that even a relatively low proportion of modules with power loss (relatively low a_{MM}) translates into noticeable string power loss. Hence, the energy yield in larger solar generators can be maximized by using modules with the lowest possible manufacturing tolerances (P_{max} and I_{MPP} tolerance $\leq 5\%$).

4.4.3 Mismatch Loss Attributable to String Inhomogeneity

Mismatch loss can also be induced by unshaded modules with identical solar module characteristic curves if a string module does not yield exactly the same insolation as the result of a factor such as unequal diffuse radiation or orientation differences – in other words, in the presence of solar generator insolation inhomogeneity. This phenomenon is particularly prevalent in stacked solar generator fields (see Figures 2.37 and 2.39). The discrepancies that occur within a solar generator in such cases can range up to several percentage points (see Example 2 in Section 2.5.8). These diffuse-radiation differences of course also lead to V–I characteristic curve differences for solar modules in a string and thus to mismatch loss. In addition, cumulative characteristic curve discrepancies attributable to manufacturing tolerances can result in a disproportionate increase in aggregate mismatch loss, as noted in Section 4.4.2 based on Figure 4.43.

Such mismatch losses can be minimized by creating strings comprising modules wired in series with the same or very similar diffusion radiation characteristics (e.g. at the lower or upper edge of a solar generator) (see Figures 2.37–2.39).

4.5 Solar Generator Structure

This section provides some practical advice concerning solar generator realization and the various options in terms of mounting, materials and wiring. In lieu of simply indicating the numbers of the relevant standards (as is done in most books), the emphasis here is on elucidating the relevant technical matters in accordance with the state of the art as of 2009. Needless to say, the relevant national standards apply in each individual country, and these standards may deviate slightly from those indicated here.

4.5.1 Solar Generator Mounting Options

In view of the steady decrease in solar module prices, in Europe, with its high proportion of diffuse radiation, the use of solar tracking systems generally does not pay, particularly in view of the higher investment and upkeep expenses entailed by these devices. That said, the use of solar tracking systems could be justified on the grounds that such systems allow for a higher number of full-power hours. The most prevalent mounting modalities for solar generators (which are normally mounted in a fixed position) are specially designed mounting racks in open spaces and mounting on building façades or pitched/flat roofs. In some cases solar generators are installed on existing terrain elements such as highway or railway sound barriers.

Framed modules are as a rule more mechanically robust than laminates. A metal frame also provides good edge protection and is beneficial for lightning protection as well. Framed modules are particularly suitable for installation in open spaces and for systems that are made for and installed on existing buildings. Framed modules are often secured to mounting racks using bolts and to a lesser extent module clips. Myriad mounting systems are available that allow solar modules to be attached with relatively little effort. However, framed modules have drawbacks relative to laminates in that grit tends to accumulate at their lower edges and snow takes a relatively long time to slide off them.

Moreover, laminates are more aesthetically appealing for building integration purposes. Most laminates weigh about the same as framed modules owing to the thicker (6–10 mm) glass that is needed for the requisite sturdiness. Laminates are often integrated into façades or roofs in the manner of plate glass, although it needs to be borne in mind that laminates are exposed to more extreme temperature fluctuations than is the case with standard building glazing. Laminates can also be installed using special module clips (with inserts made of elasticized plastic such as neoprene) or can be glued in place using

Figure 4.44 Ground-based installation comprising Siemens M55 laminates (53 Wp) for the Mont Soleil system (560 kWp) at 1270 m above sea level. The tilt angle is $\beta = 50°$ so as to allow snow to slide off (the space below the installation allowing). The laminates were glued onto the galvanized steel mounting rack using a special silicone adhesive (Photo: Siemens)

silicone adhesive. However, owing to the absence of a frame, laminates must be handled with extreme care in order to avoid breakage.

For mounting solar modules on pitched roofs, various manufacturers sell special solar roof tiles that are used in lieu of standard roofing tiles. However, if the weight per surface unit of these solar roof tiles is less than that of standard roofing tiles, additional fastening elements must be used to prevent the solar tiles from being blown off by the wind.

4.5.1.1 Ground-Based Installations

Figure 4.44 displays a ground-based PV installation on Mont Soleil (1270 m above sea level) comprising Siemens M55 laminates glued to metal carriers installed on a specially built mounting rack. This installation method allows for the realization of PV plants at sites with optimal insolation and relatively high wind energy production, although additional terrain is needed for installations of this type. The mounting rack, foundation and any grid connections that may be needed entail relatively high installation costs that cannot be shared with other construction projects. However, one advantage of this installation method is that the modules are readily accessible from both sides. For ground-based installations, mounting systems are also available for which no concrete foundation is needed and that can be bolted into the ground.

Since (a) the Mont Soleil installation is composed of laminates with glass of normal thickness and (b) the installation is not very far off the ground, some of the laminates have incurred snow pressure damage during very snowy winters (see Figures 4.67–4.70).

4.5.1.2 Solar Module Installation on a Flat Roof

Figure 4.45 displays solar modules installed using solar flat roof elements (SOFRELs), which are mass-produced prefabricated concrete elements for mounting solar modules on flat roofs. These elements are not secured to the building in any way, so as to avoid sealing problems. The SOFREL system can be used for both laminates and framed modules. However, to avoid unduly severe wind load problems, the tilt angle is only adjustable from 18° to 25°.

Figure 4.45 Flat-roof installation for framed 120 Wp solar modules using SOFRELs. Owing to the shallower tilt angle $\beta = 18°$, the wind forces are lower than for the installation shown in Figure 4.47, thus reducing (a) the load on the roof, (b) the need for additional weight and (c) the amount of spacing needed to avoid reciprocal shading Nonetheless, solar generators installed on flat roofs tend to become dirtier and produce less energy during the winter (Photo: Tritec)

Figure 4.46 displays another flat-roof solar generator with somewhat larger solar modules (approximately as in Figure 4.4). The spacing between the concrete mounting elements and the roof is somewhat larger, which makes it easier for winter snow to slide off the modules.

Figure 4.47 displays the rear of a solar generator comprising framed modules mounted on a specially made and relatively heavy structure on a flat roof. Inasmuch as the only compensatory elements for the wind forces that the installation is exposed to are frictional force and specific gravity, the roof in such cases needs to exhibit the requisite load-bearing properties – namely, around 100 kg per square metre of solar generator area. In realizing such an installation, it must be ensured that the wind forces cannot tip over the mounting rack. This mounting method renders both the front and rear of the installation readily accessible.

Figure 4.46   Relatively large solar generator (total of about 100 kWp) on the flat roof of an apartment block in Geneva. The framed approximately 215 Wp solar modules (as in Figure 4.4) were mounted using elements similar to those shown in Figure 4.45. However, the spacing between the modules and roof is somewhat larger, thus making it easier for snow to slide off the modules (Photo: Sunpower Corp./Suntechnics Fabrisolar AG)

The wind forces that the installation shown in Figure 4.47 is exposed to are counteracted by specific gravity and frictional force. The following additional features of this installation are also visible in the photo:

- The module frames and racking are connected to lightning protection installations.
- The modules are combined into series-connected strings via short wires, which exhibit a minor anomaly in that they are dangling rather than being fastened down.
- The installation contains junction boxes, where the strings are wired in parallel.

Figure 4.47   Rear view of a flat-roof installation for a solar generator comprising framed Siemens M55 modules (55 Wp, tilt angle $\beta = 35°$) that are partly held in place by the force of gravity. The modules are mounted in groups of four on aluminium sections, which are secured to the prefabricated concrete elements on the roof and weighted down at the rear of the installation to avoid possible tipping induced by wind forces

Figure 4.48 Flat-roof installation of a solar generator composed of 120 Wp Siemens M55 laminates that were mounted using Solgreen elements. Such laminates are less prone to grit accumulation and the consequent power loss, but are more vulnerable to lightning strikes (Photo: Basler & Hofmann AG)

PV systems are often installed on landscaped flat roofs. Figure 4.48 displays such an installation using Solgreen elements.

4.5.1.3 Pitched-Roof PV Installation on a Purpose-Built Roof Support System

This method is particularly suitable for pitched roofs on existing buildings, where the solar generator is installed on a purpose-built metal roof support system (see Figures 1.10, Figures 4.49 and 4.50). The requisite sealing against precipitation is provided by the existing tile roofing underneath the solar generator.

The fireproof and electrically isolating roof between the solar generator wiring and the mainly wooden roof substructure provides effective fire protection in the event of a serious problem with the solar generator.

Figure 4.49 Pitched-roof installation of ∼2.2 kWp solar generator mounted on a purpose-built roof support system. The spacing between the roof and modules allows for good underside ventilation (Photo: Siemens)

Figure 4.50 Pitched-roof installation of ∼30 kWp solar generator mounted on the purpose-built substructure of a farm building in Schwarzwald, Switzerland. Solar power is an attractive source of income for farmers if the feed-in tariffs are high enough (Photo: Sunpower Corp./SunTechnics Fabrisolar AG)

The underside of the modules, as well as the tiles under the solar generator, are extremely difficult to access once such a system is installed. In order to achieve adequate rear module cooling (which reduces power loss), sufficient clearance of the order of 10 cm should be allowed between the roof and the solar generator. Moreover, the mounting system needs to be properly sized for snow load, wind pressure and wind suction. Such systems are available from various vendors.

4.5.1.4 Pitched-Roof Installation Integrated into the Roof

Direct integration of laminates or solar tiles into a roof is a more aesthetically pleasing solution. To do this, however, the solar generator needs to carry out all of a normal roof's weather protection functions, which for new buildings obviates the cost of conventional roofing. In the interest of avoiding accidents on pitched roofs, it is advisable to use wiring connector systems that make it impossible to confuse one connector with another so that a roofer can mount and wire in series the various strings, and an electrician is only needed to hook up the string wiring and power line.

Wind suction needs to be taken into account for roof-integrated installations as well. In some cases, solar roof tiles whose area specific weight is less than that of standard roof tiles need special pressure bars or clips. It is also essential to provide sufficient rear ventilation for such systems. Figure 4.51 displays a small tile roof-integrated solar generator. Figure 4.52 displays a large laminate solar generator integrated into a pitched roof using large Megaslate solar roof tiles. This generator covers nearly half the roof.

Figure 4.53 displays a house with Solrif solar roof tiles, which are laminates with a special aluminium frame that frames the left, right and upper side of the laminate on both sides of the laminate, but at the lower edge is only arranged below the laminate so as to obviate the frequently observed problem of grit accumulating on the underside of the module frame. This solution allows the optimized lightning protection characteristics of a framed module to be combined with the lesser tendency to accumulate grit offered by laminates (see Section 6.7.7). The fact that each side of the Solrif profile exhibits a different form allows for the manufacture of laminate solar roof tiles and thus the realization of a hermetically sealed roof.

Figure 4.51   A small laminate solar generator integrated into a pitched roof (Photo: SunTechnics Fabrisolar AG)

4.5.1.5   Façade Mounting and Integration

Solar generators can also be integrated into building façades, although the projected energy yield of such installations in flat areas is around 20 to 40% lower than for optimally sized roof installations (see Sections 2.4 and 2.5). However, in mountainous areas the façade installation energy yield is on a par with that of optimally oriented roof installations. Moreover, the effects of snow loads can be almost completely obviated by mounting the solar generator vertically. Many façade installations are mounted vertically (tilt angle 90°; see Figures 4.54–4.56), although installations that serve additional purposes such as window shading and that exhibit somewhat shallower tilt angles are also realizable (see Figure 4.57). Good underside ventilation is also essential for façade-integrated installations.

Figure 4.52   Pitched-roof installation of a roof-integrated 9.3 kWp solar generator consisting of Megaslate laminate roof tiles (Photo: 3S Swiss Solar Systems AG)

Figure 4.53 A 12 kWp Solrif solar generator integrated into a pitched roof. This installation combines the advantages of framed and laminate modules (Photo: Ernst Schweizer AG)

Figure 4.54 Polycrystalline module solar generator integrated into two façades of a building in Riehen, Switzerland (Photo: SunTechnics Fabrisolar AG)

Figure 4.55 A 9.35 kWp solar generator mounted on the façade of SLF (Eidgenössischen Instituts für Schnee- und Lawinenforschung) on Weissfluhjoch Mountain (elevation 2690 m) in Davos, Switzerland (Photo: WSL Institute for Snow and Avalanche Research SLF, www.slf.ch)

Figure 4.56 Solar generator mounted 2486 m above sea level at the valley station of the Piz Nair ski lift above St Moritz (Photo: SunTechnics Fabrisolar AG)

Efficient sealing is more readily obtainable for façade installations than on roofs. Laminates (often custom made) or large modules (of which up to 300 W versions are commercially available) are mainly used for façade installations. Customized modules can cost up to 100% more than standard products. As many laminate or module configurations entail extensive exposure to the elements, it is advisable to use connector systems for the wiring that make it impossible to confuse one connector with another so that a roofer can mount and wire in series the various strings and an electrician is only needed to hook up the strings and power line.

The cost of a façade solar generator can be reduced by using it as an integral façade element, instead of merely installing the generator on the façade, particularly in cases where doing so would replace an expensive decorative material such as marble.

Figure 4.57 A polycrystalline module solar generator mounted on the façade of the US Embassy in Geneva. The modules are mounted at a 90° angle on the left side, whereas on the right side they are mounted at a smaller tilt angle so as to provide summer shading for the windows below (Photo: SunTechnics Fabrisolar AG)

Figure 4.58 As the first solar generators to be installed on a highway sound barrier, this 110 kWp installation went into operation in late 1989. An unusual but bothersome problem with it has been repeated thefts of modules, which have had to be replaced time and time again (Photo: © TNC Consulting AG, Switzerland)

4.5.1.6 Mounting a Solar Generators on an Existing Structure

In some cases, a solar generator can be mounted at little additional cost on existing structures such as highway or railway sound barriers. Figure 4.58 displays one of the first such installations near Chur, Switzerland. Recently, hybrid solutions have also been realized involving PV and acoustic attenuation elements, although such approaches inevitably necessitate certain compromises. Other structures such as bridges, retaining walls, dams, highway overpasses, parking lot roofs and so on can also potentially be used for PV installations.

4.5.1.7 Solar Tracking Systems

As noted in Section 2.4, the energy yield can be increased above that obtained with a fixed installation using single- or dual-axis trackers. Relative to peak nominal power, solar generator energy yield can be increased by around 10 to 25% with a single-axis tracker and by around 25 to 40% with a biaxial system; of course, the number of full-load hours increases accordingly (see Section A8.3, Figure A.11 and the examples in Sections 10.1.9–10.1.12). However, solar trackers are subject to size limitations for reasons of wind load. Hence larger-scale installations are subdivided into a series of around 1 to 50 kWp tracked arrays into each of which an inverter can be integrated. Various vendors sell mounting systems for biaxial solar trackers.

Single-axis trackers have less complex mechanisms, but also provide less in the way of additional energy. Solar trackers with rods that are horizontally mounted in the north–south direction and that rotate from east to west over the course of a day allow for an approximately 10 to 12% annual energy gain in Central Europe [4.16]. Such a tracker was integrated into a 6.3 MWp ground-based PV installation in Mühlhausen, Germany (see Figure 4.59), where tracking is realized in a relatively straightforward manner, using rods that are attached to the various arrays.

Single-axis trackers yield better power gains if their rotation rods face southwards. A rod oriented 30° to the south provides a 20% energy gain in Central Europe and somewhat more in Southern Europe [4.16]. Figure 4.60 displays a rooftop solar tracking system.

Figure 4.61 displays a mechanically robust construction for the dual-axis trackers integrated into the solar generators at the Berlin R&D Centre of Solon AG. In this solution each of the tracked 6.5–9 kWp solar generators (area ≈ 50 m^2) integrates an inverter. A 12 MWp installation that integrates 1500 such elements (known as Solon Movers) has been in operation since 2006 at Solarpark Erlasee near Arnstein,

Figure 4.59 Each of the arrays that make up this 6.3 MW solar generator in Mühlhausen, Germany, integrates a single-axis tracker. These mechanisms are mounted in a north–south orientation on horizontal rods that rotate from east to west over the course of the day in accordance with the path of the Sun. This also helps to prevent winter snow accumulation (Photo: Power Light GmbH)

Germany (see Figure 4.62), and numerous other Solon Movers have been installed since (see, *inter alia*, the installations described in Sections 10.1.9–10.1.11).

4.5.2 Mounting Systems

4.5.2.1 Mechanical Sizing of Mounting Systems

The factors that need to be taken into consideration when sizing a PV installation mounting system are the specific gravity of the modules or laminates (mass about 10 to 20 kg/m^2 and thus weight around 100 to 200 N/m^2) and particularly the main sporadic snow and wind load forces that the installation is exposed to. As these loads vary from one region to another, only a general idea of these loads can be provided here. To calculate such loads precisely, the relevant national standards for the site in question should be consulted.

Figure 4.60 Single-axis solar tracking system on a building roof. The modules are mounted in a north–south orientation on horizontal rods that rotate from east to west over the course of the day in accordance with the path of the Sun (Photo: Solon AG)

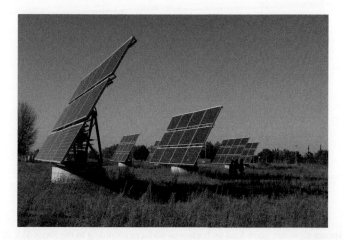

Figure 4.61 Dual-axis tracking system for solar generators, each of which integrates an inverter. The installation is used by the Solon R&D Centre (Photo: Solon AG, Norbert Michalke)

According to the applicable Swiss SIA standard, a minimum snow load amounting to $p_S = 900\,\mathrm{N/m^2}$ needs to be taken into account for the horizontal plane, although this value increases sharply with increased height above sea level (see Figure 4.63) and for very high elevations constitutes the predominant load. If a solar generator is mounted at tilt angle β relative to the horizontal plane, snow load pressure is reduced accordingly, perpendicular to the module plane. The following snow pressures need to be taken into account for a solar generator at tilt angle β:

$$p_{S\beta} = p_S \cdot \cos \beta \qquad (4.14)$$

IEC 61215 [4.11] stipulates that solar generators must be rated for a snow load of $2400\,\mathrm{N/m^2}$. The value for heavy-duty models is $5400\,\mathrm{N/m^2}$ for very snowy areas. However, the homogeneous test load that is used for tests in this regard often does not square with the real-world loads that inclined solar generators are subjected to (see Figure 4.64).

In the case of a solar generator with a tilt angle β to the horizontal plane, the uppermost snow on the generator often begins melting rapidly, whereupon some of it slides off the solar generator and piles up at

Figure 4.62 Solarpark Erlasee (12 MWp) near Würzburg, Germany. This installation is composed of 1500 dual-axis 6.5–9 kWp (depending on the technology used) Solon Movers (Photo: Sunpower Corp.)

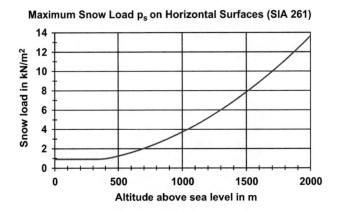

Figure 4.63 Maximum allowable snow load on horizontal surfaces according to the Swiss SIA 261 standard. The curve shown here roughly equates to snow load zone III in Germany in accordance with DIN 1055-5, Table 2 [DGS05] See [DGS05] for more detailed information concerning Germany

the bottom edge of the generator. As a result the solar module and frame loads are unequal (see Figure 4.64). The weight $F = m \cdot g$ exerted on the module consists of two components: normal force F_N which is at a right angle to the module surface and descending force F_A on the solar module plane. At equilibrium, F_N is offset by an equal and opposite reaction force F_R and F_A that stems from the opposing frictional force R, whereby $R \leq \mu_o \cdot F_N$. Since μ_o between the snow and glass is relatively small (particularly if the glass is wet), the snow can slide off during a thaw if the tilt angles are not abnormally steep.

For solar generators installed on pitched roofs, and particularly for laminates, the friction coefficient should be lower (for snow on a roof with the same pitch) than for pitched roofs with roof tiles. The energy yield is better the sooner snow slides off the roof, but also entails the risk of a roof avalanche that can cause personal injury or property damage if wet snow accumulates on a frozen roof surface and abruptly slides off. The risk of avalanches is high with roofs whose entire surface is covered with laminates, as in Figures 4.52 and 4.53. But roof avalanches can also occur with framed modules; in such cases the higher the building, the greater the risk. Hence precautionary measures aimed at preventing possible personal

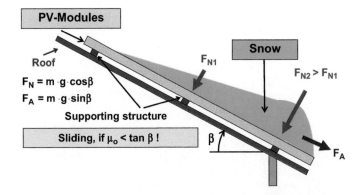

Figure 4.64 Typical snow load (shortly after snowfall) that a solar generator with tilt angle β is exposed to. If the adhesive friction coefficient $\mu_o < \tan \beta$ (e.g. with an abnormally steep tilt angle β or on formation of a film of melted water) the snow can slide off the modules – although this can result in property damage or personal injury in the case of rooftop PV installations

Figure 4.65   House roofs in Bavaria, Germany, covered with snow in spring 2006. Some rooftop PV plants incurred snow damage during this year (Photo: Schletter Solar-Montagesysteme GmbH)

injury and/or property damage should be taken for solar generators that are installed on pitched roofs. Such precautions include erecting barriers in the hazard zone after a snowfall and the use of snow collectors.

Figure 4.65 provides an impression of the possible mean snowfall in Central Europe (the figures given here are from the German state of Bavaria in the winter of 2005).

In addition to the risk of snow sliding off roofs, normal force F_N can also induce module and/or rack damage, while descending force F_A can damage the bottom edges of framed modules (see Figure 4.66). Modules can also be displaced by such forces.

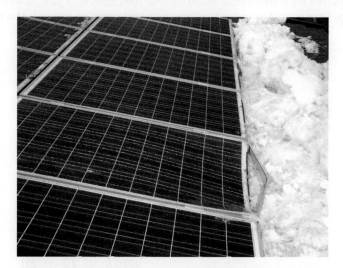

Figure 4.66   Solar module (in Germany) damaged by excessive snow load. One module was bent by an abnormally strong normal force, and in a number of modules the lower edge of the frame was torn off by descending force. This picture was kindly provided by the Schletter Company to illustrate the kind of snow damage that solar modules can incur (the system is not composed of Schletter products) (Photo: Schletter Solar-Montagesysteme GmbH)

Figure 4.67 View of 109 arrays at the Mont Soleil PV system on 8 March 2009. The installation was almost buried under a deep snow covering, some of which had slipped downwards by the time this picture was taken. This snow (see Figure 4.64) increased the snow pressure on the installation, particularly at the very bottom. The storage capacity for the numerous snowfalls during the snowy winter of 2008–2009 was insufficient. Some of the arrays in the bottom row of modules were still buried in snow at the end of March 2009 (see Figure 4.44 for a picture of an array with little snow on it) (Photo: Pierre Berger, Mont Crosin)

These pictures show that the load tests called for by module testing standards [4.11] do not accurately simulate the actual snow loads that occur on pitched roofs, and that snow damage can occur even if a PV system passes such a test. A test involving unequally distributed loads and a test where module edges are subjected to shearing force with constant force per unit of length should also be conducted.

Modules destined for use in snowy areas should integrate sufficiently thick glass, as well as frames that can withstand the force exerted by snow. A sufficiently robust mounting system is also indispensable.

Figure 4.68 A snow-free Mont Soleil installation array in late March 2009. Two bottom-row modules were damaged (snapped off) and one module in the centre was unable to withstand the snow pressure

Figure 4.69 Close-up of a Mont Soleil laminate module that was damaged by snow pressure. Like most of the damaged modules, this one was on the bottom row, where the snow pressure is highest (see Figure 4.64). The laminate is bent in the middle and the two outer elements above the rack bars simply snapped off

Modules whose edges are not robust enough can be reinforced by adding an additional support element to the bottom edge of the solar generator.

Some PV plants incurred damage during the snowy winter of 2008–2009, when, at the 1270 m elevation of the Mont Soleil installation (see Sections 4.5.1.1 and 10.1.2), 35 laminates incurred snow pressure damage that also induced a solar generator grounding fault (see Figures 4.67–4.70). It should also be borne in mind that, snowfall aside, wind can carry deep snow from higher to lower elevations and thus provoke local drifting.

Evidently the glass in these laminates was insufficiently robust for the peak snow pressure that the Mont Soleil installation was subjected to at this tilt angle ($\beta = 50°$), thus resulting in damage during the snowy

Figure 4.70 Close-up of a Mont Soleil laminate module that was damaged by snow pressure that in turn induced a grounding fault (the damaged area is at the lower right). This module was also on the bottom row, where the snow pressure is strongest. Both outer segments of the laminate, as well as its middle portion, were pushed back and snapped off

Figure 4.71 Alpine PV system module damaged by snow load (tilt angle 50°, elevation 1800 m). In this installation, both the lower modules and upper modules were damaged, i.e. apparently the entire solar generator was covered with snow for a time (Photo: Alpha Real AG/IUB)

winter of 2008–2009. Until such time as better snow load tests become available, modules installed in snowy areas should integrate glass that is thick and robust enough to withstand 5400 N/m^2.

Figures 4.71 and 4.72 illustrate the kind of damage that massive snow pressure can cause at still higher mountain sites. These pictures show the damage induced by snow pressure at a ground-based solar generator ($\beta = 50°$, around 1800 m above sea level) that is used for lighting in the Sommeregg Tunnel on the Grimsel Pass and has been in operation since 1987.

For tilt angles of greater than 60° it is safe to assume that snow will slide off the modules, which means that snow load need not be factored into the static sizing for such installations. For installations in high

Figure 4.72 Metal rack element (from the PV system in Figure 4.71) deformed by snow load (Photo: Alpha Real AG/IUB)

mountain areas, upwards of 60° tilt angles should be realized whenever possible so as to avoid the effects of snow pressure and so that snow load can be disregarded for sizing purposes. But it is essential to leave sufficient clearance for the snow to slide off modules, i.e. the modules should be mounted at a sufficient height above the ground since otherwise snow pressure damage could occur. However, at sites exposed to high winds a tilt angle of upwards of 60° does not guarantee that snow will immediately slide off the modules.

Tilt angles ranging from 75° to 90° have been shown to work well at extremely high Alpine locations. For example, as at December 2009 the installation realized in October 1993 on the Jungfraujoch (3454 m above sea level; see Section 10.1.3) had incurred no damage despite its extremely high altitude, owing to the 90° tilt angle of the system modules.

Wind load also needs to be taken into account. Wind forces, which rise four-fold as wind speed increases, are determined by the form, size and arrangement of nearby structures and by the direction of flow. Depending on the solar generator tilt angle and form, the windward side of the installation is exposed to wind pressure while the leeward side is exposed to an equal measure of wind suction, which should particularly be taken into account for roof installations.

Using Bernoulli's law, dynamic pressure can be determined as follows for a specific wind velocity v:

$$\text{Dynamic pressure } q = \tfrac{1}{2} \cdot d_L \cdot v^2 \qquad (4.15)$$

where d_L is air density (which at 0 °C at sea level is 1.29 kg/m^3).

According to the SIA 160 standard, a maximum of $q = 900\,\text{N/m}^2$ can normally be presupposed for dynamic pressure q, which equates to a wind velocity of about 135 km/h at 0 °C at sea level. The q value can be more than twice as high as this at sites that are exposed to strong winds (e.g. in high mountains, in coastal areas, or on very tall buildings), where higher wind velocities are observed. See [DGS05] for the relevant information in this regard concerning Germany.

Wind force can be estimated as follows using the value for dynamic pressure q, solar generator area A_G and a direction of flow coefficient c_W:

$$\text{Wind force } F_W = c_W \cdot q \cdot A_G \qquad (4.16)$$

Depending on the direction of flow and the form of nearby structures, the solar generator c_W ranges from about 0.4 to 1.6. The lowest c_W values apply to factors such as wind suction at roof-integrated solar generators that are either on the leeward side at the centre of a pitched roof or completely in the wind shade of other buildings or generator elements. The highest c_W values occur with ground-based PV installations that are inclined into the wind and that may be exposed to strong lifting forces. In such cases, the wind forces are at a right angle to the solar generator surface. Lifting wind forces are particularly dangerous for solar generators, and counteracting these forces usually entails greater expense than is the case with pressure forces.

The above values allow for a rough estimation of the forces that a solar generator will be exposed to. Hence the sizing of a solar generator mounting system should be mainly based not on gravity but rather on the wind and snow forces that the solar generator will be exposed to. More detailed information concerning solar generator wind load can be found in [Her92] and [DGS05].

4.5.2.2 Suitable Mounting System Materials

Aluminium alloys, galvanized steel, V2A and V4A stainless steel, and concrete (including precast concrete elements) are the materials that are most suitable for solar generator mounting systems, which are also referred to as racking systems, racks and so on. As with standard tiled roofs, the support system for roof-integrated pitched-roof solar generators can be made of wood. Wood is also often used as a mounting system material for ground-based installations with framed modules in rural Southern European areas. For example, a large 5 MWp grid-connected system was recently realized in the Leipziger Land region (Figures 1.15 and 10.46–10.48) using a wooden mounting system. For ground-based installations, however, metal mounting systems are available with extremely large ground bolts that can be screwed right into the ground without a foundation. The main fastening elements for mounting systems are bolts and clips constructed of V2A or V4A stainless steel. It has been found that galvanized bolts often begin rusting after just a few years.

Inasmuch as aluminium alloys are normally corrosion resistant and are also easy to work on following installation without inducing corrosion in the areas where work was performed, they are the ideal material for module frames and mounting systems. Unfortunately, aluminium manufacturing is energy intensive, thus making it unpopular among potential PV system owners, who tend to steer clear of it. But since aluminium rack elements can be readily recycled with only negligible energy expenditure, the grey energy integrated into this material is recouped – which means that the purported 'problem' entailed by the use of aluminium racks is greatly exaggerated. In cases where robust corrosion protection is a key factor for an installation, the use of aluminium alloys is definitively justifiable.

In a temperate climate, galvanized steel with an intact zinc layer is well protected against corrosion for many years, but in most cases, after 10–20 years it begins to show signs of corrosion. However, this is normally only an aesthetic drawback that has no impact on sturdiness. Rainwater in conjunction with module-frame aluminium can accumulate locally and thus provoke contact corrosion. If the water is able to run quickly off such contact points and if these points are well ventilated, the corrosion that is observed at contact points between the aluminium and galvanized steel is of no real importance.

When combined with either aluminium alloys or galvanized steel, joining elements made of stainless steel exhibit no corrosion and are thus ideal for use with both materials. On the other hand, owing to the far thinner zinc layer and the mechanical stress engendered by bolt tightening, which often damages the zinc layer, galvanized nuts and bolts are less well suited for mounting systems.

A more critical problem, and one that should be avoided wherever possible, is direct transitions between aluminium/galvanized steel and copper lightning protection installations. In such cases, anti-corrosion measures such as the following should be realized in the contact area: use of suitable bimetal joining elements, or nickel/tin-plated transition elements that fall within the scope of the electrochemical series; and protecting joints from direct exposure to the elements, wherever possible.

More problematic than corrosion induced by local element formation is the perennial problem of DC installation electrolyte corrosion that occurs at the metal to electrolyte current transition point. Such corrosion can be avoided via proper insulation of all DC components and good protection of all bare module plug-in connectors and junction box elements against moisture. For systems with high DC voltages, in addition to these precautions insulation monitoring measures should also be implemented.

4.5.3   Electrical Integration of Solar Generators

This section discusses the integration of a classic solar generator (see Section 4.3.3) comprising n_{SP} parallel strings composed of n_{MS} modules wired in series. The parallel strings are interconnected using generator junction boxes. For many larger systems, the various arrays are first wired in array junction boxes, whereupon the wires from these boxes are then wired into generator junction boxes.

Special types of PV plants with string or module inverters which greatly reduce the wiring workload are discussed in Section 5.2.

4.5.3.1   Risks Entailed by a Continuous Voltage Load in PV Plants

During the realization of a series connection of PV modules it is essential that the whole string wiring is ungrounded and completely decoupled from any other system component. In the interest of system

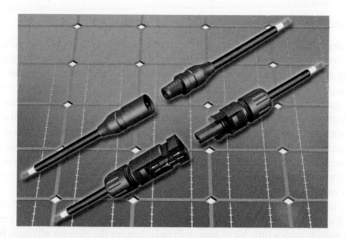

Figure 4.73 Fully insulated, non-confusable special PV connectors and the attendant coupling accessory for PV system string wiring. These devices far exceed protection class II requirements (Photo: Multi-Contact AG)

realization and servicing safety, it is useful if each string can be subdivided into separate substrings, whose maximum open-circuit voltage is 120 V (extra low voltage; such DC voltages are non-hazardous in most cases) and which are only wired in series when the system is commissioned. Moreover, string wiring safety can be substantially enhanced through the use of specially insulated PV plug-in connectors for solar module wiring in lieu of conventional screw connections in module junction boxes (shown in the system in Figure 4.47; also see Figure 4.73; the lower device is latchable and the upper one is not). Many manufacturers sell modules that integrate clearly marked wire connectors that are compatible with system plug-in connectors (see Figure 4.74), in lieu of conventional screw connections. This greatly simplifies the mounting process and reduces its cost.

Although it takes considerable force to open such a plug-in connector, in recent years it has become a requirement to be able to latch these connectors in such a way that they are locked, so as better to avert arcing in the event of the connector opening adventitiously. Such latchable PV plug-in connectors are available from various vendors (see Figure 4.73 (bottom), Figure 4.74 and Figure 4.79).

Figure 4.74 Laminate with special jacks that enable non-electricians to perform wiring in a trouble-free manner. The jacks integrate two connection wires and a special plug-in connector, which is latchable (see Figure 4.73) (Photo: Multi-Contact AG)

4.5.3.2   PV Installation Fire Hazards

When under a voltage load, PV installation wiring creates an electric field to which the following applies: the stronger the voltage, the stronger the field; and the larger the interconductor clearance, the weaker the field. Unduly strong electric fields between two conductors ionize the ambient air, resulting in a disruptive discharge and a superheated arc flash that can cause a fire. In AC systems, arcing current and hence the power generated by it exhibit a zero crossing 100 times per second, which means that the arc is likely to be extinguished. But no such zero crossing occurs in DC systems, and thus the arcing power remains constant. Owing to the power source characteristics of a PV system, an arc flash in such a system can pose a far greater hazard than in a standard AC installation. And as a PV installation's operating current is only marginally weaker than its short-circuit current, fuses are unable to protect such a system against short circuits. Moreover, string fuses cannot reliably avert excessive string reverse current or protect modules and string wiring against such a current.

Short circuits are not the only cause of arcing between conductors (parallel arc flashes). A loose connection at a loose terminal or in a faulty connector can also provoke an arc flash (referred to as a series arc flash), which, with a little bit of luck, will only damage the element affected. However, such arc flashes can also cause a fire in adjacent terminals, or even in the building that the PV system is mounted on.

An arc flash or fire can also occur if a fuse that is not rated for direct current is tripped or if a switch that is not rated for direct current is activated. Devices that are rated solely for 250 V DC are in most cases only compatible with very weak direct current and often fail under relatively low direct currents such as more than 24–48 V. Fuses and circuit breakers should only be removed or opened with the system completely powered down and should never be used to switch direct current under load. The use of devices that are not rated for the relevant system current and voltage is negligent and extremely hazardous.

Worldwide, various fires have been caused by PV system arcing, including in Switzerland (a PV installation fire at a farmhouse and in a Mont Soleil junction box). In principle, hazardous arc flashes can be detected when they first arise and before they can cause a fire. A device that does this was developed at the Bern Technical University PV Lab in collaboration with the Alpha Real Company between 1994 and 1998 and underwent extensive field testing in various PV installations under the auspices of an EU project. The device can detect arcing at a distance of up to 100 m, but owing to a lack of vendor interest the concept was not pursued further at the time. Arc-flash-induced module damage that came to light in the autumn of 2006 [4.17] may well spur renewed interest in this already developed and tested arcing detector, which could potentially be integrated into inverters at low cost with a view to providing additional protection [4.18].

Conditions are far more propitious today than in the past for the use and low-cost realization of arcing detectors. Moreover, interference voltage limit values are now available for the DC side of PV inverters (e.g. EN 61000-6-3), which means that devices where interference is efficiently suppressed now need no additional suppression elements. In addition, virtually all inverters now integrate a controller (microprocessor or microcontroller), which by dint of being under only partial load for the most part can reliably perform specific additional arcing detection tasks via a software add-on, thus under certain circumstances easing the workload of a greater or lesser portion of the detection hardware; sophisticated detection methods are also realizable for these devices. FurtherFurthermore, inverter controllers can precisely detect which devices are activated and deactivated and can thus simply disregard any arcing signal that has been erroneously generated in this phase. Figure 4.75 displays the structure of an arcing detector based on the concept that was developed up until 2008. See [4.18] for a detailed description of this device.

Arcing detectors mainly serve to protect the PV system (element 1 in Figure 4.75) solar generators and DC wiring against hazardous module or wiring arcing and against the fires that arcing can potentially cause in roofs or other adjacent structures. The possible locations where such arcing could occur are displayed in Figure 4.75. An arcing detector can detect arcing not only in a PV system's main DC wiring, but also in its individual strings. Using resonance cycle technology, the arcing

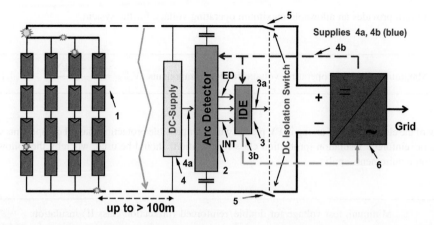

Figure 4.75 Schematic of an arcing detector that allows for reliable remote detection and deactivation of hazardous arcing on the DC side of PV plants. Elements 4 and 5 are dispensed with for integrated arcing detectors [4.18]

detector (2) detects the high-frequency oscillations generated by arcing in the installation. The intelligent detection unit (3) analyses the signals detected by the arcing detector and only generates an output signal in the presence of hazardous arcing on the DC side, and can differentiate between series and parallel arcing. A DC power supply (4) and (to extinguish series arcing) a DC-compatible switch (5) are only needed with a stand-alone arcing detector, which is considerably more expensive. Only elements 2 and 3 are necessary for the integration of an arcing detector into an inverter (6), in which case elements 4 and 5 can be dispensed with. An arcing detector can also detect parallel arcing between the positive and negative poles (relatively long arcing). This type of arcing is extinguished using a short-circuit switch (not shown in Figure 4.75).

The intelligent detection unit (3) can be realized in the system hardware or solar modules via frequent output signal scanning (2) by the inverter controller. Hybrid solutions are also an option, in which case the intelligent detection unit is hardware based to one degree or another, thus allowing for a far lower inverter controller scanning rate, which in turn reduces processor load. Integration of an arcing detector into an inverter obviates the cost of the most expensive components (4) and (5) and allows for a far more refined and reliable arcing detection process. This controller configuration for inverter power electronics in many cases would probably also allow for input-circuit short-circuiting in order to extinguish parallel arcing, and in many cases this function can be implemented at no additional cost.

4.5.3.3 Dielectric Strength and Weather Resistance Requirements for PV System Components

All DC components in a PV installation must be able to withstand a test voltage V_P that is appreciably higher than the installation's maximum open-circuit voltage V_{OCA}. Moreover, modules, junction box housings and outdoor wiring must be resistant to sunlight (UV) as well as rain, snow, ice, high ambient temperatures behind poorly ventilated modules, and so on.

This minimum test voltage is as follows in accordance with the applicable national regulations (e.g. [4.4], [4.11], [4.12]):

$$\text{Minimum test voltage } V_P \text{ for PV components: } V_P = 2 \cdot V_{OCA} + 1\,\text{kV} \qquad (4.17)$$

where V_{OCA} is the maximum open-circuit voltage for the entire PV system.

This in turn provides an allowable maximum operating voltage for the system:

$$\text{Maximum allowable operating voltage for PV components: } V_{OCA} = \frac{V_P - 1\,\text{kV}}{2} \qquad (4.18)$$

In view of the elevated risk of arcing in PV plants, wherever possible protection class II components with double or reinforced insulation (protection class II insulation) should be used, whereby the following insulation requirements apply:

$$\text{Minimum test voltage for double/reinforced (protection class II) insulation:}$$
$$V_P = 4 \cdot V_{OCA} + 2\,\text{kV} \qquad (4.19)$$

A surge voltage test (1.2 μs/50 μs) should also be conducted, as in Table 4.3.

This in turn provides an allowable maximum operating voltage for a system with protection class II insulation:

$$\text{Allowable operating voltage with protection class II insulation: } V_{OCA} = \frac{V_P - 2\,\text{kV}}{4} \qquad (4.20)$$

Although use of protection class II components with protection class II insulation does not reduce the risk of arcing, it does provide better protection against adventitious skin contact with the installation and thus enhances electrical safety.

4.5.3.4 Tips on Which Components to Use, as well as PV Generator Realization

Arcing risk can be greatly reduced and installation safety can be appreciably enhanced if the right components are used and suitable measures are taken during the system planning and installation phases. The following are some tips in this regard.

4.5.3.4.1 Systems with a High Operating Voltage (more than (120 V)

Only protection class II modules and components should be used for such installations, i.e. components that integrate protection class II insulation and whose surge voltage resistance is in accordance with Table 4.3.

Table 4.3 Peak surge test voltage (1.2 μs/50 μs) for PV modules according to [4.12]

System voltage (V)	Minimum requirement (V)	Protection class II requirement (V)
150	1500	2500
300	2500	4000
600	4000	6000
1000	6000	8000

4.5.3.4.2 Cables and Wires

Only double-insulated cables and wires that are UV resistant and rated for high operating temperatures should be used for outdoor wiring or wiring behind modules/laminates. The insulation should be halogen free, self-extinguishing and barely combustible. Inasmuch as solar cell temperature can exceed ambient temperature by up to 50 °C behind poorly ventilated modules or laminates that are mounted directly on thermal insulation, cables and wires with a rated operating temperature of at least 85 °C (but preferably upwards of 100 °C) should be used wherever possible. Wire colour coding (e.g. red for positive, blue for negative, black for all other wires) greatly eases the system realization and servicing process. Special PV cables and wires are available from many vendors.

The string wiring gauge should be at least 2.5 mm^2, although with modules comprising larger cells, 4–6 mm^2 gauges can be used in order to minimize power loss.

String wiring using a product such as double-insulated Huber + Suhner Radox 2.5 mm^2 125 PV wires, which are rated for an operating temperature of 125 °C, has been shown to be highly effective. Up to 16 mm^2 solar wires are also available for PV plants with two or more non-fused parallel strings. Shielded cable made of this same material is available for DC wiring for gauges of up to 2 · 10 mm^2 and 4 · 10 mm^2, as well as common shielding measuring up to 13 mm^2 that allows for optimal PV installation lightning protection (see the examples in Figures 4.76 and 4.77). Shielded individual conductors with bilateral, grounded and double-insulated shielding can also be used in such settings. Wires should always be: (a) connected to terminals or screws using wire pins and cable lugs; and (b) led into housings and module junction boxes and housings from below so as to prevent the wires from bringing in any water.

4.5.3.4.3 Grounding Fault- and Short-Circuit-Proof Wiring

The use of double-insulated cables and wires automatically provides grounding fault- and short-circuit-proof wiring, which should be used in PV installations without fail. If double-insulated cables are unavailable, or if an extra measure of safety is desired, such wiring can be realized by installing positive and negative cables in separate insulating raceways.

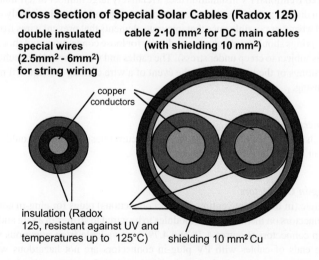

Cross Section of Special Solar Cables (Radox 125)

double insulated special wires (2.5mm^2 - 6mm^2) for string wiring

cable 2·10 mm^2 for DC main cables (with shielding 10 mm^2)

copper conductors

insulation (Radox 125, resistant against UV and temperatures up to 125°C)

shielding 10 mm^2 Cu

Figure 4.76 Cross-section of a Huber + Suhner Radox 125 PV cable with special (double) insulation (right) and a 2.5 mm^2 wire (right; available in red, blue and black) for string wiring. The special (double) insulation is indicated by the two different shades used in the depiction of the insulation. The shielded special cable for main DC leads (right), which provides optimal lightning protection, is available for voltages of up to 900 V for the shielding and up to 1500 V between the wires

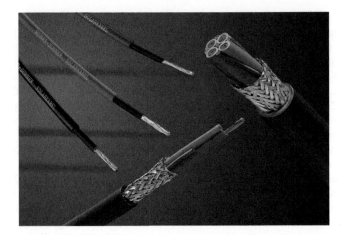

Figure 4.77   Special solar wires and shielded DC cables for optimal PV system lightning protection. Double-insulated cables should be designated as such using two different colours, as is done for solar wires (Photo: Huber+Suhner AG)

However, such raceways are more expensive and electrically less desirable when it comes to protecting a PV installation against lightning strikes in close proximity to the installation. The ideal lightning protection solution is the use of continuous metal raceways containing cables with dual insulation. If separate metal raceways are used for the positive and negative cables, it is essential that the two ends of the raceways be interconnected. This also applies to shielding for double-insulated shielded single conductors.

4.5.3.4.4   Terminals

The DC wiring of a conventional PV installation uses a relatively large number of terminals. In such cases, incorrect wiring can disable an entire string or even cause arcing and a fire. Hence only quality terminals should be used, and wherever possible ones that integrate elements that prevent wires from coming loose. Terminals without springs should be checked periodically for loose connections and should be tightened if necessary (copper is subject to creep under stress). The cables and wires should be mechanically fastened down via suitable fittings or the like so that in the event of a wire coming loose it will not touch another terminal or the housing.

4.5.3.4.5   Disconnect Terminals

Disconnect terminals should only be touched while in a currentless state and should *never* be used to switch direct current, as this could cause arcing.

4.5.3.4.6   PV Plug-in Connectors

Modules that integrate (in lieu of conventional screw connections) plugs for plug-in connectors or short wires with such connectors on them allow PV modules to be wired to strings in a safe and non-confusable manner. PV plug-in connectors (Figures 4.78 and 4.79) are available from numerous vendors.

Inasmuch as the ends of cables with PV plug-in connectors are not hazardous when plugged in, including when live, disconnect terminals are often unnecessary for systems that integrate quality PV plug-in connectors. If necessary, a string can simply be unplugged from the PV generator. Like disconnect terminals, PV plug-in connectors should only be touched with the system powered down and should never be used for switching purposes.

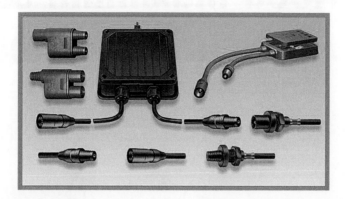

Figure 4.78 PV plug-in connectors sold by Multi-Contact, which was the first manufacturer to develop this type of connector (old, non-latched connector). The elements at the upper left are used to wire two strings to each other in parallel (Photo: Multi-Contact AG)

4.5.3.4.7 String Diodes

Inasmuch as string diodes are relatively vulnerable to power surges, which can occur in a storm even without a direct lightning strike, they are often dispensed with in today's installations (see Sections 4.3 and 4.3.1.1 for further details on the sizing of and need for these devices). Fully protecting each and every string diode against such power surges is extremely expensive and is thus hardly ever done. A certain measure of protection against such surges can be integrated into a PV system through the use of diodes whose inverse voltage is twice that of the system's open-circuit voltage V_{OCA}. As current flows continuously through string diodes, they need to be cooled adequately. Schottky diodes are available for relatively low inverse voltage and thus are normally unsuitable for use as string diodes.

4.5.3.4.8 String Fuses

String fuses protect string wiring and intact modules in a faulty string in systems containing numerous strings wired in parallel (see Sections 4.3.3.1 and 4.3.3.2) andshould be sized for around 1.4 to 2 times the string short-circuit current $I_{SC\text{-}STC}$, and upwards of 1.6 times the string short-circuit current for Alpine installations (see Section 4.3.1.2). For modules with higher reverse current loads, the

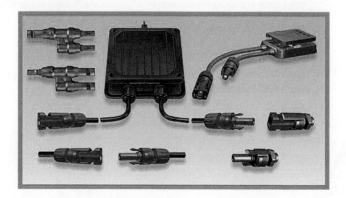

Figure 4.79 Multi-Contact's range of latchable PV plug-in connectors, which were introduced by popular demand (Photo: Multi-Contact AG)

Need maximum safety in your PV system?

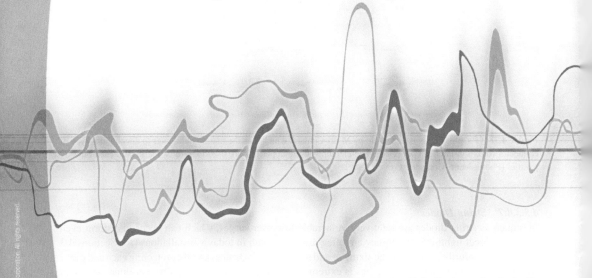

SolarMagic™ Distributed Electronics for Photovoltaic Systems Arc Detection and Safety Shut-off Solutions.

Maximum Reliability

The SolarMagic product portfolio offers a line of dedicated ICs and reference designs for reliable and cost-effective solutions in DC arc detection and PV module emergency shut-off systems. Unsurpassed reliability with Renewable Energy Grade quality, high integration and professional design-in support are setting the standard for the new generation of PV safety systems. The remote shutdown feature can be used during installation, maintenance or in emergency situations to de-energize the PV system. The SolarMagic SM73201-ARC-EV and SM3320-RF-EV are examples of National's reference solutions for arc detection and remote shutdown.

Committed to PV Safety

National Semiconductor's SolarMagic reference designs and custom ICs address the need for improved safety in PV systems. The recently published US-regulation NEC 690.11 requires arc detection and effective countermeasures in new PV installations in the US. Other markets are following with similar regulations. National Semiconductor is a member of the UL1699.11B standards body that defines the testing procedures for new PV safety devices in compliance with the US NEC 690.11 code.

For more information, and to see how you can improve the safety in your PV system, visit: **www.solarmagic.com**

SolarMagic™
by National Semiconductor

Sun protection factor

Professional protection for photovoltaic Systems

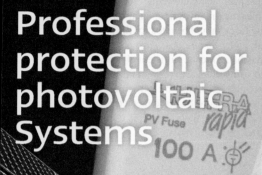

P as in professional product development

We offer product lines specially developed for photovoltaic systems:
trend-setters in cooperation with leading manufacturers of solar energy systems

Fuse Rated Voltage	Fuse Test Voltage	Product Group	Dimensions	Rated Current up to	Part no. e.g.	Approvals
V	V		mm or size	A		
400	400	GZ	6,3 x 32	8	70 065 26	-
600	600	URZ	10 x 38	20	50 225 26	ᴙ pending
1000	1000	URZ	10 x 38	20	50 215 26	ᴙ pending
900	1000	URZ	10 x 38	20	50 215 06	ᴄᴙus
1100	1100	URZ	14 x 65	25	*50 235 26	ᴙ pending
1100	1100	URZ	10/14 x 85	25	*50 238 26	ᴙ pending
1500	1500	URZ	20 x 127	25	90 081 10	-
900/1000	900/1000	URM	NH1	160	20 556 20	-
1000	1100	URM	NH1	200	20 028 20	ᴄᴙus
1000	1100	URM	NH3	400	20 031 20	ᴙ pending

* 25 A = 1000 V

SIBA GmbH & Co. KG
Borker Str. 20-22 • D-44534 Lünen
T.: +49-2306-7001-0 • www.siba.de

Our Protection.
Your benefit.

Sicherungen | Fuses

maximum value can simply be the module reverse current $I_{R\text{-}Mod}$ specified by the vendor, the wiring rating permitting. Only DC fuses or circuit breakers that are specified by the vendor for direct current should be used, and they should be rated for around 1.2 times maximum system open-circuit voltage $V_{OCA\text{-}STC}$ for the solar generator at STC power output. String fuses should be rated for the entire range, which means that gPV or possibly gR fuses should be used; under *no circumstances* should aR fuses be used. To avoid failure induced by heat build-up or cloud enhancements, sufficient clearance should be allowed between individual fuses and between fuses and string diodes, to prevent false trips. Fuses should never be used for direct current, as this could provoke arcing. Although electric cut-outs rather than fuses are a better solution in that they allow individual strings to be disconnected while under load, these devices are rarely used owing to their high cost. Circuit breakers should be used to protect the circuits between arrays and the solar generator in larger installations.

4.5.3.4.9 Main DC Load Switchgear
In order to ensure that the DC side can be shut down while under load, a DC-compatible load switchgear device should always be used whose rated direct current is higher than the maximum rated solar generator short-circuit current and whose rated DC voltage is higher than the system's maximum open-circuit voltage V_{OCA} for the solar generator. String fuses should be changed or circuit breakers opened only when this isolating switch has been opened. This switch is often accommodated in the generator junction box. For larger installations (e.g. those comprising a number of arrays) an additional DC switch should be installed near the charge controller or inverter.

4.5.3.4.10 Overload Protection
As all solar generators are mounted outdoors, they are exposed to lightning-current-induced power surges. To protect the modules and wiring against such surges, the clearance between (a) module areas that are electrified by the module wiring and (b) lighting rods and down-conductors should be as small as possible. Moreover, suitable surge diverters (varistors, if possible either sufficiently oversized or equipped with thermal monitoring devices for ageing induced leakage current increases) should be integrated into the solar generator junction boxes (in relatively close proximity to the modules) between the positive and negative conductor and ground. For further information concerning lightning and power surge protection for PV plants, see Chapter 6.

4.5.3.5 Solar Generator and Array Junction Boxes
Many of the components discussed in the previous section are accommodated in a special junction box near the solar generator and referred to as the generator junction box. The string wiring for large multi-array PV generators is initially realized in array junction boxes that exhibit the same structure as generator junction boxes. The array wires that lead out of the array junction boxes are then wired, in the superordinate generator junction boxes, to the PV generator as a whole (see Section 4.5.6 for more information concerning large PV generators). Figure 4.80 displays the basic structure of a generator junction box, which should be readily accessible to allow a technician to check the fuses and surge diverters from time to time. These boxes protect the solar generator/array components from the elements and for outdoor installations should exhibit a UV-resistant housing that meets protection class IP54 and preferably IP65 requirements. The housing should be constructed of a non-combustible or barely combustible self-extinguishing material.

The various solar generator strings, which are wired in parallel into the generator junction boxes, contain the elements necessary for such wiring, such as string connection terminals, string diodes, string fuses and disconnect terminals, as well as (in many cases) DC circuit isolating switches and varistors for power surge protection. Each generator junction box should also be rendered grounding fault- and short-circuit-proof via double-insulated terminals and positive and negative conductors, which should exhibit adequate reciprocal clearance and may also need to be insulated from each other. If desired, the positive and negative conductors can be accommodated in separate generator junction boxes, although this costs

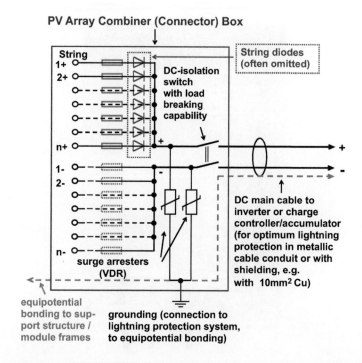

Figure 4.80   Layout of a generator junction box for a PV system for the wiring of $n_{SP} = n$ parallel strings, each of which can be measured separately if desired, and disconnected from the other solar generator devices via a disconnect terminal and fuse. The box also accommodates: (a) a DC circuit isolating switch that allows the system to be powered down before a fuse is changed or a junction box is opened; and (b) the surge diverters for the main positive and negative conductors. For systems with low n_{SP} values such as $n_{SP} \leq 3$, disconnect terminals can be used in lieu of fuses

more and precludes optimal lightning current protection. Metal housings are non-combustible, which is a major advantage in the event of a terminal fire inside the housing. But as these housings are usually grounded for reasons of safety, realization of grounding fault-proof wiring inside the housing is somewhat more difficult.

Once the strings have been wired, each string is in turn wired to the string connection terminals (with the isolating switch open, the disconnect terminals open and all fuses removed) and the open-circuit voltage at each string is then measured.

As a rule, faulty strings are immediately detectable by virtue of their open-circuit voltage reading being considerably lower than the mean value. Fuses should be changed or disconnect terminals closed (followed by closing of the isolating switch) only if an anomaly is detected in one or more strings.

As noted in Section 4.3.3.3, it is absolutely essential to avoid erroneously activating a string whose polarity is reversed. In other words, both string voltage and polarity must be verified.

Faulty fuses can be readily detected by simply measuring their voltage while the installation is running normally, i.e. while a wired inverter is operating at the MPP or if a stand-alone system is drawing current from the solar generator (i.e. when the battery bank is not fully charged). In such a case, a string with a faulty fuse will be in an open-circuit state and will exhibit considerably higher voltage than that measured at the DC isolating switch. Prior to replacement of a faulty fuse, however, the isolating switch must be opened so that the fuse can be replaced in a currentless state. Many of today's generator junction boxes for larger PV plants integrate a continuous monitoring mechanism for all string currents.

Figure 4.81   Older metal generator junction box for wiring in parallel up to 20 strings (only 10 are integrated here) with a high operating voltage (V_{OCA} up to around 800 V). The positive side of the various strings integrates large DC-compatible fuses and string diodes (the bottom fuse is shown with the dedicated fuse removal and replacement tool). The disconnect terminals for the negative wires for disconnection of all poles in the various strings, which is a desirable feature, is missing here. The two varistors can be seen at the lower left and to their right the four-pole system isolating switch. In the centre is the DC isolating switch with a special shielded $2 \cdot 10\,\text{mm}^2$ cable and $10\,\text{mm}^2$ common shielding. The wires are connected via spring-loaded terminals (Photo: Tritec)

Figure 4.81 provides a view inside a generator junction box used for parallel connection of up to 20 strings with high operating voltage (normally around 500 V). The strings integrate string diodes and large DC-compatible string fuses, although the installation around 15 kWp shown here has only 10 strings. In accordance with Swiss regulations that came into effect in the mid 1990s, the isolating switch integrates a $10\,\text{mm}^2$ common shielding that provides good lightning protection (see Section 6.9.4).

Figure 4.82 displays an SMA array junction box with an integrated string monitoring system for larger PV installations with central inverters. This box can accommodate up to 8 or 16 strings, depending on configuration. For each input, the positive and negative conductor integrates a DC-compatible fuse (not shown here for reasons of clarity). The current in the eight connecting leads (each with one or two strings) is monitored using converters from LEM.

The string output signals of the monitoring device (upper right) are received by an RS-485 interface (no connection terminals in the right centre) and are routed to a central monitoring element that immediately detects a string failure. In order for the surge diverters to work, a ground cable (as short as possible but with an adequately large gauge) must be attached to the large grounding terminal at the centre (see Chapter 6 for more on PV system lightning protection). The outbound DC wire cannot be switched, and should instead be wired into the superordinate solar generator junction boxes (preferably to a circuit breaker).

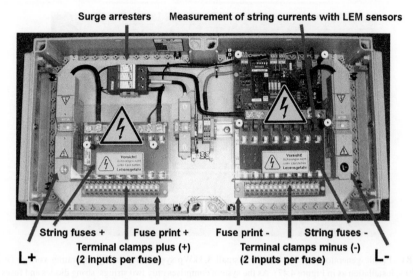

Figure 4.82 SMA array junction box for a maximum of 8/16 strings (for direct wiring in parallel of two strings with $I_{MPP} < 5.6$ A) (Photo: based on an SMA document)

Figure 4.83 displays a Sputnik array junction box for up to 16 strings, each of which integrates the following elements: a fuse and ammeter on the positive terminal side for each string; a disconnect terminal on the negative terminal side for each string; varistors for surge protection; and a common load break switch. A remote string monitoring device can also be integrated into this box if desired.

Figure 4.83 Sputnik MaxConnect Array junction box for up to 16 strings: a, surge diverter (varistors); b, string module with ammeters and string fuses (for 16 strings); c, 6 mm^2 string, positive-pole connection terminals (without string monitoring electronics); d, terminal for potential free message output; e, RJ-45 jacks for connection to the MaxComm monitoring network; f, string monitoring electronics (optional feature); g, strain-relief rail with quick wire installation element; h, 6 mm^2 positive-pole string connection terminals (with string monitoring electronics); i, 35 mm^2 grounding terminal; j, 6 mm^2 negative-pole string connection terminals and disconnect terminal; k, DC load break switch; l, connection terminal for array connection wire (50 or 150 mm^2); m, 6 mm^2 terminal for centre-point transmission (Photo: Sputnik Engineering AG)

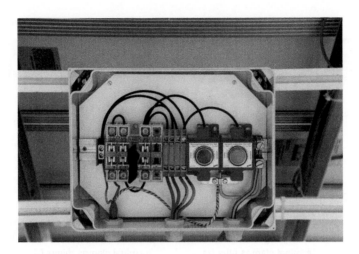

Figure 4.84 Plastic generator junction box for a small 3.3 kWp system with a high operating voltage ($V_{OCA\text{-}STC}$ around 650 V, installation as in Figure 4.47). As the system comprises only two strings, string diodes and fuses can be dispensed with. The two strings are connected (in the centre) solely via two disconnect terminals each. To the right are the thermally monitored surge diverters; the three-pole system isolating switch is on the left

Figure 4.84 displays a generator junction box for a 3.3 kWp system comprising only two strings whose nominal operating voltage is around 500 V. No string fuses are needed here, thus considerably simplifying the installation. The two strings are wired to the positive and negative conductor solely via a disconnect terminal. Here too, the isolating switch integrates a 10 mm^2 common shielding.

In the arrangement shown in Figure 4.84, the string wires are attached to the module frames by conventional white cable ties, which constitutes an egregious defect since such ties become brittle over the years, causing the wires to dangle from the box. The cable ties in the installation shown here will have to be replaced within a few years. Black cable ties are usually more durable. UV and weather-resistant cable ties should always be used to secure wires.

4.5.4 DC Wiring Power Loss

PV plants generate a considerable amount of energy. When a solar generator is operating at the MPP (as is the case, for example, with a grid-connected installation), it often pays to use a wire whose gauge is greater than the minimum required by the applicable standard, so as keep DC wiring loss to an acceptable level. The additional cost is a negligible factor in most such cases.

The ohmic resistance R of a wire with length l and gauge a is calculated as follows using specific resistance ρ of the wire material:

$$R = \rho \cdot \frac{l}{A} \tag{4.21}$$

For copper, ρ at 20 °C is around 0.0175 $\Omega\,mm^2/m$ and at 85 °C is around 0.022 $\Omega\,mm^2/m$. To use the equation above correctly with these ρ values, l must be expressed in m and A in mm^2, and for two-wire cables must be twice as large as the wire length.

To determine the ohmic loss in solar generator wiring, only one equivalent loss resistance R_{DC} should be determined. For n_{SP} strings wired in parallel with one resistance R_{STR} each and a main DC cable with resistance R_H, the ohmic loss is determined as follows:

$$\text{Equivalent DC loss resistance: } R_{DC} = R_H + \frac{R_{STR}}{n_{SP}} \qquad (4.22)$$

In calculating R_{STR} and R_H, both pure line resistance as well as additional resistance stemming from any fuses such as diode fuses and terminals must also be factored in (e.g. 1 mΩ per terminal if terminal resistance is taken into account).

In such a case, the total ohmic loss P_{VR} on the DC side is as follows for current I_{DC}:

$$\text{Ohmic loss } P_{VR} \text{ on the DC side: } P_{VR} = R_{DC} \cdot I_{DC}^2 \qquad (4.23)$$

The diode flow voltage at the string diodes is almost always $V_F \approx 0.8$ V for silicon diodes. Thus the total loss at the string diodes is as follows:

$$\text{DC loss } P_{VD} \text{ at string diodes: } P_{VD} = V_F \cdot I_{DC} \qquad (4.24)$$

$$\text{Total loss } P_V \text{ on the DC side: } P_{VDC} = P_{VR} + P_{VD} \qquad (4.25)$$

The value of interest is usually not absolute but rather relative loss, i.e. system loss that occurs in the presence of nominal direct current I_{DCn} relative to nominal DC loss P_{DCn} incurred by the inverter. If the solar generator is not oversized, I_{DCn} will roughly equate to solar generator current at the MPP and P_{DCn} will roughly equate to effective solar generator output P_{Ao} at STC power output (1 kW/m², 25 °C), which is usually somewhat lower than nominal solar generator output P_{Go} at STC. However, P_{DCn} can also be defined as a value that is somewhat lower than P_{Ao}.

With V_{DCn} as the nominal voltage on the DC side (usually solar generator MPP voltage) $P_{DCn} = V_{DCn} \cdot I_{DCn}$, the following holds true for relative DC-side power loss under nominal voltage conditions:

$$\frac{P_{VDCn}}{P_{DCn}} = \frac{R_{DC} \cdot I_{DCn} + V_F}{V_{DCn}} \qquad (4.26)$$

In the case of stand-alone installations without solar trackers, relative DC power loss of up to around 5% has little impact on energy yield since the voltage of such solar generators is usually somewhat oversized (except in very hot locations). This scenario is clearly shown in Figure 4.39, as it basically makes no difference whether voltage loss occurs at shaded modules or at diodes and resistances. On the other hand, when it comes to installations with solar trackers – that is, primarily grid-connected installations and particularly at sites with PV system feed-in tariffs – efforts should be made to keep relative DC power loss to under 1%, and at the outside under 2%, so as to avoid needless power loss. A key factor in such cases is power loss in lengthy DC wire runs and at string diodes with low nominal voltages V_{DCn}.

If a PV system's insolation and energy yield distribution across the various power stages is a known quantity, these data can be used to estimate annual DC power loss based on wiring power loss, as follows:

$$\text{Relative annual DC power loss } \frac{E_{VDCa}}{E_{DCa}} = \frac{k_{EV} \cdot R_{DC} \cdot I_{DCn} + V_F}{V_{DCn}} \qquad (4.27)$$

In Burgdorf (in the Swiss Mittelland region), with $P_{DCn} = P_{Ao}$ the energy loss coefficient $k_{EV} \approx 0.5$ is obtained.

This value can also be applied to other flat areas in Europe. If $P_{DCn} < P_{Ao}$, or for Southern European or high Alpine installations, a value higher than 0.5 (up to about 0.65) should be used for k_{EV}.

4.5.5 Grounding Problems on the DC Side

4.5.5.1   Grounding of Metal Housings and Metallic Mounting Systems

Metal generator junction boxes and mounting systems, as well as module frames, need to be grounded for safety reasons, except in cases where the insulation for active installation components meets protection class II requirements or if the PV system's open-circuit voltage V_{OCA} is lower than 120 V and the system is not connected to the low-voltage grid. For grid-connected installations with transformerless inverters, framed modules should always be grounded owing to the capacity between the inside of the module and the metal frame. Grounding is advantageous in terms of electrical safety (protection against hazardous touch voltage) and also enhances lightning protection by diverting any lightning current that may occur. However, this only holds true if an equalization bonding conductor is installed parallel to the array cable or the main DC cable. Optimal lightning protection in such cases is obtained using shielded cables (see Sections 6.6.4 and 6.9.4.2). Surge diverters in a small junction box only have the desired effect if they are grounded via the shortest possible path and if they exhibit the shortest possible equalization bonding conductor length to the solar generator's metal racking and module frames.

4.5.5.2   Grounding of Active Solar Generator Components

Grounding active solar generator components is highly problematic and rarely done. Such grounding has both advantages and disadvantages, which should be carefully weighed. The available modalities for this type of grounding are as follows: unipolar (e.g. grounding the negative pole); centre-point grounding; and ungrounded.

Unipolar grounding as shown in Figure 4.85 engenders clearly defined potential conditions across the entire solar generator and eases the task of lightning and surge protection. However, if additional precautionary measures are not taken (namely, implementation of DC fault current monitoring), a solar generator grounding fault will pose an immediate safety hazard (strong grounding fault current I_E, danger to life from touch voltage in the event of a solar generator malfunction).

Unipolar grounding is worthwhile only with extremely low operating voltages or if DC fault current circuit breakers are used. A circuit breaker (DC-FI) that reacts solely to direct current is realizable (a device with differential current sensitivity of $\Delta I = 5$ mA at 30 A operating current), but such devices need an auxiliary power supply and are not commercially available [4.5]. Such circuit breakers shut down the solar generator immediately in the event of a grounding fault and thus provide the same level of protection as classic circuit breakers in AC grids. In the interest of clarity, Figure 4.85 only displays the varistors on the appliance side at the end of the main DC cable and omits the varistors in the generator junction box near the solar generator.

Centre-point grounding (see Figure 4.86) is sometimes used for larger PV installations with high operating voltages (e.g. the Mont Soleil system). In cases of low-ohm grounding, the potential conditions are clearly defined in the manner of unilateral grounding. All solar generator insulation is subjected to only 50% of the system's open-circuit voltage V_{OCA}. And as the surge diverters need only be sized to this value, such installations provide effective surge protection despite their high operating voltage. However, centre-point grounding exhibits the same drawback as unipolar grounding and should only be used in combination with circuit breakers that react solely to direct current and that instantaneously disconnect the relevant generator element in the event of a grounding fault. The generator junction box varistors have likewise been omitted from Figure 4.86. If grounding-fault monitoring is forgone, centre-point grounding can of course be realized with DC-FI directly in the solar generator.

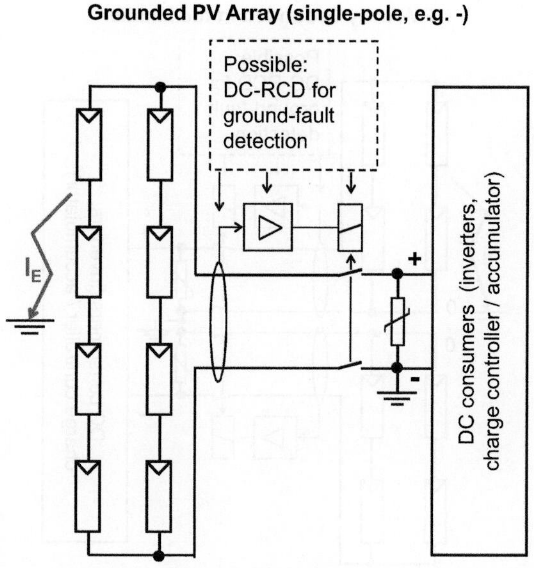

Grounded PV Array (single-pole, e.g. -)

Figure 4.85 Unipolar grounding of a solar generator, with the generator junction box varistors omitted. Here, the voltage conditions in the solar generator as a whole are clearly defined, and little in the way of lightning protection is needed on the appliance side. In the event of a grounding fault in the absence of fault current monitoring, relatively strong grounding-fault current I_E (which can range as high as the solar generator's short-circuit current) is generated, and a person touching a faulty module would be exposed to an immediate hazard if the system were under high voltage. Use of a DC-sensitive circuit breaker (DC-FI) allows for immediate disconnection of the solar generator in the event of a malfunction, thus providing the same level of safety as in FI-protected low voltage grids

In ungrounded solar generators (see Figure 4.87), the system's potential conditions under normal operation are not precisely defined. However, the varistors (usually integrated in such cases) often allow for approximate electrical symmetry, i.e. the positive and negative conductors exhibit roughly the same potential under normal operation. Owing to the very high-ohm ground connection, even low leakage current provokes a substantial and readily detectable symmetry shift. The insulation in an ungrounded solar generator can be monitored continuously using commercially available insulation monitoring devices, which are often integrated into inverters. Moreover, if need be the installation can continue in operation following an initial grounding fault, as such an event cannot result in a life-threatening situation and only becomes critical if a second grounding fault occurs.

A drawback of ungrounded operation, on the other hand, is that in the event of a grounding fault the potential can increase bilaterally to the system's full open-circuit voltage V_{OCA}, which by definition constitutes the dielectric strength of the system components and needs to be taken into account in selecting the surge diverters. Hence surge protection in ungrounded installations is unavoidably poorer and the insulation requirements more stringent than for solar generators with centre-point grounding.

4.5.6 Structure of Larger-Scale Solar Generators

In order to avoid unnecessarily long wiring runs, larger solar generators (as from 10 to 30 strings) are subdivided into arrays (see Figures 4.34 and 4.88) and the strings of these arrays are wired together using array junction boxes (TGAK) whose wires lead to a generator junction box where they are parallel connected to the solar generator as a whole. The main DC cable (DC-HL) leads from this generator

PV Array Grounded in the Middle

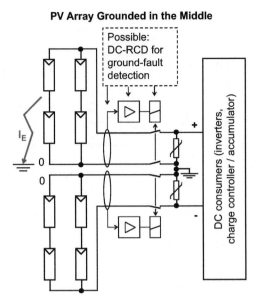

Figure 4.86 Layout of a solar generator with centre-point grounding (generator junction box varistors omitted) that is chiefly (sometimes) used for installations with high operating voltages. The voltage conditions are clearly defined across the entire solar generator, each of whose components is exposed to a maximum of half of the solar generator open-circuit voltage V_{OCA}. In the absence of fault current monitoring, the same problems arise as with a unipolar-grounded solar generator (for reasons of simplicity, the centre point of this generator was grounded directly in the field). Use of a DC-sensitive circuit breaker (DC-FI) on the positive and negative sides allows for instantaneous disconnection of the solar generator in the event of a malfunction

Floating PV Array (not grounded)

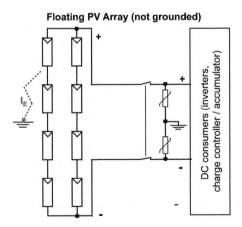

Figure 4.87 Layout of an ungrounded solar generator, with the generator junction box varistors omitted. The wiring here is relatively simple. Under normal operating conditions no component is subjected to more than about half of the solar generator's open-circuit voltage V_{OCA}. However, in the event of a malfunction entailing an unfavourably located grounding fault, full-voltage V_{OCA} may occur that does not allow for instantaneous shutdown – which means that all components including the varistors must be rated for this voltage. A major advantage in this regard is that the grounding-fault current I_E in the event of a grounding fault is very weak, which means that there is no danger to life, limb or health. This type of wiring allows for continuous insulation monitoring

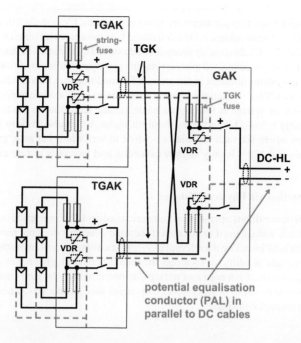

Figure 4.88 Layout of a relatively large solar generator. The array strings are wired in parallel in the array junction box (TGAK). The array wiring (TGK) leading from this junction box is connected to the solar generator via these generator junction boxes (GAK) and for larger generators should always be fused, preferably via circuit breakers (see Figure 4.34). In cases where numerous array conductors are wired in parallel in generator junction boxes, each of these conductors should be protected against reverse current from the solar generator. The ideal lightning protection solution in such cases is the use of metal raceways for bilaterally grounded shielded cables for array wiring and main DC cables (DC-HL) with a shield gauge of $\geq 10\,mm^2$ Cu

junction box to the DC appliance (normally a large grid-connected inverter). To achieve effective lightning protection, it is essential that: (a) thermally monitored varistors are integrated into both the array and generator junction boxes; (b) the equalization bonding conductor (PAL) has a sufficiently large gauge ($\geq 10\,mm^2$ Cu, [4.13]) and is in close proximity to the relevant array wire or main DC cable. This equalization bonding conductor is attached to the module's metal frame or to a metal rack and is grounded to the inverter.

In the interest of clarity, Figure 4.88 displays only the main elements. For more on PV system lightning protection, see Section 6.9.

4.5.7 Safety Protection Against Touch Voltage

In the event that a PV system component is damaged (e.g. by dint of a module having come into contact with broken front glass), electrical safety needs to be restored via suitable measures. Accidents resulting from falls (secondary to a person being frightened by non-life-threatening electric shock) can also occur in connection with roof- or façade-mounted PV installations. The requisite safety protection can be achieved in various ways, as follows.

4.5.7.1 Installation Operation Using Protective Low Voltage (Protection Class III)

Up to 12 V DC does not pose a safety hazard under normal conditions. In cases where a PV installation's open-circuit voltage V_{OCA} is less than 120 V and the installation is not connected to the low-voltage grid

($\sim$230 V), no further safety measures are necessary. However, in actual PV systems it is not always obvious for which conditions open-circuit voltage is to be determined. As most electrical data on vendor datasheets pertain to STC (1 kW/m^2, 25 °C), this readily accessible data should by all means be used. In a 12 V system such as the M55 with five series-connected monocrystalline or polycrystalline modules, 120 V will not be exceeded. Moreover, in systems installed in flat areas, STC conditions are almost never reached.

In the summer, when 1 kW/m^2 insolation can occur from time to time, module temperatures far exceed 25 °C and thus open-circuit voltage is lower. And while module temperatures are lower in the winter, insolation is considerably below 1 kW/m^2 at the usual tilt angles, which also leads to lower open-circuit voltage. Thus in cases where six such modules are wired in series in a flat area, protection class III conditions often obtain and 120 V is only slightly exceeded, and then only in rare cases.

4.5.7.2 Safety Protection via Physical Distance

In cases where a PV system exhibits open-circuit voltage exceeding 120 V or is grid connected, additional precautions against touch voltage need to be taken if the solar generator does not have basic insulation. One way to achieve this is by installing the relevant elements at a distance from each other by, for example, mounting the solar generator at a difficult-to-access site such as behind a high fence or on a normally non-walkable roof (e.g. a pitched roof, but not a roof terrace), or mounting the solar generator at an inaccessible height. It is also advisable to install other critical components such as inverters, junction boxes, battery banks and the like in an enclosed space.

4.5.7.3 Special (Double) Insulation (Protection Class II)

In cases where a solar generator and particularly its modules meet protection class II insulation requirements, in keeping with the current interpretation of the applicable standards it is considered to be unnecessary to implement further electrical safety measures against touch voltage, including for high-voltage and/or grid-connected systems. For a number of years now, solar modules that meet protection class II requirements have been available from numerous vendors.

That said, I find the above take on the applicable standards to be somewhat risky. Inasmuch as solar module glass is relatively thin and fragile and the active devices behind it, which exhibit hazardously elevated potential, are continuously exposed to the elements, these devices are not provided with the same level of long-term protection as for example a protection class II drill with a thick, robust-plastic power cord that is only used indoors, or outdoors in dry weather only. Hence it seems that the widespread use of PV systems, particularly in buildings, calls for an additional electrical safety barrier via elements such as the following:

- Ungrounded solar generators that are electrically isolated from the low-voltage grid (see Figure 4.87).
- Additional fault current monitoring (DC-FI) for grounded solar generators or in installations with transformerless inverters that are wired to the low-voltage grid (integrated into most such inverters nowadays).
- Physical separation as described in Section 4.5.7.2.

4.5.8   Factors that Reduce Solar Generator Power Yield

Both inverter quality and the long-term performance of the DC side of a PV installation are crucial factors for a PV system's ability to deliver high energy yields in a sustained fashion. In order to obtain optimal baseline conditions for the long-term degeneration to which PV installations are unavoidably subject, it is essential that modules with the lowest possible tolerances be used (e.g. minus tolerance 0% or ±3%, instead of the ±10% or even higher values obtained in the past). Low power tolerances also help to reduce solar generator mismatch loss (see Section 4.4).

Figure 4.89 Soiling (from bird droppings) of a framed horizontal Siemens M55 module with a 30° tilt angle. By looking carefully at the bottom edge of the module, it is possible also to see the striations of grit that typically form over time on installations of this type with flat tilt angles

The following are the main long-term problems that have been observed on the DC side:

- Soiling, notably on the edges of framed modules.
- Increasing local shading induced by tree growth.
- Ageing-induced degradation.
- String failure resulting from faulty string diodes, fuses or the like.
- Faults in modules or in generator junction boxes.
- Snow covering, and possible mechanical damage from snow pressure.

4.5.8.1 Partial Shading Induced by Soiling of Individual Solar Cells

Hot-spot formation and reduced solar generator output resulting from partial shading of individual solar cells are very real possibilities in PV systems. Such problems can be induced by bird droppings, fallen leaves, or unkempt natural vegetation. Figure 4.89 displays soiling on the individual cells of a solar generator resulting from bird droppings, which are observed relatively frequently in this installation owing to its having been installed below a high antenna mast and substantially reduce the output of the string affected. As a result of the 30° tilt angle, this type of module soiling is removed almost completely by heavy rain.

With the passing years, moss and lichen also often grow on module frames, substantially reducing the output of the modules affected. Figure 4.90 displays moss growing on a fouled module edge, while Figure 4.91 displays a relatively large expanse of lichen growth on a solar cell.

4.5.8.2 Dirt Accumulation on Module Frames or Laminate Edges

In either framed modules or pitched-roof-integrated laminates with sealing elements that are sized for high annual output and have relatively small tilt angles ($0° < \beta < 35°$), after rainfall a small amount of water containing dirt is always trapped behind the glass and frame of each module. A few centimetres above this area, a layer of dirt that is not completely washed away by rain tends to accumulate.

This phenomenon mainly poses a problem in cases where, to avoid shading from solar generators mounted further in front (e.g. on shed roofs; see Figure 1.11), modules are mounted horizontally, and if there is virtually no clearance between the module frames and the solar cells. This same phenomenon can

Figure 4.90   Moss growing on the underside of the fouled lower edge of a framed Siemens M55 module

also occur with solar roof tiles that exhibit high frames. Although striving to achieve optimal module efficiency by minimizing the spacing between solar cells and module frames may seem like a good idea on paper, in actual installations this practice promotes dirt soiling that reduces output. Even a few millimetres of clearance with tilt angles around 30° substantially improve the situation and also ameliorate module surge and lightning current resistance.

After manual cleaning, the output of such modules improves by several percentage points. In the case of the 60 kWp Bern University of Applied Sciences test installation, which is around 50 m from a heavily used railway line near a railway station and therefore subject to soiling in the form of a rust film, the potential output increase from cleaning was just under 10% after more than four years of operation [4.6], [4.7]. Figure 4.92 displays a cleaned and an uncleaned array at the test installation, along with the lower module edge soiling that accumulated over a five-year period. Figure 4.93 displays a close-up of such soiling in a solar generator with a 30° tilt angle.

After such modules are manually cleaned, dirt gradually re-accumulates. When evaluating the long-term performance of solar generators, the solar generator correction factor $k_G = Y_A/Y_T$ (see Chapter 7) as a function of time should be determined.

For PV installations with relatively shallow tilt angles, it is best to determine the k_G waveform for April to September, when module output is not impaired by either snow or shading – providing no unusual events such as an inverter failure, cleaning or alterations occurred during the period. Inasmuch as temperature is already factored into Y_T, ideally k_G should be as close to 1 as possible and not time dependent. Figure 4.94 displays the solar generator correction factor waveforms for the summers of 1994

Figure 4.91   Large expanse of lichen growth that has developed over the years on the edge of a framed M55 Siemens module. Slight front-contact delamination is also visible here

Figure 4.92   Solar generator for the 60 kWp PV test facility at Bern University of Applied Sciences in Burgdorf ($\beta = 30°$). The module on the left has just been cleaned. Dirt that accumulated between June 1993 and May 1998 is visible at the lower edge of the module on the right. The module surface exhibits relatively little dirt

to 2009 for the Bern University of Applied Sciences PV Lab test installation array that was monitored for the longest period.

Figure 4.94 shows that: (a) the solar generator k_G and thus energy yield decreased slowly at first, but at an increasingly rapid rate after a number of years; and (b) lengthy periods of winter snow cover slowed the decrease in k_G. The installation cleanings realized in 1998, 2002 and 2006 using a strong cleaning agent (Transsolv) successfully reversed most of the measured power loss. Prior to these cleanings, the power loss relative to the installation's baseline state was around 10%, 12% and 8% respectively. When measurements were performed in the summer of 2002, a faulty solar generator

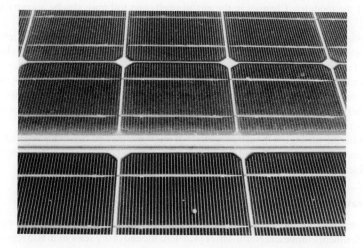

Figure 4.93   Close-up of pollen- and insect- dropping-induced soiling of the solar generator in Figure 4.92 (framed Siemens M55 modules) during a lengthy springtime dry spell. The clearance between the cells and frames is extremely small (around 0 to 3 mm). Most of the soiling occurs just above the lower edge of the module

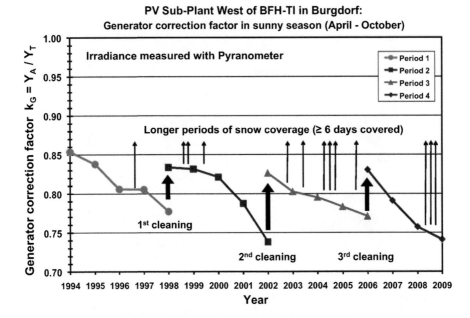

Figure 4.94 Solar generator correction factor k_G for an array at the Bern University of Applied Sciences PV Lab installation, with the installation cleaning dates. The installation has horizontal framed M55 modules with a tilt angle of $30°$, is located near a railway line and tends to become heavily fouled. The low baseline k_G is partly attributable to the elevated DC power loss in this test installation relative to that of a normal PV system

in the west array that was responsible for around 1% of the observed k_G reduction was replaced. Cleaning in 1998 and 2006 eliminated 8% of the power loss in each year, and an additional 10% was eliminated via the 2002 cleaning. Part of the approximately 3% power loss incurred over a 12-year period appears to be irreversible, however, probably on account of (a) slight soiling that occurs soon after cleaning and (b) permanent changes in the glass surface properties at the soiled lower edges of the modules. Changes in the module cells and foil may also be a contributing factor in this regard.

Soiling is less of a problem with: (a) vertically mounted modules as in Figure 2.44; (b) relatively large modules; and (c) $40°$–$50°$ tilt angles insofar as rainwater runoff is relatively robust (and thus has a stronger cleaning effect) and there is sufficient rainfall. At another installation located near a railway station and a heavily used railway line, after around 20 months of operation the installation's vertically mounted modules exhibited a soiling-induced power loss of up to 7% [4.9]. Comparable power losses have been observed at other installations (for which no operating periods are indicated), in most cases ranging from 2 to 6% and in rare cases as high as 18% [4.8].

However, at steeper tilt angles, the cleaning effect of rainwater may be somewhat negated. Figure 4.95 displays the lower edge of a $\beta = 65°$ and $\gamma = -61°$ module whose bottom solar cells exhibit appreciable expanses of deleterious moss and lichen.

4.5.8.3 Overall Installation Output Reductions Induced by Soiling

Power losses ranging up to 15–30% have been observed in horizontally mounted modules during lengthy dry periods in Southern Europe and North Africa [4.20]. At the end of dry seasons in arid regions of Africa, massive dust accumulation has resulted in power losses ranging as high as 80% in certain cases [Wag06]. Periodic solar generator cleaning is a necessity at such sites and should be factored into the installation's upkeep costs.

Figure 4.95 Soiling striations can also occur on modules with high edges and steeper tilt angles. The tilt angle of the module shown here is $\beta = 65°$. Insolation during the winter is somewhat lower here since $\gamma = -61°$

The degree of soiling and the power loss observed are strongly influenced by local conditions. The Bern University of Applied Sciences PV Lab has been monitoring the performance of a large number of PV installations in Burgdorf since 1992. While only minor permanent soiling has been observed at many of these installations after a number of years of operation, some of them have exhibited soiling-induced power reductions ranging up to 10%.

In particularly severe cases, however, elevated soiling-induced power losses of the order of 30% can also occur in PV plants in moderate climates. Over the course of 2005, a substantial decline in specific energy yield was observed in a PV installation that was installed in 1997 on the roof of a Burgdorf fast food restaurant located on a heavily trafficked thoroughfare and a railway line, and is also in close proximity to a sawmill. In October 2005, the PV Lab conducted a more detailed investigation of the installation.

During the first six months of 2005, a tall building adjacent to this installation was renovated, whereupon not only the lower edges but also the entire surface of the installation's framed modules were severely fouled (see Figures 4.96 and 4.97). In this case, the natural cleansing effect of rain was not sufficient to remove this soiling, which was probably induced by construction dust.

Figure 4.96 View of the solar generator ($\beta = 30°$, 3.3 kWp) comprising M55 modules and mounted on the roof of a fast food restaurant in Burgdorf. The front modules were cleaned just before this picture was taken. Unlike the modules shown in Figures 4.92 and 4.93, the modules in this picture exhibit the same level of soiling across their entire surface, which is why the power loss measured for them was far larger

Figure 4.97 Heavily fouled solar cell ($\beta = 30°$) from the PV system in Figure 4.96, with 30% output loss. Apart from the soiling that is also visible on the lower edge in Figure 4.93, the entire remaining surface of the cell exhibits a homogeneous degree of soiling. Taken together, these two instances of soiling result in acute output loss

The installation was cleaned in order to measure the power loss that it had incurred as the result of this soiling. Prior to and following this cleaning, measurements were realized using a characteristic curve measuring device, accompanied by synchronous measurement of irradiance G and the module temperature of the I–G characteristic curve for the installation as a whole and its individual strings; these results were then standardized based on STC (see Figure 4.98). An overall output reduction of around 29% was observed. Inasmuch as all of the installation's strings were functioning normally, this output reduction could only have been attributable to soiling, and not to the failure of specific strings. Another particularly deleterious factor that probably has a negative effect on this solar generator (comprising M55 modules) is that it is exposed to kitchen exhaust from the restaurant, in addition to the ambient emission load.

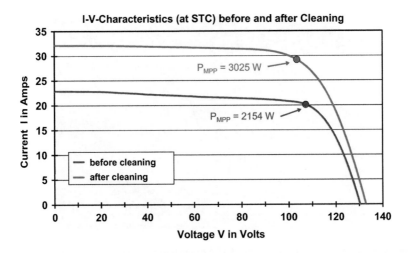

Figure 4.98 I–V characteristic curve standardized at STC for the PV installation in Figure 4.96, before and after cleaning (nominal output 3.3 kWp). The pre-cleaning output $P_{MPP} = 2154$ W was 28% lower than the 3025 W measured after cleaning

Figure 4.99 Trees can induce shading in some PV installations with the passing years. This picture taken on a late November morning displays relatively severe shading of a PV system that was installed in the mid 1990s

4.5.8.4 Shading Induced by Tree Growth

Some PV installations are subject to increasing shading with the passing years as the result of trees (sometimes in neighbouring gardens) that are small and innocuous at first, but grow into large trees that can increasingly shade the installation during the winter. We know that partial solar generator shading reduces energy yield disproportionately (see Section 4.4.1). When the PV system shown in Figure 4.99 was realized, the shadow cast by the fir tree to the right in the picture did not cast a shadow on the installation's modules. Trees planted near a PV system should (a) be located at a sufficient distance from the installation; (b) be a slow-growing species; and (c) lend themselves to cropping without disfigurement if they induce PV installation shading.

PV installations on landscaped flat roofs may also experience shading if the modules are mounted too low and the vegetation grows too high in summer.

Figures 10.2–10.4 display the progressive power loss in another installation resulting from gradually growing trees.

4.5.8.5 Electrical-Safety-Related Anomalies in Solar Modules and Generators

Although solar modules are normally far more reliable than inverters, individual modules may over time exhibit anomalies that are attributable to installation errors or substandard module quality. Figures 4.100–4.103 illustrate this type of anomaly in Siemens M55 installations, most of which had been in operation for well over 10 years when these pictures were taken. These anomalies were observed by PV Lab technicians during an inspection. Thermographs are extremely useful for detecting solar generator anomalies, including incipient problems.

The anomalies illustrated in Figures 4.100 and 4.101 are attributable to installation errors that can be avoided through the use of high-quality special PV plug-in connectors.

Figures 4.102 and 4.103 display a manufacturing-induced module anomaly.

Figures 4.102 and 4.103 show this type of hot-spot anomaly, which damaged the module insulation. These hot spots were induced by substandard contact strip transitions, where resistance increased steadily and the temperature rose so high that the backsheet was damaged and the insulation was compromised.

Anomalies can occur over time not only in individual modules, but also in generator junction boxes. Figures 4.104–4.106 display damage induced by a smouldering fire in a generator junction box at an 11-year-old installation in Burgdorf.

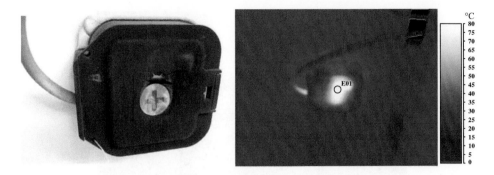

Figure 4.100 Fused connection sockets were observed in some M55 modules, whereby, with the connection socket closed, a temperature of more than 80 °C was measured at point E01 (see thermographic image on the right). The cover is also fused. Figure 4.101 illustrates two possible causes for this type of anomaly

Figure 4.101 Causes of elevated temperatures. Left: loose terminal screw. Right: partially severed connecting cable wire

In this PV installation (15 kWp, $V_{MPP} \approx 500$ V) there is more than a 100 V difference between open-circuit voltage V_{OC} and MPP voltage V_{MPP}. This is more than enough to allow for the development of stable arcing, for which the combustion voltage is 30 V or higher.

An anomaly was also observed in the generator junction box of an older PV installation (3.2 kWp, $V_{MPP} \approx 100$ V) resulting from a fused jumper that connected the two terminal blocks for five strings.

Figure 4.102 Superheated contact strip transition in a module (front views: left, visible area; right, thermographic image). It is clear from these pictures that the contact strip transition resistance was excessive and that superheating occurred here

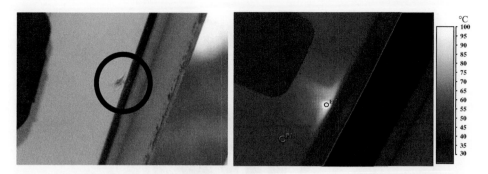

Figure 4.103 Backsheet fusing was also observed, induced by more than 90 °C hot spots. Thermographic images of some older installations reveal numerous such hot spots, which can cause problems over the long term

But no arcing occurred in this case as there was only about a 20 V difference between the open-circuit voltage V_{OC} and the MPP voltage V_{MPP}. In fact, following a brief fusing phase between the two terminal blocks induced by the burned material, the anomalous process came to a halt. Figure 4.107 displays the fused bridge (left) and the thermographic image of the generator junction box in go-live operation. This picture shows that the temperature rose substantially solely on the left, active side of the box, but hardly rose at all on the right, inactive side.

The long-term damage shown here was detected by the PV Lab as part of a long-term monitoring project that began in 1992 and now encompasses more than 60 PV installations. This installation mainly used (since discontinued) M55 modules from Siemens Solar, which was the market leader at the time. Whereas the damage shown in Figures 4.100 and 4.101 would be very unlikely to occur today owing to the predominant use of special PV plug connectors (although substandard or partly plugged-in jacks can overheat), modern modules are prone to other problems, one example being the ageing of internal connections that are on the way to failure in Figures 4.102 and 4.103. Two synchronous defects have been observed in two modules to date, with the result that the current in the module string affected flowed

Figure 4.104 Smouldering fire that broke out in a generator circuit in an 11-year-old 15 kW PV installation in Burgdorf (V_{MPP} around 500 V). An arcing detector would have presumably detected this anomaly at an incipient stage and would have shut down the installation. As a result of long-term monitoring, as well as the realization of standardized energy yield and power output diagrams and tables from the PV Lab, the anomaly was detected before it developed into a more serious problem

Figure 4.105   Close-up of the negative terminals which are seared and fused

through only one bypass diode that became severely overheated and will ultimately fail. Bypass diodes are also prone to faults, particularly if they have been damaged by a power surge induced by a nearby lightning strike (see Section 6.7.7.6). Although solar generators have proven to be more reliable than inverters, the former are nonetheless prone to ageing-induced problems and thus a certain number of spare modules should be kept on hand, particularly for integrated installations.

4.5.8.6   Power Loss Attributable to Solar Module Degradation

Solar modules are prone to other changes as well, which, though non-hazardous, can result in power loss as time goes by. A study of such changes was conducted in the autumn of 2009 under the auspices of the EU's PV Servitor project. Figures 4.108–4.110 show changes in the M55 module, which was frequently used in the 1990s and at the time was still being made by US-based Arco Solar, which merged with

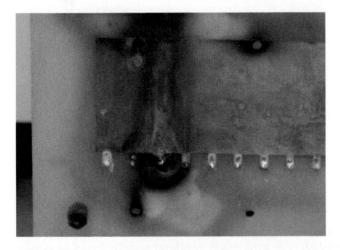

Figure 4.106   This fire apparently originated at the soldering point of string no. 3, which either was not properly soldered or underwent gradual moisture-induced ageing. Heat gradually built up over the years, until finally arcing occurred during a brief power surge secondary to a cloud enhancement event (see Figure 7.19), causing a fire to break out

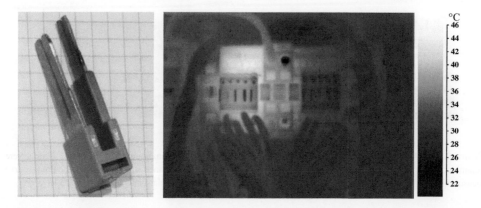

Figure 4.107 Generator junction box damage in a PV installation ($V_{MPP} \approx 100$ V) in the absence of arcing. The damage was halted by a fused bridge on the left side, whereupon installation operation resumed but at only at 50% of normal output (see thermographic image on the right). The damage here was minor as the arcing's potential combustion of 20 V was relatively low

Siemens in 1990. Siemens continued making the M55 for a few years into 2000 and it was used in numerous early PV installations in Europe. Thus relatively extensive long-term experience has been acquired with this device (Figure 4.108 displays a brand-new M55).

Initially marketed as a device with 53 W of nominal output, the M55 manufacturing process was modified in about 1988 and the M55 became a device with 55 W of nominal output. Unfortunately, as a result of this change in the production process, after a few years of operation many of these devices exhibited long-term degradation, which mainly took the form of delamination, notably around the contact strips (see Figure 4.109). Moreover, the M55s rarely exhibited their rated output. After the Siemens merger, beginning in about 1991 the manufacturing process improved and later devices were far less prone to delamination. However, such modules frequently exhibited lesser delamination after around 10 years of use.

Figure 4.108 displays a brand-new M55, Figure 4.109 displays an M55 degraded by delamination around the contact and Figure 4.110 displays a close-up of this delamination.

As Figure 4.111 shows, such delamination is not merely an aesthetic defect (as was initially claimed by the vendor) but also appreciably reduces module output.

The M55 also exhibited another type of long-term degradation, which mainly took the form of clouding that spread inwards from the edges of the cells and that likewise induced power loss, albeit to a lesser degree.

Figure 4.108 Brand-new Siemens M55 module from the small stock of backup devices at the Bern University of Applied Sciences PV Lab (129.7 cm × 32.9 cm, made in around 2000; vendor's then specified nominal output: 55 W ± 10%). Being new, this module has never been exposed to light for very long

Figure 4.109 Arco Solar M55 (made in 1988; vendor's then specified nominal output: 55 W $\pm$ 10%) with severe delamination around the contact strips. The picture shows the module in a cleaned state prior to characteristic curve measurement as in Figure 4.111. The device was operated at the MPP in a solar generator for around 18 years

Looking back at these events, it is amazing that the products of the then market leader exhibited these types of problems. Hence caution is in order when production processes are modified, and the potential impact of such changes on long-term device stability should be taken into account.

4.5.8.7   Hail Damage in Solar Modules

Among the various tests called for in [4.11] is a hail test, for which hailstones 25 mm in diameter and with an impact speed of 23 m/s are normally used. Unfortunately, these parameters are not compatible in all cases with the conditions that obtain in severe hailstorms. A severe hailstorm on 23 July 2009 in the Swiss Mittelland region damaged some buildings, and hailstones measuring up to 5 cm in diameter were observed. Figures 4.114 and 4.115 display the damage that this storm inflicted on two solar modules. Although PV modules are indisputably more hail resistant than solar collectors used for heat reclamation, in areas with frequent hail, modules with greater hail resistance should be used such as devices that have been tested with hailstones that are 35 mm in diameter.

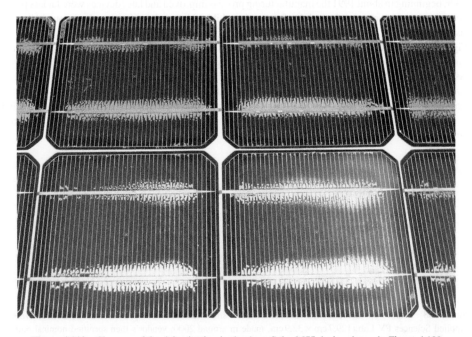

Figure 4.110 Close-up of the delamination in the Arco Solar M55 device shown in Figure 4.109

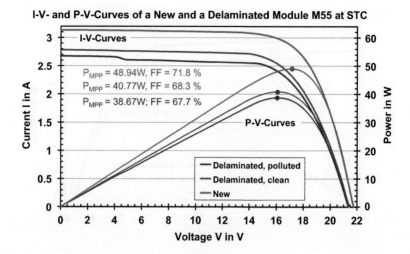

Figure 4.111 *I–V* and *P–V* characteristic curves for a brand-new M55 solar module and for a delaminated M55. The new device stems from a small stock of spare modules, while the delaminated module is the delaminated device shown in Figures 4.109 and 4.110 in a fouled state (after not having been cleaned for a number of years) and a cleaned state. This module was used for around 18 years in a solar generator operated at the MPP

4.5.8.8 Snow Covering on Solar Generators

Solar generators with relatively shallow tilt angles are often covered with snow for varying periods as the result of heavy winter snowfalls. Even a few centimetres of snow are sufficient to prevent virtually any light from arriving at the solar module surface, which means that the solar generator will hardly produce any power. The situation is even worse in cases where wet snow falls on a cold solar generator and then freezes. Figure 4.116 displays a solar generator ($\beta = 30°$) after such a snowfall involving a blanket of around 25 cm of freshly fallen snow. In such cases, relatively deep snow will not slide off modules until the temperature goes above freezing and a thaw begins, thus enabling a film of water to develop under the snow so that it can slide off. If the layer of snow has holes in it for one reason or another, the exposed portions of the solar generator warm up very quickly, thus allowing the remaining snow to melt and slide off. Light, powdery snow at low temperatures is less of a problem as it is less adhesive and is readily blown away by the wind.

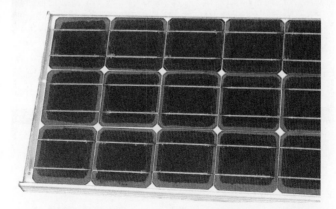

Figure 4.112 A cleaned M55 module with solar cell clouding that extended from the edges of the cells inwards and that likewise induced output loss (see Figure 4.113)

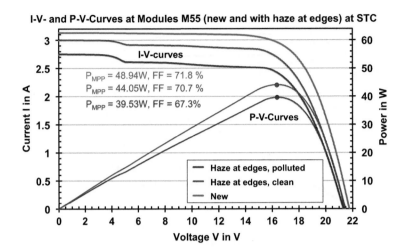

Figure 4.113 *I–V* and *P–V* characteristic curves for a brand-new M55 solar module and for a delaminated M55 with edge clouding. The new device stems from a small stock of spare modules, while the clouded module is the device shown in Figure 4.112 in a fouled state (after not having been cleaned for a number of years) and a cleaned state. This module was likewise used for around 18 years in a solar generator operated at the MPP. The power loss exhibited by this module in a fouled state amounted to more than 10%

Figure 4.114 Hail damage incurred by a solar generator (top left module) © Urs Muntwyler (Solarcenter Muntwyler, Zollikofen)

Figure 4.115 Hail damage incurred by a solar module. The deformation of the lower panel is indicative of the force of the hail. © Urs Muntwyler (Solarcenter Muntwyler, Zollikofen)

Figure 4.116 Snow-covered solar generator at Bern University of Applied Sciences in winter ($\beta = 30°$, snow depth around 25 cm). The leading edges of the horizontally mounted modules (see Figures 4.92 and 1.11) act as snow traps. In most such cases the snow does not slide off until the temperature reaches 0° and a film of water forms under the snow. At this tilt angle, even when a thaw sets in and the Sun shines, this kind of snow remains on the solar generator for quite some time, i.e. until the first bare patches begin appearing that allow the Sun to warm the generator to a greater extent

Shallow tilt angles, framed modules, solar tile edges and leading laminate edges prevent snow from sliding off because the leading edges act as mini snow traps. Lengthy 'snowed-in' periods for solar generators can be avoided using the same measures as those discussed above for reducing soiling in the vicinity of the lower module edges.

Large amounts of snow result in power loss while snow remains on the solar generator. However, it was also observed during the Bern University of Applied Sciences' long-term solar generator monitoring projects that snow remaining in place for lengthy periods counteracts module soiling and that the snow sliding off has a de-soiling effect to some degree.

The Mont Soleil installation shown in Figure 4.44 (1270 m above sea level) enables snow to slide off the solar generator owing to the installation's steep tilt angle ($\beta = 50°$) and glued-down laminates without leading edges; however, this only holds true if enough clearance remains in front of the solar generators for the snow to slide off – which is not always the case during snowy winters (see Section 4.5.2.1). IIn most cases the snow slides off after being exposed to direct sunlight for a few hours. To avoid blankets of snow on solar generators in high mountain areas, the solar modules should be mounted vertically (as in Figure 2.44) with 75°–90° tilt angles.

Several maps showing the duration of winter snow covering in various regions of the world can be found in Section 8.1.3.

4.6 Examples

Problem 1 (solar cell temperature determination for a solar module)

Determine the following for a solar module whose nominal cell temperature is 48 °C and that operates at an ambient temperature of $T_U = 40$ °C at open circuit:

(a) Cell temperature T_Z with insolation at 1 kW/m^2
(b) Cell temperature T_Z with insolation at 200 kW/m^2

Solution: Use Equation 4.1, which yields the following results:

(a) $T_Z = 40\,°C + (48 - 20\,°C) \cdot (1000/800 \text{ W/m}^2) = 75\,°C$
(b) $T_Z = 40\,°C + (48 - 20\,°C) \cdot (200/800 \text{ W/m}^2) = 47\,°C$

Problem 2 (energy yield of a partly shaded string in a stand-alone PV installation)

The solar generator of a stand-alone PV installation uses M55 solar modules with characteristic curves as shown in Figures 4.12 and 4.13. The generator is exposed to 1 kW/m^2 insolation in the AM1.5 spectrum and its cell temperature is $55\,°C$. Of the n_{MS} modules wired in series per string, n_{MSB} are completely shaded (0 W/m^2) and $(n_{MS} - n_{MSB})$ are unshaded. The string is connected to a battery. Each module is bypassed by a bypass diode, each of which exhibits ideal barrier behaviour and a constant voltage drop of $V_F \approx 750 \text{ mV}$ in the forward direction (this represents an idealization). Two operating states with battery voltage V_{A1} (battery almost completely discharged) and two with V_{A2} (battery almost completely charged) will be investigated, as follows:

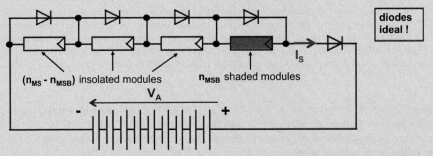

(a) A 48 V stand-alone installation without partial shading: $n_{MS} = 4$, $n_{MSB} = 0$, $V_{A1} = 48$ V, $V_{A2} = 56$ V. Determine I_{S1} for V_{A1} and I_{S2} for V_{A2}.
(b) A 48 V stand-alone installation with partial shading: $n_{MS} = 4$, $n_{MSB} = 1$, $V_{A1} = 48$ V, $V_{A2} = 56$ V. Determine I_{S1} for V_{A1} and I_{S2} for V_{A2}.
(c) A 120 V stand-alone installation with partial shading: $n_{MS} = 10$, $n_{MSB} = 1$, $V_{A1} = 120$ V, $V_{A2} = 140$ V. Determine I_{S1} for V_{A1} and I_{S2} for V_{A2}.
(d) A 120 V stand-alone installation with partial shading: $n_{MS} = 10$, $n_{MSB} = 2$, $V_{A1} = 120$ V, $V_{A2} = 140$ V. Determine I_{S1} for V_{A1} and I_{S2} for V_{A2}.

Solution: The voltage at an insolated module is as follows: $V_M = [V_A + (n_{MSB} + 1)V_F]/(n_{MS} - n_{MSB})$. Use the relevant I_S value from Figure 4.13.

(a) $V_{A1} = 48\,V \Rightarrow V_{M1} = 12.2\,V \Rightarrow I_{S1} \approx 3.4\,A$, $V_{A2} = 56\,V \Rightarrow V_{M2} = 14.2\,V \Rightarrow I_{S2} \approx 3.3\,A$
(b) $V_{A1} = 48\,V \Rightarrow V_{M1} = 16.5\,V \Rightarrow I_{S1} \approx 2.65\,A$, $V_{A2} = 56\,V \Rightarrow V_{M2} = 19.2\,V \Rightarrow I_{S2} \approx 0\,A$
(c) $V_{A1} = 120\,V \Rightarrow V_{M1} = 13.5\,V \Rightarrow I_{S1} \approx 3.35\,A$, $V_{A2} = 140\,V \Rightarrow V_{M2} = 15.7\,V \Rightarrow I_{S2} \approx 3.0\,A$
(d) $V_{A1} = 120\,V \Rightarrow V_{M1} = 15.3\,V \Rightarrow I_{S1} \approx 3.1\,A$, $V_{A2} = 140\,V \Rightarrow V_{M2} = 17.8\,V \Rightarrow I_{S2} \approx 1.5\,A$

Problem 3 (calculation for a solar generator under low-DC-voltage conditions)

A solar generator is composed of $n_{SP} = 12$ strings with $n_{MS} = 6$ M55 series-connected modules each (characteristic curves as in Figures 4.12 and 4.13). The cross-sectional area of the string wiring is $A_{STR} = 2.5 \text{ mm}^2$ Cu and the gauge of the main-cable conductors is $A_H = 10 \text{ mm}^2$ Cu. No string diodes are used, thus $\rho_{Cu}=0.02\,\Omega\,\text{mm}^2/\text{m}$ (at around 50 °C). Nominal DC output P_{DCn} of the connected device (grid-connected inverter) is 4 kW.

Determine the following:

(a) Solar generator open-circuit voltage V_{OCA} at 1 kW/m² and $T_Z=10$, 25 and 55 °C.
(b) Short-circuit current I_{SCA} for the entire solar generator at STC power output.
(c) Maximum possible return current I_R for an intact solar module in a string under short-circuit conditions across four modules at STC power output and in the absence of string fuses.
(d) Whether or not to use string fuses, and, if so, which value and type should be used.
(e) Solar generator P_{MPPA}, V_{MPPA} and I_{MPPA} values at 1 kW/m² and $T_Z=25$ and 55 °C under ideal conditions (i.e. identical modules, which means no module mismatch, no ohmic loss, no soiling and so on).
(f) Relative DC loss at STC power output with the system at the MPP, string wire length $l_{STR}=10$ m and two-string main-cable length $l_H=20$ m. Additional resistance per string: 50 mΩ for the fuses and terminals; 10 mΩ for the main cable.
(g) Using the data from task (f), determine the approximate relative DC-side annual energy loss for this system at a flat site in Central Europe; also determine whether your suppositions are on the optimistic or the pessimistic side.

Solution:

(a) $V_{OCA}=n_{MS}\cdot V_{OC}$, whereby the following is determined based on Figure 4.13: for $T_Z=10$ °C, $V_{OCA}=138.6$ V; for $T_Z=25$ °C, $V_{OCA}=130.2$ V; for $T_Z=55$ °C, $V_{OCA}=114.6$ V.
(b) $I_{SCA}=n_{SP}\cdot I_{SC\text{-}STC}=40.8$ A ($I_{SC\text{-}STC}$ as in Figures 4.12 and 4.13).
(c) $I_R=(n_{SP}-1)\cdot I_{SC\text{-}STC}=37.4$ A.
(d) Yes, $n_{SP}>3$; options: 5 A or 6 A DC-compatible fuses.
(e) $T_Z=\mathbf{25\,°C}$: Module: $V_{MPP}=17.45$ V, $I_{MPP}=3.15$ A, $P_{MPP}=55$ W. Installation: $V_{MPPA}=n_{MS}\cdot V_{MPP}=104.7$ V, $I_{MPPA}=n_{SP}\cdot I_{MPP}=37.8$ A, $P_{MPPA}=n_{MS}\cdot n_{SP}\cdot P_{MPP}=3.96$ kW.
   $T_Z=\mathbf{55\,°C}$: Module: $V_{MPP}=15.1$ V, $I_{MPP}=3.15$ A, $P_{max}=47.6$ W. Installation: $V_{MPPA}=n_{MS}\cdot V_{MPP}=90.6$ V, $I_{MPPA}=n_{SP}\cdot I_{MPP}=37.8$ A, $P_{MPPA}=n_{MS}\cdot n_{SP}\cdot P_{MPP}=3.42$ kW.
(f) With Equation 4.21, $R_{STR}=130$ mΩ, $R_H=90$ mΩ; with Equation 4.22, $R_{DC}=100.8$ mΩ, $I_{DCn}=I_{MPPA}=37.8$ A, $R_{DC}\cdot I_{DCn}=R_{DC}\cdot I_{MPPA\text{-}STC}=3.81$ V, $\Rightarrow$ with Equation 4.26, $P_{VDC}/P_{DCn}=(R_{DC}\cdot I_{DCn})/V_{DCn}=(R_{DC}\cdot I_{MPPA\text{-}STC})/V_{MPPA\text{-}STC}=3.64\%$.
(g) For flat-site installations in Central Europe with $P_{DCn}\approx P_{Ao}$, $k_{EV}\approx 0.5$; with Equation 4.27, $E_{VDCa}/E_{DCa}\approx (k_{EV}\cdot R_{DC}\cdot I_{DCn})/V_{DCn}=1.82\%$ (overly optimistic as the module temperature generally exceeds 25 °C).

Problem 4 (calculations for a solar generator under high-DC-voltage conditions)

The solar generator in the installation in Figures 4.47 and 4.84 is composed of $n_{SP}=2$ strings with $n_{MS}=30$ M55 modules wired in series, $P_{DCn}\approx P_{Ao}$. String wiring: cross-sectional area $A_{STR}=2.5$ mm² Cu, length $l_{STR}=20$ m, total of 64 terminals with 2 mΩ per string. Cable (two-wire): gauge $A_H=2.5$ mm² Cu, length $l_H=20$ m, resistance for terminals and main switch 20 mΩ. Specific copper wire resistance: $\rho_{Cu}=0.02\,\Omega\,\text{mm}^2/\text{m}$ (at around 50 °C). As $n_{SP}=2$, string fuses are dispensed with.

Determine the following:

(a) Solar generator open-circuit voltage V_{OCA} at STC power output (1 kW/m², AM1.5, $T_Z=25$ °C).

(b) Solar generator P_{MPPA}, V_{MPPA} and I_{MPPA} values at STC power output under ideal conditions (identical modules, which means no mismatch, very little ohmic loss, no soiling and so on).
(c) Relative DC loss at STC power output with the system operating at the MPP.
(d) Using the data from task (c), determine approximate relative DC-side annual energy loss for this system at a flat site.

Solution:

(a) $V_{OCA} = n_{MS} \cdot V_{OC} = 651$ V.
(b) Module: $V_{MPP} = 17.45$ V, $I_{MPP} = 3.15$ A, $P_{MPP} = 55$ W. Installation: $V_{MPPA} = n_{MS} \cdot V_{MPP} = 523.5$ V, $I_{MPPA} = n_{SP} \cdot I_{MPP} = 6.3$ A, $P_{MPPA} = n_{MS} \cdot n_{SP} \cdot P_{MPP} = 3.3$ kW.
(c) With Equation 4.21, $R_{STR} = 288$ mΩ, $R_H = 340$ mΩ; with Equation 4.22, $R_{DC} = 484$ mΩ, $V_{DCn} = V_{MPPA} = 523.5$ V, $I_{DCn} = I_{MPPA} = 6.3$ A, $R_{DC} \cdot I_{DCn} = 3.05$ V ⇒ with Equation 4.26, $P_{VDC}/P_{DCn} = (R_{DC} \cdot I_{DCn})/V_{DCn} = 0.58\%$.
(d) At a flat Central European site with $P_{DCn} \approx P_{Ao}$, $k_{EV} = 0.5$ ⇒ with Equation 4.27, $E_{VDCd}/E_{DCa} \approx (k_{EV} \cdot R_{DC} \cdot I_{DCn})/V_{DCn} = 0.29\%$.
A comparison of Problems 3 and 4 shows that wiring loss is appreciably lower with higher system voltage.

Problem 5 (fuse sizing for a large solar generator)

A solar generator with a total $n_M = 1000$ modules (SW225 poly; see the data in Table 4.1) is composed of $n_{TG} = 10$ arrays with $n_{SP\text{-}TG} = 5$ parallel-connected strings each.

Determine the following:

(a) Open-circuit voltage $V_{OCA\text{-}STC}$ at STC power output.
(b) Lowest possible nominal current I_{SN} for a suitable commercial string fuse ($k_{SN} = 1.4$).
(c) Lowest possible $I_{SN\text{-}TG}$ for a suitable circuit breaker for this array wire ($k_{SN} = 1.4$).
(d) Possible long-term reverse current I_R in a string with no string fuses ($k_f = 1.45$).
(e) How the calculated values would change if the system were installed in a high Alpine location and thus $k_{SN} = 1.7$ ($k_f = 1.45$) were used in order to be on the safe side.

Solution:

(a) $n_{MS} = n_M/(n_{TG} \cdot n_{SP\text{-}TG}) = 1000/50 = 20 \Rightarrow V_{OCA\text{-}STC} = n_{MS} \cdot V_{OC\text{-}STC} = 20 \cdot 36.6$ V $= 732$ V.
      Note that, as open-circuit voltage can be somewhat higher at extremely low temperatures, the maximum allowable inverter input voltage V_{DCmax} at a flat site should be around 1.15 times as high, i.e. more than 842 V(also see Section 5.2.6.8).
(b) With Equation 4.6 and $k_{SN} = 1.4 \Rightarrow I_{SN} \geq 1.4 \cdot 8.17$ A $= 11.4$ A ⇒ with Table 4.2: use $I_{SN} = 12$ A.
(c) With Equation 4.10 and $k_{SN} = 1.4 \Rightarrow I_{SN\text{-}TG} \geq k_{SN} \cdot n_{SP\text{-}TG} \cdot I_{SC\text{-}STC} = 1.4 \cdot 5 \cdot 8.17$ A $= 57.2$ A ⇒ next standard value $I_{SN\text{-}TG} = 63$ A.
(d) Using Equation 4.1: $I_R \approx k_f \cdot I_{SN\text{-}TG} + (n_{SP\text{-}TG} - 1) \cdot I_{SC\text{-}STC} = 1.45 \cdot 63$ A $+ 4 \cdot 8.17$ A $\Rightarrow I_R \approx 91.3$ A $+ 32.7$ A $= 124.0$ A.
(e) With Equation 4.6 and $k_{SN} = 1.7 \Rightarrow I_{SN} \geq 1.7 \cdot 8.17$ A $= 13.9$ A with Table 4.2: use $I_{SN} = 16$ A.
      With Equation 4.10 and $k_{SN} = 1.7 \Rightarrow I_{SN\text{-}TG} \geq k_{SN} \cdot n_{SP\text{-}TG} \cdot I_{SC\text{-}STC} = 1.7 \cdot 5.8 \cdot 17$ A $= 69.4$ A ⇒ next standard value $I_{SN\text{-}TG} = 100$ A (or if available of course, 80 A as well).
      Using Equation 4.1: $I_R \approx k_f \cdot I_{SN\text{-}TG} + (n_{SP\text{-}TG} - 1) \cdot I_{SC\text{-}STC} = 145$ A $+ 32.7$ A $= 177.7$ A, with $I_{SN\text{-}TG} = 80$ A a maximum of 1.45 · 80 A $+ 32.7$ A $= 148.7$ A.
      The fuses or circuit breakers used for array wiring should exhibit the finest possible gradations so as to ensure that these devices can be adapted to the specific situation to the greatest extent possible.

4.7 Bibliography

[4.1] Ch. Isenschmid: 'Neue DC-Verkabelungstechniken bei PV-Anlagen'. Diplomarbeit Ingenieurschule Burgdorf (ISB), 1995 (internal report).

[4.2] R. Hotopp: 'Verzicht auf Rückstromdioden in Photovoltaik-Anlagen'. *etz*, Band 114 (1993), Heft 23–24, pp. 1450ff.

[4.3] R. Minder: 'Engineering Handbuch für grosse Photovoltaik-Anlagen'. Schlussbericht PSEL-Projekt No. 23, March 1996.

[4.4] Eidgenössisches Starkstrominspektorat: 'Provisorische Sicherheitsvorschrift für photovoltaische Energieerzeugungsanlagen', Ausgabe Juni 90, STI No. 233.0690 d. Available from SEV, CH-8320 Fehraltorf (out of print).

[4.5] H. Häberlin: 'Das neue 60kWp-Photovoltaik-Testzentrum der Ingenieurschule Burgdorf'. *Bulletin SEV/VSE*, 22/1994, pp. 55ff.

[4.6] H. Häberlin, J. Graf: 'Gradual Reduction of PV Generator Yield due to Pollution'. Proceedings of the 2nd World Conference on Photovoltaic Energy Conversion, Vienna, 1998.

[4.7] H. Häberlin, Ch. Renken: 'Allmähliche Reduktion des Energieertrags von Photovoltaikanlagen durch permanente Verschmutzung und Degradation'. *Bulletin SEV/VSE*, 10/1999.

[4.8] H. Becker, W. Vassen, W. Herrmann: 'Reduced Output of Solar Generators due to Pollution'. Proceedings of the 14th EU PV Conference, Barcelona, 1997.

[4.9] M. Keller: 'Netzgekoppelte Solarzellenanlage in Giubiasco'. *Bulletin SEV/VSE*, 4/1995.

[4.10] Norm IEC 60364-7-712:2002: *Electrical installation of buildings – Part 7-712 Special installations or locations – Solar photovoltaic (PV) power supply systems.*

[4.11] IEC 61215:2005: *Crystalline silicon terrestrial photovoltaic modules – Design qualification and type approval.*

[4.12] IEC 61730-2:2005: *Photovoltaic module safety qualification – Part 2 Requirements for testing.*

[4.13] STI 233.1104d (2004): *Solar-Photovoltaik (PV) Stromversorgungssysteme*, Schweizerische Adaptation der IEC 60364-7-712. Available from Schweizerischen Starkstrominspektorat, CH-8320 Fehraltorf.

[4.14] H. Laukamp, M. Danner, K. Bücher: 'Sperrkennlinien von Solarzellen und ihr Einfluss auf Hot-Spots'. 14. Symposium PV-Solarenergie, Staffelstein, 1999.

[4.15] H. Laukamp, K. Bücher, S. Gajewski, A. Kresse, A. Zastrow: 'Grenzbelastbarkeit von PV-Modulen'. 14. Symposium PV-Solarenergie, Staffelstein, 1999.

[4.16] H. Gabler, F. Klotz, H. Mohring: 'Ertragspotenzial nachgeführter Photovoltaik in Europa – Anspruch und Wirklichkeit'. 20. Symposium PV-Solarenergie, Staffelstein, 2005.

[4.17] A. Schlumberger, A. Kreutzmann: 'Brennendes Problem – Schadhafte BP-Module können Feuer entfachen'. *Photon*, 8/2006, pp. 104–106.

[4.18] H. Häberlin, M. Real: 'Lichtbogendetektor zur Ferndetektion von gefährlichen Lichtbögen auf der DC-Seite von PV-Anlagen'. 22. Symposium PV-Solarenergie, Staffelstein, 2007.

[4.19] P. Kremer: 'Sind Sicherungen in PV-Anlagen unsicher?'. 24. Symposium PV-Solarenergie, Staffelstein, 2009.

[4.20] M. Ibrahim, B. Zinsser *et al.*: 'Advanced PV test park in Egypt for investigating performance of different module and cell technologies'. 24. Symposium PV-Solarenergie, Staffelstein, 2009.

Additional References

[DGS05], [Häb07], [Hag02], [Hum93], [Wag06].

5

PV Energy Systems

Inasmuch as insolation varies over time, in most cases a specific load cannot be directly allocated to PV electricity, which must instead be processed or stored in a suitable form. Direct use of PV electricity via stand-alone installations without storage systems is conceivable: for example, for facilities such as irrigation systems (see Figure 1.6), fan operation (e.g. to cool down cars during the summer) or in settings where it is of lesser importance for electricity to be available at all times. Figure 5.1 displays a breakdown of various types of PV systems according to their basic structure. Figure 5.2 displays the typical power ranges of today's PV systems.

A stand-alone system is an electricity installation that supplies one or more appliances with energy completely independently of a power grid. The possible output of such a system ranges from milliwatts to watts for power supplies for portable devices (e.g. watches, calculators, small radio devices and so on) to tens of kilowatts (for transport infrastructure facilities, large buildings, radio transmitters and so on). Appliances that depend on a steady supply of electricity need an accumulator, usually a battery or battery bank – although such components appreciably drive up electricity costs, in most cases by around €0.66 per kWh.

In the interest of avoiding electricity storage costs, relatively large PV systems with outputs of 1 kW or more are hooked up to the electricity grid. In areas with readily available grids such as Central Europe and the USA, an accumulator can be dispensed with in most cases. During periods when grid-connected PV systems produce more energy than needed, the surplus energy is fed into the grid and vice versa, i.e. energy is drawn from the grid at times when the PV system produces too little energy, namely in inclement weather during daylight hours, and after dark. Hence the electricity grid is in a sense the storage 'device' for a grid-connected PV system.

In the sections that follow, in view of the overriding practical importance of stand-alone installations with storage systems, as well as grid-connected installations without storage systems, these two types of PV installations will be discussed.

5.1 Stand-alone PV Systems

Since in most settings a continuous supply of electricity is needed, PV electricity needs to be stored for use during periods when there is little or no Sun, i.e. during inclement weather or at night. Nowadays, mainly storage batteries (mostly lead batteries, and nickel–cadmium batteries to a lesser extent) are used, although other storage methods are available. For example, short-term storage can be realized using large-capacitance supercapacitors or very high-RPM flywheel gears constructed of extremely robust materials. Long-term storage of PV energy can be realized using storable hydrogen produced via electrolysis that can be converted back into electricity using fuel cells.

Figure 5.3 displays the layout of a stand-alone PV system that integrates a battery bank for energy storage purposes. The *I–V* characteristic curves of solar generators (1) are very suitable for battery

Photovoltaics: System Design and Practice. Heinrich Häberlin.
© 2012 John Wiley & Sons, Ltd. Published 2012 by John Wiley & Sons, Ltd.

Photovoltaic plants
Stand-alone plants
Grid-connected plants
With storage
Without storage
With storage
Without storage

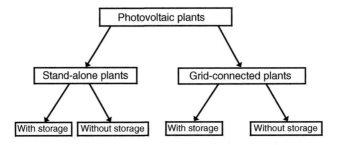

Figure 5.1 Breakdown of PV system types according to structure

charging owing to their power source characteristics. The charge controller (2) keeps the battery from being overcharged, since otherwise the aqueous electrolyte (diluted sulphuric acid) would break down into hydrogen and oxygen, resulting in battery gassing. This process (which forms a detonating gas): (a) to some extent creates a safety hazard, which can be kept under control with good ventilation; and (b) reduces the battery's life span owing to water loss, which means that the distilled water in the battery needs to replenished manually from time to time. With relatively large installations, in lieu of a charge controller,

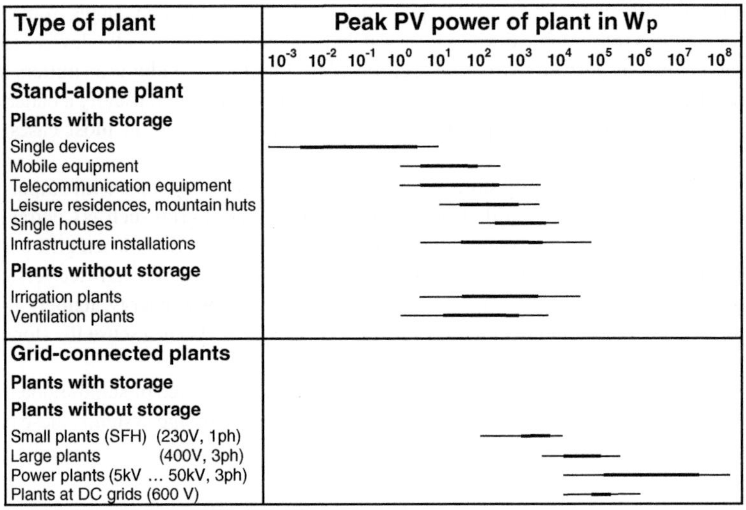

Type of plant
Peak PV power of plant in Wp
10^-3 10^-2 10^-1 10^0 10^1 10^2 10^3 10^4 10^5 10^6 10^7 10^8
Stand-alone plant
Plants with storage
Single devices
Mobile equipment
Telecommunication equipment
Leisure residences, mountain huts
Single houses
Infrastructure installations
Plants without storage
Irrigation plants
Ventilation plants
Grid-connected plants
Plants with storage
Plants without storage
Small plants (SFH) (230V, 1ph)
Large plants (400V, 3ph)
Power plants (5kV ... 50kV, 3ph)
Plants at DC grids (600 V)

Figure 5.2 Typical PV system power ranges

H
8
1
2
3
4
5
6
7a
7b
7c

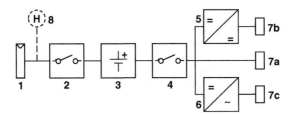

Figure 5.3 Layout of a stand-alone PV system with a battery bank (the energy flow is from left to right): 1, solar generator; 2, charge controller/solar tracker; 3, battery bank; 4, discharge controller/deep-discharge protection device; 5, DC converter; 6, inverter; 7, consumer; 8, auxiliary generator, e.g. a diesel or wind generator in a hybrid system

a device called a solar tracker can be used, which ensures that the solar generator is operated at its maximum power point (MPP) at all times and prevents overcharging.

The actual energy storage process occurs in the battery bank (3), whose nominal voltage in a stand-alone system is the same as the installation's system voltage. The other possible system voltages are 12 V for small installations, and 24 V, 48 V or even higher for larger installations, with the highest voltages being used in installations in the kilowatt range. System voltage determines the operating voltage of directly connected DC appliances (7a). The 12 V systems, which are used for auto accessories, camping equipment and portable devices, provide the highest operating voltages, followed by 24 V systems, which provide considerably less, and by upwards of 24 V systems, which provide virtually none. Apart from system voltage, battery capacity (storable electrical charge as expressed in ampere-hours (Ah)) is also an important factor as it is the key determinant of battery life during inclement weather.

Inasmuch as battery life span is affected by both overcharging and deep discharging, the discharge controller or deep-discharge protection device (4) shuts down the connected appliances if a (possibly current-dependent) minimum battery voltage is undercut. In large installations, the connected appliances can be broken down into groups that are assigned various priorities. Less important groups are shut down first so that electricity is available for longer for critical appliances.

It is often necessary to hook up appliances using different operating voltages (7b), particularly in cases where system voltages exceeding 12 V come into play. DC converters (5) carry out the requisite voltage conversion, which usually takes the form of a voltage reduction. As many appliances run on 230 V AC (7c), a stand-alone PV system must integrate an inverter (6). These devices simplify the appliances-side installation process, but they also add to the system cost and lower overall installation efficiency and exhibit a certain amount of open-circuit loss. Hence inverters should only be activated when an AC appliance is in use. For more on the use of inverters in stand-alone PV systems, see Section 5.1.3.

AC devices that use DC internally (mainly consumer electronics products) can often be converted into DC devices via a minor modification, which often greatly reduces inverter energy consumption by eliminating transformer open-circuit loss. In such cases PV DC is more useful than AC.

In small PV installations without a solar tracker, a charge controller (2) and discharge controller (4) are often combined into a single device known as a charge controller, system controller or battery controller. Such devices are available in various sizes and for various system voltages, feature connectors for the solar generator, battery and appliances, and in some cases integrate a DC distribution board. Figure 5.4 displays a holiday-home stand-alone PV system that integrates a charge controller. Inverters with integrated charge controllers and deep-discharge protection devices are now available that allow for the simple realization of stand-alone systems for 230 V AC after connecting a battery and a solar generator.

5.1.1 PV System Batteries

Apart from the fact that batteries store energy for use at night or during inclement weather, they are also a key cost factor for stand-alone PV systems. Table 5.1 lists the key technical characteristics of PV installation batteries.

Batteries account for around 10 to 40% of the capital investment and as much as 30 to 60% of the operating costs of a PV installation, owing to the need to replace the batteries periodically. Unlike solar cells, battery costs are unlikely to decrease appreciably at any time in the foreseeable future. Lead batteries that are suitable for PV systems now cost (depending on quality) anywhere from €130 to €530 per kWh (storable energy calculated using K_{100}).

5.1.1.1 Use of Nickel–Cadmium Batteries in PV Systems

Nickel–cadmium (NiCd) batteries are chiefly used in PV installations in special cases (applications calling for an extremely long life span, for miniature devices, for situations where absolute gas-tightness is needed or at very low temperatures) as they are around three to five times more expensive than lead batteries and exhibit the same energy storage capacity. Moreover, many NiCd batteries (particularly small gas-tight devices with sintered electrodes) exhibit a memory effect if not fully discharged, as often

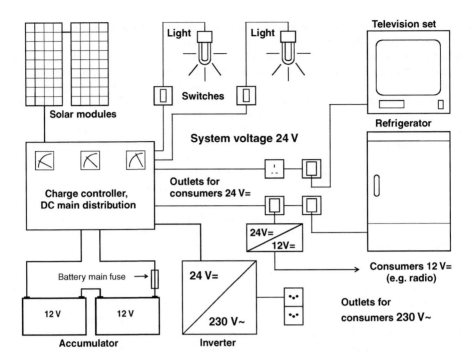

Figure 5.4 Stand-alone PV system for a holiday home

happens with stand-alone PV systems; as a result, their charging capacity diminishes if they are recharged before being completely discharged. This memory effect is extremely disadvantageous for stand-alone PV systems. Hence in cases where NiCd batteries are used, only those that integrate pouch-plate or earth electrodes should be integrated, as the memory effect is much less severe in such batteries. In the sections

Table 5.1 Key technical characteristics of PV installation batteries. A fully charged lead battery with an aqueous electrolyte can be operated at around −20 °C. In cases where a full charge cannot be guaranteed at low temperatures (e.g. for stand-alone PV systems that are operated all year round and that may be operated in a partially charged state in winter), the minimum operating temperature should be around 5 °C (and no lower) to prevent the battery electrolyte from freezing

Battery type	Cell voltage (V)		Energy density (referred to K_{20})		Operating temperature	Cycle lifetime (typical)	Energy efficiency η(Wh)	Relative cost
	V_{nom}	V_{gas}	Wh/kg	Wh/l	°C	(DOD 60%)	%	
Lead acid (vented, liquid electrolyte)	2	≈ 2.4	15–45	30–90	min. −5 (−20) max. 55 opt. 10–20	250–2000	70–85	1
Lead acid (sealed, gel–electrolyte)	2	≈ 2.35	15–35	25–90	min. −20 (−30) max. 45 (50) opt. 10–20	400–1600	70–85	1–2
NiCd	1.2	≈ 1.55	15–45	40–90	min. −40 (−50) max. 50 (60) opt. 0–30	up to 5000	60–75	3–5

here, only stand-alone PV systems that use lead batteries as energy storage devices will be discussed in detail as these are by far the most practical batteries for such systems.

5.1.1.2 Key Electrical Characteristics of Lead Batteries

Figure 5.5 displays the structure of a lead battery, the reactions that occur at the two electrodes during the charging and discharging process, as well as the overall reaction.

When a lead battery is discharged, lead oxide (PbO_2) is converted into lead sulphate ($PbSO_4$) at the positive electrode and absorbs electrons from the electric circuit, while at the negative electrode metallic lead (Pb) is likewise converted into lead sulphate ($PbSO_4$) and emits electrons. At the same time, the sulphate ions that are responsible for this conversion process are removed from the electrolyte (diluted sulphuric acid, H_2SO_4), whereby additional water is formed secondary to the decrease in electrolyte acid concentration. Inasmuch as sulphuric acid prevents the electrolyte from freezing at low temperatures, at such temperatures a partially or fully discharged battery is far more prone to damage than a fully charged battery. The battery should not be discharged too deeply as this can result in battery plate damage (secondary to sulphation and the formation of large-grained $PbSO_4$ crystals, particularly in a deeply discharged state) that cannot be fully reversed by charging and that reduces the battery's active mass and thus its capacity. Hence a deep-discharge protection device should be integrated that avoids undercutting of the allowable final discharging voltage and shuts down the load if necessary.

The chemical reactions above occur in reverse during the charging process, such that lead sulphate is again converted into lead oxide at the positive electrode and metallic lead at the negative electrode, and the electrolyte acid concentration reverts to its previous level. Once the battery is fully charged, oxygen gas (O_2) forms at the positive electrode and hydrogen gas (H_2) forms at the negative electrode. As this mixture (known as detonating gas) exhibits around 4 to 75% explosiveness, the charging voltage should be limited and the battery bank room should be adequately ventilated so as to keep the H_2 concentration sufficiently low and avoid an explosion [Köt94], [5.5]. As batteries containing aqueous electrolyte (sealed batteries) are operated in close proximity to the end-of-charge voltage, a small amount of gas unavoidably forms and water is gradually lost, which means that the battery electrolyte needs to be topped up with

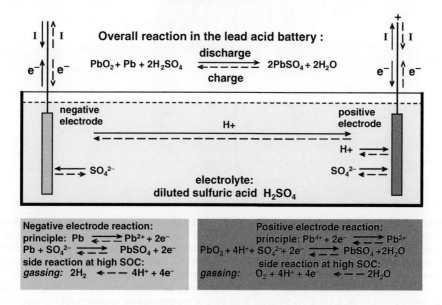

Figure 5.5   Structure and functional principle of a lead battery

distilled water once a year or so. Continuous overcharging expedites water loss, promotes electrode corrosion and reduces battery life span.

In the case of sealed gel-electrolyte batteries (Figure 5.6), which are portable to some extent and are leak-proof, normally no gas is formed during the charging process and thus there is no water loss. However, the maximum end-of-charge voltage for such batteries must be strictly observed, since even a slight overcharge will ramp up pressure inside the bypass diodes, cause the surplus gas to bubble out of the safety valve, and thus cause irreversible water loss that can appreciably shorten the battery's life span.

Figures 5.7–5.16 display the typical characteristics of a lead battery that is well suited for a stand-alone PV system and whose capacity is $K_{10} = 100$ Ah (e.g. OPzS battery featuring a special lead alloy containing Se–1.6% Sb). The technical data for the aforementioned illustrations stem from Varta, Exide, BAE, Hoppecke and SWISSsolar documentation, as well as from [5.1]–[5.5], [Köt94], [Sch88] and [Lad86].

Needless to say, higher-capacity batteries are used in practice. As rounded-off capacity figures have been used in the illustrations, it is very easy to determine the main characteristics of larger batteries as

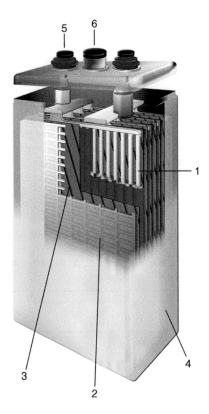

Figure 5.6 Configuration of a sealed lead–gel cell (OPzV). The gel (not shown) contains the electrolyte H_2SO_4 in a bonded form: 1, positive plates: lead–calcium alloy liner plates that are particularly robust owing to the use of stabilizing tubes and are optimized for corrosion resistance; 2, negative plates: grid plates made of a lead–calcium alloy; 3, separator: microporous and robust, the separator allows for electrical insulation of the plates and is optimized for low internal resistance; 4, casing: made of SAN (Styrol–Acrylnitril), also available in flame-retardant ABS (Acrylnitril–Butadiene–Styrene); 5, pole: screw connection for simple and secure mounting, maintenance-free joints and outstanding conductivity; 6, valve: briefly opens in the event of overpressure and keeps atmospheric oxygen out of the battery. (Illustration: Exide Technologies)

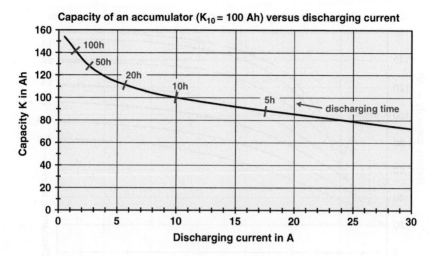

Figure 5.7   Usable capacity C of a lead battery with nominal capacity $C_{10} = 100$ Ah as a function of discharge current at 20 °C

well, whereby the following principle applies: current and charge are proportional to each other, and internal resistance is inversely proportional to battery capacity.

5.1.1.3   Key Battery Parameters and Characteristics

5.1.1.3.1   Voltage

The nominal voltage of a lead battery cell is $V_{cell} = 2$ V. Higher voltages can be obtained by series connecting various cells. The most common nominal battery voltage is 12 V, followed by 6 or 4 V. Most large batteries are composed of 2 V cells. Series connecting cells or batteries has no effect on their capacity.

Nominal voltage of a battery with n_Z series-connected cells: $V_{battery} = n_Z \cdot V_{cell}$ (5.1)

Battery voltages apart from nominal voltage are also important and are discussed separately.

5.1.1.3.2   Capacity (C)

Battery capacity C, which means a battery's charge storage capacity Q in ampere-hours (Ah), is determined by discharge time, discharge current (see Figure 5.7) and operating temperature (see Figure 5.8). Hence battery capacity and discharge current are often indicated in conjunction with a subscript for discharge time in hours. Hence C_{10} means battery capacity C for a discharge time of 10 h, while I_{10} means the volume of discharge current that flows during this period. The higher the discharge current, the lower the battery capacity. C_{10} is normally used to designate a battery's nominal capacity.

These considerations are crucially important for comparisons of various manufacturers' battery capacity data, which should always be compared head to head for the *same* discharge period. As Figure 5.7 shows, battery capacity C_{100} is considerably higher than capacity C_{20}, and the latter is in turn higher than capacity C_{10} for the same battery. Hence manufacturers of stand-alone PV system batteries tend to indicate C_{20} or C_{100} capacity in their datasheets, although the achievable cycle number is also a key factor for sizing such systems. Moreover, the manufacturer's tests in this regard are normally based on nominal capacity C_{10} or even on 60% of this value (in accordance with IEC 896-1 and 896-2). The relevant curves are displayed in Figure 5.12. Deep discharging to C_{100} should occur in exceptional cases only, as it shortens battery life span, and many manufacturers (including those that indicate C_{100}, C_{120} and even C_{240} capacity) expressly advise against it.

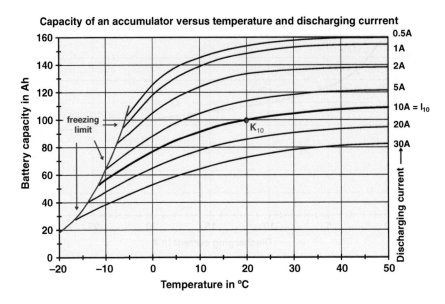

Figure 5.8 Usable capacity of a lead battery (nominal capacity $K_{10} = 100$ Ah) as a function of cell temperature, using discharge current as the value of merit. Usable capacity decreases markedly at less than around $-5\,°C$, because the battery freezes when in a low-charge state. As temperature increases, usable capacity also increases but so does self-discharge (see Figure 5.15). Life span also decreases at high temperatures by virtue of the far quicker rate of internal corrosion (see Figure 5.13). Hence PV installation batteries are best operated at $10\text{--}20\,°C$, which is the usual temperature in the basement of a building

If C_{20} and C_{100} are the only known values, the following approximate result is obtained for C_{10}:

$$\text{Approximate equation for nominal capacity } C_{10}\text{: } C_{10} \approx 0.85 \cdot C_{20} \approx 0.7 \cdot C_{100} \qquad (5.2)$$

5.1.1.3.3 Final Discharging Voltage

The final discharging voltage is the voltage that is not to be exceeded in order to avoid shortening the battery's life span. It is determined by discharge current (see Figure 5.9) and is between 1.7 and 1.85 V per cell at ambient temperature. The final discharging voltage decreases at higher voltages since internal battery resistance exhibits a greater voltage decrease. In order to ensure that the final discharging voltage cannot be undercut, the discharge controller or deep-discharge protection device in stand-alone PV systems shuts down the connected appliances if need be.

5.1.1.3.4 Charge Limiting Voltage During Cycle Operation

In order to prevent battery electrolytes from forming a detonating gas secondary to the breakdown of hydrogen and oxygen, the charge limiting voltage should not be exceeded during the battery charging process. So-called gassing voltage, which is induced by elevated gas formation, is around 2.4 V per cell at ambient temperature. In most cases, the recommended charge limiting voltage for stand-alone PV systems during cycle operation at ambient temperature falls within the range of gassing voltage. As Figures 5.10 and 5.11 show, charge limiting voltage is temperature dependent and increases as temperature falls. Many manufacturers specify an allowable constant charge limiting voltage in roughly the $10\text{--}15\,°C$ and $30\text{--}35\,°C$ temperature ranges (see Figure 5.11).

In a stand-alone PV system, the charge controller limits the charging current once the charge limiting voltage has been reached in the battery, so as to ensure that this voltage cannot be exceeded. Following

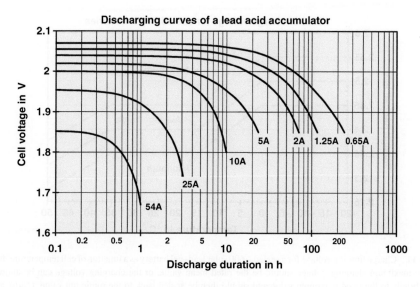

Figure 5.9 Final discharging voltage curve for a lead battery: cell voltage as a function of time for various discharge currents for a battery with $C_{10} = 100$ Ah at $20\,°C$

deep discharge of aqueous electrolyte batteries, the formation of corrosion-inducing acid stratification secondary to elevated acid concentrations at the bottom of the battery can be avoided by briefly charging the battery using a somewhat higher charging voltage, so as to allow the ensuing brief gassing period to better homogenize the electrolyte, thus eliminating the acid stratification. Modern charge regulators register the discharge depth of the previous cycle and automatically launch the brief gassing period following a deep-discharge event. However, such gassing periods provoke permanent battery water loss in most cases, and thus only charge regulators that do not enable such gassing periods or that allow for their

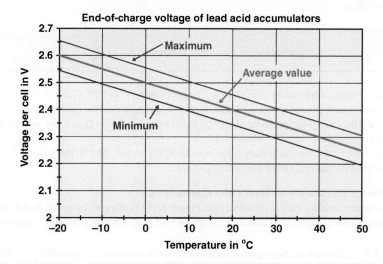

Figure 5.10 Characteristic charge limiting voltage for a sealed lead battery as a function of cell temperature during PV installation cycle operation. Following deep discharge of such batteries, it is advisable to charge them with a somewhat higher voltage for a time in order to avoid acid stratification

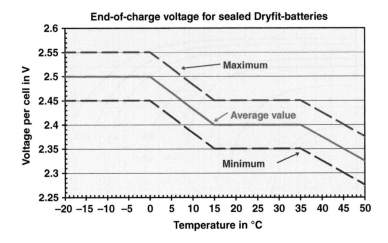

Figure 5.11 Charge limiting voltage for a sealed A600 solar lead–gel battery as a function of cell temperature. In such cases, the maximum charging voltage should be the rated mean value, or the charging voltage can be allowed to increase briefly to the rated maximum value and should then be scaled back to the minimum value. (*Data source*: Exide Technologies)

deactivation should be used. This also means that, for sealed batteries, the manufacturer's recommended charging voltage should be strictly adhered to.

5.1.1.3.5 *Ampere-Hour Efficiency η_{Ah}/Coulomb Efficiency/Charging Factor*

A minute portion of a battery's charge (particularly in non-gas-tight batteries) is used not to power the battery, but rather for secondary chemical processes that do not promote energy storage but are necessary for gas formation and related processes. Under deep-discharge conditions (and thus relatively low charging voltage), virtually all of the charging voltage is used for charge storage, whereas in the case of gassing voltage a portion of the current is used for gas formation purposes. The ratio of the available charge Q_E during discharge to the charge Q_L needed for battery charging is known as ampere-hour or Coulomb efficiency:

$$\text{Ampere-hour or Coulomb efficiency: } \eta_{Ah} = \frac{Q_E}{Q_L} \qquad (5.3)$$

η_{Ah}, which is determined by battery charge state, ranges from 80% to 98% (Table 5.2).

$\eta_{Ah} = 90\%$ is a mean figure of merit (somewhat higher for gel batteries) for flashover calculations. The charging factor is the ratio of the incoming charge Q_L to the available charge Q_E, i.e. the reciprocal value of η_{Ah}. The charging factor is typically around 1.11.

If a battery is operated in the gas voltage range over a lengthy period, Ah efficiency η_{Ah} decreases since the current is used for gassing rather than energy storage.

5.1.1.3.6 *Energy Efficiency/Watt Hour (Wh) Efficiency η_{Wh}*

Inasmuch as battery voltage during charging (see Figure 5.16) is always somewhat higher than mean discharge voltage (see Figure 5.9), a battery's energy or watt hour efficiency is always somewhat lower

Table 5.2 η_{Ah} for various charging states (in percentage of nominal capacity) according to [DGS05]

State of charge (SOC)	90%	75%	50%
η_{Ah}	>85%	>90%	>95%

than its Ah efficiency, and for lead batteries is usually between 70 and 85%. $\eta_{Wh} = 80\%$ is a mean figure of merit for flashover calculations.

5.1.1.3.7   Depth of Discharge t_Z

Depth of discharge is the ratio of discharged battery capacity Q_E to nominal battery capacity C_{10} and thus indicates the depth of battery discharge during a cycle relative to nominal capacity C_{10}:

$$\text{Depth of discharge } t_Z = \frac{Q_E}{K_{10}} \tag{5.4}$$

In the interest of maximizing battery life span, in some batteries the allowable depth of discharge is limited (e.g. for certain lead–calcium batteries it is limited to $t_Z \leq 50\%$).

5.1.1.3.8   Cycle-Based Life Span n_Z

Batteries in stand-alone PV systems are usually cyclical charging and discharging devices, respectively, during the day and at night/during periods of inclement weather. Cycle-based life span means the number of cycles that a battery can carry out before its capacity reaches 80% of nominal capacity. Cycle-based life span is chiefly determined by battery type and depth of discharge t_Z. Figure 5.12 displays the cycle-based life span for selected batteries as a function of depth of discharge, based on manufacturers' data. A battery's life span in a PV installation can be estimated based on (a) cycle-based life span for the installation's anticipated depth of discharge and (b) cycle frequency in the installation (e.g. one cycle daily).

In the interest of keeping cycle-based life span calculation time within reasonable bounds, discharging during such tests is realized using relatively high current (e.g. $2 \cdot I_{10}$) and the batteries are fully recharged immediately after being discharged. However, such a procedure is a typical for actual stand-alone PV systems (whose mean discharge currents are usually considerably lower in winter and inclement weather), and cycling sometimes occurs in the presence of lower battery charge states without full charging in the interim. However, inasmuch as battery operation for lengthy periods in a low-charge state promotes sulphation [5.1],[5.2], the attainable cycle counts for stand-alone PV systems may be somewhat lower.

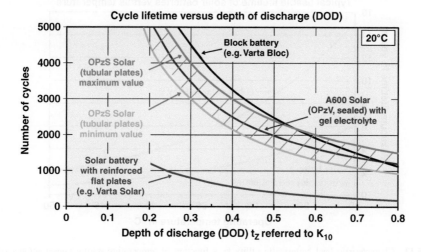

Figure 5.12   Cycle-based life span for selected batteries as a function of depth of discharge, based on manufacturers' data. The relevant intermediate values have been interpolated in cases where the manufacturer only indicated the values for a 30% and 70% depth of discharge

Stand-alone PV system battery life span can be optimized by sizing system batteries in such a way that, during the envisaged periods of autonomy, they are not discharged by more than around 75 to 80% or 50 to 60% of C_{10} or C_{100} respectively.

5.1.1.3.9 Life Utility

For low depths of cycle counts, life utility, which is determined by chemical ageing processes, can be a limiting factor for battery life span, which is determined by battery type and operating temperature. Figure 5.13 displays the life utility for two types of batteries as a function of temperature during conservation charging. If suitable types of batteries are used, according to manufacturers' data at temperatures of around 20 °C life utility ranging from 15 to 20 years is attainable. Inasmuch as a battery is subject to greater strain during cycle operation than during conservation charging, a life utility of 15–20 years is the upper limit for the projected battery life span in PV systems.

5.1.1.3.10 Full-Cycle Life Span n_{VZ}/Available Total Capacity C_{Ges} and Energy Throughput E_{Ges}

Although some manufacturers indicate a cycle-based life span n_Z for a specific depth of discharge t_Z, usually based on C_{10} (see Figure 5.12), these life spans always vary from one PV installation to another and normally bear little resemblance to those indicated in the manufacturers' datasheets. But in many batteries the product of cycle-based life span and depth of discharge is virtually constant. Full cycle-based life span n_{VZ} can be determined as follows using the datasheet information:

$$\text{Full-cycle life span: } n_{VZ} = n_Z \cdot t_Z \qquad (5.5)$$

Full-cycle life span n_{VZ} is an arithmetic value, which like t_Z is usually keyed to C_{10}. Needless to say, in practice it must be ensured that maximum allowable depth of discharge t_{Zmax} (e.g. 80%) is not exceeded. Total available battery capacity during a battery's lifetime is roughly as follows:

$$\text{Total available capacity: } C_{Ges} = n_{VZ} \cdot C_{10} \qquad (5.6)$$

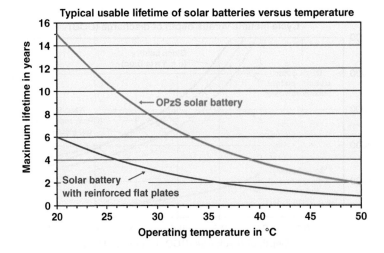

Figure 5.13 Characteristic lead battery life utility as a function of temperature during conservation charging, with around 2.25 V per cell. These values, which are roughly indicative of battery corrosion rates, constitute an upper limit for battery life span since in cycle operation batteries are subject to somewhat greater strain than during conservation charging

If C_{10} is not indicated on the manufacturer's datasheet, an approximate value can be derived from C_{20} or C_{100} using Equation 5.2.

Total storable energy E_{Ges} during a battery's life span can be determined based on full-cycle life span n_{VZ}, battery voltage $V_{battery}$ and total available capacity $C_{Ges} \cdot E_{Ges}$, which is also referred to as energy throughput, is pivotal for determining energy storage costs:

$$\text{Energy throughput: } E_{Ges} = V_{battery} \cdot C_{Ges} = V_{battery} \cdot n_{VZ} \cdot C_{10} \qquad (5.7)$$

5.1.1.3.11 Internal Resistance

Internal battery resistance, which is important for some applications, rises during the discharging process, is proportional to the number n_Z of series-connected cells and is inversely proportional to battery capacity. R_i can be readily determined from the internal resistance $R_{i100\,Ah}$ of a 100 Ah cell:

$$\text{Internal resistance } R_i \text{ of a battery with capacity } C: R_i = \frac{n_Z \cdot R_{i100\,Ah}}{K/100\,Ah} \qquad (5.8)$$

Figure 5.14 displays the main waveform of inner resistance as a function of available battery capacity for a 100 Ah OPzS battery with 2 V cells. The values indicated by various manufacturers for a fully charged 2 V OPzS battery at 20 °C range from 1.5 to 3 mΩ.

5.1.1.3.12 Self-discharge

A battery discharges at a very slow pace, even with no appliances connected. The higher the temperature, the more rapid this process is (see Figure 5.15).

Hence the selected battery operating temperature should be as low as possible. Self-discharge is also heavily influenced by the battery's internal structure. The self-discharge rate for batteries that are well suited for PV systems should range from 2 to 5% of nominal capacity per month (see Figure 5.15). However, PV installation batteries are also available whose monthly self-discharge rate ranges up to 10% at 20 °C.

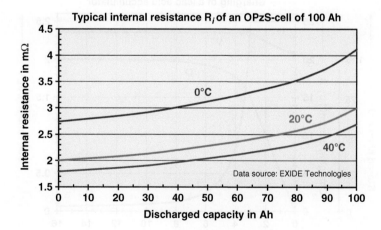

Figure 5.14 Characteristic internal resistance R_i of an OPzS lead battery (2 V) capacity $C_{10} = 100$ Ah at 0, 20 and 40 °C, as a function of discharged battery capacity, according to the manufacturer's data. R_i is lower than the manufacturer's values (e.g. 50%) for tubular or grid plate batteries. However, cycle-based life span is shorter for grid-plate batteries. (*Data source:* [5.5])

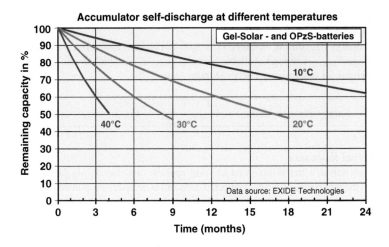

Figure 5.15   Lead battery self-discharge as a function of temperature during conservation charging, according to the manufacturer's data. The self-discharge rate is considerably faster at higher temperatures. The diagram here displays the curves for solar gel batteries, but also roughly applies to OPzS batteries. (*Data sources*: [5.4] and [5.5])

5.1.1.3.13   Conservation Charging
Conservation charging is a process that compensates for self-discharging and that keeps a fully charged battery fully charged at all times (around 2.25 V per cell at 20 °C for a lead battery).

5.1.1.3.14   Lead Battery Charging
A fully discharged lead battery can consume an extremely large amount of charging current. In practice, charging current is limited not by a battery's properties but rather by the battery charger or (in PV systems) the solar generator. The voltage increases slowly during charging until the charge limiting voltage is reached (see Figure 5.16).

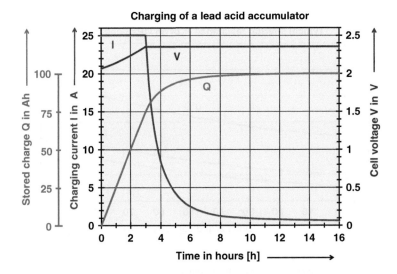

Figure 5.16   Charging a lead battery whose capacity is $C_{10} = 100$ Ah in accordance with the I–V characteristic curve. Waveform for charging current I, cell voltage V and battery charge Q, as a function of time

Once this voltage has been reached, charging does not come to an end, otherwise the battery would not be charged to full capacity. During this process a charge regulator circuit ensures that this voltage is not exceeded, which means that the charging current needs to be reduced accordingly. The battery does not reach its full capacity until it has been charged for a number of hours in accordance with the relevant charge limiting voltage. The so-called *IV* charging process described here is widely used.

5.1.1.3.15 Battery Interconnection

Only batteries of the same type, with the same capacity, of the same age and with the same charge state should be interconnected.

When a number n_{AS} of the same type of battery are interconnected, their voltages and internal resistance aggregate, while their capacity C remains the same:

$$\text{Voltage for series-connected batteries: } V'_{battery} = n_{AS} \cdot V_{battery} \tag{5.9}$$

$$\text{Internal resistance for series-connected batteries: } R'_i = n_{AS} \cdot R_i \tag{5.10}$$

When n_{AP} batteries are parallel connected, the battery voltage remains unchanged and the battery capacity and internal conductance are concatenated, which means that internal resistance decreases:

$$\text{Capacity for parallel-connected batteries: } C' = n_{AP} \cdot C \tag{5.11}$$

$$\text{Internal resistance for parallel-connected batteries: } R'_i = \frac{R_i}{n_{AP}} \tag{5.12}$$

When batteries are parallel connected, particularly in low-voltage systems, a fuse should be integrated into each string (see Figure 5.17) so as to avoid damage resulting from a plate short circuit and the extremely high equalization currents that can be induced by such an event. For parallel-connected configurations, the wire runs between the nodes and the various positive and negative poles should be exactly the same length so as to avoid unequal loads on the various batteries.

5.1.1.3.16 Target Characteristics for PV System Batteries

These are as follows:

- High capacity (in most cases for use over a period of days).
- High cycle-based life span with a relatively low depth of discharge (t_Z in most PV systems is relatively low).

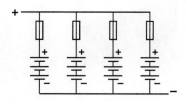

Figure 5.17 In parallel-connected batteries (which *must* all be of the same type and exhibit the same capacity and age), a fuse should be integrated into each string to prevent damage in the event of a plate short circuit. Moreover, the wiring should be realized in such a way that all wires exhibit exactly the same line resistance

- Occasional deep discharge cycles with 80–100% depth of discharge are allowable, without any life span reduction.
- Available extremely low charging currents (e.g. I_{1000}) should be leveraged for energy storage purposes.
- Low self-discharge rate (maximum 2–5% of nominal capacity per month).
- Good ampere-hour efficiency η_{Ah} and low charging factor.
- Little or no maintenance necessary (topping up with distilled water at intervals of one year or more). No maintenance at all should be necessary during the entire life span of sealed gas-tight batteries such as the Dryfit A600 PV installation battery.

5.1.1.4 Lead Batteries for PV Systems

5.1.1.4.1 Solar Batteries with Reinforced Positive and Negative Grid Plates

This type of battery is a modified auto battery with reinforced grid plates for the positive and negative electrodes. Essentially these are standard car batteries that have been optimized for improved cycle performance and lower self-discharge (example see Figure 5.18).

These batteries contain an aqueous electrolyte and are mainly suitable for use in settings such as weekend homes, i.e. smaller PV systems with a low number of cycles, whose full cycle-based life span n_{VZ} is around 150 to 250 cycles and that during cycle operation require maintenance about once a year by topping up the battery with distilled water. The monthly self-discharge rate of such batteries is around 5%. These batteries are a good buy in terms of their available energy (as calculated using C_{100}), with prices ranging from €130 to €200 per kWh.

5.1.1.4.2 OPzS Batteries with Positive Liner Plates and Negative Grid Plates

In this type of battery, the positive plate is composed of numerous lead rods (usually in a chamber-like configuration) around each of which a special mesh pouch is arranged that aggregates the active mass and prevents premature loss thereof. This allows for a far longer cycle-based life span. According to the manufacturer's data, a full cycle-based life span n_{VZ} can number between 900 and 1200 cycles.

For such OPzS batteries (Figures 5.19 and 5.20), in most cases a special lead alloy containing a minute amount of selenium and around 1.6% of antimony is used. The maintenance interval for cycle operation is around two years, and monthly self-discharge at $20\,^\circ$C is less than 4%. Such batteries are particularly suitable for (a) heavily used medium to large-sized PV systems and (b) connecting inverters. Multi-cell monoblock batteries with capacity C_{100} up to 430 Ah, as well as 2 V individual cells with C_{100} up to 4500 Ah, are available. In terms of their available energy (as calculated using C_{100}), these batteries range in price from €230 to €360 per kWh. In addition to aqueous-electrolyte models, sealed so-called OPzV batteries with gel electrolyte are available for a somewhat higher price.

Figure 5.18 SWISSsolar sealed solar battery with $V_{battery} = 12$ V and $C_{100} = 150$ Ah, with reinforced grid plates. (Photo: SWISSsolar)

Figure 5.19 Sealed OPzS solar aqueous-electrolyte battery (6 V monoblock battery with series-connected $K_{100} = 280$ Ah cells) with positive liner plates or tubular plates around each of which a mesh pouch is arranged. This type of battery has an extremely long cycle life. However, the internal resistance of OPzS batteries is relatively high (see Figure 5.14) and thus they are less suitable for discharge cycles involving very high currents, which are in any case extremely rare in PV systems. (Photo: Exide Technologies)

Figure 5.20 Aqueous-electrolyte batteries of 110 V with 2 V OPzS cells ($C_{10} = 600$ Ah) in a battery room with an acid-resistant clinker tile floor. (Photo: Banner Batterien Schweiz/BAE)

Figure 5.21   Some 2 V single cells of varying sizes with positive liner plates, for the realization of large batteries up to $C_{100} = 3000$ Ah. These Varta-Bloc (Vb) batteries, which have been on the market for years, are apparently no longer available directly from Varta but are sold by various wholesalers, in some cases under other brand names such as SWISSsolar. (Photo: Varta/SWISSsolar)

5.1.1.4.3   *Batteries with Positive Liner Plates and Negative Grid Plates*

In this type of battery (Figure 5.21), the positive plate is composed of numerous lead rods which, unlike those in OPzS batteries, also possess a diagonal mesh (as with a wire mesh), thus creating a positive liner plate with lower internal resistance. As with the mini-tubes in OPzS batteries, around each of these liner plates a special glass-fleece pouch is arranged that aggregates the active mass and prevents the latter's premature loss. This allows for a longer cycle-based life span than that obtained with OPzS. According to manufacturers' data, these batteries exhibit a full cycle-based life span n_{VZ} numbering from 1000 to 1350 cycles. The maintenance interval is around three years, and the monthly self-discharge rate is less than 3%. These batteries are particularly suitable for (a) larger and extremely heavily used PV systems whose capacity is upwards of 500 Ah and (b) connecting inverters. In terms of their available energy (as calculated using C_{100}), these batteries range in price from around €300 to €480 per kWh.

The cells integrated into large batteries should not be unduly tall (i.e. the height should not be more than double the width), since tall cells tend to exhibit deleterious acid stratification. If relatively tall cells are used, the electrolyte should be homogenized from time to time by pumping air into the battery to prevent acid stratification.

5.1.1.4.4 Sealed Lead–Calcium Batteries with Grid Plates

These batteries are normally gas-tight, electrolyte-tight and maintenance free, and exhibit a relatively low self-discharge rate of around 2% per month. However, they are not suitable for deep discharges, which can provoke irreversible grid passivation, and are incompatible with inverter operation as well. The (arithmetic) full cycle-based life span n_{VZ} of these batteries numbers anywhere from 200 to 300 cycles. They are mainly used for small PV systems with low battery use (e.g. for lighting, measuring device power supply and so on). In terms of their available energy (as calculated using C_{100}), these batteries cost anywhere from €140 to €200 per kWh.

5.1.1.4.5 Sealed Gel-Electrolyte Lead Batteries

These batteries (see Figure 5.22) are normally gas-tight, electrolyte-tight and maintenance free; depending on internal structure, they exhibit self-discharge rates of up to 4% per month at 20 °C. Models are available with reinforced positive and negative grid plates. The life span of these batteries in PV systems is comparable with that of standard grid plate batteries with aqueous electrolyte. However, special long-lasting models that integrate positive liner plates (OPzV) and are specially designed for PV systems are also available (e.g. Dryfit A600 Solar). Such batteries deliver a cycle-based life span comparable with that of the counterpart aqueous electrolyte batteries (see Figure 5.12).

As with all sealed batteries, in the above batteries the charge limiting voltage must not be exceeded so as to avoid irreversible water and capacity loss. These batteries cost more than comparable aqueous-electrolyte products. In terms of their available energy (determined using C_{100}), the price of these batteries, depending on whether they integrate grid or tubular (OPzV) plates, ranges from around €230 to €530 per kWh.

Figure 5.22   Sealed gas-tight, gel-electrolyte solar battery made by Sonnenschein. Foreground: batteries with reinforced grid plates. Background: A600 solar cells with $K_{100} = 1200$ Ah (OPzV battery with positive liner or tubular plates). (Photo: Exide Technologies)

5.1.1.4.6 Specimen Energy Storage Cost Calculations

1. A 12 V grid plate battery has a capacity $C_{100} = 150$ Ah, a cycle-based life span $n_Z = 800$ and a depth of discharge $t_{Z10} = 30\%$ of C_{10}, i.e. a full cycle-based life span $n_{VZ} = 240$ of C_{10}. According to Equation 5.2, $K_{10} \approx 0.7 \cdot K_{100}$, i.e. the battery's total available energy during its service life is as follows according to Equation 5.7:

$$E_{Ges} = V_{battery} \cdot n_{VZ} \cdot K_{10} \approx V_{battery} \cdot n_{VZ} \cdot 0.7 \cdot K_{100} = 302\,\text{kWh}$$

   At a battery price of around €300, this works out to an energy storage cost of around 1 € per kWh, not taking interest into account.

2. A 6 V sealed OPzS solar battery has a capacity $C_{10} = 300$ Ah, a cycle-based life span $n_Z = 3000$ and a depth of discharge $t_Z = 30\%$ of C_{10}, i.e. a full cycle-based life span $n_{VZ} = 900$ of C_{10}. Hence this battery's total available energy during its life span is as follows according to Equation 5.7:

$$E_{Ges} = V_{battery} \cdot n_{VZ} \cdot K_{10} = 1620\,\text{kWh}$$

   At a battery price of around €600, this works out to an energy storage cost of around €0.37 per kWh, not taking interest into account.

Note: Inasmuch as a lead battery's price is largely determined by the price of lead on the international market, which can fluctuate considerably according to the economic situation, the prices indicated in Section 5.1.1.4 are rough estimates only.

5.1.2 Structure of Stand-alone PV Systems

5.1.2.1 Definition of System and Module Voltage

One of the key factors for defining system voltage (i.e. nominal battery voltage) for a PV installation is the amount of voltage made available to DC appliances: 12 V systems deliver by far the greatest amount of voltage, whereas 24 V systems provide considerably less and 48 V systems provide hardly any. However, owing to lead line power loss, which increases in direct proportion to I^2, higher system voltages should be used in cases where appliance power consumption is relatively high.

A simple rule of thumb concerning the appliance power threshold at which the use of higher system voltage should be considered is the 1 Ω rule. If the resistance of the connected appliances goes far below 1 Ω, the power loss in wiring, switches, plugs, circuit breakers and contact terminals increases to such an extent that it makes sense to transition to a higher system voltage, since use of a larger wire gauge would appreciably ramp up material costs. The only case where substantial undercutting of this value can be considered (up to about 0.1 Ω) is if an inverter is directly connected to the battery terminal. However, inverters with higher operating voltages exhibit the same output, but with better efficiency.

Table 5.3 displays the 1 Ω theoretical power limits that arise from such considerations. Owing to the large number of DC appliances that come into play, for low system voltages the tendency is to exceed these limits to some degree. Hence an additional practical power limit should be defined that also takes account of the available materials.

Table 5.3 Power threshold figures of merit for DC appliances and for inverters that are directly connected to the battery, as a function of system voltage

System voltage	12 V	24 V	48 V
Power limit (1Ω-rule)	144 W	576 W	2304 W
Power limit (practical)	300 W	1 kW	3 kW
Power limit inverter	1.2 kW	5 kW	10 kW

Needless to say, solar generator voltage needs to be commensurate with the defined system voltage and should also be at least high enough to allow the battery to be fully charged. In other words, it should be possible to reach at a minimum the battery's end-of-charge voltage at all anticipated ambient temperatures. For installations in temperate climates, one solar module with 32 or 33 series-connected monocrystalline or polycrystalline cells is sufficient for each 12 V of system voltage. In such regions, solar modules with 36 series-connected cells provide a small amount of additional energy in summer only. On the other hand, in tropical or desert regions solar modules with 36 series-connected cells should be used owing to the higher ambient temperatures.

However, at low ambient temperatures such modules cannot be operated at the MPP without the use of a solar tracker since the battery voltage shifts the module operating point down to lower voltages, thus reducing output.

5.1.2.2 Reverse Current Diode; Deep Discharge Protection; Overcharge Protection

Solar generators produce no voltage at night. In order to prevent the battery from solely discharging into the solar generator, a reverse current or blocking diode is used that allows current to flow from the solar generator to the battery but not vice versa (see Figure 5.23). Under normal operating conditions, this diode allows for a 0.3–0.8 V voltage drop, according to the type of diode used.

In the interest of avoiding deep-discharge events that reduce battery life span, if the final discharging voltage goes above around 1.75 to 1.85 V per cell, the connected appliances should be disconnected from the battery (see Figure 5.9). This task is performed by a deep-discharge protection device, discharge controller or battery monitoring device (see Figure 5.23).

Most commercial charge controllers integrate a deep-discharge protection device (via a relay, solid state transistor or power MOSFET) that automatically reconnects system appliances following a sufficiently large battery voltage rise. However, in such cases it is essential that the maximum rated load current not be exceeded. The function of a deep-discharge protection device and a main fuse for connected appliances can also be combined by using a circuit breaker that integrates an operating current trip. A comparator trips the circuit breaker if the battery voltage goes lower than around 10.5 V in a 12 V installation and below around 21 V in a 24 V installation. Although this device provides a relatively simple solution, it needs to be manually reset after being tripped and is thus mainly suitable for use as a backup deep-discharge protection device in the event that the primary device fails.

A backup surge protector can be realized in a similar manner in such a way that the battery is protected against catastrophic overcharging in the event that the electronic charge controller fails. A comparator can trip a solar-generator-side circuit breaker via an operating current trip that is tripped if the gas voltage exceeds 2.4 V per cell, or 14.4 V for 12 V installations or 28.8 V for 24 V installations. For smaller installations, it is presumed that the charge controller will function reliably and thus a stand-alone emergency overcharge protector is forgone in most such cases. However, as a safety precaution larger installations should integrate an overcharge protector that operates independently of the charge controller.

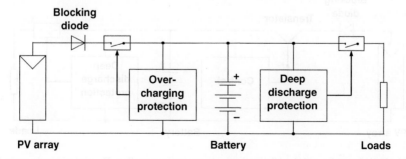

Figure 5.23 Diagram displaying the functional principle of reverse current diodes, deep-discharge protection and overcharge protection in a PV system

Needless to say, the overcharge protector switching threshold should be set high enough that it does interfere with normal operation of the charge controller or solar tracker.

5.1.2.3 Installations with Series Controllers

Installations that integrate series controllers have a current limiting element, in the guise of a solid state transistor, power MOSFET or relay contact, between the solar generator and battery bank. Figure 5.24 displays the basic layout for such installations.

In a linear or continuous controller, the transistor or (power MOSFET) remains fully connected as long as the defined charge limiting voltage V_G has not been reached. When V_G is reached, the transistor's excitation is reduced in such a way that the charge current is throttled down to the level needed to maintain V_G. This current reduction process produces relatively high transistor voltage, whereby the resulting power loss heats up the transistor, which must then be cooled. However, contrary to the operating principle of other controllers, the maximum power loss in such cases is considerably lower than total solar generator power. But with series controllers, even in a fully connected state a moderate voltage drop occurs via the transistor, thus somewhat reducing the solar generator's usable output.

Switching series controllers for the most part avoids the drawbacks entailed by transistor warming. In such cases the controller is realized as a comparator that exhibits a certain amount of hysteresis or is designed as a so-called on–off controller. The transistor is fully connected until the charge limiting voltage V_G is reached, i.e. the battery is operated under maximum current. When V_G is reached, the transistor is completely shut down, whereupon the battery voltage gradually declines (a battery that is charged to V_G has an effect comparable with that exhibited by many high-capacity farads). The transistor is not fully reactivated until a switching threshold slightly under V_G is undercut. Figure 5.25 displays the battery voltage and charging current waveform for this process.

Owing to their steep switching edges, poorly designed switching charge controllers may cause electromagnetic compatibility problems, i.e. they may interfere with nearby sensitive electronic devices such as radios, radio appliances and so on. Such problems can be mitigated via measures such as reducing the steepness of the switching edges, which will provoke elevated power loss in the semiconductor elements, or disturbance suppression mechanisms may be needed as in quality inverters.

In addition, transistors can be substituted for relay contacts, a solution that reduces connected-state loss but also shortens the contact life span. Substituting transistors for relays in high-altitude installations that are extremely prone to lightning strikes can be beneficial in that relays are far less sensitive to power surges.

5.1.2.4 Installations with Parallel Controllers

In installations with parallel controllers (also referred to as shunt controllers), any solar generator current not needed by the battery is absorbed by the controllers (Figure 5.26). The element that shunts the

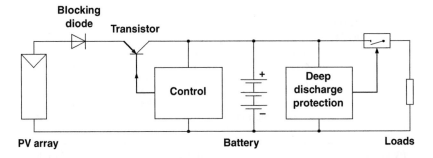

Figure 5.24 Basic layout for a PV installation that integrates series controllers. Detailed wiring diagrams for such configurations can be found in [Jäg90], [Köt94] and elsewhere in the literature. Many types of switching series controllers are commercially available

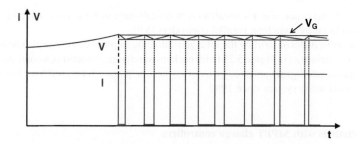

Figure 5.25   Waveform of battery voltage V and charging current I in a switching charge controller

superfluous current (power transistor, power MOSFET or relay contact) is usually located just across the solar generator connecting leads, i.e. in front of the blocking diode.

A major advantage of a parallel controller lies in the fact that it only absorbs power in cases where surplus energy is available. So long as the battery is not fully charged, the solar generator's full output is usable.

With a linear or continuous controller, the transistor cuts out if the defined charge limiting voltage V_G has not been reached. Once V_G is reached, the transistor is connected in such a way that the solar generator current not needed by the battery or appliances can flow through the transistor. This of course results in considerable transistor power loss, which means that the transistor needs to be cooled. In the worst-case scenario, the parallel controller must absorb virtually all of the solar generator power. Maximum transistor power loss can be reduced by around 25% if an appropriately sized power resistance R is series connected to the transistor. This in turn appreciably reduces the amount of cooling needed since resistances can withstand far higher temperatures than semiconductors.

The use of a switching parallel controller can greatly reduce transistor heat build-up. A solar generator can in principle be shorted as it exhibits the characteristics of a power source. In such a case, the controller is likewise realized as a comparator that exhibits a certain amount of hysteresis or is designed as an on–off controller. The transistor remains shut down until the charge limiting voltage V_G is reached, i.e. the battery is operated under full current. Once V_G is reached, the transistor is fully switched and only receives the saturation current, thus minimizing transistor power loss. The transistor is not fully shut down again until a switching threshold slightly under V_G is undercut, whereupon the battery's charging current and its voltage behave according to Figure 5.25. It should be noted, however, that at various PV congresses in

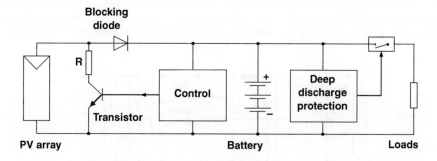

Figure 5.26   Basic layout for a PV installation that integrates parallel controllers. The use of linear controllers allows for a substantial reduction of transistor power loss via a resistance R. If switching controllers are used, R can be forgone ($R = 0$) and thus virtually no power loss occurs in the controllers. For detailed wiring layouts, see [Jäg90] and [Köt94], among others

2005, there were reports concerning PV installation module damage in warm climates resulting from hot-spot formation induced by switching parallel controllers [5.9].

Rather than shutting down or shorting the solar generator, the surplus current can also be put to a useful purpose. In an installations as in Figure 5.23 (with backup overcharge protection), a continuous controller can also be used as an intelligent appliance for auxiliary heating purposes. I myself have been using such a controller in a stand-alone system since 1989.

5.1.2.5   Installations with MPPT charge controllers

An MPPT charge controller is a specialized DC converter whose purpose is to ensure that a solar generator is always being operated at the MPP. From a design engineering standpoint, an MPPT charge controller is of course a highly elegant solution – one that, according to MPPT charge controllers manufacturers, increases power output by around 15 to 30% in a properly sized stand-alone PV system that integrates a battery.

MPPT charge controllers pay off in large stand-alone systems as from 100 W. However, it is important to determine on a case-by-case basis whether an MPPT charge controller is worth the extra cost or whether this money might not be better spent on additional solar modules. Use of an MPPT charge controller also ramps up system complexity, and it is highly unlikely that the life span of the device's power electronics will be as long as that of the solar generator into which it is integrated. As with switching charge controllers, MPPT charge controllers can cause electromagnetic compatibility problems, i.e. they may interfere with nearby sensitive electronic devices such as radios, radio appliances and so on. Such problems can be mitigated via disturbance suppression measures, as is done with quality inverters.

When an MPPT charge controller is used for battery charging purposes, in view of the battery's constant voltage the simplest approach is to ensure that charging current is I_L is maximized.

Figure 5.27 displays the basic layout of a downconverting enabled MPPT charge controller ($V_{PV} > V_L$; deep-discharge protection and appliances not shown for reasons of space). For purposes of explaining the main functional principle that comes into play here, it is assumed below that L, C_1 and C_2 are very high and that the diode D and transistor T are ideal.

A voltage drop that occurs at shunt R_{Sh} is proportional to current I_L at the MPPT charge controller output. The controller monitors this value continuously and generates the duty cycle t_E/T (the ratio of transistor turn-on time t_E to time period T) that is optimal for maximum I_L under the current insolation conditions. The charging current I_L fluctuates minimally around the maximum value, whereby the solar generator current I_{PV} equals the mean transistor current value I_T and is roughly equal to the current I_{MPP} in the MPP (see Figure 5.28).

Figure 5.29 displays the voltages that are produced in this wiring layout. During transistor turn-on time t_E, the transistor T is fully connected and the voltage at diode D is $v_D = V_{PV}$ (diode blocked). The transistor conducts current I_L during t_E. During the remaining time, the transistor cuts out, the inductance voltage is negative and the current I_L flows through the now conductive diode, where the

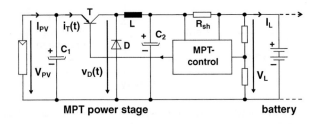

Figure 5.27   Basic circuit configuration for a charge controller with an MPPT charge controller and downconverting (output voltage lower than input voltage). Power MOSFETs are often substituted for bipolar transistors

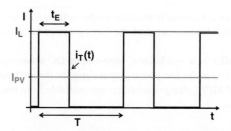

Figure 5.28 Waveform of solar generator current I_{PV}, current $i_T(t)$ in the transistor T and battery charging current I_L (red) as a function of time for the wiring layout in Figure 5.27, where t_E = transistor turn-on time, T = time period, t_E/T = duty cycle (assumption: $L, C_1, C_2 \approx \infty$)

voltage is $v_D \approx 0$ (ideal diode). The output voltage V_L comprises the mean value of the voltage $v_D(t)$ at the diode.

$$\text{The following applies to solar generator current: } I_{PV} = I_L \frac{t_E}{T} \qquad (5.13)$$

$$\text{Conversely, the following applies to charging current: } I_L = I_{PV} \frac{T}{t_E} \qquad (5.14)$$

$$\text{The following applies to output voltage } V_L: V_L = V_{PV} \frac{t_E}{T} \qquad (5.15)$$

$$\text{Conversely, the following applies to solar generator voltage: } V_{PV} = VB_L \frac{T}{t_E} \qquad (5.16)$$

In view of the negligible power loss here, the following applies, in keeping with the law of conservation of energy:

$$V_{PV} \cdot I_{PV} = V_L \cdot I_L \qquad (5.17)$$

Thus charging current I_L is higher than the current I_{PV} delivered by the solar generator. The entire circuit has the effect, so to speak, of a DC transformer, i.e. the battery-side power output is virtually the same as the solar-generator-side input power.

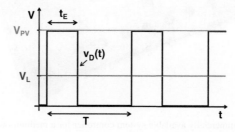

Figure 5.29 Solar generator voltage V_{PV} and voltage $u_D(t)$ via (*ideal*) diode D, and voltage V_L via the battery as a function of time for the wiring layout in Figure 5.27, where t_E = transistor turn-on time, T = time period, t_E/T = duty cycle (assumption: $L, C_1, C_2 \approx \infty$)

The MPPT charge controller continuously monitors output voltage so as to avoid battery overcharging. Once the charge limiting voltage has been reached, the duty cycle is reduced in such a way that this voltage is not exceeded.

An MPPT charge controller is a specialized, high-output DC converter with around 90 to 99% efficiency, some of which is lost under low power output since the device's open-circuit loss needs to be covered first. Mass-produced MPPT charge controllers are available from manufacturers such as Phocos and Outback Power Systems.

5.1.2.6   Examples of Commercially Available Solar Controllers (Charge Controllers with Deep-Discharge Protection)

Steca, Phocos, Morningstar (Figure 5.30) and many other manufacturers sell charge controllers for charging current ranging from 4 to 140 A, for load currents ranging from 4 to 70 A and for 12, 24 and 48 V system voltages (often with automatic recognition of and switching to the relevant voltage). Some manufacturers provide such devices for up to 300 A by special order (Table 5.4).

Many switching charge controllers use pulse width modulation, i.e. they connect the solar generator until the power limit voltage is fully reached at the battery. Following a deep-discharge event in a sealed, aqueous-electrolyte battery, equalizing charging can be realized briefly at a somewhat higher voltage, whereupon the following occurs: (a) the charging current is shunted to the battery as short current impulses, and thus the charging is pulsed (f_{pulse}, e.g. 300 Hz); and (b) via variation of the ratio of turn-on time t_E to time period T (as in Figure 5.28), the battery is continuously maintained at the (now somewhat lower) maintenance charging voltage (e.g. 2.25 V per cell).

5.1.3   PV Installation Inverters

Inasmuch as not all appliances are available as small-DC-voltage devices (particularly appliances with powerful motors such as washing machines and power tools), an inverter is often used for stand-alone

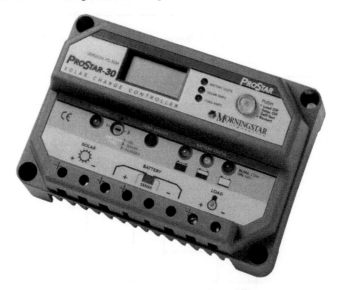

Figure 5.30   Example of a commercially available system controller for a medium-output PV installation: Prostar 30 switching pulse width modulation series controller for 12 V/24 V installations with solar generator current up to 30 A and load current up to 30 A. The device, which can be switched between various types of batteries (using a screwdriver) at the LED on the far left (battery type 1, 2, 3), integrates LED lamps and displays that provide general information concerning PV installation operating status and battery charging status. (Photo: Morningstar datasheet)

Table 5.4 Characteristics of selected PV switching system controllers from various manufacturers (P = parallel controller; S = series controller). Apart from the products listed here, many others are available that have different characteristics

Type	Steca Solarix	Steca Solarix	Steca Power TARON	Phocos	Phocos	Morningstar	Morningstar
	Alpha	Omega	4140	CA06-2	CX40	Sunsaver 6L	Prostar-30
Nominal voltage	12 V/24 V	12 V/24 V	48 V	12 V	12 V/24 V	12 V	12 V/24 V
Mode of operation	P	P	P	S	S	S	S
Charging current	8 A	30 A	140 A	5 A	40 A	6 A	30 A
Load current	8 A	30 A	70 A	6 A	40 A	6 A	30 A
Self-consumption	<5 mA	<5 mA	<14 mA	<4 mA	<4 mA	6–10 mA	25 mA
Temperature (°C)	−25 to 50	−25 to 50	−10 to 60	−40 to 50	−20 to 50	−40 to 85	−40 to 60

systems so that 230 V/50 Hz DC appliances can also be operated. For stand-alone system inverters, different characteristics come into play than for appliances that are used with grid-connected PV systems (see Section 5.2).

The requirements for stand-alone system inverters are as follows:

- A DC power source that exhibits a stable, internally generated frequency (generally 50 Hz).
- Good efficiency, including in the low-load range.
- No open-circuit loss due to low open-circuit current.
- The capacity to provide the reactive power needed for inductive appliances, without undue efficiency loss.
- The capacity to deliver high starting currents (around 2–3.5 times higher than nominal current) for around 300 ms.
- Short-term overload capacity (e.g. 50–100%) for periods ranging from several seconds to several minutes.
- Start-up current limit on the DC side (no unnecessary battery fuse tripping on start-up).
- Power surge protection via shutdown of inductive appliances.
- Protection against overload and short-circuit events on the AC side.
- For large appliances, remote operation or automatic start-up when the 230 V output is under load (standby switching).
- Sinusoidal output signal with low total harmonic distortion.
- Low level of high-frequency interference in the output signal.

The current state of the art comprises inverters with a sinusoid or, lacking that, a nearly sinusoidal output signal. Although devices with other types of output signals are commercially available and are widely used in older installations, the lower price and higher overload capacity they offered have since been cancelled out for the most part by technical advances in sine wave inverters.

Needless to say, meeting all of the above requirements entails extremely high costs. The inverters that meet these requirements in various ways fall into the following four categories according to their output signal curve characteristics (see Figure 5.31):

- rotary converter (device where a DC motor drives an AC generator)
- square-wave oscillator
- pulse-width-controlled square-wave oscillator
- sine wave inverter

the last three comprising a static inverter.

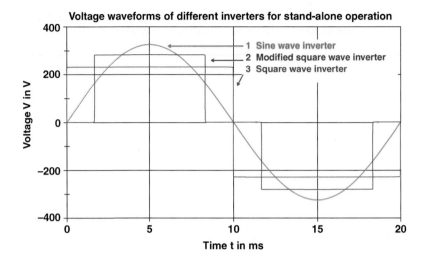

Figure 5.31   Output voltage waveform for various types of PV installation inverters

5.1.3.1   Rotary Converter

In a rotary converter a DC motor drives an AC generator, an arrangement that is advantageous by virtue of the sinusoidal output voltage and the device's capacity to provide high start-up currents (owing to the energy stored in the rotary elements). The drawback of these appliances is that they exhibit relatively high open-circuit loss and extremely poor efficiency (maximum 70–85%) since the energy needs to be converted twice over; further, the device is subject to mechanical and electrical wear.

5.1.3.2   Square-Wave Oscillators

Square-wave oscillators are quite inexpensive owing to their simple design (see Figure 5.32), are highly efficient and can withstand brief periods of moderate overload. However, their square-wave output voltage (see Figure 5.31) exhibits elevated harmonics (total harmonic distortion around 44%), which can provoke elevated heat build-up in appliances such as motors and transformers, and in appliances with DA power modules, bidirectional thyristors or silicon-controlled rectifiers can even cause malfunctions or premature failure.

Moreover, owing to the transformer's fixed transformation ratio, its output voltage is mainly determined by the battery voltage. For example, effective 230 V output voltage for 24 V battery decreases to 201 V with 21 V battery voltage, and increases to 276 V with 28.8 V (gassing voltage). Apart from the elevated harmonics entailed by such voltage fluctuations, they also cause problems in many other appliances. Hence square-wave oscillators should be used solely for *occasional* operation of single-phase collector motors (in appliances such as drills, vacuum cleaners and electric mixers) or with simple power supply units that are devoid of electronics.

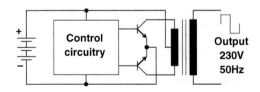

Figure 5.32   Block diagram of a simple square-wave oscillator

5.1.3.3   Pulse-Width-Controlled Square-Wave Oscillators

Pulse-width-controlled square-wave oscillators (also sometimes called modified sine wave inverters) keep output voltage and load constant, regardless of battery voltage. This is realized via a third voltage step at 0 V whose duration can be adjusted via a rapid regulating switch in such a way that the effective output voltage is kept at a steady 230 V (see Figure 5.31).

The third voltage step at 0 V also reduces harmonics in the output voltage, particularly in the odd-numbered integers of the third harmonic (i.e. 3, 9, 15, 21 and so on); this in turn reduces total harmonic distortion to just under 30%. Hence pulse-width-controlled square-wave oscillators can be used in a broad range of appliances over relatively lengthy periods. The efficiency of these oscillators is on a par with that of simple square-wave oscillators, reaching maxima ranging from 85 to 95%. The open-circuit loss of these appliances, which at 1–4% is extremely low, can be reduced to less than 1% using special automatic power-saving switches that do not fully power up the inverter until it is under load.

Pulse-width-controlled square-wave oscillators can withstand brief overload events (e.g. 100% overload for a few minutes and 200–300% for a few seconds) and are thus also compatible with appliances that exhibit high start-up current. Moreover, a suitable reduction of the steepness of the switching edges (e.g. rise and fall time > 50 µs) does not significantly reduce efficiency and at the same time obviates radio reception problems induced by the high-frequency interference voltages often exhibited by inverters. This procedure also makes square-wave pulses somewhat trapezoidal in shape.

Problems can arise when appliances are connected that are sensitive to the residual low-frequency interference voltages in supply voltages in devices such as stereos. Sine wave inverters are a better choice for such applications.

5.1.3.4   Sine Wave Inverters

Sine wave inverters generate a practically sinusoidal output voltage (see Figure 5.31) that just about any device can be connected to. As a result of considerable technical progress, these appliances now represent the state of the art. Like pulse-width-controlled square-wave oscillators, modern sine wave inverters can briefly withstand high overload (e.g. 200–300%) and thus exhibit no start-up current problems. Despite their now greater technical complexity, sine wave inverters are only marginally more expensive than in the past, and their efficiency is on a par with that of pulse-width-controlled square-wave oscillators. Thus sine wave inverters or at a minimum quasi-sine-wave inverters are today the gold standard when it comes to purchasing new appliances.

Modern sine wave inverters realize a sinusoidal output signal by generating a pulse-width-controlled signal with a frequency ranging from around 10 to 50 kHz inside the device. This relatively high frequency can be readily filtered out (often a simple inductor suffices). Most of a sine wave inverter's low-frequency harmonics are suppressed via switching. Figure 5.33 displays a block diagram for a sine wave inverter, while Figure 5.34 illustrates how a sinusoidal output signal is generated via pulse width modulation.

This method allows for maximum efficiency of well over 90% (e.g. 92–96%; see Figure 5.35 and Table 5.5) with sine wave inverters as well. However, the high frequencies entailed by pulse width modulation accompanied by the extremely steep edges that are necessary to attain high efficiency result in high-frequency interference voltages at the output that cannot be satisfactorily suppressed using a low-frequency optimized output filter. Hence, good high-frequency sine wave inverters should be fitted with additional *bilateral* high-frequency filters (see Figure 5.33) so as to avoid interference with other appliances (e.g. radio reception interference).

Owing to low-loss standby switching, modern sine wave inverters need only a small fraction (usually less than 1%) of nominal capacity at open circuit and can thus remain in operation continuously. Some of today's inverters are also available with an integrated charge controller, or with an integrated rectifier charging device which, on connection of an auxiliary generator, charges the battery following a deep-discharge event induced by a lengthy stretch of inclement weather. Table 5.5 lists the key characteristics for a number of sine wave inverters from various manufacturers.

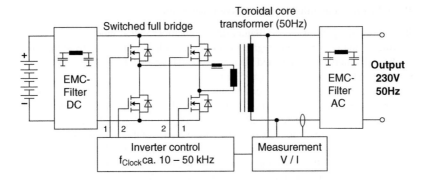

Figure 5.33 Block diagram of a modern sine wave inverter that integrates a power MOSFET. The two control signals (1 and 2) each power up an FET located diagonally opposite, whereas the other two FETs remain blocked. The device approximates a sine curve waveform by means of varying start-up and shutdown times (pulse width, pulse duration and pulse width modulation)

Modern sine wave inverters exhibit an excellent sine wave (see Figure 5.36) and relatively low harmonics and are considerably more technically advanced than square-wave or modified sine wave inverters. While the sine curves for each and every one of these appliances is not perfect, most of them match or even exceed grid quality.

The difference between sine wave and modified sine wave inverters is most pronounced in the harmonics spectrum. Figure 5.37 displays this spectrum for a TC13/24 sine wave inverter and an Atlas 24/1200 pulse-width-controlled square-wave oscillator, relative to the EN 61000-2-2 threshold values, which define the maximum anticipated harmonics in the grid.

In selecting an inverter for a stand-alone installation, the importance of start-up current should not be underestimated. Despite the appreciable overload capacity of today's inverters, it is often necessary to use an inverter with a nominal capacity considerably exceeding that of appliances that are connected for

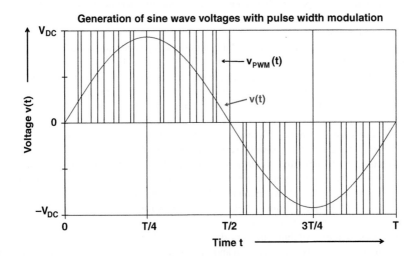

Figure 5.34 Representing the functional principle of a sine wave inverter as in Figure 5.33. The device generates a sinusoidal voltage via pulse width modulation and frequent activation and deactivation within a half cycle (current duty cycle in accordance with the transient value of the desired sine voltage). The frequency is considerably higher than that of the generated voltage and can thus be readily eliminated using a low-pass filter

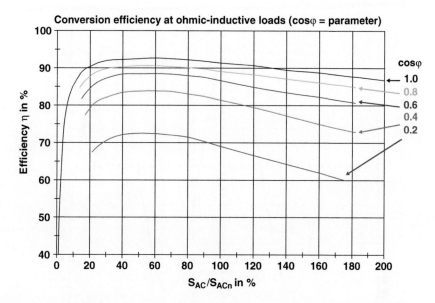

Figure 5.35 Characteristic sine wave inverter efficiency curve for stand-alone installations under ohmic–inductive load conditions. This graph shows the efficiency, for various cos φ values, as a function of standardized apparent AC power, i.e. apparent AC-side power S_{AC} relative to nominal apparent power S_{ACn}. Efficiency for low cos φ values is considerably lower than for nearly ohmic loads. The following device scenario was used as an example here: TopClass 08/24 (nominal DC voltage 24 V, $S_{ACn} = 800$ VA) [5.6]

stationary operation, since otherwise it may not even be possible to power up the inverter. Figure 5.38 displays the powering-up voltage and current waveforms for an inverter (nominal capacity 1300 W, nominal DC voltage 24 V) in a compression refrigerator with only 90 W of continuous output. Figure 5.39 displays the maximum available output for a sine wave inverter under transient overload conditions.

Table 5.5 Selected PV system 230 V/50 Hz sine wave inverters from various manufacturers. In addition to these products, numerous other models with different characteristics and features, such as an integrated DC battery recharger or a solar charge controller, are available from other manufacturers. All of the data indicated here are based on manufacturers' data

Type	Aton 2.3/12	Allegro 10/24	TopClass TC35/48	AJ 275-12	AJ 1300-24	XPC 2200-24	HPC 8000-48
Manufacturer	ASP	ASP	ASP	Studer	Studer	Studer	Studer
Continuous power	160 VA	1000 VA	3200 VA	200 VA	1000 VA	1600 VA	7000 VA
Maximum efficiency	94%	94%	93%	93%	94%	95%	96%
Nominal voltage	12 V	24 V	48 V	12 V	24 V	24 V	48 V
Input V_{DC}	10.5–16 V	21–32 V	42–64 V	10.5–16 V	21–32 V	19–32 V	38–68 V
cos φ	0.3–1	0.3–1	0.3–1	0.1–1	0.1–1	0.1–1	0.1–1
Self-consumption Standby/AC on	0.6 W/2 W	0.5 W/10 W	0.5 W/12 W	0.3 W/1.9 W	0.4 W/10 W	0.9 W/7 W	3 W/30 W
Temperature (°C)	−25 to 50	−25 to 50	−25 to 50	−20 to 50	−20 to 50	−20 to 55	−20 to 55
Charge controller (I_{max})	Yes (20 A)	No	No	Opt. (10 A)	Opt. (25 A)	Opt. (30 A)	No
Charger (I_{max})	No	No	No	No	No	Yes (37 A)	Yes (90 A)

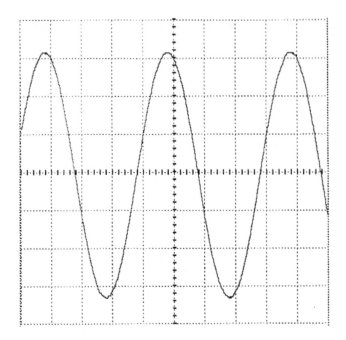

Figure 5.36 Output voltage readings for an SI1224 inverter (1200 W, 24 V) under 300 W ohmic load. The controller circuit performs optimally and the voltage exhibits an extremely good sine curve. Dimensions: 100 V/div. vertical, 5 ms/div. horizontal

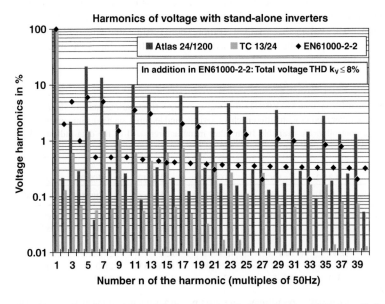

Figure 5.37 Higher harmonic voltage relative to first-harmonic oscillation for a pulse-width-controlled square-wave oscillator (Atlas 24/1200) and a sine wave inverter (TC13/24) at around 30% of nominal capacity (ohmic load), compared with the tolerability levels in the low-voltage grid in accordance with EN 61000-2-2 [5.7]. The pulse-width-controlled inverter greatly exceeds the limit values for numerous frequencies, whereas the sine wave inverter provides close to grid quality in terms of harmonics

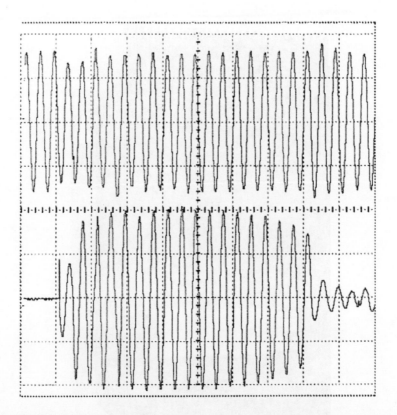

Figure 5.38 Waveforms on powering up a 90 W compressor refrigerator, for a TC13/24 at open circuit. Upper curve: voltage *V*, scale 200 V/div. Lower curve: current *I*, scale 5 A/div. Scale for time (*x*-axis): 50 ms/div. During the roughly 350 ms powering-up process, the refrigerator used nine times the device's nominal capacity, and the inverter used around *1.3 times* its nominal apparent power

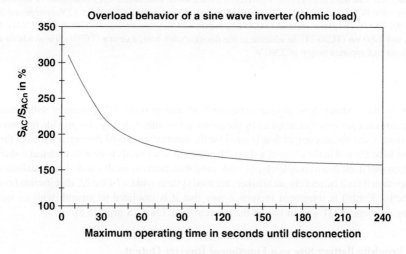

Figure 5.39 Performance of an 800 VA sine wave inverter (TC08/24) under overload conditions. The maximum possible operating time under a specific excess load is determined by pre-overload inverter temperature and by ambient temperature. The curve shown here was registered at an ambient dry mass of 20 °C and cooling element temperature of around 30 °C [5.6]

Figure 5.40   ASP sine wave inverter (230 V/50 Hz) for a stand-alone installation. The picture shows a Top Class 22/48 for 2000 VA with 48 V system voltage. This type of device is also available for: (a) a 12 V system and apparent output of 1000 VA (TC13/12) and 1800 VA (TC20/12); and (b) a 24 V system voltage for 1200 VA (TC15/24), 2000 VA (TC22/24) and 2700 VA (TC30/24). In addition to the device shown here, a device (TC35/48) is available for 48 V system voltage and apparent output of 3200 V

If properly sized, stand-alone inverters (Figure 5.40) can provide AC appliances with voltage and current quality on a par with that supplied by the public grid – although they also provide far lower short-circuit current. If the inverter output fuse is rated for the inverter's nominal current, it will take far longer for the fuse to be tripped in the event of a short. Hence an inverter needs to be able to detect a short on its own and then shut itself down completely. AC-side insulation faults in small stand-alone installations pose less of a problem if fault current circuit breakers are used system-wide, or if the AC distribution (except for PE (Protective Earth)) is integrated in such a way that it is insulated to ground, with an insulation monitoring mechanism (IT grid in lieu of the conventional TN–C–S grid) [5.8].

5.1.3.5   Requisite Battery Size as a Function of Inverter Output

It is important that a sufficiently large battery be used for a stand-alone installation that integrates an inverter, since: (a) internal resistance increases in small batteries; and (b) batteries are discharged continuously by the omnipresent 100 Hz alternating component on the AC side.

Under higher loads, the effective value of this alternating current in actual installations amounts to around 50 to 70% of DC input. Some battery manufacturers indicate an allowable AC component of around I_{10}. Hence the following applies to the DC-side nominal current I_{DCn} of the inverter:

$$(0.5 - 0.7) \cdot I_{DCn} = I_{10} \qquad (5.18)$$

It therefore follows that the requisite minimum battery capacity during inverter operation is as follows:

$$\text{Minimum inverter operation battery capacity: } C_{10} = 10\,\text{h} \cdot I_{10} = (5 - 7\,\text{h}) \cdot I_{DCn} \qquad (5.19)$$

Hence the battery should be sized large enough that its stored charge is sufficient for at least 5–7 hours of operation with nominal inverter capacity.

With this sizing, the voltage drop at the internal resistance of a fully charged OPzS battery (see Figure 5.14) will be around 40 mV per cell or around 2% of battery voltage, which is acceptable.

5.1.3.6 DC Waveform of Stand-alone Installation Sine Wave Inverters

It needs to be borne in mind for one-phase stand-alone inverters that under heavy load conditions the transient value of the current $I_{DC}(t)$ on the DC side will vary, according to bilateral AC-side frequency, between close to 0 and around twice the mean DC value I_{DC}, if a $\cos \varphi = 1$ load is connected on the AC side (see Figure 5.41). The figure of merit I_{DCeff} for this current is therefore somewhat *higher*. The following applies under ideal conditions:

$$\text{Figure of merit } I_{DCeff} \text{ for AC loads where } \cos \varphi = 1\text{: } I_{DCeff} = I_{DC}\sqrt{1.5} \approx 1.22 \cdot I_{DC} \qquad (5.20)$$

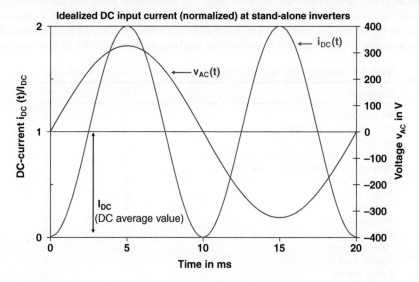

Figure 5.41 DC-side transient figure of merit $i_{DC}(t)$ for a stand-alone installation sine wave inverter (normalized representation referred to mean I_{DC} of the direct current), where a $\cos \varphi = 1$ load is connected on the inverter side. The figure of merit of this current would then be 1.22 times I_{DC}. The current $i_{DC}(t)$ is the sum total of I_{DC} and an AC component whose fluctuation equates to double the output frequency and with amplitude I_{DC}. Hence inverter input voltage $i_{DC}(t)$ will Exhibit 100 Hz fluctuation in an inverter that generates an output voltage with a 50 Hz frequency

If $\cos \varphi < 1$, the DC-side figure of merit increases even further. Using Equation 5.21, in this case as well I_{DCeff} can be *approximated* as follows:

$$\text{Figure of merit } I_{DCeff} \text{ for AC loads where } \cos \varphi < 1: I_{DCeff} \approx \sqrt{I_{DC}^2 + \left(\frac{S_{AC}}{V_{DC}\sqrt{2}}\right)^2} \qquad (5.21)$$

where:
I_{DC} = mean value of the current absorbed by the DC-side inverter
S_{AC} = apparent output on the AC side
V_{DC} = mean voltage on the DC side

This factor needs to be taken into account for sizing the wiring and fuses between the battery and inverter. However, as the input voltage V_{DC} is virtually constant, the DC-side input power is computed as $P_{DC} = V_{DC} \cdot I_{DC}$.

5.1.4   Stand-alone Installation DC Appliances

It is essential that only highly efficient appliances that exhibit the lowest possible energy consumption be used for stand-alone installations. In view of the unavoidable power loss exhibited by inverters, whenever possible it is best to integrate appliances directly into the installation's DC side, particularly small- and medium-output appliances such as energy-saving light bulbs for which special DC adaptors are available.

Only appliances that exhibit relatively high power consumption and for which no DC version is available should be powered by an inverter, in which case inverter operating times should be kept to a minimum. Moreover, in the interest of protecting batteries, insofar as possible such large appliances should be run around noon or during other periods when solar power is abundantly available.

Table 5.6 lists the DC appliances that are currently commercially available for 12 and 24 V PV systems. Such appliances (e.g. refrigerators and freezers) often cost considerably more than the counterpart AC device as they are produced in small series, or they are less expensive (e.g. auto accessory appliances) and are unsuitable for continuous operation. The stand-alone installation battery voltage fluctuates during operation by around 10–15 V for 12 V system voltage and by around 20–30 V for 24 V system voltage.

Table 5.6   Commercially available DC appliances for stand-alone PV installations

Consumers	12 V	24 V
Incandescent lamps	X	X
High-efficiency LED lamps	X	X
Adaptors for normal FL lamps	X	X
Adaptors for FL compact lamps (e.g. PLC)	X	X
Adaptors for radios, pocket calculators, etc.	X	X
Television sets	X	X
Pumps	X	X
Ventilators	X	X
Refrigerators	X	X
Deep freezers	X	X
Vaccum cleaners	X	X
Electric irons		X
Coffee makers		X
Handheld mixers		X
Small immersion heater	X	
Small fan heaters	X	
DC–DC converter fors 13.8 V radio equipment		X
Outboard motors	X	X

5.1.5   Stand-alone 230 V AC PV Installations

With today's sine wave inverters, which exhibit low open-circuit loss and allow for partial load operation, stand-alone PV installations can be realized whose appliance sides are compatible with standard 230 V AC technology, thus allowing for the use of standard 230 V appliances. The structure of such an installation is considerably less complex than that of a conventional PV system.

Figure 5.42 displays the block diagram for such AC-compatible installations, which are coming into increasing use. However, the power loss in such installations exceeds that of conventional systems, particularly for small-appliance outputs, and the inverters used in these installations exhibit continuous open-circuit loss. This factor needs to be taken into account during the sizing phase. Furthermore, there is a major risk with such installations that, as in a standard grid, increasing numbers of appliances will be connected despite the fact that PV installation output is *always* limited by insolation conditions and solar generator size – even if the system integrates a sufficiently large battery. Hence it is essential that only highly efficient AC appliances that exhibit the lowest possible energy consumption be used for stand-alone installations.

Inverters with integrated charge controllers and deep-discharge protection devices are available for which only the solar generator, batteries and 230 V AC appliances need be connected (see Figure 5.43). This in turn reduces the realization costs for such systems. In addition, for small PV installations, solar home systems are now available that integrate a sealed lead battery to which only a few solar modules need be connected, and to which standard 230 V/50 Hz appliances can then be connected.

5.1.6   Stand-alone PV Installations with Integrated AC Power Busses

In this type of installation, energy is distributed on the appliance side while at the same time, insofar as technically feasible, AC power is distributed to all system conductors. This is done via a common

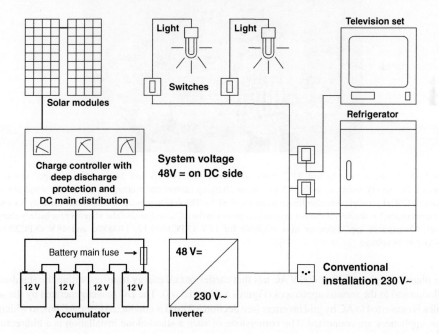

Figure 5.42   Block diagram for a 230 V AC stand-alone PV installation. In such cases, the appliance side can be realized using standard AC technology, thus greatly simplifying the system layout. Moreover, inasmuch as directly connected DC appliances need not be taken into account, DC system voltage can be higher for larger installations and a lower-capacity battery comprising lighter-weight 2 V cells can be used. In addition, many inverters exhibit somewhat higher efficiency with higher DC voltages

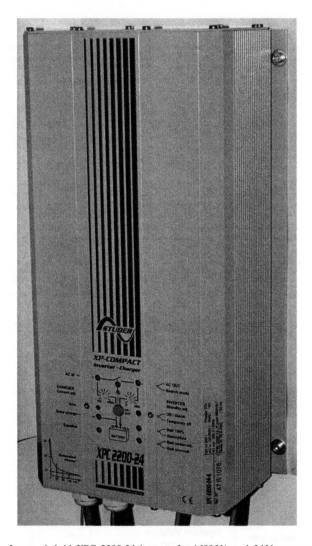

Figure 5.43 Studer-Innotec hybrid XPC 2200-24 inverter for 1600 VA and 24 V system voltage. The device integrates a DC battery recharger with 37 A maximum charging current and a charge controller for stand-alone PV installations (PWM parallel controller for a maximum of 30 A). The device uses up to its nominal capacity for fourth-quadrant operation, i.e. the device can not only deliver power to the AC side, but also absorb power for battery charging purposes. Comparable appliances are also available for 12 V (XPC1400-12, 1100 VA) and 48 V (XPC2200-48, 1600VA) system voltage

single-phase or triphase 230 or 400 V AC bus that carries out all energy transfers within the stand-alone installation and to the various appliances (Figures 5.44 and 5.45). The DC power generated by the solar modules is converted to AC by grid inverters (see Section 5.2) and is conducted to the AC bus, to which the various appliances are connected. The cornerstone of such a stand-alone installation is a bidirectional stand-alone inverter (with a battery connected to it) that continuously supplies the correct voltage and frequency to the AC bus and if need be converts DC to AC and also converts surplus AC to DC for battery charging purposes. To do this, the inverter also needs to take over the function of a charge controller. If the AC bus exhibits surplus energy that cannot be used for battery charging purposes, the bus frequency increases slightly such that the connected grid inverter slightly reduces its output. Appliances of this type,

Stand-alone plant with AC bus

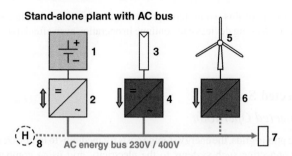

Figure 5.44 Stand-alone installation with AC energy bus (energy flow direction indicated by the arrows): 1, battery; 2, bidirectional inverter; 3, solar generator; 4, grid inverter; 5, wind turbine; 6, wind turbine inverter; 7, appliance; 8, auxiliary petrol/diesel generator

known as Sunny Island off-grid inverters, have been available for a number of years from SMA (power range 3.3–220 kVA).

In order for the system to maintain its stability, the connected grid inverters need to be designed in such a way that they do not shut down in the event of a slight frequency increase, but instead reduce their output. Another option would of course be for the system to power up certain appliances during surplus energy periods.

The following elements can potentially be connected to such AC busses using compatible inverters: (a) other renewable energy sources such as wind turbines; (b) petrol or diesel generators in hybrid systems; (c) relatively large long-term energy storage appliances in the guise of fuel cells; or (d) the public grid in areas with inadequate electricity grids.

This concept renders stand-alone installations substantially more complex, which would undoubtedly complicate installation, operation and maintenance tasks in remote areas in developing countries and the like. However, the advantage of this approach is its ready expandability, since all installations are realized in standard AC technology and standard inverter appliances can be used. This is a great advantage for the electrification of large buildings or remote villages that may be connected to the public grid after it has

Figure 5.45 Triphase stand-alone PV installation (230 V/400 V, 3 · 4.5 kW) with AC bus. Left: battery room. Centre: directional Sunny Island off-grid inverter (one device per phase). Far right: grid inverter that shunts PV power to the AC bus. (Photo: SMA)

been expanded. Investments of this nature have proven to be highly worthwhile, since all appliances, installations and power generating devices can continue in operation even after having been hooked up to the grid.

5.2   Grid-Connected Systems

5.2.1   Grid-Connected Operation

In areas with extensive public grids, the energy storage costs for medium- to large-sized PV systems can be reduced by directly connecting such systems to the electricity grid using compatible inverters. The components available today allow for grid connection of installations with output amounting to upwards of 100 Wp.

Figure 5.46 illustrates how grid-connected systems work. Inasmuch as, by Swiss law, at any given time the sum total of the generated power in an electricity grid must equal the sum total of the power being consumed in the grid, an electricity grid cannot act as a storage platform per se. Hence the power fed into a grid-connected system by a PV installation (or for that matter a wind turbine) is used by appliances somewhere in the grid and need not be generated by any other power plant. Hence the energy thus fed into the grid reduces the amount of primary energy production at a power plant, which can quickly adjust its output to fluctuations in demand. In Switzerland, this arrangement saves a small amount of water at a hydro power plant somewhere in the grid, and thus this water can be kept in a storage lake for use when PV systems are producing too little energy or even none at all.

Grid-connected systems are small power plants that feed their energy into the electricity grid, but that by law must not interfere with normal grid operation and thus must obtain a permit that authorizes the PV system to connect to the relevant electrical power plant. Inasmuch as most energy is used in buildings, it

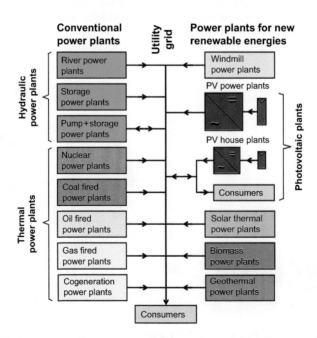

Figure 5.46   Integration of PV systems into grid-connected systems (the energy flow directions are indicated by the arrows)

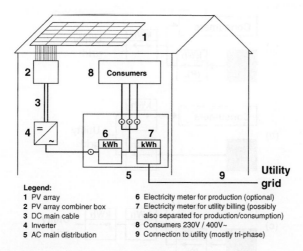

Legend:
1 PV array
2 PV array combiner box
3 DC main cable
4 Inverter
5 AC main distribution

6 Electricity meter for production (optional)
7 Electricity meter for utility billing (possibly
  also separated for production/consumption)
8 Consumers 230V / 400V~
9 Connection to utility (mostly tri-phase)

Figure 5.47 Layout of a PV system mounted on a building. Although a production meter is usually not mandatory, it is extremely useful for monitoring proper operation of the system. Meters can be wired into the system in other ways depending on the accounting and metering arrangement with the local power company (see Figure 5.48)

makes sense to mount PV systems on building roofs and façades so that at least some of the building's energy can be produced without using additional land for PV installation; also, a power lead that can be used free of charge is usually close at hand.

Figure 5.47 displays the layout of a grid-connected system that is installed on a building via a classic wiring configuration using a central inverter. The solar generator (1) on the roof converts solar energy into direct current. The solar generator's strings are wired in parallel into the generator circuit box (2). The direct current thus generated is conducted to the inverter (4) via the DC cable (3), whereupon the inverter converts the direct current into 230 V/50 Hz alternating current. The energy produced by a PV installation should be measured using a separate production meter (6). The energy thus generated can be used to power standard AC appliances (8), and any surplus energy is fed into the public grid. If the PV installation is producing an insufficient amount of electricity or none at all, the requisite energy is drawn from the grid. The energy interchange between the PV system and the grid is metered via one or possibly two electric meters (7).

5.2.1.1 Possible Meter Wiring Configurations for Grid-Connected Systems

Energy meters can be wired into grid-connected systems in various ways, depending on the accounting and metering arrangement with the local power company (see Figure 5.48).

Figure 5.48a displays the simplest such arrangement using only one meter (account metering), which runs forward when the building draws energy from the grid and runs backward when the building feeds energy back into the grid. This is a very simple circuit configuration entailing only one meter and hence the power loss associated with only one meter, and over the long term is probably the best solution. However, in this configuration the system owner is paid the exact amount per kilowatt hour (kWh) that the owner pays for the electricity consumed by the owner.

Figure 5.48b displays a wiring configuration that is used in cases where PV systems are subsidized via electricity prices, i.e. the PV system owner receives a cost-adjusted payment for the electricity sold to the local utility. In such cases the local power company purchases all the energy produced at the higher (subsidized) price, whereas the owner is billed for electricity use at the standard rate, which is of course lower than the subsidized price. This type of circuit arrangement was first introduced in Burgdorf, Switzerland, in 1991 via the so-called Burgdorf model, and has been in use throughout Germany since the country's Renewable Energy Act (EEG) was adopted into law in 2000.

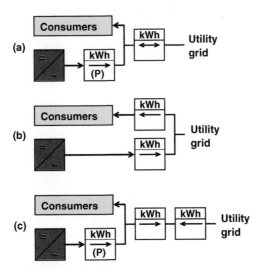

Figure 5.48 Possible meter circuit arrangements for accounting and metering between PV system owners and power companies (P = production meter). (a) Account measurement (1:1) with only one meter: when the PV system owner uses grid power the meter runs forward and vice versa. The local power company credits or debits the user's account as the case may be. (b) Circuit arrangement for scenarios involving energy price subsidies or scenarios or emoluments that cover electricity costs. Such a system was in operation in Burgdorf from 1991 to 1996 and has been in use in Germany since the country's EEG was adopted into law in 2000. In such cases the local utility purchases all energy produced at the higher (subsidized) price, whereas the owner is billed for electricity use at the standard rate, which is of course lower than the subsidized price. (c) Separate measurement and accounting for energy drawn from and fed into the grid, using two meters, each of which integrates an anti-reversing device

Figure 5.48c displays the standard circuit arrangement involving two meters, each of which integrates an anti-reversing device for separate metering and accounting of the energy fed into and drawn from the grid. This arrangement is mainly used by power companies that are leery of photovoltaics and that in most cases pay the PV system owner a lower rate than the rate this owner is billed for the electricity used.

5.2.1.2 Possible System Configurations for Grid-Connected Systems

Grid-connected systems can be configured in any of a number of ways (see Figures 5.49–5.51). In the classic configuration with a central inverter, the solar generator is composed of n_{SP} strings that are parallel connected to n_{MS} modules. This type of system, which entails extensive DC wiring and a special generator junction box (see Section 4.5.3.5), is sensitive to partial shading, as well as mismatch loss within the installation.

For more than a decade now, smaller one-string inverters have been on the market and are advantageous in that (a) wiring is limited to series connecting a number of modules (e.g. 5–40) and (b) a generator junction box can be dispensed with. For larger installations of this type, a series of such single-string

PV plant with central inverters

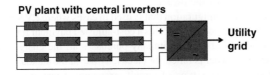

Figure 5.49 Classic PV system configuration involving relatively extensive DC wiring, a number of parallel-connected strings, a central inverter and a generator junction box

PV plant with string inverters

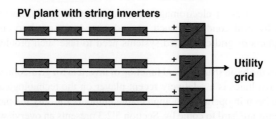

Figure 5.50 Installation with greatly simplified DC wiring (only the module of one string needs to be series connected). One string inverter per string is used

inverters is used, although these devices need to be parallel connected on the grid side. Inasmuch as each string inverter can adjust to the specific insolation conditions at the module to which the inverter is connected, such systems are less sensitive to partial shading. Moreover, DC power loss is usually lower than with central inverter systems owing to the absence of a DC cable and the generally shorter string wiring runs. Based on these advantages, installation efficiency is on a par with or even superior to that exhibited by central inverter systems, despite the fact that the DC–AC conversion efficiency of smaller inverters is usually lower than that of larger inverters.

This configuration can be optimized through the use of module inverters, which are available for power output ranging from around 100 to 400 Wp. Installations with module inverters require either no wiring (for inverters that are attached with adhesive) or very little wiring (e.g. plugging in two connectors) on the DC side, are very sensitive to partial shading and power mismatches, and exhibit virtually no DC-side loss. But since solar modules become quite hot when operated, it is advisable to isolate module inverters thermally from their modules so as to reduce inverter operating temperature, which also increases their life span. And as the life span of a module inverter is considerably shorter than that of a solar module on account of the module inverter's electronics, it is also advisable to connect each module to its inverter using a detachable element so that the inverters can be replaced if necessary.

Module inverters, as well as (in many cases) string inverters, need to be installed adjacent or in close proximity to a PV system's solar modules. Moreover, like solar modules, module inverters are exposed to the elements (Sun, rain, snow, major temperature fluctuations, wind and in some cases melted snow) and thus need to be designed in such a way that they can durably withstand the attendant loads without malfunctioning or corroding. Central inverters, on the other hand, are usually installed indoors, or in a weatherproof container or protective structure, and are thus considerably less exposed to the elements.

5.2.1.3 Basic Problems in Grid-Connected PV Systems

The inverter is the cornerstone and at the same time the critical element of any grid-connected system. Operation of such inverters can interfere with other nearby electrical or electronic appliances secondary to

PV plant with module inverters

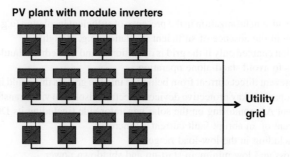

Figure 5.51 Installation with one module inverter per module. No DC wiring is needed for this configuration, which is devoid of hazardous high-DC voltages and virtually rules out the possibility of arcing

grid power surges or harmonics, or electromagnetic radiation. Conversely, inverter malfunctions can potentially be induced by grid conditions such as grid impedance or short-circuit capacity at the connection point. Designers of grid-connected systems need to take such problems into account and resolve them.

Section 5.2.2 discusses the basic structure and function of inverters for grid-connected systems. In view of the fact that (a) photovoltaics is a relatively recent player on the technological stage and (b) grid-connected systems have been in operation in appreciable numbers for only the past few years, specific standards for these systems are hard to come by. Section 5.2.3 presents an overview of the standards and regulations that apply to grid-connected systems, most of which stem from general electrical standards. Section 5.2.4 centres around the problem of islanding (undesired stand-alone installation operation following a power failure in a disconnected sector of the grid). Section 5.2.5 briefly describes the Bern University of Applied Sciences PV Lab tests of 1.5–100 kW inverters, as well as the failures observed during these tests and the causes of these failures. Section 5.2.6 takes a closer look at voltage and power sizing for grid-connected systems, and notably addresses the issue as to which PV output can be hooked up to a specific grid point without causing problems. Section 5.2.7 briefly describes the long-term regulation and stability problems that could potentially occur in electricity grid systems as a whole if grid-connected systems come into extremely wide use.

5.2.2 Design Engineering and Operating Principles of PV System Inverters

In order for solar generator power to be fed into the grid, it must be converted to 50 Hz by an inverter. Apart from this central task, most PV inverters that are specially designed for grid-connected systems also support a solar tracker, i.e. they automatically track the MPP (see Section 3.3.3) in the solar generator's characteristic curve and automatically switch themselves on and off according to the available solar generator power. Inverters also need to be compatible with the specific conditions that obtain in the micro-grid system to which they are connected, and to this end need to have the capacity to detect and terminate hazardous operating states such as post-power-failure stand-alone operation.

In most cases the characteristics that are important for grid-connected system operation differ from those that apply to stand-alone installation operation (see Section 5.1.3). For example, reactive power production and overload capacity are of secondary importance in grid-connected systems. The key requirements for grid-connected inverters are described in Section 5.2.2.1, and each such requirement is discussed in detail in the ensuing sections.

5.2.2.1 Requirements for Grid-Connected Inverters

These are as follows:

- AC power source whose frequency is determined by the *grid*. Fully synchronous operation with the grid system.
- Automatic start-up and synchronization in the presence of sufficient insolation (e.g. in the morning) and automatic shutdown in the absence of sufficient insolation (e.g. at night).
- Start-up and operation enabled only if the grid is operational, and immediate shutdown in the event of a power failure, so as to avoid stand-alone operation.
- The inverter must prevent direct current from being fed into the grid, as this could interfere with normal grid operation or jeopardize key protective devices. Electrical isolation via a transformer or continuous adaptive fault current ΔI monitoring on the solar generator side (including its DC components), and shutdown in the event of an abrupt fault current ΔI increase.
- Good efficiency, including in the low-load range.
- Low open-circuit loss and low minimum start-up and shutdown power.
- Control units should be fed from the DC side. Minimum power absorption from the grid (if possible, zero) in a shutdown state (at night) to avoid unnecessary loss.

- High level of reliability (several years or more of trouble-free operation between inverter malfunctions) and a life span of 15–20 years, i.e. on a par with the life span of major household appliances.
- Trouble-free maximum power tracking for a large power range within a sufficiently large MPP tracking range from (at a minimum) $0.8 \cdot V_{MPPA\text{-}STC}$ to $1.25 \cdot V_{MPPA\text{-}STC}$ ($V_{MPPA\text{-}STC} =$ solar generator voltage at the MPP under STC power output conditions, i.e. at a cell temperature of 25 °C and with $G_o = 1\,\text{kW/m}^2$).
- The inverter should not incur any damage when exposed to input DC voltages of at least $1.4 \cdot V_{MPPA\text{-}STC}$ (see definition in the previous item) and should be able to start up without any difficulty (for more on optimal voltage sizing of PV installations, see Section 5.2.6.8).
- Adequate DC-side filtering: in single-phase inverters, the frequency of the energy fed into the grid is twice that of the grid. Hence a sufficiently large electrolytic condenser is needed on the DC side to prevent the input voltage's 100 Hz ripple from becoming unduly large, otherwise the resulting fluctuations around the MPP will reduce the power obtained from the solar generator.
- The inverter needs to do the following in the event of surplus DC-side power resulting from brief cloud enhancements or the like: limit the power fed into the grid to AC-side nominal output (operating point shifts towards open-circuit voltage V_{OC}), but without shutting down.
- Power surge protection on both the AC and DC sides.
- Immunity against grid commands (ripple control signals) in the 110 Hz to 2 kHz range.
- Low reactive power consumption from the grid (power factor $\cos \varphi = P/S = 0.85\text{–}1$).
- Current curve that is as sinusoidal as possible, i.e. minimal generation of current harmonics in accordance with EN 61000-3-2 (formerly EN 60555-2) for up to 16 A AC appliances, and conforming to EN 61000-3-12 for higher-power appliances with up to 75 A per phase.
- For PV inverters used in residential buildings: the device should not interfere with neighbouring electronic appliances such as radios, i.e. low generation of high-frequency interference voltages on the DC and AC sides (conformance with the limit values of the electromagnetic interference standard EN 61000-6-3 [5.23] on the AC and DC sides).
- Solar generator insulation monitoring for inverters that integrate a transformer.

In the interests of simplifying grid regulation in the grid system, it is preferable if the inverter limits the power fed into the grid in cases where the grid voltage reaches its upper limit value, i.e. 230 V + 10%. The PV system should be shut down only in cases where the voltage is too high when no power is being fed into the grid. If grid-connected systems come into wider use, over the long term the cause of grid stability would be greatly served if power fed into the grid were limited successively in cases where grid frequency goes above 50.2 Hz, and if a full PV system shutdown only occurred at, say, 51.5 Hz.

However, these characteristics would increase the likelihood of stable stand-alone installation operation in a shutdown sector of the grid, and should only be allowed if adequate precautions against islanding have been taken (see Section 5.2.4).

5.2.2.2 Line-Commutated Inverters

Line-commutated inverters are simple appliances that are relatively inexpensive as they are used for a host of other applications, and are nowadays used in some large grid-connected systems. However, these inverters also have drawbacks in that during operation they generate square- or trapezoidal wave current with high harmonics in the grid and use reactive power. Both of these phenomena are undesirable from the grid operator's standpoint. In addition, a large 50 Hz transformer is needed to isolate these devices electrically from the grid.

The functional principle of a line-commutated inverter will now be discussed for a single-phase scenario. Figures 5.52 and 5.53 display a circuit arrangement of this type, which essentially comprises a bridge connection composed of four thyristors, as well as high inductance L on the DC side. The functional principle of these appliances is easier to grasp if rectifier operation, which is illustrated in Figure 5.52, is considered first.

Rectifier operation ($0° < \alpha < 90°$):

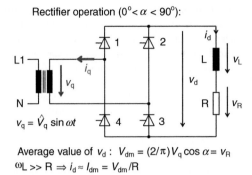

Average value of v_d : $V_{dm} = (2/\pi) V_q \cos \alpha = V_R$

$\omega L \gg R \Rightarrow i_d \approx I_{dm} = V_{dm}/R$

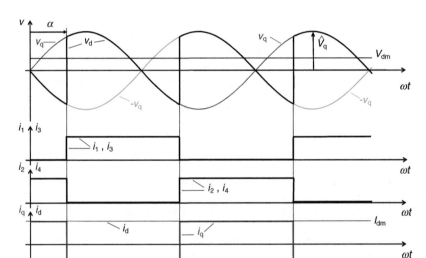

Figure 5.52 Rectifier operation of a single-phase line-commutated inverter with a four-thyristor bridge circuit. The grid can be represented on the secondary side of the transformer by a nearly ideal voltage source $v_q(t) = \hat{V}_q \sin \omega t$. The i_q direction is defined for the appliance metering system at the source v_q. Thyristors 1 and 3 are each ignited in accordance with triggering angle α following the v_q zero-crossing increase, whereas thyristors 2 and 4 are ignited a half period later

A thyristor is essentially a diode that only begins conducting current in the flow direction (a) on receipt of a positive ignition pulse from an auxiliary electrode known as the gate and (b) after blocking this pulse when the thyristor current is 0. In the circuit arrangement shown here, thyristors 1 and 3 are each ignited in accordance with angle α following the v_q zero-crossing increase, whereas thyristors 2 and 4 are ignited a half period later. The resulting voltage v_d via the R–L series connection is shown in black in Figure 5.52. When $\omega L \gg R$, the alternating component portion of v_d is filtered out via the inductance L such that the current i_d is virtually constant. Thyristor current flow is enabled by inductance, where because $v_L = L \, di/dt$ the voltage can also be negative and can also remain so following the v_q sign digit change, until the next two thyristors are started up. Thus resistance R exhibits virtually constant voltage V_{dm}, whereby the following applies:

$$\text{Linear mean } v_d: V_{dm} = \frac{2}{\pi} \cdot V_q \cdot \cos \alpha \qquad (5.22)$$

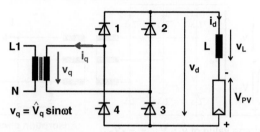

Inverter operation ($90° < \alpha < 180°$):

Average value of v_d: $V_{dm} = (2/\pi)V_q\cos\alpha = -V_{PV}$

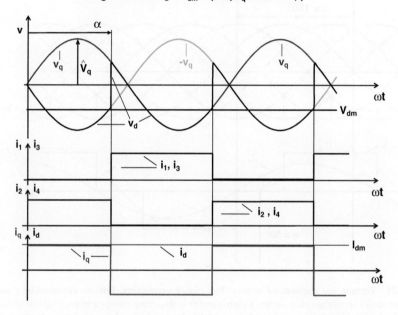

Figure 5.53 Inverter operation of the single-phase line-commutated inverter circuit from Figure 5.52 ($\alpha = 150°$ in the diagram). The power i_q fed into the grid is rectangular in form and thus exhibits high total harmonic distortion. Its first-harmonic oscillation is only slightly leading (about 30°) relative to v_q, i.e. active power is fed into the grid, but the inverter uses some inductive reactive power from the grid

Hence according to Equation 5.22, the voltage at resistance R can be adjusted via triggering angle α. In addition to the relevant voltages, Figure 5.52 also displays currents $i_1 - i_4$ that are conducted through the thyristors, as well as source current i_q, which exhibits a symmetrical rectangular curve and thus also displays high harmonics.

According to Equation 5.22, negative V_{dm} values could also occur for $\alpha > 90°$, although this could not occur at a passive resistance R as the voltage and current there always need to be flowing in the same direction. However, voltage and current at a DC power source can flow in opposite directions. This scenario is illustrated in Figure 5.53, where a solar generator is substituted for the resistance such that the circuit arrangement shown operates with an inverter and power flows from the DC side to the inverter side.

A triggering angle α far exceeding 90° is usually defined for inverter operation, but should be well below 180° as commutation failures at the inverter can occur at such an angle. Inasmuch as L is always very high, i_d is nearly constant. Under inverter operation, thyristor currents $i_1 - i_4$ are considerably lagging compared with rectifier operation. As V_{PV} and i_d flow in opposite directions, the solar generator injects power into the

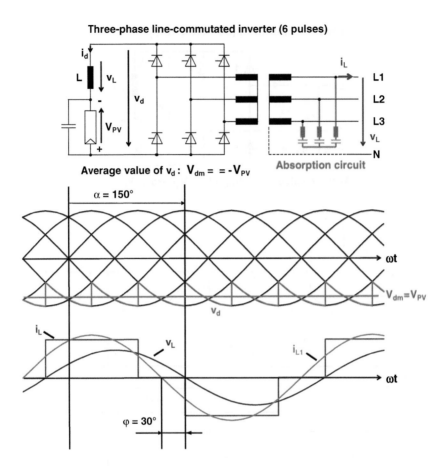

Figure 5.54 Triphase line-commutated inverter. Top: circuit arrangement. Bottom: characteristic *v* and *i* curves (without harmonic current absorption circuits, shown in green). In this arrangement, a triphase bridge circuit is substituted for a single-phase bridge circuit, and the solar generator is bypassed via a large condenser. As the first-harmonic oscillation i_{L1} of the grid current is not in phase with the grid voltage, the inverter uses inductive blind current

circuit. Current i_q on the AC side is always rectangular in form and thus exhibits high total harmonic distortion, but at this juncture is nearly in phase with v_q such that the power is absorbed by the source v_q. Hence in this circuit arrangement energy flows from the DC to the AC side, which means that this arrangement acts as an inverter. The energy flow can be regulated by varying the ignition angle α. As i_q is not fully in phase with v_q, depending on the size of α this circuit arrangement always consumes a certain amount of reactive power, whereby the following applies to the first-harmonic oscillation: $\cos \varphi_1 = \cos(\alpha + \pi) = -\cos \alpha$ with our i_q metering direction.

Line-commutated inverters are usually realized as triphase devices. Figure 5.54 displays the circuit arrangement for a six-pulse line-commutated inverter. Owing to the triphase design, the phase conductors exhibit square-wave current with a 120° lead angle per half wave (in lieu of the single-phase 180° angle), i.e. the current is trapezoidal with an interim stage at 0. The curve for this current equates to that of a pulse-width-controlled sine wave inverter (see Figure 5.31). The odd-numbered integers of the third harmonic (i.e. 3, 9, 15, 21 and so on) are absent from such currents. The triphase version is thus a far better solution from the standpoint of harmonics, which can be further reduced by integrating a filter into the harmonic current absorption circuits (shown in green) and through the use of higher-pulse circuit arrangements (e.g. 12 pulses).

Inasmuch as the relatively high inductance L in all of these circuit arrangements keeps the current I_D at a nearly constant level, the inverters used in the scenarios illustrated in Figures 5.52–5.54 are also referred to as current source inverters. For more on such devices, see [5.11] among others.

Owing to the fact that the price of self-commutated inverters (whose harmonic characteristics are far superior) has come down considerably in recent years, no single-phase line-commutated inverters are used nowadays and high-output three-phase line-commutated inverters are used only rarely. However, self-commutated inverters are coming into increasing use at the expense of line-commutated inverters, including those with high output.

5.2.2.3   Differences Between Self-commutated and Line-Commutated Inverters

The trend nowadays is towards the use of self-commutated inverters owing to the advent of evermore stringent harmonics regulations and the recent major price decreases for these appliances.

Line-commutated inverter thyristors are switched on and off once per period. However, these thyristors can only be shut down indirectly in that the ignition of additional thyristors at the opportune moment in a formerly conductive thyristor triggers a current zero crossing and thus an extinguishing process. Inasmuch as the grid is needed for commutation, line-commutated inverters have very little tendency to operate under self-commutation conditions (i.e. post-power-failure stand-alone operation in a disconnected sector of the grid), owing to the switching procedure that is used.

Self-commutated inverters integrate electronic switches that are switched on and off directly via the gate a number of times per period, thus greatly reducing reactive power consumption and current harmonics. In lieu of thyristors, self-commutated inverters integrate the following elements that act as rapid electronic switches:

- Gate turn-off thyristors (GTOs), which as their name suggests are thyristors that can be turned off via a gate.
- Bipolar solid state transistors.
- Power MOSFETs (power FETs with insulated gates).
- Insulated gate bipolar transistors (IGBTs).

Nonetheless, the switching procedure used in self-commutated inverters tends to induce stand-alone operation following a power failure. But as this is undesirable for inverters during grid-connected system operation for reasons of safety, appropriate precautions need to be taken in this regard.

To attain high efficiency, the semiconductor switches that are used need to be either fully connected or completely cut off. Moreover, extremely steep-edged pulses should be used in inverters that generate high-frequency interference, which is particularly pronounced in self-commutated inverters that exhibit numerous switching cycles per period. Without adequate filtering on both the DC *and* AC sides, such inverters induce extremely pronounced high-frequency interference in the connected lines, and this in turn can interfere with radio reception and the operation of other electronic appliances.

Sections 5.2.2.4–5.2.2.6 describe some of the key circuit arrangements that are used in commercially available PV inverters for grid-connected systems. Needless to say, not all available appliances of this type are discussed here, and other circuit arrangements that differ to varying degrees from those described (e.g. push–pull stages with a tapped primary transformer winding arrangement in lieu of bridge circuits; PWM signal generation with only one switching element (boost or buck converter) and an attached 50 Hz flip-over device; and a DC regulator with a solar tracker function upstream of the actual inverter circuit).

5.2.2.4   Self-commutated Pulse-Width-Modulated Inverters with LF Transformer

Figure 5.55 displays the basic circuit arrangement for a single-phase self-commutated and pulse-width-modulated inverter. The condenser-smoothed solar generator voltage is located at the inverter input, whereby such a device is referred to as a voltage-controlled inverter. The main circuit arrangement here is similar to that used for the stand-alone installation inverter in Figure 5.33. The sinusoidal output voltage is

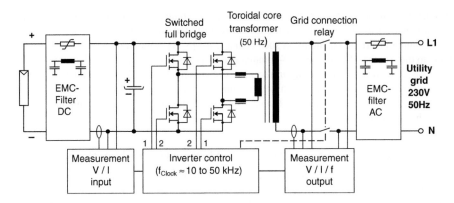

Figure 5.55 Single-phase self-commutated and pulse-width-modulated inverter with a 50 Hz transformer

realized in a bridge circuit comprising solid state transistors, power MOSFETs or (for high-output systems) IGBTs with pulse width modulation as in Figure 5.34. A relatively high pulse rate of n per period is used in most cases. The higher the value n, the more the low-frequency harmonics can be eliminated directly without the use of additional filters. Varying pulse width allows for the adjustment of the effective first-harmonic oscillation value to that needed to adapt to grid voltage and to control the energy flow. In most cases a simple inductor adequately suppresses the relatively high PWM pulse frequency (high kilohertz). Electrical isolation is realized using a low-loss 50 Hz toroidal core transformer at whose output, however, a grid relay is also necessary in order for the inverter to be connected to the grid in parallel.

Regulating an inverter that is used to feed electricity into the grid is of course a more complex proposition than for a stand-alone installation inverter. In addition to actual inverter operation, it is also necessary to implement trouble-free MPP tracking that automatically regulates the switching cycles and reliably obviates untoward operating states such as islanding, operation with unduly low or high voltage, impermissible grid frequency and so on. Single-phase inverters with this circuit arrangement are available from ASP, SMA and other manufacturers.

If a triphase bridge circuit is used in lieu of a single-phase bridge circuit, self-commutated pulse-width-modulated triphase inverters can be realized in a similar manner. Figure 5.56 displays the circuit arrangement for such an inverter. In such cases IGBTs are usually used in lieu of power MOSFETs owing to the large amount of power that needs to be converted. Pulse width modulation for this relatively large semiconductor switch is generally realized using a lower frequency than for small single-phase inverters; 20–500 kW inverters that exhibit this circuit arrangement are available from Sputnik, SMA, Siemens and other manufacturers.

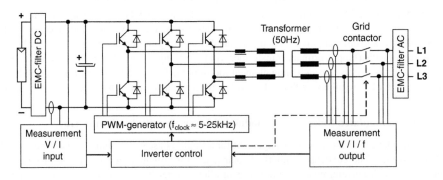

Figure 5.56 Self-commutated triphase and pulse-width-modulated inverter with a 50 Hz transformer

Electrically isolating the DC and AC sides from each other via a transformer is advantageous from both an operational and safety standpoint. With such an arrangement, any voltage desired can be used on the DC side (for small installations, e.g. from 50 to 250 V; for installations with protective low voltage less than 120 V), thus keeping the load on the module insulation within reasonable bounds and minimizing risk to life and limb in the event of a solar generator malfunction. Moreover, in the event of an inverter malfunction, the grid voltage is not automatically shunted to the solar generator. In addition, use of an insulation monitoring device on the DC side allows for continuous monitoring of solar generator insulation, which means that incipient problems can be detected and corrected before they develop into more serious problems.

5.2.2.5 Self-commutated Pulse-Width-Modulated Inverters Without Transformers

Although many power companies no longer insist on electrical separation of solar generators from the AC grid, they nonetheless do require that precautions be taken to ensure that direct current cannot be fed into the grid in the event of a solar generator malfunction. This objective can be attained not only via the electrical isolation of the solar generator from the AC grid (an arrangement that always entails a certain amount of power loss), but also via measures such as adaptive fault current monitoring. Dispensing with a transformer can enhance an inverter's efficiency.

Figure 5.57 displays the circuit arrangement for a transformerless self-commutated pulse-modulated inverter. In such a case, in lieu of a transformer only a low-pass filter and adaptive fault current monitoring are necessary to respond in the event of an abrupt change in fault current ΔI (e.g. if someone touches a faulty module); however, within certain limits these elements do not react to gradual leakage current changes occasioned by events such as the solar modules being wetted by rain or snow. This circuit logic necessitates a relatively high input-side input voltage that should be a minimum of 400 V, although actual PV systems exhibit higher voltages of the order of 450–700 V. These higher voltages also minimize DC wiring power loss, with the result that the specific energy yield of grid-connected systems that integrate such inverters can potentially be several percentage points higher than conventional installations with a DC cable and a central inverter that integrates a transformer.

One example of a commercially available inverter that uses this circuit arrangement is the Sputnik AG Solarmax 4000C (previously the Solarmax S).

Owing to the high operating voltage, as well as the AC components that are sometimes integrated into the DC side (see Figure 5.58), module and cable insulation is subjected to a relatively high load, with the result that after a number of years or decades this insulation is likely to break down. Figure 5.58 displays

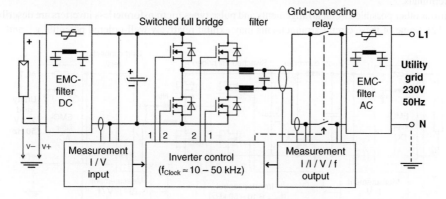

Figure 5.57 Single-phase self-commutated and pulse-width-modulated inverter without a transformer. Solar generator fault current ΔI is monitored continuously (i.e. the DC and AC components). In the event of an abrupt change in ΔI induced by an event such as someone touching a defective module, the inverter is shut down in a process known as adaptive fault current monitoring. This circuit arrangement requires DC input in excess of 400 V

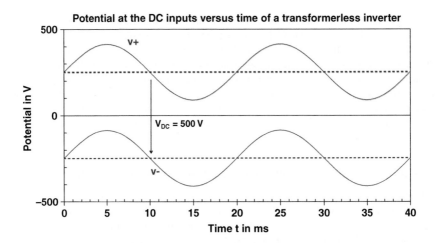

Figure 5.58 The v^+ and v^- potential for a transformerless inverter with a symmetrical pulse (the two switches diagonally opposite each other are always closed at the same time) for 500 V DC operating voltage. Both terminals Exhibit 50% of nominal DC operating voltage as well as 50% of nominal grid voltage ($f = 50\,\text{Hz}$)

the characteristic waveform for the potential v^+ and v^- to earth that occurs at the solar generator terminals. Each of the two terminals Exhibits 50% of nominal DC operating voltage and 50% of nominal grid voltage. However, for an inverter with a transformer as in Figure 5.55 or 5.56, the counterpart voltage curve for a ungrounded solar generator roughly corresponds to the dashed line (some single-phase appliances integrate a relatively weak 100 Hz component with only a few volts of power).

To offset the disadvantage of high-input DC voltage and at the same time facilitate greater variation for such voltage (e.g. for a string inverter), upstream of the actual inverter circuit a boost converter can be integrated that also has a solar tracker (see Figure 5.59). Such an arrangement reduces efficiency somewhat on account of the additional loss in the boost converter. Examples of commercial appliances that employ this circuit arrangement are Sputnik's Solarmax 6000C, as well as SMA's Sunny Boy SB2100TL and SB3300TL models.

Figure 5.60 displays the characteristic waveform for the grounded potential v^+ and v^- for such an inverter/boost converter that occurs at the solar generator terminals. In this arrangement each of the two terminals Exhibits 50% of nominal grid voltage, but DC voltage is unevenly divided between the two terminals.

Various other possible circuit arrangements and pulse types for transformerless inverters are described in [5.12]. Some of these circuit arrangements may exhibit relatively well-ordered potential v^+ and v^-,

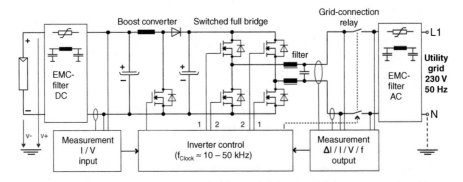

Figure 5.59 Self-commutated pulse-width-modulated inverter (transformerless) with a boost converter for low input voltages

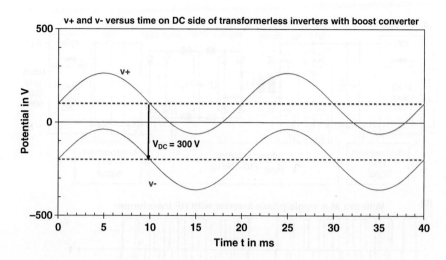

Figure 5.60 Potential v^+ and v^- for a transformerless inverter/boost converter and symmetrical pulse (the two switches diagonally opposite each other are always closed at the same time) with 300 V DC operating voltage. Although each terminal Exhibits 50% of nominal grid voltage, DC voltage distribution is asymmetrical

while in others this potential is far more chaotic – for example, nearly square jumps from 0 to V DC with v^+ and from 0 to $-V$ DC with v^- for so-called single-phase choppers. Such circuit arrangements exhibit strong electromagnetic disturbances and place considerable long-term strain on solar generator insulation and wiring in a manner for which these elements are neither rated nor tested. Hence the use of single-phase choppers should be avoided at all costs.

5.2.2.6 Self-commutated Pulse-Width-Modulated Inverters with High-Frequency Intermediate Circuits

In small installations with a nominal power of a few kilowatts, conventional 50 Hz transformers exhibit relative power loss (e.g. 5% of nominal output) and are also quite heavy. Moreover, open-circuit loss considerably reduces the installation's efficiency in the low-load range that the inverter mainly operates in.

One way to scale back this transformer loss and to realize lighter-weight transformers is to use high-frequency intermediate circuits (see Figure 5.61). In such cases, solar generator direct current is converted into alternating current using a self-commutated high-frequency inverter whose frequency ranges from around 10 to 100 kHz. Hence, based on the high frequency in this arrangement, electrical isolation can be realized using a very slow transformer that exhibits far lower power loss. In most cases the high-frequency alternating current is also pulse modulated (see Figure 5.61).

Downstream of the high-frequency transformer, the high-frequency alternating current is rectified and filtered, resulting in an output current in the form of a rectified sinusoidal half wave. In a final step, each half wave is flipped over in a bridge circuit composed of thyristors, solid state transistors or power MOSFETs in such a way as to generate a nearly sinusoidal output current that can be fed into the grid. The first mass-produced, self-commutated, 1–3 kW inverters for single-phase grid-connected systems, which were marketed in Germany, Austria and Switzerland from 1998 to 1991, mainly used the functional principle mentioned above (e.g. the SI-3000 from Photoelectric; Solcon 3000, 3300 and 3400 from Hardmeier; and PV-WR 1500 and 1800 from SMA). Owing to the fact that early flip-over appliances and high-frequency rectifiers were located directly on the grid and had little in the way of protective series impedance, they often failed. Many of these initial appliances also suffered from inadequate filtering of high-frequency interference voltages that induced extreme radio reception interference and interference with neighbouring electronic appliances.

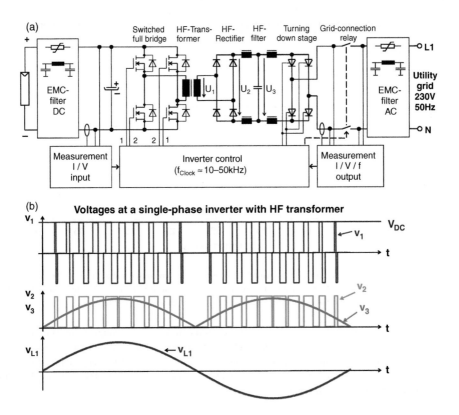

Figure 5.61 Circuit arrangement and voltage curves for a single-phase self-commutated inverter with a high-frequency intermediate circuit; electrical isolation is realized via a high-frequency transformer. Possible variants: push–pull stage in lieu of a bridge circuit (as in Figure 5.32) at the input; power MOSFETs in the flip-over device at the output, in lieu of thyristors

The circuit arrangement described here is essentially a single-phase solution. Hence three such single-phase high-frequency inverters are needed, which greatly increases the complexity and cost of the systems into which these appliances are integrated.

5.2.2.7   Recent concepts and Optimizations for Grid-Connected Inverters

The key basic forms of modern grid-connected inverters are described in Sections 5.2.2.2–5.2.2.6. However, numerous inverters that integrate various optimizations that build on these concepts are now commercially available, but cannot all be described in detail here owing to space limitations. The most important of these new concepts are as follows:

- An internal DC bus is fed by a series of independent DC–DC converters without electrical isolation, each of which integrates a solar tracker for two to four arrays. The power in this DC bus is converted to AC power using a shared inverter. Appliances of this type are available from SMA, Mastervolt and others.
- Electrical isolation on the DC side via DC–DC converters that integrate a high-frequency transformer. In this solution, the AC conversion is realized via a transformerless inverter. Such appliances are available from Fronius, SMA and other manufacturers.
- A team configuration in lieu of a master–slave arrangement, as in the past. Under low-insolation conditions, a solar generator's entire power output is handled solely by one of the team inverters, thus

allowing for higher low-load efficiency. When insolation is higher, two or more inverters are run simultaneously. Such appliances are available from SMA and other manufacturers.

- New circuit arrangements for even greater efficiency increases of transformerless inverters. Such solutions are available from Sunways (HERIC technology) [5.12]. In this concept, in the circuit configuration in Figure 5.57 two switchable anti-parallel diodes are used (two strings consisting of a diode in series with a semiconductor switch, both strings in anti-parallel configuration). Each of these diodes is activated continuously during a half period by closing its series switch bypasses the choke current during the freewheeling phases (switch in the bridge open), with the result that fewer power losses occur.
- Transformerless inverter with a grounded negative pole (for thin-film solar cell installations).
- Triphase transformerless inverters are also available. Such devices generally integrate a series of electrically non-isolated solar trackers with or without a shared DC bus.

5.2.2.8   Central Inverter Stations with Direct Feeding into the Medium-Voltage Grid

The output of extremely large PV systems is often too high for connection to the low-voltage grid and thus energy is fed into the medium-voltage grid instead. In such cases, twice an independent electrical isolation between the DC and medium-voltage sides is not absolutely necessary, and some power loss can be avoided by forgoing electrical isolation between the input on the DC side and the output on the AC side in the inverters at one transformer. With this method, the absolutely necessary electrical isolation vis-à-vis the medium-voltage grid can be realized using a medium-voltage transformer.

In the interest of (a) avoiding reciprocal disturbances on the part of the large transformerless triphase inverters that are used and (b) easing the task of detecting anomalies in the solar generator fields of the various inverters, it is advisable to use a separate triphase undervoltage winding for each inverter. Manufacturers such as ABB, SMA, Sputnik and Voltwerk sell special transformerless inverters for use in such installations. In order to ensure that this relatively large input range (e.g. $V_{MPP} = 450–800$ V) does not need a boost converter at the input (which would lower system efficiency somewhat), these inverters often use, at the input, a somewhat lower phase-to-phase voltage ranging from 280 to 300 V. However, this is no longer relevant as in any case a special medium-voltage transformer must be used for conversion of the medium voltage (e.g. 16 or 20 kV). According to manufacturers' data, the peak conversion efficiency attained by these devices ranges up to 99%, while European standard efficiency is up to 98%. Inasmuch as the peak efficiency of relatively large medium-voltage transformers likewise ranges up to 99%, overall conversion efficiency is far greater for feeding electricity into the medium-voltage grid.

Thus to hook up two such inverters, a transformer with three triphase windings is needed, while to connect three such inverters a transformer with four triphase windings is needed. Figure 5.62 displays a block diagram for such a central inverter station, for feeding electricity directly into the medium-voltage grid.

Containers that can accommodate a medium-voltage transformer (which often possesses a small switching device for two outputs) as well as two or three 250–630 kW central inverters are available for relatively large ground-based PV installations.

5.2.2.9   Inverters for Grid-Connected Systems

For these inverters, see Figure 5.63–5.65.

5.2.3   Standards and Regulations for Grid-Connected Inverters

Before a grid-connected system is connected to the public grid, it is absolutely essential to obtain a permit from the relevant local power company. Grid-connected inverters are not allowed to interfere with normal grid operation or with devices connected to the grid, and thus must conform with various standards and regulations.

**Central inverter station for direct power
injection into medium voltage grid**

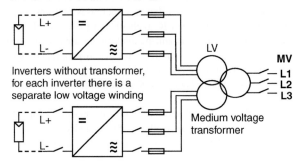

Figure 5.62 Block diagram of a central inverter station for feeding electricity directly into the medium-voltage grid. The medium-voltage transformer integrates a number of undervoltage windings that are electrically isolated from each other (MS = medium voltage; US = undervoltage)

Instead of following the usual practice of listing the relevant standard designation numbers (which in any case often change), this section summarizes the key content and purpose of the most important standards, regulations and limit values which, to the best of my knowledge, apply to PV systems that are coupled to the low-voltage grid in Switzerland. For reasons of space, these matters have been simplified to some extent, without sacrificing coherence. Although installations realized in accordance with the information in this section should function properly, I cannot of course guarantee that the information provided is either complete or accurate. The aim of this section is to provide electricians and electrical engineers whose work involves grid-connected systems with a convenient summary of the relevant information. For reasons of space, it was simply not possible to go into detail concerning all applicable standards, which are in any case not always fully consistent with each other. For full details concerning these standards, see the texts of the actual standards [4.10], [4.13], [5.14]–[5.23], which are available from Electrosuisse, Postfach, CH-8320 Fehraltorf, Switzerland.

In recent years, many European standards have been harmonized, and there are now harmonized standards for Germany, Austria and Switzerland such as [5.19], although the applicable German and

Figure 5.63 Single-phase SB3800 inverter with transformer as in the circuit arrangement of Figure 5.55. The underside integrates positive and negative PV plug connector jacks (not shown) that allow for direct connection to three strings. The grid is connected via a special coupling (not shown). The nominal AC power for this device is $P_{ACn} = 3.8\,kW$

Figure 5.64   Triphase Solarmax 25C inverter with transformer as in the circuit arrangement of Figure 5.56 The DC cable and the grid are interconnected via terminals inside the device. The nominal AC power of this device is $P_{ACn} = 25\,\text{kW}$

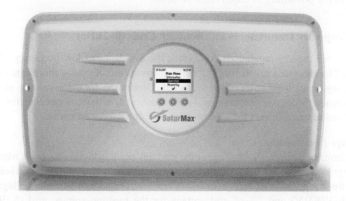

Figure 5.65   Single-phase Solarmax 6000S transformerless inverter as in the circuit arrangement of Figure 5.59. The underside integrates positive and negative PV plug connector jacks that allow for direct connection to three strings. The grid is connected to this device via a special coupling. The nominal AC power for this device is $P_{ACn} = 4.6\,\text{kW}$

Austrian regulations differ somewhat from the Swiss regulations. Moreover, the rules promulgated by local power companies also need to be taken into account in many cases. Hence, if need be, it is advisable to consult the actual texts of the applicable standards and regulations, which are available from local power companies and from national power company associations (VDEW in Germany, VEÖ in Austria). For general information concerning the current situation in this regard in Germany, also see [DGS05].

Like other electrical installations, PV systems must exhibit trouble-free performance in their electromagnetic environment, where they are not allowed to generate any disturbance via inadmissibly elevated electromagnetic emissions. This property is known as electromagnetic compatibility.

5.2.3.1 Provisional Safety Regulations for PV Systems

The safety regulations issued in 1990 by the Swiss regulatory organization known as Eidgenössischen Starkstrominspektorat (ESTI) [4.4] contain stipulations concerning the following: lightning protection; the mechanical and electrical structure of solar generators; and wiring and grounding on both the DC and AC sides. The ESTI regulations also stipulate the following concerning the inverters used in grid-connected systems: they must shut down within five seconds after a power failure; they are not to feed any direct current into the grid; and they are not to generate any inadmissible harmonics or conducted disturbances. The ESTI regulations further state that small single-phase PV systems up to 3.3 kVA and triphase 10 kVA building installations are the equivalent of each other and that such installations need not be submitted to ESTI prior to realization. This relatively permissive regulation (which has since been superseded), along with various local subsidy programmes, were undoubtedly largely responsible for the spate of grid-connected systems that were realized in Switzerland from 1990 to 1995.

5.2.3.2 Provisional Safety Regulations for PV System Inverters

The safety regulations issued in 1993 by the Swiss regulatory organization known as Schweizerischen Elektrotechnischen Verein (SEV) [5.13] contain the following, apart from general safety regulations concerning PV system elements: information about the applicable electromagnetic compatibility standards for the low- and high-frequency ranges; information about the required self-tests; and a stipulation to the effect that overload tests must be conducted for any inverter that is operated for up to two hours under 140% of the solar generator's nominal output at STC power output, with a view to ensuring that the inverter will continue to function properly during brief cloud enhancements (see Figure 5.66). Even though this regulation has since been superseded, such tests are extremely important for grid inverters owing to the possibility of cloud enhancements.

5.2.3.3 Low-Frequency Disturbances in Electricity Grids (0–2 kHz)

This section discusses the key regulations concerning the avoidance of low-frequency disturbances in electricity grids. The measures called for by these regulations are of course more important to power companies than, say, regulations concerning conducted disturbances for the avoidance of radio and TV reception interference. Moreover, monitoring low-frequency disturbances necessitates a completely different fleet of instruments than is the case with high-frequency disturbances.

5.2.3.3.1 Voltage Fluctuations

Operation of an inverter increases the voltage at the grid link point, which is the point where a device is connected to the grid – for example, the point where a building's main distribution board is connected to a grid meter. The limit value for very rarely occurring voltage changes is 3% of grid voltage for the low-voltage grid (230 V/400 V) and 2% of grid voltage for the medium-voltage grid [5.19]. According to [5.15] and [5.16] (both of which are standards), values of up to 4% are also allowable for rarely occurring voltage fluctuations.

In PV systems, output can vary from around 10 to 100% of nominal power several times a minute under windy conditions with scattered clouds. The above standards stipulate that the allowable voltage

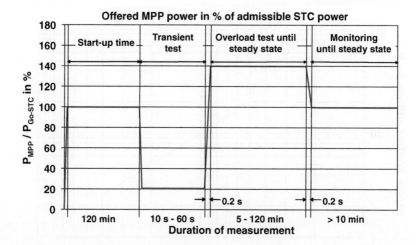

Figure 5.66 Overload test in accordance with [5.13] involving a DC-side power overload $P_{MPP} > P_{Go}$, for the purpose of ensuring that the inverter will function properly during a cloud enhancement (P_{Go} = maximum allowable solar generator nominal capacity at STC power output). Cloud enhancements can provoke voltage peaks of up to 1400 W/m² in flat areas and up to 1900 W/m² at high altitudes

fluctuations can potentially be reduced via higher repetition rates (see Figure 5.67). However, inasmuch as the above weather conditions rarely occur, and if the number of grid-connected systems remains within reasonable bounds, it would pose no problem if the power leads of such PV systems were sized in such a way that, when the inverters are operated at nominal voltage, the voltage at the low-voltage grid link point does not exceed the voltage of a shutdown inverter by more than 3% (from both L to N (live to neutral) and L to L (live to live)). This requirement applies to stand-alone PV systems. If grid-connected systems come into greater use, it may be necessary to reduce the voltage increase further to, say, around 1%. In addition to adhering to the maximum allowable voltage fluctuation, PV systems must of course also conform with the maximum allowable grid voltage (230 V/400 V + 10%) at the grid link point.

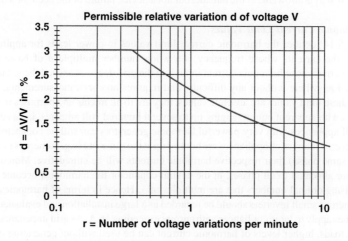

Figure 5.67 Allowable relative voltage fluctuation d in the low-voltage grid (230 V/240 V) as a function of repetition rate according to [5.19]. According to [5.15] and [5.16], d values of up to 4% are also allowable for sporadic voltage fluctuations

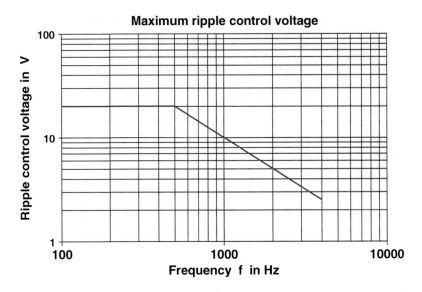

Figure 5.68 Allowable ripple signal voltages (L–N) in low-voltage grids [5.19]. Higher signal levels ranging up to 10 V can occur at frequencies greater than 1 kHz [5.18]

5.2.3.3.2 Immunity to Ripple Control Signals
Figure 5.68 displays the figures of merit for the maximum anticipated ripple control signal (a low-frequency control signal of a few volts, superimposed on the line voltage by the utility to block, enable or disable heavy loads of consumers) in the 100 Hz to 4 kHz range. Although the level of the signals emitted by power companies is usually far below the maximum values [5.19], owing to reactive power compensation systems the grid impedance ratios at higher frequencies are not clearly defined and thus can vary greatly. Hence, resonance phenomena may result in local levels that at certain times differ considerably from the values indicated in Figure 5.68. This in turn means that PV inverters should be sized in such a way that they can withstand such levels without incurring serious failure, i.e. a brief fault followed by an automatic restart is allowable at the outside, but not a device failure or the need for a manual restart.

5.2.3.3.3 Harmonic Current Limit Values
EN 61000-3-2 [5.14] defines the harmonic current limit values in power leads for appliances $\leq$ 16 A (Table 5.7), i.e. the currents whose frequency is a whole-number multiple n of basic frequency. A stationary PV inverter whose harmonic current input into the grid does not exceed the values in the table can be connected anywhere without any difficulty and requires no check measurements.

Inasmuch as these limit values are absolute values, they are to all intents and purposes meaningless for small appliances with apparent output ranging up to several hundred volt amperes, i.e. a *relatively small number* of small appliances with very powerful harmonic generators can still be connected without any difficulty. However, if many such appliances are hooked up at the same site (e.g. numerous module or string inverters in the same phase) their respective harmonic currents will be cumulative. Moreover, if single-phase devices are allocated to all phases, in the neutral conductor the harmonics become a cumulative quantity comprising ordinal numbers that are multiples of 3. Hence in terms of harmonics a PV system comprising numerous small inverters should be regarded as a large installation and evaluated accordingly.

Different values apply to large appliances with current exceeding 16 A. As grid impedance is lower with high connected loads, higher levels of harmonic current can be used without generating impermissibly high harmonic voltages. A limit value for total harmonic current distortion is often defined in addition to the limit values for discrete current harmonics I_n. A distinction is made between (a) total harmonic distortion THD$_I$, which is derived from the total figure of merit I for nominal device current, and (b) total

Table 5.7 Effective harmonic current limit values according to EN 61000-3-2 [5.14]

Harmonic number n	Maximum permissible harmonic current I_n (in A)
odd harmonics:	
3	2.3
5	1.14
7	0.77
9	0.4
11	0.33
13	0.21
$15 \leq n \leq 40$	0.15–15/n
even harmonics:	
2	1.08
4	0.43
6	0.3
$8 \leq n \leq 40$	0.23–8/n

EN61000-3-2

harmonic distortion THD_{I1}, which is derived from the (somewhat lower) figure of merit for the first-harmonic portion of nominal device current:

$$\text{Total harmonic current distortion } k_I = \text{THD}_I (\text{based on total current}): k_I = \sqrt{\frac{\sum_{n=2}^{40} I_n^2}{I}} \qquad (5.23)$$

where I_n = the figure of merit of the nth harmonic and I = the figure of merit of total current; $k_{I1} = \text{THD}_{I1}$ is determined in a similar manner, except that I_1 is substituted for I in this equation. Hence this value constitutes the ratio of the figure of merit of harmonic current and the figure of merit of total current I to first-harmonic current I_1.

In some cases, partially weighted total harmonic distortion (PWHD) is important, whereby higher harmonics that generate disturbances are given greater weight:

$$\text{PWHD} = \frac{\sqrt{\sum_{n=14}^{40} n \cdot I_n^2}}{I_1} \qquad (5.24)$$

In many larger installations, the allowable harmonic current is often determined by the apparent short-circuit capacity of the grid at the grid link point (for more on calculating S_{KV}, see Section 5.2.6.1.5).

According to [5.17], the limit values in Table 5.8 apply to PV systems whose apparent short-circuit capacity S_{KV} is at least 33 times larger than apparent nominal capacity, which is almost always the case with correctly sized power leads [5.17].

Also indicated in [5.19] are allowable harmonic current limit values as a function of the S_{KV}/S_{WR} ratio; these values are mandatory in Germany, Austria, Switzerland and the Czech Republic. However, according to [5.19], *only 50% of the harmonic current allowed for appliances is allowable for grid-connected power generating appliances owned and operated by consumers* (i.e. the p_n values in Table 5.9):

$$\text{Limit values for ordinal number } n \text{ harmonics (for inverters): } \frac{I_n}{I} \leq 0.5 \cdot p_n \sqrt{\frac{S_{KV}}{S_{WR}}} \qquad (5.25)$$

Table 5.8 Harmonic current limit values (derived from first-harmonic oscillation I_1) according to EN 61000-3-12 for $S_{KV} \geq 33 \cdot S_{WR}$. Somewhat higher harmonic currents are allowable for short-circuit capacity $S_{KV} \geq 120 \cdot S_{WR}$ [5.17]

Harmonic number n	Maximum permissible harmonic current I_n/I_1 in %
odd harmonics:	
3	5
5	10.7
7	7.2
9	No indication
11	3.1
13	2
even harmonics:	
$2 \leq n \leq 40$	16/n
In addition:	THD $\leq$13%, PWHD $\leq$ 22%

EN61000-3-12 für Rsce = 33

where:
S_{KV} = apparent short-circuit capacity S_{KV} of the grid at the grid link point
S_{WR} = apparent nominal capacity of the inverter that is to be connected
  I = nominal inverter current (total current)

0.5 is the factor for in-plant generation appliances such as inverters, according to [5.19]
  The limit value for total harmonic distortion THD_I as derived from total current, according to [5.19], is

$$\text{Total harmonic current distortion (for inverters): } k_1 = THD_I \leq 0.5 \cdot 2\% \sqrt{\frac{S_{KV}}{S_{WR}}} \qquad (5.26)$$

Figure 5.69 exhibits the [5.19] limit values as a function of S_{KV}/S_{WR}. It should be noted hat higher harmonic currents are allowable in higher-capacity grids with lower grid impedance and higher short-circuit capacity. Owing to the 0.5 factor for generation installations, these values are somewhat more restrictive than the limit values in Table 5.8 [5.17], which mainly apply to appliances. If an installation exhibits high harmonic current, suitable measures should be taken to ensure that the [5.19] limit values are adhered to. These measures include intake circuit realization, grid bolstering, using a different type of inverter and so on.

5.2.3.4 High-Frequency Interference (150 kHz to 30 MHz)

PV inverters integrate a rapid electronic switch that allows for high-efficiency operation with the steepest possible switching edges. However, steep-edged high currents and voltages exhibit an elevated proportion of high frequencies that can interfere with radios and other sensitive electronic appliances in the environs if countermeasures are not taken. The solar generator wiring and the power leads in grid-connected systems are high-radiation elements that act as transmitting antennas for such interference. Inasmuch as

Table 5.9 Limit values for total harmonic distortion (THD), referred to total current, according to [5.19]

n	3	5	7	11	13	17	19	>19
p_n in %	0.6	1.5	1	0.5	0.4	0.2	0.15	0.1

Grenzwerte für die p_n-Werte nach D-A-CH-CZ-2004

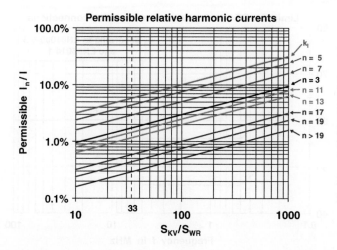

Figure 5.69 Allowable harmonic current for generation installations in the low-voltage grid, as derived from total current I, in accordance with [5.19], as a function of the S_{KV}/S_{WR} ratio. The $S_{KV}/S_{WR} = 33$ value has also been inserted so as to facilitate comparisons with the values in Table 5.8. Note that twice the values indicated here are allowable for appliances

the internal switching frequency of most PV inverters is considerably below 100 kHz, such appliances in the 30 MHz range normally generate little interference.

There is a device-specific standard for PV inverters concerning emissions of conducted high-frequency interference. As grid-connected systems are often installed on buildings, European electromagnetic radiation standard EN 61000-6-3 [5.23] should be used. This standard, to which mandatory DC standards were recently added, applies to all electrical appliances that are used in residential areas and for which no specific device standard exists. The de facto counterpart to this standard is the household appliance standard EN 55014 [5.21], which specifies limit values for conductors other than power cables (in a PV system this would be DC conductors); these values are extremely important for the realization of PV installations.

Conformance with the limit values indicated in these standards does not necessarily rule out radio reception interference in cases where the distance between the source of the interference and the device affected by it is less than 10 m. Unlike most small electrical appliances, grid-connected systems operate continuously from dawn to dusk. Hence in the interest of trouble-free use of PV systems in residential areas, the somewhat stricter limit values in the now superseded German standard VDE 871B should be applied if possible.

5.2.3.4.1   Conducted Disturbances in Power Leads
As stipulated in [5.21] and [5.23], power leads that exhibit conducted disturbances in the 150 kHz to 30 MHz range in a 50 Ω grid model are not to exceed the limit values indicated in Figure 5.70.

5.2.3.4.2   Other Types of Leads, Notably DC Connecting Leads
Whereas the measurement procedure and limit values are identical in the [5.21] and [5.23] standards, there are differences in terms of measuring interference in other types of leads.

Unlike many other electromagnetic compatibility standards, for many years now the household appliance standard EN 55014-1 [5.21] has prescribed the conducted-disturbance limit values for other types of connected leads (including PV inverter DC leads) indicated in Figure 5.71. While the standard stipulates that these voltages are to be measured with a 1500 Ω high-ohm sensor, it does not stipulate that any particular grid simulation needs to be realized in these wires while the measurements are being taken. This can make it difficult to reproduce PV inverter measurements. Inasmuch as PV inverters, like

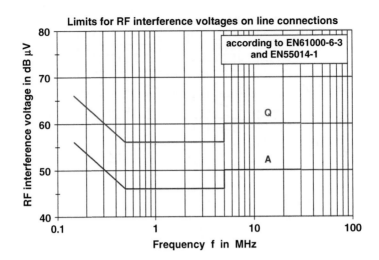

Figure 5.70   Conducted disturbance limit values in power leads (according to EN 61000-6-3/EN 55014-1) that apply to PV inverters [5.1],[5.23]. Q: limit values for quasi-peaks and for sources of broadband interference; A: limit values for mean/sources of broadband interference

household appliances, are often used in residential buildings, it makes sense to apply EN 55014-1. However, since inverters are not common household appliances, many inverter manufacturers have tended to disregard this standard.

Realistic electromagnetic compatibility readings are best obtained if characteristic conditions are used. To this end, the PV inverter being tested should be fed by a solar generator or solar generator simulator that is in turn fed by a simulated grid that exhibits a defined high-frequency impedance to earth. In this regard, [5.24] and [5.25] recommend that a 150 Ω simulated grid (according to EN 61000-4-6) with relatively low capacity to ground be used for the current levels that occur on the DC side. However, conventional 50 Ω simulated grids for the AC side are unsuitable for such tests as they tend to falsify the readings owing to the fact that their capacity to ground is far too high [5.75].

The EN 55014-1 limit values apply to load conductors that are no more than a few metres in length. However, PV system DC wiring runs are mostly more extended and can easily be in the range of quarter- and half-wave antenna dimensions for short waves. This in turn means that, under unfavourable conditions, solar generator wiring can exhibit relatively good radiation characteristics; thus in some cases, adhering to the prescribed limit values does not suffice. The electric field strength radiation readings observed at a number of PV installations during two EU projects resulted in the recommendation that somewhat more restrictive limit values be adopted, as indicated by the dashed lines in Figure 5.71. Inverters that exhibit good electromagnetic compatibility should not exceed these values.

Since 1992, the Appendix to EN 50081-1 has recommended interference current limit values as in Figure 5.72 in lieu of limit values for conducted disturbances. In 2004, these limit values were incorporated into the new EN 61000-6-3 standard, which applies to appliances used in residential and commercial settings. To obtain defined and replicable conditions, in realizing such measurements it is necessary to connect a conductor to a suitable grid model that exhibits relatively low earth capacitance, using high frequency and a terminal impedance of 150 Ω to ground [5.24]. The current limit value readings can if necessary be readily converted to voltage limit values using the known terminal impedance (150 Ω) of the grid model.

The differences between the various standards do not exceed more than a few decibels. The prescribed 150 Ω terminal impedance (see Figure 5.72) can be used to determine the conducted disturbances that arise from these differences. The counterpart DC conducted disturbances in dBµV can be determined by increasing the current limit values in dBµA by 43.5 dB. A comparison of these conducted disturbances with the limit values in EN 55014-1 (see Figure 5.71) shows that above 500 kHz the limit values thus

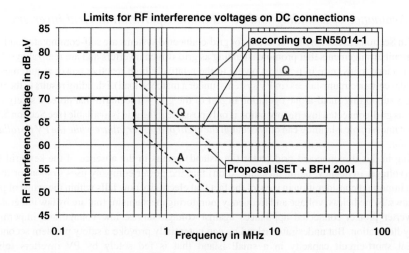

Figure 5.71 Limit values for conducted disturbances in other leads in accordance with EN 55014-1 [5.21]. Q: limit values for quasi-peaks and for sources of broadband interference; A: limit values for mean/sources of broadband interference. The dashed lines indicate the recommended limit values based on the findings of EU projects [5.24],[5.25]

obtained are *virtually identical*, which means that PV inverters must adhere to electromagnetic compatibility limit values on the DC side as well.

5.2.3.5 Other Standards

The draft version of EN 61000-6-1 lays out additional electromagnetic compatibility requirements for appliances that are used in residential areas. Experience has shown that measures aimed at reducing the kinds of high-frequency interference discussed in the previous section also improve immunity to rapid and transient power surges, and this in turn enhances the operational reliability of such inverters.

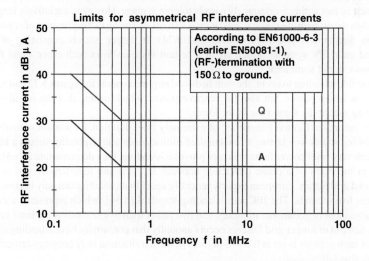

Figure 5.72 Limit values for asymmetrical current for other conductors (particularly DC conductors) measured using high-frequency current sensors and line terminations to ground with 150 Ω, in accordance with EN 61000-6-3 [5.23]. Q: limit values for quasi-peaks and for sources of broadband interference; A: limit values for mean/sources of broadband interference

5.2.4 Avoidance of Islanding and Stand-alone Operation in Grid Inverters

As noted in Section 5.2.2, many PV inverters for grid-connected systems are self-commutated and tend to exhibit circuit arrangements that promote islanding as grid-driven inverters and are realizable solely for high-capacitance loads (rare in practice) owing to the relatively high reactive power requirements that apply. Under certain circumstances (in PV systems, under a matched load), islanding results in the inverter or inverters supporting stand-alone operation in part of the grid – a scenario that poses a safety hazard. However, as grid-connected inverter operation is only possible if the grid is available (see Section 5.2.2.1), this type of undesirable islanding can only occur *in the event of a power failure while the PV installation is in operation*.

Islanding in this disconnected grid sector (i.e. island) occurs in the absence of the key grid figures of merit (voltage, frequency, harmonics and so on) that the power company uses to monitor the grid. This is an important consideration in view of the fact that electricity now falls within the scope of product liability laws. Nonetheless, voltage and frequency monitoring mechanisms that are by law integrated into every inverter protect connected appliances against voltage surges, low voltage and impermissible frequency fluctuation. But undesirable islanding can potentially provoke a safety problem secondary to insufficient short-circuit capacity in a small island that is fed solely by PV inverters (also see Section 5.1.3.4) – although a short is likely to induce termination of stable islanding. Moreover, such islanding could only be hazardous for a technician who disregarded the most basic safety procedures – namely, a post-shutdown voltage check.

Hence the *matched-load conditions for islanding* are met insofar as *all* of the following conditions obtain *at the same time*:

1. The installation must be disconnected from the grid via a disconnect planned by the power company or via a safety disconnect following a malfunction.
2. In the disconnected grid sector (the resulting island), the connected appliances' effective and reactive power requirements are met by one or more inverters during each phase.

If these criteria are met, it is also admissible for intra-island appliance voltage and input to be provided solely by the inverter or inverters following disconnection from the grid. In order to ensure that PV inverter voltages and frequencies remain within the limits that are continuously monitored by the inverters, condition 2 must be met within an extremely narrow tolerance range. However, a relatively lengthy period greatly decreases the probability that stable islanding will occur, owing to the presence of continuous load and insolation fluctuations (see Figures 5.73 and 5.74). In larger islands consisting of numerous consumers and many PV systems, which are at a certain distance from each other, these fluctuations may tend to smooth out somewhat.

In order for stable islanding to occur, the matched-load requirements as in Figures 5.73 and 5.74 must be met not only for active power, but also for reactive power, which means that the likelihood of both conditions being met at the same time is far lower.

Most power failures in Central Europe are of extremely brief duration, a common scenario being arc flash extinguishing shutdowns lasting a few hundred milliseconds after lightning strikes a high-voltage line. While such very brief power failures do not provoke islanding, they do constitute an additional load for inverters in the event of a phase difference between the grid and inverters (if still in operation) following a rapid grid restart. European experts generally agree that islanding can only become a problem after more than five seconds. The IEC anti-islanding standard [5.34], which represents a compromise resulting from years of negotiations, stipulates that two rather than five seconds is the limit value in such cases. Only a handful of longer grid failures occurs annually that potentially have islanding implications (the number of such outages is set at 0.2 to 2 in [5.29]). Such a situation only becomes critical if inverter safety devices also fail.

Grid-connected inverters also integrate additional safety functions that detect and shut down such islanding (which is in any case a rare occurrence), thus further reducing the likelihood of islanding events.

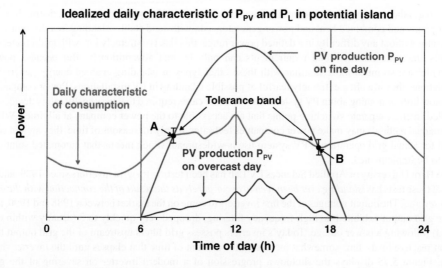

Figure 5.73   Idealized PV installation power output P_{PV} and load power consumption P_L progression in a potential island, for a single day. The matched-load requirements are met only at points A and B, and then only temporarily. Figure 5.74 displays a detailed example of a possible waveform for point A or B

All inverters must at a minimum monitor grid voltage and frequency, for which the applicable limit values vary from one country to another. According to the latest German pre-standard [5.27], the allowable voltage ranges from 80 to 115% of nominal voltage, while the allowable frequency ranges from 47.5 to 50.2 Hz. In the presence of voltages or frequencies that lie outside of these ranges, the inverter is not to start up and if in operation must shut down within 0.2 s. In addition, [5.27] stipulates that an inverter must also shut down if the 10-minute mean grid voltage at the grid link point is more than 110% of nominal voltage. Inverters exhibit still other safety functions to allow these appliances to pass the requisite islanding recognition tests, whereby these functions include grid impedance monitoring, frequency shifting procedures and triphase voltage monitoring for single-phase inverters.

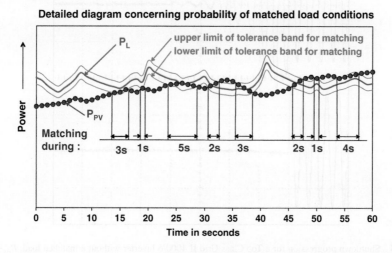

Figure 5.74   Detail of the waveform in the matched-load range (point A or B in Figure 5.73). The mandated matched-load requirements are met if PV installation power output P_{PV} coincides with the tolerance range for power consumption P_L. However, this state rarely lasts for more than five seconds

Whereas other energy generation installations such as small-scale hydro power plants, district heating power plants and combined heat and power generators can withstand stable islanding in a disconnected grid sector without any difficulty in a defined power range, this risk is extremely low with grid-connected systems since the matched-load requirements can only be met adventitiously. But because power companies are so much more familiar with these other types of islanding-enabled energy generation installations, they are often extremely fearful of possible islanding in PV installations. Power companies that know little or nothing about PV systems have in some cases required that very costly safety devices be installed, such as separate switching points that are accessible to the power company at all times. While such manual methods may make power companies feel safer, if only for reasons of time, they are not very useful for actual grid operation if PV system use is widespread, which means that automated solutions need to be implemented.

The Bern University of Applied Sciences PV Lab has been testing PV grid inverters since 1989, and in none of these tests has *islanding ever been seen to occur purely as the result of the connection with the grid being severed*. The output voltages of the first inverters to come on the market between 1988 and 1990, and whose grid output exhibited no high-frequency filtering, dropped very quickly to 0 (usually within one period) following a power outage. Today's inverters possess grid filters upstream of the grid output that exhibit reactive loads, thus somewhat increasing the amount of time that elapses until the inverter shuts down. Figure 5.75 displays the shutdown progression of a modern inverter on severing of the grid connection and with the matched-load conditions not met ($R = \infty$ in the test circuit arrangement as in Figure 5.76).

5.2.4.1   Islanding under Matched Load Conditions

In order to induce islanding for the lab inverter tests above, it was necessary to hook up the inverter to a load that absorbs the device's active power P (see Figure 5.76). As a result, the current between the grid

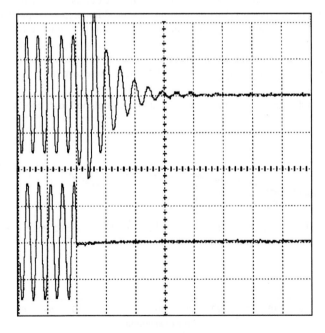

Figure 5.75   Shutdown progression for a Top Class Grid II 4000/6 inverter without a matched load, $P_{AC} \approx 900$ W. Oscilloscope settings: vertical, 200 V/div.; horizontal, 50 ms/div. Upper curve: inverter output voltage; lower curve: grid voltage. Following grid shutdown, the output voltage rises substantially for around two periods before the inverter shuts down

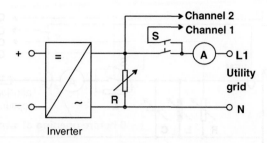

Figure 5.76 Simple test circuit for islanding tests with a matched load for cos $\varphi \approx 1$ inverters. Prior to the opening of the two switches, R was set in such a way that minimal current was flowing through the ammeter. Channel 1 was used to determine switching time

and the parallel connection between the grid and load prior to opening of the connection switch was minute (0 for 50 Hz components, in which case the current consisted almost entirely of harmonics). Inasmuch as self-commutated inverters with high-frequency transformers (see Figure 5.61), which were in wide use at one time, exhibit cos $\varphi \approx 1$, this active power matching was usually sufficient to trigger islanding of unlimited duration, particularly in new appliances from inexperienced manufacturers. Such islanding is shown in Figure 5.77 for the first Solcon appliances that came on the market.

The inverters manufactured in around 1991, which were structured as shown in Figures 5.55 and 5.56 and integrated an NF transformer such as a low-loss 50 Hz toroidal core transformer, always needed to draw some (usually relatively little) reactive power Q from the grid. To meet the matched-load conditions for such appliances, it was also necessary to compensate for this reactive power. To this end, additional parallel-connected reactances were added to the circuit arrangement shown in Figure 5.76 [5.13], resulting in the arrangement shown in Figure 5.78.

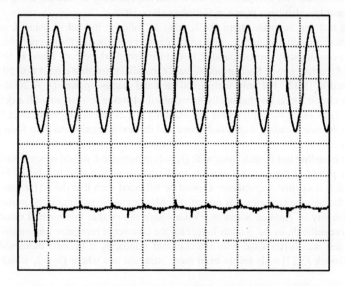

Figure 5.77 Islanding in the oldest-model Solcon inverter with $P_{AC} \approx 300$ W and a matched load, using the circuit shown in the previous illustration. Oscilloscope settings: vertical, 200 V/div.; horizontal, 20 ms/div. Upper curve: inverter output voltage; lower curve: grid voltage. This scenario resulted in islanding of indeterminate duration. The output voltage curve was virtually sinusoidal during the islanding phase (low harmonics)

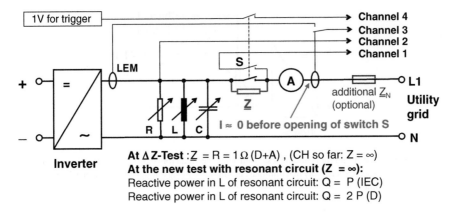

Figure 5.78   Universal test circuit for inverter island testing (also with cos $\varphi \neq 1$) with and without ENS. Before the two switches are opened, R, L and C are set in such a way that minimal current A (but only its 50 Hz portion) is flowing through the ammeter. Channel 1 (and channel 4 for ENS) allows for the determination of switching time; channel 2 measures the output voltage; and channel 3 measures the inverter output current (and power lead current prior to opening of S). For inverter tests without ENS, $Z = \infty$ was set, and L and C were set in accordance with the minimum required quality Q_S. For ENS tests with Z, according to [5.27], in Germany $R = 1\,\Omega$ should be used in lieu of $0.5\,\Omega$ [5.26] as in the past, and $1\,\Omega$ should be used in Austria; [5.27] also calls for a certain amount of grid impedance in series with power leads (blue)

In Germany, the so-called ENS, which was based on *continuous grid impedance monitoring*, was introduced in the mid 1990s (see Section 5.2.4.5). Function testing of this ENS was initially conducted using a relatively complex test circuit [5.32]. It was shown in [5.33], via the relevant grid conversions, that this circuit arrangement equates to the test circuit shown in Figure 5.78 if the switch that establishes the direct connection with the grid is bypassed by a low-ohm impedance Z. Something along the lines of this circuit arrangement has also been in use in Germany since 1999.

The islanding test can be made more challenging as follows: instead of using L or C solely to compensate for random inductive or capacitive reactive power absorbed by the inverter, this power can be used to realize a continuously present parallel resonance circuit where both L and C always absorb a certain amount of reactive power. Such an additional LC parallel resonance circuit with 50 Hz resonance was first proposed in [5.32], but only with $\pm 100\,$var fixed reactive power Q. A 2008 IEC standard described such a circuit for an internationally harmonized islanding test circuit with a defined factor of merit (e.g. $Q_S = 1$) [5.34]. $Z = \infty$ allows for the islanding test circuit arrangement as in Figure 5.78 in accordance with the new standard mentioned above and the Swiss regulations have been in effect for a number of years.

To perform an islanding test in such cases with $Q_S = 1$, inductance L should be set in such a way that it absorbs reactive power Q whose level equates to the active power absorbed by R. Following this, before the switch S is opened, capacitance C must be balanced such that the 50 Hz component of the current flowing through the ammeter A is minimized. Measurements realized using numerous inverters at the Bern University of Applied Sciences PV Lab have shown that this method usually makes the islanding test more difficult, in that it takes longer for the inverter to recognize impermissible islanding. Once this new test was harmonized with international standards, it became more widely accepted in Germany, although [5.27] calls for an even more stringent test where $Q_S = 2$, which would mean that $Q = 2P$.

The universal test circuit shown in Figure 5.78 thus allows for the realization of the currently standard test with a balanced load and with much less effort. Triphase inverter tests using this circuit should be performed phase by phase, whereby the latter two phases should be directly connected to the grid. However, [5.34] indicates that one resonance circuit per phase is needed for triphase tests.

5.2.4.2   Main Methods for Detecting Undesired Islanding

According to standard power company practice, responsibility for averting undesired islanding lies with the operator and/or manufacturer of the grid-connected PV system – although in principle power companies also have resources that would certainly ease the task of detecting such events. The companies could, for example, send out a ripple control pilot signal when a PV installation is connected to the grid. PV inverter islanding can be detected without difficulty using the following methods, all of which have been used successfully (also see [5.31] among other sources):

- Passive methods (monitoring key grid parameters):
  - Grid voltage monitoring (inverter shuts down in the presence of a voltage surge or low voltage).
  - Frequency monitoring (inverter shuts down if the frequency is out of tolerance).
  - Harmonics monitoring (inverter shuts down if harmonics are unduly high).
  - Inverter shuts down if any of the aforementioned parameters changes abruptly.
  - Inverter shuts down if the size of the phase angle between V and I increases abruptly
- Active methods:
  - Active frequency shift when the inverter is not connected to the grid and inverter shutdown if the tolerance limit is reached.
  - Output power adjustment for active and/or reactive power.
  - Grid impedance measurement and inverter shutdown if this parameter is unduly high.

Since use of any one of these methods on its own does not allow for completely reliable islanding detection, it is best to employ a combination of methods, which greatly increases the likelihood of successful detection and thus makes it highly unlikely that islanding will occur. But of course there is no magic bullet here either. Specific appliances can detect islanding very efficiently, depending on the method used, but problems can always arise if numerous appliances are installed adjacent to each other.

When a relatively serious anomaly occurs, such as a voltage surge or low voltage or high/low frequency, the inverter should shut down quickly (e.g. within 200 ms) so as to avoid jeopardizing adjacent appliances. Reliable differentiation between normal and abnormal operation takes more time in the cases of brief islanding accompanied by voltages and frequencies that lie within the normal grid tolerances. The regulations in Switzerland, Germany, Austria and many other countries call for detection within five seconds in such cases, although a mere two seconds was recently promulgated by [5.34].

Problems can arise in cases of brief inverter islanding under balanced-load conditions following a rapid restart if, for example, the grid goes down briefly during a storm and immediately restarts (e.g. after 250 ms) during an islanding event that exhibits a frequency slightly different from grid frequency secondary to a phase shift. No such event has been observed to date (including in inverters that exhibit a frequency shift procedure) for the following reasons: (a) as noted, islanding where matched-load conditions are adventitiously met rarely occurs; and (b) during storms, where such rapid restarts tend to occur, PV system inverters usually shut down secondary to unduly low insolation. That said, the cause of optimal inverter reliability would be greatly served by research on inverter characteristics during brief grid outages in the presence of balanced loads.

5.2.4.3   Triphase Voltage Surge and Low-Voltage Monitoring

The advantage of single-phase inverters is that all three phases always shut down simultaneously on initiation of a pre-announced power outage. Simultaneous monitoring (for low voltage) of the concatenated voltage between the three phases via a so-called asymmetry relay greatly increases the likelihood of detection, particularly if a single-phase inverter is used. A voltage surge can be detected by merely monitoring the phase that power is being fed into (see Figure 5.79). For this method, which was frequently used for up to 4.6 kVA single-phase installations in Germany and Austria in the early 1990s, the voltage surge or low voltage must be shut down within 200 ms. This relatively simple method offers the advantage that shutdown occurs almost instantaneously following a power outage, thus avoiding phase shifting

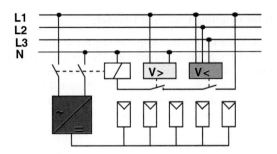

Figure 5.79 Anti-islanding arrangement for a single-phase inverter, with triphase low-voltage monitoring of concatenated voltage as in [5.35]. The break contact opens (at $V>$) if the voltage rises to more than 15% above nominal voltage while energy is being fed into the inverter. If any of the concatenated voltages fall below 80% of nominal grid voltage, the make contact (which is normally closed) opens (at $V<$). The inverter is automatically disconnected from the grid in both such cases

problems between inverter output voltage and restart grid voltage following storm-induced outages. This method also obviates reciprocal failure of adjacent installations. One drawback of this method, however, is that connecting a single-phase inverter necessitates a three-phase power lead and two special relays, which use energy. Although many German manufacturers now use ENS, a few, such as Kaco and Sunways, still use triphase grid monitoring. Testing at three-year intervals is no longer required [5.27].

Although in three-phase PV systems the likelihood that matched-load conditions will be met in all three phases during a power failure is far lower than for the single-phase systems above, such an event can nonetheless occur if, for example, a newly formed island is exposed solely to symmetrical triphase loads. Hence it would seem that sole use of triphase voltage surge and low-voltage monitoring with triphase inverters is somewhat less reliable than with single-phase inverters. This is why in such cases some power companies require that an additional switching point be established that is accessible at all times to the company's technicians (see Figure 5.80). However, such switching points will be impracticable if a large number of grid-connected systems go on line.

That said, measurements realized at PV systems show that triphase inverters likewise do not always exhibit precisely symmetrical power distribution in all phases. In view of the observations in Section 5.2.4.7 concerning the likelihood of all matched-load conditions being met, it would seem that if an additional simple measure such as a frequency shifting procedure is used, the additional switching point can be dispensed with for triphase appliances used for triphase voltage surge and low-voltage monitoring.

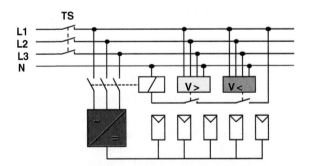

Figure 5.80 Anti-islanding arrangement for triphase inverters according to [5.35]. In the presence of more than 115% of nominal grid voltage, or if the voltage falls below 80% of nominal grid voltage, the make contact (which is normally closed) opens (at $V<$). The inverter is automatically disconnected from the grid in both such cases. In addition, for triphase inverters, power companies require a separate switching point TS that is accessible round the clock

5.2.4.4   Frequency Shifting Procedures

In view of the relatively tight tolerances for grid system frequencies in Western Europe (around 49.8 to 50.2 Hz), very low-tolerance frequency monitoring can also be used. When used in conjunction with voltage surge and low-voltage monitoring (e.g. $1.15 \cdot V_n$ or $0.8 \cdot V_n$, where V_n = nominal voltage), this type of frequency monitoring is a highly suitable solution for averting damage to connected customer installations in the event of undesired islanding. Moreover, if, in the presence of islanding, an inverter automatically shifts its frequency to a level that is outside of the grid frequency tolerance range – which means that the inverter's frequency is only available within this tolerance range if the grid is available (a mechanism referred to the frequency shift method) – this then constitutes a highly efficient islanding detection function for a balanced load that works well for individual appliances under single-phase conditions. Figure 5.81 displays islanding detection for a balanced load via a connected single-phase inverter that uses the frequency shifting method. The frequency change is so minute in such a case that it is indiscernible in the oscillogram. Nonetheless, the frequency changes so drastically after a few hundred milliseconds that the device's continuous frequency monitoring mechanism is activated and the device shuts down.

The islanding rules of the Swiss power grid watchdog agency (ESTI) have been relatively lax for as long as anyone can remember. In the provisional safety regulations from 1990, ESTI merely stipulated that: (a) when the power fails a grid-connected inverter must shut down in every case (i.e. including in the presence of a balanced load) within five seconds [4.4]; and (b) voltage and frequency must be monitored and the inverter must be disconnected from the grid in the event of an impermissible deviation from the relevant nominal values. Hence the choice of anti-islanding method is left up to the inverter or installation manufacturer. This standard has since been superseded by [4.13], which unfortunately is devoid of

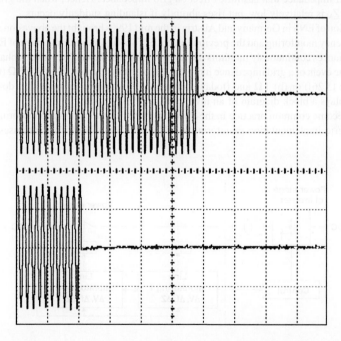

Figure 5.81   Shutdown of a Solarmax S inverter during an islanding test with a balanced load where $P_{AC} \approx 380$ W. This device detects islanding using the frequency shifting method. Oscilloscope settings: vertical, 200 V/div.; horizontal, 100 ms/div. Upper curve: inverter output voltage; lower curve: grid voltage. Around 380 ms following the power failure, the frequency has shifted so drastically that the device's continuous frequency monitoring mechanism is triggered and the device shuts down. The time this procedure takes (far less than five seconds) conforms with current Swiss regulations

islanding or islanding detection rules. It is for this reason that Swiss manufacturers still always use the easy-to-realize frequency shifting method, which is also used in The Netherlands and other countries.

However, in order for the frequency shifting method to be used for numerous inverters operated in close proximity to each other without engendering reciprocal interference, an international standard should stipulate the direction in which an inverter that is not connected to the grid should shift its frequency. For example, Dutch regulations recommend a downward shift [5.36].

But as grid-connected system use grows (it already amounted to around 10 GW of installed capacity in Europe as at December 2008), these frequency limits should be liberalized somewhat so as to avoid exacerbating the frequency stability problems that could potentially arise if all grid-connected installations abruptly shut down (also see Section 5.2.7).

5.2.4.5 Continuous ENS Monitoring of Grid Impedance

The triphase voltage surge and low-voltage monitoring of single-phase inverters that was required in Germany and Austria during the early 1990s was burdensome for both power companies and installation owners because of the requirements concerning mandatory triphase connections and testing at three-year intervals.

A promising concept in this regard, which was first proposed by [5.32], is continuous grid impedance Z_N monitoring via the detection of undesired islanding. Grid impedance Z_N mainly consists of relatively low transformer impedance, which converts medium voltage into low voltage, and line impedance, which is usually higher (see Section 5.2.6). On the other hand, connected appliances in islands of moderate size exhibit far higher impedance that has little effect on grid impedance. Hence, when the grid is available, grid impedance Z_N is relatively low, but rises abruptly if islanding suddenly occurs.

The introduction of ENS in Germany and Austria in the mid 1990s brought the above concept to fruition along with frequency monitoring and the previous practice of voltage monitoring. Each of ENS's two grid monitoring appliances continuously monitors the level of grid impedance Z_{1N} between phase and neutral conductors. In the event of a grid impedance jump ΔZ_{1N} amounting to more than (a) $0.5\,\Omega$ (the previously applicable value [5.26]) or (b) $1\,\Omega$ and (today's value) [5.27], the inverter must shut down within 5 s. Figure 5.82 displays a block diagram of an inverter that exhibits ENS.

It has since become common practice in the PV industry to refer to both this monitoring method and the concept of grid implementation monitoring as ENS, which this book therefore uses in this sense of the term.

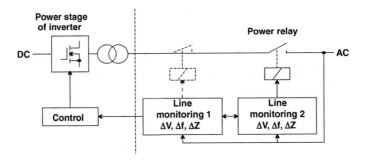

Figure 5.82 Block diagram of an inverter that exhibits ENS [5.26] for anti-islanding purposes. In addition to grid voltage and frequency monitoring, grid impedance is also measured periodically using two separate and redundant measuring circuits that act upon separate disconnection devices. However, in cases where ENS is integrated into an inverter, one disconnection device can be dispensed with if one of the two grid monitoring devices disables the power phase controller in the event of an anomaly. But if ENS is realized as a separate device, two isolating relays are needed (the second relay is indicated by the dashed line)

The triphase connection can be dispensed with and ENS safety can be strengthened by integrating two separate grid monitoring appliances, each of which has a switching element. If ENS is realized as a stand-alone device, two separate switching elements should be used. Integration of such a stand-alone device in the main distribution board immediately adjacent to the automatic cutout to which the inverter is connected is by far the safest method, particularly in cases where the inverter exhibits a frequency shifting procedure and/or other anti-islanding measures. But if ENS is integrated into the inverter itself, one of the two grid monitoring devices can act directly on the power phase controller such that only one isolating relay is needed, thus reducing costs.

Current German and Austrian regulations allow for the connection of up to single-phase 4.6 kVA inverters using ENS and no longer require periodic testing. The key requirements that ENS must meet are as follows:

- Shutdown within 0.2 s following a serious deviation from either of the following key grid parameters:
  - Grid voltage $> 1.15 \cdot V_n$ or $< 0.8 \cdot V_n$ ($V_n =$ nominal voltage).
  - Frequency > 50.2 Hz or < 47.5 Hz.
- Shutdown within 5 s following an islanding event in the absence of a serious deviation from either of the following figures of merit:
  - Grid impedance change $\Delta Z_{1N} > 1\,\Omega$ (as from the last measurement; formerly $0.5\,\Omega$ in Germany).
  - Grid impedance $Z_{1N} > 1.75\,\Omega$ (in Austria; this also used to apply in Germany).

In this context, impedance jump tests are to be performed using various supplementary ohmic–inductive impedances (shown in blue in Figure 5.78) and values ranging up to around $1\,\Omega$.

Grid impedance monitoring can *only* be used for triphase inverters and *not* for single-phase appliances, i.e. triphase ENS that monitors triphase grid impedance $\underline{Z}_{3N}$ is also realizable.

For a number of years now, ENS has been used in Germany for triphase devices up to 30 kVA as well, and according to [5.27] even this limit no longer applies. However, it would make sense to adapt the aforementioned fixed impedance values to the rated power of the appliances for high-output installations, i.e. to use lower impedance values for devices with higher rated power. Otherwise, reliable islanding detection is unrealizable for high-output installations as the impedance of all connected appliances in a large island can be around $1\,\Omega$ or even less.

In cases where the inverter's operating current is used to measure impedance, minimum current or power is necessary in order to obtain a detectable signal. To this end, [5.26] stipulates that ENS testing need only be realized with around 25, 50 and 100% of nominal inverter capacity. Although this method simplifies ENS integration, it lacks the requisite consistency. This holds true for all capacity ranges in cases where islanding could pose a serious problem, and thus islanding should be suppressed in all cases. This is more readily realizable with a fully stand-alone ENS in the main distribution board.

Figure 5.83 illustrates how grid impedance is measured using an inverter that integrates ENS. In this device, a relatively strong, approximately 2.5 ms pulse current is used near the grid voltage zero crossing. In most appliances, a reference measurement is realized before the device shuts itself down, so as to attain reasonable immunity to interference induced by events such as grid voltage fluctuation.

As power companies mainly use 50 Hz impedances, the values indicated above pertain to the amount Z_{1N} of impedance $\underline{Z}_{1N}$ at 50 Hz. Grid impedance Z_{1N} at 50 Hz is usually fairly well defined but increases at higher frequencies. Resonance events could also potentially occur if reactive power compensation appliances are in the vicinity of the inverter. Figure 5.84 displays the frequency dependence of grid impedance Z_{1N} at the connection point of a single-phase inverter, in the absence of such resonance. An inverter that measures impedance using a current impulse containing a specific level of higher harmonics (see Figure 5.83) will measure impedance not at 50 Hz but rather at an undefined higher frequency, and in most cases the impedance reading will be higher than would otherwise be the case. In the interest of better measurement reproducibility, impedance should be measured using 50 Hz current impulses or thereabouts.

Although ENS works quite well in Central European urban grids, less favourable conditions (smaller transformers, greater distance between inverters and transformers, longer low-voltage overhead power

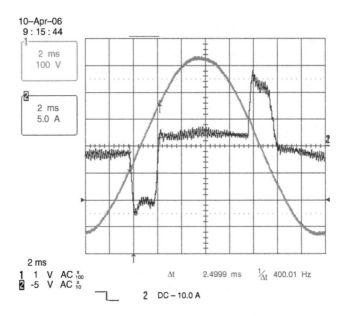

Figure 5.83 Grid impedance measurement in an SMA SB3800 inverter that integrates ENS. This curve was produced at relatively low power ($P_{AC} \approx 450$ W) so that the current impulse used for the measurement process could be shown: curve 1, grid voltage (100 V/div.); curve 2, device output current (5 A/div.). Time scale: 2 ms/div. Impedance was measured using a virtually square current impulse (duration around 2.5 ms; peak value around 13 A) near the grid voltage zero crossing

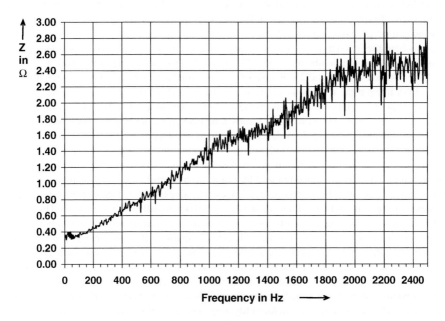

Figure 5.84 Waveform of single-phase inverter grid impedance at the inverter connection point Z_{1N} as a function of frequency in the Gfeller 3.18 kWp PV system in Burgdorf. Impedance rises with increasing frequency owing to the skin effect and the effect of line inductance

lines and so on; also see Section 5.2.6) can give rise to problems, depending on the type of inverter used. In Austria, the required impedance jump was raised from 0.5 to 1 Ω many years ago, and this same step was taken in Germany a few years ago. Excessive inductance in grid impedance can mimic excessive grid impedance (depending on the inverter measurement method) that prevents the inverter from switching on despite the presence of sufficiently low impedance at 50 Hz. The operation of ENS inverters can also be impaired by ripple signal or grid voltage fluctuations.

One basic drawback of ENS is that it causes additional grid voltage fluctuations, which in turn result from the current impulses needed for grid impedance measurement. Hence the number of active ENS mechanisms in an electricity grid should be kept to a minimum.

Active grid monitoring measures tend to interfere with each other. Inasmuch as grid impedance is usually measured transiently (for a few periods at the most) and very sporadically (e.g. at more than 100 period intervals), hardly any such reciprocal disturbance occurs if only a few ENS inverters are connected at the same grid link point. On the other hand, problems can arise in a large installation containing a great many module or string inverters, all of which exhibit ENS. In such cases a central ENS integrated into the main distribution board is definitely a better solution.

Nonetheless, following a power failure, large islands with more than around 50 kW of power per phase exhibit a grid impedance of less than 1 Ω, which undermines the reliability of grid impedance monitoring.

In view of these problems with ENS (which has yet to catch on internationally) and issuance of the new IEC standard [5.34], many manufacturers have begun using methods other than ENS that will need to pass the resonance circuit test described in this standard.

5.2.4.6 New Automatic Switching Point Between Energy Generation Installations and the Public Grid

Inasmuch as the original ENS [5.26] did not catch on internationally, in Germany a new and less restrictive standard titled 'Automatic switching points between the public low voltage grid and an energy generation installation that runs in tandem with the public low voltage grid' was recently devised that lifts the grid impedance measurement requirement and allows for additional methods in the guise of triphase voltage monitoring (*without* periodic retesting) and a resonance circuit test (see Section 5.2.4.1 and [5.34]). Unlike its predecessor, which applied solely to PV systems, this standard applies to all energy generation installations that run in tandem with the grid and stipulates no power limits.

Like [5.26], the new standard: (a) requires that two stand-alone switching elements be integrated; (b) stipulates that a single irregularity in the monitoring system should not provoke the loss of system safety functionality; and (c) indicates that one such switching element can be the system inverter (see Figure 5.85).

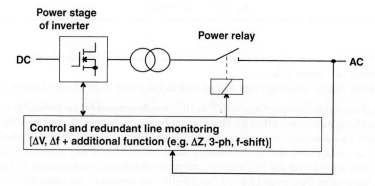

Figure 5.85 Block diagram of an automatic switching point for use in PV inverters, in accordance with the standard that came into force in February 2006 [5.27]. This method is referred to as a bidirectional safety interface (BISI)

The main properties of a BISI are as follows:

- Shutdown within 0.2 s following a major deviation from either of the following key grid figures of merit:
  - Grid voltage $> 1.15 \cdot V_n$ or $< 0.8 \cdot V_n$ ($V_n =$ nominal voltage)
  - Frequency > 50.2 Hz or < 47.5 Hz.
       The inverter must also shut down if the 10 min grid voltage mean at the grid link point is more than 110% of nominal voltage, although the standard allows a certain amount of leeway to compensate for a voltage drop in the interconnecting line.
- The following applies to the remaining tests, which can be realized at the operator's discretion, but with a balanced load as in Figure 5.78:
  - For impedance measurement as in Section 5.2.4.5: shutdown within 5 s following a change in grid impedance $\Delta Z_{1N} \geq 1\,\Omega$.
  - For the resonance circuit test: shutdown within 5 s for the resonance circuit test with a balanced load and $Q_S = 2$.
  - For triphase grid voltage monitoring: shutdown within 0.2 s in the event that the voltage in at least one outer conductor is $< 0.8 \cdot V_n$ or $> 1.15 \cdot V_n$.

For transformerless inverters, DC fault current must also be monitored. Leakage current > 300 mA must trigger a shutdown within 0.3 s; the same holds true for abrupt changes in leakage current $\Delta I > 30$ mA. The required shutdown time for higher ΔI is even shorter (as low as 40 ms depending on the ΔI value).

5.2.4.7 Probability of PV installation Inverter Islanding

At an International Energy Agency (IEA) conference on grid system installations held in Zurich in 1997, no islanding-induced problems or damage were reported, despite the fact that thousands of grid-connected PV systems had been in operation for many years worldwide.

At an IEA Task 5 Conference that was held in Arnhem in 2002 following years of exhaustive research on islanding and that was attended by many international experts on the subject, the following statements were made: (a) the likelihood of a grid-connected system provoking an islanding event is extremely small [5.28],[5.29]; and (b) as most appliances also use reactive power, in the interest of avoiding islanding events inverters should be designed in such a way that they generate active power only [5.28]. However, in view of the ever-growing use of PV inverters and in the interest of minimizing voltage increases, in Germany PV inverters must be designed in such a way that they feed not only active power into the grid, but also inductive or capacitive reactive power, depending on the amount of stored energy available in the PV system in question [5.71],[5.72].

The probability p_{IB} of hazardous islanding occurring in any given PV system is calculated as the product of the probability for specific islanding events, as follows:

$$p_{IB} = p_{AN} \cdot p_{NA} \cdot p_{ND} \tag{5.27}$$

where:
$p_{AN} =$ probability that matched-load conditions will be met (within a defined tolerance)
$p_{NA} =$ probability of a power failure
$p_{ND} =$ probability that an islanding event will go undetected owing to an inverter fault or the like

In [5.28], a characteristics range of $p_{AN} = 10^{-5}$ to 10^{-6} was determined for the probability of matched-load conditions being met for more than 5 s. However, probability p_{AN} is largely determined by the scope of the presumed tolerance range for meeting matched-load conditions for effective and reactive power. In [5.28], it was presupposed that, in determining these values, only a very narrow tolerance range comprising 5% for active power and 2% for reactive power needs to be taken into account. But appreciably higher p_{AN} values ranging from around $0.8 \cdot 10^{-4}$ to $2.3 \cdot 10^{-3}$ are obtained if the values likewise indicated in this study for a broader tolerance range comprising 15% for active power and 10% for reactive power are taken into account.

The characteristic probability of a power failure affecting individual PV systems in Central Europe is around two failures per year [5.29], although this figure can be higher in rural locations with less favourable grid conditions. Inasmuch as an inverter shuts down immediately after a power failure if matched-load conditions are not met (see Figure 5.75 for example), the key factor is not power failure count but rather duration. And since only daytime power failures are critical for a PV system, $p_{NA} \approx 1$ annually for a typical PV system.

However, this is far from being the case when it comes to maintenance activities (mainly realized during the day), which involve $p_{NA} \approx 1000$ scheduled shutdowns per year of portions of the grid. Such shutdowns are necessary in order for technicians to service the grid safely.

The probability p_{ND} that an islanding event will go undetected is determined by grid figures of merit such as island size and matched-load quality that obtain at the time of a given power failure, as well as by inverter characteristics such as the detection criterion applied, inverter installation quality, and the extent to which hardware and software operations exhibit anomalies. The minimum of $p_{ND} = 0.001$ posited in [5.29] is probably quite realistic for small islands with a small inverter fleet that exhibits one of the detection methods described in Sections [5.2.4.2]–[5.2.4.6], e.g. during electrical contracting work on a building with a small fleet of small- to medium-output inverters. Hence stable islanding is highly unlikely to occur in such settings.

However, each detection method has a blind spot, i.e. a range in which stable islanding is likely to go undetected. For example, grid impedance monitoring (see Section 5.2.4.5) is extremely efficient in small islands with a small inverter fleet but is inherently unreliable in large islands that are subject to loads of more than around 50 kW per phase. Hence in sizing such installations, allowance must be made for the worst-case scenarios and the attendant higher values.

5.2.4.7.1 Possible Scenarios Involving High PV Penetration where Installed PV Capacity in a Grid Sector Exceeds Minimal Capacity in that Sector

Conservative estimate of islanding event probability p_{IB} for individual PV systems:
   Suppositions:

> Probability that matched-load conditions will be met, $p_{AN} = 10^{-3}$.
> Power failure probability (per year), $p_{NA} \approx 1$.
> Probability of an islanding event going undetected $p_{ND} = 10^{-2}$.

It then follows that the probability of an islanding event > 2 s going undetected in a specific PV system following a power failure under absolute worst-case conditions is $p_{IB} \leq 10^{-3} \cdot 1 \cdot 0.01 = 10^{-5}$ per year (≤ 1 undetected islanding event > 2 s in 100 000 *years*).

5.2.4.7.2 Possible Scenarios Involving Work Being Performed on Relatively Large Grid Sectors and the Consequent Risk of Large-Scale Islanding

Suppositions:

> Probability that matched-load conditions will be met, $p_{AN} = 10^{-3}$.
> Power failure probability (per year), $p_{NA} \approx 1000$ (planned shutdowns).
> Probability of an islanding event going undetected $p_{ND} = 0.1$ (a higher value
> is presupposed here as such work can often result in large islands).

It then follows that the probability of an islanding event of more than 2 s duration going undetected in a specific PV system following grid shutdown for servicing is, under the absolute worst-case conditions, $p_{IB} \leq 10^{-3} \leq 1000 \leq 0.1 = 0.1^{-5}$ per year (≤ 1 severe islanding event > 2 s in 10 *years*).

If the relevant technicians adhere to the mandated safety procedures and if the required voltage test is adventitiously omitted for 1 in every 1000 installations, the probability of such a technician risking injury or death secondary to an islanding event is around 10^{-4} per year.

If such an extremely rare islanding event does in fact occur while grid maintenance is being performed, normally it will resolve very quickly on its own by virtue of the continuous insolation and load fluctuations to which islands are subject.

However, in scenarios involving an extremely large appliance fleet, elevated PV penetration and numerous widely scattered PV installations, such fluctuations may be smoothed out to some extent, which means that in exceptional cases longer-lasting islanding events may be acceptable insofar as they occur at all. It is of course potentially hazardous if a grid sector that has been disconnected for maintenance work is not instantaneously voltage free. However, such hazardous situations can be avoided if the technician verifies (according to the safety rules) that the relevant grid sector is dead. To this end, technicians can modify the island load by, for example, disabling specific outbound branches of the low-voltage transformer busses and then restarting these branches once the branch in question is dead. It is highly unlikely that all of the sub-islands engendered by the shutdown of the medium-voltage side will also exhibit stable islanding. In any case, the aforementioned action, which is performed centrally at the transformer, is far simpler and far more practicable for many large PV systems than the previous requirement, to the effect that a continuously accessible switching point had to be provided for each grid-connected system – an approach that was impractical for many such installations.

Members of the IEC Photovoltaic Standards Committee reported in June 2009 that stable islanding of up to 13 minutes' duration had apparently occurred during maintenance work in large portions of the Spanish grid that exhibit high PV penetration and capacity ranging from 600 kW to 2.5 MW.

There are also fears that islanding is more likely to occur under matched-load conditions in PV installations with a number of parallel-connected inverters. Such a scenario would involve spontaneous load matching secondary to staggered shutdown of only some of the system's converters while the remaining appliances remain in operation using only part of the system's total inverter capacity. In an extensive PV installation field study where such a scenario was induced using various makes of up to 18 parallel-connected inverters, not a single impermissible islanding incident was observed and in each instance all of the devices tested shut down within less than 5 s.

The ideal anti-islanding mechanism would of course be for power companies to transmit in the grid: (a) a continuous pilot signal that would only allow a converter to connect to the grid in the presence of this signal; or (b) a shutdown command to all decentralized energy generation installations in the disabled grid sector (apart from PV systems) such as district heating power plants, biogas plants and the like. This signal could be transmitted via a classic ripple signal or over the Internet. With increasing penetration of the low-voltage grid by energy generation installations, which like district heating power plants and biogas plants can exhibit islanding over a far larger capacity range than is the case with grid-connected systems, this issue will take on greater importance and will need to be addressed.

In this regard, recent German regulations already stipulate that, in the event of exceptional capacity or other problems in a grid sector, operators of large-scale grid-connected energy generation installations – including of course PV installations – can be required to reduce system capacity or shut down their installations altogether. Needless to say, such a procedure presupposes sufficient communication with such installations and their operators to allow for the transmission of such commands [5.71],[5.72]. Hence newly developed inverters should enable reduction of the power being fed into the grid or inverter shutdown on receipt of the relevant external command.

5.2.5   Operating Performance and Characteristics of PV Grid Inverters

Using a series of cases as illustrative examples, this section discusses the evolution and current state of the art of PV grid inverters and problems that can potentially arise with these devices. However, a full account of all currently available PV grid inverters could not be provided here for reasons of space and practicality. Updated information concerning such appliances is available in the relevant trade and research publications.

Various PV inverters have been undergoing tests since 1989 under the auspices of the Bern University of Applied Sciences PV Lab and the technology training institution Ingenieurschule Burgdorf (ISB), within the framework of term papers and diploma theses, as well as a number of research projects [5.30],

[5.37]–[5.52]. Other institutions have been conducting such tests as well [5.53]–[5.56], *inter alia*. Inasmuch as the discussion in this section centres around the most important characteristics of inverters, the findings of the PV Lab tests have been used for the most part as they were the most accessible to me and were the most complete. For reasons of space, it was only possible to provide a brief account of the test findings discussed here. In most cases, only one device of each type was tested. I cannot guarantee that the information presented here is either complete or accurate. Further information can be found in the publications referenced in the present chapter, many of which, together with the test reports, are available at www.pvtest.ch.

In the following, a brief overview of the key data and test results concerning the inverters investigated thus far at the PV Lab will be presented. Table 5.10 contains the test results for 1989–2000. The development of suitable semi-automatic solar generator simulators allowed for more detailed lab tests, which had heretofore been cost prohibitive and whose results from 2004 to 2006 are listed in Tables 5.11 and 5.12.

The European standard efficiency η_{EU} value indicated in Tables 5.10–5.12 constitutes a mean efficiency scale for Central European insolation conditions that is useful for energy yield calculations for solar generators that are not unduly oversized and was determined using the following equation, where the subscripted values are the percentages of nominal direct current:

$$\eta_{EU} = 0.03 \cdot \eta_5 + 0.06 \cdot \eta_{10} + 0.13 \cdot \eta_{20} + 0.1 \cdot \eta_{30} + 0.48 \cdot \eta_{50} + 0.2 \cdot \eta_{100} \tag{5.28}$$

The weighting factor used in this equation was first proposed in [5.35], since when other such factors have been discussed [5.78]–[5.81].

A standard definition of what constitutes nominal capacity in a PV installation has yet to emerge. Some manufacturers indicate nominal DC capacity, which is useful for solar generator sizing, while others indicate nominal AC capacity, which is useful for power lead sizing. In addition, newer inverters are prone to brief overloading. Table 5.10 lists manufacturers' data concerning AC-side nominal capacity for inverters. Reactive power consumption and harmonic current for self-commutated inverters are relatively low, while the power factor $\lambda = P/S$ roughly equates to first-harmonic oscillation $\cos \varphi$ and is generally close to 1 for medium and high capacities, such that apparent output $S \approx P_{AC}$. Line-commutated inverters, on the other hand, consume a relatively large amount of reactive power and in some cases generate elevated harmonic current, such that $\lambda < 1$ and $S > P_{AC}$ (see Figures 5.53 and 5.128).

A comparison of the data in these tables reveals the following:

- **DC–AC conversion efficiency η:** Conversion efficiency has greatly improved since PV grid inverters were first introduced in 1989, from which year until 1991 the peak efficiency exhibited by these devices (all of which integrated transformers) ranged from 89 to 92% and European standard efficiency η_{EU} ranged from 85 to 90%. These values have improved appreciably in the interim. Medium-capacity inverters with transformers today exhibit peak efficiency ranging from 93 to 96%, while European-standard efficiency η_{EU} is between 92 and 95%. Transformerless appliances normally exhibit values that are 1–2% higher.
- **Maximum power point tracking (MPP tracking):** As measuring MPP tracking is relatively difficult and cannot be done without highly accurate solar generator simulators, manufacturers often fail to specify it. Some simulators are particularly challenged by MPP tracking measurement of low-capacity appliances.
- **Harmonic current:** The harmonic current generated by most of today's self-commutated inverters is below the EN 61000-3-2 and EN 61000-3-12 limit values. However, problems can arise with such devices if a great many string or module inverters are connected at the same grid link point. Problems can also arise with line-commutated inverters (e.g. EGIR 10 from 1991).
- **Islanding:** Making inverters that reliably shut themselves off after a power failure was often something of a headache for novice manufacturers in the past, but today is no problem as a rule for ENS and non-ENS devices. But unfortunately, the attendant methods have only been standardized internationally to a limited extent (see Section 5.2.4).

Table 5.10 Key data and test results for some of the inverters that were tested between 1989 and 2000 at the Bern University of Applied Sciences PV Lab (the accuracy and completeness of these data cannot be guaranteed)

Type	Test year	S_N [kVA]	V_{DC} (typ.) [V]	η_{EU} [%]	Transformer	Current harmonics (0.1–2 kHz)	EMV AC	EMV DC	Ripple control sensitivity	Islanding
SI-3000	89	3	48	90	HF	0	−[1]	−	0[3]	−/++[3]
SOLCON	90/91	3.3	96	90	HF	+	−[1]	−	+[3]	−/++[3]
EGIR 10	91	1.7	165	89	LF	−	−	−	n.t.	n.t.
PV-WR-1500	91	1.5	96	85.5	HF	++	0	−	0	++[5]
ECOVERTER	91/92	1	64	92	HF	++	0	0	+	++
PV-WR-1800	92	1.8	96	86.5	HF	+	++	0	0	++[5]
TCG 1500	92	1.5	64	89.5	LF	+	+[1]	0[1]	++	−/++[3]
TCG 3000	92	3	64	91.5	LF	0	+[1]	0[1]	++	−/++[3]
EcoPower20 *	94/95	20	760	92.6	LF[6]	0	0/+[1]	++	++	0
Solcon3400	94/95	3.4	96	91.9	HF	0	0/+[1]	0	++	++
NEG 1600	95	1.5	96	90.4	LF	+	++	0	++	++
SolarMax S	95/98	3.3	550	91.7	TL	+	−/+[7,8]	−/0[1]	++	0/++[3]
SolarMax20 *	95	20	560	89.4	LF	0	+	0	++	++
TCG II 2500/4	95	2.2	64	91.9	LF	0	+	−	++	++
TCG II 2500/6	95	2.2	96	90.4	LF	0	+	−	++	++
TCG II 4000/6	95	3.3	96	90.2	LF	0	0/+[2,8]	−/++[2]	++	++
Edisun 200	95/96	0.18	64	90.7	HF[6]		++	0[4]	++	++
SPN 1000	95/96	1	64	89.8	LF	++	++	++	0	++
Sunrise 2000	96	2	160	89.3	LF	0	++	+	0	++
SWR 700	96	0.7	160	90.8	LF	0	0[8]	++	+	++
TCG III 2500/6	96	2.25	96	91.5	LF	+	+[8]	++	++	++
TCG III 4000	96	3.5	96	91.9	LF	++	+[8]	++	++	++
TC Spark	98/99	1.35	180	90.6	LF	++	+[8]	−[4]	++	++
OK4E-100	98/99	0.1	32	90.3	HF	++		−[4]	++	0
Solcolino	99/00	0.2	64	90.6	HF[6]	++	0		++	++
Convert 4000	99/00	3.8	550	92.5	TL	++	+[8]	++	++	++
SWR1500	99/00	1.5	400	94.4	TL	++	+[8]	++	++	++

++ very good, much lower than limits
+ good, lower than limits
0 sufficient, limits nearly respected
− insufficient, higher than limits
− − bad, much higher than limits
n.t. not tested
* three-phase device
HF/LF/TL high/low frequency/without transformer

1) after modification by HTI Burgdorf
2) with optional DC choke
3) with new control software
4) sufficient for module inverters (PV array small)
5) only with three-phase connection
6) without galvanic separation DC–AC
7) new, improved model
8) small trespassing of limits for < 300 kHz

Table 5.11 Key data and test results for a selection of the new inverters that were tested between 2004 and 2006 at the Bern University of Applied Sciences PV Lab, using semi-automatic solar generator simulators (the accuracy and completeness of these data cannot be guaranteed). Unlike the previous tests, device characteristics under various DC voltage conditions were tested, including in conjunction with static and dynamic MPP tracking. *Note:* In some of the tests, the measurement conditions were too stringent, i.e. the voltage ranges between low- and high-power stages were unduly large, and are marked in red

Inverter type	Test year	S_N [kVA]	Transformer	MPP-voltage [V]	η_{EU} [%]	$\eta_{MPPT,EU}$ [%]	$\eta_{tot,EU}$ [%]	Dyn. MPPT-behaviour	Harmonics of current (0.1-2 kHz)	EMC AC	EMC DC	Ripple control sensitivity	Monitoring of frequency	Monitoring of voltage	Islanding
Sunways NT4000	04	3.3	TL	400	95.4	99.5	94.9	+	+	0	+	++	−	++	++
				480	94.9	99.0	94.0								
				560	94.6	98.0	92.6								
Fronius IG30	04	2.5	HF	170	91.0	99.8	90.8	0	++	+	+[4)]	+	++	+	+
				280	92.1	99.7	91.8								
				350	91.6	99.5	91.2								
Fronius IG40	04	3.5	HF	170	91.1	99.9	91.1	−	++	++	+[4)]	+	++	+	+
				280	92.5	99.6	92.2								
				350	91.8	99.5	91.3								
Sputnik SM2000E	05	1.8	TL	180	92.4	99.9	92.3	0*	++	0[1)]	+[4)]	++	++	++	+
				300	93.4	99.7	93.1								
				420	94.0	99.2	93.2								
Sputnik SM3000E	05	2.5	TL	250	93.5	99.5	93.0	0*	+	0[1)]	++	++	++	++	+
				330	94.0	99.4	93.4								
				420	94.7	99.7	94.4								
Sputnik SM6000E	05	5.1	TL	250	94.3	99.8	94.1	0*	−	0[1)]	++	+[3)]	++	++	++
				330	94.8	99.9	94.6								
				420	95.2	99.6	94.9								
Sputnik SM6000C*	05	4.6	TL	250	94.5	99.7	94.2	+	+	0[1)]	++	+	++	+	+
				330	95.1	99.6	94.7								
				420	95.4	99.5	95.0								

Table 5.11 (*Continued*)

Inverter type	Test year	S_N [kVA]	Transformer	MPP-voltage [V]	ηEU[%]	ηMPPT,EU[%]	ηtot,EU[%]	Dyn. MPPT-behaviour	Harmonics of current (0.1–2 kHz)	EMC AC	EMC DC	Ripple control sensitivity	Monitoring of frequency	Monitoring of voltage	Islanding
Sputnik SM25C[9]	05	25	NF	490	93.6	99.6	93.2	+	++	0[1]	+[6]	++	+[7]	++	+[8]
				560	93.3	99.5	92.8								
				630	93.0	99.7	92.7								
ASP TC Spark	05	1.4	NF	160	90.0	99.7	89.8	++	++	0[1]	++	0	+	+	0[5]
				190	90.4	99.8	90.3								
SMA SB3800*	05	3.8	NF	200	94.8	99.6	94.4	+	++	++	++	++	++	++	+
				280	94.2	99.7	93.9								
				350	93.5	99.7	93.2								
SMA SMC6000	05	5.5	NF	280	94.7	99.6	94.3	0**	++	++	++	+	++	++	+[2]
				350	94.1	99.6	93.8								
				420	93.7	99.7	93.4								

++   very good
+   good
0   sufficient
−   insufficient
−−   bad
*   { η measured with new, more precise power meter }
**   measurement a little too stringent (ΔV too high)

1) Higher than limits for frequencies < 300 kHz
2) Operation only with activated ENS
3) Shutdown at relatively low ripple control voltages for $f \leq 200\,\text{Hz}$
4) Higher than limits for frequencies < 200 kHz
5) Complies only with earlier ENS standard (not with actual version)
6) Higher than limits for frequencies < 400 kHz
7) Minor violation of new VDE 126-1-1 at overfrequency (> 50.2 Hz)
8) Test power somewhat too low, a little lower than in VDE 126-1-1
9) Efficiency measurements at 630 V only to 86%, at 560 V only to 76%, at 490 V only to 66% of rated power. For European efficiency calculations extrapolation of curves with trend line to rated power.

- **Sensitivity to ripple signals:** In previous years, novice manufacturers in particular found it difficult to make inverters that exhibited robust ripple signal immunity, owing to occasional device faults when ripple signals occurred in tandem with elevated grid voltage. Most of today's devices are fault free, although ENS inverters tend to be more sensitive to ripple signals and sometimes shut down unnecessarily when ripple signals occur.
- **Electromagnetic compatibility:**
  - *AC side:* The AC connecting leads in the first inverters to come on the market, in the 1990s, exhibited extremely elevated high-frequency interference voltages that far exceeded the prescribed limit values. Inasmuch as the AC-side standards were clearly defined, experienced manufacturers quickly recognized and eliminated this defect in their subsequent products. Today, some devices $< 400\,\text{kHz}$ and devices from novice manufacturers still exhibit such problems.
  - *DC side:* Impermissible high-frequency emissions need to be obviated on the DC side of PV systems as they are high-radiation elements. The filters used for this radiation also enhance immunity to injected transient overvoltage during thunderstorms. Until the 2004 adoption of the addendum to EN 61000-6-3 [5.23], there were no legally binding standards for high-frequency emissions on the DC side of PV installations, with the result that for a long time many less conscientious manufacturers integrated inadequate filtering mechanisms or none at all into their inverters. Today the standards are clearly defined and devices from seasoned manufacturers conform with the applicable limit values.
- **Reliability:** Devices from experienced manufacturers nowadays exhibit between 0.1 and 0.2 failures per inverter operating year.

5.2.5.1 Conversion Efficiency

The most important figure of merit for PV installations is inverter conversion efficiency η_{UM} (often referred to as efficiency η), which indicates how efficiently an inverter converts solar generator direct current into alternating current and is defined as follows:

$$\text{Inverter conversion efficiency } \eta_{UM} = \eta = \frac{P_{AC}}{P_{DC}} \tag{5.29}$$

where P_{DC} is DC consumption and P_{AC} is alternating current fed into the grid, which in practice equates to first-harmonic oscillation capacity.

The inverter conversion efficiency for stationary stand-alone operation can be readily measured as the devices used for this purpose do not track MPP.

However, such measurements are more difficult to perform for grid-connected inverters owing to the continuous MPP tracking algorithms exhibited by most such devices (also see Section 5.2.5.2). To perform such measurements, it is necessary to ensure that the alternating and direct currents are measured during at least one basic frequency period and if possible synchronously. As most inverters integrate energy storage components, during a defined measurement time T_M a series of discrete measurements should be performed at constant insolation, under similar capacity conditions and at a defined interval such as one or two seconds. Moreover, total measurement time T_M should be appreciably longer than the highest MPP control algorithm time constants. If power remains relatively stable during the measurement time, efficiency can be expressed as the quotient of the AC energy E_{AC} fed into the grid during the measurement time T_M and DC energy consumed E_{DC}, and can then be allocated to mean capacity during the measurement time T_M. The remaining measurement uncertainty can be mitigated via iteration of this measurement procedure for the same output and mean output derived from these measurements.

This efficiency is of course not constant, but is instead determined by the power and DC voltage used, i.e. $\eta = f(P_{DC}, V_{DC})$ [5.30], [5.45–5.51], [5.55], [5.58], [5.59], [5.60]. Efficiency is relatively low in small devices owing to open-circuit loss in the device electronics and transformer, but rises dramatically as power increases. Efficiency in electrically isolated inverters is usually highest at medium power and

Table 5.12 Key data and test results for a selection of the new inverters that were tested between 2006 and 2009 at the Bern University of Applied Sciences PV Lab, using semi-automatic solar generator simulators (the accuracy and completeness of these data cannot be guaranteed). Here, too, device characteristics under various DC voltage conditions were tested, including in conjunction with static and dynamic MPP tracking. New ramped capacity variants as in prEN 50530 were used to measure dynamic performance in lieu of the standard square capacity variants

Inverter type	Test year	S_N[kVA]	Transformer	MPP-voltage [V]	ηEU[%]	ηMPPT.EU[%]	ηtot.EU[%]	Dyn. MPPT-behaviour	Harmonics of current (0.1-2 kHz)	EMC AC	EMC DC	Ripple control sensitivity	Monitoring of frequency	Monitoring of voltage	Islanding
Convert 6T[1]	06	6.0	TL	170	87.6	99.7	87.4	+	++	++	+	+	+	+	0
				400	93.0	99.9	92.9								
				630	94.7	99.8	94.5								
SMA SMC8000TL	06	8.0	TL	350	97.4	99.7	97.1	n.t.	++	++	++	n.t.	n.t.	n.t.	n.t.
				400	97.3	99.8	97.1								
				490	97.0	99.7	96.7								
Sunways AT5000	08	5.0	TL	250	94.9	99.9	94.8	+	++	++	++	++	n.t.	n.t.	n.t.
				370	95.1	99.9	95.0								
				480	95.1	99.9	95.0								
Sputnik SM100C[4]	08	100	NF	440	94.9	100	94.9	0*	++	0	++	++[2]	0[2]	++[2]	0[3]
	09			560	94.3	100	94.3								
				680	93.9	100	93.9								
Sputnik 6000S	08	5.0	TL	220	94.2	100	94.1	++*	++	++	++	+	++	++	++
	09			320	95.1	100	95.1								
				420	95.3	99.9	95.3								

++ **very good**[1]

+ **good**

0 **sufficient**

-- **insufficient**

-- **bad**

n.t. **not tested**[2]

* *Stringent test with ramps according to newest draft standard prEN50530*

At all devices efficiency η measured with new, more precise power meter

1 Efficiency measurement at 400 V only up to 68%, at 170 V to 29% of rated power. Calculation of European efficiency referred to these correspondingly reduced maximum power values. Reason: DC supply during test only possible with one (out of three) MPP tracker, current limitation at 9 A per tracker.

2 Test performed only at reduced power (about 12 kW).

3 Test performed only at reduced power (about 25 kW).

4 Efficiency measurements at 680 V only to 96%, at 560 V only to 79%, at 440 V only to 62% of rated power. For European efficiency calculations extrapolation of curves with trend line to rated power.

declines slightly at high power owing to increased transformer copper loss. No such power loss occurs under high-power conditions in transformerless devices or devices with autotransformers. Efficiency is also determined by AC voltage, but to a lesser extent.

Inverter efficiency η_{UM} or η can be expressed as a function of either DC output P_{DC} or AC output P_{AC}. For head-to-head comparisons of the efficiency of inverters with differing power output, efficiency should be expressed as a function of standardized AC or DC power, as follows:

$$\text{Standardized DC output} = P_{DC}/P_{DCn} \qquad (5.30)$$

$$\text{Standardized AC output} = P_{AC}/P_{ACn} \qquad (5.31)$$

where P_{ACn} is nominal AC output and P_{DCn} is nominal DC output or P_{ACn}/η_n (η_n = nominal efficiency under nominal output conditions).

Figures 5.86–5.94 display selected conversion efficiency values for various PV grid inverters. Figure 5.86 displays the efficiency of selected transformer inverters from the early 1990s. Figure 5.87 displays the efficiency of two popular transformer inverters from the mid 1990s. Figures 5.88–5.94 display the efficiency curves for selected modern devices for three or four different MPP voltages. These diagrams clearly show that transformer inverter efficiency has improved by several percentage points over the years and that the efficiency of transformerless devices has improved even more than this.

As Figures 5.88–5.94 show, the correlation between conversion efficiency and DC voltage varies from one device to another. The only quasi-general principle that emerges here is that in most cases efficiency peaks under either low- or high-voltage conditions, although some devices also exhibit peak efficiency under medium-voltage conditions (see Figure 5.91).

As this Solarmax inverter's maximum AC power is 8 kW it should only be operated in groups of three at each of the various phases of the devices that are connected to the triphase grid, so as to avoid substantial asymmetry.

Note that, as far as measurement conditions are concerned, all of the efficiency curves shown in Figures 5.88–5.106 are derived from the I–V characteristic curves of highly stable solar generators that exhibit a 75% filling factor.

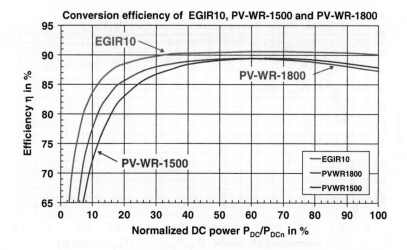

Figure 5.86 Efficiency as a function of standard AC output (based on nominal output) for selected older single-phase transformer inverters: EGIR10 (1.75 kW), PV-WR-1500 (1.5 kW), PV-WR-1800 (1.8 kW). The SI-3000 and Solcon curves can be found in [Häb91]

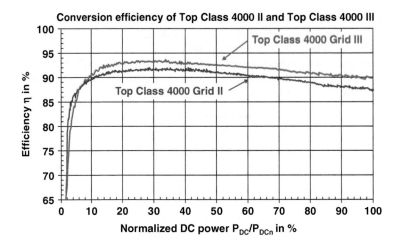

Figure 5.87 Efficiency as a function of standardized AC output of the ASP Top Class 4000 Grid III inverter relative to the efficiency of the predecessor device Top Class 4000 Grid II, both of whose nominal output was $P_{ACn} = 3.3\,\text{kW}$. The more recent model is considerably more efficient (both devices are transformer inverters)

5.2.5.2 MPP Tracking Efficiency and MPP Control Characteristics

Any given PV installation solar generator exhibits a specific I–V characteristic curve (depending on current insolation and module temperature) that in turn exhibits maximum power P_{MPP} at MPP (see Section 3.3.3) at voltage V_{MPP}. In the interest of attaining optimal energy yield, an inverter should always operate at this MPP.

But since inverters track the MPP at most times, they do not usually operate at the exact MPP but rather only somewhere near it, which results in a certain amount of power loss depending on the quality of the MPP control algorithm.

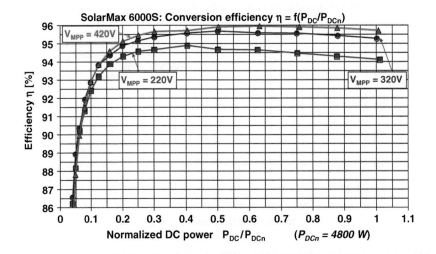

Figure 5.88 Efficiency of the transformerless Solarmax 6000S as a function of standardized DC output for three DC voltages. This device exhibits peak efficiency under high-DC-voltage conditions

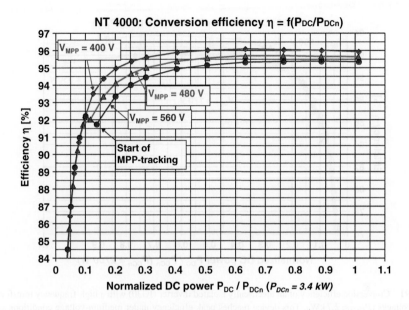

Figure 5.89 Conversion efficiency of a transformerless NT4000 inverter for three DC voltages ($P_{DCn} = 3.4\,\text{kW}$). This device exhibits peak efficiency under low-DC-voltage conditions

To define a measure for (a) the precision of MPP matching achieved by the inverter control and (b) the energy loss associated with the remaining deviation from the MPP, MPP matching efficiency η_{AN} or MPP tracking efficiency η_{MPPT} can be defined by comparing, under constant insolation conditions for a defined period, inverter DC energy consumption with the energy that the solar generator feeds into a load that matches the MPP perfectly.

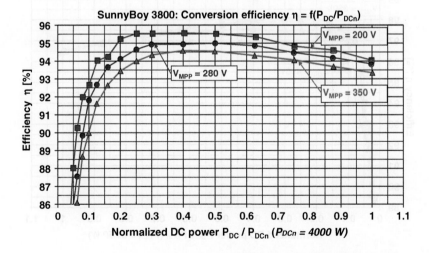

Figure 5.90 Conversion efficiency of an SB3800 ($P_{DCn} = 4\,\text{kW}$) with a 50 Hz transformer for three different DC voltages. This device, which achieves peak efficiency with the lowest voltage, exhibits extremely high efficiency for a transformer inverter. Under high-output conditions, this efficiency exhibits little correlation with DC voltage

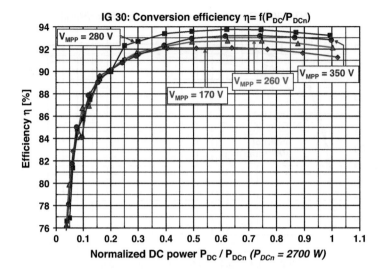

Figure 5.91 Conversion efficiency of an electrically isolated inverter (IG30) with a high-frequency transformer for four DC voltages ($P_{DCn} \approx 2.7\,\text{kW}$). This device reaches peak efficiency under medium-voltage conditions

Under static operating conditions, the characteristic curve and MPP power P_{MPP} exhibited by the solar generator are constant. Hence static MPP tracking efficiency η_{MPPT} is defined as follows:

$$\eta_{MPPT} = \frac{1}{P_{MPP} \cdot T_M} \int_0^{T_M} v_{DC}(t) \cdot i_{DC}(t) \cdot dt \tag{5.32}$$

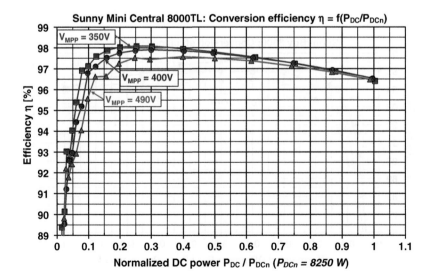

Figure 5.92 Efficiency of the single-phase transformerless Solarmax SMC8000TL inverter as a function of standardized DC output for three DC voltages. This device exhibits peak efficiency under low-DC-output conditions. Its efficiency, which correlates very weakly with voltage, ranges as high as 97.5–98%

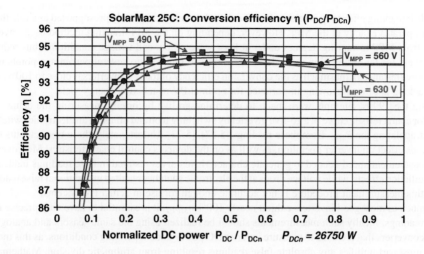

Figure 5.93 Conversion efficiency of the triphase Solarmax 25C transformer inverter as a function of standardized DC output for three DC voltages. This device exhibits peak efficiency under low-DC-output conditions

where:

$v_{DC}(t)$ and $i_{DC}(t)$ = voltage and current at the inverter's DC input, respectively

P_{MPP} = maximum available solar generator power at the MPP, depending on irradiance G and cell and module temperature Tz

T_M = measurement time as from $t = 0$. Recommended measurement time: > 60 s up to 600 s per power stage (although the new draft standard prEN 50530 recommends *600 s* per power stage)

For inverters that exhibit good MPP control characteristics, static MPP tracking efficiency from Equation 5.32 should be close to 100%, at least for higher output.

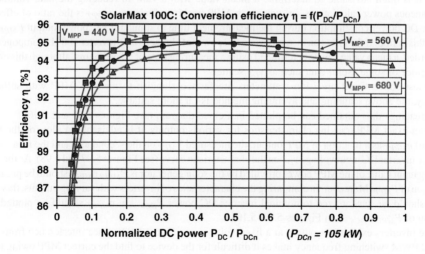

Figure 5.94 Conversion efficiency of the triphase Solarmax 100C transformer inverter as a function of standardized DC output for three DC voltages. This device exhibits peak efficiency under low-DC-output conditions and, owing to its higher nominal voltage, is far more efficient than the smaller Solarmax 25C

As MPP tracking efficiency is relatively difficult to determine, it is tacitly presupposed as a rule that an inverter operates exactly at the MPP. However, depending on the MPP tracking method used, inverters deviate from MPP to varying degrees under at least some output and voltage conditions, thus reducing energy yield for the installation as a whole (this loss can range up to several percentage points under certain circumstances). Inasmuch as the conversion efficiency of today's inverters is close to 100%, good MPP tracking is of ever-growing importance for optimal PV installation energy yield.

During the 1990s, attempts were made to determine go-live by briefly measuring the actual solar generator characteristic curve of a go-live inverter by switching it on and off using an electronic switch and then extrapolating the P_{MPP} from these readings [5.57], [5.73]. Unfortunately, these brief cutouts had a deleterious effect on inverter operation, as well as on the device's control characteristics. Moreover, the high measurement speed needed to generate characteristic curves resulted in false readings induced by RLC oscillation effects, while voltage drops at the electronic switches resulted in additional false readings. Hence this measurement procedure is not accurate enough for use in actual PV systems.

Inasmuch as most good inverters exhibit nearly 100% η_{MPPT}, it is essential to have very precise P_{MPP} output readings. Ideally such measurements should be realized using the same sensors and analogue to digital converters that are used to measure inverter DC power under operating conditions, as this method for the most part nullifies any absolute false readings resulting from arithmetic division. Mathematical models for P_{MPP} measurement based on irradiance G and cell and module temperature readings T_Z (for go-live solar generators) or on control data (for solar generator simulators) are far less accurate by virtue of the various absolute errors that are folded into the result.

To obtain precise and replicable measurements of static MPP tracking efficiency, rapid and highly stable solar generators are needed [5.46], [5.60]; the diode chain simulators used by many test labs are less well suited for this purpose owing to the inherent thermal stability anomalies exhibited by these devices. PC-controllable solar generator simulators, on the other hand, allow for: (a) synchronous measurement of numerous variables (e.g. η, η_{MPPT}, $\cos \varphi$ and harmonics) during the same power phase; and (b) automatic measurement by gradually modulating the current for a specific characteristic curve.

For static MPP tracking efficiency η_{MPPT} measurement, after a new power phase has been set and prior to the actual measurement phase, a brief stabilization period should be observed (e.g. 120 s, but more for inverters that track MPP sluggishly). During the ensuing measurement time T_M, direct current I_{DC} and DC voltage v_{DC} should be measured as synchronously as possible using a high scan frequency (e.g. 1000 to 10 000 measuring points per second), whereupon the results are used to determine instantaneous DC power. It is often advisable to determine a mean value with a view to reducing the data volume for instantaneous power calculations. In such cases, MPP tracking efficiency η_{MPPT} is the ratio of effective inverter DC power consumption during the measurement time to simulator DC power output $P_{MPP} \cdot T_M$ during this period. If the effect on instantaneous DC power of characteristic inverter 100 Hz components is disregarded, a mean value can be determined using a measuring time of 50 or 100 ms, although this entails the sacrifice of some useful data such as v_{DC} scatter.

As most conventional precision wattmeters are far too sluggish to measure MPP with sufficient accuracy, the above scanning and averaging method is far more suitable.

The readings thus obtained can be plotted in a cloud diagram (see Figures 5.95 and 5.96).

To plot exact MPP tracking characteristics for various outputs, it is advisable to plot (on the same diagram) η_{MPPT} as a function of MPP output, the measured V_{MPP} on the second axis and the mean of the effective inverter DC input voltage v_{DC} on the characteristic curve (see Figures 5.97 and 5.98). As the MPP tracking input value of the MPP power provided by the solar generator is P_{MPP}, η_{MPPT} should be plotted as a function of P_{MPP}. Moreover, in comparing the characteristics of inverters with varying outputs, this MPP power should be standardized to nominal inverter DC power P_{DCn}, i.e. η_{MPPT} should be plotted as a function of P_{MPP}/P_{DCn} as in Figures 5.97–5.101.

Some inverters exhibit low output at a fixed voltage (see Figure 5.97), since interference from their internal PWM switching frequency makes it difficult for the device to find the correct MPP owing to the device's inability to detect weak electrical signals under low-power conditions.

This device strategy allows for reliable operation even under low-power conditions, although under such conditions it also induces varying degrees of power loss depending on effective MPP voltage V_{MPP}

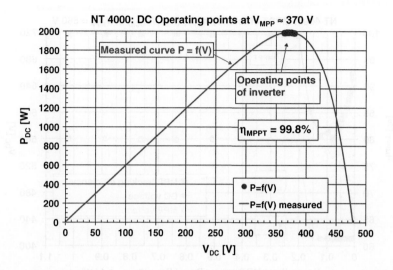

Figure 5.95 Cloud diagram for an NT4000 inverter where $P_{MPP} \approx 2$ kW and $V_{MPP} \approx 370$ V. This device exhibits measured η_{MPPT} efficiency of around 99.8%, which means that here it exhibits very good MPP tracking characteristics

(see Figure 5.99) since solar generator power is not fully used, particularly when the generator exhibits a fixed voltage (see Figure 5.97) that differs greatly from the effective V_{MPP}. Hence, under low-power conditions it would presumably be preferable either to use 0.8 times the previous open-circuit voltage V_{OC} reading or to make this fixed voltage adjustable under low-power conditions. On the other hand, the inverter illustrated in Figure 5.98 exhibits far superior static MPP tracking characteristics across all voltages (see Figure 5.100) and its operating voltage V_{DC} is only slightly below V_{MPP}, including under low-power conditions, resulting in considerably less power loss and thus considerably higher η_{MPPT}.

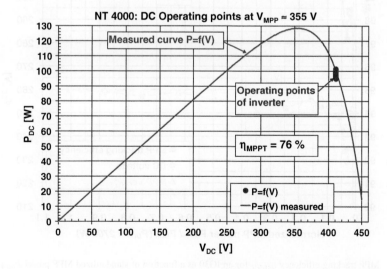

Figure 5.96 Cloud diagram for an NT4000 inverter where $P_{MPP} \approx 130$ kW and $V_{MPP} \approx 355$ V. This device operates at $V_{DC} \approx 410$ V and very close to the MPP, whereby its η_{MPPT} is around 76%, i.e. its MPP tracking characteristics here are very poor

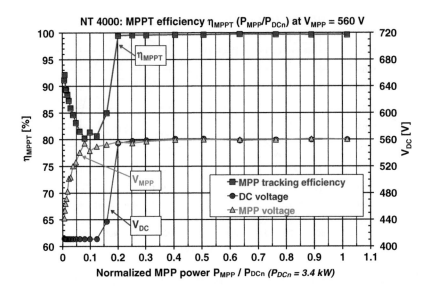

Figure 5.97   MPP tracking efficiency η_{MPPT} for an NT4000 as a function of standardized MPP power P_{MPP}/P_{DCn} for $V_{MPP} \approx 560$ V.

In as much as with low output and the lower voltage the device exhibits $V_{DC} \approx 410$ V, η_{MPPT} will also be lower under these conditions, but rises to close to 100% in the presence of higher P_{MPP}, as $V_{DC} \approx V_{MPP}$ under these conditions

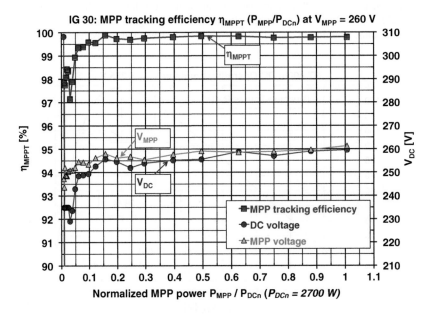

Figure 5.98   MPP tracking efficiency η_{MPPT} for an IG30 as a function of standardized MPP power P_{MPP}/P_{DCn} for $V_{MPP} \approx 260$ V.

V_{DC} differs only slightly from V_{MPP}, even under lower output conditions, and the device's static MPP tracking characteristics are very good under these conditions

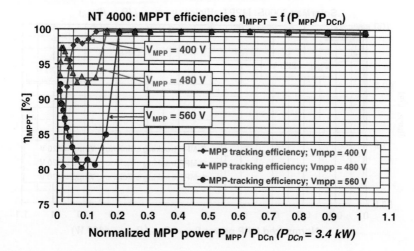

Figure 5.99 Static MPP tracking efficiency η_{MPPT} for an NT4000 as a function of standardized MPP power $P_{MPP}/$ P_{DCn} for three MPP voltages. Inasmuch as the device exhibits fixed $V_{DC} \approx 410$ V under lower output conditions, η_{MPPT} is more or less lower than 100% under these conditions depending on V_{MPP} position

If measurement time T_M is long enough (following a defined stabilization period such as 300 s), static MPP tracking efficiency η_{MPPT} measurements are readily replicable. However, in inverters whose MPP tracking process periodically occurs far from the current operating point, discrepancies between the various readings may occur, particularly with low T_M values (see Figure 5.102). These discrepancies reduce as T_M increases. With $T_M = 600$ s, the measured η_{MPPT} for a device as shown in Figure 5.102 would range from 98.71 to 99.28%, depending on device clock status.

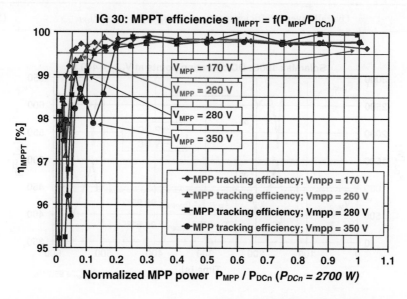

Figure 5.100 Static MPP tracking efficiency η_{MPPT} for an IG30 as a function of standardized MPP power P_{MPP}/P_{DCn} for four MPP voltages. Relative to Figure 5.99, the η_{MPPT} scale is considerably elongated, the static MPP tracking characteristics are far better, and the characteristics under low-output conditions are also very good

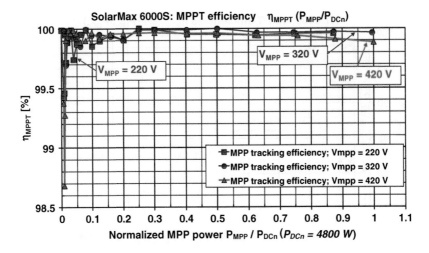

Figure 5.101 Static MPP tracking efficiency η_{MPPT} for a Solarmax 6000S as a function of standardized MPP power P_{MPP}/P_{DCn} for three MPP voltages. Relative to Figure 5.100, the η_{MPPT} scale is considerably elongated. This graph displays MPP tracking characteristics across the board

5.2.5.3 Overall Inverter Efficiency

The method for determining overall inverter efficiency η_{tot} will now be described (see Figure 5.103).

A solar generator provides specific power P_{MPP} depending on current irradiance G and temperature T. Under stationary operating conditions, the inverter only uses $P_{DC} = \eta_{MPPT} \cdot P_{MPP}$ of this power, which it uses to produce $P_{AC} = \eta \cdot P_{DC}$ (where η = conversion efficiency with $P_{DC} \leq P_{MPP}$). This in turn allows for the definition of a new efficiency value as follows:

$$\text{Overall inverter efficiency: } \eta_{tot} = \eta \cdot \eta_{MPPT} = P_{AC}/P_{MPP} \qquad (5.33)$$

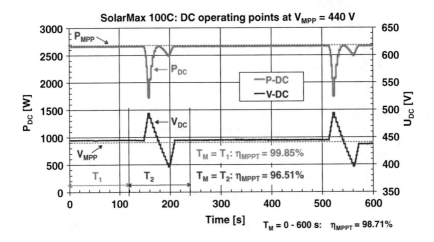

Figure 5.102 P_{DC} and V_{DC} for a 100 kW inverter under low-power conditions ($P_{MPP} \approx 2.65$ kW). The device tracks MPP at around six-minute intervals not only directly, but also over a broader range. This results in different η_{MPPT} values ranging from 93 to 100% depending on measurement period T_M duration and start time

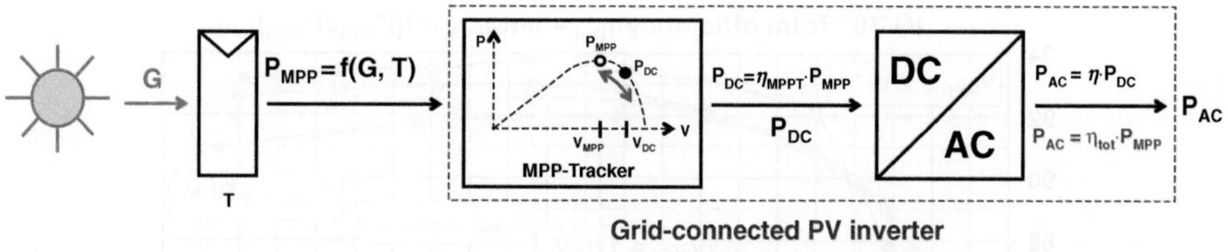

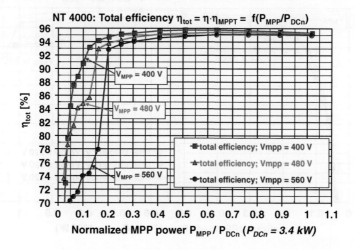

Grid-connected PV inverter

Figure 5.103 Determination of overall grid inverter efficiency. A grid inverter is composed of two main elements: its solar tracker, which enables the solar generator to deliver maximum possible power at all times; and the inverter component per se, which converts the solar generator's direct current into alternative current as efficiently as possible. Note that η is the conversion efficiency for $P_{DC} \leq P_{MPP}$, but *not always* for P_{MPP}

The following applies under stationary conditions:

$$P_{AC} = \eta \cdot P_{DC} = \eta \cdot \eta_{MPPT} \cdot P_{MPP} = \eta_{tot} \cdot P_{MPP} \tag{5.34}$$

Hence the overall inverter efficiency is a direct quality characteristic that is more relevant in practice than inverter conversion efficiency η. Like η and η_{MPPT}, η_{tot} is of course also determined by P_{MPP} and V_{MPP} and should be computed using suitable measurements.

The advantage of using the German term for overall efficiency (*totaler Wirkungsgrad*) and the symbol η_{tot} lies in the fact that 'total' has the same meaning in many languages, thus rendering the η_{tot} symbol intuitively decipherable.

Figures 5.104, 5.105 and 5.106 display overall efficiency η_{tot} thus computed for the NT4000, IG30 and 6000S inverters, respectively, as a function of standardized MPP power P_{MPP}/P_{DCn} for three and four different MPP voltages.

Although the central concept that allows for the determination of overall efficiency was apparently first put forward in 1992 [5.73], the concept fell into oblivion for a time because, when applied to actual measurements, it failed to provide sufficiently accurate MPP tracking efficiency η_{MPPT} results. The concept first came into use with the advent in 2004 of highly stable and sufficiently rapid solar generator simulators that allowed for the highly precise η_{MPPT} measurements that were needed [5.49], [5.50], [5.51].

Figure 5.104 Overall efficiency η_{tot} of an NT 4000 inverter as a function of P_{MPP} for three MPP voltages. On account of the device's relatively poor η_{MPPT} under low-power conditions and its relatively high V_{MPP}, it exhibits relatively low η_{tot} under these conditions despite its relatively high efficiency η

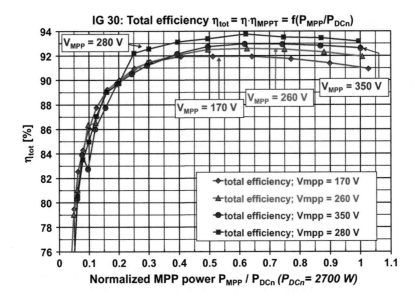

Figure 5.105 Overall efficiency η_{tot} of an IG 30 inverter as a function of P_{MPP} for four MPP voltages. Under low-power conditions at η_{tot}, the device's good MPP tracking compensates for its lower conversion efficiency η

The term overall efficiency has come into increasing use since 2007 largely because the journal *Photon* began publishing monthly inverter test reports at that time containing converter efficiency values. The term also appears in the provisional inverter test standard that was issued in 2008 [5.83].

Wherever possible, conversion efficiency η, MPP tracking efficiency η_{MPPT} and overall efficiency η_{tot} should be measured synchronously using a highly stable solar generator simulator, otherwise false readings may be obtained. If a solar simulator's solar tracker is not working properly, it means that the

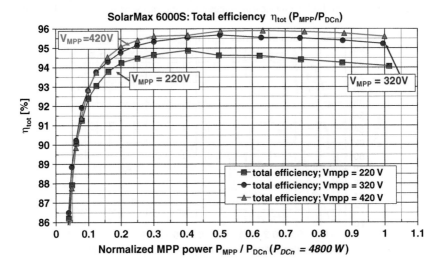

Figure 5.106 Overall efficiency η_{tot} of a Solarmax 6000S as a function of P_{MPP} for three MPP voltages. Owing to the device's outstanding MPP tracking efficiency (see Figure 5.101), its overall efficiency curves are virtually the same as those for conversion efficiency η (see Figure 5.88)

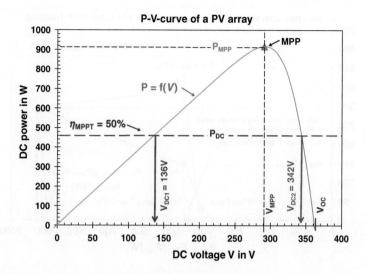

Figure 5.107 *P–V* curve of a solar generator with $P_{MPP} = 920$ W at $V_{MPP} = 290$ V, whereby the generator integrates an inverter with $\eta_{MPPT} = 50\%$, i.e. with $P_{DC} = 460$ W. The possible operating points under these conditions are $V_{DC1} = 136$ V and $V_{DC2} = 342$ V. Unless V_{DC} is defined, the inverter's operating voltage v_{DC} is not defined solely by the η_{MPPT} value

simulator inverter's operating voltage V_{DC} differs from MPP voltage V_{MPP} for the simulator's characteristic curve and that DC power consumption P_{DC} is lower than MPP power consumption P_{MPP} (see Figures 5.96, 5.97 and 5.103). This phenomenon is elucidated in greater detail in the hypothetical example displayed in Figures 5.107, 5.108 and 5.109.

The operating voltage of a solar generator as in Figure 5.107 is at $\eta_{MPPT} = 50\%$, in which case the generator can potentially exhibit two completely different v_{DC} values and exhibits conversion efficiency as in Figure 5.108.

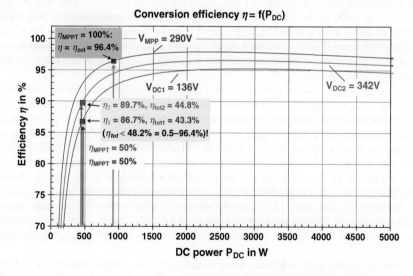

Figure 5.108 Conversion efficiency as a function of DC power P_{DC} for the inverter in Figure 5.107 with DC voltage as a parameter for $V_{DC} = V_{MPP} = 290$ V, whereby $V_{DC1} = 136$ V and $V_{DC2} = 342$ V. The efficiency levels for these three voltages differ considerably

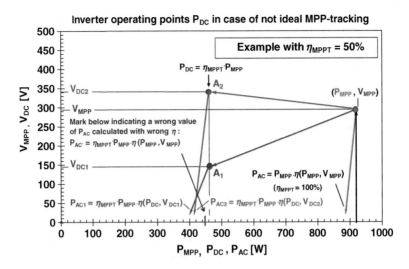

Figure 5.109 Operating the inverter as in Figure 5.108 on the *P–V* curve as in Figure 5.107 results in operating points A_1 and A_2 at the *P–V* level and the consequent AC power P_{AC1} and P_{AC2}

Figure 5.109 displays the resulting *P–V* operating points, as well as the AC outputs P_{AC1} and P_{AC2}. For $P_{MPP} = 920$ W and $V_{MPP} = 290$ V, conversion efficiency $\eta = 96.4\%$; for $P_{DC} = \eta_{MPPT} \cdot P_{MPP} = 0.5 \cdot 920$ W $= 460$ W and $V_{MPP} = 290$ V, $\eta = 92.8\%$. If η_{tot} is determined using the first η value, $\eta_{tot} = 48.2\%$, and $\eta_{tot} = 46.4\%$ for the second η value. However, both of these values are incorrect. The correct value for $V_{DC1} = 136$ V is $\eta_1 = 86.7\%$ and thus $\eta_{tot1} = 43.3\%$. Similarly, for $V_{DC2} = 342$ V, $\eta_2 = 89.7\%$ and thus $\eta_{tot2} = 44.8\%$.

To sum up, in a scenario where $\eta_{MPPT} < 100\%$, multiplying separately measured η values (whereby $P_{DC} = P_{MPP}$ and $V_{DC} = V_{MPP}$) by another set of separately measured η_{MPPT} values (for P_{MPP} and V_{MPP}) produces a systematic error due to the fact that: (a) in the case of $\eta_{MPPT} < 100\%$, $V_{DC} \neq V_{MPP}$ and $P_{DC} < P_{MPP}$; and (b) η is also still determined by DC voltage to some extent. (I was apparently insufficiently aware of this phenomenon, as described in earlier standards such as EN 61683 and EN 50524.) If a relatively small set of measured η curves for many different V_{DC} values is available, as a rule η_{tot} cannot be accurately determined from separately measured η_{MPPT} and η values, otherwise the data concerning V_{DC} status are lost.

In as much as η_{tot} is the product of η and η_{MPPT} that can be plotted on the abscissa of separate graphs as a function of P_{DC}/P_{DCn} and P_{MPP}/P_{DCn}, an untoward situation arises in $\eta_{MPPT} < 100\%$ scenarios whereby the measurement readings derived from the same measuring point for the $\eta = f(P_{DC}/P_{DCn})$ and $\eta_{MPPT} = f(P_{MPP}/P_{DCn})$ diagrams for a specific V_{MPP} are assigned slightly differing P_{DC}/P_{DCn} and P_{MPP}/P_{DCn} values that need to be correctly multiplied for Equation 5.34 (also see Figures 5.89, 5.97, 5.99, 5.104, etc.). To avoid this problem, conversion efficiency η and η_{MPPT} can of course also be plotted directly as a function of P_{MPP}/P_{DCn} as has been done for inverter tests in the journal *Photon* and elsewhere. In such a case, the measurement readings derived from the same power phases P_{MPP} are then assigned the same abscissa value. This somewhat unusual plotting method has a certain educational value and is fully accurate. Another approach is to plot $\eta_{tot} = f(P_{AC}/P_{ACn})$ and $\eta = f(P_{AC}/P_{ACn})$ using V_{MPP} as a parameter instead, although the parameter derived from $\eta_{MPPT} = f(P_{AC}/P_{ACn})$ is less useful in this regard.

To describe an inverter's characteristics using a small value set, a mean efficiency (i.e. European standard efficiency) can also be determined for η_{tot} and η_{MPPT}. The values resulting from this calculation are indicated in Tables 5.11 and 5.12.

Note that $\eta_{tot\text{-}EU}$ is not exactly the same as $\eta_{EU} \cdot \eta_{MPPT\text{-}EU}$ since η_{EU} is not based on P_{MPP}.

5.2.5.4 Dynamic MPP Tracking Test

Apart from static operating characteristics, which can be satisfactorily described using η, η_{MPPT} and η_{tot}, of similar practical interest for MPP tracking are dynamic characteristics, i.e. those that occur during periods of rapidly fluctuating insolation. From the standpoint of PV installation operators, efficient dynamic MPP tracking is a highly desirable system characteristic. The first measurement readings for dynamic MPP tracking using simple square test patterns were published in [5.52].

The extensive measurement data obtained during the long-term monitoring projects realized by the Bern University of Applied Sciences PV Lab conclusively demonstrate that go-live PV installations do in fact exhibit MPP tracking problems. Figure 5.110 displays an example of such a problem, which was caused by repeated insolation fluctuations.

In conducting dynamic solar tracking tests, which simulate days with fluctuating cloud conditions, it is necessary to create artificial discrepancies between the various power stages using known P_{MPP} values.

Similar to Equation 5.32, dynamic MPP tracking efficiency $\eta_{MPPTdyn}$ can be defined as follows for dynamic power fluctuations:

$$
\text{Dynamic MPP tracking efficiency } \eta_{MPPTdyn} = \frac{\displaystyle\int_0^{T_M} v_{DC}(t) \cdot i_{DC}(t) \cdot dt}{\displaystyle\int_0^{T_M} P_{MPP}(t) \cdot dt} \tag{5.35}
$$

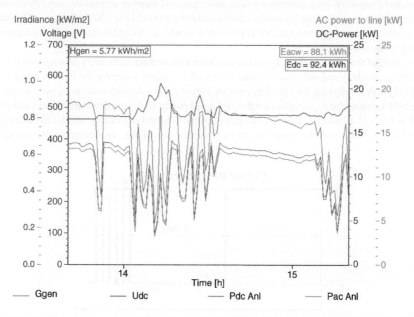

Figure 5.110 Problem with dynamic MPP tracking at a PV installation on 24 September 2007. Between 2.00 and 2.35 p.m. on this date, the DC input power P_{DC} absorbed by the inverter was at times unduly low compared with insolation, because, owing to insolation fluctuations, the device's DC operating voltage V_{DC} was sometimes considerably higher than actual MPP voltage V_{MPP}

where:

$v_{DC}(t)$ and $i_{DC}(t)$ = voltage and current at the inverter's DC input, respectively

$p_{MPP}(t)$ = maximum available PV generator simulator power for each momentary MPP

T_M = measurement time as from $t = 0$

The denominator integral constitutes overall MPP energy that can be absorbed by the inverter under optimal conditions during measurement T_M.

Often during PV generator simulator tests, $p_{MPP}(t)$ is discontinuous and is instead broken down into numerous stages that exhibit transient and constant P_{MPPi}. The following applies in such a case:

$$\int_0^{T_M} p_{MPP}(t) \cdot dt = \Sigma P_{MPPi} \cdot T_{Mi} = P_{MPP1} \cdot T_{M1} + P_{MPP2} \cdot T_{M2} + \dots + P_{MPPn} \cdot T_{Mn} \qquad (5.36)$$

(the sum total of all MPP power that the inverter can potentially absorb under optimal conditions during the various power stages), where T_{Mi} is the time during which the solar generator simulator provides power P_{MPPi}:

$$\text{Total measurement time } T_M = \Sigma T_{Mi} = T_{M1} + T_{M2} + T_{M3} + \dots + T_{Mn} \qquad (5.37)$$

5.2.5.4.1 *Simple Dynamic MPP Tracking Test Using Quasi-square Test Patterns*

Tests involving relatively rapid (i.e. quasi-square), artificially induced discrepancies between two different power stages are relatively easy to perform (see Figure 5.111). Small-scale PV installations exhibit steeper power variation edges than is the case with large installations. For installations with up to a few kilowatts of capacity, solar generator power under unusual weather conditions (notably following a cold front during the spring or early summer) can be varied by around 15 to 120% of nominal DC output in under two seconds. A good inverter should not shut down under such conditions, and should certainly not exhibit any other anomaly.

One example of this test is a virtually square variation with around 20 to 100% of nominal current and power with steep edges and only one to three intermediate stages, whereby the current and power are absorbed only very briefly (e.g. for 100–200 ms) (see Figure 5.111). As with static tests, prior to the start of dynamic MPP tracking the P_{MPP} values for the envisaged power stages are determined and a stabilization period of 1–5 min is allowed. Six or so test cycles are then carried out during which effective dynamic MPP tracking is measured. Inasmuch as most inverters are unable to determine the actual MPP

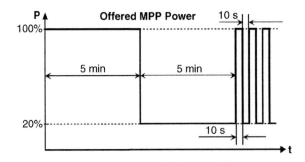

Figure 5.111   Solar generator simulator power profile during a simple dynamic MPP tracking test, prior to which MPP output at both power stages was precisely measured and a 5 min device stabilization period was allowed. The actual test begins with rapid low–high–low toggling for 10 s per power stage

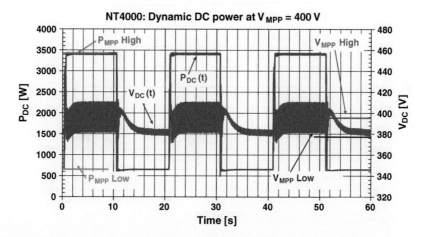

Figure 5.112 Dynamic power $P_{DC}(t)$ and DC voltage $V_{DC}(t)$ for operation on a characteristic curve with $V_{MPP} = 400\,V$ (with maximum power) for an NT4000 inverter and power output as in Figure 5.111. This device exhibits excellent dynamic characteristics at this voltage, whereby dynamic MPP tracking efficiency is $\eta_{MPPTdyn} = 99.4\%$ under these conditions

instantaneously, the entirety of the available MPP power P_{MPPi} is not absorbed immediately following a current/power change. The test cycle time T during which high and low current are absorbed can vary between 2 and 60 s, which equates to 12 minutes of maximum total test time $T_M = \sum T_{Mi}$ for the defined power and voltage range.

The advantage of such quasi-square dynamic tests (see Figure 5.111) lies in their simplicity and the opportunity they afford to measure precisely the high and low stage output as is done with static measurements. Such tests also provide a clear overview of the main characteristics of an inverter's MPP controller, although these data reflect go-live characteristics to a very limited extent only. Figures 5.112 and 5.113 display the results of such simple dynamic MPP tracking tests.

The results of other tests of this type can be found in [5.52] and [5.70].

5.2.5.4.2 *Dynamic MPP Tracking Test with Ramping in Accordance with prEN 50530*

Although tests involving current ramping (proportional insolation) and MPP power ramping reflect real operating conditions more realistically than the above tests, they are more difficult to perform to the requisite high level of accuracy. As with static tests, prior to a dynamic MPP tracking test it is necessary to allow up to a five-minute stabilization period. Six or so test cycles are then carried out during which effective dynamic MPP tracking is measured. As inverters somewhat lower the actual MPP, the available MPP power is not fully absorbed following a change.

In the interest of devising a standard overall inverter efficiency figure of merit [5.83], in 2008 various test patterns for such overall inverter efficiency tests involving ramping were investigated (see Figures 5.114 and 5.115). During these tests, extreme variations between n and t_1–t_4 were applied with a view to determining whether an inverter's MPP tracking algorithm can successfully keep pace with rapidly fluctuating insolation.

The above tests were defined via a characterization code that was composed of the following elements (see Table 5.13): number n of repetitions, followed by x and (in brackets) the duration t_1, in seconds, of the increase, followed by dwell t_2 at a high level, in seconds (number + H), followed by the duration of the decrease t_3 in seconds, followed by dwell at low level t_4 in seconds (number + L). According to the characteristic curve model (single-diode model) used in the simulator, P_{MPP} is roughly proportional to G, while V_{MPP} varies slightly according to current **irradiance** G. The measurements are extremely time consuming, e.g. a measurement in Table 5.13 (including a 300 s stabilization time at the baseline stage) took nearly 6.5 hours.

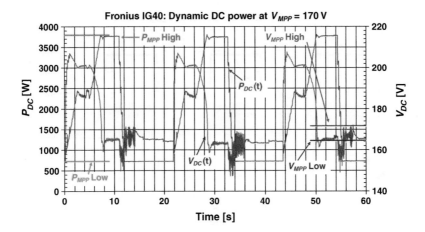

Figure 5.113 Dynamic power $P_{DC}(t)$ for $V_{MPP} = 170$ V (for maximum power) for an IG40 and output as in Figure 5.111. The dynamic characteristics here are suboptimal in that the upper-stage power is absorbed only gradually following a power ramp-up, and dynamic MPP tracking efficiency is only $\eta_{MPPTdyn} = 82.5\%$

On some occasions, many of the tests realized until October 2008 exhibited test ramps that were steeper or less steep than indicated in Table 5.13, whose test ramps represent a compromise that: (a) was reached after lengthy discussions and many test measurements; and (b) has since been folded into the draft of the provisional European standard prEN 50530. Finely calibrated measurements (in seconds), previously realized in Freiburg, Burgdorf and Vienna for solar generators [5.61, 5.78], reveal that the insolation change slopes in go-live solar generators vary considerably and that such changes become more infrequent as the change slope grows steeper.

A broad measurement spectrum ranging from 0.5 to 100 W/m²/s was defined that encompasses all possible values likely to occur in practice with reasonable frequency. Although steeper slopes lying outside this spectrum may occur, they are rare.

Exact measurement replicability is not always attainable for devices whose solar trackers perform poorly, in that identical tests on the same inverters sometimes yield differing dynamic MPP tracking efficiency results depending on current inverter DC voltage, the current V_{MPP} tracking direction (upwards or downwards) and device clock status.

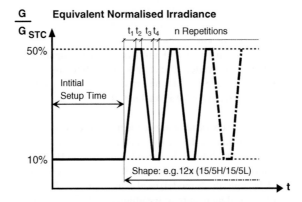

Figure 5.114 Test of fluctuations between various low- and medium-output levels (10% and 50% of G_{STC} with appreciable V_{MPP} variation)

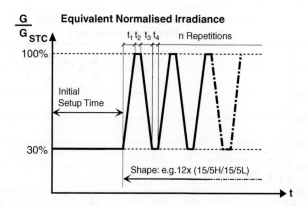

Figure 5.115 Test of fluctuations between various medium- and high-output levels (30% and 100% of G_{STC} with appreciable V_{MPP} variation)

During the test development phase, analogue tests were also performed that began at a high rather than a low power level, and these tests of course took twice as long as the previous tests. However, as these analogue tests were not very informative, they were excluded from the definitive test patterns so as to keep the amount of work entailed by the tests within reasonable bounds.

5.2.5.4.3 Overview Diagram Containing the Dynamic Ramp-Test Results as in Table 5.13

Using the ramps from these tests, seven different inverters (numbered 1–7) were tested until mid 2009. The tests showed that the manufacturer-defined MPP tracking algorithm has a major impact on dynamic

Table 5.13 Envisaged ramp tests, iteration counts and power variation curve steepness according to pr EN50530 [5.83]. A 300 s stabilization period is allowed at the baseline stage prior to each measurement cycle

Irradiance Variation 10% ⇒ 50% of G_{STC}

Number	Slope [W/m²/s]	Mode	Duration [s]
2	0.5	(800/10H/800/10L)	3540
2	1	(400/10H/400/10L)	1940
3	2	(200/10H/200/10L)	1560
4	3	(133/10H/133/10L)	1447
6	5	(80/10H/80/10L)	1380
8	7	(57/10H/57/10L)	1374
10	10	(40/10H/40/10L)	1300
10	14	(29/10H/29/10L)	1071
10	20	(20/10H/20/10L)	900
10	30	(13/10H/13/10L)	767
10	50	(8/10H/8/10L)	660
Total	Time for Setup + Measurement		15939

Irradiance Variation 30% ⇒ 100% of G_{STC}

Number	Slope [W/m²/s]	Mode	Duration [s]
10	10	(70/10H/70/10L)	1900
10	14	(50/10H/50/10L)	1500
10	20	(35/10H/35/10L)	1200
10	30	(23/10H/23/10L)	967
10	50	(14/10H/14/10L)	780
10	100	(7/10H/7/10L)	640
Total	Time for Setup + Measurement		6987

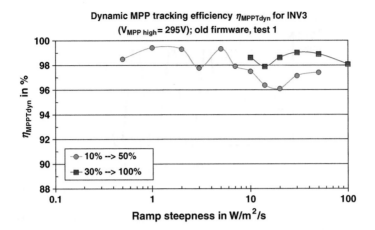

Figure 5.116 Results of a complete ramp test as in Table 5.13 for inverter 3 using older firmware from 2005 (first test cycle). The measured $\eta_{MPPTdyn}$ values ranged from 96 to 100%, but no curve exhibited pronounced 'resonance points' where the device became completely disoriented

MPP tracking. Particularly noteworthy in this regard were tests involving two inverters using older MPP tracking firmware for one device and optimized firmware of this type for the other.

Inasmuch as these tests involved basic research and were not intended to evaluate specific inverters, in a departure from customary practice the inverter types and manufacturers were anonymized.

Figures 5.116, 5.117 and 5.118 display the results of these tests for inverters 3 and 6 using older firmware. As a comparison of Figures 5.116 and 5.117 shows, not all test results proved to be replicable (also see Figures 5.123 and 5.125). The devices also experienced problems with certain ramp curves, and some of the other measured $\eta_{MPPTdyn}$ values are well below 100%.

Figures 5.119, 5.120 and 5.121 display the results of analogue tests using the exact same devices as above, but with improved firmware in lieu of the older versions. Both devices exhibited outstanding test results and reproducibility, which were also observed in all other devices that passed this test with flying colours. Reproducibility problems were only observed so far in devices that exhibited substandard

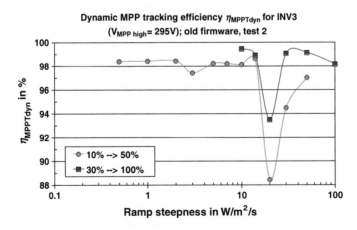

Figure 5.117 Results of a complete ramp test as in Table 5.13 for inverter 3 using older firmware from 2005 (second test cycle). With a 20 W/m²/s ramp curve, pronounced 'resonance points' were observed that resulted in the device becoming completely disoriented. Device characteristics with this ramp curve were far poorer than in Figure 5.116, and no exact replicability was observed

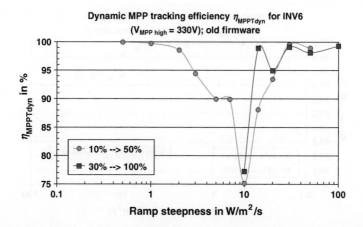

Figure 5.118 Results of a complete ramp test as in Table 5.13 for inverter 6 using older firmware. This device handled extremely sluggish ramps well, but lost its footing in roughly the 3–20 W/m²/s range, where it exhibited pronounced 'resonance points' at a ramp curve of 10 W/m²/s

dynamic characteristics. Hence simply improving the firmware can greatly improve dynamic MPP tracking in inverters.

5.2.5.4.4 Dynamic Test Timeline Diagram for Analytic Purposes

To obtain greater insight into the characteristics exhibited during dynamic tests, it is useful to plot as a function of time not only the power P_{DC} absorbed by the test device, but also its working voltage V_{DC} (see Figures 5.122–5.127), and in some cases using different test patterns as in Table 5.13.

Figures 5.122 and 5.123 display the characteristics for inverter 3 during such a ramp test. The cause of the non-replicable characteristics of the MPP tracking algorithm used is very clearly shown for these tests, whereby a 10 W/m²/s curve appears to be absolutely critical. However, the 20 W/m²/s curve appears to be even more critical for the low to high tests shown in Figures 5.116 and 5.117.

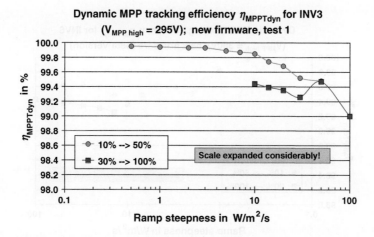

Figure 5.119 Results of a complete ramp test as in Table 5.13 for inverter 3 using the new firmware (first test cycle). The scale is greatly elongated here. Dynamic MPP tracking with this firmware was outstanding and the measurement results were readily replicable (also see Figure 5.120)

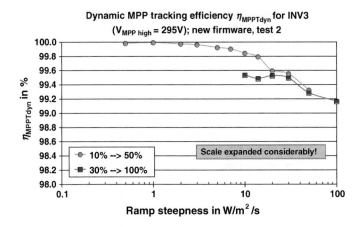

Figure 5.120 Results of a complete ramp test as in Table 5.13 for inverter 3 using the new firmware (second test cycle). The scale is greatly elongated here. Dynamic MPP tracking with this firmware was outstanding and the measurement results were readily replicable (also see Figure 5.119)

In Figures 5.122 and 5.123 (elongated section), the device begins (as it is supposed to) at the high V_{MPP}, but it then increases V_{DC} steadily during increasing P_{MPP} ramps, with the result that the device runs on the characteristic I–V curve towards open circuit and therefore P_{DC} decreases at an ever-faster rate. The tracking direction does not reverse until the open-circuit voltage state is attained on the low level.

As Figure 5.123 clearly shows, V_{DC} increases immediately before the beginning of the rising P_{DC} ramp. The MPP tracking algorithm now assumes that this increase of P_{DC} is the result of the previous increase of V_{DC}, and that therefore it has simply to continue in the same direction and increase V_{DC} further. However, if P decreases or remains constant, V_{DC} will remain roughly constant as well.

Figures 5.124 and 5.125 (elongated section) display a detailed timeline diagram of a ramp test as in Table 5.13 with 20 W/m²/s (10% ⇒ 50%) for inverter 3 with the old firmware. However, since the internal clock was set differently at the beginning of the test, the device now steadily lowers V_{DC} with the result that it moves along the characteristic curve towards the short-circuit point and P_{DC} decreases gradually. Inasmuch as output does not decrease as quickly when voltage decreases as it does when the device is moving towards an open-circuit state, P_{DC} decreases less rapidly and $\eta_{MPPTdyn}$ does not decrease as much.

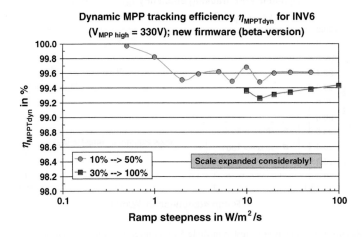

Figure 5.121 Results of a complete ramp test as in Table 5.13 for inverter 6 using the new firmware. The scale is greatly elongated here. Dynamic MPP tracking with this firmware was outstanding

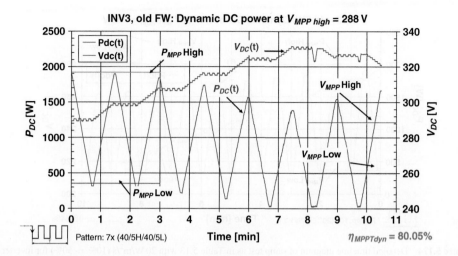

Figure 5.122 Timeline diagram for a ramp test of inverter 3 using the old firmware and beginning at a high-power stage. If P increases or remains constant, V_{DC} also remains constant; if P increases, V_{DC} also increases, although it is well above V_{MPP}. A detailed analysis, with commentary, can be found in the next illustration

Figures 5.126 and 5.127 display the same test as in Figures 5.124 and 5.125 with 10 ramps (20/10H/20/10L, 10% ⇒ 50%, curve 20 W/m²/s) for inverter 3 with the new firmware. The device exhibits far better characteristics in this case.

The MPP tracking algorithm comes into play here. Inasmuch as a change in P_{DC} is not necessarily attributable to a previous change in V_{DC} but can be prompted by an insolation change instead, from time to time the algorithm tracks in the opposite direction and if necessary remains at this voltage.

The tests realized to date show that the dynamic MPP tracking characteristics of inverters can be greatly improved through the use of suitable and intelligent software, without lowering the quality of static MPP tracking. In view of the fact that the dynamic MPP tracking characteristics of two inverters were drastically improved by simply optimizing the MPP tracking software and without modifying the

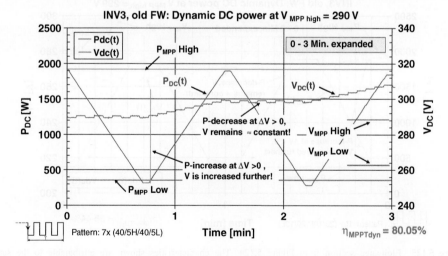

Figure 5.123 Elongated section of Figure 5.122. As long as power P_{DC} continues to rise, the MPP tracking algorithm simply assumes that this power increase was caused by its previous change in V_{DC}; thus, for reaching MPP, V_{DC} has simply to be changed in the same direction. Observation interval ≈ 5 s

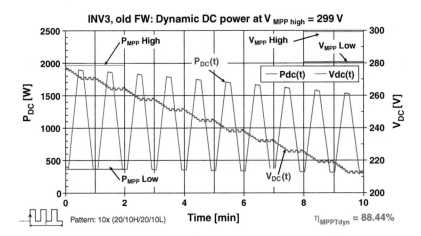

Figure 5.124 Detailed timeline diagram of ramp test as in Table 5.13 with 20 W/m²/s (10% ⇒ 50%) for inverter 3 with the old firmware. If *P* increases or remains constant, V_{DC} also remains constant; if *P* decreases, $\eta_{MPPTdyn}$ is 88.4% here. A detailed analysis, with commentary, can be found in the next illustration

hardware, any manufacturer should be able to effect comparable optimizations and thus attain good dynamic MPP tracking.

However, some devices that exhibit substandard dynamic MPP tracking characteristics unfortunately also exhibit problems when it comes to measurement value replicability at the various stages. But as was shown above, this is a basic problem in that MPP tracking validation is always realized via internal clock-controlled periodicity and a test device, and its inner clock simply cannot be synchronized. That said, replicability for devices with good dynamic MPP tracking characteristics has been outstanding during all of the tests conducted to date.

The impact of dynamic MPP tracking on energy yield should not be underestimated. The main inverter selection criterion should be possible overall efficiency η_{tot} in the range of the envisaged MPP voltage

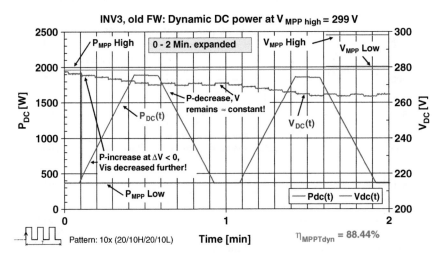

Figure 5.125 Elongated section from Figure 5.124. The characteristics shown are attributable to the same phenomenon as in Figures 5.122 and 5.123: as long as power P_{DC} continues to rise, the MPP tracking algorithm simply assumes that the reason for this power increase was caused by its previous change in V_{DC} and thus, to reach the MPP, V_{DC} has simply to be changed in the same direction

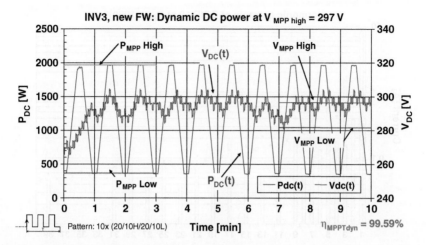

Figure 5.126 Detailed timeline diagram of a ramp test as in Table 5.13 with 20 W/m²/s (10% ⇒ 50%) for inverter 3 with new firmware (same test pattern as in Figure 5.124). V_{DC} begins just below the V_{MPP} of the low level and quickly follows the correct generator V_{MPP} value in each case. $\eta_{MPPTdyn}$ attains an excellent level of 99.6%

V_{MPP}, whereby dynamic MPP tracking efficiency is of only secondary importance. A good MPP tracker should be able to handle power fluctuations without exhibiting abrupt voltage jumps, and should instead adjust the voltage relatively slowly but also accurately, without moving away from the MPP [5.82].

5.2.5.5 Harmonic Current

Figure 5.128 displays the current spectrum for the EGIR10 line-commutated single-phase inverter (1.75 kW) at 1 kW of power. First-harmonic current is considerably higher than it should be, in view of the converted active power with $\cos \varphi = 1$, and thus the apparent output is far higher than active power, causing the device to use a relatively large amount of reactive power. Moreover, under these output conditions, the EN 61000-3-2 limit values for harmonic current are exceeded (see Section 5.2.3.3.3).

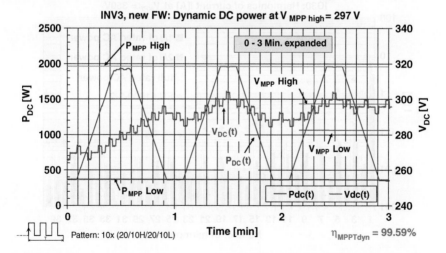

Figure 5.127 Elongated section of Figure 5.126. This diagram clearly shows the functional principle of the new and improved MPP tracking algorithm. As the observation interval is only around 2 s here, the MPP tracking algorithm reacts much more quickly to changes

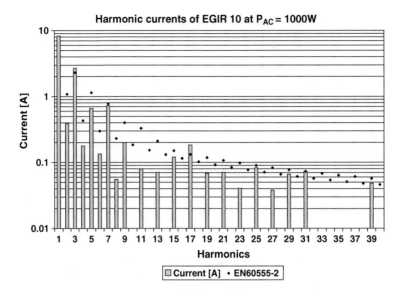

Figure 5.128 Harmonic current in the EGIR10 line-commutated inverter(1.75 kW) at $P_{AC}=1$ kW. The bars in the diagram display effective current for the various harmonics, and the EN 61000-3-2 limit values are indicated by ◆. The limit value for the 3rd and 17th harmonics is already exceeded at a mere 1 kW

Inasmuch as all of the other inverters tested are self-commutated and exhibit high-frequency pulse width modulation, their harmonic currents should normally induce no problems in go-live settings if grid impedance is not unusually high.

Figure 5.129 displays the harmonic current readings for a self-commutated single-phase inverter, all of whose harmonics for the harmonic currents measured are far below the allowable limit values. Virtually all newer single-phase inverters adhere to the EN 61000-3-2 limit values for devices up to 16 A at full power.

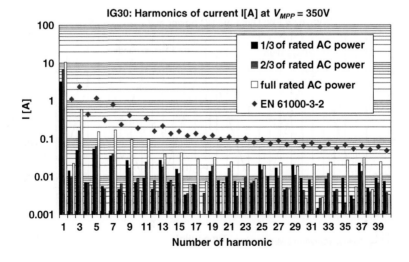

Figure 5.129 Harmonic current in the IG30 inverter ($P_{ACn}=2.5$ kW) for $V_{MPP}=350$ V and output of 33, 66 and 100% of nominal AC power, relative to EN 61000-3-2 limit values. (Ordinal numbers = 400% of basic frequency 50 Hz.)

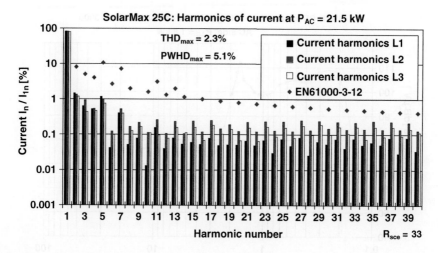

Figure 5.130 Standardized harmonic current for the SolarMax 25 triphase inverter ($P_{ACn} = 25$ kW) for $V_{MPP} = 630$ V and power of $P_{AC} = 21.5$ kW relative to the EN 61000-3-12 limit values for $S_{KV} \geq 33 S_{WR}$ (see Table 5.8)

Figure 5.130 displays harmonic current readings for a Solarmax 25C ($P_{ACn} = 25$ kW) triphase inverter, which conforms with EN 61000-3-12 limit values [5.17] for 16–75 A devices, and with the limit values in [5.19] for $S_{KV} \geq 33 S_{WR}$ (see Section 5.2.3.3.3). Hence this device should be compatible with any PV system from a harmonic current standpoint.

5.2.5.6 High-Frequency Interference Voltage Emissions

Many manufacturers developed inverters between 1987 and 1992, often on shoestring budgets, with the result that in many cases these devices exhibited only the most rudimentary functions and important details such as sufficient suppression of high-frequency interference voltages in power leads were disregarded; this in turn often caused problems in PV installations. Back then, many manufacturers were simply unaware of the fact that good suppression of high-frequency emissions almost automatically translates into improved immunity against transient external interference voltages and thus optimizes device reliability.

Such filtering as was implemented was often limited to the AC side. The manufacturers were fond of citing in their datasheets standards such as EN 55011 that only indicated limit values for the AC side. But for medium and large inverters, which in some cases have extensive DC wiring on the solar generator side, it is essential to implement adequate disturbance suppression measures on both the AC *and* DC sides (see Section 5.2.3.4 for the relevant limit values).

Figure 5.131 displays the high-frequency interference voltages on the AC side for a particularly serious offender, namely the original version of the first Solcon inverter to be mass produced in Switzerland. This device, whose conducted disturbances ranged up to 56 dBμV above the allowable limit values, not surprisingly also caused extreme interference in nearby electronic devices.

Although the Bern University of Applied Sciences PV Lab managed to reduce these interference emissions considerably, it was not possible to get them below the allowable limit values. In most cases, such disturbances cannot be remedied by simply integrating a grid filter somewhere or other in the device.

Manufacturers with their eyes on future trends soon realized that electromagnetic compatibility is a key green-technology issue and began modifying their devices accordingly, at least on the AC side, whose standards had been clearly defined for quite some time. However, for many years manufacturers failed to realize that disturbance suppression is also necessary on the DC side, and to some extent such measures are still lacking for inverters in the range up to 500 kHz. SMA made major advances in the field of

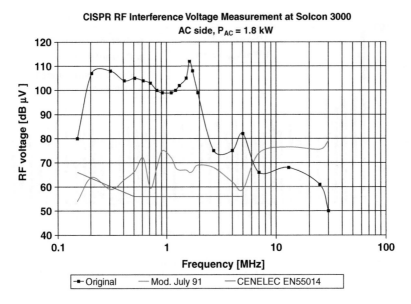

Figure 5.131 High-frequency disturbances induced by an old Solcon 3000 inverter on the AC side before and after modification in July 1991, relative to the EN 55014 limit values for quasi-peak measurements. At around 1.8 MHz, the original version exceeded the limit value by a massive 56 dBµV

electromagnetic compatibility early on, and the high-frequency emissions of the company's latest inverters are far below the allowable limit values (see Figures 5.132 and 5.133).

Figure 5.134 displays the high-frequency interference voltages for a triphase Solarmax 25C inverter, which is challenged by disturbance suppression below 400 kHz (also on the AC side, not shown here). However, the emission characteristics of power cables and solar generators in this frequency range are

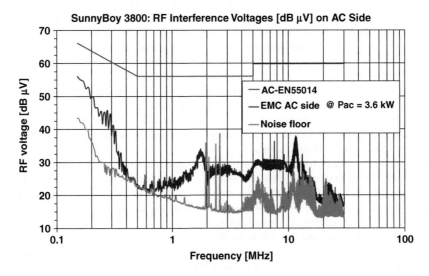

Figure 5.132 High-frequency interference voltages produced on the DC side by an SB3800 inverter at nominal output, relative to the EN 55014 limit values for quasi-peak measurements. The basic disturbance level measured for this device when it is shut off is extremely low, and when in operation it undercuts the limit values by a truly impressive 10–40 dBµV

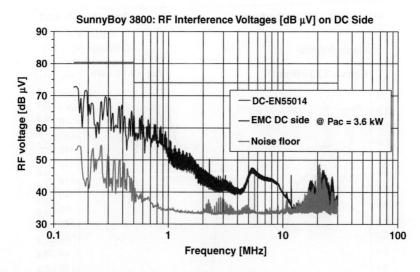

Figure 5.133 High-frequency interference voltages produced on the AC side by an SB3800 inverter at nominal output, relative to the EN 55014 limit values for quasi-peak measurements. Here too the EN 55014 limit values are undercut by 10–40 dBμV, and the high-frequency interference voltages are even lower than the more stringent recommended limit values as in Figure 5.71

poor, and this range is hardly ever used for radio reception in residential areas; thus limit value exceedance in this frequency range is of little importance.

5.2.5.7 A Brief History of Inverter Reliability (Part 1)

The first inverters (SI-3000 and Solcon 3000) to be tested at the Bern University of Applied Sciences between 1989 and 1991 exhibited an average of around three inverter failures per operating year.

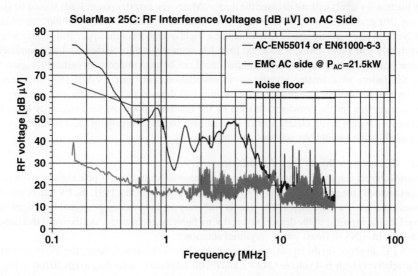

Figure 5.134 High-frequency interference voltages produced on the DC side by a Solarmax 25C inverter at nominal output, relative to the EN 55014 and EN 61000-6-3 limit values for quasi-peak measurements. The limit values are egregiously overshot for frequencies up to 400 kHz (including on the AC side)

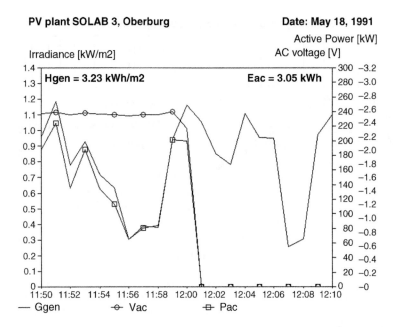

Figure 5.135 Failure of a Solcon inverter at noon on 18 May 1991 triggered by a 317 Hz power company ripple signal under relatively high grid voltage conditions

Tests conducted at the PV Lab revealed that many of these failures were attributable to ripple signals (particularly in conjunction with relatively high grid voltage) and high grid voltage at the inverter connection point (particularly during cloud enhancements under low grid load conditions). Figure 5.135 displays this type of failure in a Solcon inverter attributable to a 317 Hz ripple signal under relatively high grid voltage conditions. Other failures of this type are described in [5.37] and [5.38]. Inverters for grid-connected operation need to be sized in such a way that exposure to grid ripple signals with sometimes strongly fluctuating levels will not damage the device. Moreover, inverter power leads should be sized in such a way that grid impedance and the voltage rises that occur under go-live conditions are taken into account. Such inverters should also be able to handle a slight overshoot of the upper grid voltage tolerance value (currently 230 V and 400 V + 10%, up from 6% some years ago) without failing completely.

The exact cause of other such inverter failures could not be determined with certainty. Figure 5.136 displays the failure of a master inverter in an installation with four PV-WR-1800 inverters in a master–slave configuration. The device manufacturer had devised this particular circuit arrangement partly to compensate for the drawback entailed by the relatively poor low-load efficiency of large PV installations. As Figure 5.136 clearly shows, the inverters began powering up successively prior to the failure, i.e. while the installation was still working properly.

5.2.5.7.1 *Research on Ripple Signal Sensitivity*

Once manufacturers learned that their inverters were prone to failure from ripple signals, most of them took steps to remedy the problem by improving the device circuitry. In late 1991, the PV Lab began testing all relevant inverters via simulated ripple signals with varying frequencies generated by a ripple signal simulator. Despite exposure of the devices up to 20 V, no failure was observed, and the worst that happened (particularly with ENS devices) was a very brief shutdown.

Figure 5.137 displays: (a) the ripple signal sensitivity that was measured (using the above set-up) in an SB3800 at relatively high grid voltage (242 V); and (b) the limit curve according to EN 50160 [5.18]. With the lowest ripple signal frequencies tested and ripple signal levels of more than around 16 V, the device shut down briefly a number of times but did not fail. Like many of today's inverters, this one exhibits excellent immunity to ripple signals.

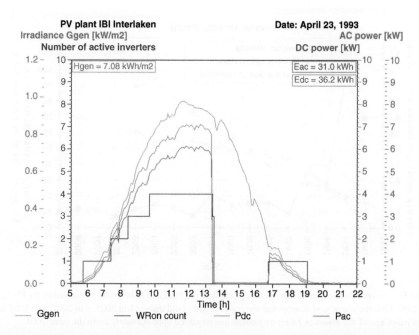

Figure 5.136 Inverter failure (master device) in a 9 kW installation with four PV-WR-1800 inverters in a master–slave configuration. The inverters began switching on successively prior to the failure. This failure was detected at the end of the day and the installation was manually reconfigured for operation with the devices that were still working

5.2.5.7.2 A Brief History of Inverter Reliability (Part 2)

The mean number of inverter failures per device operating year has decreased exponentially owing to: (a) the elimination of various initial problems such as the above sensitivity to ripple signals and high voltage; (b) the manufacturer's learning curve; and (c) measures aimed at reducing high-frequency

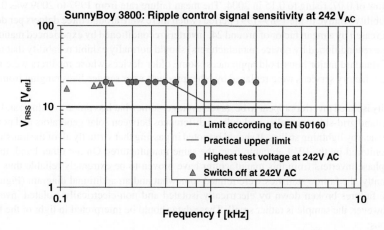

Figure 5.137 Sensitivity of an SB3800 to ripple signals with $P_{AC} = 3.8$ kW and $V_{AC} = 242$ V. Only in the lowest ripple signal frequencies in the 110–220 Hz range at ripple signals exceeding 16 V did the device briefly shut down, but without failing. Moreover, at frequencies of upwards of 283 Hz and with maximum voltage (20 V) there were no shutdowns at all

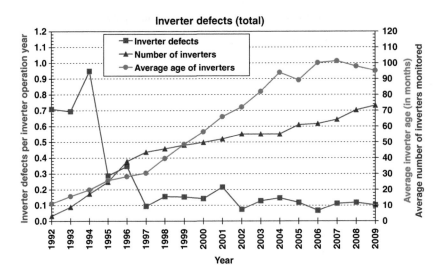

Figure 5.138   Inverter failures per device operating year and mean number of inverters monitored by the PV Lab from 1992 to 2009. With the use of new devices and the replacement of old ones (11 in 2005, 9 in 2007, 8 in 2008 and 5 in 2009) the mean age of the inverters being monitored has remained approximately constant since 2004

interference voltages, which have also improved inverter DC- and AC-side immunity to transient voltage surges from external devices.

As Figure 5.138 shows, the various PV Lab measurement projects revealed that inverter failures per device operating year decreased from around 0.71 in 1992 to less than 0.1 in 1997 (with a transient increase to 0.94 in 1994 owing to the commissioning of numerous new and untested types of inverters). Since 1997, this figure has varied between 0.07 and 0.21 inverter failures per inverter operating year, reaching a relative maximum of 0.21 in 2001 due to the failure of, for the most part, older devices. As for the 2001 failures, the failure of three devices was probably partly attributable to damage resulting from lightning-induced voltage surges, since these failures occurred in quick succession at the same site. In 2002, when no serious lightning was observed, the failure rate decreased to a record low of 0.07, rising to 0.15 in 2004. The mean failure rate from 1997 to 2009 was 0.24. Thus inverter reliability relative to the 1989–1991 period (when there were around three defects per device and year) has increased by a mean factor of around 24. Inverters reconditioned by experienced manufacturers (but not those reconditioned by novice manufacturers) should normally exhibit reliability that is on a par with that of standard major household appliances. Some older devices whose repairers were no longer providing satisfactory services were replaced by other products after service lives ranging from 4.5 from 12 years.

Electrically isolated inverters appear to be: (a) on average, somewhat less prone to failure; and (b) better able to withstand grid-side anomalies and voltage differences between solar generators and power leads secondary to nearby lightning strikes. It should be noted in passing that virtually all of the transformerless inverters monitored by the PV Lab came from the same manufacturer. On the other hand, the higher-capacity triphase inverters from this manufacturer have proven to be extremely reliable thus far.

Consequently, the failure statistics were refined somewhat and an additional diagram (Figure 5.139) showing the failures broken down by electrically isolated and non-electrically isolated inverters was realized. However, the sample is rather small. These data should be interpreted in light of the following considerations:

• The high failure rate of non-electrically isolated inverters in 1996 was mainly attributable to a series of failures of EcoPower20 devices, which were subsequently replaced and were phased out many years ago.

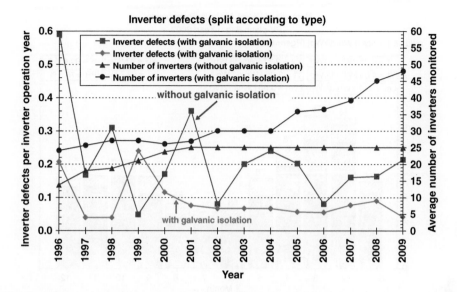

Figure 5.139 Inverter failures per device operating year (1996–2009), broken down by devices with and without electrical isolation. The data appear to indicate that electrically isolated inverters exhibit better long-term reliability. However, the sample used here was relatively small

- The high failure rate of electrically isolated devices in 1999 was occasioned by failures of a number of older Solcon devices that were then replaced by other devices. These Solcon devices were phased out many years ago as well.
- The record high number of transformerless device failures in 2001 was probably due to voltage surges induced by lightning strikes near an installation containing a number of inverters.

Needless to say, the impact of an inverter failure varies greatly depending on timing and the size of the device in question. The energy loss in the installations affected may well have been around 10% in some years (see Figures 5.140 and 5.141).

On the other hand, the *mean* energy yield loss attributable to inverter failure was considerably lower (long-term mean around 1.1%). Figure 5.142 displays the mean energy yield loss occasioned by these failures for PV installations in Burgdorf from 1996 to 2008.

5.2.5.8 Power Limitation of Oversized Solar Generators

The first grid-connected inverters to come on the market (SI-3000, Solcon and others) simply shut down if they were oversized or exposed to a cloud enhancement. However, this device characteristic is completely unacceptable for PV installation owners. It makes much more sense for an inverter to respond to excess power on the DC side by allowing the MPP to drift towards open-circuit voltage and then simply scale back the nominal AC voltage. This is the current state-of-the-art method. The PV-WR-1500 was the first mass-produced inverter to exhibit this characteristic (see Figure 5.143), thus allowing the solar generator to be up to 50% oversized. Under the same peak AC power conditions, the inverter is able to feed far more energy into the grid than is possible with a solar generator whose sizing is fully congruent with nominal DC power. Power companies find this characteristic highly desirable as it somewhat mitigates the peak power problems associated with grid-connected systems (see Section 5.2.7).

Inasmuch as power-limited inverters are now available, solar generators in flat areas should be slightly oversized, i.e. power P_{Go} at STC power output conditions (1 kW/m^2, cell temperature 25 °C, AM1.5) should be somewhat higher than the inverter's nominal DC power P_{DCn}, although the maximum

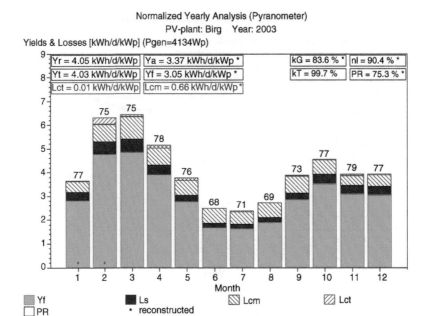

Figure 5.140 Standardized 2003 statistics for the Birg installation (2670 m above sea level); the energy loss attributable to inverter failure in January and February 2003 is also shown (see Figure 5.141). Excluding this failure, solar energy production for the year was 1113 kWh/kWp (also see Chapter 7)

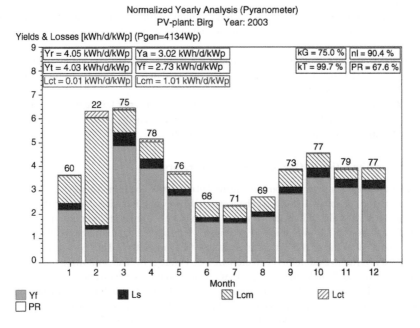

Figure 5.141 Standardized 2003 statistics for the Birg installation (2670 m above sea level). Owing to a failure in late January, the installation produced active power amounting to only 996 kWh/kWp, i.e. 10.5% less than normal (also see Chapter 7)

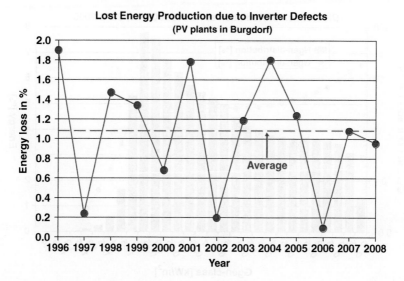

Figure 5.142   Energy yield loss (in %) attributable to inverter failure as tracked by a long-term PV Lab monitoring project. The long-term mean loss was 1.07%

connectable generator power as in the manufacturer's datasheet should also be observed. This method can potentially ramp up the installation's low-load efficiency and reduce costs related to installation size.

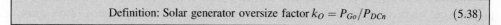

Definition: Solar generator oversize factor $k_O = P_{Go}/P_{DCn}$ (5.38)

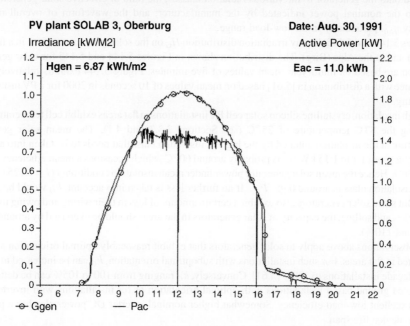

Figure 5.143   Limitation of AC power fed into the grid in a PV-WR-1500 inverter at around 1.4 kW in the presence of excess DC power and an oversized 2.8 kWp solar generator. The brief shutdowns at noon and 4 p.m. were attributable to ripple signals

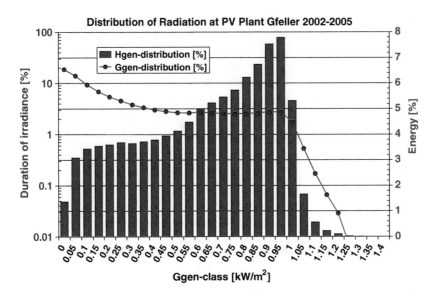

Figure 5.144 Multi-year distribution of energy irradiated onto the solar generator plane H_G (based on five-minute mean values) for a flat-roof PV installation in Burgdorf (540 m above sea level) with $\beta = 28°$ and $\gamma = -10°$. Despite occasional cloud enhancements ranging up to 1.4 kW/m², the values above around 1.15 kW/m² are irrelevant from an energy standpoint

The following factors should be taken into consideration in determining the factor k_O by which a solar generator can be oversized without inducing excessive energy loss: the distribution of the energy irradiated onto the generator to the various radiation classes; the ratio of effective solar generator STC power to the nominal power indicated by the manufacturer; and the waveform of overall inverter efficiency η_{tot}, particularly in the low-load range.

Figures 5.144 and 5.145 display irradiation distribution H_G on the solar generator surface in a flat area (Gfeller) and in the Alps (Birg). The distribution shown in Figures 5.144 and 5.145 for solar generator insolation are based on multi-year mean values of five minutes. Figure 5.144 exhibits relatively good congruence with a distribution in [5.61] based on mean values of 10 seconds in 2000 for a PV installation in Freiburg.

At high insolation, crystalline silicon solar cell PV installations in flat areas exhibit cell temperatures far exceeding the STC temperature of 25 °C (see Sections 3.3.3 and 4.1). The mean solar generator temperature at the insolation entailed by the highest insolation class that needs to be taken into account and ranging from 1.1 to 1.15 kW/m² is probably around 60 °C, which occasions a mean efficiency loss of around 16%. Hence the mean solar generator power under mean insolation conditions ($G = 1.125$ kW/m²) for the insolation class is around $0.95 \cdot P_{Go}$. If no further loss is taken into account, k_O would be around 105%. But since solar generators also exhibit a certain amount of loss in their wiring and owing to power mismatches and soiling, the capacity of solar generators in flat areas should be oversized by around 110% (k_O around 110%).

The observations above apply to solar generators that exhibit reasonably optimal orientation and that are located in flat areas. For such installations with suboptimal orientation, k_O can be increased to 120%, and for façade installations to 130% [5.61]. Conversely, k_O ranging from 100 to 105% can be defined for well-cooled ground-based installations with optimal orientation or for installations with inverters that exhibit excellent low-load efficiency. Somewhat higher nominal inverter DC power also has a positive effect on device life span.

Although k_O ranging from around 0.7 to 1.25 was recommended at one time [5.62], [5.63], it should be borne in mind that the active power P_{Ao} of a 1990s' solar generator at STC power output was normally around 10 to 15% lower than nominal power P_{Go} as derived from aggregate STC nominal module power

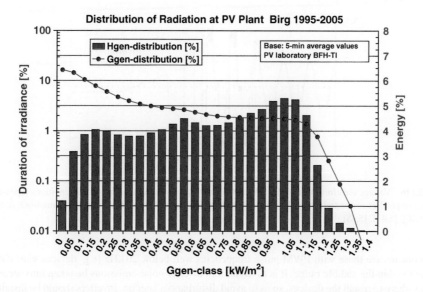

Figure 5.145 Multi-year distribution of energy irradiated on the solar generator plane H_G (based on five-minute mean values) for a flat-roof PV installation in Birg (2670 m above sea level) with $\beta = 90°$ and $\gamma = -5°$. There were occasional cloud enhancements ranging up to 1.45 kW/m^2, but the values above around 1.30 kW/m^2 are irrelevant from an energy standpoint

P_{Mo}. In other words, manufacturers often provided devices whose active power fell several percentage points short of the minimum power indicated on the device datasheet. Nominal power tolerance in 1990s' devices was often $\pm 10\%$. These devices also exhibited a low-load efficiency (see Figure 5.86) that fell far short of today's inverters (see Section 5.2.5.1) and their inverter power costs per kilowatt were appreciably higher. Thus around 10 to 20% higher k_O values made sense during this era. But nowadays, modules with considerably lower tolerances are available (even with a tolerance of -0%!), and these devices also exhibit considerably better low-load efficiency.

However, the situation is different for high-Alpine installations (Figure 5.145), which are exposed to lower ambient temperatures and more frequent cloud enhancements (mostly in the winter when the ground in front of the solar generators is covered with snow), which means that higher-power spikes need to be factored into the planning process. The mean solar generator temperature under the insolation conditions entailed by the highest insolation class that needs to be taken into account and ranging from 1.25 to 1.3 kW/m^2 is probably around 50 °C, which occasions a mean efficiency loss of around 12%. Hence the mean solar generator power under mean insolation conditions ($G = 1275$ kW/m^2) for this insolation class is around $1.13 \cdot P_{Go}$. If no further loss is taken into consideration, k_O should thus be around 90%, i.e. P_{Go} at STC power output should be defined as an even lower value than the inverter's nominal DC power.

Figures 5.144 and 5.145 show that transient irradiance G spikes (i.e. measured in seconds) far exceeding 1 kW/m^2 also occur. Such spikes ranging as high as 1.7 kW/m^2 have been observed at the Jungfraujoch installation (3454 m above sea level) secondary to cloud enhancements. Although such insolation spikes are rare, PV installation components and notably their inverters should nonetheless be able to withstand such events without being damaged. Hence inverter tests involving 140% of nominal DC power P_{DCn} should be routinely conducted so as to ensure that these devices limit their power and incur no damage during cloud enhancements (see Figure 5.66).

5.2.5.9 Inverter Noise

Inverter noise emissions vary greatly from one device to another. Some devices exhibit virtually no disturbing noise, while others make noise that would be unacceptable indoors. The inverters most prone to

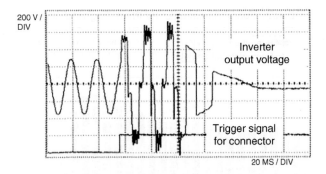

Figure 5.146 Voltage waveform in an older SWR700 following a power failure under load conditions (open-circuit shutdown, no parallel-connected load). The device shuts down as required but exhibits a maximum amplitude of around 760 V [5.46], [5.47], [5.48]

noise problems are those with PWM pulse frequencies well below 20 kHz (e.g. devices with IGBTs), which are within the audible range. It is essential that inverter noise emissions be taken into account in deciding where to install the devices, so as to avoid disturbances later on. Inverters should be installed as far as possible in an attic or cellar.

5.2.5.10   Shutdown under Load Following a Power Failure

A power failure without a balanced load and/or a load parallel to the inverter can induce elevated output voltages in some inverters across a number of periods. Figure 5.146 displays the characteristics of an older SWR700 with voltage spikes up to 760 V, which is more than 2.3 times the normal amplitude. Lower-power devices that are connected when such spikes occur can potentially be ruined. This effect is considerably less pronounced in the SWR1500, which is a more recent device from the same manufacturer (see Figure 5.147). When tested under these conditions, other inverters also exhibit similar characteristics, albeit with far lower amplitudes (see Figure 5.75).

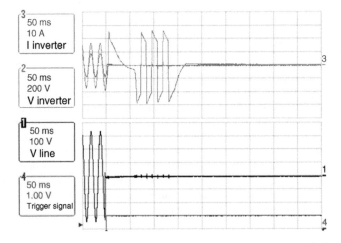

Figure 5.147 Open-circuit shutdown (no parallel-connected load) characteristics of a Sunny Boy SWR1500 for nominal AC power ($P_{ACn} = 1.5$ kW). In such a case, the voltage at the inverter output increases to about 500 V for three periods [5.46]

5.2.6 Problems That Occur in Grid-Connected Systems and Possible Countermeasures

This section discusses some of the key problems that have caused difficulties in PV installations. First, there is an introduction to the topic for electricians and electrical engineers, along with detailed descriptions of the relevant calculation methods. This is followed by practical tips, which will by and large obviate the key problems. The section concludes with a series of illustrative examples.

Grid-connected systems are small power plants that feed their energy into the electricity grid, but that by law must not interfere with normal grid operation and thus must obtain a permit that authorizes them to connect to the relevant power plant.

Figure 5.148 displays a single-phase equivalent circuit for a grid-connected system, whose grid at the grid link point (i.e. the point at which other users are also connected) is illustrated here by an equivalent voltage source V_{1N} connected in series with source impedance Z_N (R_N in series with L_N). The interconnecting line exhibits impedance Z_S (R_S in series L_S). In lieu of inductances L_N and L_S, it is possible (as is customary in the field of energy technology) also to indicate the 50 Hz values of X_N and X_S. Thus in investigating the problems associated with harmonics, it should be borne in mind that the X values for the nth harmonics are n times as high as for 50 Hz. As grid-connected PV inverters are realized either as single-phase devices (L–N connection) or for higher outputs as symmetrical triphase devices, the relevant single-phase equivalent circuit should be used.

The following equations apply to reactances X_N and X_S:

$$X_N = 2\pi f \cdot L_N; \; X_S = 2\pi f \cdot L_S \tag{5.39}$$

As shown in the sections below, the key calculation values are as follows: for the voltage increase, R_N and R_S; and for harmonics, L_N, L_S, X_N and X_S.

5.2.6.1 Determination of Grid Impedance Z_N at the Link Point

Many power companies now have computer databases containing the key characteristics of grid impedance, short-circuit capacity and so on for the entire grid. For the PV installation designer, it is

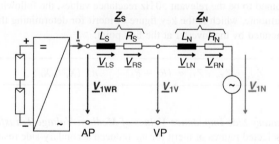

Figure 5.148 Single-phase equivalent circuit for a grid-connected system for calculation of voltage fluctuations, short-circuit capacity and harmonics. Cable capacities and reactive power compensation device capacities aside, this circuit arrangement is approximately valid for low frequencies up to around 2 kHz [5.64]. The following additional numbers can be used for R, L and X if necessary: 1, single phase; 3 = triphase.

Legend:

AP = inverter connection point V_{1WR} = connection point phase voltage
VP = link point V_{1V} = grid link-point phase voltage
V_{1N} = single-phase equivalent voltage source for the grid (ideal part)
R_N = grid resistance L_N = grid inductance
R_S = interconnecting line resistance L_S = interconnecting line inductance

of course very convenient and labour saving to be able to obtain such data for the envisaged link point from the relevant power company.

Grid impedance (amount and phase) is determinable not only for 50 Hz but also for up to 2 kHz frequencies, providing that suitable measuring instruments are available. Such an instrument was used for Figure 5.84, which displays the grid impedance Z_{1N} waveform as a function of inverter connection point frequency.

Described in [5.64] is a method for calculating low-voltage grid impedance involving a reasonable amount of computation. Inasmuch as very high-output PV installations in any case require the power company and installation owner to conduct collaboratively a detailed investigation of grid characteristics and conditions, for reasons of space the discussion here will be confined to grid impedance calculations for the low-voltage grid – although it should be borne in mind that the voltage increase calculation methods described here apply in principle to the medium-voltage grid as well (see [5.19] for detailed information on medium- and high-voltage grid calculations).

Grid impedance in the low-voltage grid is mainly determined by: (a) the size of the transformer that is interposed between the medium-voltage and low-voltage grid; and (b) the low-voltage line, whereby the characteristics of the superordinate medium-voltage grid are of minor importance in this regard. Hence it can be assumed, for example (albeit somewhat reductively), that only one transformer feeds into the grid (i.e. the low-voltage side has no loop network) and that impedance Z_N is the sum total of medium-voltage transformer impedance $Z_T = R_T + jX_T$ and impedance Z_L in the line from the transformer to the link point VP. As the neutral conductor's impedance is irrelevant for triphase connections to a symmetrical inverter at L1, L2 and L3, a distinction needs to be made between the single-phase and triphase scenarios.

The following equations apply to 5 Hz impedance, which is the main parameter for calculating voltage fluctuations and short-circuit capacity S_{KV} at the link point:

$$\underline{Z}_N = R_N + jX_N = Z_N \angle \psi = \underline{Z}_T + \underline{Z}_L = (R_T + R_L) + j(X_T + X_L) \tag{5.40}$$

$$\text{Amount of grid impedance } Z_N = \sqrt{R_N^2 + X_N^2} \tag{5.41}$$

$$\text{Grid impedance angle } \psi = \arctan(X_N/R_N) \tag{5.42}$$

If X_T and X_L are assumed to be the relevant 50 Hz reactance values, the following grid impedance is obtained for the nth harmonic, which is the key figure of merit for determining the harmonic-current-induced harmonics generated by the inverter at the link point:

$$\underline{Z}_{Nn} = \underline{Z}_{Tn} + \underline{Z}_{Ln} = (R_T + R_L) + j \cdot n \cdot (X_T + X_L) \tag{5.43}$$

5.2.6.1.1 Mean Secondary-Side Impedance Values of Medium-Voltage Transformers

Table 5.14 [5.64] lists selected figures of merit for the reduced secondary-side resistance and reactance values of 16 kV and 0.4 kV transformers (including a small-impedance portion of a 16 kV, 8 mm gauge

Table 5.14 Figures of merit for 16 kV/0.4 kV transformers according to [5.64]

S_n in kVA	63	100	160	200	250	300	400	500	630	800	1000	1600
R_T in mΩ	42.1	24.8	13.7	10.9	8.6	7.1	5.4	4.4	3.5	3	2.5	1.8
X_T in mΩ	104	69	45	37	30	26	20	16	13	11	9.1	6.5
Z_T in mΩ	112	73.3	47	38.6	31.2	27	20.7	16.6	13.5	11.4	9.4	6.7
ψ_T	68°	70.2°	73.1°	73.6°	74°	74.7°	74.9°	74.6°	74.9°	74.7°	74.6°	74.5°
S_{KS} in MVA	1.43	2.18	3.4	4.15	5.13	5.94	7.72	9.64	11.9	14	17	23.7

wire run that is 5 km in length) as a function of nominal transformer output S_n. These values stem from [5.64] but exhibit very high congruence with the values in [5.19] that are obtained using more labour-intensive calculation methods. X_T is considerably higher than R_T in medium-voltage transformers (the higher the transformer voltage, the more pronounced this phenomenon). Table 5.14 also indicates the values for transformer impedance Z_T, Z_T phase angle ψ_T and short-circuit power S_{KS} under short-circuit conditions at the bus bar of this type of transformer.

5.2.6.1.2 Approximate Values for Conductor Reactance Layers

Although ohmic resistance is often the predominant factor in conductors, reactance should also be taken into account, although normally an approximate calculation method suffices. The reactance layers X'_L (reactance per metre of wire run) are indicated in [5.19] and [5.64] for various types of wires. As is known from field theory, the inductance layer L' and thus X'_L are mainly determined by whether a cable or overhead line (with insulation by air) is involved, whereby the wire gauge is of secondary importance. The mean values indicated in Table 5.15 can be used if the relevant data are not available from the manufacturer. Line reactance is particularly important for overhead lines.

5.2.6.1.3 Line Resistance and Reactance

The wire run that interconnects a transformer and link point is composed of various sections. Resistance R_{Li} and reactance X_{Li} are determined using the ith segment wire gauge A_i and length l_i. To be on the safe side, a line temperature of around 70 °C should be presupposed for resistance calculation purposes, i.e. copper wire resistance can be presumed to exhibit a specific resistance $\rho = 0.022\,\Omega\,mm^2/m$; for aluminium wires the figure is $\rho = 0.036\,\Omega\,mm^2/m$.

Resistance and reactance in segment i of the outer conductor are determined as follows:

$$\text{Resistance in segment } i \text{ of the outer conductor: } R_{Li} = \rho \frac{l_i}{A_i} = R'_{Li} \cdot l_i \qquad (5.44)$$

$$\text{Reactance in segment } i \text{ of the outer conductor: } X_{Li} = X'_{Li} \cdot l_i \qquad (5.45)$$

The resistance layers R'_L and exact reactance layer values X'_L are indicated in [5.19] and [5.64] for various types of underground and overhead cables. The mean X'_L values from Table 5.15 and the actual figures of merit for a given installation are virtually the same.

In low-voltage grids, resistance R_L often far exceeds reactance X_L. This holds true in particular for underground cables, i.e. the ratio is exactly the inverse of the transformer ratio.

In the **triphase scenario** (L1–L2–L3), the line impedance is the sum total of all impedances in the various segments of the outer conductor, i.e.

$$\underline{Z}_{3L} = R_{3L} + jX_{3L} = \sum_{i=1}^{m} R_{Li} + j\sum_{i=1}^{m} X_{Li} \qquad (5.46)$$

Table 5.15 Reactance layers in low-voltage lines according to [5.19] and [5.64]

Type of line	Reactance per metre of line $X_L' = X_l/l$		
	Average value for calculations	Minimum	Maximum
Overhead line	0.33 mΩ/m = 0.33 Ω/km	0.27 mΩ/m	0.37 mΩ/m
Single-conductor cable	0.18 mΩ/m = 0.18 Ω/km	0.15 mΩ/m	0.22 mΩ/m
Four-conductor cable	0.08 mΩ/m = 0.08 Ω/km	0.065 mΩ/m	0.1 mΩ/m

In the **single-phase scenario** (L–N), in most cases the gauge of the outer and neutral conductors is the same along most of the wire run, whose total impedance is almost double the outer conductor impedance, i.e.

$$\underline{Z}_{1L} = R_{1L} + j x_{1L} = \sum_{i=1}^{m} 2 \cdot R_{Li} + j \sum_{i=1}^{m} 2 \cdot X_{Li} = 2 \cdot \underline{Z}_{3L} \tag{5.47}$$

5.2.6.1.4 Overall 50 Hz Grid Impedance $\underline{Z}_N$

Using Equation 5.40, grid impedance $\underline{Z}_N = R_N + j X_N$ for 50 Hz can be determined as follows.

In the triphase scenario:

$$\underline{Z}_N = \underline{Z}_{3N} = Z_{3N} \angle \psi_3 = \underline{Z}_T + \underline{Z}_{3L} = (R_T + R_{3L}) + j(X_T + X_{3L}) \tag{5.48}$$

In the single-phase scenario:

$$\underline{Z}_N = \underline{Z}_{1N} = Z_{1N} \angle \psi_1 = \underline{Z}_T + \underline{Z}_{1L} = (R_T + R_{1L}) + j(X_T + X_{1L}) \tag{5.49}$$

where ψ_3 and ψ_1 are the arguments of the triphase and single-phase grid impedance.

In cases for larger distances from the transformer, the grid impedance in the single-phase scenario is nearly double that in the triphase scenario. Triphase grid impedance is usually around 60% of single-phase grid impedance.

5.2.6.1.5 Link-Point Short-Circuit Capacity

Short-circuit capacity S_{KV} is a key figure of merit in today's electricity grids. The new standards concerning allowable harmonic current define triphase inverter limit values that are determined by the ratio of apparent nominal capacity to link-point short-circuit capacity, which is also useful for determining the voltage increase induced by an inverter at the connection and link point.

Link-point short-circuit capacity S_{KV} is the apparent output that the grid needs to provide for a triphase short circuit at the link point. The following then applies as in Figure 5.148:

$$\text{Triphase short-circuit capacity } S_{KV} = 3 \cdot \frac{V_{1N}^2}{Z_{3N}} = \frac{V_N^2}{Z_{3N}} \tag{5.50}$$

where V_{1N} is the phase voltage and $V_N = \sqrt{3} V_{1N}$ is the concatenated grid voltage.

The link point of the transformer that feeds energy into the grid exhibits a short-circuit capacity S_{KS} (see Table 5.14). The greater the distance between the transformer and the link point, the lower the short-circuit capacity. Hence in practice S_{KS} constitutes an upper short-circuit power limitation for the portion of the grid that the transformer is feeding energy into.

A single-phase short-circuit capacity S_{1KV} can be similarly defined that occurs in conjunction with a single-phase short between an outer conductor and the neutral conductor:

$$\text{Single-phase short-circuit capacity } S_{1KV} = \frac{V_{1N}^2}{Z_{1N}} \tag{5.51}$$

In close proximity to the transformer, S_{1KV} equates to approximately $S_{KV}/3$ and decreases successively to around $S_{KV}/6$ as the distance to the transformer grows larger.

5.2.6.2 Voltage Increases at the Connection and Link Point

The low-voltage grid voltage in Central Europe is 230 V/400 V, give or take 10%. Connected devices exhibit trouble-free operation under these voltage conditions. To guarantee this voltage throughout the grid (including for the most remote devices), despite frequently fluctuating loads, the transformer stations that convert medium voltage of, say, 16 kV to 230 V/400 V low voltage are configured such that the transformer open-circuit voltage is close to the upper tolerance limits, which average (for example) 240 V.

Voltage can be increased by several volts at the grid connection point without any difficulty under full-capacity conditions. The PV system of a building located near a transformer will often increase the building voltage and notably the inverter voltage still further. But if this is done while the grid load is low and the transformer station voltage is close to the upper limit value, the grid voltage in the building in question may be excessively high, which can potentially cause damage to connected devices resulting from overheating or the like. Hence, inverters should automatically limit their grid input power on reaching the upper voltage limit.

The scope of such voltage surges can be reduced by configuring a PV system in certain ways. A PV inverter is always connected directly to the building's main distribution board via a special, separately fused line so as to minimize impedance at the grid link point and at the same time reduce the attendant harmonics. In addition, voltage fluctuations are reduced through the use of a building conductor with a sufficiently large gauge. High-capacity inverters (around 30 kW and upwards) should be directly connected to the transformer using a separate line.

This problem is bound to worsen in the coming years as the number of grid-connected systems rises. If one grid sector contains a relatively large number of PV systems that collectively generate peak power of a magnitude that is fairly substantial relative to maximum capacity, these systems will occasion major voltage fluctuations not only in individual buildings, but across the entire grid. In such a case, the power company needs to locate the open-transformer station's open-circuit voltage in the centre of the tolerance band rather than at its upper end, and, in order to avoid major downward voltage fluctuations, must either use more robust power lines or optimize transformer voltage regulation – both of which measures involve considerable additional cost.

The following symbols are used for the discussion below of the problems associated with voltage increases (also see Figure 5.148):

V_N = concatenated grid voltage; V_{1N} = grid phase voltage = $V_N/\sqrt{3}$

V_V = concatenated voltage at the link point VP; V_{1V} = phase voltage at the link point VP

V_{WR} = concatenated voltage at the inverter; V_{1WR} = phase voltage at the inverter

P = total active power fed into the grid by the inverter

Q = total (inductive) reactive power drawn from the grid by the inverter

S = apparent inverter capacity, i.e.

$S = \sqrt{P^2 + Q^2} \approx V_{1N} \cdot I$ (single-phase) and $\sqrt{3} \cdot V_N \cdot I$ (triphase)

$\cos\varphi = P/S$ = inverter power factor

I = power fed into the grid by the inverter

$I_W = I \cdot \cos\varphi$ = active power fed into the grid by the inverter

$I_B = I \cdot \sin\varphi$ = reactive power drawn off by the inverter

n (subscript) = nominal; under nominal operating conditions

All the remaining designations are as defined in Section 5.2.6.1.

In cases where inverter $\cos\varphi$ is close to 1, which is desirable for the power companies, voltage V_{WR} must always be higher than grid open-circuit voltage V_N. Moreover, link-point voltage V_V is always lower than V_{WR}, but is also somewhat higher than V_N (see Figure 5.148).

Figure 5.149 is a vector diagram for the $\cos\varphi = 1$ scenario. Inasmuch as a correctly sized power lead will exhibit voltage decreases at the circuit resistances and inductances as in Figure 5.148 relative to V_{1N}, the voltage decreases at the reactances are irrelevant for calculation of the absolute voltage increases ΔV_{1WR} and ΔV_{1V}. Thus the following equations yield approximate figures of merit in such a case:

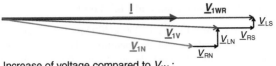

Increase of voltage compared to V_{1N} :
$\Delta V_{1V} \approx R_N \cdot I; \quad \Delta V_{1WR} \approx (R_N + R_S)I$

Figure 5.149 Vector diagram for the circuit in Figure 5.148 for connection of an inverter with $\cos \varphi = 1$

$$\Delta V_{1V} = V_{1V} - V_{1N} \approx R_N \cdot I \tag{5.52}$$

$$\Delta V_{1WR} = V_{1WR} - V_{1N} \approx (R_N + R_S) \cdot I \tag{5.53}$$

To avoid impermissible voltage fluctuations, R_N should be defined such that Equation 5.52 yields ΔV_{1V} that is no more than 3% of phase voltage V_{1N} (see Section 5.2.3.3.1 and [5.19]).

Conversely, if R_N is known, it can be used to determine the maximum nominal inverter power that can be connected at this link point for $\cos \varphi = 1$. In the low-voltage grid, this nominal voltage is as follows for a maximum allowable voltage increase of 3%:

$$\text{For a single-phase connection: } P_{AC1} = 0.03 \frac{V_{1N}^2}{R_{1N}} \tag{5.54}$$

$$\text{For a triphase connection: } P_{AC3} = 0.09 \frac{V_{1N}^2}{R_{3N}} = 0.03 \frac{V_N^2}{R_{3N}} \tag{5.55}$$

The following also applies in such a case: $P_{AC3} \leq S_n$ and $P_{AC1} \leq S_n/3$ (S_n = nominal apparent transformer capacity).

In cases where a separate transformer is directly connected to the medium-voltage grid, only two-thirds of the values above are allowable, since the maximum allowable voltage change in the medium-voltage grid is only 2%.

As R_{3N} usually ranges from 50 to 60% of R_{1N}, five to six times more power can usually be fed into the grid at the same triphase grid connection point than is the case with a single-phase connection.

The phase voltage increase is smaller for inverters that use reactive power Q; this occurs for example with a line-commutated inverter, where $Q > 0$ (see Figure 5.150).

Voltage increases can be reduced by drawing off regulated amounts of inductive reactive power, while voltage can be increased by drawing off capacitive reactive power.

The following equations apply to the single-phase scenario:

$$\Delta V_{1V} \approx R_{1N} \cdot I_W - X_{1N} \cdot I_B = R_{1N} \frac{P}{V_{1N}} - X_{1N} \frac{Q}{V_{1N}} \tag{5.56}$$

$$\Delta V_{1WR} \approx (R_{1N} + R_{1S}) \cdot I_W - (X_{1N} + X_{1S}) \cdot I_B = (R_{1N} + R_{1S}) \frac{P}{V_{1N}} - (X_{1N} + X_{1S}) \frac{Q}{V_{1N}} \tag{5.57}$$

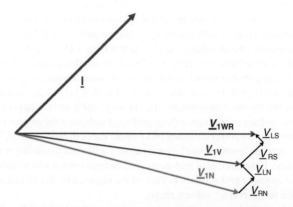

Figure 5.150 Vector diagram for the circuit in Figure 5.148 for connection of an inverter that draws reactive power from the grid (cos $\varphi \approx 0.7$ in the example). In such a case, the grid absorbs active power and capacitive reactive power produced by the inverter. As the voltage drops at R and X partly offset each other, the voltage increase is considerably lower in this case

The following equation applies to the relative voltage increase d_V at the link point:

$$d_V = \frac{\Delta V_{1V}}{V_{1N}} \approx \frac{P}{S_{1KV}} \cos\psi_1 - \frac{Q}{S_{1KV}} \sin\psi_1 \tag{5.58}$$

In cases where single-phase power is fed into the grid in several phases, the resulting voltage increase is smaller owing to the fact that the currents partly in the neutral conductor offset each other partly or (in the symmetrical scenario) fully. Hence the voltage increase determined using Equations 5.56, 5.57 and 5.58 constitutes the worst-case scenario.

The following equation applies to symmetrical feed-in for the triphase scenario:

$$\Delta V_{1V} \approx R_{3N} \cdot I_W - X_{3N} \cdot I_B = R_{3N} \frac{P}{3V_{1N}} - X_{3N} \frac{Q}{3V_{1N}} \tag{5.59}$$

$$\Delta V_V = \sqrt{3} \times \Delta V_{1V} \approx R_{3N} \frac{P}{V_N} - X_{3N} \frac{Q}{V_N} \tag{5.60}$$

$$\Delta V_{1WR} \approx (R_{3N} + R_{3S}) I_W - (X_{3N} + X_{3S}) I_B = (R_{3N} + R_{3S}) \frac{P}{3V_{1N}} - (X_{3N} + X_{3S}) \frac{Q}{3V_{1N}} \tag{5.61}$$

$$\Delta V_{WR} = \sqrt{3} \cdot \Delta V_{1WR} \approx (R_{3N} + R_{3S}) \frac{P}{V_N} - (X_{3N} + X_{3S}) \frac{Q}{V_N} \tag{5.62}$$

With S_{KV}, the following equation applies to the relative voltage increase d_V at the link point for a symmetrical triphase feed-in scenario:

$$d_V = \frac{\Delta V_{1V}}{V_{1N}} = \frac{\Delta V_V}{V_N} \approx \frac{P}{S_{KV}} \cos\psi_3 - \frac{Q}{S_{KV}} \sin\psi_3 \tag{5.63}$$

Owing to the dramatic increase in the number of PV systems in Germany since 2000 (spurred by the Renewable Energy Act), the most recent German regulations stipulate that PV installation inverters must also contribute to grid regulation and voltage stability. In this regard, [5.71] and [5.72] require new grid-connected installations to contribute to voltage regulation via regulated draw-off of inductive reactive power, i.e. via the use of capacitive or inductive reactive power when, respectively, small or large amounts of power are being fed into the grid. The new rules also stipulate that: (a) power companies should to some extent be allowed to define the relevant characteristic curve $Q = f(P)$; and (b) larger PV installations that are connected to the low-voltage or medium-voltage grid must scale back the amount of power fed into the grid on request from the relevant power company (via a ripple signal or other means). These requirements will entail additional costs for inverter manufacturers as they will need to use larger semiconductors for the same nominal voltage. If the number of grid-connected systems rises, it will be indispensable for these systems to support the requisite grid services in view of the importance, for all connected users, of grid stability and adherence to the correct voltage range.

Voltage V_{1V} is not permitted to exceed the upper tolerance level of $230\,\text{V} + 10\% = 253\,\text{V}$, while concatenated voltage V_V must adhere to a $400\,\text{V} + 10\% = 440\,\text{V}$ limit. Inasmuch as power companies for the most part try to keep transformer station open-circuit voltage (i.e. V_{1N}) in the upper half of the tolerance band and if possible at least a few percentage points below the upper tolerance limit (e.g. $V_{1N} = 240\,\text{V}$), if the power fed into the grid exhibits less than the maximum capacities in Equations 5.54 and 5.55 and if grid-connected system concentration is kept within reasonable bounds, there will be no link-point voltage surge problems whatsoever.

For $\cos\varphi \approx 1$, the voltage V_{1WR} at the inverter is still around $R_S \cdot I$ higher than the voltage V_{1V} at the link point. If around 30 m of clearance is allowed between the link point and inverter, the grid voltage upper tolerance limit, particularly for single-phase devices, can be overshot by a few percentage points without any problem. Such unavoidable voltage increases should normally not interfere with inverter functionality or cause device failure. But it makes little sense for an inverter to shut itself down on attaining the upper grid voltage tolerance limit, because if V_{WR} has reached the tolerance limit, it means that V_V has long since fallen below it.

The voltage tolerance of a well-designed grid-connected inverter should be a few percentage points higher than grid voltage tolerance, while the device's operating voltage V_{1WR} should be compatible with the approximately 230 V (with a $+15\%$ and -20% tolerance). It would of course be ideal if the upper voltage limits of such devices could be adjusted upward by 10–15% in accordance with local conditions.

In the interest of limiting possible voltage increases, it would be preferable if: (a) the inverter limited the power fed into the grid in cases where the grid voltage reaches its upper limit value; and (b) the inverter's nominal voltage could be increased by, say, 110–115% to compensate for voltage decreases in the interconnecting line. Thus an inverter should only shut itself down if its voltage reaches the upper grid voltage limit value (including in the event where no power is fed in). However, as noted, this characteristic would increase the likelihood of stable stand-alone operation in a shutdown sector of the grid, and should only be allowed if adequate precautions against islanding have been taken (see Section 5.2.4). Insofar as allowed or requested by the power companies, reactive power draw-off in a grid sector with a high concentration of grid-connected systems can reduce the voltage increase induced by large amounts of power being fed into the grid, and thus under certain circumstances limits on such infeeding could perhaps be suspended or lowered.

The scope of the resulting voltage surges can be reduced via intelligent grid-connected system configuration strategies that increase overall system efficiency. For example, mounting the inverter adjacent to the main distribution board, or running a dedicated inverter interconnecting line with a sufficiently large wire gauge (if possible, larger than required by the building code, e.g. 4 or even 6 mm² gauge in lieu of 2.5 mm²) from the building's main distribution board, can minimize R_S. Use of a building power line with a sufficiently large gauge reduces R_N, which is likewise beneficial. It makes little sense for manufacturers to go to great expense to ramp up inverter efficiency by 1, 2 or even 3%, only to turn around and authorize use of an interconnecting line that exhibits unduly elevated resistance loss of the same magnitude – loss that could readily have been avoided.

Calculation examples for grid impedance, allowable connected loads and voltage increases in go-live settings can be found in Section 5.2.6.7.

5.2.6.2.1 Interconnecting Line Loss

Calculation methods similar to those used to determine DC wiring loss (see Section 4.5.4) also come into play for AC line (interconnecting line) loss, as follows:

The following equations apply to the single-phase scenario:

$$\text{AC line loss } P_{vAC} = R_{1S} \cdot I^2 = R_{1S} \frac{P^2}{V_{1N}^2 \cos^2 \varphi} \qquad (5.64)$$

As interconnecting line loss is inversely proportional to $\cos^2\varphi$, long interconnecting lines for line-commutated inverters should be avoided at all costs.

Relative AC line loss for nominal power:

$$\frac{P_{v_{ACn}}}{P_{ACn}} = R_{1S} \frac{P_{ACn}}{V_{1N}^2 \cos^2 \varphi} = R_{1S} \frac{I_n}{V_{1N} \cos \varphi} \qquad (5.65)$$

Relative annual energy loss can be derived from ohmic interconnecting line loss (also see Section 4.5.4) if the insolation distribution and energy production in the various power stages are known.

Relative annual AC energy loss:

$$\frac{E_{v_{ACa}}}{E_{ACa}} \approx k_{EV} \frac{P_{v_{ACn}}}{P_{ACn}} = k_{EV} \cdot R_{1S} \frac{P_{ACn}}{V_{1N}^2 \cos^2 \varphi} = k_{EV} \cdot R_{1S} \frac{I_n}{V_{1N} \cos \varphi} \qquad (5.66)$$

In Burgdorf (in the Swiss Mittelland region), with $P_{DCn} = P_{Ao}$ (effective solar generator power at STC power output) the energy loss coefficient $k_{EV} \approx 0.5$ is obtained. This value can also be applied to other flat areas in Europe. If $P_{DCn} < P_{Ao}$ for an oversized solar generator, or for Southern European or high Alpine installations, a value higher than 0.5 (up to about 0.7) should be used for k_{EV}.

The following equation applies to symmetrical feed-in for the triphase scenario:

$$\text{AC line loss } P_{vAC} = 3 \cdot R_{3S} \cdot I^2 = 3 \cdot R_{3S} \left(\frac{P}{3V_{1N} \cos \varphi} \right)^2 = R_{3S} \frac{P^2}{V_N^2 \cos^2 \varphi} \qquad (5.67)$$

Relative AC line loss for nominal power:

$$\frac{P_{v_{ACn}}}{P_{ACn}} = R_{3S} \frac{P_{ACn}}{V_N^2 \cos^2 \varphi} = R_{3S} \frac{I_n}{V_{1N} \cos \varphi} \qquad (5.68)$$

Relative annual AC energy loss:

$$\frac{E_{v_{ACa}}}{E_{ACa}} \approx k_{EV} \frac{P_{v_{ACn}}}{P_{ACn}} = k_{EV} \cdot R_{3S} \frac{P_{ACn}}{V_N^2 \cos^2 \varphi} = k_{EV} \cdot R_{3S} \frac{I_n}{V_{1N} \cos \varphi} \qquad (5.69)$$

(where 0.5–0.7 should be applied for k_{EV}; see comment on Equation 5.66).

The inverter and main distribution board should be interconnected at the production and billing meter using a dedicated conductor (see Figures 5.47 and 5.48), whose line loss is assumed entirely by the installation operator. Unnecessary energy production loss can be avoided through use of an interconnecting line that is designed in such a way that its line loss does not exceed 1 or 2% of nominal power.

5.2.6.3   Harmonics

5.2.6.3.1   Interference from Harmonic Currents

As grid current contains harmonics in addition to 50 Hz first-harmonic oscillation, inverters distort grid current to one degree or another depending on the circuit arrangement used. Line-commutated inverters, despite being relatively simple and expensive, generate relatively strong harmonic current.

The higher the harmonic current and inductivity (as in Figure 5.148), the higher the link-point harmonic current at the relevant frequency. Devices nearby may experience interference if the grid voltage exhibits excessive harmonics.

Using Equation 5.43, grid impedance $\underline{Z}_{Nn}$ for the nth harmonic can be approximately derived from the 50 Hz grid impedance $\underline{Z}_N$ (see Section 5.2.6.1).

Equation 5.43 does not consider: (a) a resistance increase due to the skin effect at higher-frequency conditions; and (b) the effect of reactive power compensation condensers without small series inductors, which can cause resonance points. Assuming that the connected inverter is at the nth harmonic of a harmonic current source I_n, the harmonic voltage $\underline{Z}_{Nn} \cdot \underline{I}_n$ that generates this current in the grid impedance $\underline{Z}_{Nn}$ can be determined. In the worst-case scenario, these additional harmonics will be in phase with the harmonics already exhibited by the grid at this frequency and will be added to them.

5.2.6.3.2   Assessment of Inverter Compatibility and Conformance Based on Measured Harmonic Currents

An inverter can be used without dedicated harmonic reduction measures if the harmonic current generated by the inverter is relatively low under all power conditions, i.e. lower than the limit values indicated for single-phase inverters in Table 5.7, or for larger triphase inverters in Table 5.7 and Figure 5.69 (as the case may be). All harmonic current limit values can be found in Section 5.2.3.3.3.

In cases where a PV installation integrates a series of smaller inverters, the inverter assessment should be based on the sum total of all inverter harmonic currents in the installation. If the inverter assessment is not based on aggregate harmonics, serious problems could arise in a PV installation that uses a large number of small inverters (e.g. module inverters) whose harmonics are just below the limit values indicated in Table 5.7.

5.2.6.3.3   Preliminary Assessment for Line-Commutated Inverters

During PV installation planning for a large triphase installation, a preliminary assessment of line-commutated inverters can be realized based on apparent inverter capacity S_{WR} and grid short-circuit capacity S_{KV} at the link point [5.19] – bearing in mind, however, that only half of aggregate appliance output is allowable in such cases.

In terms of inverter harmonics in a high-voltage grid with $S_{KV}/S_{WR} \geq 300$, any device can be connected regardless of operating principle.

The maximum allowable inverter output for six-pulse inverters is

$$\text{Maximum apparent output for six-pulse inverters: } S_{WR6} \leq S_{KV}/300 \qquad (5.70)$$

and the maximum allowable inverter output for devices with 12 or more pulses is

$$\text{Maximum apparent output for higher-pulse inverters: } S_{WR12} \leq S_{KV}/75 \qquad (5.71)$$

Only half of the apparent output from Equations 5.70 and 5.71 may be directly connected to the medium-voltage grid. For line-commutated inverters, it is also necessary to determine whether the commutation notches exhibited by the device are impermissibly large [5.19].

In the event of non-conformance with any of the aforementioned criteria, the harmonic current generated by other installation appliances should also be taken into account by means of a detailed calculation as in [5.19] that factors in all appliances that are connected at the same connection point.

5.2.6.3.4 Mitigation of Harmonic Current Effects

In the event of non-conformance with any of the criteria referred to in Sections 5.2.6.3.2 and 5.2.6.3.3, the power company concerned will not issue an installation permit or will issue a permit with reservations that: (a) requires the owner to carry out post-commissioning monitoring; and (b) may call for the realization of remedial measures at the owner's expense. Such measures may include reducing the harmonic current fed into the grid via, for example, intake circuits or use of a different and possibly smaller inverter, or increasing the allowable limit values, e.g. by beefing up the grid or installing dedicated inverter power leads at the grid connection point that exhibit higher short-circuit capacity. The greater the overshoot of the allowable limit values, or the lower the short-circuit capacity S_{KV} relative to apparent inverter output S_{WR}, the greater the scope of the untoward events that can potentially arise and the expense entailed by averting them.

The use of triphase inverters connected solely to L1, L2 and L3 for the installation's entire output is far preferable to the use of star-connected single-phase inverters. Unlike 50 Hz current, which increases in the neutral conductor under symmetrical load conditions, harmonic currents whose ordinal numbers are multiples of 3 are always in phase in the neutral conductor and thus aggregate, including in cases where the device output is symmetrically distributed across the various phases. Hence it is essential that single-phase inverters at these frequencies exhibit only weak harmonic current. The use of small triphase inverters (down to string and module inverters) can also reduce the grid harmonic load. Such devices are available with output down to a few kilowatts.

Neutral-conductor harmonic voltages can provoke anomalies, as well as interference with other single-phase devices and with system inverters themselves. In cases where a grid sector has an extremely large number of single-phase appliances, such as computers, electronic devices, TVs and the like that generate elevated harmonics, the neutral-conductor current may chiefly consist of harmonic currents that can potentially be stronger than outer conductor current. Neutral-conductor impedance and thus such harmonic currents can be reduced through the use of a neutral-conductor wire gauge that is larger than the regulatory standard.

5.2.6.4 Electromagnetic Compatibility

Many PV installations with first-generation inverters provoke radio reception disturbances (notably in short-, medium- and long-wave radio transmission), as well as disturbances in other devices – one example being the case where an inverter that exhibited electromagnetic radiation caused a doctor's X-ray machine to malfunction. In view of the fact that low-output solar generators in conjunction with their wiring runs (which can be quite extensive) in effect constitute an electromagnetic generator, and since PV installations are in continuous operation all day long, the AC side of building PV installations should at least adhere to the limit values as in Figure 5.70 and the DC side should conform with the limit values as in Figure 5.71 or 5.72.

If an inverter overshoots these limit values or repeatedly causes disturbances in nearby devices, the inverter's high-frequency interference voltages should be further reduced by attaching external line filters mounted directly on the (metallic) inverter housing, and on the DC and AC sides of the device if necessary. Such disturbance suppression measures are far more effective if the inverter has a metal housing. In some cases, residual disturbances on the DC side can be mitigated by placing on or in the inverter a high-frequency choke (L around 50 to 500 μH) consisting of a few bifilar windings for the positive and negative power leads mounted on a suitable toroidal core.

5.2.6.5   Inverter Failure

In the early days of inverter use, failures were often induced by low-frequency voltage surges and/or ripple signals accompanied by elevated grid voltage. These anomalies were eliminated through: (a) V_{WR} reduction measures (see Section 5.2.6.2), which also enhance overall installation efficiency; and (b) the use of inverters that exhibit ripple signal immunity. Most newer inverters from experienced manufacturers exhibit no failures attributable to ripple signals.

Similarly important in this regard is adequate protection against transient power surges, through the use of varistors with sufficient current immunity at all DC and AC power leads. Voltage surge damage is not always readily apparent immediately after the relevant event and may take some time to manifest itself.

Power failures that occur under unfavourable circumstances such as high-ohm grid impedance resulting from overcurrent trip deactivation can also cause some older inverters to fail, since under such conditions they exhibit characteristics similar to those of a power source. In such cases, a power failure can result in a transient voltage surge that can damage inverter electronics. For inverters that exhibit these problems, the main DC-side switch should always be opened before the AC side is shut down. Hence, wherever possible, inverters should be used that can also withstand a grid power lead cutout without incurring damage. It is for this reason that all inverter evaluations realized at the Bern University of Applied Sciences PV Lab include an AC-side shutdown test under full load conditions.

5.2.6.6   Grid Power Lead Sizing Tips

If the calculations described in Sections 5.2.6.1–5.2.6.3 seem too labour intensive, or if no detailed grid data are available, observing the following guidelines can help to avoid many of the problems that arise with inverters:

1. Only use inverters with certified immunity to ripple signals.
2. Only use inverters whose grid-side operating voltage is at least $230\,V \pm 10\%$ or $400\,V \pm 10\%$ (but preferably $+15\%, -20\%$). At voltages exceeding $230\,V + 10\%$ or $400\,V + 10\%$, the inverter should have the capacity to shut itself down or limit the power fed into the grid without exhibiting a failure anywhere in the nominal voltage range – including in cases where a voltage surge and ripple signal co-occur.
3. Only use inverters that can withstand a grid power lead cutout without incurring damage.
4. Wherever possible, use inverters that conform to the EN 61000-3-2 and EN 61000-3-12 harmonics standards or with the limit values in [5.19] (Figure 5.69; see Section 5.2.3.3.3). Do not use single-phase line-commutated inverters.
5. Wherever possible, use inverters that conform to the electromagnetic compatibility limit values in EN 55014 [5.21] or EN 61000-6-3 [5.23] for the AC side (Figure 5.70) and the DC side (Figure 5.71 or 5.72). If the inverters do not conform to these limit values, install additional grid or other filters.
6. Connect inverters directly to the building's main distribution board using a dedicated cable that exhibits the lowest possible ohms, i.e. the maximum voltage decrease should not exceed the grid voltage by more than around 1 or 2%. The main distribution board should be fused for an output that is at least three times higher than the inverter's nominal output.
7. Inverters whose output exceeds around 100 kW should be connected directly to the transformer station using a dedicated wire that exhibits the lowest possible ohms, i.e. the maximum voltage decrease should not exceed 2 or 3%. Nominal apparent transformer output S_n should be at least three times higher than maximum inverter output.
8. Use only triphase line-commutated inverters, if possible only in wired grids, and keep the distance between these devices and the transformer as short as possible. For six-pulse inverters, the transformer's apparent nominal output should be at least 16 times higher than the PV installation's output, and for 12 or more pulse inverters four times higher. In addition, wherever possible a direct interconnecting line should be used between inverters and transformers. If any of the foregoing conditions are not met, a critical harmonic intake circuit should be integrated as described in Section 5.2.6.3.4.

9. Line-commutated inverters whose output is around 100 kW or higher should be connected directly to the medium-voltage grid using a dedicated transformer.

5.2.6.7 Specimen Power Lead Sizing Procedure

Four inverters (A, B, C and D) are to be connected at three grid connection points (1, 2 and 3), whereby it can be presupposed that, apart from these inverters, very few appliances that exhibit harmonics will be connected at these points.

The grid impedances and short-circuit capacities, as well as the allowable voltage-increase-related connected load, are determined first. Based on these calculations, it is then determined whether or not the voltage increases and harmonics exhibited by the inverters allow them to be connected. Finally, the relative voltage increase with the inverters connected under nominal voltage operating conditions is determined:

Inverter A: Single-phase self-commutated inverter that exhibits high-frequency pulse width modulation ($P_{ACn} = 3$ kW, $V_{1N} = 230$ V, $\cos \varphi \approx 1$) and is EN 61000-3-2 conformant.

Inverter B: Triphase self-commutated inverter that exhibits high-frequency pulse width modulation ($P_{ACn} = 20$ kW, $V_N = 400$ V, $\cos \varphi \approx 1$) and is EN 61000-3-2 conformant for all current harmonics except the 17th and 19th ($I_{17} = 250$ mA, $I_{19} = 200$ mA).

Inverter C: Triphase, six-pulse grid-connected inverter ($S_{ACn} = 30$ kVA, $V_N = 400$ V, $P_{ACn} = 27$ kW, $Q_{ACn} = 13$ kvar).

Inverter D: Triphase, 12-pulse grid-connected inverter ($S_{ACn} = 30$ kVA, $V_N = 400$ V, $P_{ACn} = 27$ kW, $Q_{ACn} = 13$ kvar).

Example 1: High-Quality Municipal Grid

			R (mΩ)	X (mΩ)	Z_N (mΩ)	ψ
Transformer	$S_n = 1600$ kVA	As in Table 5.14	1.8	6.5		
Conductor	Four-lead	$A_1 = 240$ mm^2,	6.88	6.0		
	conductor 1	$l_1 = 75$ m				
	Four-lead	$A_2 = 95$ mm^2,	2.32	0.8		
		$l_2 = 10$ m				
	conductor 2					
Total		Grid impedances Z_{3N}	11.0	13.3	17.26	50.4°
		Z_{1N}	20.2	20.1	28.5	44.9°

Short-circuit capacities: $S_{KV} = 9.27$ MVA, $S_{1KV} = 1.86$ MVA

(Conductor material: Cu, $\rho = 0.022 \,\Omega\,$mm^2/m, X'_L as in Table 5.15.)

Voltage increase:
Connectable outputs as in Equations 5.54 and 5.55: $P_{AC1} = 78.6$ kW, $P_{AC3} = 436$ kW.
The voltage increase exhibited by all four inverters allows for their connection.

Harmonics:
Inverter A: EN 60555-2 conformant, no connection restrictions apply.
Inverter B: $S_{KV}/S_{WR} = 463 \Rightarrow$ connectable as in Section 5.2.6.3, as the inverter's apparent output is
$$S = 20 \text{ kW} < S_{KV}/300 = 30.9 \text{ kVA}.$$
Validation: $I_{ACn} = 28.9$ A $\Rightarrow$ Allowable as in Equation 5.25: $I_{17zul} = 622$ mA $> I_{17} = 250$ mA.
$\Rightarrow$ Allowable as in Equation 5.25: $I_{19zul} = 467$ mA $> I_{19} = 200$ mA.
$\Rightarrow$ Connecting inverter B is allowable.

Inverter C: $S_{WR6} = S_{KV}/300 = 30.9\,\text{kVA} > 30\,\text{kVA} \Rightarrow$ Connecting inverter C is allowable.
Inverter D: $S_{WR12} = S_{KV}/75 = 124\,\text{kVA} > 30\,\text{kVA} \Rightarrow$ Connecting inverter C is allowable.

Relative link-point voltage increase d_V as in Equations 5.58 and 5.63:
Inverter A: $d_V = 0.11\%$; inverter B: $d_V = 0.14\%$; inverter C: $d_V = 0.08\%$; inverter D: $d_V = 0.08\%$.

Example 2: Substandard Rural Grid

			R (mΩ)	X (mΩ)	Z_N (mΩ)	ψ
Transformer	$S_n = 63\,\text{kVA}$	As in Table 5.14	42.1	104		
Conductor	Overhead line	$A_1 = 50\,\text{mm}^2$,	308	231		
		$l_1 = 700\,\text{m}$				
	Four-lead	$A_2 = 25\,\text{mm}^2$,	26.4	2.4		
	conductor 2	$l_2 = 30\,\text{m}$				
Total		Grid impedances Z_{3N}	377	337	506	41.9°
		Z_{1N}	711	571	912	38.8°

Short-circuit capacities: $S_{KV} = 316\,\text{kVA}$, $S_{1KV} = 58.0\,\text{kVA}$

(Conductor material: Cu, $\rho = 0.022\,\Omega\,\text{mm}^2/\text{m}$, X'_L as in Table 5.15.)

Voltage increase:
Connectable outputs as in Equations 5.54 and 5.55: $P_{AC1} = 2.23\,\text{kW}$, $P_{AC3} = 12.7\,\text{kW}$.
Thus the envisaged inverter outputs are too high and connecting the devices is impermissible from a voltage increase standpoint.

Harmonics:
Inverter A: EN 60555-2 conformant, no harmonics-related connection restrictions apply.
Inverter B: $S_{KV}/S_{WR} = 15.8 \Rightarrow S = 20\,\text{kW} > S_{KV}/300 = 1.05\,\text{kVA} \Rightarrow$ more exact investigation needed.
    $I_{ACn} = 28.9\,\text{A} \Rightarrow$ Allowable as in Equation 5.25: $I_{17zul} = 115\,\text{mA} < I_{17} = 250\,\text{mA}$.
        $\Rightarrow$ Allowable as in Equation 5.25: $I_{19zul} = 86\,\text{mA} < I_{19} = 200\,\text{mA}$.
        $\Rightarrow$ Connecting inverter B is not allowable.
Inverter C: $S_{WR6} = S_{KV}/300 = 1.05\,\text{kVA} < 30\,\text{kVA} \Rightarrow$ Connecting inverter C is not allowable.
Inverter D: $S_{WR12} = S_{KV}/75 = 4.22\,\text{kVA} < 30\,\text{kVA} \Rightarrow$ Connecting inverter D is not allowable.

Relative link-point voltage increase d_V as in Equations 5.58 and 5.63:
Inverter A: $d_V = 4.0\%$; inverter B: $d_V = 4.7\%$; inverter C: $d_V = 3.6\%$; inverter D: $d_V = 3.6\%$.

None of these inverters should be connected at this weak grid connection point. However, a connection would be just barely possible if the solar generator power for inverter A were reduced (e.g. to $P_{Go} \leq 2.5\,\text{kWp}$).

Connecting the six-pulse line-commutated inverter C would provoke particularly serious harmonics problems resulting from substantial overshooting of the apparent output. Connecting the 12-pulse inverter would also provoke serious harmonics problems. On the other hand, despite the somewhat higher output (30 kVA) attributable to the use of inductive reactive power, the resulting voltage increase in both line-commutated inverters C and D is somewhat lower than for the self-commutated inverter B, which exhibits only 20 kW and cos $\varphi = 1$.

The Example 2 *scenario can be considerably improved by installing a medium-voltage transformer near the feed-in point. For* Example 3 *(see below), it is presupposed that the medium-voltage transformer would also supply other appliances, i.e. the link point VP would still be on the low-voltage side.*

> ## Example 3: Improved Rural Grid Owing to Installation of a Dedicated Transformer
>
> | | | | R (mΩ) | X (mΩ) | Z_N (mΩ) | ψ |
> |---|---|---|---|---|---|---|
> | Transformer | $S_n = 160$ kVA | As in Table 5.14 | 13.7 | 45 | | |
> | Conductor | Four-lead cable | $A_1 = 25$ mm^2, $l_1 = 30$ m | 26.4 | 2.4 | | |
> | Total | | Grid impedances $\underline{Z}_{3N}$ | 40.1 | 47.4 | 62.1 | 49.8° |
> | | | $\underline{Z}_{1N}$ | 66.5 | 49.8 | 83.1 | 36.8° |
>
> Short-circuit capacities: $S_{KV} = 2.58$ MVA, $S_{1KV} = 637$ kVA
>
> (Conductor material: Cu, $\rho = 0.022\,\Omega\,$mm^2/m, X'_L as in Table 5.15.)
>
> *Voltage increase:*
> Connectable outputs as in Equations 5.54 and 5.55: $P_{AC1} = 23.9$ kW, $P_{AC3} = 120$ kW.
> The voltage increase exhibited by all four inverters allows for their connection.
>
> *Harmonics:*
>   Inverter A: EN 60555-2 conformant, no connection restrictions apply.
>   Inverter B: $S_{KV}/S_{WR} = 129 \Rightarrow S = 20$ kW $> S_{KV}/300 = 8.59$ kVA $\Rightarrow$ more exact investigation needed.
> $\quad I_{ACn} = 28.9$ A $\Rightarrow$ Allowable as in Equation 5.25: $I_{17zul} = 328$ mA $> I_{17} = 250$ mA.
> $\quad\quad \Rightarrow$ Allowable as in Equation 5.25: $I_{19zul} = 246$ mA $> I_{19} = 200$ mA.
> $\quad\quad \Rightarrow$ Thus the findings of the detailed investigation indicate that connection of inverter B is permissible.
>   Inverter C: $S_{WR6} = S_{KV}/300 = 8.59$ kVA < 30 kVA $\Rightarrow$ Connecting inverter C is not allowable.
>   Inverter D: $S_{WR12} = S_{KV}/75 = 34.4$ kVA > 30 kVA $\Rightarrow$ Connecting inverter C is allowable.
>
> *Relative link-point voltage increase d_V as in Equations 5.58 and 5.63:*
> Inverter A: $d_V = 0.38\%$; inverter B: $d_V = 0.50\%$; inverter C: $d_V = 0.29\%$; inverter D: $d_V = 0.29\%$.
>   Thus inverters A, B and D can be connected without any problem. Connection of the six-pulse inverter C poses no voltage-increase-related problem, but is problematic in terms of the device's harmonics; thus additional harmonics reduction expense will be incurred for measures such as installing an intake circuit at the critical harmonics. However, in view of the fact that the S_{WR6} overshoot is far smaller than in Example 2 (only a factor of around 3.5), averting harmonics problems on possible connection of inverter C will be considerably less expensive than for this example.

5.2.6.8 Optimal DC Operating Voltage for Grid-Connected Systems

For most inverters, the manufacturer specifies an allowable voltage range V_{MPP} at the MPP ($V_{MPPmin} - V_{MPPmax}$) within which the device will operate properly and at which the device will always be able to find an MPP on the solar generator's I–V characteristic curve. Often indicated as well are a minimum and maximum operating voltage (V_{Omin}, V_{Omax}) at which the device works, but not always in the correct MPP, in which case $V_{Omin} < V_{MPPmin}$ and $V_{Omax} > V_{MPPmax}$. A maximum voltage V_{DCmax} is also sometimes given that is not to be overshot at open circuit at the lowest allowable module temperature T_{Cmin}. In properly sized inverters, $V_{DCmax} > V_{Omax} > V_{MPPmax}$ and V_{DCmax} is considerably higher than V_{MPPmax}.

Unfortunately, some manufacturers indicate the same value for V_{MPPmax}, V_{Omax} and V_{DCmax}, which of course makes no sense. In planning a grid-connected system, the questions arise as to (a) the MPP voltage $V_{MPPA\text{-}STC}$ range that should be defined for the solar generator at STC power output and (b) which voltage is optimal for the inverters being used.

5.2.6.8.1 Voltage and Output Parameters

In the interest of clearly differentiating between the various types of voltage and output that come into play, it is necessary to use specific parameters, some of which are associated with one or more indices (i.e. subscripts). These parameters are as follows (for reasons of space, only some of them appear in the list of symbols at the end of the book).

Main symbols: V = voltage; G = irradiance; P = power/output (and in some cases capacity); T = temperature

Subscripts:		
	MPP	Maximum power point
	STC	Standard test conditions ($G = 1\,\text{kW/m}^2$, cell temperature $25\,^\circ\text{C}$)
	OC	Open circuit
	A	Installation/system
	B	Operation
	C	Cell/module temperature (T_C = cell/module temperature)
	LI	At low insolation such as $0.1 \cdot G_{STC}$
	min	Minimum
	max	Maximum

A closer look at the most important symbols:

Symbols related to solar generators

G_{STC}	Irradiance at STC power output ($1\,\text{kW/m}^2$)
G_{LI}	Lowest irradiance at which the installation is still operational (e.g. $0.1 \cdot G_{STC}$)
T_{STC}	STC reference temperature ($25\,^\circ\text{C}$) at which nominal solar generator power P_{Go} is defined
T_{Cmax}	Maximum solar generator cell temperature
T_{Cmin}	Minimum solar generator cell temperature

Symbols related to PV installations

$V_{MPPAmin}$	Minimum PV installation MPP voltage at G_{STC} and T_{Cmax}
$V_{MPPAmin\text{-}STC}$	Minimum PV installation MPP voltage at STC power output
$V_{MPPA\text{-}STC}$	PV installation MPP voltage at STC power output ($V_{MPPAmin\text{-}STC} < V_{MPPA\text{-}STC} < V_{MPPAmax\text{-}STC}$)
$V_{MPPA\text{-}GLI}$	PV installation MPP voltage at G_{LI} and T_{STC} ($G_{LI} \ll G_{STC}$)
$V_{MPPAmax}$	Maximum PV installation MPP voltage at G_{STC} and T_{Cmin} (usually at STC power output, i.e. $V_{MPPAmax} \approx V_{MPPAmax\text{-}STC}$, as this condition only occurs in extremely cold locations with G_{STC}, $T_C < T_{STC}$)
$V_{MPPAmax\text{-}STC}$	Maximum PV installation MPP voltage at STC power output
$V_{OCAmin\text{-}STC}$	Minimum solar generator open-circuit voltage at STC power output
$V_{OCA\text{-}STC}$	Solar generator open-circuit voltage at STC power output ($V_{OCAmin\text{-}STC} < V_{OCA\text{-}STC} < V_{OCAmax\text{-}STC}$)
$V_{OCAmax\text{-}STC}$	Maximum solar generator open-circuit voltage at STC power output
$V_{OCA\text{-}TCmin}$	PV installation open-circuit voltage at G_{STC} and T_{Cmin}

Inverter-related symbols:

V_{Omin}	Minimum operating voltage specified by the manufacturer (shutdown voltage)
V_{MPPmin}	Minimum inverter MPP voltage specified by the manufacturer
V_{Ein}	Inverter activation voltage ($V_{MPPmin} < V_{Ein} < k_{LI} \cdot V_{OCAmin\text{-}STC}$)
V_{MPPmax}	Maximum MPP voltage specified by the manufacturer
V_{Omax}	Maximum operating voltage specified by the manufacturer (inverter operation without MPP tracking)
V_{DCmax}	Maximum allowable input voltage under open-circuit conditions (inverter not in operation)

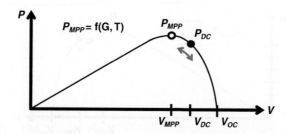

Figure 5.151 $P = f(V)$ characteristic curve for a solar generator. This curve is determined by solar generator and module irradiance G and by solar generator cell temperature T_C. This in turn means that maximum output P_{MPP} at the MPP is determined by G and T_C. However, an inverter sometimes operates at voltages other than the MPP and at voltage $V_{DC} \neq V_{MPP}$ uses somewhat less power $P_{DC} < P_{MPP}$

5.2.6.8.2 Solar Generator Characteristics under Various Insolation and Temperature Conditions

Curves showing operational characteristics $P = f(V)$ can be generated in lieu of more familiar I–V characteristic curves. As with I–V characteristic curves, P–V characteristic curves are of course also determined by solar generator and module irradiance G and by solar generator cell temperature T_C (see Figure 5.151).

Figure 5.152 displays the P–V characteristics of a solar generator under standard test conditions ($G = G_{STC} = 1\,\text{kW/m}^2$, cell temperature $T_C = 25\,°\text{C}$). Under the same insolation conditions G_{STC} but at lower cell temperatures, P_{MPP} and V_{MPP} increase (see Figure 5.153). Conversely, P_{MPP} and V_{MPP} decrease under the same insolation conditions but at higher cell temperatures (see Figure 5.154).

In the event of an abrupt insolation drop (e.g. overcast, $G_{LI} \approx 0.1 \cdot G_{STC}$) under the highest envisaged cell temperature conditions, P_{MPP} and V_{MPP} go even lower (see Figure 5.155). Lower T_{Cmax} values are likely to occur in this setting since the cell temperatures under these conditions are somewhat lower than the highest absolute temperatures.

5.2.6.8.3 Definition of the Relevant Voltage Factors (with Indication of the Characteristic Values)

The data obtained using Figures 5.151–5.155 can be used to define various voltage factors that are each related to the relevant voltage at STC power output. As G_{STC}voltage at the MPP is lower than under open-circuit conditions, the following is defined:

$$\text{MPP voltage factor: } k_{MPP} = V_{MPP\text{-}STC}/V_{OC\text{-}STC} \tag{5.72}$$

Characteristic values for k_{MPP}:
Crystalline module where FF $\approx 75\%$: $k_{MPP} \approx 0.8$
Amorphous module where FF ≈ 55–60%: $k_{MPP} \approx 0.7$

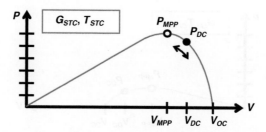

Figure 5.152 Characteristic curve for output P as a function of voltage V for a solar generator under standard test conditions ($G = G_{STC} = 1\,\text{kW/m}^2$, cell temperature $T_C = T_{STC} = 25\,°\text{C}$)

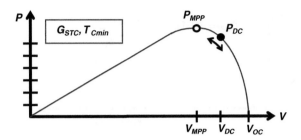

Figure 5.153 Characteristic curve $P = f(V)$ for a solar generator with $G_{STC} = 1\,\text{kW/m}^2$ and thus minimum envisaged cell temperature T_{Cmin} (e.g. $-10\,°C$ in installations in flat areas)

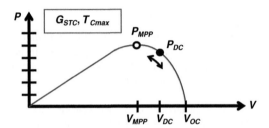

Figure 5.154 Characteristic curve $P = f(V)$ for a solar generator with $G_{STC} = 1\,\text{kW/m}^2$ and thus maximum envisaged cell temperature T_{Cmax} (e.g. $55–65\,°C$ in installations in flat areas)

At low temperatures (T_{Cmin}), open-circuit voltage V_{OC} for G_{STC} is higher than at STC power output:

$$\text{Low-temperature voltage factor: } k_{TCmin} = V_{OC\text{-}TCmin}/V_{OC\text{-}STC} \qquad (5.73)$$

Characteristic values for k_{TCmin} for installations in flat areas

$(T_{Cmin} \approx -10\,°C \text{ and } 1\,\text{kW/m}^2)$: $k_{TCmin} \approx 1.15$

Realistic value for Alpine installations (somewhat lower T_{Cmin},

   somewhat higher G_{max}): $k_{TCmin} \approx 1.2$

Similarly, realistic values for high-Alpine installations: $k_{TCmin} \approx 1.25$

Exact manufacturer's data are often lacking for amorphous modules. However, inasmuch as in such modules the temperature correlation is weaker but the voltages are somewhat higher (at least during the initial degradation phase), in the absence of exact data approximately the same factor can be used as for crystalline modules.

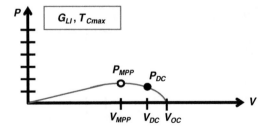

Figure 5.155 Characteristics $P = f(V)$ for a solar generator on a hot cloudy day following an insolation drop to $G_{LI} < G_{STC}$

Thus at high temperatures, MPP voltage V_{MPP}at G_{STC}decreases relative to STC as follows:

$$\text{High temperature voltage factor: } k_{TCmax} = V_{MPP\text{-}TCmax}/V_{MPP\text{-}STC} \qquad (5.74)$$

Characteristics value for crystalline modules at $T_{Cmax} \approx 60\,°C$ for G_{STC}: $k_{TCmax} \approx 0.84$

Exact manufacturer's data are often lacking for amorphous modules. However, inasmuch as in such modules the temperature correlation is weaker but the voltages are somewhat higher (at least during the initial degradation phase), in the absence of exact data approximately the same factor can be used as for crystalline modules.

Under low-insolation conditions $G_{LI}$and at T_{STC}, the MPP voltage $V_{MPPA\text{-}GLI}$is lower than for G_{STC}:

$$\text{Low-light temperature voltage factor } k_{LI} = V_{MPPA\text{-}GLI}/V_{MPPA\text{-}STC} \qquad (5.75)$$

Characteristics value for $G_{LI} \approx 0.1 \cdot G_{STC}$: $k_{LI} \approx 0.91$

5.2.6.8.4   DC-Side Voltage Sizing for PV Installations

Unfortunately, many manufacturers fail to provide complete specifications for DC operating voltages. To be on the safe side, the missing specifications should be defined as follows:

If V_{DCmax} is missing, use $V_{DCmax} = V_{Omax}$.
If V_{Omax} is missing, use $V_{Omax} = V_{MPPmax}$.

Manufacturers should specify inverter operating voltages as precisely as possible so that installation planners have the broadest possible input voltage range to work with.

A PV installation can be considered to be optimally sized if its inverter:

- can withstand the worst-case scenario (usually with $G_{STC} = 1\,kW/m^2$ and minimum cell temperature T_{Cmin}, but also at higher-altitude installations with higher G values) for the theoretically highest possible installation open-circuit voltage $V_{OCA\text{-}TCmin}$ without failing;
- performs reliably when exposed to the maximum V_{OC} figures of merit;
- performs reliably under all anticipated MPP V_{MPP} conditions.

In order to attain trouble-free dynamic MPP tracking, the highest possible cell temperature T_{Cmax} exhibited by the installation at minimum MPP voltage $V_{MPPAmin}$ should be higher than inverter V_{MPPmin}:

$$V_{MPPAmin} = V_{MPPmin}/k_{LI} \qquad (5.76)$$

Hence in order for inverter testing to yield reliable and relevant results, appliance characteristics under these minimum realistic solar generator voltage conditions should be investigated via a series of tests.

Using Equation 5.74, the minimum PV installation MPP voltage at STC power output can be determined as follows:

$$V_{MPPAmin\text{-}STC} = V_{MPPAmin}/k_{TCmax} \qquad (5.77)$$

Using the typical values specified in Equation 5.75 and below for k_{LI} and k_{TCmax}, the following applies approximately:

$$V_{MPPAmin-STC} \approx 1.3 \cdot V_{MPPmin} \text{ (in practice } 1.25 \cdot V_{MPPmin} \text{ often suffices)} \qquad (5.78)$$

Under variable G conditions, T_{Cmean} is usually lower than T_{Cmax}, and thus 1.25 is still acceptable.

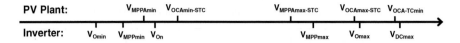

Figure 5.156   Key voltages on the installation and inverter side.

The following applies to the installation's maximum open-circuit voltage at STC power output:

$$V_{OCAmax\text{-}STC} = V_{DCmax}/k_{TCmin} \tag{5.79}$$

Hence the maximum allowable installation MPP voltage at STC power output is as follows:

$$V_{MPPAmax\text{-}STC} = V_{OCAmax\text{-}STC} \cdot k_{MPP} \tag{5.80}$$

Hence in order for inverter testing to yield reliable and relevant results, appliance characteristics under these minimum realistic solar generator voltage conditions should be investigated via an additional series of tests. Apart from this, a series of tests should be conducted to determine installation characteristics under other voltage conditions such as for the mean of $V_{MPPAmin}$ and $V_{MPPAmax}$.

The following approximately apply to installations in flat areas based on the characteristic values for k_{MPP} and k_{TCmin}:

$$\text{For crystalline modules: } V_{MPPAmax\text{-}STC} \approx 0.7 \cdot V_{DCmax} \tag{5.81}$$

$$\text{For amorphous modules: } V_{MPPAmax\text{-}STC} \approx 0.6 \cdot V_{DCmax} \tag{5.82}$$

To attain sizing that allows for optimal PV installation energy yield, installation $V_{MPPA\text{-}STC}$ should be defined such that the highest possible efficiency $\eta_{tot} = \eta \cdot \eta_{MPPT}$ is attained (see Section 5.2.5.3), whereby the following applies:

$$V_{MPPAmin\text{-}STC} < V_{MPPA\text{-}STC} < V_{MPPAmax\text{-}STC} \tag{5.83}$$

Figure 5.156 gives an overview of the position of the different important voltages discussed in this section.

5.2.6.8.5   Specimen DC-Side Voltage Sizing for PV Installations

Fronius IG30: $V_{DCmax} = 500\,V$, $V_{MPPmax} = 400\,V$, $V_{MPPmin} = 150\,V$
With Equation 5.76, $V_{MPPAmin} \approx 165\,V$
With Equation 5.78, $V_{MPPAmin\text{-}STC} \approx 195\,V$
With Equation 5.72, the following is obtained for crystalline modules: $V_{OCAmin\text{-}STC} \approx 244\,V$
With Equation 5.72, the following is obtained for amorphous modules: $V_{OCAmin\text{-}STC} \approx 278\,V$
With Equation 5.79, $V_{OCAmax\text{-}STC} = V_{DCmax}/k_{TCmin} \approx 435\,V$ for installations in flat area
With Equation 5.81, the following is obtained for crystalline modules in installations in flat areas:
    $V_{MPPAmax\text{-}STC} \approx 350\,V$
With Equation 5.82, the following is obtained for amorphous modules in installations in flat areas:
    $V_{MPPAmax\text{-}STC} \approx 300\,V$
For inverter tests, a series of tests for characteristics under the following conditions should be
    conducted: $V_{MPPAmin} \approx 170\,V$, $V_{MPPAmax} \approx 350\,V$, and at a mean of $260\,V$.

Sunways NT4000: $V_{DCmax} = 850\,V$ (previously $800\,V$), $V_{MPPmax} = 750\,V$, $V_{MPPmin} = 350\,V$
With Equation 5.76, $V_{MPPAmin} \approx 385\,V$
With Equation 5.78, $V_{MPPAmin\text{-}STC} \approx 455\,V$
With Equation 5.72, the following is obtained for crystalline modules: $V_{OCAmin\text{-}STC} \approx 569\,V$
With Equation 5.72, the following is obtained for amorphous modules: $V_{OCAmin\text{-}STC} \approx 650\,V$
With Equation 5.79, for $V_{OCAmax\text{-}STC} = V_{DCmax}/k_{TCmin} \approx 739\,V$ for installations in flat areas
With Equation 5.81, the following is obtained for crystalline modules in installations in flat areas:
$V_{MPPAmax\text{-}STC} \approx 595\,V$
With Equation 5.82, the following is obtained for amorphous modules in installations in flat areas:
$V_{MPPAmax\text{-}STC} \approx 510\,V$
For inverter tests, a series of tests for characteristics under the following conditions should be conducted using the currently available data: $V_{MPPAmin} \approx 400\,V$, $V_{MPPAmax} \approx 590\,V$, and at a mean of $465\,V$. Inasmuch as V_{DCmax} was previously defined as $800\,V$, the measurements referred to in Section 5.2.5 were conducted for 400, 480 and 560 V [5.49,5.50].

5.2.6.9 Possible DC-Side Problems Resulting from the Use of Latest-Generation Cell Technologies

Recently, irreversible degradation problems have been observed in some modules that possess thin-layer cells and crystalline back-side contact cells, in settings where transformerless inverters have been used. These problems can often be eliminated by grounding the negative terminal of thin-layer cells and grounding the positive terminal of back-side contact cells of the type made by Sunpower and other manufacturers. Hence, to be on the safe side, electrically isolated inverters should be used in installations that exhibit the aforementioned technologies.

5.2.6.10 Chapter Recap; Outlook for the Future

This section discussed some of the key problems that arise with grid-connected systems. Some of the observations made here, which were previously published elsewhere based on the then-valid standards [5.38], have been validated in practice.

The calculation methods discussed in this chapter allow possible problems to be identified during the planning phase and the relevant countermeasures to be taken early on. This definitely represents major progress compared with what used to be done: an installation was hooked up to the grid without further ado and any problems that arose were contended with in the field using cost-intensive ad hoc measures.

Voltage increases pose less of a problem in medium- and high-voltage grids since resistances relative to conductor reactances are lower in these grids than in low-voltage grids. Although experts will undoubtedly be conversant with the material presented in this chapter, PV installation planners need to be familiar with the relevant information in specialized electrical engineering fields such as AC technology, solar cell theory, 50 Hz grids, low-frequency harmonics and high-frequency technologies. In this chapter, I have attempted to present all of the essential information from these various disciplines in a reasonably simplified manner.

Needless to say, the recommendations presented are strictly my own and are based on my analysis of numerous problems that have arisen in a number of PV installations (my own included), in light of the applicable standards. My recommendations make good technical sense and can help to eliminate many of the inverter problems that have been observed time and time again. However, I cannot of course guarantee that following these recommendations will completely eliminate all problems or will translate into an installation that fully conforms with the valid national or power company standards, which are in any case in a constant state of flux.

Nonetheless, it seems to me that improved planning methods going forward will allow for the realization of largely trouble-free grid-connected systems.

5.2.7   Regulation and Stability Problems in Grid Systems

The main purpose of a grid-connected system is to obviate the need for a battery bank and thus eschew storage costs in connection with solar generator power, which is an inherently fluctuating energy resource (also see Figure 5.46 in Section 5.2.1).

This section briefly discusses (based on the situation in Switzerland) the technical limits of grid system operation, i.e. how much solar power can be stored in the system pretty much free of charge without the need for major expenditures on the build-out of pump storage systems or the like. *Needless to say, this limit has not been reached by a long shot, which means that it will be many years before a massive build-out of grid-connected systems takes on importance in this regard.* Production spikes under very high-insolation conditions can be brought down by means of both technical and tariff-related measures, although the latter are probably more cost effective. The current situation in the European grid system will also be briefly discussed, including ways to increase the amount of storable PV energy.

5.2.7.1   Output Regulation in Grid Systems

As previously noted (see Section 5.2.1), for reasons related to the laws of physics the amount of energy being generated and consumed in an electricity grid at any given time must be exactly the same. Thus the amount of electricity produced by power plants is determined not by the installations per se, but rather by the totality of devices that use grid energy. Needless to say, today's power plants cannot regulate their output from one moment to the next on an ongoing basis.

The amount of time a power plant needs to reach full production capacity varies from a matter of minutes for hydro power or gas turbine power plants to a number of hours for large nuclear power plants and other thermal power plants, whose efficiency and life span suffer if the facility has to power up and down repeatedly. Hence such power plants need to be operated continuously for base-load power production and to cover the basic grid load. Run-of-river plants are likewise best suited for base-load electricity generation, as they often exhibit relatively low storage capacity; further, from an energy management standpoint it would be pointless to allow available river water to flow through dams unused solely for the purpose of grid output regulation. Only in run-of-river plants that exhibit low to medium water throughput and relatively large dams can production be adjusted to demand at short notice so as to ease the task of grid regulation.

In contrast, the kinetic energy stored in the rotating elements comprising all turbines and generators allows for indispensable second-by-second adjustment of production to fluctuating grid load. All such generators operate in absolute lockstep with a grid system. In the event of a sharp and abrupt rise in demand relative to output, the deficit is filled transiently by the kinetic energy exhibited by the totality of the rotating elements, which slows them somewhat, thus reducing grid frequency. This results in the activation of so-called primary regulation, with the result that power plants that are already on line but are not producing to full capacity (so-called regulating power plants, but also of course, in Switzerland, storage hydroelectric power stations, or pump storage systems, and also in many cases gas turbine power plants in flat areas) increase their output until the nominal 50 Hz is reached, thus bringing energy production and demand back into equilibrium. Around 3 GW of power is available for primary regulation in the European grid system as a whole.

If the regulation range of power plants is not sufficient to cover a large load jump, secondary regulation kicks in automatically, whereby one of two things happens: either power plants that are already running immediately begin feeding additional power into the grid (this is known as rotating reserve); or additional reserve power plants that lend themselves to a quick start-up are activated. In Switzerland these are often storage hydroelectric power stations, although gas turbine power plants are often used in flat areas. Secondary regulation is activated within a few seconds and is deactivated 15 minutes later.

When secondary regulation is used, the central grid control system quickly authorizes the activation of reserve power plants with relatively short start-up times, so as to allow for restoration of the normal regulation reserve within a maximum of 15 minutes (this is known as the tertiary regulation reserve). This procedure comes into play in cases where a large nuclear power plant or other major facility fails.

In the reverse case, i.e. an abrupt drop in demand, the above procedure unfolds in reverse, whereby grid frequency briefly increases, the regulating power plants scale back their output, and any storage hydroelectric power stations or gas turbine power plants that are on line are shut down.

All Western and Central European power plants are integrated into the European grid system. The advantage of this arrangement is that it provides considerably greater insurance against grid collapse, but at a low cost for reserve retention in all participating countries – which means that the requisite primary regulating power can be shared among all participating states. The allowable frequency fluctuation in this grid is minute, however. The start-up and shutdown processes of many power plants are based on the so-called critical RPM range. If the turbines and generators in these power plants operate within this range for very long, they will incur severe damage from resonance effects. Hence all generator and turbine power plants shut down automatically in the event of a major deviation from the standard 50 Hz grid frequency. If this phenomenon occurred synchronously in numerous power plants, the entire European grid system would collapse, causing a power failure of continental magnitude. Thus proper frequency and output regulation is supremely important in the European grid system, and under no circumstances should grid-connected PV systems (or for that matter grid-connected wind turbines) be allowed to interfere with this regulation mechanism. Under normal operating conditions, the European grid system is regulated within an extremely narrow tolerance range, namely 50 ± 0.2 Hz.

A frequency drop in the European grid system can be induced not only by power plant failures, but also by an abrupt rise due to a severe disturbance in the high-voltage grid (e.g. power outages induced by the failure of key power lines in areas with relatively few power plants and a high concentration of connected appliances) resulting from the disappearance of the power that would otherwise be supplied to this grid.

As PV electricity is roughly proportional to the insolation on the generator surface, it is subject to major fluctuations over the course of a day and as the result of seasonal and weather changes. Thus PV installations are best suited as a substitute, on sunny days, for the electricity produced by power plants that lend themselves to fairly rapid to very rapid output regulation (referred to respectively as medium or peak load power plants, which in Switzerland are storage hydroelectric power stations and pump storage systems). On the other hand, the steady supply of electricity from nuclear power plants, conventional thermal power plants and run-of-river plants, whose output does not lend itself to regulation, can only be directly replaced by PV installations during the day and to a very limited extent only, since only a minute fraction (a few percentage points) of maximum PV installation output is guaranteed to be available under any given weather conditions.

The power plants in the 24 states that make up the European grid system (UCTE) exhibit an overall capacity of around 640 GW and generate around 2600 TWh of energy annually [5.84]. On the other hand, as at December 2008 total installed PV system power in Europe amounted to around 9.6 GW [1.17] and the installed power of all wind turbines was 66 GW. Hence the aggregate maximum grid-infeed capacity exhibited by these two technologies is considerably higher than the European grid system's approximately 3 GW of regulating power.

There have long been calls for wind turbines to help mitigate the effects of grid system disruptions instead of shutting down immediately, as is now the case. With growing overall output, this procedure is sure to take on ever-growing importance for PV installations as well. In view of the fact that grid failure can cause not only frequency drops but also the reverse, the 2006 [5.27] regulation requiring PV installations to shut down immediately if grid frequency overshoots 50.2 Hz no longer makes sense. For if upwards of 10 GW of output abruptly disappears, the grid frequency can also drop precipitously, causing frequency swings that could potentially jeopardize grid stability. In Germany, which currently has the world's highest installed PV capacity relative to maximum grid capacity (see Figure 5.158), a recently adopted regulation stipulates that PV and wind turbine installations that are directly connected to the medium-voltage grid need not be shut down immediately if the 50.2 Hz mark is overshot, but: (a) must instead (in keeping with the characteristic curve in Figure 5.157) reduce their power infeed at an increasing rate as the grid frequency climbs and thus promote grid stability; and (b) need only shut down if the grid frequency rises above 51.5 Hz [5.71].

Inasmuch as aggregate European capacity of small and medium-sized PV systems that are installed directly in the low-voltage grid will probably exceed 3 GW before long, the regulation method as in

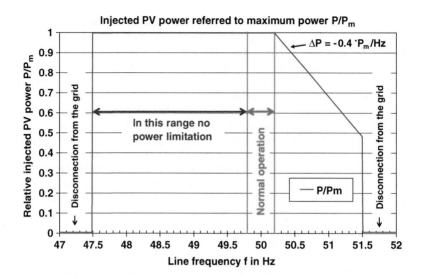

Figure 5.157 Characteristic infeed restriction curve for inverters as a function of grid frequency pursuant to Germany's recently adopted medium-voltage grid regulation [5.71]. According to this regulation, full possible maximum capacity P_m can be fed into the grid when it Exhibits 47.5–50.2 Hz, whereas above this frequency the power infeed is to be reduced at an increasing rate as grid frequency climbs (40% per Hz reduction) and is to be shut down if 51.5 Hz is exceeded

Figure 5.157 should also be adopted for small PV installations, and inverter manufacturers should be authorized to integrate the attendant characteristics into their devices. Inasmuch as an inverter can in any case scale back its output in the event of an anomaly (e.g. to protect itself against cloud enhancements), this function could be readily implemented via a minor software modification. Also, the question arises here as to whether a shutdown at 51.5 Hz is necessary and advisable; perhaps the characteristic curve that comes into play here should be ramped up somewhat.

5.2.7.2 Grid Load over the Course of a Day

Figure 5.158 displays electricity demand in Switzerland, Austria and Germany (where a different scale applies) on the third Wednesday of March, June, September and December 2005.

The advantage of PV installations over wind turbines (whose scope has been massively expanded over the years, particularly in Germany) is that all of their electricity is produced during the day when demand is highest, whereas wind power can decrease at any time. However, weather reports can be used to render wind turbine production predictable to a certain degree, thus allowing for its inclusion in power company electricity production planning.

In Germany, Austria and Switzerland demand is highest during the day, peaking shortly before noon, and is lowest in the early morning. This elevated daytime demand must be met by power plants whose output can be regulated fairly quickly (i.e. medium and peak load power plants) and can thus only be replaced by solar power during daylight hours. But as PV installation production is extremely low in inclement weather, particularly in the winter in flat areas (see Chapter 10), the lion's share of these regulatable power plants needs to be kept in reserve for such cases.

5.2.7.3 Duration Curves, Energy Use Efficiency, Full-Load Periods and Capacity Factor

In the interest of shedding greater light on the factors that come into play here, the actual output exhibited by PV installations will now be discussed, via duration curves, which are extremely useful in this regard (see the diagrams on the left in Figures 5.159 and 5.160).

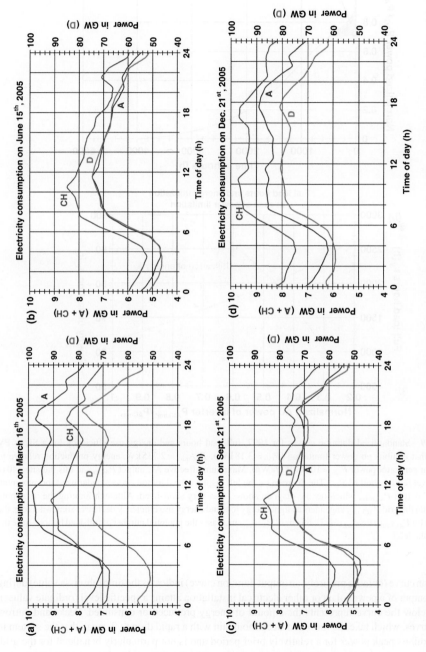

Figure 5.158 Electricity demand on a typical spring, summer, autumn and winter day in Switzerland (CH), Austria (A) and Germany (D; subject to a separate scale) in 2005. The load diagram displays overall output for all grid users over the course of the day on the third Wednesday of March, June, September and December 2005. (*Data source:* UCTE, www.entsoe.eu)

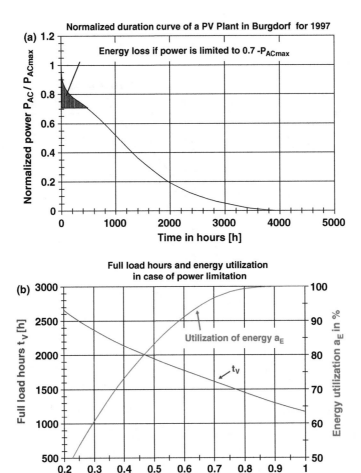

Figure 5.159 Standardized duration curves for 1997, full-load hours and energy use efficiency for a Swiss PV installation that exhibits no power limitation ($P_{Go} = 3.18$ kWp, $P_{ACmax} = 2.735$ kW, energy production relative to nominal solar generator power $P_{Go} = 1004$ kWh/kWp). Maximum effective AC output P_{ACmax} at this flat site 540 m above sea level fell far short of P_{Go}. The diagram on the left displays the duration curve standardized for maximum effective AC output P_{ACmax}, which (as the diagram shows) was very short-lived. Although limiting the maximum energy fed into the grid $P_{ACLimit}$ engenders a slight energy loss and energy use efficiency decreases somewhat (e.g. the hatched area for $P_{ACLimit}/P_{ACmax} = 0.7$), this limit $P_{ACLimit}$ increases the potential number of full-load hours t_V (see the diagram on the left)

A duration curve (or more precisely, an output duration curve) indicates the time (abscissa value) during which the output of a power plant or other electrical installation attains a specific level (ordinate values). The area below the curve represents the amount of energy produced or consumed. Peaks and narrow duration curves, which take the form of a steep mountain with a rapid drop to 0, mean that the system in question exhibits peak power for a relatively brief period and is not particularly beneficial for the grid. Blunt and trapezoidal duration curves, which take the form of a flat cliff and ideally never drop all the way to 0, are more beneficial for the grid, on the other hand. Duration curves for which the output P_{AC} of the installation concerned is standardized to maximum capacity P_{ACmax} are useful for comparing the output of various installations. The area below such curves represents the number of full-load hours t_{Vm} exhibited by the installation, i.e. the number of hours during which the installation needs to generate maximum power P_{ACmax} in order to produce or absorb energy E_{AC} during the observation period (e.g. one year). The following equation thus applies to an electrical installation:

$$\text{Full-load hours at maximum output: } t_{Vm} = \frac{E_{AC}}{P_{ACmax}} \tag{5.84}$$

If a power limitation $P_{ACLimitation} < P_{ACmax}$ is imposed (e.g. if a lower-output inverter is used or if output is reduced by a ripple signal from the grid control centre), a small amount of generated energy is lost, i.e. the effective energy fed into the grid E'_{AC} is lower than it otherwise would be without the limit E_{AC}. However, this effect is relatively minor in non-tracked PV installations in temperate climates, which exhibit relatively peaked duration curves. For purposes of illustration, this energy loss is shown in the duration curves in Figures 5.159 and 5.160 as a hatched area for $P_{ACLimit}/P_{ACmax} = 0.7$, which is extremely low relative to the overall area below the curve. Thus energy use efficiency E'_{AC}/E_{AC} for such a power limitation scenario is still nearly 100%.

This power limitation to $P_{ACLimit}$ substantially increases the number of possible full-load hours t_V for the new, lower maximum output $P_{ACLimit}$, thus improving grid load balance. The situation resulting from such a power limitation is illustrated on the right of Figures 5.159 and 5.160 for the $P_{ACLimit}/P_{ACmax} = 0.2$–1 range. The following values are shown here:

$$\text{Energy use efficiency } a_E = \frac{E'_{AC}}{E_{AC}} = \frac{\text{Energy with power limitation}}{\text{Energy without power limitation}} \tag{5.85}$$

$$\text{Full-load hours with a power limitation: } t_{Vb} = \frac{E'_{AC}}{P_{ACLimit}} \tag{5.86}$$

t_V can also be used to define the capacity factor of the installation for time T as follows:

$$\text{Capacity factor CF} = t_V/T \tag{5.87}$$

The mean annual capacity factor is obtained by dividing annual full-load hours t_V by $T = 365 \cdot 24\,\text{h} = 8760\,\text{h}$.

Figure 5.159 displays the standardized annual duration curve for 1997 for a PV installation in Burgdorf (540 m above sea level). This curve is relatively peaked, i.e. this installation attained peak power only briefly. During the observation year, insolation was slightly above average and the installation produced $E_a = E_{AC} = 3193\,\text{kWh}$. Relative to nominal solar generator power $P_{Go} = 3.18\,\text{kWp}$, the installation's specific annual energy yield was $Y_{Fa} = 1004\,\text{kWh/kWp}$, while its maximum registered peak AC output P_{ACmax} was only 2.735 kW and thus fell far short of P_{Go}. Thus, according to Equation 5.84, $t_{Vm} = 1167$ full-load hours would be necessary to generate this amount of energy. Thus the mean annual capacity factor was $CF_m = 13.3\%$.

If the output of this installation were limited to $P_{ACLimit} = 0.7 \cdot P_{ACmax} = 1.915\,\text{kW}$, E'_{AC}/E_{AC} would equal 96.6%, i.e. 3.4% of the installation's installed capacity would go unused but its annual energy yield would still be $E'_{AC} = 3085\,\text{kWh}$. According to Equation 5.86, this output $t_{V0.7}$ would allow for 1611 full-load hours and the annual capacity factor, as in Equation 5.87, would rise to $CF_{0.7} = 18.4\%$.

If the output of this installation were limited to $P_{ACLimit} = 0.4 \cdot P_{ACmax} = 1.094\,\text{kW}$, E'_{AC}/E_{AC} would equal 73.1%, i.e. 26.9% of the installation's installed capacity would go unused but its annual energy yield would still be $E'_{AC} = 2335\,\text{kWh}$. This output $t_{V0.4}$ would allow for 2135 full-load hours and the annual capacity factor would rise to $CF_{0.4} = 24.4\%$.

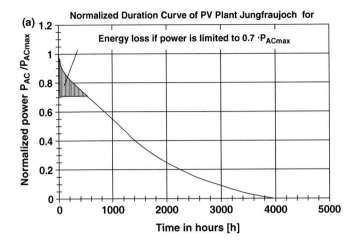

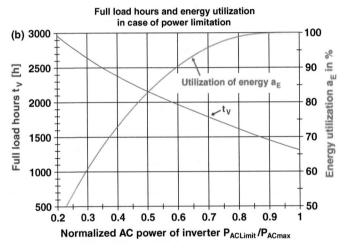

Figure 5.160 Similar diagrams to Figure 5.159 for the PV installation on the Jungfraujoch (3454 m above sea level) for 1997 ($P_{Go} = 1.15$ kWp, $P_{ACmax} = 1.303$ kW; energy production relative to nominal solar generator power $P_{Go} = 1478$ kWh/kWp). This installation, which likewise exhibited no power limitation, produced considerably more energy and its maximum effective AC power P_{ACmax} was considerably higher than P_{Go} owing to the thinner atmosphere, the absence of fog and the additional insolation engendered by light bouncing off the glacier in front of the installation. The duration curve is wider and full-load hours t_V higher than for the flat-site installation. However, on account of the higher maximum output P_{ACmax}, the increase in t_V was considerably lower than the increase in annual energy yield

For purposes of comparison, Figure 5.160 displays the standardized annual (1997) duration curve for a high-Alpine PV system on the Jungfraujoch (3454 m above sea level). Although this curve is somewhat wider, it is still quite peaked, i.e. peak power was short-lived here as well. During the observation year, insolation was somewhat above average and the installation produced $E_a = E_{AC} = 1700$ kWh. Relative to nominal solar generator power $P_{Go} = 1.15$ kWp, the installation's specific annual energy yield was $Y_{Fa} = 1478$ kWh/kWp, while its maximum registered AC-side peak power P_{ACmax} was only 1303 kW and thus fell far short of $P_{Go} = 1.152$ kWp. Thus, according to Equation 5.84, $t_{Vm} = 1305$ full-load hours would be necessary to generate this amount of energy. Thus the annual capacity factor was $CF_m = 14.9\%$.

If the output of this installation were limited to $P_{ACLimit} = 0.7 \cdot P_{ACmax} = 912$ W, E'_{AC}/E_{AC} would equal 95.7%, i.e. 4.3% of the installation's installed capacity would go unused but its annual energy yield would still be $E'_{AC} = 1628$ kWh. According to Equation 5.86, this output $t_{V0.7}$ would allow for 1783 full-load hours and the annual capacity factor, as in Equation 5.87, would rise to $CF_{0.7} = 20.4\%$.

If the output of this installation were limited to $P_{ACLimit} = 0.4 \cdot P_{ACmax} = 521$ W, E'_{AC}/E_{AC} would equal 72.9%, i.e. 27.1% of the installation's installed capacity would go unused but its annual energy yield would still be $E'_{AC} = 1239$ kWh. Thus output $t_{V0.4}$ would allow for 2378 full-load hours and the annual capacity factor would rise to $CF_{0.4} = 27.1\%$.

A comparison of the two installations shows that, despite the relatively sizeable difference between their respective specific annual energy yields Y_{Fa} for maximum AC power relative to full-load hours t_{Vm}, their respective annual capacity factors and energy use efficiency under the same conditions were far less dissimilar.

These examples show that keeping the energy fed into the grid below the maximum level can considerably increase the number of annual full-load hours, if the installation owner is willing to accept a certain amount of energy loss. Such losses will prove to be acceptable in the event that PV installations come into exponentially greater use, providing that the energy loss induced by such a limit is on a par with the loss associated with energy storage.

5.2.7.4   Maximum Energy Fed into the Grid

5.2.7.4.1   Scenario Involving the Use of a Fixed Solar Panel Installation in Switzerland

This section discusses power, and the amount of energy, that can be fed into a Swiss electricity grid without destabilizing it.

Figure 5.161 displays overall electricity generation in Switzerland on a typical summer day, 16 June 2004. The energy production of the base-load power stations varied little over the course of the day, ranging from 4.6 to 5.4 GW; this fluctuation was attributable to slight regulation by run-of-river power plants, which could also be modified for somewhat higher production during the night instead of during the day. On the other hand, storage hydroelectric power station production exhibited extreme fluctuations in order to adjust to demand over the course of the day. A comparison of Figures 5.161 and 5.158 reveals

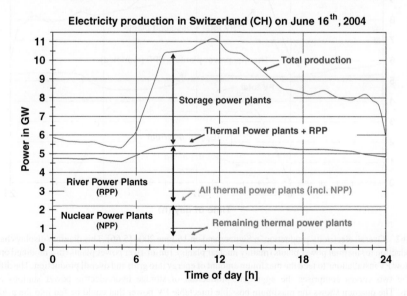

Figure 5.161   Overall electricity production in Switzerland on 16 June 2004 [5.66]. Production for the following elements is shown here: base-load energy from thermal power plants (chiefly nuclear power plants); run-of-river plants; and overall production. The difference between the curves comprises the electricity produced by regulatable storage hydroelectric power stations that can be readily replaced by electricity from other installations

that electricity production on summer days in most cases appreciably overshoots demand in Switzerland. This is attributable to the great abundance, by and large, of hydro power in Switzerland during the summer, some of which is exported to Switzerland's neighbours, thus allowing them to throttle down or even shut down their fossil fuel power plants. On the observation day, exported power ranged from around $-900\,MW$ (i.e. imported energy) to $+2600\,MW$, whereby total exports amounted to around 24 GWh.

Run-of-river plant production was not entirely constant over the course of the day. Some of these power plants that exhibit limited storage capacity could potentially increase their daytime production, although in principle this would also be possible at night (see Figure 5.162).

Figure 5.162 presupposes that the relevant run-of-river stations and storage hydroelectric power stations are operated in such a way that the grid can absorb the maximum output of PV installations, some of which are presumed to be operating with and some without a power limitation. The maximum difference between (a) overall production and (b) aggregate grid production of thermal power plants and run-of-river stations was just under 6 GW at noon and was continuously just under 4 GW from 9 a.m. to 5 p.m. It was possible to replace all of this output with PV installation output. Because CET was in effect here, PV power production peaked shortly after 1 p.m. rather than at noon. The diagram also displays a putative daily curve in dark green for PV power production that could potentially be absorbed by the grid without destabilizing it. This curve is composed of the energy output of PV installations that: (a) exhibit no power limitation and peak AC output of $P_{ACu} = 0.66\,GW$; and (b) limit output to 50% of maximum capacity via $P_{ACb} = 3.5\,GW$ ($P_{ACmax} = 7\,GW$). PV output peaked at 4.16 GW at around 1 p.m. The aggregate output of the PV installations, thermal power plants and run-of-river stations for the entire day

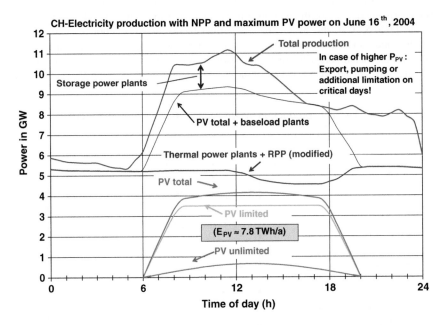

Figure 5.162 Overall electricity production in Switzerland on 16 June 2004 [5.66]. This diagram displays base-load energy production for thermal power plants (mainly nuclear plants), run-of-river power plants (in a modified operating mode to allow PV installations to feed the maximum amount of energy into grid) and overall production. The difference between the two curves comprises the aggregate production of storage hydroelectric power stations and PV installations. The diagram shows the maximum possible injectable PV power that could be fed into the grid in this scenario without regulation problems, provided that storage hydroelectric power stations and run-of-river stations adjusted their output accordingly. This PV production is composed partly of the output of (a) PV plants that exhibit no power limitations and (b) PV plants that exhibit a power limitation, which here is presumed to be $0.5 \cdot P_{ACmax}$. As daylight savings Central European Time (CET) is used here, maximum PV production occurred at around 1 p.m

never exceeded overall grid production. The remaining output (less than 3 GW for the entire day) was produced by storage hydroelectric power station without any difficulty.

The remaining, approximately 27 GWh of energy that the storage hydroelectric power stations needed to produce on this sunny day is considerably less than in Figure 5.161, without PV plants, where the storage plants had to produce 73 GWh. The resulting, roughly 46 GW difference was produced by the PV installations and remained in the storage lakes as an additional water reserve for inclement weather days.

Inasmuch as, for reasons of grid stability, all of the storage hydroelectric power stations could not be shut down during the day, it was necessary to run some of these facilities as regulating power plants. And since a running storage hydroelectric power station needs to exhibit a minimum production level at all times, zero overall production is only attainable if some of the station's output is absorbed by the pump storage systems, which is unsatisfactory from an energy standpoint. This situation would have arisen from approximately 4.30 to 5.30 p.m. on the observation day. However, normally the international grid system that comes into play here would obviate the need for such use of the pump storage systems to maintain grid stability, whereby the minimum storage hydroelectric power station output could be exported or could be temporarily down-regulated slightly by suitable run-of-river stations, water throughput levels permitting.

For purposes of comparison, analogous data were compiled for another summer day where electricity production and exports were higher (see Figures 5.163 and 5.164). On this day, exported power ranged from around 900 to 2700 MW, total exported energy was around 50 GWh and storage hydroelectric power station production was 96 GWh.

Figure 5.164 again presupposes that the run-of-river stations were operated in such a way that the grid could absorb a maximum amount of PV installation output. This was attainable in the present case without any major change in run-of-river station operating modality.

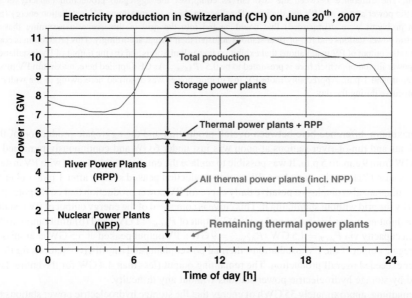

Figure 5.163 Overall electricity production in Switzerland on 20 June 2007 [5.85]. Production for the following elements shown here: base-load energy from thermal power plants (chiefly nuclear power plants); run-of-river plants; and overall production. Production is substantially higher than in Figure 5.161 at night in particular, since both thermal power plant and run-of-river station output were also considerably higher. The afternoon output of the storage hydroelectric power station was also far higher. The difference between the two curves comprises the electricity produced by regulatable storage hydroelectric power stations that can readily be replaced by electricity from other installations. This type of production curve allows appreciably more PV installation energy to be fed into the grid without any problem

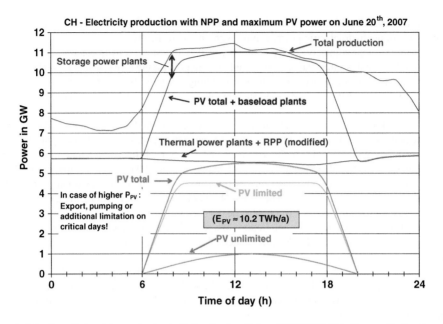

Figure 5.164 Overall electricity production in Switzerland on 20 June 2007 [5.85]. This diagram displays base-load energy production for thermal power plants (mainly nuclear plants), run-of-river power plant production (in a readily modified operating mode to allow PV installations to feed the maximum amount of energy into grid) and overall production. The difference between the two curves comprises the aggregate production capacity of storage hydroelectric power station and PV installations. The diagram shows the maximum PV installation energy (possible regulation problems were disregarded) that could be fed into the grid in this scenario assuming that storage hydroelectric power stations and run-of-river stations adjusted their output accordingly. This production is composed partly of (a) the output of PV installations that exhibit no power limitations and (b) the output of PV installations that exhibit a power limitation, which here is presumed to be $0.5 \cdot P_{ACmax}$. As CET applied here, maximum PV production occurred at around 1 p.m. Appreciably more PV installation power was absorbed here owing to the wider overall production curve during the day

The maximum difference here between overall production and the aggregate production of thermal power plants and run-of-river stations at noon was just under 6 GW and continuously remained at just under 5 GW from 9 a.m. to 5 p.m. It was possible to replace the entirety of this output with PV installation output. Because CET was in effect here, PV power production peaked shortly after 1 p.m. rather than at noon. The diagram also displays a putative daily curve for PV power production that could potentially be absorbed by the grid without stabilizing it. This curve is composed of the energy output of PV installations that: (a) exhibit no power limitation and peak AC output of $P_{ACu} = 1$ GW; and (b) limit output to 50% of maximum capacity via $P_{ACb} = 4.5$ GW ($P_{ACmax} = 9$ GW). PV output peaked at 5.5 GW at around 1 p.m. The aggregate output of the PV installations, thermal power plants and run-of-river stations for the entire day never exceeded overall production. The remaining output (less than 4.4 GW for the entire day) was produced by storage hydroelectric power stations without any difficulty.

The remaining, approximately 35 GWh of energy that the storage hydroelectric power stations needed to produce on this sunny day is considerably less than in Figure 5.163, which omits the roughly 96 GWh produced by the PV installations. The resulting, roughly 61 GWh difference was produced by the PV installations and remained in the storage lakes as an additional water reserve for inclement weather days.

Annual PV energy capacity can be determined using the possible annual number of full-load hours t_V with and without a power limitation $P_{ACLimit} = 0.5 \cdot P_{ACmax}$. Hence, if the mean of all grid-connected systems in Switzerland (i.e. in the Alps, Tessin, Wallis and in foggier Mittelland) for installations with no power limitation $t_{Vu} = 1200$ h and with a power limitation $t_{Vb} = 2000$ h is determined, the following is obtained for the annual PV energy that can be produced in Switzerland without destabilizing the Swiss electricity grid:

$$E_{ACa} = P_{ACu} \cdot t_{Vu} + P_{ACb} \cdot t_{Vb} \tag{5.88}$$

For these suppositions and using Equation 5.88 and the values $P_{ACu} = 0.66$ GW and $P_{ACb} = 3.5$ GW that were determined using Figure 5.162, possible annual PV production $E_{ACa} = 7800$ GWh $= 7.8$ TWh is obtained, which amounts to around 13% of Switzerland's total electricity demand (around 60 TWh) in 2005.

Using (a) the values $P_{ACu} = 1$ GW and $P_{ACb} = 4.5$ GW that were determined using Figure 5.164 and (b) Equation 5.88, $E_{ACa} = 10\,200$ GWh $= 10.2$ TWh, or around 17% of Switzerland's total annual electricity demand (around 60 TWh) in 2005 is obtained.

The scenario will now be discussed that would result from shutting down all nuclear power plants, as has been advocated by environmentalists. This would of course enable the grid to absorb a higher level of peak power.

Figure 5.165 displays a scenario of this nature based on the overall production curve for 16 June 2004, which is composed of: (a) the energy produced by PV installations that exhibit a power limitation and peak AC output $P_{ACu} = 0.66$ GW; and (b) the energy produced by PV installations that exhibit a power limitation amounting to 50% of maximum possible output at $P_{ACb} = 5.3$ GW ($P_{ACmax} = 10.6$ GW). Using Equation 5.88 for installations that exhibit no power limitation at $t_{Vu} = 1200$ h and installations that

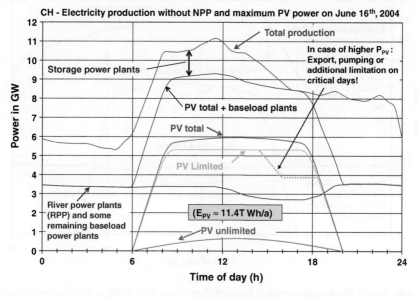

Figure 5.165 Overall electricity production in Switzerland on 16 June 2004 [5.66]. In the absence of nuclear power, the output of the remaining base-load power stations would be considerably reduced (blue curve). Here again, run-of-river station operation has been modified to allow for the maximum possible amount of PV installation to be fed into the grid. Overall production is represented by the red curve. The difference between the two curves comprises the aggregate production capacity of storage hydroelectric power stations and PV installations. The diagram shows the maximum PV installation energy (possible regulation problems were disregarded) that could be fed into the grid in this scenario assuming that storage hydroelectric power stations and run-of-river stations adjusted their output to allow for maximum PV installation input into the grid. This production is composed of a portion of (a) the output of PV installations that exhibit no capacity limits and (b) the output of PV installations that exhibit a capacity limit, which here is presumed to be $0.5 \cdot P_{ACmax}$. As CET applied here, maximum PV production occurred at around 1 p.m

exhibit a power limitation at $t_{Vb} = 2000$ h, annual PV energy that can be fed into the Swiss grid without destabilizing it, $E_{ACa} = 11\,400$ GWh $= 11.4$ TWh, is obtained, which amounts to around 19% of 2005 national electricity demand (around 60 TWh).

The energy balance for the scenario in Figure 5.165 will now be briefly discussed. Most PV installations generate around 68 GWh on a sunny day (in lieu of the 46 GWh as in Figure 5.162), while base-load power stations still produce 78 GWh in the wake of the shutdown of all nuclear power plants, which produce around 44.5 GWh. This means that storage hydroelectric power stations need to produce around 50 GWh on this sunny summer day.

Figure 5.166 displays a scenario of this nature derived from the overall production curve on 20 July 2007, when production was higher. This curve is composed of the maximum energy output of PV installations that: (a) exhibit no power limitation and peak AC output of $P_{ACu} = 1.2$ GW; and (b) limit output to 50% of maximum capacity with $P_{ACb} = 6.5$ GW ($P_{ACmax} = 13$ GW).

Using Equation 5.88 for installations that exhibit no power limitation at $t_{Vu} = 1200$ h and installations that exhibit a power limitation at $t_{Vb} = 2000$ h, the annual PV energy that can be fed into the Swiss grid without destabilizing it amounts to $E_{ACa} = 14\,400$ GWh $= 14.4$ TWh, which equates to around 24% of 2005 national electricity demand (around 60 TWh).

The energy balance for the scenario in Figure 5.166 is as follows. Most PV installations generated around 87 GWh on this sunny day (in lieu of only 68 GWh in Figure 5.165), while the base-load power

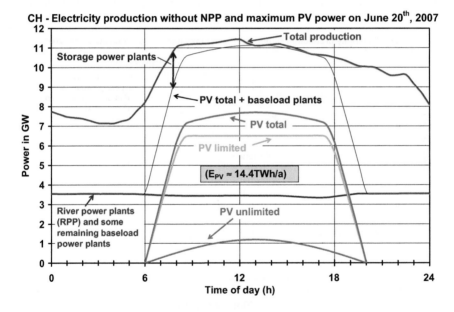

Figure 5.166 Overall electricity production in Switzerland on 20 June 2007 [5.85]. In the absence of nuclear power, the output of the remaining base-load power stations would be considerably reduced (blue curve). Here again, run-of-river station operation has been modified to allow for the maximum possible amount of PV installation to be fed into the grid. Overall production is represented by the red curve. The difference between the two curves comprises the aggregate production capacity of storage hydroelectric power stations and PV installations. The diagram displays maximum PV installation energy (possible regulation problems were disregarded) that could be fed into the grid in this scenario, assuming that storage hydroelectric power stations and run-of-river stations adjusted their output to allow for maximum PV installation input into the grid. This production is composed of a portion of (a) the output of PV installations that exhibit no capacity limits and (b) the output of PV installations that exhibit a capacity limit, which here is presumed to be $0.5 \cdot P_{ACmax}$. As CET applied here, maximum PV production occurred at around 1 p.m. Appreciably more PV installation power was absorbed here owing to the wider overall production curve during the day

stations still produced 83 GWh in the wake of the shutdown of all nuclear power plants (which produced around 52 GWh) and the storage hydroelectric power stations produced 62 GWh.

As the calculations derived from Figures 5.165 and 5.166 show, the grid can absorb a considerably greater amount of PV installation power in the absence of nuclear power plants. However, unless nuclear power plant production is not replaced at certain times by energy sources such as biomass power stations, geothermal power plants, wind turbines, gas turbine power plants, pump storage systems and imported electricity, a reliable electricity supply cannot be guaranteed at all times. In 2007 Switzerland's nuclear power plants generated more than 26 TWh or around 40% of the country's total energy production. This issue is addressed in the sections that follow.

5.2.7.4.2 Possible Energy Production Optimization via Further Technical and Price Measures

Although tracked PV installations cost more, they also exhibit greater annual energy capacity, and in particular increase by 25–40% the number of full-load hours obtainable with the same maximum AC peak output at the same site (see Sections 10.1.9–10.1.12). Of course, such PV installations also increase, to the same extent, the amount of energy that can be fed into the grid without destabilizing it. Tracked PV installations are particularly advantageous in southern and desert regions, where t_{Vu} values of around 2000 are presumably obtainable for fixed solar panel installations, and around 3000 full-load hours are presumably obtainable with dual-axis trackers.

Moreover, additional technical measures can ramp up still further the maximum energy that the grid can absorb. For example, in lieu of programming a fixed power limitation into individual inverters, by using ripple signals or the Internet power companies could strategically down-regulate or even shut down individual PV installations, but only in cases where grid stability is genuinely jeopardized. Implementation of this technology in new installations is already required in Germany (which has a high concentration of PV installations) [5.71], [5.72] and could considerably reduce the energy loss induced by infeed restrictions.

An even more useful reform would be to institute technical and price measures that would allow loads such as washing machines, refrigerators/freezers, heat pumps and hot water heaters to be shifted from the low-tariff evening and night period to peak PV production times [5.67], [5.68]. To this end, programmes could be instituted aimed at modifying electricity customers' behaviour through appeals to ecologically conscious individuals, and by adjusting electricity prices to anticipated solar energy production levels (and informing customers accordingly ahead of time). Such programmes have already been piloted successfully [5.67]. An even more productive approach might be to combine price and technical measures with bidirectional energy management interfaces (BEMIs) that allow power companies to activate and deactivate loads on line according to current grid load and electricity rates [5.68].

In addition, electric vehicles connected to the grid at suitably equipped parking spaces would help to mitigate grid power spikes by storing energy in car batteries. This in turn would increase the maximum amount of energy that PV installations can feed into the grid in a manner that would obviate infeed restriction loss and avoid further increasing the grid load.

A constellation of such measures would allow for a sustainable increase of least 15–25% in the amount of PV installation energy that can be fed into Central European electricity grids without destabilizing them and without undue energy generation loss. But this would necessitate a major shift in the power companies' current grid regulation paradigm, which would need to exhibit greater technical complexity.

5.2.7.4.3 Status of the European Grid System

Inasmuch as the daily electricity demand curve is essentially the same in all European countries (see Figure 5.158), the observations made here concerning Switzerland apply to the rest of Europe, assuming that PV installation concentration is roughly the same in all countries. Although not all European countries have the kind of rapidly regulatable power plants that Switzerland has, all countries nonetheless have a certain number of regulatable power plants, e.g. gas turbine power plants in lieu of storage hydroelectric power stations. Moreover, owing to the electricity interchange options afforded by the European grid system, local electricity surpluses can be readily interchanged. In view of the fact that the weather is rarely sunny everywhere in Western Europe at the same time, if the scope of electricity interchange is enlarged in

Europe there are probably not very many occasions when limits would need to be placed on the power that PV installations feed into the grid in order to maintain grid stability; this in turn means that energy loss induced by such restrictions will likewise be a rarity. But increased electricity exchange in Europe would also necessitate a build-out of our high-voltage grid.

In addition, the many air-conditioners used during the summer in Southern Europe dovetail ideally with daily PV installation production curves in this region, whose high number of full-load hours allows PV installations to cover a higher proportion of electricity demand there.

Hence, if a combination of the above technical and price-related measures were implemented in the European grid system, within a few decades PV installations could readily meet 25–30% of European electricity demand.

5.2.7.5   Possible Substitutes for Base-Load Energy Sources

A modicum of base-load power stations is indispensable for an uninterrupted electricity supply. And while ideally base-load energy should be provided by run-of-river stations, they are unfortunately not ubiquitously available in sufficient numbers. The most suitable substitutes for non-renewable base-load energy resources (e.g. energy from coal-fired, natural gas, oil-fired or nuclear power plants) are newer carbon-neutral technologies such as biomass and geothermal power stations, which have great potential but are currently in very short supply. Solar thermal power stations in very sunny regions could also be used in conjunction with suitable thermal storage technologies (e.g. liquid salt and the Andasol project; see Chapter 11) to produce electricity at night and thus supply a modicum of base-load power after dark as well as during lengthy inclement weather periods, which are relatively rare. Biomass power stations can also be used for grid regulation purposes, which would be highly desirable if the use of PV installation electricity becomes more widespread. Nonetheless, the potential offered by biomass technology is extremely limited on account of the relatively low efficiency entailed by converting solar energy into biomass.

If an appreciable amount of base-load energy were to be replaced by PV or wind turbine energy, a portion of the energy produced would need to be kept in reserve for low-power periods. Wind energy in Germany has been developed to a relatively major extent, with expansion far beyond the present level in the works (see Figure 5.167); thus wind energy storage problems will arise long before the challenges of PV energy storage need to be faced.

Some of the same issues arise in connection with wind and PV energy storage. Stepped-up use of solar and wind energy will increase the need for rapidly regulatable power and for power stations that exhibit both short- and medium-term storage capabilities. It is with this end in mind that far-sighted power companies have already begun ramping up these capacities.

The cost of PV energy storage varies greatly from one site to another. For PV installations that are mainly used for winter energy production and are located in low-fog areas such as the Alps or Southern Europe, storage for a number of days is sufficient to make it through inclement weather periods and to balance daytime and nocturnal energy demand. These storage needs could be largely met by expanding all existing storage hydroelectric power stations into pump storage systems. But when it comes to rooftop PV systems in Northern and Central Europe that are optimized for maximum annual energy production, a certain amount of energy needs to be stored seasonally, which is both far more expensive and unlikely to be unattainable using pump storage systems alone.

As noted, wind energy (which is currently available in far greater supply than solar power) is beset with similar peak load and energy storage problems – although the fact that wind turbine production is higher during the lower insolation winter months than in summer provides a very welcome counterweight to PV energy production.

5.2.7.5.1   Electricity Storage in Pump Storage Systems
Pump storage systems are a proven electricity storage technology that is already up and running (see Figures 5.168 and 5.169), but whose overall efficiency (which unfortunately entails converting electricity into water and then *back* into electricity) is only around 75 to 80%, i.e. a 20–25% energy loss.

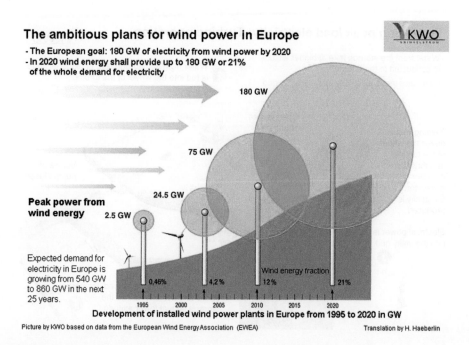

Figure 5.167 Envisaged expansion of wind energy capacity in Europe. Wind energy development between 2004 and 2008 unfolded essentially as forecast in the diagram. European installed wind power capacity as at December 2008 was around 66 GW, according to the European Wind Energy Association. (Illustration: Oberhasli power station; *data source*: European Wind Energy Association (EWEA))

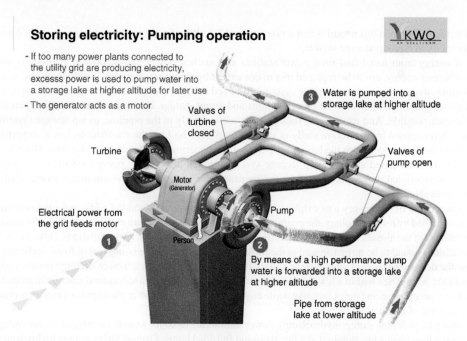

Figure 5.168 Operating principle of pump storage systems (pumping phase). At times when electricity supply exceeds demand, this (momentarily) inexpensive energy is used to pump water from lower-lying lakes into storage lakes at a higher elevation. (Illustration: Kraftwerke Oberhasli (KWO))

Producing peak load electricity: Turbine operation

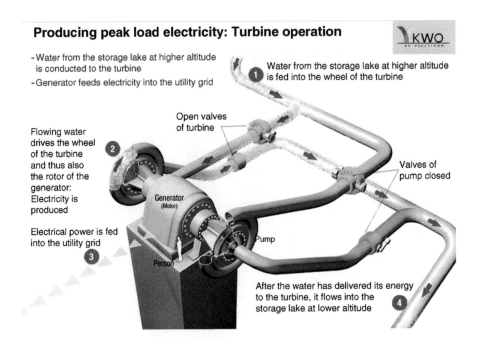

Figure 5.169 Operating principle of pump storage systems (turbine generation phase). At times when electricity supply falls short of demand (e.g. around noon, or if PV installations or wind turbines are underproducing), water from higher-elevation storage lakes is shunted to power-generating turbines. (Illustration: Kraftwerke Oberhasli (KWO))

The rule of thumb in this regard is that it takes around 1.3 kWh of pumping energy to produce 1 kWh of electricity in a pump storage system.

If energy from fossil-fuel-fired power stations and nuclear power plants, as well as biomass and geothermal energy, are to be replaced to a major extent by electricity from PV installations and wind turbines, the number of pump storage systems will need to be increased considerably and the capacity of existing storage hydroelectric power stations and pump storage systems will need to be expanded wherever possible. And while many such projects are currently in the pipeline, pump storage systems are often opposed by environmentally minded politicians who lack the requisite technical expertise and are not conversant with the basic technical issues concerning electricity grid stability. Moreover, for the reasons stated above, pump storage systems, which are already being used to store energy from conventional power plants, will be indispensable for an electricity system that is based wholly on renewables.

On account of the energy loss exhibited by pump storage systems, and the additional infrastructure costs entailed by this technology, the energy stored in these systems is more expensive, which also drives up the cost of the already cost-intensive electricity from wind turbines and PV installations. However, this situation is bound to be turned around in the medium to long term by the rising cost of fossil fuel energy and the decreasing costs of wind turbine and PV energy. And as for nuclear power, uranium reserves will probably be depleted within a few decades. Moreover, nuclear power is a suboptimal solution in terms of political acceptance, and the radioactive waste engendered by nuclear power plants poses a major problem on account of international terrorism.

Many large Alpine storage hydroelectric power stations are rated for seasonal storage of winter energy for anywhere from a few hundred to a few thousand full-load hours. Overall Swiss storage hydroelectric power station capacity is around 8.5 TWh, which represents around 14% of annual Swiss energy demand in 2005. Most pump storage systems exhibit a storage capacity ranging from 10 MW to around 1 GW.

Inasmuch as their attainable full-load hour capacity in most cases falls far short of that exhibited by storage hydroelectric power stations, ranging from less than 10 hours to several hundred hours, pump storage systems can only store energy for a few days. But pump storage systems also have the virtue of being by far the least-expensive storage technology available today.

5.2.7.5.2   *Short-Tem Storage Using Battery Banks*

Short-term storage systems based on large batteries and bidirectional inverters can be used to cover grid load spikes or underproduction for several hours at a time for output ranging up to around 30 MW and stored energy ranging up to several hundred megawatt hours. A range of battery technologies such as Pb, NiCd, NaS, lithium ion and vanadium redox flow can be used for this purpose [5.86], [5.92]. Batteries are charged during periods of surplus output, whereas during low-output periods battery power is discharged back into the grid. Large appliances mainly use batteries for the following purposes: to provide spinning reserves (i.e. meeting sudden demands for power) so as to reduce electricity costs for peak load output use; for grid load stabilization in relatively low-output stand-alone power systems; and to provide readily useable regulating power for frequency regulation purposes [5.92]. Most batteries are rated for relatively few full-load hours (2–10 or the like), are hooked up to the medium-voltage grid, and exhibit AC-power-to-battery-to-AC-power cycle energy efficiency ranging from around 70 to 80% [5.92].

As early as several decades ago, a 17 MW system of this type with 14 MWh of storage capacity was in use in Berlin [5.88], and an ABB-provisioned battery bank based on 6 MW sodium sulphide (NaS) batteries with 49 MWh of storage capacity has been in operation in Japan for more than 10 years. A 34 MW system of this type with 200 MWh of storage capacity was in the works as of 2008, at which time more than 80 NaS battery banks with varying outputs were in operation worldwide. Such systems exhibit energy cycle efficiency of around 75% [5.86], although this efficiency is put at only 72% for a short-term storage lead battery bank in California (output 10 MW, storage capacity 40 MW) that is likewise hooked up to the medium-voltage grid [5.86].

An obvious alternative to hooking up large-scale battery banks to the medium-voltage grid would be for all grid-connected systems to integrate a battery bank for decentralized storage of part of the energy produced by grid-connected inverters, for use at night or during inclement weather. This would allow for a reduction of the maximum peak PV output fed into the grid and thus for a grid connection of higher overall PV output. Brief power outages could also be buffered with such systems, whereby those described in Section 5.1.6 in principle already exhibit this capability. However, hooking up such decentralized storage system PV installations to the public grid is prohibited under current regulations and would necessitate a major shift in the mindset of power companies.

As previously noted (see Section 5.1.1), batteries have limited cycle lives and need to be replaced from time to time. A battery's cycle life is determined by its technology and discharge depth. Manufacturing batteries and the electronics needed for rectifiers and inverters necessitates the use of so-called grey energy (see Section 9.2), whose cost is depreciated over the life of the battery and needs to be factored into the overall energy balance calculation. Hence the effective energy efficiency of batteries is lower than the value indicated for an individual battery cycle. Battery manufacturers are not overly concerned about such matters and tend to indicate the per-cycle efficiency only.

Reported in [5.90] is a study involving a grid support lead battery that exhibited 450 kWh of storage capacity and underwent a 150 kWh daily cycle (discharge depth t_Z, i.e. 33.3%) for five years, during which time the battery achieved 75% cycling efficiency (i.e. discharged AC energy divided by AC energy used for charging purposes). This efficiency level can be attained with, for example, rectifier efficiency amounting to $\eta_{GR} = 95\%$, battery energy efficiency amounting to $\eta_{-battery} = 83\%$ (DC energy to DC energy) and inverter efficiency amounting to $\eta_{WR} = 95\%$. It was determined in [5.90] that this system's effective efficiency was 66.4%, including the grey energy used for battery cycling, whereby this grey energy amounted to around 1.5 times the amount of energy obtained from the battery. This calculation presupposed that 50% of the lead used in the battery was recycled. If 99% recycled lead is presupposed instead, the battery would Exhibit 68% efficiency. Needless to say, these efficiency levels fall far short of those exhibited by a pump storage system.

5.2.7.5.3   *Short-Term Energy Storage Using Batteries from Electric Cars*
One way to reduce vehicle carbon emissions is to run electric vehicles (BEVs) and plug-in hybrid electric vehicles (PHEVs) using electricity from PV installations or other low-carbon technologies. Most electric vehicles contain a battery bank whose storage capacity typically ranges from 5 to 40 kWh and is in most cases not fully used over the course of a day.

Hence the idea is to use a portion of electric vehicle storage capacity to store surplus output from PV installations at noon, or from wind turbines and so on, and then to provide this energy back to the grid at opportune times such as in the evening or at night. As the vast majority of vehicles are parked most of the time, this system would work if parking lots equipped with stations that could return power to the grid from the vehicles were widely available [5.87], [5.88], [5.89].

To do this, however, the following problems would need to be solved:

- The grid infrastructure would have to be rendered compatible with such a system and would need to be fitted with the relevant accounting platforms.
- Battery cycling expedites the battery ageing process, which is of course undesirable for vehicle owners; however, this problem could be solved via a battery loan system run by power companies.
- Vehicle owners would no longer have full battery capacity at their disposal, thus partly eliminating one of the key benefits of owning a car (i.e. uninterrupted availability), and rapid charging facilities would be few and far between, assuming they were available at all.

Only a minute fraction of the overall battery capacity needed to institute such a system is currently available, since in the morning vehicle storage batteries are never dead and should in any case never be fully discharged. A very useful and relatively simple approach in this regard would be to charge fully all car batteries over the course of the day, PV installation production permitting – although frequent deep discharge in the evening or at night to promote grid stability would presumably be unacceptable to car owners.

That said, it seems to me that something along the lines of the following system could be instituted in Switzerland.

Assuming a 2 million vehicle fleet of electric vehicles (BEVs) and plug-in hybrids (PHEVs) with a mean storage capacity of 10 kWh, connected load amounting to 4 kW and available grid regulation storage capacity amounting to 4 kWh, these vehicles could provide a total of around 8 GWh back to the grid if necessary (e.g. up to 8 GW for 1 hour or 1 GW for 8 hours). Charging this fleet's battery banks during the day would require around 10 GWh of power, on account of the presumably higher energy cycle efficiency of the new lithium-ion batteries that would probably be used. Useful though this system would undoubtedly be, it would not be nearly sufficient to replace the nocturnal energy generated by thermal power plants; furthermore, it would only work on sunny days. In any case this solution, like other battery storage systems, could only be used for short-term energy storage and a very low number of full-load operating hours.

As noted in Section 5.2.7.5.2, owing to the grey energy used for battery cycling by such systems, their effective energy efficiency falls short of that of cycling efficiency alone and thus needs to be taken into account for energy efficiency evaluations of the system as a whole.

5.2.7.5.4   *Other Energy Storage Systems*
The following additional systems can also be used for energy storage purposes:

- Compressed air energy storage (CAES). Wherever possible, such systems should also allow for compression of the attendant thermal energy as latent heat or the like, in order to compensate for the thermal energy needed to reheat the air secondary to air expansion, otherwise this energy must be obtained from outside sources, which of course reduces energy cycle efficiency.
- Short-term storage as kinetic energy in ultrahigh-RPM flywheel gears, ideally with friction-free bearings.

- Short-term storage in electrically heated high-temperature accumulators powered by conventional steam turbines and generators whose electricity is relatively inexpensive. For direct electrical heating using resistances, energy cycle efficiency ranging up to 40% (and up to 60% if heat pumps are used), and storage periods of up to several days, should be attainable.

5.2.7.5.5 Hydro Power
PV electricity can also be used for water electrolysis to produce hydrogen, which can be converted back into electricity in fuel cells.

A solar–hydrogen economy would allow for the production of hydrogen using solar energy in desert areas. After being transported to devices in a fluid, gaseous or metal hydride form, this hydrogen would be converted into fuel-cell-based electricity that could be fed into the grid or used as motor vehicle or aircraft fuel.

However, the currently attainable efficiency for converting electricity into hydrogen and back into electricity is only around 30 to 50% and thus falls far short of the efficiency exhibited by classic pump or battery storage systems. The only viable use of this fuel cell technology that can currently be contemplated is for hydrogen storage to power emission-free cars and as a substitute for the batteries now used in mobile devices. The relevant industries are currently hard at work on fuel cell development and optimization, but this technology is still in its infancy and costs far more than pump storage systems. Hence the use of hydrogen storage in the electricity industry is not a major priority at present. For further information on hydrogen storage technologies, see [Win89], [Bas87], [Sche87], [Web91], [Kur03] and [5.69].

5.2.7.5.6 The Supergrid and Global Grid Solutions
A supergrid with sufficient capacity would go a long way towards solving the energy storage conundrum. East–west power transmission lines would allow for energy interchange between regions in different time zones, while north–south links would facilitate electricity exporting and importing on a seasonal basis. However, currently available technologies do not allow for the realization of such a grid since transporting electricity over thousands of kilometres would entail excessive transmission loss, which with today's high-voltage lines ranges from around 5 to 10% per thousand kilometres. This loss could be reduced to around 2.5 to 4% per 1000 kilometres through the use of high-voltage direct current (HVDC) transmission lines, which today only support point-to-point transmission.

A supergrid of this type that would link Europe, North Africa and the Middle East was proposed in mid 2009 in the guise of the Desertec project. This grid would be composed of the following elements: hydro power plants in Northern Europe, the Alps, the Pyrenees and elsewhere in Europe; wind turbines in all suitable coastal areas from Northern Europe to North Africa; PV installations in Europe; geothermal power stations in suitable locations; and solar thermal power stations in desert regions (see Figure 5.170 and www.desertec.org).

Although this project could be brought to fruition using currently available technologies, it would entail investments on a monumental scale. In addition to the existing high-voltage AC grid, an extremely far-flung HVDC transmission grid would be needed whose energy would have to be fed into the conventional high-voltage AC grid in close proximity to the relevant appliances. To do this, the transmission line capacity that currently allows for international energy exchange via the largely AC-based high-voltage grid would have to be quadrupled, i.e. such a supergrid would entail realization of an extremely large number of new high-voltage lines.

With the technical advances in the field of superconductors, a global supergrid might be realizable at some point in the future, although an even greater number of new high-voltage lines would be needed for such a project than for the Desertec supergrid. In order for such large-scale electricity grids to be brought to fruition, all of the countries involved would need to be politically stable – a desirable but highly unrealistic aim.

5.2.7.6 Economic Impact on the Operation of Other Power Plants

Only the technical aspects of grid stability have been considered thus far. But power plants are also business enterprises that aim to sell their electricity at the highest possible price so as to depreciate the

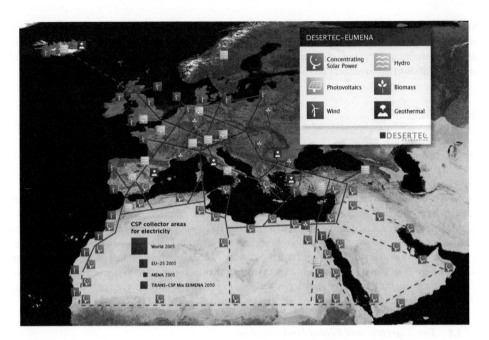

Figure 5.170 Desertec supergrid composed of large power plants that would allow for the use of renewable energy in Europe, North Africa and the Middle East, would be interconnected by HVDC lines, and would be able to provide basic loads owing to the grid's integrated accumulators and wide geographical reach. (Illustration: Desertec Foundation, www.desertec.org)

extremely high capital investments entailed by electricity production. It is for this reason that Switzerland's rapidly regulatable power plants produce as much electricity as possible during peak demand periods when rates are highest. Other countries also have oil-fired plants, gas turbine power stations and the like that specialize in the production of energy at peak demand times and which, on account of their shorter operating times, need to charge higher rates per kilowatt hour than do other power plants. If these power plants were required to scale back their output during such peak times to allow the grid to absorb the output of numerous PV installations and wind turbines, and had to generate electricity during low-load periods, their electricity would be sold for lower prices and their sales revenue would decline. As noted, a similar effect would occur if the grid possessed a large proportion of grid-connected systems, in that daytime electricity rates would tend to fall while nocturnal prices would rise.

Owing to deregulation of the energy industry, power companies are today subject to extreme cost pressures and would thus understandably resist any such rearrangement of their production times. However, in this context a clear distinction needs to be made between economic and technical considerations. Power company reluctance to solve technical problems simply to protect their bottom line is totally unacceptable. But jeopardizing the survival of environmentally-friendly hydro power plants and storage hydroelectric power stations solely on economic grounds would not of course be acceptable either, particularly since such peak load power stations would be essential at night and during low-wind periods in a grid that obtains considerable amounts of its power from wind and solar energy. If our leaders are able to muster the political courage needed to institute extremely large-scale use of PV, solar thermal and wind power, surely ways can and will be found to compensate power companies for the consequent financial losses. For in any case, resource scarcity is ultimately going to force us to transition to renewables whether we like it or not, because all non-renewable energy resources are finite. If this transition is postponed for too long, we will face severe problems when push finally comes to shove and such a transition is forced on us by circumstances that are likely to be more than just economic in nature.

That said, energy sector deregulation and the severing, prescribed by law, of the production and distribution domains are also advantageous for the PV sector in that power companies can no longer oppose the concept of PV installations providing electricity to the grid or refuse to allow such provisioning. Many European countries have already adopted or plan to adopt laws subsidizing PV installation and wind turbine development by increasing feed-in tariffs, as was recently done in Germany by passage of that country's Renewable Energy Act (EEG).

On 1 January 2009 in Switzerland, such a renewable energy subsidy framework (albeit a relatively modest one from the PV installation standpoint) came into effect that is financed by up to 0.6 Rp per kWh (0.4 eurocents per kWh) for energy that is fed into the high-voltage grid by small hydro power stations, wind turbines, biomass power stations, geothermal energy facilities and PV installations.

However, in contrast to Germany the funds available for these subsidies are severely limited: so long as PV system costs remain relatively high, the slice of the energy subsidy pie earmarked for renewables will be extremely small. For more on how this works for PV installations, see Section 1.4.7. For more on compensatory feed-in tariffs for renewables, see www.swissgrid.ch, which contains the currently valid regulations and remuneration in accordance with the current market situation, all of which are subject to change.

5.2.7.7 The Way Forward

In Sections 5.2.7.1–5.2.7.6 it was shown that there are absolutely no technical obstacles to around 15 to 20% of European electricity demand being covered by grid-connected systems in the existing grid, and that such an increase would engender no insurmountable problems. However, a reform of this scope would have a major impact on grid operation and on the economic fortunes of existing power plants. The institution of specific technical and rate-related measures in conjunction with pan-European energy interchange would probably allow PV installations to meet 25–30% of European energy demand – although this would necessitate expansion of the European high-voltage transmission grid. To exceed this level, additional accumulators would need to be built and the grid would have to be expanded still further. This could be done by the build-out of existing storage hydroelectric power stations, building new pump storage systems, or through the use of new storage technologies that have yet to be fully developed. Increasing the number of PV installations in areas that are sunny in winter, such as the Alps or other mountain regions, would help to redress the current imbalance between summer and winter PV installation energy production in Central Europe. Wind turbines could also make an analogous contribution to the European energy supply. All such scenarios would, however, involve additional capital costs for whose financing viable economic models are needed – although to some extent the rising cost of conventional energy may render the renewables financing issue moot. It is now presupposed in most quarters that, by 2020, the cost of PV energy in most of Europe will be on a par with end-customer electricity prices (see Section 9.1.2).

5.3 Bibliography

[5.1] D. Berndt (ed.): *Bleiakkumulatoren (Varta)*, 11th edition.VDI-Verlag, Düsseldorf, 1986, ISBN 3-18-400534-8.

[5.2] D.U. Sauer, H. Schmidt: 'Speicher für elektrische Energie in Photovoltaik-Anlagen', in [Rot97].

[5.3] EXIDE Technologies: 'Handbuch für verschlossene Gel-Blei-Batterien – Teil 1: Grundlagen, Konstruktion, Merkmale', Ausgabe 3, March 2005.

[5.4] EXIDE Technologies: 'Handbuch für verschlossene Gel-Blei-Batterien – Teil 2: Montage, Inbetriebsetzung und Betrieb', Ausgabe 9, March 2005.

[5.5] EXIDE, Technologies: 'Handbuch für geschlossene Classic-Batterien – Teil 2: Montage, Inbetriebsetzung und Betrieb', Ausgabe 1, March 2005.

[5.6] J. Jost, S. Wagner: 'Aufbau eines Messplatzes für Inselbetriebs-Wechselrichter'. Ingenieurschule Burgdorf (ISB) Diploma Thesis, 1997 (in-house report).

[5.7] EN 61000-2-2 (2002): *Elektromagnetische Verträglichkeit (EMV). Teil 2: Umweltbedingungen. Abschnitt 2: Verträglichkeitspegel für niederfrequente leitungsgeführte Störgrössen und Signalübertragung in öffentlichen Niederspannungsnetzen.*

[5.8] K. Preiser, G. Bopp: 'Die Wahl der geeigneten Netzform in photovoltaischen Inselsystemen', in [Rot97].

[5.9]   A. Neuheimer, P. Adelmann, G. Rimpler et al.:'Hot-Spots durch Shuntregler – Schäden in der Praxis dokumentiert'. 20. Symposium PV Solarenergie, Staffelstein, 2005.

[5.10] H. Häberlin, M. Kämpfer.'Optimale DC-Betriebsspannung bei Netzverbundanlagen'. 20. Symposium PV Solarenergie, Staffelstein, 2005 (see www.pvtest.ch).

[5.11] F. Zach: *Leistungselektronik – Bauelemente, Leistungskreise, Steuerungskreise, Beeinflussungen*. Springer-Verlag, Vienna, 1988, ISBN 3-211-82028-0.

[5.12] H. Schmidt, B. Burger, Ch. Siedle:'Gefährdungspotenzial transformatorloser Wechselrichter–Fakten und Gerüchte'. 18. Symposium PV Solarenergie, Staffelstein, 2003.

[5.13] Schweizerischer Elektrotechnischer, Verein: 'Provisorische Sicherheitsvorschriften für Wechselrichter für photovoltaische Stromerzeugungsanlagen – 3.3kVA einphasig/10kVA dreiphasig'. Ausgabe 1, March 1993 (rescinded).

[5.14] EN, 61000-3-2 (2000) *Elektromagnetische Verträglichkeit (EMV), Teil 3-2: Grenzwerte für Oberschwingungsströme (Geräte-Eingangsstrom ≤ 16A je Leiter)*.

[5.15] EN, 61000-3-3 + A1 (2000): *Elektromagnetische Verträglichkeit (EMV), Teil 3-3: Grenzwerte – Begrenzungen von Spannungsänderungen, Spannungsschwankungen und Flicker in öffentlichen Niederspannungsnetzen für Geräte mit einem Bemessungsstrom ≤ 16A je Leiter*.

[5.16] EN, 61000-3-11 (2001): *Elektromagnetische Verträglichkeit (EMV), Teil 3-11: Grenzwerte – Begrenzungen von Spannungsänderungen, Spannungsschwankungen und Flicker in öffentlichen Niederspannungsnetzen für Geräte mit einem Bemessungsstrom ≤ 75A, die einer Sonderanschlussbewilligung unterliegen*.

[5.17] EN, 61000-3-12 (2004): *Elektromagnetische Verträglichkeit (EMV), Teil 3-12: Grenzwerte für Oberschwingungsströme in öffentlichen Niederspannungsnetzen für Geräte mit einem Bemessungsstrom ≤ 75A, die einer Sonderanschlussbewilligung unterliegen*.

[5.18] EN, 50160 (1999): *Merkmale der Spannung in öffentlichen Stromversorgungsnetzen*.

[5.19] 'Technische Regeln zur Beurteilung von, Netzrückwirkungen'., Ausgabe 2, 2007. Obtainable from VSE, CH-5001 Aarau, VDN, D-10115 Berlin, or from VEÖ, A-1040 Vienna.

[5.20] IEC, 61727 (2004): *Photovoltaic (PV) Systems – Characteristics of the utility interface*.

[5.21] EN, 55014, Aenderung, 1/Oktober 1988: *Grenzwerte und Messverfahren für Funkstörungen von Elektro-Haushaltgeräten u. ä.*

[5.22] EN, 50081-1 (February 1991): *Electromagnetic compatibility – Generic emission standard. Generic standard class: Domestic, commercial and light industry* (rescinded).

[5.23] EN, 61000-6-3 + A11 (2004): Elektromagnetische Verträglichkeit (EMV). Teil 6-3: Fachgrundnormen – Fachgrundnorm Störaussendung – Wohnbereich, Geschäfts- und Gewerbebereiche sowie Kleinbetriebe.

[5.24] H. Häberlin:'New DC-LISN for EMC-Measurements on the DC side of PV Systems: Realisation and first Measurements at Inverters'. Proceedings of the 17th EU PV Conference, Munich, 2001.

[5.25] N. Henze, G. Bopp, T. Degner, H. Häberlin, S. Schattner:'Radio Interference on the DC Side of PV Systems: Research Results and Limits of Emission'. Proceedings of the 17th EU PV Conference, Munich, 2001.

[5.26] DIN, VDE, 0126 (1999-04): *Selbsttätige Freischaltstelle für Photovoltaikanlagen mit einer Nennleistung ≤ 4,6 kVA und einphasiger Paralleleinspeisung über Wechselrichter in das Netz der öffentlichen Versorgung*.

[5.27] DIN, V, VDE, V, 0126-1-1 (2006-02): *Selbsttätige Freischaltstelle zwischen einer netzparallelen Eigenerzeugungsanlage und dem öffentlichen Niederspannungsnetz*.

[5.28] B. Verhoeven:'Probability of Islanding in Utility Networks due to Grid-Connected PV Power Systems'. Report *IEA PVPS T5-07: 2002*, September 2002.

[5.29] N. Cullen, J. Thornycroft, A. Collinson:'Risk Analysis of Islanding of Photovoltaic Power Systems within Low Voltage Distribution Networks'. Report *IEA PVPS T5-08: 2002*, March 2002.

[5.30] J.D. Graf, H. Häberlin:'Qualitätssicherung von Photovoltaikanlagen'. Schlussbericht des BFE-Projektes DIS 2744/61703, ENET Nr. 200023, July 2000.

[5.31] W. Bower, D. Ropp:'Evaluation of Islanding Detection Methods for Photovoltaic Utility-Interactive Power Systems'. Report *IEA PVPS T5-09: 2002*, March 2002.

[5.32] U. Lappe:'Selbsttätige Freischaltstelle für Eigenerzeugungsanlagen einer Nennleistung ≤ 4,6kVA bzw. bei Photovoltaikanlagen ≤ 5kWp mit einphasiger Paralleleinspeisung in das Netz der öffentlichen Versorgung'. 10. Symposium Photovoltaische Solarenergie, Staffelstein, March 1995.

[5.33] H. Häberlin, J. Graf:'Islanding of Grid-connected PV Inverters: Test Circuits and some Test Results'. Proceedings of the 2nd World Conference on Photovoltaic Energy Conversion, Vienna, 1998.

[5.34] IEC, 62116: 'Test procedure of islanding prevention measures for utility-interconnected photovoltaic inverters'. September 2008.

[5.35] R. Hotopp: 'Private Photovoltaik-Stromerzeugungsanlagen im Netzparallelbetrieb', RWE Energie AG, Essen, October 1990.

[5.36] S. Verhoeven: 'New Dutch Guidelines for Dispersed Power Generators'. Proceedings of a Workshop on Grid Interconnection of PV Systems, IEA PVPS Task V, Zurich, September 1997.

[5.37] H. Häberlin, H.P. Nyffeler, D. Renevey: 'Photovoltaik-Wechselrichter für Netzverbundanlagen im Vergleichstest'. *Bulletin SEV/VSE,* 10/1990.

[5.38] H. Häberlin: 'Photovoltaik-Wechselrichter für Netzverbundanlagen – Normen, Vorschriften, Testergebnisse, Probleme, Lösungsmöglichkeiten'. *Elektroniker,* 6/1992, 7/1992.

[5.39] H. Häberlin, H.R. Röthlisberger: 'Neue Photovoltaik-Wechselrichter im Test'. *Bulletin SEV/VSE,* 10/1993.

[5.40] H. Häberlin, H.R. Röthlisberger: 'Vergleichsmessungen an Photovoltaik-Wechselrichtern'. Schlussbericht BEW-Projekt EF-REN(89)045, 1993.

[5.41] H. Häberlin: 'Vergleichsmessungen an Photovoltaik-Wechselrichtern'. 9. Symposium Photovoltaische Sonnenenergie, Staffelstein, March 1994.

[5.42] H. Häberlin, F. Käser, Ch. Liebi, Ch. Beutler: 'Resultate von neuen Leistungs- und Zuverlässigkeitstests an Wechselrichtern für Netzverbundanlagen'. 11. Symposium Photovoltaische Sonnenenergie, Staffelstein, March 1996.

[5.43] H. Häberlin, F. Käser, Ch. Liebi, Ch. Beutler: 'Resultate von neuen Leistungs- und Zuverlässigkeitstests an Photovoltaik-Wechselrichter für Netzverbundanlagen'. *Bulletin SEV/VSE,* 10/1996.

[5.44] C. Liebi, H. Häberlin, Ch. Beutler: 'Aufbau einer Testanlage für PV-Wechselrichter bis 60kW'. Schlussbericht BEW-Projekt DIS 2744, ENET Nr. 9400561, January 1997.

[5.45] Ch. Renken, H. Häberlin: 'Langzeitverhalten von netzgekoppelten Photovoltaikanlagen 2 (LZPV2)'. Schlussbericht BFE-Projekt DIS 39949/79765, September 2003.

[5.46] H. Häberlin: 'Fotovoltaik-Wechselrichter werden immer besser – Entwicklung der Fotovoltaik-Wechselrichter für Netzverbundanlagen 1989–2000'. *Elektrotechnik,* 12/2000.

[5.47] H. Häberlin: 'Entwicklung der Photovoltaik-Wechselrichter für Netzverbundanlagen 1989–2000'. 16. Symposium Photovoltaische Solarenergie, Staffelstein, 2001.

[5.48] H. Häberlin: 'Resultate von Tests an neueren Photovoltaik-Wechselrichtern für Netzverbundanlagen'. *Bulletin SEV/VSE,* 10/2001.

[5.49] H. Häberlin: 'Wirkungsgrade von Photovoltaik-Wechselrichtern – Bessere Charakterisierung von Netzverbund-Wechselrichtern mit den neuen Grössen "Totaler Wirkungsgrad" und "Dynamischer MPPT-Wirkungsgrad"'. *Elektrotechnik,* 2/2005.

[5.50] H. Häberlin, L. Borgna, M. Kämpfer, U. Zwahlen: 'Totaler Wirkungsgrad – ein neuer Begriff zur besseren Charakterisierung von Netzverbund-Wechselrichtern'. 20. Symposium Photovoltaische Solarenergie, Staffelstein, 2005.

[5.51] H. Häberlin, M. Kämpfer, U. Zwahlen: 'Neue Tests an Photovoltaik-Wechselrichtern: Gesamtübersicht über Testergebnisse und gemessene totale Wirkungsgrade'. 21. Symposium Photovoltaische Solarenergie, Staffelstein, 2006.

[5.52] H. Häberlin, L. Borgna, M. Kämpfer, U. Zwahlen: 'Measurement of Dynamic MPP-Tracking Efficiency at Grid-Connected PV Inverters'. 21st EU PV Conference, Dresden, September 2006.

[5.53] W. Vassen: 'Technische Begleitung des 1000-Dächer Programms – Messungen an netzgekoppelten Wechselrichtern'. 8. Symposium Photovoltaik, Staffelstein, March 1993.

[5.54] W. Knaupp: 'Wechselrichter-Technik, Kenngrössen und Trends'. 8. Symposium Photovoltaische Sonnenenergie, Staffelstein, March 1993.

[5.55] Ch. Bendel, G. Keller, G. Klein: 'Ergebnisse von Messungen an Photovoltaik-Wechselrichtern'. 8. Symposium Photovoltaik, Staffelstein, March 1993.

[5.56] F. Hummel, H. Müh, R. Wenisch: 'Sieben Wechselrichter im Test'. *Sonnenenergie & Wärmetechnik,* 1/95.

[5.57] M. Jantsch, M. Real, H. Haeberlin et al.: 'Measurement of PV Maximum Power Point Tracking Performance'. 14. EU PV Conference, Barcelona, 1997.

[5.58] H. Haeberlin: 'Evolution of Inverters for Grid connected PV-Systems from 1989 to 2000'. 17th EU PV Conference, Munich, 2001.

[5.59] F. Baumgartner, A. Breu, S. Roth, H. Scholz et al.: 'MPP Voltage Monitoring to Optimise Grid-Connected PV Systems'. 19th EU PV Conference, Paris, 2004.

[5.60] H. Haeberlin, L. Borgna: 'A New Approach for Semi-Automated Measurement of PV Inverters, Especially MPP Tracking Efficiency, Using a Linear PV Array Simulator with High Stability'. 19th EU PV Conference, Paris, 2004.

[5.61] B. Burger: 'Auslegung und Dimensionierung von Wechselrichtern für netzgekoppelte PV-Anlagen'. 20. Symposium Photovoltaik, Staffelstein, 2005.

[5.62] A. Woyte, S. Islam, R. Belmans, J. Nijs *et al.*: 'Unterdimensionieren des Wechselrichters bei der Netzkopplung – Wo liegt das Optimum?'. 18. Symposium Photovoltaik, Staffelstein, 2003.

[5.63] K. Kiefer: 'Erste Auswertungen aus dem Intensiv-Mess- und Auswerteprogramm (I-MAP) des 1000-Dächer-PV-Programms'. 9. Symposium Photovoltaik, Staffelstein, 1994.

[5.64] SEV-Norm, SEV3600-2., 1987: *Begrenzung von Beeinflussungen in Stromversorgungsnetzen (Oberschwingungen und Spannungsänderungen), Teil 2: Erläuterungen und Berechnungen* (rescinded).

[5.65] H. Schmidt, B. Burger, K. Kiefer: 'Welcher Wechselrichter für welche Modultechnologie?'. 21. Symposium Photovoltaik, Staffelstein, 2006.

[5.66] 'Schweizerische Elektrizitätsstatistik, 2004'., Available from www.bfe.admin.ch; also published in *Bulletin SEV/VSE*, 12/2005.

[5.67] S. Gölz, G. Bopp, B. Buchholz, R. Pickham: 'Waschen mit der Sonne – Direkter Verbrauch von lokal erzeugtem PV Strom durch gezielte Lastverschiebung in Privathaushalten'. 21. Symposium Photovoltaik, Staffelstein, 2006.

[5.68] Ch. Bendel, M. Braun, D. Nestle, J. Schmid, P. Strauss: 'Energiemanagement in der Niederspannungsversorgung mit dem Bidirektionalen Energiemanagement Interface (BEMI) – Technische und wirtschaftliche Entwicklungslösungen'. 21. Symposium Photovoltaik, Staffelstein, 2006.

[5.69] Elektrowatt Ingenieurunternehmung, AG: 'Alternative Energie Wasserstoff'. EGES-Schriftenreihe Nr. 5, EDMZ Bern, July 1987.

[5.70] H. Häberlin: 'Neue Tests an Netzverbund-Wechselrichtern unter spezieller Berücksichtigung des dynamischen Maximum-Power-Point-Trackings'. *Elektrotechnik*, 7-8/2006.

[5.71] BDEW: 'Technische Richtlinie Erzeugungsanlagen am Mittelspannungsnetz'. Bundesverband der Energie- und Wasserwirtschaft e.V., Berlin, June 2008.

[5.72] Forum Netztechnik/Netzbetrieb beim, VDE: 'Erzeugungsanlagen am Niederspannungsnetz'. Draft, June 2008.

[5.73] V. Weeber, A. Bleil: 'Messungen an Wechselrichtern'. 7. Symposium Photovoltaische Solarenergie, Staffelstein, March 1992.

[5.74] H. Häberlin: 'Bau eines Solargenerator-Simulators von 100 kW'. 23. Symposium Photovoltaische Solarenergie, Staffelstein, 2008.

[5.75] C. Bendel, P. Funtan, T. Glotzbach, J. Kirchhof, G. Klein: 'Ergebnisse aus dem Projekt OPTINOS – Defizite und Unsicherheiten bei Prüfprozeduren von Photovoltaik-Stromrichtern'. 23. Symposium Photovoltaische Solarenergie, Staffelstein, March 2008.

[5.76] B. Burger: '98,5% Wechselrichterwirkungsgrad mit SiC MOSFETs'. 23. Symposium Photovoltaische Solarenergie, Staffelstein, March 2008.

[5.77] H. Häberlin, L. Borgna, D. Gfeller, Ph. Schärf, U. Zwahlen: 'Development of a Fully Automated PV Array Simulator of 100 kW'. 23rd EU PV Conference, Valencia, September 2008.

[5.78] B. Bletterie, R. Bründlinger, H. Häberlin *et al.*: 'Redefinition of the European Efficiency – Finding the Compromise Between Simplicity and Accuracy'. 23rd EU PV Conference, Valencia, September 2008.

[5.79] H. Häberlin, Ph. Schärf: 'Verfahren zur Messung des dynamischen Maximum-Power-Point-Trackings bei Netzverbundwechselrichtern'. 24. Symposium Photovoltaische Solarenergie, Staffelstein, March 2009.

[5.80] H. Häberlin, L. Borgna, D. Gfeller *et al.*: 'Resultate von ersten Tests von 100kW-Wechselrichtern mit dem neuen Solargenerator-Simulator von 100 kW'. 24. Symposium Photovoltaische Solarenergie, Staffelstein, March 2009.

[5.81] B. Burger, H. Schmidt, B. Bletterie *et al.*: 'Der Europäische Jahreswirkungsgrad und seine Fehler'. 24. Symposium Photovoltaische Solarenergie, Staffelstein, March 2009.

[5.82] H. Schmidt, B. Burger, U. Bussemas, S. Elies: 'Wie schnell muss ein MPP-Tracker wirklich sein?'. 24. Symposium Photovoltaische Solarenergie, Staffelstein, March 2009.

[5.83] CENELEC: 'Final Draft for European Standard prEN 50530:2008'.

[5.84] UCTE: *Statistical Yearbok 2007*. UCTE, Brussels (www.ucte.org).

[5.85] 'Schweizerische Elektrizitätsstatistik, 2007'. Available from www.bfe.admin.ch; also published in *Bulletin SEV/VSE*, 7/2008.

[5.86] D., Chartouni:, 'Battery Energy Storage Systems for Electric Utilities'. Paper presented at Powertagen, Zurich, 2008.

[5.87] D.U., Sauer:, 'Stromspeicher in Netzen mit hohem Anteil erneuerbarer Energien'. 23. Symposium Photovoltaische Solarenergie, Staffelstein, March 2008.

[5.88] R., Horbaty:, 'Netz mit Autobatterien regulieren'. *Bulletin SEV/VSE*, 3/2009.

[5.89] A., Vezzini:, 'Lithiumionen-Batterien als Speicher für Elektrofahrzeuge (2 Teile)'. *Bulletin SEV/VSE*, 3/2009, 6/2009.

[5.90] C.J., Rydh:, 'Environmental assessment of vanadium redox and lead-acid batteries for stationary energy storage'. *Journal of Power Sources*, 1999, 80, 21–29.

[5.91] D. Chartouni, T. Bühler, G. Linhofer:'Wertvolle Energiespeicherung'. *Elektrotechnik,* 1/09.

[5.92] A. Oudalov, T. Buehler, D. Chartouni:'Utility Scale Applications of Energy Storage'. IEEE Energy Conference, Atlanta, GA, November 2008.

Additional References

[Häb07], [Köt94], [Luq03], [Mar03], [Qua03], [Rot97], [Sch00], [Wag06].
More recent publications of mine are available for download at www.pvtest.ch.

SHIFTING THE LIMITS

**RELIABLY HIGH YIELDS ARE
ONLY POSSIBLE USING
TOP-OF-THE-RANGE-TECHNOLOGY.
WE AIM HIGHER.**

6

Protecting PV Installations Against Lightning

PV installations need to occupy a fairly large amount of space on account of the relatively low power density of solar radiation. Lightning strikes about one to four times per year and square kilometre in Germany and about three to six times per year and square kilometre in Switzerland (see Section 6.1) [Has89], [Deh05]. The Austrian rate is on a par with that of Switzerland. These rates are likely to increase in the coming years, owing to global warming, and even today can range even higher at extremely exposed sites such as those in the Alps. In subtropical and tropical regions, lightning strikes from 30 to 70 times per year and square kilometre [Pan86]. Many solar module mounting systems are good lightning conductors by virtue of their being constructed of metal, although laminates with non-metal frames are just as prone to lightning strikes since their wiring contains metal. A lightning strike on a solar generator can damage not only solar modules, but also the PV installation's charge controllers, inverters and other elements; of course, lightning can also cause personal injury. *It should be noted, however, that the presence of a PV installation on a building does not increase the likelihood of the building being struck by lightning.*

There are no magic bullets to protect against direct lightning strikes, no matter how costly a lightning protection system may be. However, affordable lightning protection can be realized for PV installations that rules out the possibility of fire or personal injury and that can limit the scope of any lightning damage that a PV installation may sustain.

PV systems are at risk not only from direct lightning strikes on the installation itself or the building on which it is installed, but also from the indirect effects of nearby lightning strikes – for example, on an adjacent building. Moreover, grid-connected systems can be damaged by lightning-induced voltage surges from the grid. Such indirect effects of lightning are far more common than direct lightning strikes, but the cost of averting damage from nearby strikes is also far lower than for direct strikes. Hence all PV installations should at a minimum exhibit sufficient protection against nearby lightning strikes.

For PV installation owners who wish to apply more than just generic solutions and would instead like to delve deeper into the issues that come into play here, PV installation lightning protection needs to be seen in its larger context. To this end, the following is an introduction to the general principles of lightning protection.

6.1 Probability of Direct Lightning Strikes

The probability of a particular construction being struck by lightning can be estimated based on the construction's dimensions and the structures that are in its immediate vicinity [Deh05], [6.2,6.21]. All constructions attract lightning not only to their own surfaces, but also a portion of the lightning that strikes the immediate vicinity of the construction, whereby the likelihood of the latter event is positively

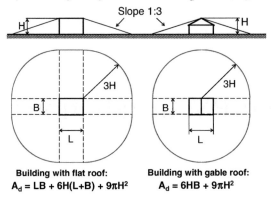

Equivalent lightning collection area A_d of buildings

Building with flat roof:
$A_d = LB + 6H(L+B) + 9\pi H^2$

Building with gable roof:
$A_d = 6HB + 9\pi H^2$

Figure 6.1 Equivalent intercepting area A_d of a square/rectangular building with a flat or pitched roof

correlated with construction height. To determine the contour of a construction's equivalent intercepting surface A_d, find the construction's sight line from an 18.4° angle, which equates to a 1:3 slope (see Figure 6.1). A_d can be determined for simple rectangular or square buildings as follows.

Equivalent intercepting surface for a square/rectangular building with a flat roof:

Intercepting surface of a building with a flat roof: $A_d = L \cdot B + 6 \cdot H \cdot (L+B) + 9 \cdot \pi \cdot H^2$

(6.1)

Equivalent intercepting surface for a square/rectangular building with a pitched roof with an incline of more than 18.4°:

Intercepting surface for a building with a pitched roof: $A_d = 6 \cdot H \cdot B + 9 \cdot \pi \cdot H^2$ (6.2)

Equivalent intercepting surface of a mast of height H:

Intercepting surface of a mast: $A_d = 9 \cdot \pi \cdot H^2$ (6.3)

where:
L = building length (perpendicular to a gable on a pitched roof with an incline of more than 18.4°)
B = building width
H = building or mast height

These values can then be used to determine the mean annual number of direct lightning strikes N_D on a construction as follows [6.2,6.21]:

Mean annual number of direct lightning strikes: $N_D = N_g \cdot A_d \cdot C_d \cdot 10^{-6}$ (6.4)

where:
N_g = mean number of lightning strikes per year and square kilometre at a particular site: for Germany, around 1–4 (lower rate in the north, higher rate in the south and in hilly areas); for Switzerland and

Table 6.1 Ambient coefficients C_d [6.21]

Construction location relative to its surroundings (particularly at distance $3H$)	C_d
Constructions surrounded by buildings or trees of the same or greater height	0.25
Constructions surrounded by other buildings or trees of the same or lesser height	0.5
Free-standing construction with no nearby element	1
Free-standing construction located on a mountaintop or cliff	2

Austria, around 3–4 in flat areas and just north of the Alps, and 5–6 in the Alps and just south of the Alps

A_d = the equivalent intercepting surface in square metres for the construction in question, as calculated using Equations 6.1, 6.2 and 6.3.

C_d = the ambient coefficient as in Table 6.1, which takes account of ambient features.

The values from Equation 6.4 are too low for buildings that are more than about 100 m high. Radio or TV transmission towers exceeding 150 m in height may be struck by lightning up to 30 or 40 times a year [Has89].

6.1.1 Specimen Calculation for the Annual Number of Direct Lightning Strikes N_D

The values obtained for N_D vary greatly according to location and local characteristics.

The following are specimen calculations of N_D values for two different buildings A and B at four locations (see Table 6.2).

Building A: Single-family dwelling with pitched roof; $L = 10$ m, $W = 8$ m, $H = 8$ m, $A_d = 2194$ m^2

Building B: Factory building with flat roof; $L = 30$ m, $W = 20$ m, $H = 20$ m, $A_d = 17\,910$ m^2

Locations for which calculations were realized:

1. **Northern Germany, $N_g = 1.5$ lightning strikes/km^2/a**; numerous nearby buildings of the same height as the building in question
2. **Central Germany, $N_g = 3$ lightning strikes/km^2/a**; all nearby buildings lower than the building in question
3. **Switzerland (just north of the Alps), $N_g = 4$ lightning strikes/km^2/a**; free-standing building
4. **Switzerland (just south of the Alps), $N_g = 6$ lightning strikes/km^2/a**; free-standing building on a cliff

Table 6.2 Mean annual number of direct lightning strikes at two different buildings at four different locations

Location	N_g Ground strokes per km^2 and year	C_d	Building A (family home)		Building B (factory)	
			N_D per year	Years between two strokes	N_D per year	Years between two strokes
1	1.5	0.25	0.00082	1215	0.0067	149
2	3	0.5	0.00329	304	0.0269	37.2
3	4	1	0.00877	114	0.0716	14
4	6	2	0.0263	38	0.215	4.7

For low N_D values, for reasons of cost it may be advisable to forgo protection against the effects of direct lightning strikes and simply insure the building adequately. This would apply, for example, to single-family dwellings in flat residential areas, particularly if they are densely populated. On the other hand, protection against the effects of direct lightning strikes is advisable for relatively large and tall buildings in flat areas, particularly those with overhead power lines or cable railways.

6.2   Lightning Strikes: Guide Values; Main Effects

Lightning current is a transient but extremely strong surge current. Figure 6.2 illustrates the main lightning current waveform, which very rapidly increases to a very high peak value i_{max}. The ensuing drop to 0 takes far longer, however. There are three different types of lightning, which are discussed in the next section.

6.2.1   Types of Lightning

These are as follows:

- **Type 1 lightning (positive or negative charge):** Type 1 lightning exhibits the highest peak current values and generates the largest amount of heat in electrical wires. The maximum value i_{max} can range from 100 to 200 kA in extreme cases, although the average value is around 30 kA. The maximum rate of current increase di/dt_{max} is usually lower than with type 2 lightning. The wavefront time T_1 is typically around 10 μs, while the wave-tail half-value time T_2 is around 350 μs. Of all the transient lightning currents, type 1 lightning transports the largest loads Q_S.
- **Type 2 lightning (negatively charged):** As this type of lightning uses a lightning channel that has been ionized by a previous direct lightning strike, it can reach its peak value far more quickly and thus attains the highest values for the rate of current increase di/dt_{max} (maximum value between 100 and 200 kA/μs, which for an average type 2 lightning strike is around 25 kA/μs). Hence type 2 lightning generates the highest induced voltages in conductor loops. The maximum values i_{max} of type 2 lightning are considerably lower than for type 1 lightning and in extreme cases can go as low as 25–50 kA. The wavefront time T_1 is typically around 0.25 μs, while the wave-tail half-value time T_2 is around 100 μs.
- **Type 3 lightning (positively or negatively charged):** This type of lightning exhibits relatively low current ranging from a maximum of 200 to 400 A, but is of relatively long duration (several hundred milliseconds, typically 500 ms). Of all of the types of lightning, type 3 lightning can transport the largest loads Q_L, but is otherwise relatively harmless and can occur either immediately before or after a type 1 lightning or type 2 lightning event.

A lightning bolt often comprises up to 10 type 1 lightning events, which occur within no more than 1 s and use the same lightning channel and impact point.

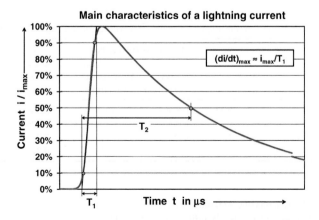

Figure 6.2   Main lightning current waveform: T_1 = wavefront; T_2 = wave-tail half-value time

6.2.2 Effects of Lightning

A lightning strike on a building or an electrical installation can provoke a number of hazardous events, the best known undoubtedly being a fire when lightning strikes a flammable structure such as a farmhouse that has no lightning rod. Other important effects of lightning include the following:

- **A sharp rise in potential in the affected structure, relative to the surroundings:** The key guide value in this regard is maximum current i_{max}.
- **Elevated voltages induced in adjacent conductor loops:** The key guide value in this regard is maximum rate of current change: $(di/dt)_{max} \approx (i/t)_{max} = i_{max}/T_1$.
- **Calefaction in/force exerted on wires that conduct lightning current:** The key guide value in this regard is specific energy $W/R = \int i^2 \, dt$.
- **Fusing at the lightning strike point; explosive effect of lightning strikes:** The key guide value in this regard is $Q = \int i \, dt$, of which there are two types, namely surge current Q_S and long-term current load Q_L

As with all electrical installations, in PV installations the most important effects of lightning strikes are increased potential and voltage, which are discussed in separate sections later in this chapter.

6.2.3 Lightning Protection Installation Classes and Efficiency

In order for a lightning protection installation to be effective, it should ensure that the lion's share of current from a direct lightning strike is deflected in such a way that the structure affected is not damaged. Lightning protection installation efficiency means the ratio of (a) the number of lightning strikes on a protected structure that cause no damage to (b) the total number of lightning strikes sustained by the structure. The higher the required efficiency, the higher the relevant guide values, whereby the four protection or requirements classes listed in Table 6.3 come into play.

6.2.4 Use of Approximate Solutions for Lightning Protection Sizing

In as much as the applicable lightning guide values vary greatly, the exact waveform exhibited by lightning-induced voltages and currents is unpredictable. Thus lightning protection installations are sized using the limit values indicated in Table 6.3, according to the protection class selected.

Since exact figures concerning lightning-induced voltages and currents are usually unnecessary for purposes of lightning protection installation sizing, such sizing is based on an accuracy of $\pm 30\%$ or the like. Hence it is similarly unnecessary to perform lengthy mathematical calculations to arrive at the exact values for mutual inductances and inductances. Approximate equations can also be used if the error relative to the exact solution is not unduly large.

Table 6.3 The protection or requirements classes for lightning protection installations. Also indicated is the efficiency attained for the various lightning protection classes if a lightning sphere with radius r_B is sized in accordance with the prescribed limit values [Deh05], [6.20], [6.32]

Requirement	Efficiency (%)	i_{max} (kA)	$(di/dt)_{max}$ (kA/µs)	W/R (MJ/Ω)	Q_S (A s)	Q_L (A s)	r_B (m)
Low (class IV)	84	100	100	2.5	50	100	60
Normal (class III)	91	100	100	2.5	50	100	45
High (class II)	97	150	150	5.6	75	150	30
Extreme (class I)	99	200	200	10.0	100	200	20

6.3   Basic Principles of Lightning Protection

6.3.1   Internal and External Lightning Protection

The purpose of external lightning protection, which cannot avert a lightning strike, is to deflect the superheated lightning channel from flammable or otherwise vulnerable structures and to conduct the lightning current safely to ground via a conductor. To this end, lightning conductors or lightning rods that have a sufficiently large gauge (e.g. more than 25–35 mm² Cu) are mounted directly in front of the relevant structure. These conductors or rods then trap the lightning and conduct it to the ground installation via special conductors, and from there to the ground. A foundation can be used for grounding by connecting the grounding cable to the foundation reinforcement. Another method is to install a sufficiently corrosion-resistant circuit (e.g. made of 8 mm round copper wire) around the building about 50–100 cm underground. The grounding installation exhibits a specific grounding resistance to remote ground R_E that usually ranges from around 1 to 10 Ω. As efficient grounding is difficult to attain on mountain cliffs, grounding resistance may be somewhat higher at such sites. Further information concerning the realization of external lightning protection can be found in [Has89], [Deh05], [6.2] and [6.22].

Internal lightning protection comprises all measures aimed at averting damage that could be incurred by the internal elements of the relevant elements as the result of current flowing through the external lightning protection installation conductors secondary to a lightning strike.

6.3.2   Protection Zone Determination Using the Lightning Sphere Method

In a cloud-to-ground lightning strike (which is the most common type), a flash of lightning moves from the electrically charged clouds in steps. The point towards which a lightning flash moves in such a case is random at first and is not determined by any ground structure. When a lightning flash comes within around 30–100 m of the ground, the insulation capacitance of the air near the ground is overshot. As a result, an arresting discharge arising from the ground moves towards the lightning flash, provoking the return stroke. The so-called geometric–electric model presupposes that the return stroke always occurs across the shortest possible distance, i.e. the lightning strikes the location that is nearest to the leader stroke (see Figure 6.3). The return stroke radius (in the form of a lightning sphere around the leader stroke) that needs to taken into account depends on the protection class selected (see Table 6.3; for further details see [Has89], [Deh05] and [6.22]).

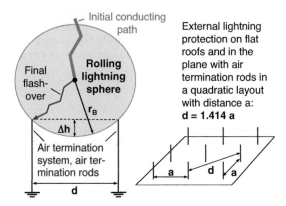

Figure 6.3   The lightning sphere method: The return stroke occurs at the structure that is nearest to the leader stroke. Lightning conductors can avert a lightning flash below the lightning sphere. The maximum penetration depth Δh between two lightning conductors is determined by d and the lightning sphere radius r_B ($r_B = 20$–60 m, depending on protection class; see Table 6.3)

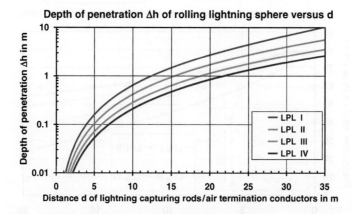

Figure 6.4 Penetration depth Δh of the lightning sphere into the surface that the lightning rod or lightning conductor is attached to, as a function of distance d, for the four lightning protection classes

The lightning sphere method for determining the size of the area that needs protection against lightning is particularly suitable for installations on flat roofs, on level sites and for complex and unusual situations. The method involves rolling (either putatively or via a model) a lightning sphere (with radius r_B and in accordance with the protection class selected) over the structure that has the envisaged lightning conductors, in such a way that the lightning sphere only touches the relevant structure at its juncture with the lightning conductor or lightning rod.

In the case of vertical lightning rods at distance d, the lightning penetrates a specific depth Δh into the surface that the lightning rod is attached to (see Figures 6.3 and 6.4), in accordance with the following equation:

$$\text{Lightning sphere penetration depth } \Delta h = r_B - \sqrt{r_B^2 - \left(\frac{d}{2}\right)^2} \qquad (6.5)$$

For lightning protection on level surfaces, the so-called mesh procedure is used, whereby a square mesh element composed of lightning conductors with a maximum distance a is presupposed (see Table 6.4; penetration depth Δh relatively shallow with a relatively small d).

6.3.3 Protection Zone for Lightning Conductors and Lightning Rods

The protection angle method is very suitable for buildings with pitched roofs and metal lightning rods. In such cases, it can be assumed that a vertical or horizontal lightning rod will protect a specific zone from lightning strikes. This zone can be characterized simply by the protection angle α relative to the vertical [Deh05], [6.22].

Protection angle α is determined by height h and the protection class selected from Table 6.3. Figure 6.5 displays the zone protected by a lightning rod along with the relevant protection angle, as a function of h and the protection class selected. In such cases, the lightning rod protects a spherical zone and the lightning conductor protects a conical zone that is defined by angle α relative to the vertical.

Table 6.4 Maximum mesh distance a for the mesh method

Lightning protection class	I	II	III	IV
Maximum mesh distance a (m)	5	10	15	20

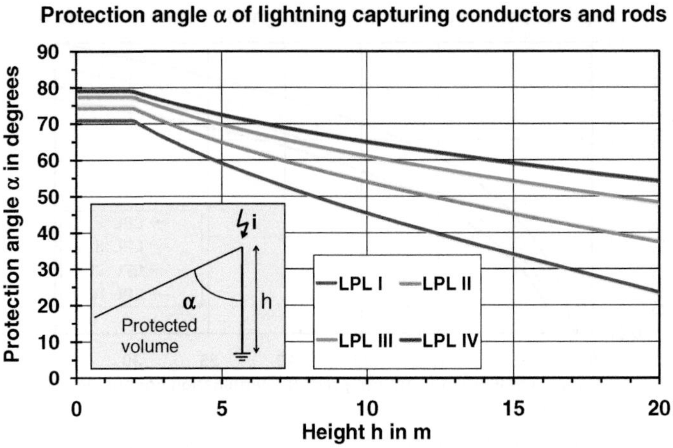

Protection angle α of lightning capturing conductors and rods

Figure 6.5 Protection angle for lightning rods and horizontal lightning conductors. It is assumed that no lightning will strike the zone defined by protection angle α, which is spherical for an individual lightning rod and conical for horizontal lightning conductors [Deh05], [6.22]

6.3.4   Lightning Protection Measures for Electricity Installations

These are as follows:

- Direct lightning strikes on electricity installations should be avoided at all costs, through the use of efficient outer lightning protection elements.
- Wires that conduct lightning current should be of a sufficiently large gauge ($\geq$ 16–25 mm^2 Cu), and the walls of the relevant containers and housings should be sufficiently thick.
- Wherever possible, lightning current should be broken down into a series of sub-currents that are shunted to ground via a series of parallel conductors, with a view to mitigating the effects of lightning, which are proportional to i or i^2.
- Equipotential bonding should be realized from outside of the installation for all conductors that feed current into the installation.
- Sufficient clearance should be allowed between lightning current conductors and adjacent conductor loops, which should extend over the smallest possible area.
- For conductors, shielded conductors should be grounded *bilaterally* and surge protection devices should be integrated on either side of each conductor.

6.4   Shunting Lightning Current to a Series of Down-conductors

One of the most effective ways to mitigate the effects of lightning current is to break it down instantaneously into smaller current flows that are then conducted to various down-conductors. This greatly reduces the untoward effects of lightning current (which are proportional to i^2, i or di/dt), although the consequent magnetic fields become stronger. Since, as a rule, only the portion i_A of the lightning current flowing through the nearest down-conductor determines the induced-voltage strength, it is important to ascertain the ratio of the current flowing through the shunt i_A to the overall lightning current for various configurations.

This the following is defined:

Proportion of lightning current in a down-conductor relative to overall current: $k_C = \dfrac{i_A}{i}$ (6.6)

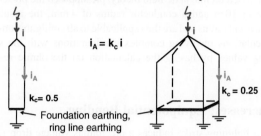

Figure 6.6   Lightning current sectioning for a symmetrical shunting configuration

Here, k_C is determined by the reciprocal geometry of the lightning conductors and down-conductors, as well as by the juncture point of the lightning strike.

In the case of a symmetrical lightning strike relative to the lightning conductors and down-conductors, the lightning current is distributed evenly to all down-conductors (see Figure 6.6), whereby the following applies:

Relative proportion of lightning current for a symmetrical scenario for n down-conductors:

$$k_C = \frac{1}{n} \qquad (6.7)$$

If the down-conductors leading away from the lightning juncture point are of unequal length, the lightning current will not be distributed equally to all such conductors. The strongest current will flow through the shortest connection to ground. The greater the distance between a down-conductor and the lightning juncture point, the weaker the current in the conductor. Figure 6.7 displays the k_C values for the shortest

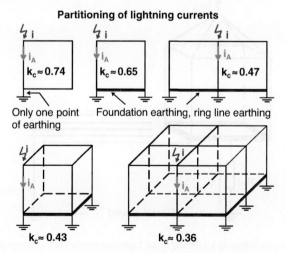

Figure 6.7   Specimen lightning current segmentation for asymmetrical lightning protection device configurations. The effective k_C value secondary to a lightning strike is determined not only by the type of mesh and the number of down-conductors used, but also by the lightning juncture point

distance to ground for selected configurations that are used in actual installations. Although these calculations, which are based on magnetic field theory, presupposed the presence of a square mesh with a conductor clearance of 10 m and a conductor radius of 4 mm, the results barely correlate at all with conductor clearance and radius and are thus applicable to all configurations. A labour-saving method in this regard for calculations involving complex configurations with the same conductor radius is to approximate the k_C values by basing the calculation on the ohmic resistances for each of the various elements.

6.5 Potential Increases; Equipotential Bonding

As Figure 6.8 shows, a lightning strike induces a precipitous voltage drop $v_{max} = V_{max}$ relative to the remote surroundings (remote ground) at the building or installation grounding resistance R_E, whereby the peak potential increase occurs at i_{max}:

$$\text{Peak potential increase } V_{max} = R_E \cdot i_{max} \qquad (6.8)$$

In as much as grounding resistance R_E usually ranges from around 1 to 10 Ω, extremely large potential increases ranging from around 300 kV to several megavolts can occur.

In view of the relatively large distance between metallic lines and the like (electrical wires, phone wires, water pipes and so on) leading into the building/installation and zero potential, these conductors should be regarded as being connected to remote ground. Hence voltage $v_{max} = V_{max}$ is also conducted between the metallic lines leading into the building/installation and the ground installation; fortunately, in most cases this only occurs in some of these conductors.

Hence, in the interest of minimizing the voltage differences induced by a lightning strike, equipotential bonding should be realized between the lightning protection installation and all metallic conductors

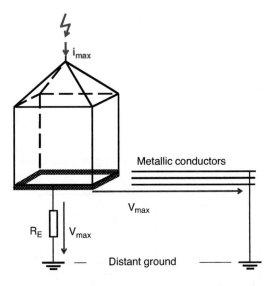

Figure 6.8 Potential increase following a lightning strike. Lightning current i flows through grounding resistance R_E to ground, which causes a voltage drop, as well as a maximum potential increase $V_{max} = R_E \cdot i_{max}$. Voltage V that equates to the potential increase flows between the structure affected and remote ground. As all conductors that lead to the structure affected are in effect connected to remote ground, in the absence of equipotential bonding an extremely high voltage flows between the structure affected and these conductors

leading into the building in question. In the event of a lightning strike, all conductors integrated into the equipotential bonding element will conduct a portion of the lightning current.

6.5.1 Equipotential Bonding Realization

Equipotential bonding can be realized for elements such as water pipes that are non-conducting during operation of the installation in question, by simply wiring in the grounding installation (e.g. a foundation grounding device) using wires with a sufficiently large gauge. Electrical conductors must be shielded in order to be directly connected to ground, whereas active conductors cannot of course be grounded for operational reasons. Active conductors should be incorporated into the equipotential bonding installation via suitable lightning protection devices that are rated for the lightning current that the wires would be conducting. In order to facilitate inspections (among other reasons), in many cases all conductors are connected to the equipotential bonding installations via an equipotential bonding bus bar, which is in turn connected to the grounding installation via a low-inductance ribbon-shaped conductor or the like which, as with all equipotential bonding conductors, is kept as short as possible in order to minimize inductive voltages. To this end, a series of equipotential bonding bus bars should be used for larger installations or buildings. Equipotential bonding installations are generally mounted in an attic or basement, although in tall buildings mounting an additional equipotential bonding device on the upper floors is sometimes advisable. Figure 6.9 shows the configuration of a typical equipotential bonding installation in a building.

6.5.2 Lightning Current in Conductors that are Incorporated into the Equipotential Bonding Installation

On completion of the equipotential bonding phase, lightning current is distributed at the grounding resistance R_E and across the conductors that are part of the equipotential bonding installation, while at the same time the potential increase induced by the lightning strike decreases accordingly. Described in [6.20] is a method for distributing lightning current across the various conductors and leads. It can be assumed that around half of the lightning current will be dissipated by the grounding resistance and the remaining

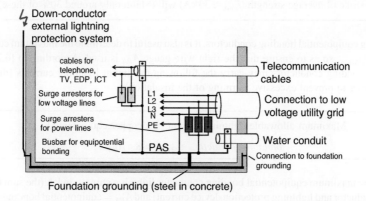

Foundation grounding (steel in concrete)

Figure 6.9 Equipotential bonding installation (usually mounted in an attic or cellar). All metallic conductors leading into the building are connected to the building's grounding installation either directly, if this can be done without interfering with building activities or if there is no risk of corrosion, or indirectly, via the relevant protection devices or spark gaps so as to avoid hazardous voltage differences in the building. To this end, in the event of a lightning strike, each of the lines entering the building, as well as the lightning protection devices used, conduct a portion of the lightning current

half by the building's n_L in equal shares. In a conductor comprising n_A leads, lightning current i_L is distributed equally to all leads. In the case of bilaterally shielded conductors, however, it can be assumed that most of their lightning current will flow to their shielding, providing that this current can be conducted there without causing damage. This scenario can be expressed as follows:

$$\text{Lightning current in one conductor: } i_L \approx 0.5 \frac{i}{n_L} \qquad (6.9)$$

$$\text{Lightning current in one lead: } i_{LA} = \frac{i_L}{n_A} \approx \frac{i_L}{n_A} \qquad (6.10)$$

where:

$i = $ lightning current

$n_L = $ number of conductors connected to the building

$n_A = $ number of conductor leads

Using the values in Table 6.3, the maximum anticipated lightning current i_{max} can be inserted into the two equations above according to the desired protection class, whereupon the maximum current i_{LAmax} for the various leads and conductors can be calculated and then specified accordingly.

Example

A galvanized iron water pipe, a four-lead high-voltage cable (230/400 V), a two-lead phone line and a cable TV line are to be connected to a house that has a lightning protection installation. If the protection class III limit values in Table 6.3 are applied ($i_{max} = 100$ kA), according to Equation 6.9 each of these lines is likely to exhibit a maximum lightning current of $i_{Lmax} = 12.5$ kA, while according to Equation 6.10 each high-voltage cable conductor will exhibit a maximum current of $i_{LAmax} = 3.1$ kA. A lightning strike of average strength ($i_{max} \approx 30$ kA) will exhibit only around 30% of these values.

When sizing equipotential bonding conductors, it is also useful to determine the maximum current they will exhibit, i_{PAmax}, so as to ensure that the right wire gauge A_{PA} is used. According to [6.2], copper equipotential bonding conductors may carry the following maximum lightning currents (this limit is imposed in order to prevent excessive warming of the insulation):

$$\text{Maximum allowable lightning current in a copper equipotential}$$
$$\textit{bonding conductor: } i_{PAmax} = 8 \cdot A_{PA} \qquad (6.11)$$

where $i_{PAmax} = $ maximum equipotential bonding conductor lightning current, in kA (the sum total of all connected conductor and lightning protection device current) and $A_{PA} = $ equipotential bonding conductor gauge in mm^2.

Hence, if a copper equipotential bonding conductor exhibits a specific current i_{PAmax} (in kA), the following minimum gauge A_{PAmin} (in mm^2) should be used:

$$\text{Minimum gauge for a copper equipotential bonding conductor: } A_{PAmin} = 0.125 \cdot i_{PAmax} \quad (6.12)$$

As the two equations above are so-called adjusted quantity equations, they only apply to numerical values if i_{PA} is given in k A and A_{PA} is given in mm².

The equations above can also be used to estimate whether a conductor can handle a specific lightning current.

In the interest of attaining satisfactory mechanical robustness, national lightning protection standards for equipotential bonding conductors often specify a 6 mm² minimum gauge for copper wires (5 mm² according to [6.22]). For bus bars that interconnect equipotential bonding conductors and the grounding installation, the minimum gauge for copper wires is 16 mm² (14 mm² according to [6.22]).

6.5.3 Lightning Protection Devices

Two main types of lightning protection devices are commercially available. Lightning current lightning protection devices (type 1) allow conductors that are not grounded for go-live operation to be incorporated into the equipotential bonding installation. These conductors: (a) provide coarse lightning protection; (b) are mainly composed of specialized spark gaps which, under nominal current conditions, typically reduce voltage that is 10–15 times higher than nominal operating voltage; (c) are rated for lightning current (up to a maximum i_{max} of 50–100 kA for power lines and 2.5–40 kA for telecommunication lines); and (d) are rated for wavefront times of $T_1 = 10\,\mu s$ and wave-tail half-value time T_2 amounting to 350 μs. However, in the main they are available only for AC voltages ranging from 250 to 280 V, which means that they are often unusable for lightning current protection on the DC side of PV installations. However, a type 1 lightning protection device specially designed for DC-side applications has come onto the market, is rated for 1000 V and can shunt a 50 kA surge current that exhibits a 10/350 μs waveform (see Section 6.9.1).

Many surge diverters for currents that Exhibit 8 μs wavefront time T_1 and 20 μs wave-tail half-value time T_2 are also available (type 2 devices for medium protection and type 3 for fine protection), which possess the following characteristics: (a) they are mainly composed of zinc oxide (ZnO) varistors, whose resistances are voltage independent; (b) they are chiefly intended for inductive surge diversion (see Section 6.6); and (c) they are mainly available for the currents $I_{8/20}$ (around 0.1 to 100 kA) that are allowable for this type of waveform. Under nominal current conditions, surge protection devices typically reduce voltage that is three to five times that of nominal DC voltage V_{VDC} and are thus often also used for medium and fine protection downstream from a lightning down-conductor and current limiting device (inductance, conductor several metres long). Varistor pick-up voltage tends to decline as the result of overloading and/or ageing. In an extreme case, leakage current may provoke varistor calefaction and severe damage at operating voltage, although under certain circumstances secondary grid current can provoke further damage. This problem has prompted some manufacturers to offer thermally monitored varistors that exhibit $I_{8/20}$ of around 5 to 20 kA and a cutout device that shuts off the varistor if it exhibits non-conformant calefaction, so as to avoid further damage.

Surge protection devices containing varistors are available for voltages ranging from 5 to 1000 V. Figure 6.10 displays specimen *V–I* characteristic curves for selected surge protection devices that integrate ZnO varistors.

When varistors are used as surge protection devices, it should be borne in mind that they exhibit inherent capacitance (usually around 30 pF up to a few nanofarads) that can sometimes cause signal line problems. As gas-filled surge protection devices exhibit inherent capacitance amounting to only a few picofarads, they are compatible with signal lines but not with PV installations, since their relatively low potential combustion prevents them from extinguishing arc flashes, and they could thus be knocked out by secondary solar generator current.

Owing to the paucity of suitable lightning protection devices on the market, surge protection devices are occasionally used for this purpose. The following can be estimated based on the energy exhibited by a lightning protection device secondary to a surge current load: a surge protection device that is subjected to a 10/350 μs lightning current curve can only withstand a load amounting to around 6 to 12% (depending on type) of the surge protection device's nominal current $I_{8/20}$; however, this is probably sufficient for

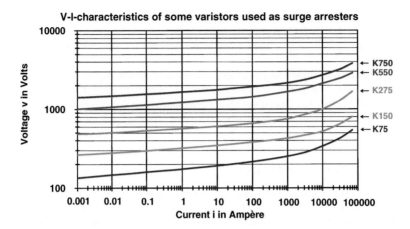

Figure 6.10 Characteristic curves for selected Siemens Matsushita SIOV-B60K series varistors that can be used as surge protection devices for up to 70 kA (8/20 μs) surge current. The numbers next to the letter K indicate the effective value of the allowable operating voltage for alternating current. The allowable direct current is around 30 to 35% higher in each case, which means that the attendant values for the various models would be as follows: K75, 100 V; K150, 200 V; K275, 350 V; K550, 745 V; K750, 1060 V. *Source*: Siemens Matsushita

conductor leads that exhibit only moderate lightning current. With stronger current, a series of such lightning protection devices can be used.

6.6 Lightning-Current-Induced Voltages and Current

Like all current, lightning current generates an ambient magnetic field which, like lightning current itself, changes very rapidly. A long, straight conductor through which current i flows will exhibit a magnetic flow density of $B = \mu_0 i / 2\pi r$ at interval r.

A magnetic field induces a flow Φ in a closed conductor loop whose characteristics can be determined based on loop surface area via integration of magnetic flow density B. If a conductor loop flow changes, induced voltage $v = d\Phi/dt$ results, in keeping with the law of induction.

Mutual inductance M is an extremely useful parameter for determining voltage in adjacent conductor loops, as it allows for calculation of this value without detailed knowledge of electromagnetic field theory. Its application for simple configurations is discussed in Section 6.6.1.

A variable current i induces a voltage in an adjacent conductor loop according to the following equation:

$$\text{Voltage induced by current } i \text{ in a conductor loop: } v = \frac{d\Phi}{dt} = M\frac{di}{dt} \qquad (6.13)$$

where M is the mutual inductance between (a) a conductor that exhibits current i and (b) the conductor loop affected, as determined solely by the geometry between the loop and current, and di/dt is the mathematical derivative of current i over time, often expressed as $di/dt \approx \Delta i/\Delta t$, the change in current per unit of time.

The larger the surface area occupied by a conductor loop and the nearer it is to the current, the stronger the mutual inductance will be, by virtue of the strong magnetic field that obtains in close proximity to the conductor through which the current flows. Conductor loops that integrate part of a down-conductor exhibit particularly high mutual inductance M (see Figure 6.11).

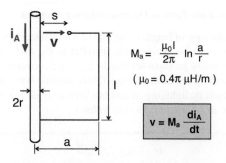

Figure 6.11 Mutual inductance between a lightning down-conductor and a rectangular conductor loop that also integrates part of the lightning protection device (length l) (See Section 6.6.2 for a discussion of the s section.)

Current and thus also di/dt can be reduced in individual down-conductors by distributing lightning current to the various down-conductors (see Section 6.4). To calculate the voltage induced in a conductor loop, the various voltages induced by two or more lightning current conductors in the loop are tallied. In view of the fact that the strength of the magnetic fields in some of a building's lightning protection devices often increases, to be on the safe side it is usually advisable only to take account of the rough calculations for the down-conductor that exhibits lightning current $i_A = k_C \cdot i$ and that is closest to the loop. In such a case, the conductor loop voltage induced by lightning current will be as follows:

$$\text{Voltage induced by lightning current } i_A: \quad v = M \frac{di_A}{dt} = M \cdot k_C \frac{di}{dt} = M_i \frac{di}{dt} \qquad (6.14)$$

where $M_i = k_C \cdot M$ is the effective mutual inductance between overall lightning current i and the relevant conductor loop. In this equation, M_i factors in not only the geometry between the lightning current i_A in the lightning conductors, but also the previous distribution of overall lightning current i to the various down-conductors and the consequent magnetic field strength reduction. M_i is a particularly useful factor for discussions of the damage that can potentially be caused by induced voltage and current. M_i can also be used (via Equation 6.37) to take into account any other factors such as metallic module frames that weaken the magnetic field still further.

The maximum of di_A/dt is $(di_A/dt)_{max} = k_C \cdot (di/dt)_{max}$, i.e. owing to the distribution of lightning current i to various down-conductors, the maximum loop voltage $v_{max} = M \cdot di_A/dt_{max}$ will be lower by a factor k_C.

Lightning-current-induced voltage is mainly of importance during the lightning current increase phase, which, on account of the extremely elevated maximum rate of current change di/dt_{max} (100 kA/μs up to 200 kA/μs, see Table 6.3) that may be exhibited, can result in extremely elevated induced voltages ranging up to many kilovolts or (if the conductor loop integrates some of the available down-conductors) even several megavolts.

It can be assumed that induced voltage only occurs during the lightning current increase phase, whose waveform is roughly $v_{max} \cdot e^{-t/\tau}$, whereby $v_{max} = M \cdot di_A/dt_{max}$ (see Figure 6.51 for an example of this waveform) and which equals 0 both before and after this phase. After lightning falls away from a previously attained maximum, di/dt is extremely low, reaching a maximum of only around 0.2% of the anticipated maximum current increase; thus voltage-induced current drops are of little relevance in actual practice.

6.6.1 Mutual Inductance and Induced Voltages in a Rectangular Loop

In essence, the configurations commonly found in practice can either be deduced from the basic scenarios illustrated in Figures 6.11 and 6.13, or be obtained from the computations for these scenarios.

6.6.1.1 Rectangular Loops That Possess Down-conductor Sections

Figure 6.11 shows a rectangular loop of length l, one side of which integrates a straight lightning current conductor whose radius is r and whose other side (parallel to the first side) is distance a from the conductors axis. The equation below presupposes that the lightning current conductor is of infinite length, but only a relatively minor anomaly occurs in practice if this condition is not met. The conductor radius in the part of the loop that conducts no lightning current has been disregarded. Mutual inductance M_a for a conductor loop of this nature as in Figure 6.11 is as follows:

$$M_a = 0.2 \cdot l \cdot \ln\frac{a}{r} \qquad (6.15)$$

where M_a is expressed in μH and all lengths (l, r, a) are expressed in metres.

In as much as most of the flow that is responsible for induced voltage occurs in very close proximity to the current conductor, in some cases a conductor is also directly allocated a specific – albeit of course somewhat random – inherent inductance L_a that equates to M_a for a specific and relatively large distance such as 10 m. However, for a large distance a, the waveform of the natural logarithm is relatively flat, which means that no major error can occur in the event of reasonable deviations. This approach can be used, for example, to estimate the inductive voltage drop exhibited by a grounding conductor.

If M_a is used in Equation 6.14 and if di_A/dt is known, the voltage induced in a loop of this nature can also be determined.

In the interest of obtaining an optimally generic waveform, in Figure 6.12 a mutual inductance per unit of length $M_a' = M_a/l$ and induced voltage per unit of length v/l for a 100 kA/μs di_A/dt are shown for this configuration instead of using M_a directly. This maximum di_A/dt is a useful supposition for an initial estimate. By multiplying by l, the voltage and mutual inductance that will be exhibited by a specific configuration can be readily derived from M_a' and u/l.

Although the numerical values of M_a' are relatively low and do not overshoot 1.5–2 μH/m even in the presence of extremely elevated a/r ratios, the voltages induced in these types of loops for the value a are extremely elevated on account of the very sizeable rate of current change that occurs. Voltage v/l per unit of length very quickly attains values ranging from 10 to 200 kV/m.

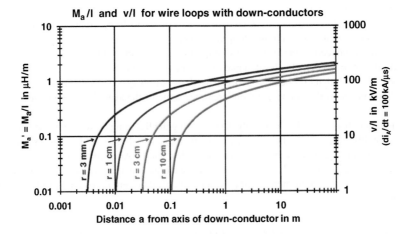

Figure 6.12 Mutual inductance $M_a' = M_a/l$ per unit of length, and, for $di_A/dt = 100$ kA/μs, induced voltage v/l per unit of length for a rectangular loop (as in Figure 6.11) that is derived from a down-conductor whose length is l and other installations

Example Based on the Configuration in Figure 6.11

Installation loop as in Figure 6.11 between (a) a down-conductor ($l = 7$ m and $r = 3$ mm) that conducts the entirety of the lightning current and (b) a PV installation DC interconnecting line at a distance of $a = 10$ m.
What are the values for (a) M_a and (b) maximum induced voltage v_{max} with $(di/dt)_{max} = 100$ kA/µs?

Solution: $k_C = 1 \Rightarrow (di_A/dt)_{max} = (di/dt)_{max} = 100$ kA/µs. Using Equation 6.15, $M_a = 11.4$ µH. Using Equation 6.14, it then follows that the voltage is $v_{max} = 1.14$ MV during the lightning current increase phase.

6.6.1.2 Rectangular Loops Not Connected to Down-conductors

Figure 6.13 illustrates a rectangular loop of length l and width b that is at distance d from a straight lightning current conductor. The equation below presupposes that the lightning current conductor is of infinite length, but only a relatively minor anomaly occurs in practice if this condition is not met.

The mutual inductance for a conductor loop of this nature as in Figure 6.13 is as follows:

Mutual inductance for a down-conductor that is not connected to the loop:

$$M_b = 0.2 \cdot l \cdot \ln\frac{b+d}{d} \tag{6.16}$$

In this equation, M_b is the mutual inductance in µH if all lengths (l, b, d) are expressed in metres and d is the distance between the conductor loop and i_A.

B is often far smaller than d, whereby the following constitutes a good approximation in this case:

$$M_b \approx 0.2\frac{l \cdot b}{d_S} = 0.2\frac{A_S}{d_S} \tag{6.17}$$

In this equation, M_b is the mutual inductance in µH if all lengths (l, b, d_S) are expressed in metres and A_S is expressed in square metres. $d_S = d + b/2$ represents the distance between the down-conductor axis and the conductor loop centre of gravity S, and A_S is the surface area of the conductor loop. Equation 6.17 is applicable from $d > 2b$. If b is defined as the maximum width of the loop perpendicular to the down-conductor, then Equation 6.17 for $b \ll d$ can also be applied to triangular, circular and other loop configurations.

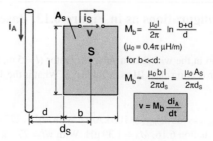

Figure 6.13 Mutual inductance between a down-conductor and a rectangular loop of length l and width b and at distance d from the lightning protection device axis

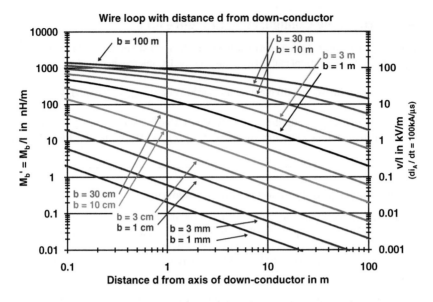

Figure 6.14 Mutual inductance $M_b' = M_b/l$ per unit of length and for induced voltage v/l ($w\ di_A/dt = 100\,\text{kA/µs}$) per unit of length for a rectangular loop as in Figure 6.13 for various loop widths b as a function of gap d

If M_b is used in Equation 6.14 and if di_A/dt is known, the voltage induced in a loop of this nature can also be determined.

In the interest of obtaining an optimally generic waveform, in Figure 6.14 mutual inductance per unit of length $M_b' = M_b/l$ and induced voltage per unit of length v/l for a 100 kA/µs di_A/dt are shown for this configuration as in Figure 6.13 instead of using M_b directly. By multiplying by l, the voltage and mutual inductance that will be exhibited by a specific configuration can be readily derived from M_b' and v/l.

The ratio of $(b + d)/d$ in Equation 6.16 is usually far smaller than a/r in Equation 6.15, such that M_b is usually far lower than M_a for the same loop length. However, the extremely high voltages in such loops induced by very elevated rates of current change secondary to lightning current can nonetheless easily inflict severe damage on unprotected charge controllers, inverters and the like. To reduce such voltages, the area occupied by unavoidable installation loops such as solar module wiring should be kept to a minimum and should be installed as far as possible from lightning down-conductors. Moreover, simply twisting conductor leads can considerably reduce inherent voltages. Somewhat more precise but far more complex equations for scenarios involving relatively short lightning current conductors are described in [Has89].

Examples for the Configurations as in Figure 6.13, Based on Protection Class III

Problem 1: Installation loop in the wiring of a solar generator ($l=5$ m, $b=3$ m, $d=1$ m); $k_C=1$, $(di/dt)_{max} = 100\,\text{kA/µs}$ for protection class III as in Table 6.3. What is the M_b value and the maximum induced voltage v_{max}?

Solution: $k_C=1 \Rightarrow (di_A/dt)_{max} = (di/dt)_{max} = 100\,\text{kA/µs}$. With $M_b' = 277\,\text{nH/m}$ from Figure 6.14 and multiplying by l, or using Equation 6.16, $M_b = 1.39\,\text{µH}$. With $v/l = 27.7\,\text{kV/m}$ from Figure 6.14 and multiplying by l, or using Equation 6.14, $v_{max} = 139\,\text{kV}$ during the lightning current increase phase.

Problem 2: Two-lead conductor, length 10 m (axis clearance 1 cm) at a distance of 10 cm and parallel to the lightning down-conductor as in Figure 6.13, whereby $k_C = 0.5$ and $(di/dt)_{max} = 100$ kA/µs from Table 6.3 for protection class III. What is the M_b value and the maximum induced voltage v_{max}?

Solution: $k_C = 0.5 \Rightarrow (di_A/dt)_{max} = k_C \cdot (di/dt)_{max} = 50$ kA/µs. In this case, $l = 10$ m, $d = 0.1$ m, $b = 0.01$ m $\Rightarrow M_b' = 19.1$ nH/m as in Figure 6.14l, or using Equation 6.16 or 6.17, $M_b = 191$ nH. Using Equation 6.14, it then follows that the voltage is $v_{max} = 9.5$ kV during the lightning current increase phase.

Problem 3: Two-lead conductor, length 10 m (axis clearance 1 cm) at a distance of 10 cm and perpendicular to the lightning down-conductor as in Figure 6.13, whereby $k_C = 0.5$ and $(di/dt)_{max} = 100$ kA/µs from Table 6.3 for protection class III. What is the M_b value and the maximum induced voltage v_{max}?

Solution: $k_C = 0.5 \Rightarrow (di_A/dt)_{max} = k_C \cdot (di/dt)_{max} = 50$ kA/µs for standard requirements. In this case, $l = 10$ m, $d = 0.1$ m, $b = 0.01$ m, where $M_b' = 923$ nH/m from Figure 6.14, and multiplying by l, or using Equation 6.16, $M_b = 9.23$ nH. With $v/l = 92.3$ kV/m from Figure 6.14 and multiplying by l and k_C, or using Equation 6.14, $v_{max} = 462$ kV during the lightning current increase phase.

6.6.2 Proximity Between Down-conductors and other Installations

The voltages induced in lightning current loops as in Figure 6.11 can be extremely high. If the loop is open in such a case and the safety clearance s is unduly small, a disruptive discharge can occur, which needs to be taken into account if a lightning down-conductor and other installations in the same building, such as the DC cable of a PV installation, are in close proximity to each other.

The leader of a negatively charged lightning flash exhibits maximum $(di/dt)_{max}$ for up to 0.25 µs. According to [Has89], the breakdown voltage for rod-to-rod spark gaps with clearance s is as follows for this type of transient load:

Mean breakdown voltage between lightning rods and with clearance s:

$$V_D = s \cdot k_m \cdot 3000\,kV/m \qquad (6.18)$$

where k_m is a material factor as follows: $k_m = 1$ for air distances; $k_m \approx 0.5$ if a proximate distance covers wood, concrete or brick.

As in Figure 6.12, the length-specific mutual inductance for large conductor loops in Figure 6.11 ranges up to around $M_a' = 1.2\,\mu H/m$ in buildings for $a \approx 2$ m where $a \gg r$.

On account of the flat increase in the natural logarithm for high hypothetical values, this inductance does not increase appreciably for higher a/r ratios and thus can be regarded as a representative guide value for large clearances a.

If the voltage from Equation 6.14 induced in a loop as in Figure 6.11 is set as equal to breakdown voltage V_D according to Equation 6.18, the result is as follows:

$$v = M_a' \cdot l \cdot k_C \cdot di/dt = 1.2\,\mu H/m \cdot l \cdot k_C \cdot di/dt = s \cdot k_m \cdot 3000\,kV/m \qquad (6.19)$$

If $(di/dt)_{max} = 100$ kA/µs is set for standard requirements (i.e. protection class III) as in Table 6.3 and if the calculation is performed based on s, the minimal required safety clearance s_{min} pursuant to the latest standard is obtained so as to ensure that no disruptive discharge occurs in the event of proximity as in Figure 6.11:

$$\text{Minimum safety clearance for proximities: } s_{\min} = 0.04 \frac{k_C}{k_m} l \qquad (6.20)$$

where k_m is a material factor as follows: $k_m = 1$ for air distances; $k_m \approx 0.5$ if the proximity distance covers wood, concrete or brick. l is the length of the loop down-conductor section through which the lightning current flows.

For protection class II, 0.06 should be used in lieu of 0.04 in Equation 6.20; for protection class I, 0.08 should be used [6.22]. According to [Deh05] and [6.2], the higher values applied in the past were as follows: for protection classes III and IV, 0.05; for protection class II, 0.075; for protection class I, 0.1.

If an installation is realized such that these minimum safety clearances are adhered to, and if it is ensured through conscientious planning and realization that no concealed proximities creep in, a solar generator can be mounted in the protection zone of a lightning protection installation.

The following is a specimen calculation for the required minimum safety clearances.

Example for Figure 6.15

A solar generator is to be installed in the lightning protection zone of a house that has a lightning protection installation and a steeply pitched roof (also see Figure 6.5). The task here is to determine the minimum required safety clearances s_1 relative to the lightning conductor, the clearances s_2 relative to the roof gutter attached to the lightning protection installation, and the relevant clearances s_3 and s_4 through the roof for the DC cable that will be taken through it (or for the power lead if the inverter is installed in the attic). Here, $l_1 = 13$ m, $l_2 = 6$ m, $l_3 = 7$ m, and standard lightning protection requirements apply (protection class III).

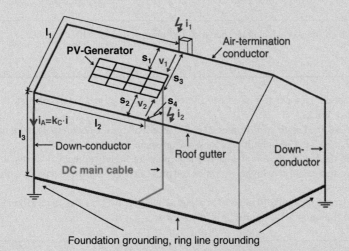

Figure 6.15 Minimum required safety clearances s_1 and s_2 on a roof and s_3 and s_4 through a roof, as well as for lightning-induced voltages v_1 and v_2 for a house with a steeply pitched roof and a lightning protection installation, in a scenario where a solar generator is to be installed in the protected area of this installation

Two different scenarios come into play:

Scenario 1: The maximum allowable voltage v_1 for s_1 and s_3 occurs if lightning i_1 strikes the chimney. In this case, $k_C \approx 0.5$ and the required down-conductor length is $(l_1 + l_3) = 20$ m. Using Equation 6.20, $s_{1min} = 40$ cm ($k_m = 1$, secondary to air flashover) and $s_{3min} = 80$ cm ($k_m = 0.5$, secondary to a disruptive discharge through a solid). In the absence of the second down-conductor (right), $k_C = 1$, $s_{1min} = 80$ cm and $s_{3min} = 1.6$ m would apply.

Scenario 2: The maximum allowable voltage v_2 for s_2 and s_4 is induced by a lateral lightning strike i_2 in close proximity to the extremity of the roof gutter. In this case $k_C \approx 0.83$ for section l_3 (estimated on the basis of the ratio of the resistance between the two down-conductor paths) and $k_C = 1$ for section l_2; for s_2, $k_m = 1$ (disruptive discharge in the air) and for s_4, $k_m = 0.5$ (disruptive discharge through a solid). The following is then obtained using Equation 6.20:

$s_{2min} = 0.04 \cdot 6$ m $+ 0.04 \cdot 0.83 \cdot 7$ m $= 47$ cm and $s_{4min} = 2s_{2min} = 94$ cm. In the absence of the second down-conductor (right), $s_{2min} = 52$ cm and $s_{4min} = 1.04$ m would apply.

For protection class I, the aforementioned safety clearances would need to be doubled. On a steeply pitched roof, the roof gutter might still be within the protection zone of the roof crest lightning conductor (see Figure 6.5), thus ruling out the possibility of the roof gutter being struck by lightning.

If a minimum safety clearance s_{min} is not adhered to, or if a connection to the lightning protection installation is required, a metallic connection should be realized between the lightning protection installation and the module frame and/or the solar generator frame that allows for diversion of part of the lightning current; in addition, shielded lightning-current-resistant conductors should be used (see Section 6.6.4).

6.6.3   Induced Current

When conductor loops are closed (as a result of the following, among other things: a short; a bypass diode in a module; impedance from a connected device; a disruptive discharge via an air break with insufficient clearance), the resulting induced voltage induces a current whose strength is particularly important to determine for purposes of sizing the surge protection device voltage. This section describes a method for determining the requisite nominal current in $I_{8/20}$ for varistors. First, short-circuit current I_{So} for the loss-free loop ($R_S = 0$) is determined (this is a very simple calculation). The requisite varistor current $I_{8/20}$ is determined using a correction factor k_V, which is found from M, k_C, R_S and the varistor's DC operating voltage V_{VDC}.

The elevated induced voltage usually lasts less than 10 µs. During the lightning current increase phase, the strength of the current i_S induced in a closed loop is mainly determined by the loop's inherent inductance L_S, since usually the voltage drop at the loop resistance R_S can be disregarded. Inherent inductance is often referred to as inductance.

For a two-lead conductor with a radius r_o, axis clearance b and length l as on the right of Figure 6.13, the following applies to $l \gg b$, whereby the internal inductance (which plays a very minor role) is disregarded as it is only important for low frequencies:

$$\text{Inductance for a long two-lead conductor: } L = 0.4 \cdot l \cdot \ln \frac{b - r_o}{r_o} \qquad (6.21)$$

Here, L is the inductance in µH if all lengths (conductor length l, axis clearance b and conductor radius r_o) are expressed in metres. In the interest of obtaining an optimally generic waveform, Figure 6.16 indicates the inductance layer $L' = L/l$ (i.e. inductance per unit of length) in lieu of inductance L for a long two-lead conductor. Inductance L for a specific configuration is then derived from L' by multiplying by l.

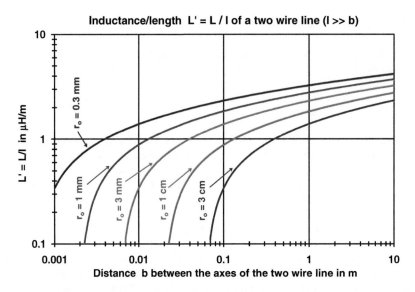

Figure 6.16 Inductance layer $L' = L/l$ of a two-lead conductor (r_o = conductor radius, b = conductor axis clearance, length $l \gg b$); see the right-hand diagram in Figure 6.13

Lightning protection installations often possess conductor loops that do not meet the $l \gg b$ requirement, in which case both length l and width b engender considerable inductance. In lieu of the exact but complex equation in [Has89], a far simpler approximation can be used that exhibits only very minor discrepancies for $l \geq b$ and $b \gg r_o$:

$$\text{Rectangular loop inductance: } L \approx 0.4 \cdot (l + b) \cdot \ln \frac{b - r_o}{r_o} - 0.55 \cdot b \qquad (6.22)$$

Here, L is the inductance in μH if all relevant dimensions (length l, width b and conductor radius r_o) are expressed in metres ($l \geq b$, $b \gg r_o$).

6.6.3.1 Short-Circuit Current Induced in a Loss-Free Loop

The calculation of short-circuit current i_{So} in a resistance-free loop that has inductance L_S is very simple. Mutual inductance $M = M_b$ engenders short-circuit current i_{So}, which is induced in a configuration as in Figure 6.13 by lightning current i_A in a down-conductor in a closed conductor loop that has $R_S = 0$ and inductance L_S:

$$\text{Induced short-circuit current in a conductor loop } (R_S = 0): \ i_{So} \approx \frac{M}{L_S} i_A = \frac{M}{L_S} k_C \cdot i = \frac{M_i}{L_S} i \quad (6.23)$$

Here $M_i = k_C \cdot M$ is the effective mutual inductance between overall lightning current i and the conductor loop.

Use of the maximum value i_{max} (100–200 kA as in Table 6.3) allows for the realization of a simple calculation of the maximum short-circuit current i_{Somax} exhibited by a resistance-free loop:

$$\text{Induced maximum short-circuit current in a conductor loop } (R_S = 0): \ i_{So\,max} \approx \frac{M_i}{L_S} i_{max} \quad (6.24)$$

6.6.3.2 Current Induced in Loops That Contain Surge Protection Devices

In a non-resistance free loop, the lower the inductance L_S, and the higher the loop resistance and varistor voltage, the more rapidly the induced current i_S falls away from the maximum value i_{Smax}.

For accurate current sizing of surge protection devices, the calculation of the current in loops that contain varistors needs to be sufficiently precise. However, this is no easy task because of the nonlinear nature of varistors and because the strength of induced current is determined by various parameters. Hence certain simplifications, as well as conservative estimates, are necessary in order to keep the calculation within reasonable bounds.

Lightning current can be determined via the following equation:

$$i(t) = I(e^{-\sigma_1 t} - e^{-\sigma_2 t}) \qquad (\sigma_2 \gg \sigma_1) \tag{6.25}$$

This equation has the following advantages: it provides a good approximation of go-live conditions without an unduly large amount of calculation; it can be readily adjusted to the maximum i_{max}, $(di/dt)_{max}$ and Q recommended by the relevant standard; and it exactly corresponds to the waveform of the surge current generated by high-voltage test lab generators.

In view of the fact that in many cases relatively little in the way of technical data is available for surge protection devices, an equivalent circuit should be developed that needs some product-specific data and that is supplied by virtually all manufacturers. The guide values that come into play here are maximum operating DC voltage V_{VDC} for the varistor and maximum allowable surge current $I_{8/20}$ for the 8/20 μs waveform. Moreover, the relevant circuits should be linearized so as to allow for the use of standard network theory tools.

In as much as the characteristics of lightning current are controlled by external current, i.e. current that exhibits the characteristics of an ideal power source, the effect of the current from the varistor loop $i_V = i_S$ on primary current i can be disregarded.

Thus the equivalent circuit in Figure 6.17 is obtained for a lightning-current-coupled loop that contains a surge protection device:

To be on the safe side, the varistor in the linearized circuit should be replaced by a real voltage source that exhibits $V_V = 2 \cdot V_{VDC}$ and inner resistance $R_V = V_{VDC}/i_{Somax}$. For relatively high mutual inductance $M_i = k_C \cdot M$ (typically M_i is more than about 1 μH), on reaching 0 the current i_V can also still be negative. In such a case, the algebraic sign of the power source used to calculate the negative range of i_V in the linearized equivalent circuit should be reversed.

6.6.3.2.1 Calculation of Current i_V Using the Laplace Transformation

Using the linearized circuit as in Figure 6.17, i_V can be analytically calculated via the Laplace transformation [6.19].

The Laplace transformation $I(s)$ of lightning current $i(t) = I(e^{-\sigma_1 t} - e^{-\sigma_2 t})$ is as follows:

$$I(s) = \frac{I}{s + \sigma_1} - \frac{I}{s + \sigma_2} = \frac{I \cdot (\sigma_2 - \sigma_1)}{(s + \sigma_1)(s + \sigma_2)} \tag{6.26}$$

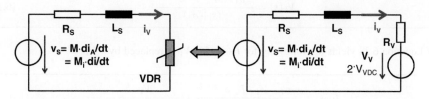

Figure 6.17 Equivalent circuit for determination of varistor current i_V in a loop that is inductively coupled with a lightning current i (left, original circuit; right, linearized variant for $i_V > 0$)

Hence the following applies to the voltage $V_S(s)$ in the loop:

$$\text{Voltage induced in the loop: } V_S(s) = s \cdot M_i \cdot I(s) = \frac{s \cdot M_i \cdot I \cdot (\sigma_2 - \sigma_1)}{(s + \sigma_1)(s + \sigma_2)} \qquad (6.27)$$

In the circuit in Figure 6.17, the following applies to varistor current $I_V(s)$ that has $\sigma_3 = (R_S + R_V)/L_S$:

$$I_V(s) = \frac{V_S(s) - \dfrac{V_V}{s}}{R_S + R_V + s \cdot L_S} = \frac{V_S(s) - \dfrac{V_V}{s}}{L_S(s + \sigma_3)} \qquad (6.28)$$

The following is obtained by using $V_S(s)$ as in Equation 6.27:

$$I_V(s) = \frac{M_i \cdot I \cdot s \cdot (\sigma_2 - \sigma_1)}{L_S(s + \sigma_1)(s + \sigma_2)(s + \sigma_3)} - \frac{V_V}{L_S \cdot s \cdot (s + \sigma_3)} \qquad (6.29)$$

Back-conversion in the time range yields the following:

$$i_V(t) = \frac{M_i \cdot I}{L_S} \left[\frac{\sigma_1 \cdot e^{-\sigma_1 \cdot t}}{(\sigma_1 - \sigma_3)} + \frac{\sigma_2 \cdot e^{-\sigma_2 \cdot t}}{(\sigma_3 - \sigma_2)} + \frac{\sigma_3 \cdot (\sigma_1 - \sigma_2) \cdot e^{-\sigma_3 \cdot t}}{(\sigma_1 - \sigma_3)(\sigma_2 - \sigma_3)} \right] - \frac{V_V(1 - e^{-\sigma_3 \cdot t})}{L_S \cdot \sigma_3} \qquad (6.30)$$

$i_V(t)$ reaches 0 at a specific time t_0, whereby $t > t_0$ when $i_V < 0$. To determine the negative range of $i_V < 0$, the algebraic sign of the source $V_V = 2 \cdot V_{VDC}$ in Figure 6.17 is reversed, i.e. $V_V' = -V_V$. Since $\sigma_2 \gg \sigma_1$ for real lightning current, the following applies to $t > t_0$ for $i(t)$:

$$i(t) \approx I \cdot e^{-\sigma_1 t} = I \cdot e^{-\sigma_1 \cdot t_0} \cdot e^{-\sigma_1 \cdot \tau}, \quad \text{where by } t = t - t_0 \qquad (6.31)$$

Hence voltage $v_S(t) = M_i \cdot di/dt$ is as follows:

$$v_S(t) = -\sigma_1 \cdot M_i \cdot I \cdot e^{-\sigma_1 \cdot t} = -\sigma_1 \cdot M_i \cdot I \cdot e^{-\sigma_1 \cdot t_0} \cdot e^{-\sigma_1 \cdot \tau} \qquad (6.32)$$

The Laplace-transformed element (as regards the time point $\tau = 0$) is as follows:

$$V_S(s) = \frac{-\sigma_1 \cdot M_i \cdot I \cdot e^{-\sigma_1 \cdot t_0}}{(s + \sigma_1)} \qquad (6.33)$$

Use of Equation 6.28 yields the following for $I_V(s)$ after V_V is replaced by $-V_V$:

$$I_V(s) = \frac{-\sigma_1 \cdot M_i \cdot I \cdot e^{-\sigma_1 \cdot t_0}}{L_S(s + \sigma_1)(s + \sigma_3)} + \frac{V_V}{L_S \cdot s \cdot (s + \sigma_3)} \qquad (6.34)$$

Back-conversion in the time range yields the following for $t > t_0$:

$$i_V(t) = -\sigma_1 \cdot M_i \cdot I \cdot e^{-\sigma_1 \cdot t_0} \frac{e^{-\sigma_1 \cdot (t-t_0)} - e^{-\sigma_3 \cdot (t-t_0)}}{L_S(\sigma_3 - \sigma_1)} + \frac{V_V(1 - e^{-\sigma_3 \cdot (t-t_0)})}{L_S \cdot \sigma_3} \qquad (6.35)$$

This solution applies to $t > t_0$ if $i_V < 0$. If i_V had reached 0, the original equivalent circuit as in Figure 6.17 would have applied. But since v_S in the lightning return stroke remains negative, there is no propulsive voltage that could again generate an $i_V > 0$, i.e. i_V perpetually remains 0.

To be on the safe side (worst-case scenario calculation), it suffices to investigate relatively low R_S and R_V values. Hence the following R_S and R_V values were used for the computer simulation, where they were derived automatically from the defined L_S and V_{VDC} values: $R_S = L_S \cdot 1\ \text{m}\Omega/\mu\text{H}$, $R_V = V_{VDC}/i_{Somax}$, i.e. overall voltage drop at the varistor when i_{Somax} is $3 \cdot V_{VDC}$.

Figures 6.18–6.23 show specimen varistor current calculations using the above suppositions and relatively high M_i and V_{VDC} values. In the interest of obtaining results that lend themselves to comparison, the ratio $M_i/L_S = {}^1/_4$ was used for all examples.

In Figures 6.18–6.20, current induced in a relatively large loop ($M_i = 5\ \mu\text{H}$ and $L_S = 20\ \mu\text{H}$) by lightning where $i_{max} = 100\ \text{kA}$ ($I = 106.5\ \text{kA}$, $\sigma_1 = 2150\ \text{s}^{-1}$, $\sigma_2 = 189\,900\ \text{s}^{-1}$) was investigated first.

In the scenarios with relatively high effective mutual inductance M_i as in Figures 6.19 and 6.20, the surge current i_S is of far longer duration than surge current $I_{8/20}$ exhibited by the 8/20 μs waveform, for which most down-conductors are specified. In order to avoid failure, down-conductors must be rated for a higher surge current $I_{8/20}$ than the maximum current i_{Smax} as in Figures 6.19 and 6.20. A conservative assumption in this regard would be as follows: the load Q flowing through the varistors secondary to i_S (sum total of the positive and negative half wave) equates to the load Q_S of a surge current $I_{8/20}$ whereby $Q_S \approx I_{8/20} \cdot 20\ \mu\text{s}$. This assumption can be deemed to be on the safe side by virtue of the fact that the varistor voltage exhibited by weaker current is lower, and thus less energy is used as well. Hence a

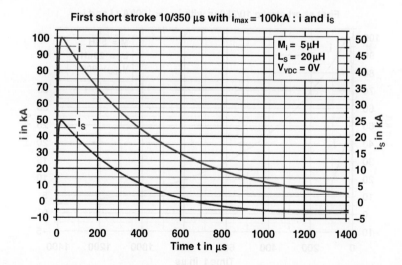

First short stroke 10/350 μs with $i_{max} = 100\text{kA}$: i and i_S

$M_i = 5\mu\text{H}$
$L_S = 20\,\mu\text{H}$
$V_{VDC} = 0\text{V}$

Figure 6.18 Short-circuit current induced in a relatively large varistor-free loop ($M_i = 5\ \mu\text{H}$ and $L_S = 20\ \mu\text{H}$) by lightning where $i_{max} = 100\ \text{kA}$ ($I = 106.5\ \text{kA}$, $\sigma_1 = 2150\ \text{s}^{-1}$, $\sigma_2 = 189\,900\ \text{s}^{-1}$) and $i_S \approx i_{So}$ (as in Equation 6.23, this applies on account of very low resistance ($R_S = 20\ \text{m}\Omega$)). The curves for i_S and hence i are thus very similar at first. However, as time goes on i_S becomes slightly negative owing to σ_3 from the third pole, although the peak negative value falls far short of i_{Smax}

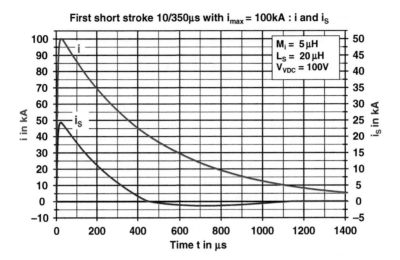

Figure 6.19   Current $i_S = i_V$ induced in the same loop (which contains a varistor with a small maximum DC operating voltage, $V_{VDC} = 100\,V$) by lightning that exhibits $i_{max} = 100\,kA$. I_S decreases more rapidly in this case because $V_{VDC} > 0$. On account of the relatively elevated M_i value and the relatively low V_{VDC} value, the post-zero-crossing negative voltage i_S induced in the lightning current tail is sufficient to generate briefly a low-level negative current

correction factor k_V for the easy-to-calculate maximum short-circuit current i_{Somax} exhibited by the loss-free loop is determinable for such cases (see Figure 6.24).

Figures 6.21–6.23 show the current induced in a relatively small loop ($M_i = 0.1\,\mu H$ and $L_S = 0.4\,\mu H$) by lightning where $i_{max} = 25\,kA$ ($I = 25.2\,kA$, $\sigma_1 = 6931\,s^{-1}$, $\sigma_2 = 3\,975\,000\,s^{-1}$).

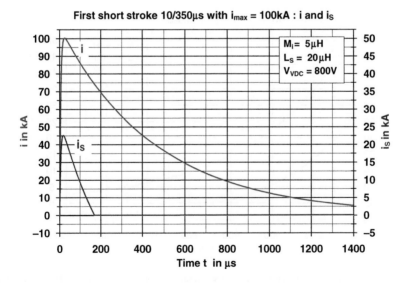

Figure 6.20   Current $i_S = i_V$ induced in the same loop (which contains a varistor with a large maximum DC operating voltage, $V_{VDC} = 800\,V$) by lightning that exhibits $i_{max} = 100\,kA$. As V_{VDC} is relatively high, in this case i_S does not attain the peak value i_{Smax} as in Figure 6.18 and drops very quickly. Once i_S reaches 0, negative voltage v_S induced in the lightning current tail is not sufficient to generate a negative current, despite the elevated M_i value induced by the elevated V_{VDC} value

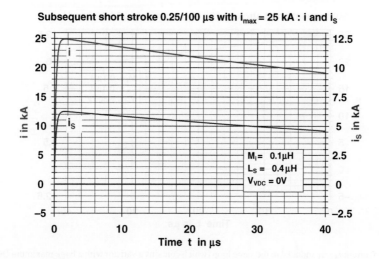

Figure 6.21 Short-circuit current induced in a relatively small varistor-free loop ($M_i = 0.1\,\mu H$ and $L_S = 0.4\,\mu H$) by negatively charged lightning where $i_{max} = 25$ kA. $I_S \approx i_{So}$ (as in Equation 6.23) applies on account of the very low resistance ($R_S = 0.4\,m\Omega$). The curves for i_S and hence i are thus very similar at first. On account of the σ_3 stemming from the third pole, i_S briefly becomes slightly negative after a time as in Figure 6.6 (not shown in the diagram), although the peak negative value falls far short of i_{Smax}

In the scenarios with relatively low effective mutual inductance M_i as in Figures 6.22 and 6.23, the surge current i_S is of far shorter duration than surge current $I_{8/20}$ exhibited by the 8/20 μs waveform, for which most down-conductors are specified. Hence the load that comes into play is far smaller than that used by the surge current $I_{8/20}$, and thus the varistor can be sized far smaller.

Since, in the presence of current exceeding nominal current, the voltage increases and damage can occur, sizing to the same load used would not be a conservative approach. On the other hand, in such cases the specified maximum current $I_{8/20}$ of the varistor being used equates to the effective maximum current,

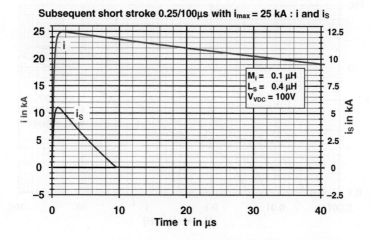

Figure 6.22 Current $i_S = i_V$ induced in the same loop (which contains a varistor with a low maximum DC operating voltage, $V_{VDC} = 100$ V) by negatively charged lightning that exhibits $i_{max} = 25$ kA. I_S decreases far more rapidly in this case because $V_{VDC} > 0$. Once i_S reaches 0, the negative voltage v_S induced in the lightning current tail is not sufficient to generate a negative current on account of the extremely low M_i and despite the low V_{VDC}

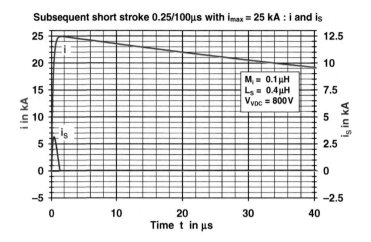

Figure 6.23 Current $i_S = i_V$ induced in the same loop (which contains a varistor with a large maximum DC operating voltage, $V_{VDC} = 800$ V) by negatively charged lightning that exhibits $i_{max} = 25$ kA. As V_{VDC} is relatively high, i_{Smax} exhibits a far lower peak value than in Figure 6.21 and drops precipitously. Hence, once i_S reaches 0, the negative voltage v_S in the lightning current tail is insufficient to generate negative current

which here is somewhat weaker than in the easy-to-calculate short-circuit current i_{Somax} of the loss-free loop (also see Figures 6.21 and 6.23). Hence a correction factor k_V for this short-circuit current i_{Somax} exhibited by the loss-free loop is determinable for such cases as well (see Figure 6.24). In scenarios involving very low M_i, the voltage generated by the somewhat negatively charged lightning (which exhibits both weaker current and higher di/dt_{max} values, i.e. 100–200 kA/μs as in Table 6.3) is only strong enough to induce varistor conductivity. But since negatively charged lightning can occur repeatedly, varistor current sizing should presuppose the same maximum current i_{max} as for 100–200 kA lightning in

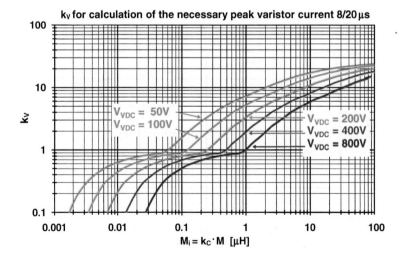

Figure 6.24 Varistor correction factor k_V for determining the requisite varistor current $I_{V8/20}$ based on the maximum short-circuit current in the loss-free loop, as in Equation 6.36. The suppositions applied to this calculation can be deemed to be conservative in that the values used here are on the safe side for scenarios involving standard lightning protection requirements

Table 6.3. This would enable a varistor to withstand four lightning flashes in quick succession with a maximum of 25–50 kA as in Table 6.3.

In the interest of devising a convenient procedure for determining the requisite nominal varistor current $I_{V8/20}$ for a 8/20 μs waveform for a given application, a varistor correction factor k_V as a function of effective mutual inductance $M_i = k_C \cdot M$ was determined, based on the presuppositions above (see Figure 6.24).

This procedure readily allows for the following determination of the requisite nominal varistor current $I_{V8/20}$ based on maximum short-circuit current i_{Somax} in the loss-free loop, which is readily derivable from M_i and L_S:

$$\text{Requisite nominal varistor current } (8/20 \text{ μs}): I_{V8/20} = k_V \times i_{Somax} = k_V \frac{M_i}{L_S} i_{max} \qquad (6.36)$$

Examples

Problem 1: Standard lightning protection requirements: lightning current where $i_{max} = 100 \text{ kA}$, $k_C = 0.5, M = 10 \text{ μH}, L_S = 20 \text{ μH}, V_{VDC} = 800 \text{ V}$. Determine the maximum short-circuit current i_{Somax} in the loss-free loop, as well as the requisite peak varistor current $I_{V8/20}$ (scenario as in Figure 6.20).

Solution: $k_C = 0.5 \Rightarrow M_i = k_C \cdot M = 5 \text{ μH} \Rightarrow i_{Somax} = i_{max} \cdot M_i/L_S = 25 \text{ kA}$ as in Equation 6.24. From Figure 6.24, $k_V = 4$ for $M_i = 5 \text{ μH}$ and $V_{VDC} = 800 \text{ V}$, thus according to Equation 6.36, $I_{V8/20} = 100 \text{ kA}$.

Problem 2: Standard lightning protection requirements: lightning current where $i_{max} = 100 \text{ kA}$, $k_C = 0.25$, $M = 0.2 \text{ μH}$, $L_S = 1 \text{ μH}$, $V_{VDC} = 400 \text{ V}$. Determine the maximum short-circuit current i_{Somax} in the loss-free loop, as well as the requisite peak varistor current $I_{V8/20}$.

Solution: $k_C = 0.25 \Rightarrow M_i = k_C \cdot M = 0.05 \text{ μH} \Rightarrow i_{Somax} = i_{max} \cdot M_i/L_S = 5 \text{ kA}$ as in Equation 6.24. From Figure 6.24, $k_V = 0.5$ for $M_i = 0.05 \text{ μH}$ and $V_{VDC} = 400 \text{ V}$. Hence $I_{V8/20} = 2.5 \text{ kA}$ from Equation 6.36.

6.6.3.3 Induced Current in Bypass Diodes

Bypass diodes can be damaged by lightning strikes near a PV installation. If a solar module is subjected to strong lightning current, the module's bypass diodes will in any case be knocked out by the elevated induced voltage and current.

Schottky diodes are mainly used in today's modules by virtue of their lower forward voltage decrease, although such diodes exhibit relatively low inverse current ranging from 40 to 100 V. However, voltage induced by lightning in a module loop can exhibit far higher values if it occurs at a specific distance from the lightning current (see Section 6.7.7). Fortunately, in the presence of moderately high voltage, many diodes can also be operated briefly in their avalanche range, which means that the problem is not as serious as might appear at first glance. Depending on the polarity of the voltage induced by the lightning current increase phase, bypass diodes can be subjected to loads in both the reverse and forward directions. Hence to calculate the possible bypass diode load for both possible polarities, this issue needs to be looked into more closely, to which end the equations in Section 6.6.3.2 can be useful.

Testing has shown that the induced voltages in laminates are commensurate with the calculated voltages exhibited by a filamentary conductor loop that traverses the solar cell's centre of gravity [6.3,6.6]. Mutual inductance M between lightning current $i_A = k_C \cdot i$ and the relevant loop can readily be determined for such module loops. In modules with metal frames, the induced voltage is reduced still further by a frame reduction factor R_R (see Section), which for individual module loops ranges from about 2.5 to 5, and for laminates is around 1. The mutual inductance M_i used in Equation 6.14 to calculate induced voltage is thus as follows, taking into consideration the frame effect k_C in Equation 6.6 and R_R in Equation 6.50:

$$M_i = \frac{k_C \cdot M}{R_R} \tag{6.37}$$

A lightning protection installation can only achieve the desired results if: (a) the solar generator is installed within the protection zone of a lightning protection device (see Sections 6.3.2 and 6.3.3); and (b) minimum clearance, which is usually around 50 cm, is allowed between lightning down-conductors and modules so as to avoid hazardous proximities (see Section 6.6.2).

Clearance b for an internal module's bypass diode conductor loop with n_Z series-connected solar cells ranges from 10 to 20 cm depending on cell size, whereby the length l of such a loop in commercial modules usually ranges from around 0.8 to 2 m. As this precludes the use of high M_i values, realistic M_i values for modules that are in relatively close proximity to lightning current i_A range from around 10 to 80 nH (see Figure 6.14).

Equation 6.22 cannot be used as it is to determine loop inductance L_S because: (a) the conductor radius r_o is unknown; (b) a different conductor radius applies to connections that are at an angle to the loop axis; and (c) certain additional bus bars come into play. However, an approximation can be formulated for L_S to provide values that are commensurate with actual go-live module readings.

Crystalline modules have a clearance b and front electrode contacts that are usually realized via two parallel-connected conductors (width $c \approx 0.02 \cdot b$) that are about $0.48 \cdot b$ apart (see Figure 6.25; for illustrations of this type of module see Figures 4.5, 4.7 and 4.9) and can be more or less regarded as a two-element bundle conductor whose equivalent radius equates to the geometric mean of conductor clearance and the radius of a bundle conductor. The latter parameter in solar cells should be assumed to be the radius of a conductor with a circular gauge that has the same surface area as a strip-like conductor.

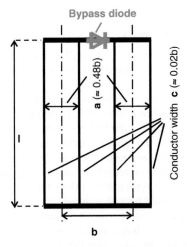

Figure 6.25   Diagram for approximate calculation of loop inductance L_S in a bypass diode loop

Hence the equivalent radius r_o is approximately as follows:

$$\text{Equivalent radius for determination of } L_S: \ r_o \approx \sqrt{0.48 \cdot b \cdot \frac{2 \cdot 0.02 \cdot b}{2\pi}} \approx 0.055 \cdot b \qquad (6.38)$$

The logarithm in Equation 6.22, which does not correlate with b for the most part, is around 18 for a typical module, thus making $\ln(18) \approx 3$. In as much as (a) connections that are at an angle to the conductor axis exhibit a smaller equivalent radius and (b) an additional conductor length ranging from $0.5 \cdot b$ to $1.5 \cdot b$ needs to be taken into account up to the module junction box containing the bypass diodes, the following approximation for loop inductance L_S (in µH) is obtained by adding 50 nH for the bypass diode and slightly modifying Equation 6.22:

$$\text{Bypass diode loop inductance } L_S \approx 1.2 \cdot (l + 2 \cdot b) + 0.05 \qquad (6.39)$$

In this equation, L_S is the loop inductance in µH that determines the bypass diode load if all lengths (length l, cell clearance b) are expressed in metres.

L_S typically ranges from around 1 to 3 µH, which has also been confirmed by lab measurements of selected modules.

As M_i is likewise proportional to l, the M_i/L_S ratio is mainly determined by clearance d, the presence of a metal frame and by k_C – but hardly at all by l.

Depending on the reciprocal locations and orientations of lightning down-conductor and internal module loops, two different scenarios can arise whose difference lies in the polarity of the voltage induced in the leader stroke, whereby the bypass and solar cell diodes are subjected to load either in the reverse direction (Figure 6.26) or in the forward direction (Figure 6.27).

The current exhibited by a bypass diode can be determined (as in Figure 6.17) by defining the relevant linearized equivalent circuit, whereupon the consequent bypass diode current can be calculated using Equations 6.26–6.35.

For operation in the avalanche range, a voltage source that exhibits forward voltage V_{ZA} and resistance R_{ZA} can be presupposed. In such a case $V_{ZA} = 20\,\text{V}$ (see Section 4.2.1) and $R_{ZA} = 5\,\text{m}\Omega$, which are conservative values, can be used as guide values. The bypass diode exhibits forward voltage V_{BA} and resistance R_{BA}. The following guide values can be presupposed for V_{BA}:

$$\text{For Schottky diodes: } V_{BA} \approx (1.5 - 2) \cdot V_{RRM} \qquad (6.40)$$

$$\text{For standard silicon diodes: } V_{BA} \approx (1.2 - 1.5) \cdot V_{RRM} \qquad (6.41)$$

where V_{RRM} = periodic inverse voltage peaks as in the diode datasheet.

The consequent bypass diode current i_{BR} under bypass operating conditions can be determined using Equations 6.62–6.35 in Section 6.6.3.2, insofar as the following V_V and R_V values are used as in Figure 6.28:

$$V_V = V_A = n_Z \cdot V_{ZA} + V_{BA} \qquad (6.42)$$

$$R_V = R_A = n_Z \cdot R_{ZA} + R_{BA} \qquad (6.43)$$

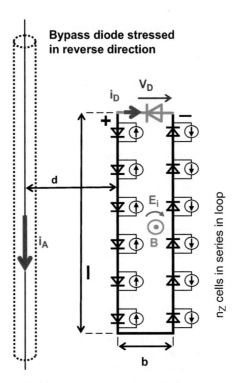

Figure 6.26 Bypass and solar cell diodes are subjected to a load in the reverse direction in the leader stroke. If the relatively high forward voltage is overshot and if the current is moderate, these diodes can operate – but only very transiently – in the avalanche range, whereupon voltage decreases precipitously on account of the elevated mutual inductance

Avalanche operation current i_{BR} was determined for selected guide values for M_i (10, 20, 40, 80 nH), for a mean value $L_S = 2\,\mu H$, and for $R_{ZA} = 5\,m\Omega$ and $R_{BA} = 50\,m\Omega$ (see Figures 6.29, 6.30 and 6.31).

Figures 6.29–6.31 show the diode loads and bypass diode energy consumption for the various current curves. Owing to the elevated mutual inductance V_A ranging from around 300 to 550 V for Schottky diodes (and up to 2 kV for silicon diodes with high breakdown voltage), bypass diode current i_{BR} decreases relatively quickly.

For silicon diodes with a breakdown voltage $V_{RRM} = 1\,kV$ for $M_i \leq 80\,nH$, the voltages induced by type 1 lightning are too low to induce an avalanche, i.e. $i_{BR} \approx 0$; even for type 2 lightning i_{BR} is still about 0 for $M_i \leq 20\,nH$ (see Figure 6.31).

Bypass diode current rises with increasing M_i, thus disproportionately increasing the risk of a diode failure. Under most circumstances diodes will not be damaged by low M_i values (about $\leq 20\,nH$), and diodes with specified avalanche characteristics can also withstand somewhat higher loads. According to the relevant datasheets, some Schottky diodes can withstand avalanche energy ranging up to 10 mJ, although the data here mostly pertain to weak current conditions. Selected correct bypass diode measurements are discussed in Section 6.7.7.6.

In as much as both M_i and L_S increase as the number of solar cells per loop increases, the number of cells n_Z per bypass diode loop is very weakly correlated with the M_i/L_S ratio. Since, according to Equation 6.42, breakdown voltage V_A increases with n_Z, higher n_Z values with relatively low V_{BA} have a beneficial effect, particularly for Schottky diodes, and help to limit the current i_{BR} in the avalanche range. Owing to voltage limitation, the externalized module voltage amounts to only about V_{BA} per bypass diode.

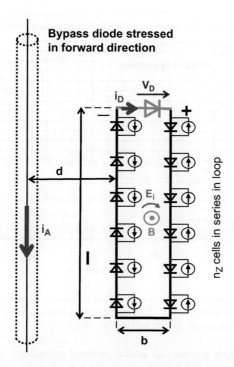

Figure 6.27 Bypass and solar cell diodes are subjected to a load in the forward direction in the leader stroke, whereby the solar cell and bypass diodes exhibit a relatively small voltage drop. The bypass diode will be knocked out if the allowable limit current is overshot. In this scenario, voltage decreases far more slowly on account of the lower mutual inductance

For the purposes of operation in the passband, a voltage source that exhibits forward voltage V_{ZA} and resistance R_{ZA} can be presupposed (see Figure 6.32). Bypass diodes likewise exhibit forward voltage V_{BF} and resistance R_{BF}. Around 4 mΩ can be used as the guide value for R_{ZF}. The following guide values can be presupposed for V_{BF} and V_{ZF}:

$$\text{For Schottky diodes: } V_{BF} \cdot 0.7 - 1 \text{ V} \tag{6.44}$$

$$\text{For standard silicon diodes and silicon solar cells: } V_{BF} \approx V_{ZF} \cdot 0.8 - 1.1 \text{ V} \tag{6.45}$$

Bypass diode and solar cells operating in avalanche mode

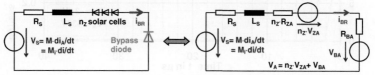

Figure 6.28 Equivalent circuit for calculation of bypass diode current $i_D = i_{BR}$ under a bypass diode load in the reverse direction as in Figure 6.26 (left, original circuit; right, linearized circuit for operation in the avalanche range)

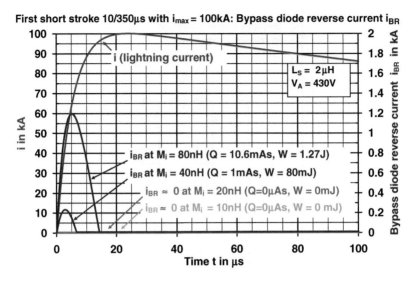

Figure 6.29 Bypass diode current i_{BR}, secondary to $i_{max} = 100$ kA lightning, under bypass operating conditions in a module loop that contains $n_Z = 18$ solar cells ($V_{ZA} = 20$ V) and a Schottky bypass diode ($V_{BA} = 70$ V), for selected guide values for effective mutual inductance M_i. $I_{BR} \approx 0$ for lower M_i values

The consequent bypass diode current i_{BF} under passband conditions can be determined using Equations 6.62–6.35 in Section 6.6.3.2, insofar as the following V_V and R_V values are used as from Figure 6.32:

$$V_V = V_F = n_Z \cdot V_{ZF} + V_{BF} \tag{6.46}$$

$$R_V = R_F = n_Z \cdot R_{ZF} + R_{BF} \tag{6.47}$$

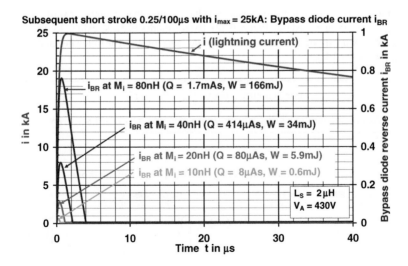

Figure 6.30 Bypass diode current i_{BR}, secondary to $i_{max} = 25$ kA negatively charged lightning, under bypass operating conditions in a module loop that contains $n_Z = 18$ solar cells ($V_{ZA} = 20$ V) and a Schottky bypass diode ($V_{BA} = 70$ V) for selected guide values for effective mutual inductance M_i

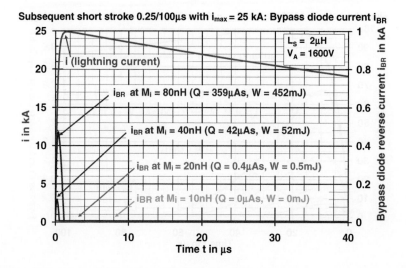

Figure 6.31 Bypass diode current i_{BR}, secondary to $i_{max} = 25\,kA$ negatively charged type 2 lightning, under bypass operating conditions in a module loop that contains $n_Z = 18$ solar cells ($V_{ZA} = 20\,V$) and a Schottky bypass diode ($V_{BA} = 1.24\,V$) for selected guide values for effective mutual inductance M_i. $I_{BR} \approx 0$ for lower M_i values

Passband current i_{BR} was determined for selected guide values for M_i (10, 20, 40, 80 nH), for a mean value $L_S = 2\,\mu H$, and for $R_{ZF} = 4\,m\Omega$ and $R_{BF} = 3\,m\Omega$ (see Figures 6.33 and 6.34).

Under passband conditions: (a) the voltage per bypass diode is just above V_{BF}, i.e. the effective externalized voltage in a module ranges up to a few volts; (b) considerably higher voltage and loads occur than in the cut-off region; and (c) a bypass diode can be knocked out by elevated peak current. The maximum allowable current for a specific type of diode should be determined experimentally (see Section 6.7.7.6). If no measurement data are available for the maximum allowable peak short-circuit current, this value can be conservatively estimated using the diode datasheet value for one-time peak current I_{FSM}, which is indicated for an 8.3 or 10 ms sinusoidal half wave and ranges from 300 to 600 A for most bypass diodes.

A failed bypass diode will generally exhibit an (imperfect) short that under certain circumstances can induce hazardous calefaction and possibly arcing as well.

6.6.4   Voltages in Lightning-Current-Conducting Cylinders

Lightning current conducted to a metal cylinder as in Figure 6.35a does not generate a magnetic field inside the cylinder, such as a metal tube with lightning-current-resistant shielding, which in turn means

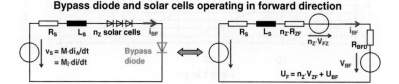

Figure 6.32 Equivalent circuit for calculation of bypass diode current $i_D = i_{BF}$ for bypass diode loads in the forward direction as in Figure 6.27 (left, original circuit; right, linearized variant). For optimally precise modelling of operation under high-current conditions, higher V_{ZF} and V_{BF} are used than for standard passband operation

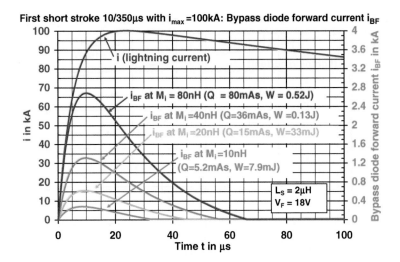

Figure 6.33 Bypass diode current i_{BR}, secondary to type 1 lightning, $i_{max} = 100\,\text{kA}$, under passband conditions in a module loop that contains $n_Z = 18$ solar cells ($V_{ZF} = 0.95\,\text{V}$) and a Schottky bypass diode ($V_{BF} = 0.9\,\text{V}$) for selected guide values for effective mutual inductance M_i

that no voltage is induced in a conductor loop inside such a cylinder. This type of two-lead conductor, which has $10\,\text{mm}^2$ Cu lightning-current-resistant shielding and is commercially available, provides ideal DC cable lightning protection for PV installations (see Figures 4.76 and 4.77). If only single-lead shielded conductors are available and, for example, dual insulation is required, a similarly effective configuration can be realized as in Figure 6.35b if the two ends of the lightning-current-conducting shielding are joined to each other and configured in a directly parallel fashion.

Although current flowing through the metal sheath does not induce voltage in the power optimizers inside the cylinder, it does induce an ohmic decrease in the direct-axis voltage current v at resistance R_M of

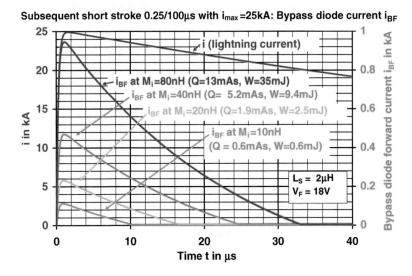

Figure 6.34 Bypass diode current i_{BR}, secondary to type 2 lightning, $i_{max} = 25\,\text{kA}$, under passband conditions in a module loop that contains $n_Z = 18$ solar cells ($V_{ZF} = 0.95\,\text{V}$) and a Schottky bypass diode ($V_{BF} = 0.9\,\text{V}$) for selected guide values for effective mutual inductance M_i

(Partial) lightning current on cylindrical metal shielding:

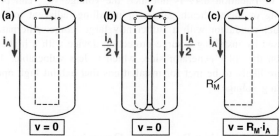

Figure 6.35 Conductor voltages in a metal shell (tube, pipe, channel, shielding) that conducts lightning current. Configurations a and b exhibit no voltage at all, while in configuration c an ohmic voltage decrease occurs solely via the resistance R_M of the metal sheath. However, this resistance can be kept acceptably low by using a large enough gauge

the metal sheath or shielding; this also occurs in the conductor loops that are connected on one side of the sheath (see Figure 6.35c):

> Ohmic voltage decrease in the metal sheath that exhibits a lightning current flow:
>
> $$v = R_M \cdot i_A \tag{6.48}$$

As a result of the skin effect in rapid processes that exhibit dense shielding, the key value for R_M may be even lower than DC resistance [Has89], a factor that should be taken into account to be on the safe side.

In lieu of individual shielded conductors, a series of conductors can be placed in a continuously enclosed metal lightning current raceway. This approach is particularly useful in cases where no cable with a suitable shielding gauge is available.

In cases where lightning current flows through the conductor shielding or a metal raceway, surge protection devices should be mounted such that they abut each lead's shielding at the beginning and end of the metal sheath, since, owing to the proximity to lightning current conductors in the shielding, an elevated voltage is induced in the non-shielded leads there. For PV systems, this should be done in the generator junction boxes and inverters. To prevent substantial lightning current from flowing through the surge protection devices during a lightning strike, which could ruin them, it should be ensured that the maximum voltage v_{max} at the metal sheath that exhibits a lightning current flow does not induce excessive conductivity in these devices. This can be accomplished by using conventional varistors if, for example, v_{max} does not exceed double the sum total of the nominal DC voltage of the two varistors (see Figure 6.10); that is, where

> Maximum voltage at a metal sheath that exhibits a lightning current flow is as follows:
>
> $$v_{max} = R_M \cdot I_{max} = R_M \cdot k_C \cdot i_{max} < 2(V_{VDC1} + V_{VDC2}) \tag{6.49}$$

where:
V_{VDC1} = maximum allowable DC operating voltage for varistor 1 on the solar generator side
V_{VDC2} = maximum allowable DC operating voltage for varistor 2 on the charge-controlled/inverter side
$i_{Amax} = k_C \cdot i_{max}$ = maximum lightning current in a metal cylinder or shielding
R_M = shielding or metal sheath resistance

If the requirement of Equation 6.49 is not met, the voltage v_{max} can be reduced by either increasing the shielding or metal sheath gauge, or installing, adjacent to the shielded line, an additional parallel relief conductor with a sufficiently large gauge. Alternatively, the allowable voltage can be increased by installing a surge protection device with a higher nominal operating voltage, particularly on the solar generator side, which is less vulnerable. The requirements of Equation 6.49 are more likely to be met for installations that exhibit high operating voltage than those with low operating voltage.

6.7 PV Installation Lightning Protection Experiments

6.7.1 Introduction

Rooftop PV modules can be struck by lightning. In 1993, a 60 kWp solar generator was installed, for PV system research purposes (particularly for inverters), on the roof of the then-new building that houses the Bern University of Applied Sciences (BFH) Electrical Engineering Department in Burgdorf. Since little was known at the time when this PV installation was designed about the effects of direct lightning strikes on PV systems, in the interest of developing an optimal lightning protection concept, from 1990 to 1993 a series of term papers and diploma theses were realized at the department's high-voltage lab involving extensive testing of solar cells, solar modules and a simulated PV installation. To this end, a surge current generator was built and a test apparatus for replicable monitoring of solar module characteristic curves under defined irradiance conditions was developed [6.4–6.7].

The initial tests revealed that solar modules with metal frames are very robust and that lightning current flows in the immediate surroundings of solar cells in such modules induced only minor damage in them. Hence, seamless lightning protection appeared to be attainable. Further experiments showed that damage from electromagnetic fields can be averted by increasing the clearance between lightning current and solar cells to a few centimetres. Moreover, high-voltage lab tests showed that lightning can be reliably deflected by lightning conductors that are only around 30 cm long and that the consequent lightning current can be shunted in a controlled manner to down-conductors that run past the solar cells at a few centimetres from them and preferably are part of the mounting system.

Extensive induced-voltage measurements were carried out on the BFH solar modules and solar generator models in 1999 and 2000 under the auspices of an EU project. Specifically, quantitative measurements were realized concerning the effects of the solar module metal frames and of various types of grounding installations on the strength of the voltages induced by lightning. These measurements allowed for experimental confirmation of theories that had been advanced in this domain.

6.7.2 The Surge Current Generator

6.7.2.1 Key Lightning Parameters for PV Installations

Lightning is a threat to PV installations on account of its high voltages, as well as from strong currents that can damage system wiring and/or surge protection devices. According to the law of induction, during the rapid current increase induced in system wiring by a lightning strike, the direct-axis voltage current ($v = L \cdot di/dt$) drops precipitously, inducing high voltage ($v = M \cdot di/dt$) in the installation loops. Hence, in order for lightning tests to be realistic, the surge current generator must have the capacity to generate di/dt values far exceeding these values for an average lightning flash. Further, since induced current in a shorted loop reaches a maximum of around $i_{max} \cdot M/L$ (see Section 6.6.3.1), the surge current generator needs to be able to generate maximum current I_{max} far exceeding these values for an average lightning flash (see Table 6.5).

Table 6.5 Guide values for lightning (also see [6.20])

	Maximum current i_{max} (kA)	Maximum curve $(di/dt)_{max}$ (kA/μs)	Load Q (A s)
Average lightning flash	30	25	9
Strong lightning flash	100	100	100

6.7.2.2 Structure and Key Characteristics of the Surge Current Generator

The surge current generator, which was used for tests between 1990 and 1998, originally comprised 10 surge condensers (1.2 μF/50 kV; also see Figure 6.39), and because of its coaxial configuration was specially designed for subjecting individual, metal-framed solar modules to acute voltage increases. The maximum module width at the time was around 50 cm.

In the EU project, the surge current generator was considerably enlarged to allow for investigations of integral modules and even solar generator models of up to 1.2 · 2.25 m. Table 6.6 lists the maximum surge current guide values that were attainable with the BFH solar generator.

Figure 6.36 displays the layout of an expanded coaxial surge current generator that allows solar modules and wired solar generators with surfaces of up to about 1.25 · 2.25 m to be subjected to the magnetic field of simulated lightning. Figure 6.37 shows three Kyocera KC60 modules mounted in the generator, while Figure 6.38 displays a surge current that was used for many of the tests.

Both generators exhibited i_{max} and di/dt_{max} values that far exceeded those of an average lightning flash, and only exhibited below-average values in this regard for load Q. The new generator was expanded in 2007–2008 to allow for longer surge currents with Q ranging up to around 100 A s.

Building and commissioning this generator was labour intensive, and owing to the installation's elevated inherent electromagnetic disturbances while in operation, getting the measurement system right (shunts, distributors, shielding, coaxial cable grounding, adjustment problems) took a great deal of effort. Figure 6.38 illustrates a characteristic surge current exhibited by the generator and the exactitude of the installation's measuring system.

6.7.3 Test Apparatus for Solar Module Characteristic Curves

To be able to quantify the effects of lightning current on solar module functionality, it is necessary to record in a precisely replicable manner the I–V characteristic curves of solar modules under defined insolation and module temperature conditions. To this end, a relatively inexpensive solar module test apparatus with 30 fluorescent lamps was developed within the framework of a term paper diploma thesis [6.14]. The light spectrum of the Osram Biolux lamps that were used is very similar to that of sunlight. This test apparatus allows for the production of insolation ($G_{max} \approx 300$ W/m^2) that is homogeneous to within a few percentage points, on a rectangular surface that is approximately 130 cm long and 50 cm wide. In 2002 an optimized version of the apparatus was built (useable surface: around 170 cm × 70 cm; $G_{max} \approx 500$ W/m^2) that allowed for measurement of solar generator I–V characteristic curves before and after exposure to simulated lightning current.

Table 6.6 Maximum guide values attainable with the BFH surge current generator

	Maximum current i_{max} (kA)	Maximum curve $(di/dt)_{max}$ (kA/μs)	Load Q (A s)
Original generator	108	53	0.6
Expanded generator	120	40	1.2

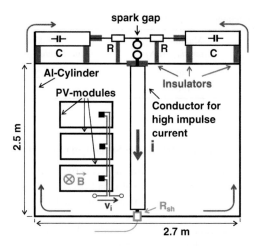

Figure 6.36 Schematic layout of the expanded coaxial surge current generator at BFH which has been in use since 1999 and is composed of 20 parallel RC elements ($R = 4.1\,\Omega$, $C = 1.2\,\mu F/50\,kV$). Guide values: $i_{max} = 120\,kA$, $di/dt_{max} = 40\,kA/\mu s$, $Q_{max} = 1.2\,A\,s$ (KFS = sphere gap). Although the maximum current curve in the expanded installation falls somewhat short of that in the original one, the new installation allows for investigations of both far larger module surfaces and the characteristics of wired solar generator models

Figure 6.37 View inside the upgraded surge current generator, with a three-module Kyocera solar generator mounted in the installation for testing purposes

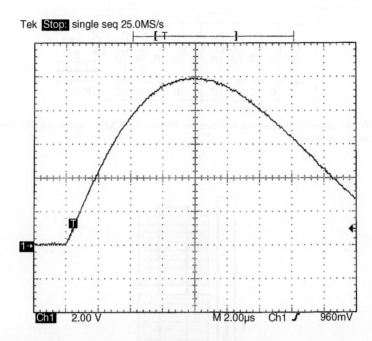

Tek ▊Stop▊ single seq 25.0MS/s

Ch1   2.00 V   M 2.00µs   Ch1 ʃ   960mV

Figure 6.38 Waveform of the surge current that was used for many of the tests ($i_{max} \approx 100\,kA$ and $di/dt_{max} \approx 25\,kA/\mu s$; unit of measure, 20 kA/div. and 2 µs/div.)

6.7.4   Solar Cell and Solar Module Damage Induced by Surge Current

For cost reasons, many of the tests during the 1990–1993 period were realized initially using individual cells, then with three-cell mini-modules, and finally with BFH's own solar modules – in all cases using the test set-up illustrated in Figure 6.39. For the tests with individual cells and three-cell mini-modules, the simulated lightning current was induced through a wire that was 1–4 mm from the edge of the cell being tested (see Figure 6.40).

For the module tests, the surge current was conducted to the shorter side of the module frame (see Figure 6.39), the centre of the module frame, or for laminates to a flat conductor that was placed just below the back of the module (see Figure 6.41). This allowed for the simulation of direct lightning strikes on module frames, or laminate mounting systems.

The primary effects that were anticipated for these tests – namely, total failure of the module resulting from direct flashovers between the module frames and the immediately adjacent solar cells – did not occur at all. The relatively high di/dt values ranging from 40 to 50 kA/µs resulted in an inductive direct-axis voltage current along a metal conductor (inductance layer $L' = L/l$ of 1 µH/m) amounting to around 40 to 50 kV/m.

6.7.4.1   Results of the Cell and Module Tests

The sole effect of the lightning current flowing through a module frame was progressive worsening of the *I–V* characteristic curve filling factor of the relevant cell or module (see Figures 6.42 and 6.43), which may have been attributable to the lightning current's rapidly changing electromagnetic field. Usually solar cells with appreciable filling factor degradation exhibit visible damage on the front and back side of the contact, which is presumably induced by lattice eddy currents. In addition to the changes in the module characteristic curves, some modules also exhibited bypass diode failure, mainly in the guise of a post-test short [6.4].

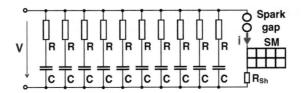

Figure 6.39 Test set-up used for load tests at BFH between 1990 and 1993. KFS = sphere gap; SM = solar module; R_{Sh} = current measurement shunt

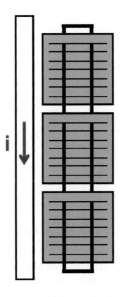

Figure 6.40 Test set-up for three-cell mini-solar modules

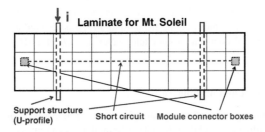

Figure 6.41 Surge current feed-in for a Siemens M55 solar laminate that is installed at Mont Soleil. For the first tests, the surge current flowed through a flat conductor that was placed just below the laminate (conductor surface cell clearance: 2 mm). Tests were then conducted in which a U-profile under the laminate was subjected to surge current

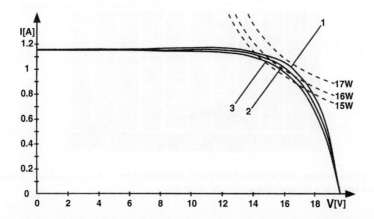

Figure 6.42 Characteristic curves for a framed MSX60 module with $G = 300$ W/m^2; this module shorted out on being subjected to surge current: 1, baseline characteristic curve; 2, post-surge current curve ($i_{max} = 53$ kA, $di/dt_{max} = 33$ kA/μs); 3, curve following a second surge current ($i_{max} = 80$ kA, $di/dt_{max} = 53$ kA/μs)

Figures 6.44 and 6.45 display the analogous damage incurred by a monocrystalline solar cell. Figure 6.46 displays a similar defect in a polycrystalline solar cell, where no damage is visible on the front as there are no shorted loops on the edge of the solar cell.

The *I–V* characteristic curve degradation was probably attributable not only to increased series resistance resulting from damage to the front- and back-contact sides, but also to semiconductor material defects induced by the rapidly changing electromagnetic field. These experiments showed that the *I–V* characteristic curve degradation in an entire solar cell is attributable to the magnetic-field-damaged area that is near the lightning current.

As a comparison of Figures 6.42 and 6.43 shows, laminates incur greater damage than framed modules under the same conditions (also see Section 6.7.7). The lightning-current-induced damage incurred by module frames is mainly attributable to the module configuration. For example, modules that have 1–2 mm

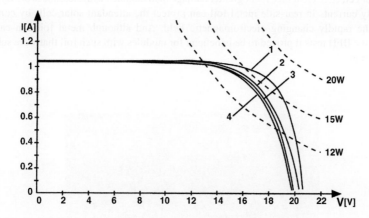

Figure 6.43 Characteristic curves for an M55 laminate (used for Mont Soleil) with $G = 320$ W/m^2; this module shorted out on being subjected to surge current: 1, baseline characteristic curve; 2, post-surge current curve ($i_{max} = 53$ kA, $di/dt_{max} = 33$ kA/μs); 3, curve following a second surge current ($i_{max} = 69$ kA, $di/dt_{max} = 43$ kA/μs); 4, curve following a third surge current ($i_{max} = 80$ kA, $di/dt_{max} = 53$ kA/μs)

Figure 6.44 Defects in a front-contact lattice of a solar cell: defects (fractures; see arrows) induced by surge current in the front-contact lattice of a Siemens monocrystalline solar cell

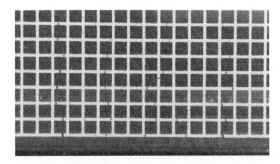

Figure 6.45 Defects in the back-contact lattice of a monocrystalline solar cell: surge-current-induced defects on the back-contact lattice of a Siemens monocrystalline solar cell. The contact lattice exhibits numerous fractures

of frame–solar cell edge clearance incur greater damage than modules with clearances of, say, 5 mm. In addition, eddy currents in rear-side metal foil can protect the attendant solar cells by considerably attenuating the rapidly changing electromagnetic field. And although metal foil can cause other problems, in the BFH tests it proved to be beneficial for modules with such foil that were subjected to lightning current.

Figure 6.46 Defects on the back-contact side of a polycrystalline solar cell: surge-current-induced cracks in a Telefunken Systemtechnik polycrystalline solar cell

In a Kyocera LA361J48 module, which by chance integrated both advantageous conditions (metal foil and a large cell–frame clearance), no change in the $I–V$ characteristic curve was observed after the module was subjected to surge current a number of times (also see Section 6.7.7). The $I–V$ characteristics changed as in Figure 6.42 in the tests of the Solarex MSX60 and Siemens M65 modules, which exhibited no metal foil and a smaller frame–cell clearance.

Lightning-current-induced damage can also be associated with module wiring. In as much as a specific lightning current induces the most serious damage in a shorted solar cell or module but induces less damage if the cell or module is at open circuit, most of the tests described above were conducted using shorted modules.

6.7.5 Improving Module Immunity to Lightning Current

The fact that the BFH tests (see Section 6.7.4) induced relatively little damage, particularly in metal-framed modules, suggests that such damage could be avoided altogether by increasing the clearance between the lightning current path and the solar cells.

In point of fact, BFH tests conducted in 1993 showed that all such damage could be eliminated in a Siemens M65 monocrystalline solar cell when lightning current (i_{max} up to 111 kA and di/dt up to 56 kA/μs) was conducted not directly into the module frame as in the previous tests, but rather into an aluminium U-profile arranged just below the frame and secured to it [6.6]. In the set-up for these tests (see Figure 6.47), the centre of gravity of the surge current was around 6 cm from the solar cell nearest to it. In earlier tests, where the frame of an M65 solar module was subjected to similar surge current, minor damage was observed (see Figure 6.42).

The results of these tests suggest that magnetic-field-induced damage to framed solar modules resulting from lightning current can be eliminated altogether by simply upping the clearance between the lightning current path and the solar cells. But of course what is needed here is to attain this minimum clearance for

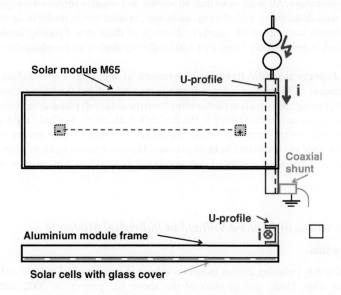

Figure 6.47 Test set-up for a framed module with increased clearance (around 6 cm) between the surge current centre of gravity and the solar cell, which was shorted during the test. The surge current was conducted directly into an aluminium U-profile (50 mm · 40 mm · 4 mm) on which a Siemens M65 was mounted. No damage was observed, including with maximum surge current ($i_{max} \approx 111$ kA and $di/dt \approx 56$ kA/μs)

all configurations (so as to avoid direct lightning strikes on the module) without shading the solar generator to any appreciable degree. This can be accomplished via apparatuses such as miniature lightning conductors or other devices (see Section 6.7.6).

Note: In the above tests, it only proved possible to afford metal-framed modules seamless lightning protection, since the shorted loop engendered by the frame weakens the magnetic field somewhat, thus mitigating the deleterious effects of lightning current (also see Section 6.7.7 and [6.3]).

M55 laminates of the type used at Mont Soleil were also tested. The lightning current centre of gravity of the laminates and their U-profiles is somewhat closer to the solar cells (around 3.5 m), and one section of the lightning-current-conducting profile is only around 2 mm below the solar cell plane. As in our earlier flat-profile tests, the use of U-profiles of the type used at Mont Soleil resulted in extreme changes in the $I–V$ characteristic curve. Hence, to completely avoid lightning-current-induced damage to laminates, the solar cell–lightning current path clearance would need to be even larger, to an extent that can only be determined via further testing.

6.7.6   Mini-lightning Conductors for PV Installations

Previous attempts to avert direct lightning strikes on PV installations have mainly involved the use of high vertical or horizontal lightning conductors that: (a) extend far above the installation's grounding cable [Ima92]; (b) at certain times of day and seasons of the year cast shadows that impinge upon PV installation energy yield; and (c) are eyesores in certain situations, e.g. when mounted on buildings. The fact that even a few centimetres of solar module–lightning current path clearance is sufficient to avert damage allows for the use of far shorter lightning conductors. To this end, mini-lightning conductors that are only 30 cm high were developed for the 60 kWp test PV installation on the building that houses BFH's Mechanical Engineering Department [6.24]. The efficiency of these devices was exhaustively tested at the Emil Haefely Company's High-Voltage Lab using lightning flashes up to 3 m long and up to 2 MV of surge current (1.2/50 μs; see Figure 6.48).

These tests showed that mini-lightning conductors mounted on either side of a solar module provide blanket lightning protection. All of the more than 40 positive and negative surge currents generated by the lightning flashes were deflected by the lightning conductors; in other words, no lightning strikes on either the modules or frames were observed. Another advantage of these mini-lightning conductors is that, owing to their minute diameter (only 1 cm), they cast hardly any shadow on the adjacent modules of large solar generators.

Figure 6.49 displays the mini-lightning conductors at the BFH PV installation, which is composed of Siemens M55 modules. A conductor is mounted above the topmost module of each array, such that lightning ends up a few centimetres from the solar cell plane after being shunted into the grounded mounting structure. Owing to the device's peak effect, not only the topmost row of modules, but also the entire array, lie within the protection zone, which means that the entire installation is protected against direct lightning strikes. However, fewer lightning conductors could have been used, as it has since emerged that spacing these devices 6 m apart is sufficient (see Section 6.3.2).

6.7.7   Measurement of Induced Voltage in Individual Modules

6.7.7.1   Introduction

As noted in Section 6.6, lightning current induces a voltage in each external module and internal solar generator conduit loop. Under the auspices of the above EU project, in 2000 further extensive measurements of the voltages induced by simulated lightning current in various types of modules were conducted using the BFH's new surge current generator. The tests were conducted on then-typical modules whose dimensions were compatible with those of the generator.

Figure 6.48 Artificial lightning flash directed at a solar module that is protected by lightning conductors: tests of the mini-lightning conductor developed at BFH's PV Lab. The tests were conducted at the Emil Haefely Company's High-Voltage Lab. A lightning flash around 3 m long and peak voltage of around 2 MV were directed at a Siemens M55 solar module that was protected by lightning conductors

As the solar cells of the crystalline PV modules tested were series connected using a meander configuration, the mutual inductance used to measure the induced voltage at the module terminals is considerably lower than the voltage that would otherwise have been induced in a rectangular loop that has the same dimensions as the module. As noted in Section 6.6.3.3, the induced voltages in laminates are commensurate with the voltages exhibited by filamentary conductor loops that traverse the solar cell's centre of gravity [6.3,6.6].

A crystalline solar module normally contains two internal module loops, whose induced voltages can (depending on reciprocal orientation) aggregate, which is the standard case, or exhibit partial or complete

Figure 6.49 View of the 60 kWp solar generator (tilt angle 30°) on the roof of BFH's Department of Solar Sciences building. The mini-lightning conductors are mounted just above the top row of the modules. Lightning is conducted directly into the mounting rack's U-profile that is arranged on the module, thus avoiding damage since the lightning current is shunted to a point that is a few centimetres away from the solar cell plane. However, this arrangement does not avert bypass diode damage at the lightning juncture point

reciprocal compensation, as occurs with certain special types of modules. Induced voltages in the latter type of module are lower.

However, this situation is complicated by the fact that the loop bypass diodes either block or conduct current. In addition, in the presence of elevated voltages that do not jeopardize the integrity of blocking bypass diodes, such diodes (if intact) may transition to the avalanche state where they limit the voltage in their respective loops to slightly above their breakdown voltage V_{BA} (see Figure 6.28 in Section 6.6.3.3). A distinction can be made between the following two types of modules, according to bypass diode effect:

- **Additive module with an even number of cell rows** (by far the most common type, one example being the Kyocera KC60): Depending on lightning current direction, induced voltages in n_B in series-connected internal module bypass diode loops reach a cumulative maximum of around $n_B \cdot V_{BA}$ (whereby the bypass diodes perform a blocking or limiting function) or are almost 0 (whereby the bypass diodes conduct current).
- **Compensating modules** (e.g. Solarex MSX60/64; Siemens SM46 and SM55): The induced-voltage polarities in the individual loops exhibit varying polarities. Externalized loop voltage always exerts either a blocking or limiting effect on the relevant loop diodes. All loops whose voltage induces reverse polarity are shorted by the bypass diodes.

In the case of framed modules, short-circuit current is induced in highly conductive frames (as in Equation 6.23) by lightning current in a nearby down-conductor. This short-circuit current weakens the loop's magnetic field, thus reducing the voltage at the solar module terminals and effective mutual inductance M_i (see Equation 6.37) shielding via a short-circuit mesh [Rod89]. The total current flow through the module is reduced to almost $\Phi = 0$, although the module's induced voltage does not decrease by this much owing to magnetic field differences.

For reasons of space, only the test results for additive modules are indicated here since they are by far the type used most frequently.

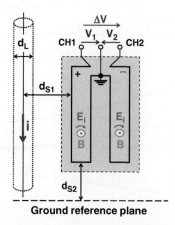

Figure 6.50   The KC60 module in a parallel configuration (lengthways sides parallel to the surge current path) with both loops at open circuit ($d_{S1} = 450$ mm, $d_{S2} = 900$ mm). KC60 module data: $P_{max} = 60$ Wp at STC power output; 751 mm long, 652 mm wide. Configuration, $9 \cdot 4$ cells; cell height, 77 mm; cell width, 154 mm. As the inner loop was nearer to the surge current path, the magnetic field and induced voltage there were higher

6.7.7.2   Voltage Induced in an Additive Module (K60) in a Parallel Configuration

The highest voltages occur when surge current first appears and di/dt is at its peak. This elevated voltage also induces the most serious problems and damage. The short-circuit current induced in a metal frame reduces the induced voltage by weakening the magnetic field. In a module that is in relatively close proximity to the lightning current path, a loop that is farther from the path may overcompensate by exhibiting a polarity reversal. As is the case with compensating modules, as a result of this reversal and on account of the bypass diodes, only the inner loop voltage has an effect on the clamps of an additive module. In the interest of avoiding needless bypass diode failures during the measurement process (these failures were particularly frequent with low cell surge current clearance), the measurements described here were realized without bypass diodes, to which end conducting diodes were replaced by shorts and blocking diodes by an open-circuit state.

Figure 6.50 displays the test configuration that was used. Figures 6.51 and 6.52 respectively display the voltages $(di/dt)_{max} = 25$ kA/μs induced by lightning current with a framed KC60 module in a parallel configuration and with 450 mm of clearance.

Since the limited usable test space in the surge current generator restricted the clearance size that could be investigated, a mathematical model for mutual inductance was developed that also took account of metal-frame lightning current and that was validated via the generator measurement readings. Figure 6.53 displays the various clearances that were determined using the model and the maximum induced-voltage readings for a KC60 module in a parallel configuration.

Tests of the worst-case scenarios were conducted. If di/dt or the module orientation is inverted, the bypass diodes are conductive and the voltage is close to 0.

6.7.7.3   Voltage Induced in an Additive Module (KC60) in a Normal Configuration

The same measurements were also performed with a module configuration rotated 90°. Figure 6.54 displays this configuration, as well as the maximum induced voltage, with various clearances, for a KC60 module in a normal configuration (i.e. with the module's lengthways sides at a right angle to the lightning current path, Figure 6.55).

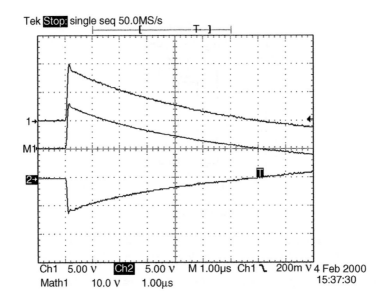

Figure 6.51 Voltage induced in a KC60 module laminate (data and configuration as in Figure 6.50) in a parallel configuration (lengthways sides parallel to the lightning current path), secondary to surge current as in Figure 6.38. The effective voltages are 100 times higher due to the fact that a 100:1 voltage divider was used. If the effect of the minor oscillation at the outset is disregarded, the approximate maximum voltages were $V_1 = 900$ V, $V_2 = 600$ V and $\Delta V = 1500$ V, which are consistent with the mutual inductances $M_{i1} \approx 36$ nH and $M_{i2} \approx 24$ nH

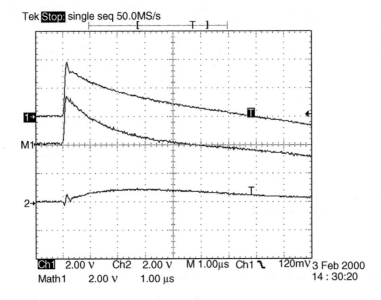

Figure 6.52 Voltage induced in an aluminium-framed KC60 module (data and configuration as in Figure 6.50) in a parallel configuration (lengthways sides parallel to the lightning current path), secondary to surge current as in Figure 6.38. The effective voltages are 100 times higher due to the fact that a 100:1 voltage divider was used. The voltage in open-voltage loop 2 is positive. In an actual PV installation, the bypass diode would be conductive and only loop 1 voltage would exert an effect. In this case, $V_1 \approx \Delta V \approx 350$ V and thus $M_{i1} \approx 14$ nH

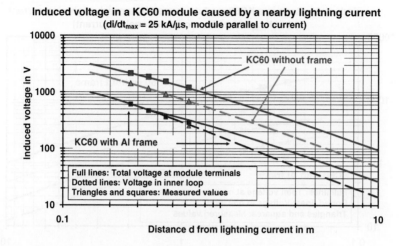

Figure 6.53 Maximum voltages ($di/dt_{max} = 25$ kA/µs) induced in an additive KC60 solar module in a parallel configuration (lengthways sides parallel to the lightning current path) with and without a metal frame (upper and lower curves respectively). The framed module exhibited a circular current, which weakened the lightning current's magnetic field, resulting in far lower induced voltages in the module

6.7.7.4 Frame Reduction Factor for Various Types of Modules

The characteristics of framed modules and laminates can be compared by folding in a frame reduction factor R_R that indicates the maximum induced voltage in the presence and absence of a frame. R_R is determined by module type and by clearance d ($= d_{S1}$ in Figures 6.50 and 6.54) between the module and lightning current path (see Figure 6.56).

Hence the frame reduction factor R_R is defined as follows:

$$R_R = \frac{\text{Maximum voltage without frame}}{\text{Maximum voltage with frame}} = \frac{V_{\text{max without frame}}}{V_{\text{max with frame}} } \qquad (6.50)$$

The frame reduction factor R_R thus defined *always* refers to a module's overall induced voltage, which is usually slightly lower in parallel-configuration modules (see Figure 6.50) for the loop nearest to the lightning current path; this loop is critical in terms of bypass diode load. When voltages in various loops

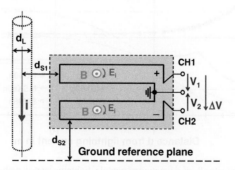

Figure 6.54 The KC60 module in a normal configuration (lengthways sides at a right angle to the surge current path) with both loops at open circuit ($d_{S1} = 450$ mm, $d_{S2} = 900$ mm). With this configuration, the magnetic fields in the two loops are identical with the same clearance, which means that $V_1 = V_2$ and thus $\Delta V = 2V_1$

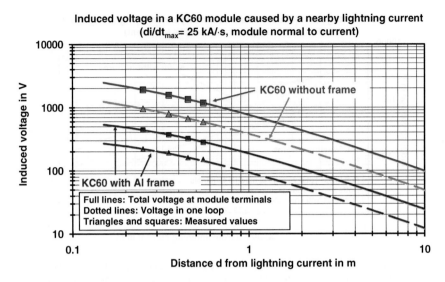

Figure 6.55 Maximum voltages ($di/dt_{max} = 25\,kA/\mu s$) induced in an additive KC60 solar module in a normal configuration (lengthways sides at a right angle to the lightning current path) with and without a metal frame (upper and lower curves respectively). The framed module exhibited a circular current, which weakened the lightning current's magnetic field, resulting in far lower induced voltages in the module

are compared with each other, the above reduction for internal loop modules with three or more parallel loops may be lower in some cases, but will always exhibit a minimum of around $R'_R = 2.2$. The following principles hold true for *all* modules: (a) a metal frame will appreciably reduce induced voltage and M_i will be lower accordingly (as in Equation 6.37); and (b) R_R will be somewhat lower for parallel-configuration than for normal-configuration modules.

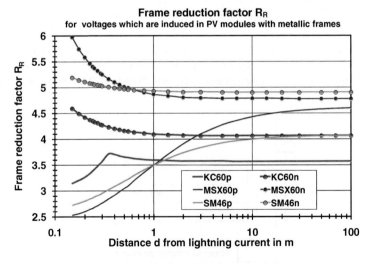

Figure 6.56 Frame reduction factor R_R as a function of module–lightning current path clearance d for the following types of modules: additive KC60 module (red curves: p = parallel configuration; n = normal configuration); four-row compensating MSX60 module; three-row compensating SM46 module. Larger clearances $d = d_{S1}$ were determined using a calculation model

Table 6.7 Recommended frame reduction factors R_R for calculation purposes

For parallel-configuration modules (see Figure 6.50)	$R_R = 3$
For bypass diode loops in parallel-configuration modules (see Figure 6.50)	$R_R = 2.5$
For normal-configuration modules (see Figure 6.54)	$R_R = 4$
For bypass diode loops in normal-configuration modules (see Figure 6.54)	$R_R = 4$

The mathematical models were also used for simulations involving characteristic square-cell modules that exhibited configurations such as $9 \cdot 4\,150$ mm cells, $9 \cdot 4\,200$ mm cells, $12 \cdot 6\,125$ mm cells and $12 \cdot 6\,150$ mm cells. Most of the frame reduction factors R_R obtained with these simulations exceeded the KC60 values, ranging from around 3.2 to 6. To be on the safe side, such calculations should be performed using the figures listed in Table 6.7.

6.7.7.5 Effect of Aluminium Foil on the Back Sides of Modules

The aluminium foil used as a moisture barrier on the back sides of solar modules exhibits eddy currents if the relevant module is struck by lightning. This weakens the magnetic current, thus substantially reducing induced voltages. In the interest of quantifying this effect, aluminium foil was glued to the back of the previously tested KC60 framed modules and laminates, such that the minimum foil–frame clearance was around 5 mm. The tests showed that the presence of this aluminium foil further reduced the induced voltage by a foil reduction factor $R_F \approx 7$–10. In other words, this reduction was above and beyond that associated with the frame reduction factor. However, it is more difficult for modules with aluminium foil to meet protection class II insulation requirements.

6.7.7.6 Measurement of Bypass Diode Lightning Current Sensitivity

As noted in Section 6.6.3.3, bypass diodes can be knocked out if lightning strikes in close proximity to a PV installation. Such diodes often exhibit an (imperfect) short circuit, which can be particularly hazardous for PV installations comprising four parallel non-fused strings. The currents engendered by such an event can be approximately calculated using the methods described in Section 6.6.3.3. The present section briefly discusses the results of tests that aimed to demonstrate the accuracy of such calculations using test-compatible waveforms (around $6/350\,\mu s$) and KC60 laminates in a parallel configuration as in Figure 6.50.

Figures 6.57 and 6.58 display two tests where solar cells and non-conducting bypass diodes were subjected to a load in the reverse direction and thus briefly exhibited an avalanche state. Owing to the voltage drop at the avalanche-state diodes, the leader-stroke current is relatively weak despite the elevated induced voltage. The diode is still intact in Figure 6.57, but is knocked out in Figure 6.58.

Hence the 80SQ045 diode used here withstood an avalanche energy of 78 mJ in an avalanche state and surge current of $i_{max} \approx 240$ A, whereby the effective mutual inductance was $M_i \approx 46$ nH. Comparing the two figures with Figures 6.29 and 6.30 reveals that in this module such diodes should normally be able to withstand $M_i \leq 20$ nH, type 1 lightning and type 2 lightning.

Figures 6.59 and 6.60 display two tests where solar cells and bypass diodes were subjected to loads in the forward direction. The diode current in the lightning current head is now considerably stronger owing to the small voltage drop. The diode is still intact in Figure 6.59, but in Figure 6.60 is knocked out by the elevated surge current.

Hence the 80SQ045 diode used withstood a surge current of $i_{max} \approx 1.95$ kA in a passband state, whereby the effective mutual inductance was $M_i \approx 52$ nH. Comparing these two figures with Figures 6.33 and 6.34 reveals that in this module such diodes should normally be able to withstand $M_i \leq 20$ nH, type 1 lightning and type 2 lightning.

With the upgrading of the surge current generator, the lightning currents used for these tests were greatly improved relative to the waveforms used in [Häb07]. The advantage of the waveform used for

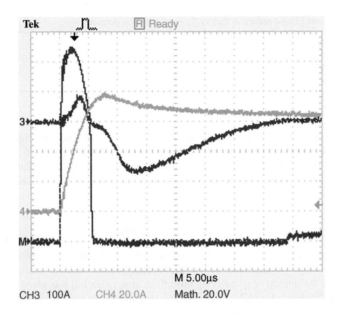

Figure 6.57 Bypass diode avalanche reverse current i_{BR} in a module loop with $n_Z = 18$ solar cells in a KC60 module in a parallel configuration ($d_{S1} = 40$ cm ($M_i \approx 46$ nH)), secondary to lightning current (green) where $i_{max} \approx 79$ kA and $di/dt_{max} \approx 20$ kA/µs. A 80SQ045 bypass diode was used (nominal inverse current $V_{RRM} = 45$ V). The diode that was used remains intact in this case. In an avalanche state, voltage (red) is limited to ≈ 90 V for 5 µs, peak voltage (purple) is around 240 A, and the calculated avalanche energy is around 78 mJ, although the datasheet indicates only 10 mJ. Inasmuch as, after peaking, the lightning current briefly exhibits a somewhat higher negative value than di/dt, the bypass and solar cell diodes in the lightning current tail are subjected to a load in the forward direction; this effect would be less pronounced with a standardized lightning current

these tests (around 6/350 µs) is that it allowed for the realization of useful tests in both the reverse and forward directions up to a maximum current of around 120 kA and di/dt_{max} of around 40 kA/µs without changing the test set-up. The test results for other diodes and modules, and the relevant theoretical investigations, can be found in [6.25].

6.7.7.7 Effect of Adjacent Framed Modules on Induced Voltage

The lightning-current-induced voltage in any given module of a solar generator that contains metal-framed modules is determined not only by the lightning current and by the current exhibited by the frame, but also by the frame current in adjacent modules. This phenomenon complicates the task of determining module voltage.

During the EU-sponsored project mentioned in Section 6.7.2.2, such voltages were investigated for relatively simple scenarios [6.15]. These investigations revealed that the voltages induced in the innermost modules (i.e. those nearest to the lightning current path) of a multi-row module configuration are normally somewhat lower (up to 40% in compensating modules) owing to the presence of more outward-lying adjacent modules. Conversely, the voltages in the outermost modules are increased by the presence of inner modules, whereby this increase can range up to 100% for compensating modules. This effect is far less pronounced in additive modules, which are by far the most common type used. However, in no case did the voltage overshoot that exhibited by an individual innermost module or a neighbouring module.

An empirical aggregation principle also came to light that allows for calculation of the voltage induced in a string comprising modules that exhibit inhomogeneous reciprocal clearances. According to this

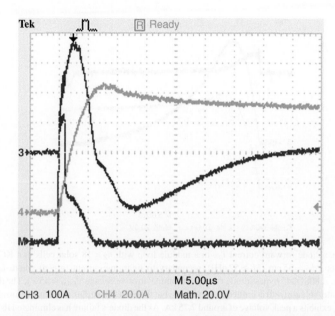

Figure 6.58 Bypass diode avalanche reverse current i_{BR} in a module loop with $n_Z = 18$ solar cells in a KC60 module in a parallel configuration ($d_{S1} = 40$ cm ($M_i \approx 46$ nH)), secondary to lightning current (green) where $i_{max} \approx 86$ kA and $di/dt_{max} \approx 22$ kA/μs. A 80SQ045 bypass diode was used (nominal inverse current $V_{RRM} = 45$ V). In this case the diode failed after around 1 μs. In an avalanche state, the voltage (red) is limited to ≈80 V for 1 μs, and the current (purple) increases to 360 A. When the avalanche range is reached, the diode fuses, engendering a bidirectional short that is, however, imperfect. As in Figure 6.57, in the tail of the lightning current some forward current is conducted through the bypass and cell diodes. However, this effect would be less pronounced with a standardized lightning current

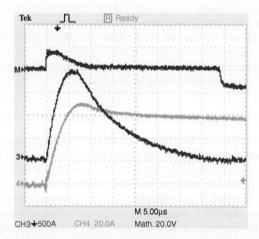

Figure 6.59 Bypass diode forward current i_{BF} in a module loop with $n_Z = 18$ solar cells in a KC60 module in a parallel configuration ($d_{S1} = 35$ cm ($M_i \approx 52$ nH)), secondary to lightning current (green) where $i_{max} \approx 72$ kA and $di/dt_{max} \approx 18$ kA/μs. A 80SQ045 bypass diode was used (nominal inverse voltage $V_{RRM} = 45$ V). The diode remains intact in this case. In a forward state, voltage (red) is limited to a few volts, but the current (purple), whose waveform is similar to that of lightning current, exhibits a peak voltage of around 1.95 kA. Following the current zero crossing, a slight load persists in the forward direction on account of the slightly negative di/dt in the lightning current tail, but this engenders only very little current

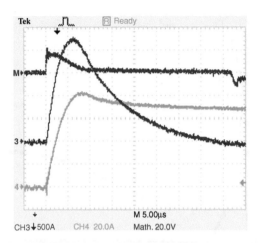

Figure 6.60 Bypass diode forward current i_{BF} in a module loop with $n_Z = 18$ solar cells in a KC60 module in a parallel configuration ($d_{S1} = 35$ cm ($M_i \approx 52$ nH)), secondary to lightning current (green) where $i_{max} \approx 84$ kA and $di/dt_{max} \approx 21$ kA/μs. A 80SQ045 bypass diode was used (nominal inverse voltage $V_{RRM} = 45$ V). The diode fails in this case. In a forward state, voltage (red) is limited to a few volts, but the current (purple), whose waveform is similar to that of lightning current, exhibits a peak voltage of around 2.25 kA. As the diode's failure has eliminated its barrier capacity and the diode exhibits an imperfect short, it continues to conduct current after the current's zero crossing, and no negative voltage occurs

principle, aggregate solar generator module voltage (including adjacent modules) roughly equates to the aggregate of all voltages in the various modules at the counterpart positions [6.15].

6.7.8   Voltage Induced in Wired Solar Generators

The measurement readings discussed in this section show that the recommendations in this chapter concerning the realization of PV installation lightning protection can and should be used (see Sections 6.8 and 6.9). Readers who are pressed for time and who are more interested in practical results than theoretical considerations and substantiating tests can skip this section. In the interest of clarity, and in order to avoid unnecessary device damage during the tests, the measurements described in this section were realized without bypass diodes or using reverse-blocking bypass diodes.

6.7.8.1   Applicability of the Superposition Theorem to Total Induced Maximum Voltage

The superposition theorem can be used to determine the maximum voltage induced by a nearby lightning strike, despite the nonlinear nature of bypass diodes and internal solar cell diodes. Maximum induced voltage occurs at the outset of a lightning current event. Because of their poor barrier properties in the reverse direction (see Section 4.2.1), blocking solar cell diodes exhibit a low-voltage drop under low-current conditions and are bypassed by a considerable junction capacitance, which at open circuit ranges up to a few microfarads per solar cell. Conducting diodes are quasi-shorts compared with relatively high induced voltage. Hence the superposition theorem still applies to maximum voltage in the absence of bypass diodes (i.e. at the outset of a lightning current event), despite the numerous nonlinear elements that come into play. On the other hand, in the presence of bypass diodes the maximum voltage per module is limited to around $V_M = n_B \cdot V_{BA}$ (see Sections 6.6.3.3 and 6.7.7.1).

In the worst-case scenario, maximum aggregate voltage V_S in a wired series string comprises the inter-module voltage V_V induced in the wiring plus the sum total of the voltages V_M induced in the various modules, as possibly limited to $n_B \cdot V_{BA}$ by bypass diodes. This voltage is a symmetrical (differential) voltage: (a) between the module's positive and negative terminals; (b) between interconnecting line conductors; or (c) between the positive and negative terminals of an inverter or other device that is connected at the end point:

$$\text{Maximum voltage induced in a string: } V_S = V_V + \sum V_M (V_M \leq n_B \cdot V_{BA}) \qquad (6.51)$$

One of the terms in this equation may be dominant in an actual PV installation, or both terms may account for a substantial portion of aggregate voltage. And it can of course happen that the wiring and module voltage have a compensatory effect, but as a rule this cannot be assumed. The lowest possible wiring voltage V_V is obtained in an installation where the space occupied by the module wiring loops is minimized, although this is not possible in the vast majority of cases.

The presence of grounded structures in close proximity to a solar generator can result in very large loops where extremely high voltages can be induced by nearby lightning current. These are asymmetrical voltages that arise between a grounded structure (e.g. a metal-module mounting rack that has been grounded intentionally or unintentionally, a metal-module frame, or an inverter housing) and the inside of a solar module or inverter, and which, if suitable countermeasures are not taken, can easily damage the module or inverter insulation and even cause a fire. Such untoward effects can be averted through proper grounding and use of suitable surge protection devices.

6.7.8.2 Reduction of Induced Wiring Voltage in Framed Modules

Tests conducted without bypass diodes on framed modules have shown that, in an extensive module wiring loop that is located on the solar generator plane and whose limit values are not overshot, induced voltages can be reduced to an extent similar to the module voltages per se. In other words, wiring voltage V_V is reduced by an analogous reduction factor R_R ranging from 3 to 5 (see Figure 6.56) that is on a par with module voltage V_M. Figure 6.61 displays a test configuration that was used to demonstrate this principle.

The following wiring scenarios were tested for each type of module:

- Wiring with the smallest possible wiring loop surface area ($b_V \approx 0$) and thus minimal V_V.
- Wiring that possesses an appreciable wiring loop surface area and thus a large V_V and whose dimensions were as in Figure 6.61.

In the minimum wiring loop surface area scenario, the maximum induced string voltage for both the framed modules and laminates was around three times the maximum induced voltage in a KC60 module with 450 mm spacing as in Figures 6.51 and 6.52 (results not shown for reasons of space).

In the tests of the scenario involving large wiring loop areas (dimensions as in Figure 6.61), the mutual inductance between the lightning current and wiring loop was $M_V \approx 93$ nH as in Equation 6.16. In the interest of not overloading component insulation, the current curve for these tests was reduced from around 14.2 to 14.6 kA/µs.

Figure 6.62 displays the voltage induced in a solar generator comprising three KC60 laminates in a parallel configuration as in Figure 6.61 ($di/dt_{max} \approx 14.6$ kA/µs).

With $M_V \approx 93$ nH and the indicated di/dt_{max}, $V_{Vmax} \approx 1.36$ kV, the voltage for a module with the same clearance where $di/dt_{max} = 25$ kA/µs is around 1.5 kV (see Figure 6.51). Hence the resulting sum total of module voltages with $di/dt_{max} = 14.6$ kA/µs is $\sum V_{Mmax} = 3 \cdot (14.6/25) \cdot 1.5$ kV $= 2.63$ kV and the sum total of V_{Smax} is 3.99 kV, which is virtually the same as the baseline value of around 4 kV as in Figure 6.62.

Figure 6.63 displays the voltage induced in a solar generator comprising three framed KC60 devices in a parallel configuration as in Figure 6.61, with $di/dt_{max} \approx 14.2$ kA/µs.

As in Figure 6.52, the voltage of a module with the same clearance is around 350 V; the resulting aggregate module voltage is $\sum V_{Mmax} = 3 \cdot (14.2/25) \cdot 350$ V $= 596$ V; and as in Figure 6.63 the maximum voltage is $V_{Smax} \approx 920$ V. The maximum voltage induced in the wiring is

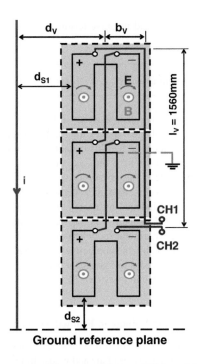

Figure 6.61 Wiring in a solar generator with a string of three KC60 modules (in a parallel configuration, as Figure 6.50) that was used to validate the applicability of the superposition theorem to maximum induced voltage. Dimensions: $d_{S1} = 450$ mm, $d_V = 735$ mm, $b_V = 255$ mm (and 0), $l_V = 1560$ mm. In the interest of not placing the test module insulation under undue strain, lower di/dt_{max} amounting to 15 kA/μs in lieu of 25 kA/μs was used for some measurements. d_{S1} was 450 mm for all of the measurements discussed here, although measurements performed using other clearances yielded comparable results. Modules with and without metal frames were used (the former are represented by the dashed lines), and all of the modules were mounted near each other. The metallic joining elements used for the modules and grounding installation had practically no effect on the induced string voltage V_S, i.e. the symmetrical voltage between CH1 and CH2, where CH1 and CH2 were connected, respectively, to the positive and negative terminals of the string. With these dimensions, the mutual inductance between the lightning current and wiring loop was $M_V \approx 93$ nH for the laminates tested

$V_{Vmax} = V_{Smax} - \sum V_{Mmax} \approx 324$ V, which falls far short of the theoretical V_{Vmax} value, for which it is presupposed that the metal frame has no effect on the voltage V_V induced in the wiring.

Application of this supposition with $M_V = 93$ H and $di/dt_{max} \approx 14.2$ kA/μs would yield $V_{Vmax} \approx 1.32$ kV, which is clearly too high. Hence this scenario has a frame reduction factor $R_R = 1320$ V/324 V ≈ 4.1 for induced maximum wiring voltage V_V and thus for effective wiring mutual inductance M_{Vi} as well. However, R_R of around 3 was observed in tests with another solar generator comprising four framed SM46 devices (compensating module).

Further measurements showed that if the module wiring is run along the entire length of the metal frame, the frame's circular current reduces the overall magnetic field effect in the wiring – and thus the voltage induced therein – to nearly zero. However, as this configuration cannot be realized in all cases, minimizing the wiring surface area is a simpler solution. It was also shown that if inter-module spacing is minimal, it makes no difference whether or not metal elements are used to join the modules to each other.

6.7.8.3 Effect of Grounding on Solar Generators

Most PV installation safety regulations stipulate that the frames of solar generators comprising metal-framed modules should be grounded. This is particularly advisable (including in installations with

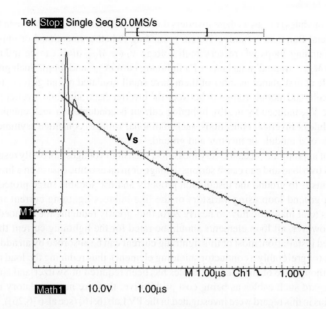

Figure 6.62 Voltage induced in a solar generator comprising three KC60 laminates in a parallel configuration as in Figure 6.61 ($di/dt_{max} \approx 14.6\,kA/\mu s$). The effective voltages are 100 times higher because a 100:1 voltage divider was used. *RLC* oscillation effects attributable to a measurement sensor input capacitance of around 100 pF and the resulting increase in loop inductance were observed during the first 500 ms. The true value for induced initial current is obtained by extending a tangent at V_S (full line) from the end of the oscillation effects up to the surge current start time

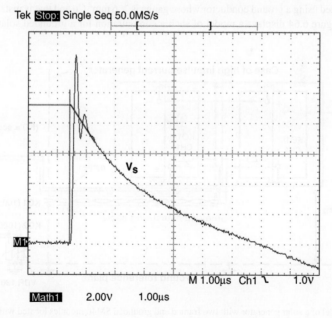

Figure 6.63 Voltage induced in a solar generator comprising three framed KC60 devices in a parallel configuration as in Figure 6.61 ($di/dt_{max} \approx 14.2\,kA/\mu s$). The effective voltages are 100 times higher because a 100:1 voltage divider was used. *RLC* oscillation effects attributable to a measurement sensor input capacitance of around 100 pF and the resulting increase in loop inductance were observed during the first 500 ms. The true value for induced initial current is obtained by extending a tangent at V_S (full line) from the end of the oscillation effects up to the surge current start time

protection class II modules) in cases where inverters that lack electrical isolation between the AC and DC sides are used. Metal-frame or mounting system grounding (whether intentional or unintentional) can result in large grounding loops where very high voltage $V_{ES} = M_{ES} \cdot di/dt$ can be induced by nearby lightning current (M_{ES} = mutual inductance of a lightning current ground loop). Such ground loops are often unidentifiable at first glance as most of them have small insulated openings (e.g. at the insulation between the solar cells and module frames or between the inverter electronics and inverter housing). All or part of the voltage V_{ES} induced by nearby lightning current is conducted to such openings.

In terms of module or inverter connections, such additional voltages constitute asymmetrical voltages between the inside of a module or inverter and ground.

In the absence of appropriate precautionary measures, such high voltages can easily result in module or inverter insulation overload and can cause serious damage or in an extreme case even a fire. The only way to avoid such events is to ground the installation properly and use suitable surge protection devices.

Closing such a ground loop using varistors or the like induces lightning current in the loop that reduces the loop's total magnetic flow to nearly zero, thus greatly reducing the induced voltage at the loop openings. However, all loop elements should be rated for the lightning current that occurs. In a suitably configured installation, most of this lightning current can be absorbed by an additional parallel ground conductor (or preferably a conductor shielding element), thus reducing the load on the varistors and DC cable. Although use of shielded DC cables has been required in Switzerland since 1990 [4.4], other countries regard such cables as being cost prohibitive; they are not mandatory in [4.10].

The key scenarios in this regard were investigated in the PV Lab [6.16] (see also [6.26]), under the aegis of the aforementioned EU project, and selected results of these investigations will now be described (geometry and measured voltages).

6.7.8.3.1 Grounded Solar Generator Without Lightning Current in the DC Cable

In those cases where solar generator conductors will normally not exhibit lightning current (because, for example, the solar generator is located in the protection zone of a lightning installation), the generator should be grounded using a ground conductor whose gauge is $\geq 6 \, mm^2$ Cu and that is installed parallel to the DC cable. Figure 6.64 displays a model of such a configuration for a grounded solar generator.

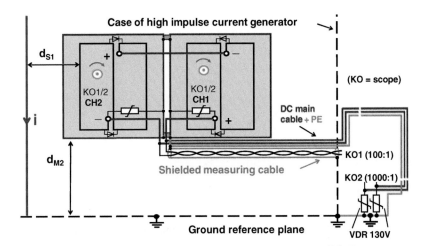

Figure 6.64 Model of a solar generator with two framed and grounded SM46 modules located within the magnetic field of a nearby lightning current i. As this configuration contains no ground loop by virtue of the ground conductor (measurement cable shielding) having been installed in very close proximity to the DC cable, no additional asymmetrical voltage V_{ES} is induced and the ground conductor exhibits no lightning current. All measurements were realized under the following conditions: $di/dt_{max} \approx 25 \, kA/\mu s$, $d_{S1} = 350 \, mm$ and $d_{M2} = 900 \, mm$; conductor length around 10 m. Maximum induced string voltage under these conditions was $V_{Smax} \approx 1.2 \, kV$

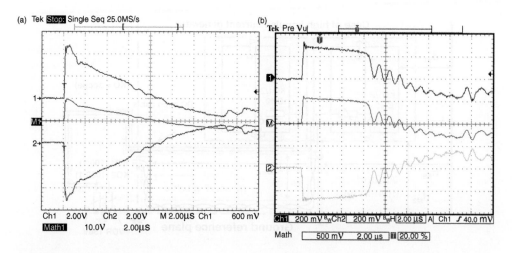

Figure 6.65 Voltage induced in the circuit of Figure 6.64 at the solar generator connections (left) and on the inverter/charge controller side (right) with two VM130 varistors ($V_{DCmax} = 130$ V) connected on the inverter/charge controller side only

The following measurements were realized with this model: with and without a ground conductor parallel to the DC cable; and with all four possible varistor circuit variants, namely (a) with a VM130 varistor on either side between the positive and negative and ground conductors, (b) varistors on the solar generator side only, (c) varistors on the inverter/charge controller side only (right) and (d) no varistors. For reasons of space, all of the results could not be shown here.

Figure 6.65 displays the results for the scenario where varistors are connected on the inverter side only.

The voltages in the positive and negative conductors are virtually symmetrical to ground. The varistors on the charge controller/inverter side limit the maximum voltage to 250–300 V, whereas, owing to the conductor inductance voltage drop, the varistors on the other side have little effect on solar generator-side voltage V_{Smax}, which is around 1 kV and thus only slightly lower than the induced maximum open-circuit voltage amounting to around 1.2 kV. However, inasmuch as solar generators with a large number of modules per string tend to exhibit far higher voltage V_{Smax} that can induce an insulation breakdown, varistors should be installed on either side of the ground conductor that limit the voltage to a value similar to that on the inverter/charge controller side.

If varistors are mounted on the solar generator side only, they limit the voltage on that side. However, in such cases RLC oscillation effects induce a voltage at the DC cable termination that is nearly twice that on the solar generator side, which means that the electronics at this termination exhibit far less effective lightning protection.

The absence of varistors on either side of the conductor induces RLC oscillation effects not only at the conductor termination, but also at its origin; this in turn results in maximum solar-generator-side voltages that exceed maximum open-circuit voltage.

The same measurements were realized without a ground conductor. As the ground conductor exhibits no lightning current, the voltage readings for this scenario were virtually the same as for the scenario with a ground conductor, providing that varistors were installed on one side of the conductors. Such discrepancies that were observed (attributable to capacitive effects) occurred only in the absence of varistors.

6.7.8.3.2   Grounded Solar Generator with Lightning Current in the Shielded DC Cable

Figure 6.66 displays the test set-up that was used here. The following measurements were realized for this model: with all four possible varistor circuit variants, namely (a) with a VM130 varistor on either side between the positive and negative and ground conductors, (b) varistors on the solar generator side only, (c) varistors on the inverter/charge controller side only (right) and (d) no varistors. In the interest of identifying anomalies entailed by improper grounding, these measurements were repeated with the

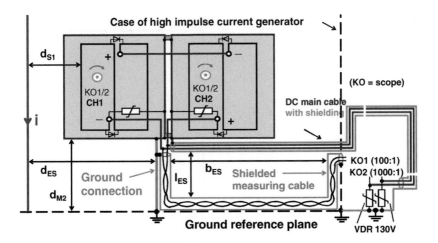

Figure 6.66 Model of a solar generator with two framed and grounded SM46 modules that contain a dedicated ground conductor in the magnetic field of a nearby lightning current i. The fact that the module frames are grounded at the lower centre parallel to the measurement conductor connection results in a large ground loop in which extremely elevated V_{ES} could potentially be induced. To avoid this, the ground loop is closed off via the bilaterally connected conductor shielding – although this would still allow lightning current to flow through the conductor. Relevant characteristics: conductor length, around 50 m; all conductor gauges for positive and negative shielding, 10 mm^2; total inductance L_{ES} for the overall ground loop, including inductance between the DC cable shielding and ground, around 40 μH; $di/dt_{max} \approx 25$ kA/μs; $d_{S1} = 350$ mm; $d_{M2} = 900$ mm. Ground loop characteristics: $d_{ES} = 560$ mm, $l_{ES} = 770$ mm, $b_{ES} = 800$ mm, $M_{ES} \approx 0.14$ μH

Note: All voltage polarities are reversed as channels 1 and 2 have been interchanged relative to their configuration in Figure 6.64

conductor termination shielded on one side only. For reasons of space, all of the results could not be shown here.

Figure 6.67 displays the results for the scenario where the shielding is grounded bilaterally and varistors are connected on the inverter side only. Figure 6.68 illustrates the same scenario with grounded shielding

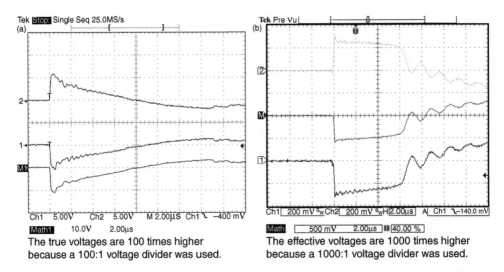

The true voltages are 100 times higher because a 100:1 voltage divider was used.

The effective voltages are 1000 times higher because a 1000:1 voltage divider was used.

Figure 6.67 Voltage induced in the circuit of Figure 6.66 at the solar generator connections (left) and on the inverter/charge controller side (right) with two VM130 varistors ($V_{DCmax} = 130$ V) connected on the inverter/charge controller side only

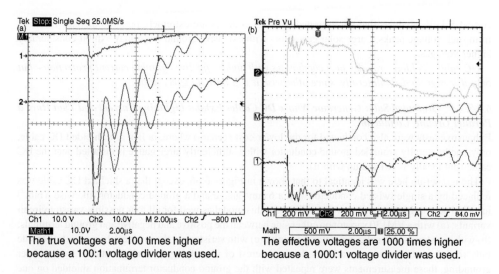

Ch1	10.0 V	Ch2	10.0V	M 2.00µs Ch2 ʃ –800 mV
Math1	10.0V	2.00µs		

The true voltages are 100 times higher
because a 100:1 voltage divider was used.

Ch1	200 mV ᴮwCh2	200 mV ᴮwH 2.00µs	A Ch2 ʃ 84.0 mV
Math	500 mV	2.00µs 25.00 %	

The effective voltages are 1000 times higher
because a 1000:1 voltage divider was used.

Figure 6.68 Voltage induced in the circuit of Figure 6.66 at the solar generator connections (left) and on the inverter/ charge controller side (right) with two VM130 varistors connected on the inverter/charge controller side only and *with the conductor shielding grounded at the conductor termination only*

at the conductor termination only and without the shorter connection shown in red in Figure 6.66 (to the right of the ground conductor arrow).

Figure 6.67 illustrates the scenario where DC cable shielding is properly grounded at both ends and the short red connection in Figure 6.66 is included.

The voltages in the positive and negative conductors are virtually symmetrical to ground. The varistors on the charge controller/inverter side limit the maximum voltage to 250–300 V, whereas, owing to the conductor inductance voltage drop, the varistors on the other side have little effect on solar-generator-side voltage V_{Smax}, which is around 1 kV and thus only slightly lower than the induced maximum open-circuit voltage amounting to around 1.2 kV. The voltages here are on a par with those shown in Figure 6.65, despite the lightning current exhibited by the shielding. (Note that, relative to the Figure 6.65 configuration, channels 1 and 2 have been interchanged and a different voltage metric has been used in the left diagram.)

In as much as solar generators with a large number of modules per string tend to exhibit far higher voltage V_{Smax} that can induce an insulation breakdown, varistors should be installed on either side of the ground conductor that limit the voltage to a value similar to that on the inverter/charge controller side.

If varistors are mounted on the solar generator side only, they limit the voltage on that side. However, in such cases *RLC* oscillation effects induce a voltage at the end of the DC cable that is nearly twice that on the solar generator side, which means that the electronics at the termination of this conductor exhibit far less effective lightning protection.

The absence of varistors on either side of the conductor induces *RLC* oscillation effects not only at the conductor termination, but also at its origin; this in turn results in maximum solar-generator-side voltages that exceed maximum open-circuit voltage.

Figure 6.68 illustrates the scenario where the DC cable shielding is *incorrectly* grounded at only one end of the conductor and the short red connection in Figure 6.66 is missing.

In this scenario, the voltage between the positive and negative conductors is asymmetrical to ground, while the varistors on the inverter/charge controller side limit maximum voltage to ground to 250–320 V. As the DC cable at the various solar generator connection points is nearly at open circuit, these points additionally exhibit nearly all of the very elevated asymmetrical voltage V_{ES} induced in the ground loop, while the difference between positive and negative voltage is virtually the same as in Figure 6.67. The very small solar generator illustrated here already exhibited a maximum voltage to ground amounting to around 5.4 kV. Modules that are not rated for such voltages are extremely prone to damage resulting from such voltage peaks. Large solar generators with extensive ground loops may exhibit considerably higher voltages.

DC cable shielding provides such conductors with highly effective lightning current protection – but only if it is grounded bilaterally, since otherwise it will have no effect at all.

6.7.8.3.3   Grounded Solar Generator Whose DC Cables Exhibit Lightning Current and That Have a Parallel Ground Conductor

Exhaustive tests were also conducted for the lower-cost scenario where, in lieu of a shielded DC cable, only one bilaterally grounded ground conductor with a sufficiently large gauge is installed near the DC cable (but ideally is twisted up with it).

The test set-up here was identical to that shown in Figure 6.66, except that a shielded DC cable was replaced by a single twisted conductor with a 6 mm^2 gauge and a ground conductor (likewise 6 mm^2) installed near it. Measurements were realized for this model as well for all four possible varistor circuit variants: (a) with a VM130 varistor on either side between the positive and negative and ground terminals; (b) with varistors on the solar generator side only; (c) with varistors on the inverter/charge controller side only (right); and (d) without varistors. In the interest of identifying anomalies entailed by improper grounding, these measurements were repeated with the ground conductor termination shielded on one side only. For reasons of space, all of the results could not be shown here.

Figure 6.69 displays the results for the scenario where the ground conductor is grounded bilaterally and varistors are connected on the inverter side only.

A comparison of Figures 6.67 and 6.68 shows that the voltages in the presence of a bilaterally grounded ground conductor are considerably higher than with a shielded conductor but fall short of the voltage induced in the absence of a ground conductor or with only bilateral shielding.

6.7.9   Conclusions Drawn from the Test Results

The tests show that, surprisingly, the only untoward effect of a simulated lightning current in a solar module frame is a change in the *V–I* characteristic curve and degradation of the filling factor. However, inadequately sized bypass diodes can induce total failure in such cases. Framed modules reduce the

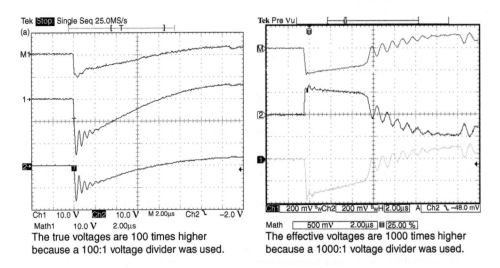

Figure 6.69   Voltage induced in the circuit of Figure 6.66 at the solar generator connections (left) and on the inverter/charge controller side (right) with two VM130 varistors connected on the inverter/charge controller side only and with, in lieu of a shielded DC cable, a bilaterally grounded ground conductor near the DC cable

voltage induced in the module itself as well as in the installation wiring (and thus also the current and load to which any down-conductors may be subjected). Aluminium foil on the back of the module exerts a similar effect. The use of surge protection devices for both the solar generator and appliances on either side of a shielded DC cable considerably reduces the transient voltage and current to which appliances are exposed.

It appears that solar generators with framed modules can be protected both efficiently and affordably against lightning insofar as: (a) lightning is diverted using suitable devices; (b) down-conductors are at least 10 cm away from the solar modules; (c) wiring is realized in a manner that minimizes mutual inductance; (d) suitable varistors are mounted at the correct locations; and (e) the grounding installation is well designed.

The sections that follow contain PV installation recommendations based on the insights that have been gained from the research discussed in this chapter. Section 6.8 defines rules for optimal sizing of solar generator protection devices. Section 6.9 provides information on how PV installations can be optimally protected against direct lightning strikes.

6.8 Optimal Sizing of PV Installation Lightning Protection Devices

To protect PV installations efficiently against lightning, it is first and foremost necessary to install an external lightning protection system that can avert direct lightning strikes on solar modules. The ideal solution in such cases is to mount the solar generator in the protection zone of the lightning protection installation such that no lightning current can flow within less than 1 m of the solar modules. Damage from a direct lightning strike on a building or installation is mainly caused by the voltage induced by the lightning's leader stroke in adjacent conductor loops. To keep this voltage within safe bounds, mutual inductance M_S between the string wiring and lightning current $i_A = k_C \cdot i$ should be minimized by installing down-conductors that prevent the induced voltage from damaging module insulation, wiring and electronics (charge controllers, inverters and so on).

However, lightning strikes on nearby structures also jeopardize the integrity of PV installations to some extent, although the hazards arising from induced voltage are not nearly as great. Although the voltage induced by lightning strikes at a distance of more than around 100 m from a solar installation is lower still, the displacement current exhibited by the solar generator surface and thus the connecting leads resulting from rapid changes in the electric field can in some such cases induce damage.

The effective mutual inductance M_{Si} of a string comprises the effective module mutual inductance M_{Mik} of the n_{MS} series-connected string modules and the effective mutual inductance M_{Vi} of the wiring:

$$\text{Effective mutual inductance in a string } M_{Si} = \sum_{k=1}^{n_{MS}} M_{Mi_k} + M_{Vi} \qquad (6.52)$$

The induced voltage exhibited by a module's intact conducting bypass diodes is extremely low, while that exhibited by its blocking bypass diodes is a maximum of around $n_B \cdot V_{BA}$ (see Section). Hence, to be on the safe side, M_{Si} should always be determined using Equations 6.52 and 6.14.

Careful planning and realization of a solar generator allow the module mutual inductance to be partly or fully offset by mutual inductance in the installation's wiring. However, in order to ensure that the worst-case scenario is taken into account, it is safer to determine the effective mutual inductance using the sum total of module mutual inductances as in Equation 6.52.

Thus in order to keep the string mutual inductance to an absolute minimum, the low module mutual inductance M_{Mi} and the lowest possible wiring mutual inductance M_{Vi} should be attained. These values are determined by the following factors: circuit geometry; string–lightning current clearance; the presence/absence of a frame; and k_C as in Equation 6.6. The calculation procedures described in Sections 6.8.1 and 6.8.2 presuppose that lightning will induce lightning current on the solar module plane. However, modules are often tilted at an angle to the horizontal plane. For vertical lightning currents at a distance of a

Table 6.8 Recommended values for a correction factor k_{MR} for derivation of module mutual inductance M_M from mutual inductance M_{MR} at the module edge

Typical values for k_{MR}: (1 overall loop per module)	Parallel mount		Normal mount	
	without frame	with frame	without frame	with frame
Additive modules (2 loops, e.g. 4×9)	0.38 … 0.42	0.1 … 0.12	0.39 … 0.41	0.1
Additive modules (3 loops, e.g. 6×12)	0.37 … 0.41	0.07 … 0.1	0.38 … 0.41	0.09 … 0.1
Recommended for additive modules	0.4	0.11	0.4	0.1
Comp. modules (3 rows, e.g. M55)	0.33	0.11	0.3	0.06
Comp. modules (4 rows, e.g. MSX 64)	0.24	0.09	0.2	0.04

few metres from a ground-based or rooftop solar generator or the like, the M_{Mi} and M_{Vi} values computed using the aforementioned methods should be multiplied by a correction factor $\sin \beta$.

6.8.1 Solar Module Mutual Inductance

Tests in the PV Lab have shown that the effective mutual inductance M_{Mi} of laminates closely equates to the filamentary conductor loop mutual inductance at the centre of gravity of the reciprocally joined solar cells. This principle allows for approximate calculation of the module mutual inductance M_{Mi} for various types of internal module circuits. Slightly different M_{Mi} values are obtained depending on the orientation of the module's internal wiring loops as regards lightning current i_A deflection.

For compensating modules such as the SM46, SM55 and MSX 60/64, in terms of overall module size and depending on module orientation, M_{Mi} can be up to half that of additive modules such as the KC60 and most other modules, since normally only half of the internal module loops are active on account of the bypass diodes. Although a solar generator of any given size will exhibit lower mutual inductance with compensating modules than with additive modules, to the best of my knowledge no manufacturers make modules that are intended to exhibit the relevant compensating functions.

In as much as determining the effective module mutual inductance M_{Mi} using all internal module loops and taking account of frame current is a labour-intensive process, it often suffices to use a readily calculated approximation of M_{Mi} as in Equation 6.53. A correction factor k_{MR} and k_C as in Equation 6.6 are derived from (readily calculated) mutual inductance M_{MR} at the module frame edge as in Equation 6.16, and lightning current $i_A = k_C \cdot i$. Correction factor k_{MR} is determined by module and module frame type:

$$\text{Effective module mutual inductance } M_{Mi} \approx k_{MR} \cdot M_{MR} \cdot k_C \qquad (6.53)$$

Using the previously realized measurements and simulations as a starting point, the k_{MR} values in Table 6.8 can be approximated.

In modules with aluminium foil, k_{MR} will be lower depending on the extent to which the magnetic field is further weakened by the foil's eddy currents. The guide values in Table 6.9 are based on the additive

Table 6.9 Recommended k_{MR} guide values for modules that contain aluminium foil and are mounted in a parallel configuration

Typical values for k_{MR} at modules with foil: (1 overall loop per module)	Parallel mount with foil	
	without frame	with frame
Additive modules (crystalline and thin film)	0.05	0.011

KC60 module onto which aluminium foil was glued. Similar values were observed during tests (in 2003) on a thin-film solar module (ST20) with integrated aluminium foil.

Note: k_{MR} can be reduced in modules with n parallel loops according to the number of such loops in the modules. The reduction factor is n for a standard configuration and somewhat lower for a parallel configuration.

Example

Standard lightning protection requirements for $di/dt_{max} = 100\,\text{kA/}\mu\text{s}$, $k_C = 0.25$, KC60 module (length $l = 0.751$ m, width $b = 0.653$ m, additive module four rows) with metal-module frames in a parallel configuration as in Figure 6.50; 0.8 m of clearance from a lightning current $i_A = k_C \cdot i$. What are the M_{MR} and M_{Mi} guide values for the maximum voltage v_{max} induced in the module?

Solution: According to Equation 6.16, $M_{MR} = 89.6\,\text{nH}$, and according to Equation 6.53, $M_{Mi} = k_{MR} \cdot M_{MR} \cdot k_C = 2.46\,\text{nH}$. Hence, according to Equation 6.14, $v_{max} \approx 246$ V, which is consistent with the measured value indicated in Figure 6.53.

6.8.2 Wiring Mutual Inductance

Examples of string wiring parallel and perpendicular to the down-converter are shown in Figures 6.70 and 6.71.

In as much as string wiring M_V is normally composed of one or more rectangular loops, its mutual inductance can be determined using Equation 6.16 or 6.17. To obtain the lowest possible M_V value, the surface area A_S occupied by wiring loops should be minimized and their distance from the down-conductors should be maximized. Twisting the feed and return conductor or use of a compensating conductor loop also reduces M_V.

It should also be noted that, as for Equation 6.37, the effective mutual inductance M_{Vi} of metal-frame module wiring can be reduced by k_C and by a frame reduction factor R_R ranging from 3 to 5 (see Section and Figure 6.56) if the wiring is on or near the module-frame plane and the relevant PV installation limit values are not overshot.

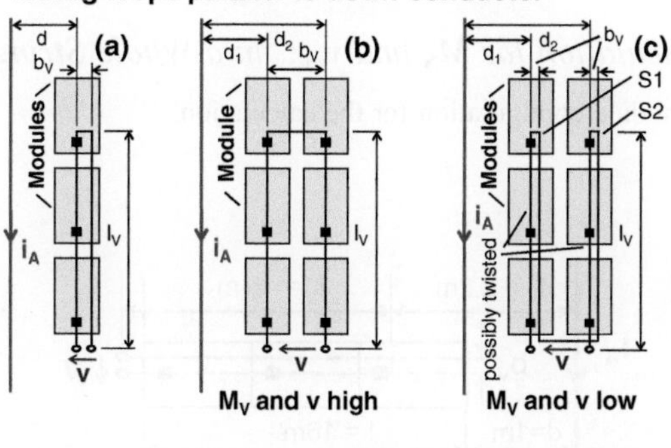

Wiring loops parallel to down-conductor

Figure 6.70   Example of string wiring, parallel to the down-conductor, with average (a), above-average (b) and below-average (c) mutual inductance M_V. To obtain the shortest possible M_V, loop width b_V should be minimized and if possible the feed and return conductor should be twisted. A compensating loop configuration can also be realized (c)

Wiring loops normal to down-conductor

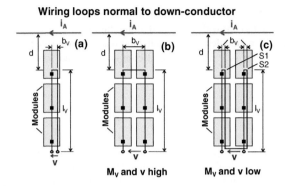

My and v high   My and v low

Figure 6.71 Example of string wiring, perpendicular to the down-conductor, with average (a), above-average (b) and below-average (c) mutual inductance M_V. To obtain the smallest possible M_V, loop width b_V should be minimized. The additional measures as in Figure 6.70 can also be realized

Example

Standard lightning current requirements: $di/dt_{max} = 100\,kA/\mu s$; $k_C = 0.25$; wiring loop as in Figure 6.70a with $l_V = 5\,m$ and clearance $d = 1\,m$ from lightning current $i_A = k_C \cdot i$.

What are the M_{Vi} and v_{max} values for (a) a scenario where $b_V = 40\,cm$ and (b) a scenario where $b_V = 0.8\,cm$ – in both cases for laminates? (c) What is the approximate v_{max} value for a framed module whose wiring is within the confines of the metal frame?

Solution:

(a) For $b_V = 40\,cm$: $M_V = 336\,nH$ and $M_{Vi} = 84\,nH$. Using Equation 6.14, $v_{max} \approx 8.4\,kV$.
(b) For $b_V = 0.8\,cm$: $M_V = 8\,nH$ and $M_{Vi} = 2\,nH$. Using Equation 6.14, $v_{max}\ 200\,V$.
(c) As noted in Section 6.7.8.2, a frame reduction factor $R_R \approx 4$ can be applied to calculations for metal-framed modules. Hence, for case (a) above, $M_{Vi} = 21\,nH$ and $v_{max} \approx 2.1\,kV$. For case (b) above, $M_{Vi} = 0.5\,nH$ and $v_{max} \approx 50\,V$.

6.8.3   Specimen Calculation for M_S and v_{max} in a Whole String

See Figure 6.72 for the module configuration for the calculation.

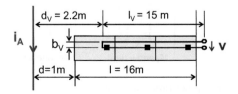

Figure 6.72   Module configuration for lightning current i_A for the specimen calculation

Example

Standard lightning current requirements: $di/dt_{max} = 100\,kA/\mu s$; $k_C = 0.25$; 10 additive modules (length $a = 1.6\,m$, width $b = 0.8\,m$) close to each other in the normal configuration. The edge of the innermost module is 1 m from a lightning current $i_A = k_C \cdot i$. The wiring loop ($l_V = 15\,m$ and $b_V = 3\,cm$) begins at a distance of $d_V = 2.2\,m$ from i_A. What is the total mutual inductance and induced maximum voltage v_{max} (a) for laminates and (b) for framed modules all of whose wiring is within the confines of the module's surface area?

Solution:

(a) As the modules are very close together, only total (aggregate) mutual edge inductance $M_{MR\text{-}tot}$ need be determined for the 10 modules. According to Equation 6.16, $M_{MR\text{-}tot} = 453\,nH$; according to Equation 6.53, $M_{Mi\text{-}tot} = k_{MR} \cdot M_{MR} \cdot k_C \approx 45\,nH$. By analogy, $M_{Vi} \approx 3\,nH$. Hence $M_{Si} = M_{Mi\text{-}tot} + M_{Vi} = 48\,nH$; and according to Equation 6.14, $v_{max} \approx 4.8\,kV$.

(b) With the frame reduction factor $R_R \approx 4$, in this case $M_{Si} \approx 12\,nH$ and $v_{max} \approx 1.2\,kV$.

Unless suitable protective measures are taken, the voltages indicated inSections 6.8.2 and 6.8.3 could damage a PV installation's inverters or charge controllers.

6.8.4 Effects of Distant Lightning Strikes

6.8.4.1 Capacitive Displacement Current and Distant Lightning Strikes

Although distant lightning (i.e. at a distance of upwards of 100 m) does not strike a building or installation per se, it induces transient and elevated electric field fluctuations which, owing to the large surface area and direct exposure involved, exerts a stronger effect on PV systems than on other electrical installations.

Electric field fluctuations induce displacement current density J_V into which a displacement current i_V flows if the displacement current density strikes a solar generator or other surface that is connected to a conductor. In a solar generator, which normally contains a plus and a minus connected conductor, half of the i_V flows into each of these conductors. And as they are usually mounted very close to each other, this divided flow has little effect on the resulting magnetic field and induced voltage (see Figure 6.73).

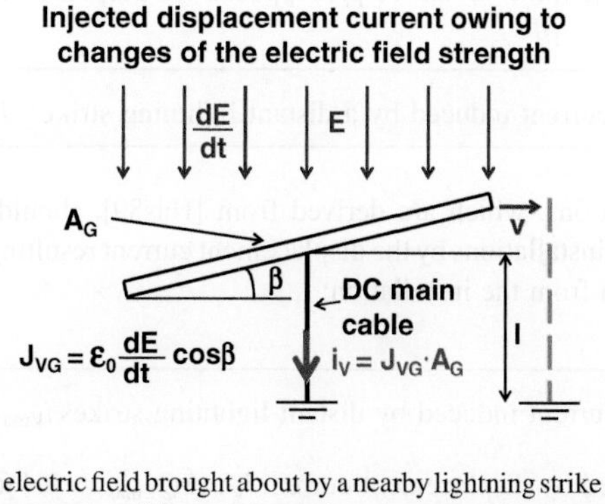

Figure 6.73 The change in the electric field brought about by a nearby lightning strike induces displacement current i_V in the connecting leads of a solar generator with surface area A_G, whereupon this current is conducted to ground via the connecting leads of the solar generator, the devices connected to it and the surge protection devices. For reasons of simplicity, only one conductor lead is shown here

The following normally applies to displacement current density J_{VG} that arrives at a solar generator surface:

$$\text{Displacement current density } (\perp \text{ solar generator surface}) \ J_{VG} = \varepsilon_0 \cdot \cos \beta \cdot \frac{dE}{dt} \qquad (6.54)$$

where:

$E =$ electric field strength
$\varepsilon_0 =$ electric field constant $= 8.854 \, \text{pF/m}$
$dE/dt =$ arithmetic lessening of electric field strength over time
$J_{VG} =$ the resulting displacement current density on the solar generator plane
$\beta =$ solar generator tilt angle

Displacement current density J_{VG} induces a current i_V in the connecting leads. The following applies to a solar generator with surface A_G in the presence of a homogeneous electrical field E:

$$\text{Displacement current } i_V = J_{VG} \cdot A_G = \varepsilon_0 \cdot \cos\beta \cdot \frac{dE}{dt} \cdot A_G \qquad (6.55)$$

In as much as this displacement current, which exhibits the characteristics of an ideal power source, strenuously seeks a path to ground, the DC side of the inverters or charge controller for each connected conductor should have a surge protection device to ground that allows for shunting of this displacement current. Also advisable in such situations is the integration of a condenser that is rated for high frequencies and whose size is derived from the allowable voltage as in Equation 6.61.

In as much as type 2 lightning induces the highest displacement currents, it suffices to estimate the hazards engendered by and the strength of such currents. (The electromagnetic fields induced in such situations have been investigated by F. Heidler [Has89].) The following considerations are important for such scenarios [Has89].

Displacement currents of this type are extremely transient ($< 0.5 \, \mu s$). Their maximum rate of current change di_V/dt, which is the determining factor for induced-voltage calculations, can range up to $i_{Vmax}/0.05 \, \mu s$. A lightning strike at a distance of 30 m can induce a maximum $(dE/dt)_{max}$ of $500 \, \text{kV/m/}\mu s$ [Has89], which is consonant with the $500 \, \text{kV/m/}\mu s$ maximum $(dE/dt)_{max}$ indicated in DIN VDE 0185 Part 103/8 for a remote lightning strike up to a distance of 100 m. As these are merely approximate values, the effect of $\cos \beta$ can be disregarded. And since most solar generators exhibit more than a 25° tilt angle, to be on the safe side the following value should be presupposed for purposes of determining J_{VG}:

$$\text{Maximum displacement current induced by a distant lightning strike: } J_{VGmax} \approx 4 \, \text{A/m}^2 \qquad (6.56)$$

Further, the following equations, which are derived from [Has89], should be used to determine the electrical load induced in PV installations by the displacement current resulting from a lightning strike at a distance of more than 100 m from the installation:

$$\text{Maximum displacement current induced by distant lightning strikes } i_{Vmax} \approx 4 \, \text{A/m}^2 \cdot A_G \qquad (6.57)$$

$$\text{Maximum load induced by displacement current } i_V: \ \Delta Q_{Vmax} \approx 0.4 \, \mu\text{A s/m}^2 \cdot A_G \qquad (6.58)$$

$$\text{Maximum rate of current change for } i_V: \ (d_{iV}/dt)_{max} \approx 80 \, \text{A/}\mu s \cdot A_G \qquad (6.59)$$

The solar generator surface area A_G should be expressed in m^2 for Equations 6.57, 6.58 and 6.59, which respectively serve the following purposes: current sizing for surge protection devices; estimation of the maximum voltage to which a condenser is exposed; and estimation of induced voltage by Equation 6.13.

The displacement current induced in a lightning protection installation by a direct lightning strike is considerably lower for buildings whose PV installations lie within the protection zone of a lightning protection installation. Assuming a mean distance of around 3 m and based on the data in [Has89], 10 times the values in Equation 6.57 and 6.59, and 20 times the values in Equation 6.58, are obtained.

6.8.4.2 Sizing Surge Protection Devices to Displacement Current

A lightning protection installation is subjected to far more distant lightning strikes than direct lightning strikes. The strength of lightning current decreases roughly in inverse proportion to the distance between the installation and the lightning strike. By extension, the greater the distance between the lightning and the installation, the greater the requisite lightning conductor surface area (see Section 6.1) and thus the greater the frequency of the consequent nearby lightning strikes. For conventional varistors that are used as down-conductors, the allowable number of surge current events with a load amounting to $1/n$ of nominal current is n^2, which means that a far greater number of weak-surge current events are allowable. In as much as displacement current is extremely transient ($< 0.5\,\mu$s), surge protection devices rated for an 8/20 µs waveform are sufficient.

The values indicated in Equations 6.57, 6.58 and 6.59 apply to distant lightning strikes of 100 m or more. To be on the safe side, the extremely conservative assumption should be applied to the effect that all lightning strikes at a distance of 100–300 m will induce the aforementioned displacement current, and that therefore each surge protection device will be subjected to around $0.25 \cdot N_g$ distant lightning strikes a year (as in Section 6.1). In this scenario, a down-conductor needs to be able to withstand around $10 \cdot N_g$ surge current events, which in Central Europe equates to around 10–60 events, assuming a PV installation life span of 40 years.

Normally the surge protection device for each solar generator cable positive and negative conductor is grounded (see Figure 6.80) such that it absorbs only half of the maximum displacement current $i_{V\text{max}}$ as in Equation 6.57. Likewise, a consideration here is that, since the current impulse is of far shorter duration, the effective load is far smaller than for the 8/20 µs waveform. Hence, the following nominal varistor current should approximately be sufficient:

$$\text{Required nominal varistor current } (8/20\,\mu\text{s}): I_{V8/20} \geq i_{V\text{max}} \approx 4\,\text{A/m}^2 \cdot A_G \qquad (6.60)$$

Here, $i_{V\text{max}}$ is the maximum displacement current (as in Equation 6.57) exhibited by the solar generator surface A_G.

Thus in such situations a so-called type 3 (formerly type D) down-conductor of the type now commonly found in commercial inverters is sufficient for a maximum current (8/20 µs) of up to a few kiloamps. However, charge controllers do not always provide sufficient surge protection.

6.8.4.3 Estimation of Displacement-Current-Induced Voltage

Like lightning current, the displacement current $i_{V\text{max}}$ that occurs at the ground resistance R_E of a solar generator induces a potential increase $V_{\text{max}} = R_E \cdot i_{V\text{max}}$ as in Equation 6.8 (see Figure 6.8) and flows into cables from external sources such as the electricity grid. However, owing to the far lower currents that come into play here, the voltages that occur are also far lower, usually from around 100 V to a few kilovolts. Larger installations can be protected against such power surges using standard type 2 (formally type C) or type 3 (formally type D) down-conductors for an 8/20 µs curve, and lightning conductors can be dispensed with.

Like lightning current, the displacement current exhibited by solar generator connecting leads in a PV installation induces magnetic field fluctuations, and by extension (as in Equations 6.13 and 6.14) the voltage in adjacent loops that is several magnitudes weaker than lightning current.

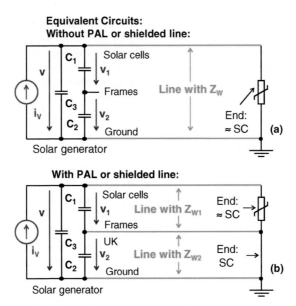

Figure 6.74 Equivalent circuit for determining the voltage induced by a distant lightning strike (KS = short circuit). Displacement current i_V feeds a parallel circuit of a series connection between C_1 and C_2 (or C_3) and a conductor with wave impedance Z_W (a). Voltage v_1 occurs between the solar cells and the module frame or mounting rack, while voltage v occurs relative to remote ground. If C_1 is large enough, v_1 generally remains within a safe range. If need be, v_1 can be further reduced (as in (b)) by wiring in an equipotential bonding conductor or through use of a shielded conductor, either of which will also reduce the maximum possible voltage via Z_{W1} (e.g. in the absence of varistors) and the reflections that occur at the varistors

But owing to the possibly very brief increase and decrease times that come into play, here we run up against the limits of induced-voltage calculations for lightning protection models, since at distances exceeding 2 or 3 m the field changes in the relevant area no longer occur in a virtually synchronous manner. Hence the displacement-current-induced voltage for larger distances should be determined using the analytical theory of simple travelling waves along parallel conductors.

Figure 6.74 displays the equivalent circuit for determining the voltage induced by a distant lightning strike. Such circuits are fed by displacement current i_V.

Two different scenarios come into play here. In the case of framed modules or metallic mounting racks that have two series-connected capacitances: C_1, between the solar cells and the module frame as a whole and the metallic mounting rack; and C_2, between the module frame and mounting rack or ground. For laminates on insulated mounting racks, only one capacitance, C_3, between the solar cells and ground comes into play, whereby a parallel conductor with wave impedance Z_W is also wired in (see Figure 6.74a). It is extremely difficult to determine the exact voltage and current that occur in such scenarios since the exact i_V curve as a function of time may vary from one scenario to another. However, usually it suffices to estimate the maximum voltage that will occur at specific points, e.g. the voltage at the beginning of the conductor to ground or between solar cells and mounting racks.

Assuming that the overall load ΔQ_{Vmax} transported by i_V (as in Equation 6.58) is conducted to a capacitance C, the maximum possible voltage v_{Cmax} will be as follows:

$$\text{Maximum condenser voltage induced by } i_V: \quad v_{Cmax} = \frac{\Delta Q_{Vmax}}{C} \tag{6.61}$$

Laminates or framed solar modules with an extensive metallic mounting system usually exhibit C_1 well above $100\,\mathrm{pF/m^2} \cdot A_G$, which is considerably higher than C_2. In such cases, $v_{1max} < 4\,\mathrm{kV}$ as in Equation 6.61, which normally poses no solar module insulation problems.

In Equation 6.61, the total capacitance to ground should be used for voltage v relative to remote ground, i.e. C_1 connected in series with C_2 or C_3. As in Equation 6.62, $C \approx C_2 \approx C_E$ ($C_1 \gg C_2$) for a building-mounted solar generator comprising framed modules or with an insulated metallic mounting system. C_2 is nearly shorted and $C \approx C_1$ for a solar generator mounted on a grounded metallic rack. $C_1 = C_2 = 0$ and thus $C = C_3 = C_E$ for a building-mounted solar generator comprising laminates without a metallic mounting system (e.g. plastic solar roof tiles as in Figure 4.9).

C_E is relatively high for solar generators that are close to the ground or are mounted on a reinforced-concrete roof (see Figures 6.85–6.90). The following guide values can be roughly estimated using Equation 6.62 (term on the left, plate capacitor capacitance; term on the right, capacitance of a half single-wire line to ground): ground capacitance of a solar generator with surface area A_G; mean height above the ground or a reinforced-concrete surface; external size l; and edge height d (framed module frame height). Thus

$$\text{Estimated solar generator ground capacitance:} \quad C_E \approx \varepsilon_0 \cdot \frac{A_G}{h} + \frac{\pi \cdot \varepsilon_0 \cdot l}{\ln(4h/d)} \qquad (6.62)$$

All surface areas should be expressed in $\mathrm{m^2}$ and all lengths in m; for the meaning of ε_0 see Equation 6.54.

On the other hand, the maximum voltage v for low C_E is mainly determined by the wave impedance Z_W in the conductor that originates at the solar generator. The voltage at the origin of such a conductor is initially determined solely by wave impedance Z_W owing to the infinite propagation rate of the electromagnetic field. The voltage induced in such cases can be estimated solely on the basis of wave impedance Z_W if this value is known for a few characteristic configurations typically exhibited by PV installations (see Figure 6.75).

The displacement current in such a case is equally divided between the two positive and negative leads, which form a coaxial conductor (as is commonly found in 230 V high-voltage lines and elsewhere) that can be regarded as being parallel connected for the purposes of calculating Z_W for this rapid asymmetrical disturbance. Hence the resulting wave impedances Z_W are lower than for a standard biaxial or coaxial conductor. The lower values apply to conductors that are installed very near each other and that exhibit

Approximate values for wave impedance Z_W of lines with two wires

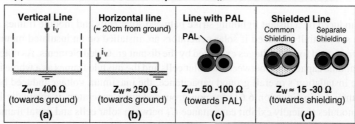

Vertical Line	Horizontal line (≈ 20cm from ground)	Line with PAL	Shielded Line	
			Common Shielding	Separate Shielding
$Z_W \approx 400\,\Omega$ (towards ground)	$Z_W \approx 250\,\Omega$ (towards ground)	$Z_W \approx 50\text{-}100\,\Omega$ (towards PAL)	$Z_W \approx 15\text{-}30\,\Omega$ (towards shielding)	
(a)	(b)	(c)	(d)	

Figure 6.75 Estimated wave impedance Z_W for conductor geometry typically exhibited by PV installations (PAL in (c) = equipotential bonding conductor). The two leads form a coaxial conductor of the type commonly found in high-voltage lines and elsewhere and can be regarded as being parallel connected for the purposes of calculating Z_W for this rapid asymmetrical disturbance. Hence the resulting Z_W is determined using the equivalent radius r_E of this two-wire cable ($r_E = \sqrt{r \cdot a}$, where r = conductor radius and a = wire clearance)

equipotential bonding or shielding and relatively large gauges. Hence the maximum voltage induced in a conductor by i_V can be determined using the following equation:

$$\text{Maximum voltage induced by iV in a conductor with } Z_W: \quad v_{Lmax} \approx k \cdot Z_W \cdot i_{Vmax} \qquad (6.63)$$

The k factor in working solar generators ranges from 1 to 1.8. The wave at the conductor termination or resulting from a disruptive charge induced by high voltage is conducted to ground via the varistors that are usually present, whereupon this wave is nearly shorted and thus is reflected to a large degree.

Hence the maximum voltage v_{max} that can be induced between solar cells and remote ground can be estimated using the following equation:

$$\text{Maximum voltage: } v_{max} \leq \min(v_{Cmax}, v_{Lmax}) = \min(\Delta V_{Vmax}/C, k \cdot Z_W \cdot i_{Vmax}) \qquad (6.64)$$

This in turn means that maximum voltage v_{max} cannot overshoot the lower of the voltages v computed with Equations 6.61 and 6.63 using the values as in Equations 6.57 and 6.58.

If need be, the voltage v_1 induced between solar cells and module frames or mounting systems can be further reduced by wiring in an equipotential bonding conductor or through use of a shielded conductor. Figure 6.74b displays the correct equivalent circuit for such a configuration, where two wave impedances are series connected to remote ground. The wave impedance Z_{W1} between the positive and negative leads (which are parallel connected for this rapid disturbance) and the incorporated equipotential bonding conductor or shielding falls far short of the conductor's overall wave impedance Z_{W2} to ground. (The approximate values for Z_{W1} are displayed in Figure 6.75.) These measures appreciably reduce the maximum voltage that can be induced in this conductor as in Equation 6.64. However, the Z_{W2} values, which determine the module frame or mounting system voltage relative to remote ground, are only slightly lower than the Z_W values in the absence of an equipotential bonding conductor or shielding.

Hence the maximum voltage v_{1max} that can be induced between solar cells and module frames or mounting systems can be estimated as follows:

$$\text{Maximum voltage } v_{1max} \leq \min(v_{C1max}, v_{L1max}) = \min(\Delta V_{Vmax}/C_1, k \cdot Z_{W1} \cdot i_{Vmax}) \qquad (6.65)$$

By extension, the maximum voltage v_{2max} (usually much greater than v_{1max}) that can be induced between module frames or mounting systems and ground can be estimated as follows:

$$\text{Maximum voltage } v_{2max} \leq \min(v_{C2max}, v_{L2max}) = \min(\Delta V_{Vmax}/C_2, k \leq Z_{W2} \leq i_{Vmax}) \qquad (6.66)$$

The voltage v_1 induced by a distant lightning strike between solar cells and module frames or mounting systems can be substantially reduced if solar generators are installed on an insulated mounting system whose insulation is rated for the voltages exhibited by the distant grounded elements. As noted, inasmuch as the framed-module capacitance between the solar cells and the module frames as a whole is usually far greater than the installation's overall ground capacitance, the maximum voltage induced between cells and module frames (as in Equation 6.61) for non-grounded solar generators is relatively low. Framed Siemens M55 modules typically exhibit capacitance C_1 between solar cells and metal frames of around $700\,\text{pF/m}^2 \cdot A_G$, which in the presence of loads as in Equation 6.58 precludes the occurrence of unsafe voltages. Hence, installing a solar generator on an insulated mounting system is the simplest way to avert excessive voltage between solar cells and modules frames induced by lightning strikes at a distance of upwards of 100 m from a PV installation.

However, a solar generator cannot be installed on an insulated mounting system in a PV system that contains transformerless inverters, as the module frames need to be grounded for reasons of contact protection.

As in Figure 6.80, V_1 can be reduced in the presence of abnormally low capacitance C_1 by installing an equipotential bonding conductor (a) parallel to and in very close proximity to the DC cable (b) that is thus well coupled to this cable and therefore exhibits low wave impedance Z_{W1}. A preferable but more expensive solution here is use of a shielded cable as in Figure 6.83.

In cases where the DC cable is connected to a string or other inverter that is in turn connected to a protective conductor but is otherwise ungrounded, the protective conductor shunts the displacement current to the building lead-in, which is usually connected to the building's ground installation. This scenario is similar to that illustrated in Figure 6.74b. The remaining conductor leads can be regarded as being parallel connected for this rapid asymmetrical disturbance and thus can be considered to be grounded. The wave impedance Z_{W1} between the protective conductor and the remaining leads is relatively low (approximately as in Figure 6.75b). V_1 between this conductor and the remaining leads can be estimated using Equation 6.65, while v_2 between the protective conductor and ground can be estimated using Equation 6.66. Voltage v_2 can be reduced by supplying current i_V via the shortest and most direct connection possible between the inverter housing and the grounding installation, as shown by the dashed line in Figure 6.80. The voltage thus obtained can then be added to the potential increase $V_{max} = R_E \cdot i_{Vmax}$ at the ground resistance.

To protect against this type of voltage surge as well as voltage surges induced by potential increases, surge protection devices with around the same surge current rating as on the DC side should be installed on the inverter grid side. Such inverters often integrate suitable varistors. With string inverters and other small devices, a good grid filter containing Y-condensers ($\geq 2.2\,\text{nF}$) provides sufficient protection against this type of rapid grid voltage surge.

6.8.4.4   Examples for Determining the Effects of Displacement Current

1. Grid-connected single-phase PV installation as in Figure 6.80 with the following additional characteristics: 60 framed Siemens M55 modules; $P_{Go} = 3.3\,\text{kW}$; solar generator installed on an insulated mounting system (no equipotential bonding conductor); external length 22.2 m; module-frame height 34 mm; mean height above ground $h = 7\,\text{m}$; surface area $A_G = 25.6\,\text{m}^2$; building ground resistance $R_E = 10\,\Omega$. In as much as the capacitance C_M between the module frames and the solar cells is 300 pF per module, and $i_{Vmax} \approx 102\,\text{A}$ as in Equation 6.57, in this case (as in Equation 6.60) two surge protection devices, each rated for nominal surge current of $I_{V8/20} > 100$ A, provide adequate protection against nearby lightning strikes; or small varistors can be used instead. Damage can occur if these surge protection devices are not installed at the conductor termination. For example, according to Equation 6.61, 51 kV can be induced at a winding capacitance $C_w \approx 200\,\text{pF}$ between the primary and secondary winding of an inverter.

   I_V as in Equation 6.8 induces a potential increase, $V_{max} = 1.02\,\text{kV}$, at the building's ground resistance R_E. If the grid side of this inverter has a good grid filter with 2.2 nF Y-condensers that exhibit $i_V/3$ in the worst-case scenario, a maximum voltage of 1.55 kV is induced as well, as in Equation 6.61. Although this voltage is unlikely to pose a problem in the presence of a good grid filter, it is safer to integrate one varistor each for L1 and N on the grid side of the inverter, upstream of the grid filter.

   $C_1 = 18\,\text{nF}$ for capacitance C_1. If it is conservatively assumed that the entire load flows from i_V to C_1, according to Equation 6.61 the maximum voltage is $v_{C1max} \approx 570\,\text{V}$, which, apart from being extremely transient, poses no problem for a 3 kV module test voltage. However, as part of the voltage i_V flows through Z_W, the actual voltage v_{1max} will be lower.

   According to Equation 6.2, the solar generator grounding capacitance C_E is around 124 pF, which means that in this case $C_1 \gg C_2 = C_E$, and thus the maximum possible voltage $v_{max} \approx 83\,\text{kV}$ as in Equation 6.61, based on the displacement current load. As in Figure 6.75a, wave impedance is around 400 Ω, and thus as in Equation 6.63, with $k \approx 1.8$, a voltage of $v_{Lmax} \approx k \cdot Z_W \cdot i_{Vmax} \approx 74\,\text{kV}$ can be expected. Hence as in Equation 6.64, a very transient maximum voltage of $v_{max} \approx 74\,\text{kV}$

occurs relative to remote ground, which usually poses no problem for solar generators installed on an insulated mounting system (e.g. a wooden rack on a tile roof).

2. Installation as in Figure 6.90 with the following characteristics: 60 850 W string inverters, $A_G = 6.3 \, \text{m}^2$ each; 10 BP585 modules per insulated mounting rack as in Figure 4.45 or 4.46; the modules are mounted behind each other; external solar generator length 25 m; module-frame height 0.3 mm (= solar cell thickness); mean height above reinforced-concrete roof $h = 20$ cm; building ground resistance $R_E = 2 \, \Omega$.

   In as much as $i_{V\text{max}} \approx 25$ A (as in Equation 6.57), in this case (as in Equation 6.60) two small surge protection devices, each rated for a nominal surge current of $I_{V8/20} > 25$ A, provide adequate protection against nearby lightning strikes. Damage can result if these devices are missing from the conductor termination. For example, according to Equation 6.61, 25 kV can be induced at a winding capacitance $C_W \approx 100$ pF between the primary and secondary winding of an inverter.

   Total displacement current at the building's ground resistance R_E is $i_{V\text{max-tot}} = 60 \cdot 25 \, \text{A} = 1.5$ kA.

   Hence, as in Equation 6.8, $i_{V\text{tot}}$ induces a $V_{\text{max}} = 3$ kV potential increase. For the building lead-in of such a large-scale PV installation, the surge protection devices for all three phases should be rated for a current load as in Equation 6.60, amounting to $I_{V8/20} \geq 1.5$ kA (8/20 μs), e.g. type 2, although type 1 is course preferable for a large building. If the grid side of this inverter possesses a good grid filter with 2.2 nF Y-condensers that exhibit $i_V/3$ in the worst-case scenario, as in Equation 6.61 the maximum voltage of 382 kV is additionally induced that normally poses no problem for a grid filter.

   In as much as, in accordance with Equation 6.62, the grounding capacitance $C_E = C_3$ of each of the 60 generators is around 367 pF, the maximum possible voltage induced by the displacement current load is $v_{\text{max}} \approx 6.9$ kV. As in Figure 6.75b, the wave impedance is around 250 Ω, and thus, from Equation 6.63, with $k \approx 1.8$ a voltage of $v_{L\text{max}} \approx k \approx Z_W \approx i_{V\text{max}} \approx 11.3$ kV can be expected. Hence, as in Equation 6.64, a very transient and non-problematic maximum voltage ($v_{\text{max}} \approx 6.9$ kV) occurs at the reinforced-concrete surface of the roof.

6.9   Recommendations for PV Installation Lightning Protection

Foolproof protection of a PV installation against direct lightning strikes is extremely expensive and often not worthwhile from a financial standpoint if the probability of a direct lightning strike is very low. Thus it is necessary to evaluate on a case-by-case basis whether comprehensive protection is needed or whether a lower level of protection would suffice. The cost of PV installation lightning protection varies considerably depending on the level of protection desired.

6.9.1   Possible Protective Measures

The simplest way to protect solar generators with metal-module frames and mounting systems against lightning is to connect conductors whose gauge is large enough (for copper, a minimum of 6 mm) to a good grounding device via the shortest possible wiring run. This method ensures that lightning current will be conducted to ground via the most direct possible path. For grid-connected systems, under no circumstances should protective conductors be used as the sole lightning protection measure; instead the installation should be connected to a suitable foundation grounding device or round circuit using 25 mm^2 gauge copper wire at a minimum. A building-mounted solar generator whose down-conductors are in close proximity to other installations and that cannot be mounted within the building's lightning conductor protection zone may need to be directly connected to the building's lightning protection installation. For such a scenario involving a large installation, the various installation elements are intermeshed. As is the case with all lightning protection installations, an equipotential bonding conductor should be connected to adjoining large metal structures so as to avoid flashovers.

Lightning strikes unleash extremely strong and rapidly changing current, which can induce extremely high voltage in adjoining conductor loops ranging from 30 to several hundred kilovolts (see Section 6.6.1); this voltage can damage PV installation electronics and insulation. Such voltage surges can be neutralized via surge protection devices that are preferably based on thermally monitored varistors (see Section 6.5.3). Conductors that originate from solar generator junction boxes should be installed at the shortest possible distance from the solar generator and should be wired, via a varistor, to the module frames and mounting system, and thus to ground. This reduces the lightning current load on internal solar module insulation.

If the solar generator and building lead-in are more than a few metres apart, lightning protection can be optimized by grounding each conductor redundantly via a varistor. If the installation's charge controller, inverter and the like are also more than a few metres from the building lead-in, each conductor leading to such a device can be grounded redundantly via a varistor. Shielded conductors should also be used if necessary. The voltage induced by lightning current between the positive and negative connections and between these connections and ground can be reduced by installing three varistors, whose rated voltage is somewhat more than half of the DC operating voltage, in a Y-circuit.

The rated varistor operating voltage should be higher than the solar generator open-circuit voltage that would be expected to occur on an extremely cold winter day under maximum insolation conditions. It is also essential that the resistor's rated current load is high enough for the estimated maximum induced current as in Section 6.6.3.2. If type 2 voltage protection varistors or the like are installed on either side of a lightning current conductor, shielding of an adequate gauge (see Section 6.6.4) should also be mounted to ensure that the varistors do not conduct current during a lightning current event as this would result in their being massively overloaded and instantaneously knocked out.

Implementation of all of these measures would be extremely expensive and may be unnecessary in some cases; thus the lightning protection measures that are adopted should be tailored to the requirements and protection goals that obtain in each individual case. As noted in Section 6.7, the cause of optimal lightning protection is in any case best served by using either framed modules or laminates that are mounted in or on enclosed metal frames.

The sections that follow indicate distances for various types of lightning protection for framed-module solar generators, though it should be borne in mind that these distances are around four times larger for laminates. The figures in these sections display various types of down-conductors, to which the following observations apply.

A type 1 surge protection device (SPD) is a lightning current arrester that can divert a minimum of $i_{max} = 25$ kA lightning current with a 10/350 μs waveform. Type 1 SPDs are mainly used for 230 or 400 V power leads that may be subjected to direct lightning strikes on the relevant building or installation (see Figure 6.76). Such a device is on the market for up to 1000 V DC-side voltage for situations where a PV installation is to be protected against lightning without the use of raceways or conductors that are shielded against direct lightning strikes (see Figure 6.77). Most down-conductors can neutralize up to around 100 A secondary grid or PV installation current. In cases where stronger currents occur, the relevant circuits must be fused.

Type 2 SPDs, which are rated for surge current $i_{max} = 12$–25 kA (8/20 μs waveform), are used on the DC side of PV installations, often in groups of three in the so-called Y-circuit (see Figures 6.78 and 6.79). Type 3 SPDs are rated for fine protection, i.e. for $i_{max} = 2$–4 kA current and an 8/20 μs waveform.

As noted in Section 6.5.3, varistor-based SPDs may gradually reduce pick-up voltage on account of overloading or ageing. Thermal cutout devices that are generally reliable for inverter loads tend to be less so for DC voltage owing to the absence of current zero crossing. This has prompted the development of new types of Y-SPDs which, in response to a thermal overload, induce a short circuit via a special DC-compatible fuse. At the same time a fault in one of the two upper varistors that has triggered a short circuit almost instantaneously induces an analogous overload and short circuit in the other varistor, thus shorting out both fuses which then fuse and are thereby transitioned to a safe state (indicated via a fault display).

6.9.2 Protection Against Distant Lightning Strikes

Foolproof protection of a PV installation against direct lightning strikes is extremely expensive and often not worthwhile from a financial standpoint if the probability N_D of a direct lightning strike is very low (see Section 6.1). However, a PV installation should at a minimum be protected against the far more common

Figure 6.76 Type 1 surge protection device (SPD) that is rated for the protection of 230 and 400 V power leads against $i_{max} = 25$ kA (10/350 μs) per phase in the power company grid. By virtue of being directly connected, the power lead PEN conductor is rated for total lightning current of $i_{max} = 100$ kA (10/350 μs). The device shown is a hybrid surge arrester comprising a type 1 SPD and a type 2 SPD for medium protection, thus allowing a limitation of the peak impulsel voltage to a relatively low value compared with rated voltage (Photo: Dehn + Söhne)

scenario involving a distant lightning strike more than 100 m from the installation. A PV installation that has reasonably low mutual inductance wiring by virtue of the fact that its wiring runs extend over the smallest possible area (see Section 6.8.2) will normally incur no damage from distant lightning-strike-induced voltage surges; thus only the displacement current protection described above (see Section 6.8.4) need be implemented.

Figure 6.77 Type 1 SPD for two DC cables (L+ and L−) up to a maximum solar generator voltage of 1 kV. Maximum surge current $i_{max} = 25$ kA (10/350 μs) for L+ to L− and 50 kA for L+ or L− to ground (protective conductor). Maximum nominal voltage between L+ and L− is 3.3 kV, and between L+ or L− and ground (protective conductor) 4 kV (Photo: Dehn + Söhne)

Figure 6.78 Type 2 PV SPD in a Y-circuit (left, positive conductor; right, negative conductor; centre, ground conductor) for a PV installation with V_{OC} up to 1 kV and $i_{max} = 20$ kA (8/20 μs); voltage limitation at nominal voltage for ≤ 4 kV. The use of other SPDs in conjunction with other voltages allows circuits in the 100 V to 1 kV range to be realized (Photo: Dehn + Söhne)

In Figure 6.80, the displacement current i_V induced in a solar generator's power leads is shunted to a protective conductor PE (or directly to ground (dashed line)) via a type 3 SPD at or in the inverter, or, in the case of stand-alone installations, to the charge controller.

Running an equipotential bonding conductor immediately adjacent to the DC cable reduces the i_V-induced voltage between internal module components and metal frames or a metal mounting system (if any). Such a conductor should also be used in scenarios that call for module-frame or mounting system grounding – for example, if transformerless inverters are used. However, this conductor can be dispensed with for smaller solar generators installed on an insulated mounting system if the insulation is sufficiently robust – for example, if it is rated for more than 50–100 kV surge voltage. This level of insulation robustness is also necessary for the cable lead-in of installations with a DC cable. For a stand-alone installation, the system controller is used in lieu of the inverter as in Figure 6.80, and of course the installation is not connected to the external utility grid. For grid-connected systems, the above type of protection needs to be provided on the DC side as well, via a suitable SPD. Many inverters integrate small varistors of this type on both the DC and AC sides, thus obviating the need for an external type 3 SPD.

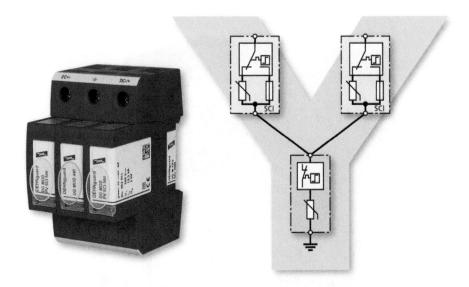

Figure 6.79 New type of type 2 PV SPD in a Y-circuit with a ground connection at the centre. A thermal overload induces a short circuit via a special PV fuse that fuses, thus transitioning the SPD to a safe state. Two models of this device are currently available for PV installations with V_{OC} up to 600 V (protection level for 2.5 kV) and 1 kV respectively (protection level for 4 kV), and $i_{max} = 12.5$ kA (8/20 μs) (Photo: Dehn + Söhne)

In many small PV installations with string inverters, the inverters are mounted near the solar modules and there is no DC cable. Such simply configured installations often possess only minimal distant lightning strike protection. The realization of cost-intensive protection measures involving the use of varistors and ground conductors partly defeats the purpose of these simple installations by

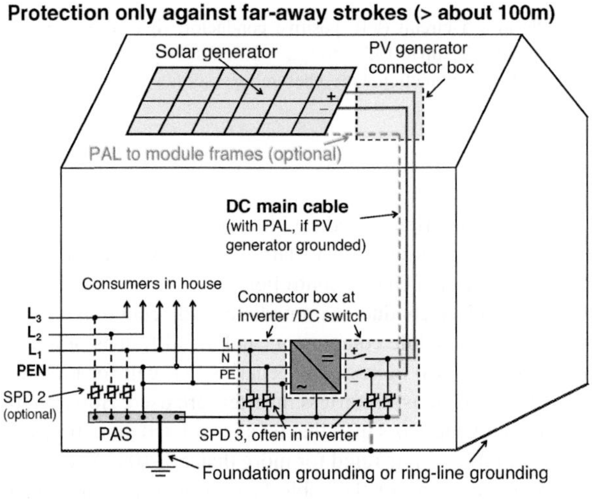

Figure 6.80 Installation that protects smaller PV systems (with low-mutual-inductance solar generator wiring) against the effects of displacement current i_V induced in solar generator power leads by distant lightning strikes. PAS = equipotential bonding bus; PAL = equipotential bonding ($\geq 6\,\text{mm}^2$, although 16 mm² is preferable); SPD3 = type 3 surge protection device. Equipotential bonding devices can be dispensed with for solar generators that are installed on an insulated mounting system

rendering them more complex. Moreover, in many settings the expense involved is not financially tenable for such a small installation, particularly if the installation can be protected against lightning relatively inexpensively.

However, it is important to bear in mind that this kind of minimum protection against distant lightning strikes affords *no protection* at all against nearby lightning strikes, much less direct ones. In other words, if a building whose PV installation thus minimally protected is hit by lightning, the installation will be damaged for sure. And a building that has no lightning protection installation is also very likely to be damaged in such a case.

6.9.3   Protection Against Both Distant and Nearby Strikes (up to about 20 m)

Protection against nearby lightning strikes involves appreciably higher induced voltages and necessitates specific countermeasures (see Figure 6.81), although not necessarily including a lightning protection installation for the building itself.

Type 2 SPDs should be used for both the solar generator junction box and the grid cable leading into the building. To limit PV installation voltage to around an equal extent to ground and between the positive and negative elements, three SPDs can be integrated using a Y- or Δ-circuit. Suitable varistors in a Y-circuit and housing are available from Dehn (see Figures 6.78 and 6.79).

Surge voltage damage would definitely result from a direct lightning strike on the solar generator in the PV installation illustrated in Figure 6.81. To prevent a fire in such a case, the gauge of the equipotential bonding conductor should be at least 16 mm^2 and these conductors should be robustly connected to an external grounding element such as a water pipe, although a round circuit or foundation grounding device is preferable.

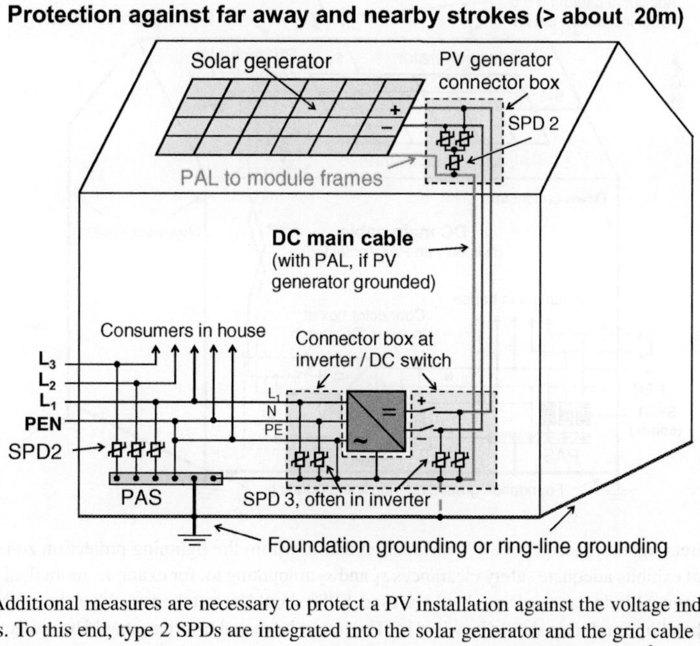

Protection against far away and nearby strokes (> about 20m)

Figure 6.81   Additional measures are necessary to protect a PV installation against the voltage induced by nearby lightning strikes. To this end, type 2 SPDs are integrated into the solar generator and the grid cable leading into the building. PAS = equipotential bonding bus; PAL = equipotential bonding conductor ($\geq$6 mm^2, although 16 mm^2 is preferable); SPD2 = type 2 surge protection device; SPD3 = type 3 surge protection device

6.9.4 Protection Against Direct Lightning Strikes on PV Installations and Buildings

6.9.4.1 Scenario Involving a Direct Lightning Strike (a) on a PV Installation Whose Solar Generator is within a Lightning Conductor Protection Zone and (b) That Induces No Lightning Current in the DC Cable

A direct lightning strike on the surface of a laminate or framed solar module will definitely knock out the module. Although a direct lightning strike on a solar module frame will in many cases not completely ruin it (see Section 6.7), the module itself will be damaged. Moreover, owing to the higher effective overall mutual inductance involved, the damage will be even greater if the frame of a module that is wired to a number of other modules incurs a direct lightning strike. In the interest of limiting the scope of the damage induced by such an event, a direct lightning strike on a solar module should be avoided, possible by mounting the solar generator in the lightning conductor protection zone (see Section 6.3.2) and by ensuring that a sufficient safety clearance s is maintained between the solar modules and lightning current conductors (see Section 6.6.2). The voltages induced by such lightning strikes can be substantially reduced through the use of framed modules and maximally low-mutual-inductance wiring that is installed on the plane and within the confines of the module frames (see Figures 6.70 and 6.71 in Section 6.8.2).

Figure 6.82 displays the lightning protection measures necessary for a building whose rooftop solar generator is within the lightning protection zone of a lightning conductor.

A minimum air clearance of approximately 60 cm on the roof should be observed for single-family dwellings with two SPDs, although a larger clearance such as 1 m would of course reduce

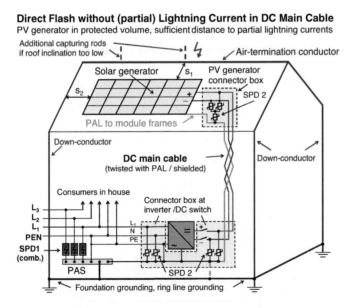

Figure 6.82 Direct lightning strike on a PV installation that is within the lightning protection zone of a lightning conductor and that exhibits adequate safety clearances s_1 and s_2 amounting to, for example, more than 60 cm. Framed modules and low-mutual-inductance wiring should be used. SPDs should be installed on either side of the DC cable and an equipotential bonding conductor should be wired in. The power leads and inverters should be protected using type 1 and type 2 SPDs respectively (PAS = equipotential bonding bus)

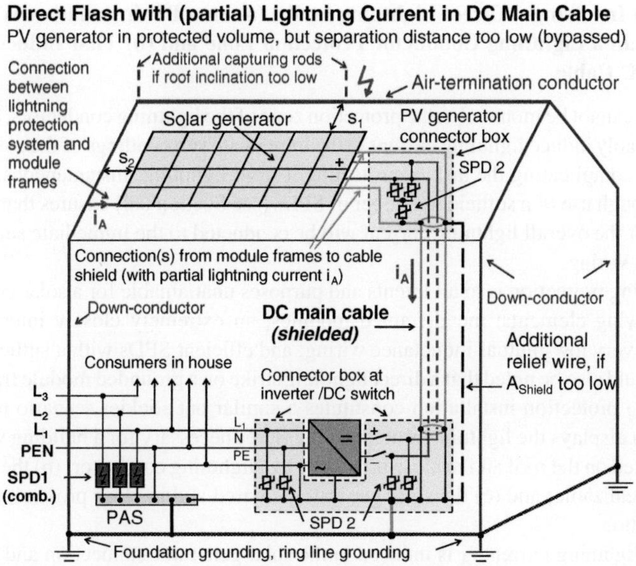

Figure 6.83 Scenario involving a direct lightning strike on a PV installation in the protection zone of a lightning conductor in the absence of sufficient safety clearance; the solar generator is wired to the lightning protection installation, which is in close proximity to other installations Owing to this connection in conjunction with DC cable shielding, the overall lightning current is broken down into sub-lightning currents i_A. The DC cable has bilateral type 2 SPDs. Effective lightning protection is to all intents and purposes unattainable without the use of framed modules and low-mutual-inductance wiring (PAS = equipotential bonding bus)

lightning-current-induced voltage more effectively. The higher the building, the larger the clearances should be (see Section 6.6.2). Use of framed modules and/or a meshed metallic mounting system is strongly recommended.

For PV installations that possess a central inverter and thus a DC cable (see Figures 6.82 and 6.83) SPDs should be installed on either side of the cable and an equipotential bonding conductor should be wired in, preferably twisted with the DC cable.

An even better solution is the use of a shielded cable as in Figure 6.83. As shown in Figure 6.15, sufficient clearances (s_3 and s_4) should be allowed between a rooftop DC cable and the remaining elements of the building's lightning protection installation. For stand-alone systems, the system controller is used in lieu of the inverter shown in Figure 6.82; as such systems are not connected to the grid, a type 1 SPD can be dispensed with.

A lightning strike on a building's lightning protection installation induces an extremely large potential increase (see Section 6.5). Grid-connected systems can be protected from the resulting and relatively long-lasting voltage surge induced by such a lightning strike via top-notch equipotential bonding; type 1 SPDs (preferably hybrid devices that keep voltage low, see Figure 6.76) should be wired into the building junction box. In cases where the SPD is more than a few metres from the solar generator, additional type 2 SPDs should be installed at the inverter power lead.

The DC cable can usually be dispensed with if string inverters are used since they are installed near the solar generator. In such cases, it is important to determine carefully (by evaluating induced current as described in Section 6.6.3) whether the internal SPDs will provide adequate protection in conjunction with the low string mutual inductance M_S that is required. If necessary, additional external type 2 SPDs should be used.

6.9.4.2 Scenario Involving a Direct Lightning Strike (a) on a PV Installation Whose Solar Generator is within a Lightning Conductor Protection Zone and (b) That Induces Lightning Current in the DC Cable

If a solar generator cannot be mounted in the protection zone of the lightning conductors, a direct lightning strike will unavoidably induce lightning current in the immediate surroundings of the solar generator (see Figure 6.83), thus complicating the task of protecting it from lightning. Instantaneous subdivision of a lightning flash through use of a sufficient number of SPDs (see Section 6.4) ensures that only a relatively small proportion of the overall lightning current will be conducted to the immediate surroundings of the solar modules and wiring.

Effective lightning protection is to all intents and purposes unattainable for a solar generator unless it exhibits the following elements: metal-framed modules; an extremely closely intermeshed metallic mounting system; very low-mutual-inductance wiring; and efficient SPDs with a sufficiently high rated current load. It should also be noted that a direct lightning strike on a grounded module frame in a building without a lightning protection installation constitutes a similar but stickier scenario relative to that in Figure 6.83, which displays the lightning protection measures necessary for a building where: (a) a solar generator is mounted on the roof and in close proximity to a lightning conductor; (b) the necessary safety clearances are unrealizable; and (c) the solar generator is wired and in close proximity to the lightning protection installation.

In such a case, lightning current i_A is induced in the solar generator connection and in the DC cable, which *must* exhibit bilateral shielding so as to minimize the effects of lightning current (see Figure 6.35 in Section 6.6.4).

As the lightning current i_A is conducted through the shielding, its gauge must be sufficiently large (e.g. 10 mm^2 Cu). The DC cable should have bilateral type 2 SPDs. As it is particularly important to ensure that the lightning current i_A to and from the DC cable shielding induces the smallest possible amount of voltage at the DC cable termination, the following measures should be taken: metal housings should be used at both ends of the DC cable; and the lightning current should be conducted directly to the shielding via the metal housing and the relevant fittings. Preferably, however, the solar generator junction box should be installed at the building lead-in. In the event that the shielded DC cable is in close proximity to other grounded structures (e.g. reinforced concrete at a building lead-in or in a ceiling) such that the safety clearance as in Equation 6.20 is undercut, this clearance should be bypassed (see Section 6.6.2).

As noted, a lightning strike on a building induces a large potential increase (see Section 6.5). Grid-connected systems can be protected from the resulting and relatively long-lasting voltage surge induced by such a lightning strike through the realization of top-notch equipotential bonding; to this end, type 1 SPDs (preferably hybrid devices that keep voltage low) should be wired into the building junction box. In cases where the SPD is more than a few metres from the solar generator, additional type 2 SPDs should be installed at the inverter power lead. For stand-alone systems, the system controller is used in lieu of the inverter; as such systems are not connected to the grid, a type 1 SPD can be dispensed with.

As noted in Section 6.6.4, the resulting shielding resistance R_M should be kept as low as possible so as to prevent the voltage drop secondary to i_A from inducing excessive conductance in the SPDs; in other words, Equation 6.49 must be fulfilled. One way to accomplish this is to reduce i_A through the use of additional SPDs. However, if the shielding gauge $A_{shielding}$ is too small, R_M can be reduced via a parallel-connected relief conductor (see Figure 6.83). If the DC cable is not shielded, type 1 SPDs should be installed on either side of it on the DC side (see Figure 6.77).

A PV installation needs not only efficient lightning protection, but also wiring that possesses the requisite short-circuit and grounding characteristics. To this end, for a number of years now special double-insulated Radox 125 wire has been used for most solar generators (see Figures 4.76 and 4.77, left) that has insulation as in Equations 4.18 and 4.19 and protection class II, as well as two distinct insulation layers.

The type of wire which is widely used in Switzerland for DC cables (2 · 10 mm^2 with 10 mm^2 shielding; see Figure 4.76 and Figure 4.77, right) also conforms to protection class II but has only single-layer Radox 125 wire insulation. Although manufacturers could make this type of wire with double-layer wire insulation, it would be prohibitively expensive and thus available by special order only. However, this type of wire is in fact the optimal solution for PV installation DC cables.

In German PV systems, positive and negative DC cable that is rated for direct lightning strikes has single shielding. This type of special coaxial cable with double insulation and two separate insulation layers is readily available in hardware stores. Although direct parallel connection of these cables and interconnecting the upper and lower shielding are likewise a good solution (see Figure 6.35b), the risk nonetheless exists that this interconnection might not be effected owing to carelessness or lack of knowledge (see [DGS05] among others); such an omission would result in sudden and extremely high induced voltage in the event of a lightning strike.

Another valid solution – particularly if wire with a sufficiently large gauge is unavailable – is to install the DC cable in a continuous and enclosed metal raceway whose lower side is grounded and whose upper side is joined to the module frame.

6.9.5 Lightning Protection for Large-Scale Ground-Based PV Installations

Large-scale ground-based PV installations also need suitable lightning conductors to protect against direct lightning strikes on the system's modules or frames. These lightning conductors – whose gauge should be as small as possible – should be installed somewhat north of the solar generators such that the rods are within the protection zone (see Sections 6.3.2 and 6.3.3) and the minimum required safety clearance s is maintained as in Equation 6.20. As ground-based PV installations generally comprise a series of staggered rows of solar generators, they should be mounted in such a way that the clearance between them precludes, as much as possible, any reciprocal shading during the winter (see Section 2.5.4.1). Figures 6.84 and 6.85 display the layout of a ground-based PV installation that is equipped with lightning rods. Although a narrow-gauge lightning rod casts a far smaller shadow than a wall, it can nonetheless reduce energy yield by a few percentage points in the module affected, and can also result in additional string mismatches (see Section 4.4.2).

Here too, optimal lightning protection is obtained using framed modules, an intermeshed metallic mounting system, very low-mutual-inductance wiring, and efficient SPDs with a sufficiently high rated current load. Figure 6.86 displays the lightning protection measures needed for such a PV installation.

If the lightning rods are installed on the module mounting system (as in Figure 6.49, which displays the less desirable scenario), in order to mitigate the lightning current induced in the array wires (TGK), as in Figure 6.83, shielded DC cable should be laid in continuously attached and bilaterally grounded metal raceways; or suitable type 1 SPDs (see Figure 6.77) should be mounted on either side of the DC cables.

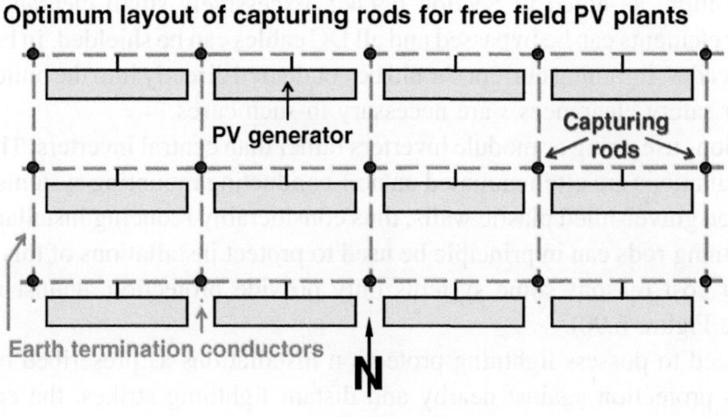

Optimum layout of capturing rods for free field PV plants

Figure 6.84 Lightning rod layout for a ground-based PV installation. The lightning rods, as well as any metallic mounting elements in the solar generators, should be wired to a suitable grounding installation (mesh width ≤ 20 m). In most cases, a clearance of around 1 m between the lightning conductors and solar generators is optimal

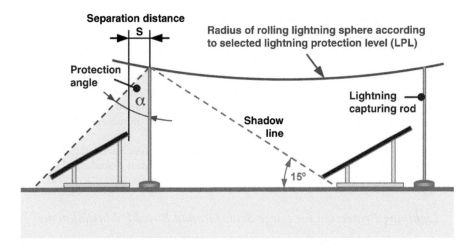

Figure 6.85 Lightning rods should be installed in such a way that they: (a) are within the solar generator protection zone; (b) exhibit the minimum required clearance *s* vis-à-vis the solar generators; and (c) do not cast excessive shadows during the winter. *Source*: Dehn + Söhne

6.9.6 Lightning Protection for PV Installations on Flat Roofs

PV installations mounted on flat roofs have been a common feature of urban landscapes for many years now. When such installations use conventional central inverters, the lightning protection scenario is somewhat similar to that for ground-based PV installations that likewise contain central inverters (see Section 6.9.5). Here too, direct lightning strikes on solar generators can in principle be deflected by lightning rods (see Figures 6.87–6.89). However, when such an installation is mounted on a large roof, it may be difficult to maintain the requisite safety clearances *s* as in Equation 6.20, since the roof substructure is usually made of steel or reinforced concrete and is normally only a few centimetres below the lightning rods and PV cables. There are a number of ways to get around this problem: a modicum of clearance can be maintained at cable junctions; SPDs and lightning rods can be mounted at a distance from each other and insulated; recently developed insulated SPDs (Dehn HVI cable) can be used; PV modules can be mounted on insulated mounting systems; and a gap can be left between the roof surface and DC cables, which can also be insulated. In any case, the use of continuously attached metal raceways always helps. In addition, as noted in Section 6.9.4.2, excessively small clearances between down-conductors and other elements can be bypassed and all DC cables can be shielded. In buildings with metal roofs and/or metal façades, lightning current should be conducted directly into the building's metal skin as considerably smaller safety clearances *s* are necessary in such cases.

Some PV installations use string or module inverters rather than central inverters. The laminate-framed modules of such installations are often mounted on non-conducting mounting systems such as SOFRELs (see Figure 4.45) or on gravel-filled plastic walls, thus considerably reducing installation and DC wiring costs. Although lightning rods can in principle be used to protect installations of this type against direct lightning strikes, for cost reasons some systems only provide protection against nearby and distant lightning strikes (see Figure 6.90).

Large buildings need to possess lightning protection installations as prescribed by law. But if a PV installation only has protection against nearby and distant lightning strikes, the entire installation is extremely vulnerable to direct lightning strikes and its power leads may therefore conduct elevated lightning current with wave-tail half-value times ranging up to 350 μs. Hence the building lead-in for such power leads should have lightning down-conductors (type 1 SPDs) at the roof as well, so as to protect the building's other electrical installations from lightning damage. Such conductors should be wired

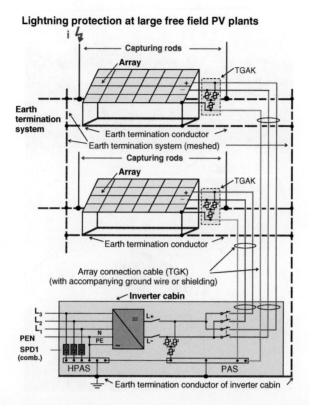

Figure 6.86 Lightning protection elements for a large ground-based PV installation with three triphase inverters. Long DC cables should be installed between the array circuit boxes (TGAK) and the generator junction boxes in the inverter cubicles. Such cables should be laid in suitable cable conduits or in above-ground metal raceways. Bilateral type 2 SPDs should be mounted at either end of these cables. Running a ground wire with the same gauge as the positive and negative wires reduces the varistor load from any lightning current that may be induced in these wires. Type 1 SPDs (preferably hybrid devices that keep surge voltage as low as possible) should be installed on the grid side. HPAS = main equipotential bonding rail; PAS = equipotential bonding rail

separately from the building's main distribution board, whose type 1 (hybrid) SPDs (which are integrated in any case) can be used for fine-protection purposes. In such situations, all wiring should be laid in metal raceways. In many cases the lightning current induced in the power leads is mitigated through the realization of spark gaps on the roof between the metallic raceways, the housings and string inverter protective conductors, and the lightning protection installation. Such measures normally help to minimize lightning-induced damage to the building's electronic installations, although at least some elements of a PV installation are bound to be damaged by a direct lightning strike.

If, however, the PV installation has suitable lightning rods for protection against direct lightning strikes, these spark gaps are unnecessary and type 2 SPDs are sufficient. The latter solution is always a better bet, since lightning current in a building's power leads always poses a problem to at least some extent.

6.9.7 PV Installation Lightning Protection as Prescribed by Swiss Law

The Swiss PV installation safety regulations that applied until 2004 [4.4] were formulated with grid-connected systems in mind, as this was the only central inverter-based concept that was known in 1990 when the regulations were written. These regulations were based on the lightning protection regulations

Figure 6.87   Lightning rods for solar generators on a flat roof; clearance between the down-conductors and lightning rods ≥ 1 m; DC cables in metal conduits (Photo: Dehn + Söhne)

that applied in Switzerland at the time but that did not call for a lightning installation protection zone [6.1]. These regulations stipulated that a building that had no lightning protection system and on which a PV installation was mounted was not required to realize such a system, since the presence of a PV installation did not increase the probability of a lightning strike. However, in line with the then-applicable antenna regulations, PV installations had to be realized in such a way that lightning current induced in the installation would not cause a fire.

The measures in [4.4] concerning grounding and voltage surge protection, which applied to some but not all of the protective measures against direct lightning strikes shown in Figure 6.83, stipulated that the DC wire between the solar generator junction box on the solar generator side and the inverter had to be shielded and have a gauge of least $10 \, mm^2$ Cu. Moreover, in buildings with a lightning protection installation the latter had to be connected to the DC cable shielding, while in buildings without a lightning protection installation a relief conductor of at least $16 \, mm^2$ Cu had to be laid alongside this cable. Thus a direct lightning strike always induced lightning current i_A in this cable. The regulations further stipulated that, in order to counteract the voltage induced in the lightning leader stroke, a down-conductor SPD was to be wired into the inverters and the circuit box DC cables if the latter were more than a few metres long. On the other hand, voltage surge protection elements were recommended for the grid side, but not required.

Unfortunately, these lightning protection measures are only sufficient for stand-alone installations, and are woefully inadequate when it comes to fully protecting grid-connected systems against direct lightning strikes. To achieve such protection, the building lead-ins need to be fitted with type 1 SPDs (preferably hybrid models), as in Figure 6.82 or 6.83, so as to avert inverter and building electrical installation damage

Figure 6.88 Lightning rods for a large PV installation on a flat roof. The DC cables are in metal conduits. The junctions of the cable conduits and down-conductors have a clearance *s* of around 30 cm (Photo: Dehn + Söhne)

induced by the increase in potential from a lightning strike. However, the lightning protection required by [4.4] can only avert fire damage resulting from a direct lightning strike on a grid-connected system, but cannot avert damage to inverter and electrical installations.

The new IEC standard issued in 2002 for building-mounted PV installations [4.10] unfortunately lacks clearly formulated lightning protection requirements. The Swiss version of this standard [4.13] issued two

Figure 6.89 PV installation on a flat roof with the solar generator in the lightning rod protection zone. In this installation, the clearance *s* between the DC cables and down-conductors is only a few centimetres, which is far too little (Photo: Dehn + Söhne)

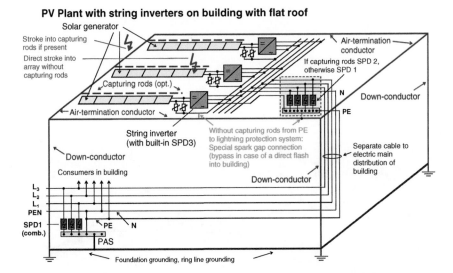

Figure 6.90 Flat-roof PV installation with string or module inverters. Air termination conductors installed at the building perimeter fully protect the building itself against direct lightning strikes. But in the absence of lightning rods and the attendant down-conductors (dark red dashed lines), the entire PV installation and its cabling would be extremely vulnerable to direct lightning strikes (red)

years later clarifies these matters somewhat and represents an improvement over [4.10] due to the lightning protection examples and advisories it offers. The version [4.13] also implicitly requires that PV installations be mounted within the protection zone of lightning rods. In keeping with the 1995 preliminary European standard [6.2] as well as the solution deemed optimal in this chapter, and that has been allowed in Germany for years, the Swiss lightning protection regulations from 2006 adopted the European standards [6.20–6.23] and stipulate that PV installations can be mounted within the protection zone of lightning rods without a connection to the lightning protection installation. Hence the recommendations in the above sections are consonant with current Swiss lightning protection standards.

6.10 Recap and Conclusions

This chapter surveys the principles of lightning protection and the applicable regulations in detail, and in so doing has extensively discussed both induced voltages (which are relatively easy to calculate) and induced currents, which are particularly important if bypass diodes and surge protection devices are to be sized accurately. The chapter also describes a straightforward method for varistor voltage sizing. The lengthy ensuing section described the results of extensive experimental investigations concerning PV installation lightning protection.

The tests on various modules showed that framed crystalline solar cell modules withstand the effects of lightning current amazingly well, and that therefore virtually foolproof lightning protection is realizable with such modules if suitable measures are taken. On the other hand, chinks in the lightning protection armour for solar modules arise with Schottky bypass diodes, which are the prevalent devices and often possess insufficient dielectric strength and surge current resistance.

Detailed experiments using solar generator models have confirmed the concept that metal-frame solar modules are considerably less expensive laminates as they greatly reduce the voltage and current induced in both the module and the solar generator cabling.

Following the recommendations laid out in Sections 6.8 and 6.9 will allow PV installations to be configured in such a way that they can withstand direct lightning strikes without incurring major damage.

6.11   Bibliography

[6.1]   SEV-Norm SEV4022.1987: *Leitsätze des SEV: Blitzschutzanlagen*, 6th edition.

[6.2]   ENV61024-1 (1995): *Blitzschutz baulicher Anlagen.*

[6.3]   H.J. Stern: 'Über die Beeinflussung von Solarmodulen durch transiente Magnetfelder'. *Fortschritt Berichte VDI Reihe* 21, Nr. 154. VDI-Verlag, Düsseldorf, 1994.

[6.4]   H. Häberlin, R. Minkner: 'Blitzschläge – eine Gefahr für Solarmodule? Experimente zur Bestimmung der Blitzstromempfindlichkeit von PV-Anlagen'. *Bulletin SEV/VSE*, 1/1993.

[6.5]   H. Häberlin, R. Minkner: 'Tests of Lightning Withstand Capability of PV-Systems and Measurements of Induced Voltages at a Model of a PV-System with ZnO-Surge-Arresters'. Proceedings of the 11th EC PV Conference, Montreux, 1992.

[6.6]   H. Häberlin, R. Minkner: 'Einfache Methode zum Blitzschutz von Photovoltaikanlagen'. *Bulletin SEV/VSE*, 19/1994.

[6.7]   H. Häberlin, R. Minkner: 'A Simple Method for Lightning Protection of PV-Systems'. Proceedings of the 12th EU PV Conference, Amsterdam, 1994.

[6.8]   M. Real: 'Optimierte Verkabelungssysteme für Solarzellenanlagenanlagen'. Schlussbericht NEFF-Forschungsprojekt 532.2, 1996.

[6.9]   R. Hotopp: 'Blitzschutz von PV-Anlagen: Stand der Technik und Entwicklungsbeitrag durch die Photovoltaik-Siedlung Essen'. 1. *VDE/ABB Blitzschutztagung*, 29.2.–1.3.1996.

[6.10] F. Vassen, W. Vaassen: 'Bewertung der Gefährdung von netzparallelen PV-Anlagen bei direktem und nahem Blitzeinschlag und Darstellung der daraus abgeleiteten Massnahmen des Blitz- und Überspannungsschutzes'. 2. *VDE/ABB Blitzschutztagung*, 6.–7.11.1997.

[6.11] H. Häberlin: 'Von simulierten Blitzströmen in Solarmodulen und Solargeneratoren induzierte Spannungen'. *Bulletin SEV/VSE*, 10/2001.

[6.12] H. Häberlin: 'Blitzschutz von Photovoltaikanlagen – Teil 1'. *Elektrotechnik*, 4/2001.

[6.13] H. Häberlin: 'Blitzschutz von Photovoltaikanlagen – Teil 2'. *Elektrotechnik*, 5/2001.

[6.14] H. Häberlin: 'Blitzschutz von Photovoltaikanlagen – Teil 3'. *Elektrotechnik*, 6/2001.

[6.15] H. Häberlin: 'Blitzschutz von Photovoltaikanlagen – Teil 4'. *Elektrotechnik*, 7–8/2001.

[6.16] H. Häberlin: 'Blitzschutz von Photovoltaikanlagen – Teil 5'. *Elektrotechnik*, 9/2001.

[6.17] H. Häberlin: 'Blitzschutz von Photovoltaikanlagen – Teil 6'. *Elektrotechnik*, 10/2001.

[6.18] H. Häberlin: 'Interference Voltages Induced by Magnetic Fields of Simulated Lightning Currents in Photovoltaic Modules and Arrays'. Proceedings of the 17th EU PV Conference, Munich, 2001.

[6.19] H. Häberlin: 'Einsatz von Surge Protection Devicen beim Blitzschutz: Notwendige Strombelastbarkeit von Varistoren in induktiv gekoppelten Schleifen'. *Bulletin SEV*, 25/2001.

[6.20] EN62305-1 (2006): *Blitzschutz. Teil 1: Allgemeine Grundsätze.*

[6.21] EN62305-2 (2006): *Blitzschutz. Teil 2: Risiko-Management.*

[6.22] EN62305-3 (2006): *Blitzschutz. Teil 3: Schutz von baulichen Anlagen und Personen.*

[6.23] EN62305-4 (2006): *Blitzschutz. Teil 4: Elektrische und elektronische Systeme in baulichen Anlagen.*

[6.24] H. Häberlin: 'Das neue 60kWp-Testzentrum der ISB'. *Bulletin SEV/VES*, 22/1994.

[6.25] H. Häberlin, M. Kämpfer: 'Measurement of Damages at Bypass Diodes by Induced Voltages and Currents in PV Modules Caused by Nearby Lightning Currents with Standard Waveform'. 23rd European PV Solar Energy Conference, Valencia, 2008.

[6.26] H. Häberlin: 'Interference Voltages Induced by Magnetic Fields of Simulated Lightning Currents in Photovoltaic Modules and Arrays'. Proceedings of the 17th EU PV Conference, Munich, 2001.

Additional References

[4.10], [4.13], [Deh05], [Häb07], [Has 89], [Pan86].

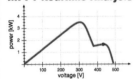
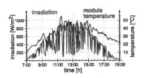
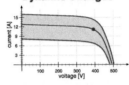
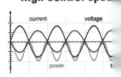
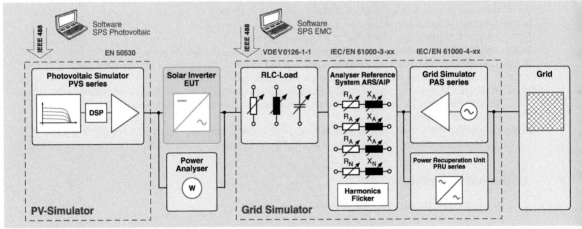

7

Normalized Representation of Energy and Power of PV Systems

In 1993, the European Commission Joint Research Centre (JRC) proposed assessment guidelines aimed at promoting impartial and valid comparisons of PV installations of various sizes and at various locations [7.1]. This chapter begins with some introductory remarks concerning such assessments, with an emphasis on grid-connected systems; these assessments have been in use for many years for other types of electrical installations.

Certain optimizations of these methods realized by the Bern University of Applied Sciences PV Lab will also be described that render the resulting diagrams more informative and that notably allow for a detailed analysis of occasional anomalies such as solar tracker malfunctions, solar generators being shaded or covered in snow, and so on [7.2]. The assessments enabled by these methods will then be illustrated via a series of examples.

Normalized assessments are useful, however, for more than just measurement data-based analyses of PV installation characteristics, as they can also be used for energy yield calculations (see Chapter 8).

7.1 Introduction

Many PV installations use data acquisition systems to document insolation and energy production data, which then allow for the generation of diagrams concerning daily, monthly or annual insolation and energy production. But since installed capacity and local insolation conditions often vary greatly from one PV installation to another, such diagrams shed light on anomalies and allow for head-to-head installation comparisons only to a limited degree. However, these comparisons are extremely useful when it comes to optimizing PV installation technology.

7.2 Normalized Yields, Losses and Performance Ratio

7.2.1 Normalized Yields

The effect of an installation's size can be disregarded if its energy yield during a reference period τ (e.g. day (d), month (mt) or year (a)) is divided by the solar generator's nominal output P_{Go} at STC power

---

Photovoltaics: System Design and Practice. Heinrich Häberlin.
© 2012 John Wiley & Sons, Ltd. Published 2012 by John Wiley & Sons, Ltd.

output ($G = 1\,\mathrm{kW/m^2}$, AM1.5 spectrum, cell temperature 25 °C), which indicates installation yield, whereby a distinction is made between final yield Y_F and array yield Y_A, as follows:

$$\text{Final yield } Y_F = E_{use}/P_{Go} \qquad (7.1)$$

$$\text{Array yield } Y_A = E_A/P_{Go} \qquad (7.2)$$

where:
E_{use} = PV installation power output during the reference period τ
E_A = solar generator DC power output during the reference period τ
P_{Go} = nominal solar generator power output (aggregate module power) under STC ($G = G_o = G_{STC} = 1\,\mathrm{kW/m^2}$, AM1.5 spectrum, cell temperature 25 °C).

The metric obtained if these values are expressed as a numerator and denominator is h/τ. Hence the Y_F and Y_A values also indicate the number of operating hours it takes solar generators with power output P_{Go} to generate the same amount of energy in time τ. This in turn allows for valid comparisons of the energy yield of PV installations of varying sizes. Figure 7.1 shows the final yield Y_F in kWh/kWp per month for 2003 of three grid-connected systems: one at Bern University of Applied Sciences, the second on Jungfraujoch Mountain [7.3] and the third in Burgdorf.

The effect of local and temporal insolation differences is taken into account by factoring in the reference yield Y_R, which is obtained by dividing irradiated energy H_G (in kWh/m²) by irradiance $G_o = 1\,\mathrm{kW/m^2}$ under STC for the reference period τ. The metric obtained by expressing these values as a

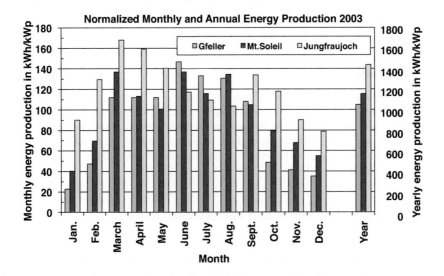

Figure 7.1 Comparison of final yield Y_F in kWh/kWp per month for three different grid-connected systems that have been in operation for 10 years or more and are at various elevations: Gfeller/Burgdorf (3.18 kWp, 540 m); Mont Soleil (555 kWp, 1270 m); Jungfraujoch Mountain (1.152 kWp, 3454 m). All of the data indicated here are based on nominal solar generator power output P_{Go} according to the relevant module datasheet

numerator and denominator is h/τ. Hence the reference yield Y_R also indicates the number of full-load solar hours – that is, the number of hours of $G_o = 1\,\mathrm{kW/m^2}$ irradiance needed to obtain energy H_G during the reference period τ on the solar generator plane. An ideal, power-loss-free PV installation whose solar generator remained at a constant 25 °C under STC would generate an exactly identical final yield $Y_F = Y_R$, hence the term reference yield:

$$\text{Reference yield } Y_R = H_G/G_o = H_G/G_{STC} \qquad (7.3)$$

where:

$$H_G = \text{irradiated energy in kWh/m}^2 \text{ during the reference period } \tau \ ([7.1]\colon H_I)$$
$$G_o = G_{STC} = 1\,\mathrm{kW/m^2} = \text{irradiance under standard test conditions (STC)}$$

A daily (d), monthly (mt) or annual (a) reference period τ can be used for such assessments. For monthly or annual reference periods, the Y value obtained is often divided by the number of days n_d in the reference period τ, which provides directly comparable normalized mean daily yields in kWh/kWp/d or h/d, with relatively little numerical variation.

In many instances, normalized yield assessments are based on large indices (F, R, A) in lieu of small ones, and H_G is replaced by H_I [7.1]. Table 7.1 below contains definitions of the various normalized PV installation yield and loss parameters.

Table 7.1 List containing the definitions of normalized PV installation yield and losses. The units presuppose that mean daily values will be indicated for relatively lengthy reference periods

Symbol	Term	Meaning / Definition	Unit	
Y_R	Reference Yield	$Y_R = H_G/G_0$ = PSH (Peak Solar Hours). Y_R is equal to the time which the sun has to shine with G_0 = 1kW/m² to irradiate the energy H_G onto the solar generator	$\dfrac{\mathrm{kWh/m^2}}{\mathrm{d\cdot 1kW/m^2}}$	[h/d]
L_C	Capture Losses	*Thermal capture losses L_{CT}:* • Losses caused by cell temperatures higher than 25 °C. *Miscellaneous capture losses L_{CM} (not temperature dependent) :* • Wiring, string diodes, low irradiance • Partial shadowing, pollution, snow covering, inhomogeneous irradiance, mismatch • Maximum power tracking errors, reduction of array power caused by inverter failures or when the accumulator is fully charged (stand alone systems) • Errors in irradiance measurements • When irradiance is measured with pyranometer: Spectral losses, losses caused by glass reflections	$\dfrac{\mathrm{kWh}}{\mathrm{d\cdot kWp}}$	[h/d]
Y_A	Array Yield	$Y_A = E_A/P_0$. Y_A is equal to the time which the PV plant has to operate with nominal solar generator power P_0 to generate array (DC-)energy E_A	$\dfrac{\mathrm{kWh}}{\mathrm{d\cdot kWp}}$	[h/d]
L_S	System Losses	Inverter conversion losses (DC-AC), accumulator storage losses (stand alone systems).	$\dfrac{\mathrm{kWh}}{\mathrm{d\cdot kWp}}$	[h/d]
Y_F	Final Yield	$Y_F = E_{use}/P_0$. Y_F is equal to the time which the PV plant has to operate with nominal solar generator power P_0 to generate the useful output energy E_{use}. For grid connected plants: $E_{use} = E_{AC}$.	$\dfrac{\mathrm{kWh}}{\mathrm{d\cdot kWp}}$	[h/d]
PR	Performance Ratio	$PR = Y_F/Y_R$. PR corresponds to the ratio of the useful energy E_{use} to the energy which would be generated by a lossless, ideal PV plant with solar cell temperature at 25 °C and the same irradiation.		[1]

$$Y_R \xrightarrow{\ -L_C\ } Y_A \xrightarrow{\ -L_S\ } Y_F \qquad Y_R \xrightarrow{\ -L_{CT}\ } Y_T \xrightarrow{\ -L_{CM}\ } Y_A \xrightarrow{\ -L_S\ } Y_F$$

7.2.2 Definition of Normalized Losses

The difference between reference yield Y_R and array yield Y_A is referred to as capture losses, $L_C = Y_R - Y_A$, which comprise thermal and miscellaneous capture losses, i.e. $L_C = L_{CT} + L_{CM}$. Hence the following applies:

$$\text{Capture losses } L_C = Y_R - Y_A = L_{CT} + L_{CM} \qquad (7.4)$$

Thermal capture losses L_{CT} are attributable to the fact that maximum solar generator power output is usually lower than P_{Go} because solar generator temperature generally exceeds temperatures at STC power output. Thermal capture losses are relatively easy to determine based on solar generator temperature readings (see Section 7.4).

Miscellaneous capture losses L_{CM} are induced by completely different factors, however. Capture losses in PV installation wiring, resistances and string diodes are normal, as are modules' efficiency loss under low insolation conditions.

However, these capture losses fold in the following additional types of occasional losses, which in some cases are only sporadic and are often not taken into account by solar module vendors:

- Capture losses due to the fact that effective module power is lower than that indicated on the module datasheet.
- Capture losses induced by solar generator shading, dirt accumulation, snow covering and insolation irregularity.
- Power mismatches between modules and strings.
- A solar tracker malfunction (e.g. in an inverter).
- Insolation measurement errors.
- In grid-connected systems, additional loss attributable to inverter failure.
- In stand-alone installations, additional infeed restriction loss in the presence of a fully charged battery.
- Spectral loss attributable to pyranometer insolation measurements; glass reflection loss occasioned by small light incidence angles.

A well-planned and well-realized grid-connected system normally should exhibit only minimal L_{CM} loss.

The difference between array yield Y_A and final yield Y_F is referred to as system or BOS losses, $L_S = L_{BOS} = Y_A - Y_F$, which comprise all losses except for the previously mentioned capture losses. L_S notably includes an inverter's DC–AC converter loss (in systems that contain one or more inverters), and for stand-alone installations, battery storage loss. Hence the following applies:

$$\text{System losses (or balance of system losses) } L_S = L_{BOS} = Y_A - Y_F \qquad (7.5)$$

Note: For grid-connected systems, losses attributable to inverter failure and solar tracker malfunctions are folded into L_{CM} but not L_S.

7.2.3 Performance Ratio

Since (a) the various components of a solar generator exhibit loss and (b) solar generator temperature usually exceeds 25 °C, the final yield Y_F of an actual PV installation will be lower than Y_R. To define a solar generator's performance relative to the ideal scenario, the performance ratio is the quotient of the final yield and the reference yield during the same reference period τ:

$$\text{Performance ratio PR} = Y_F/Y_R \qquad (7.6)$$

7.2.4 New Normalized Values of Merit

These are listed in Table 7.1.

7.3 Normalized Diagrams for Yields and Losses

7.3.1 Normalized Monthly and Annual Statistics

Converting all Y and L values to normalized mean daily yields provides a highly informative annual energy yield bar chart showing Y_F, L_S and L_C in kWh/kWp/d for any given year. This type of chart is known as a normalized annual statistics chart. A normalized monthly statistics chart can be obtained via a bar chart showing Y_F, L_S and L_C for each day of any given month [7.1].

Such charts also indirectly indicate Y_A (the boundary between L_S and L_C) and Y_R (usually the upper limit of L_C). As L_C can only be negative on extremely cold days, the maximum bar height equates to Y_A. However, the Y and L values for Y_F, L_S and L_C can be readily indicated on such charts via hatching, which preferably should be colour coded.

These compact graphical displays are far more informative if the percentage performance ratio (PR) is indicated below each bar. They then contain all of the information that used to have to be presented (according to [7.1]) in two separate graphs, where Y_F, L_S, L_C and Y_R were in one (without performance ratios) and Y_F, Y_R and the performance ratios were in the other (see Figure 7.4). These normalized graphs provide a wealth of information concerning problems that can potentially arise in connection with PV installation operation.

As can be seen in Figure 7.2 (which displays the 1994 normalized annual energy yield statistics for the PV installation on the Jungfraujoch), the PR values were considerably lower from April to June than

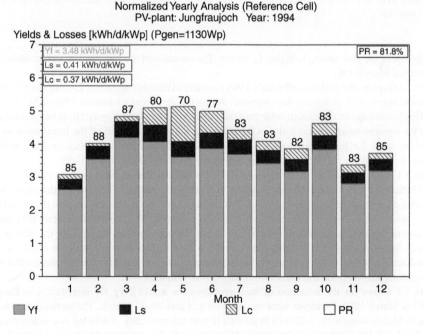

Figure 7.2 Normalized annual energy yield statistics for the grid-connected installation on the Jungfraujoch (elevation 3454 m) in 1994 [7.3], which is mounted on the façade of a research station in the high Alps. All Y and L values are expressed as mean kWh/kWp/d and h/d values. The standardization was based on the effective peak solar generator readings at STC power output amounting to $P_{Go} = 1.13$ kWp

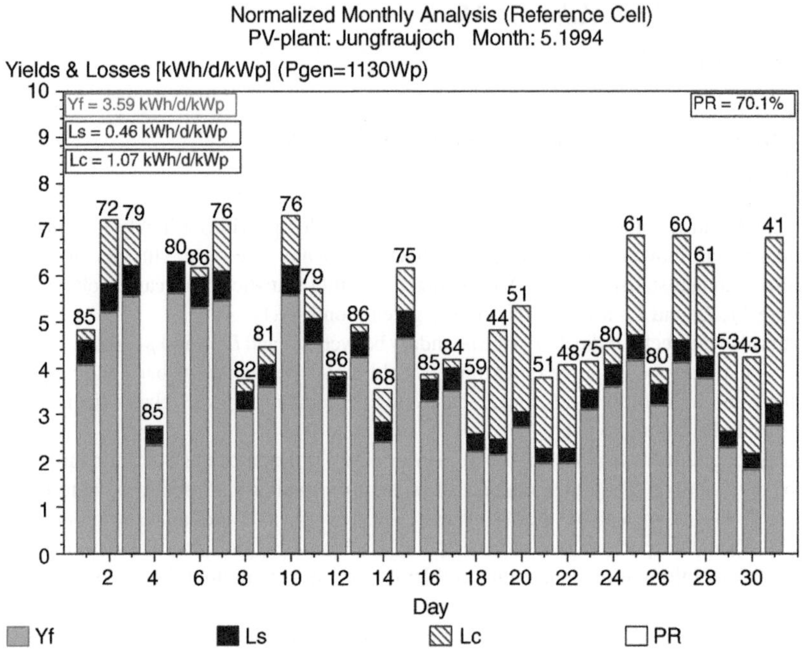

Normalized Monthly Analysis (Reference Cell)
PV-plant: Jungfraujoch   Month: 5.1994

Yields & Losses [kWh/d/kWp] (Pgen=1130Wp)

Figure 7.3 Normalized monthly statistics for the Jungfraujoch PV installation in May 1994. The energy yield was lower for the eastern half of the installation owing to a covering of snow, resulting in low PR and high L_C values. This chart contains all of the information from Figure 7.4, plus a wealth of other information such as dates, L_C, L_S and more exact PR statistics

during the rest of the year, owing to higher L_C values. The normalized monthly statistics chart allows for a more detailed assessment.

Figure 7.3 displays the same installation's 1994 normalized monthly statistics. L_C rose sharply and PR decreased to below 50% on certain days because, during the very snowy winter of 1993–1994, the snow reached the lower edge of the eastern solar generator and had to be cleared away. However, as the terrain beneath the western solar generator slopes far more, the output of this half of the installation was not impaired by snow for lengthy periods. In such cases, normalized daily statistics or daily diagrams (discussed below) allow for detailed assessments of the problems that come to light.

Figure 7.4 displays a scatter diagram (as in [7.1]) for the installation and month in Figure 7.3. This diagram indicates the Y_F and Y_R values, as well as PR, via the second axis. The readings that are far below the line that runs from the baseline to PR = 0.7 are signs of operational problems with the installation. However, this diagram displays neither L_C nor L_S, and according to [7.1] no data indicating the timing of the measurements were available. In Figure 7.4, an attempt was made to incorporate the date of each operational incident, but this is actually not possible as the diagram then becomes too cluttered.

Normalized monthly statistics also allow for the detection of occasional inverter anomalies that might well go undetected solely on the basis of monthly power statistics.

Figure 7.5 allows for the detection of such problems for a 3.18 kWp PV installation in Burgdorf ($\beta = 30°$) in March 1993. Incidents were recorded on 3, 4 and 28–31 March. The installation was shut down on 30 March owing to the failure of its inverter to start automatically. A detailed assessment based on a normalized daily diagram revealed that the solar generator had been covered with snow on the 3rd and 4th and that the inverter exhibited start-up and MPP tracking malfunctions from the 28th to 3st. The normalized daily statistics discussed in the following section can be used to investigate and pinpoint the sources of such anomalies (see Figures 7.6 and 7.7).

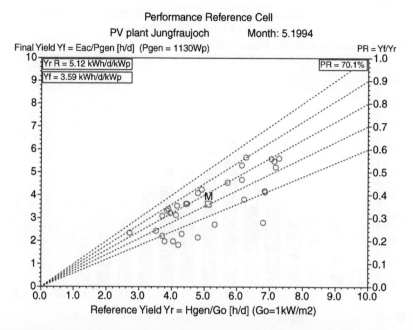

Figure 7.4   Scatter diagram [7.1] indicating reference yield Y_R, final yield Y_F and performance ratio PR $= Y_F/Y_R$ (right) for the Jungfraujoch installation in May 1994. The date is also indicated for each of the PR $< 70\%$ readings. Although this makes it possible to see the exact date of an operational incident, the diagram then becomes too cluttered

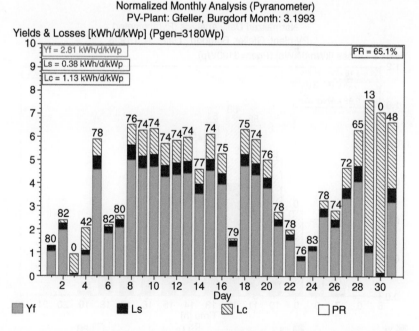

Figure 7.5   Normalized monthly statistics for the Gfeller PV installation in March 1993, which reveal that the solar generator was covered with snow on 3 and 4 March and that the inverter malfunctioned from 28 to 31 March. PR was low and L_C was high at the time of these incidents, but the two types of incidents could only be differentiated by using normalized daily statistics or daily diagrams

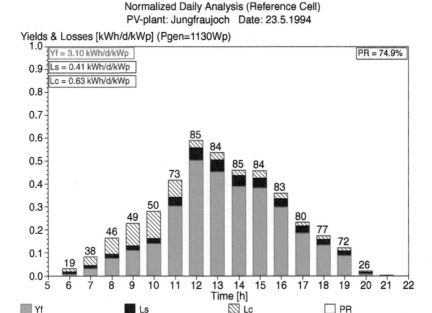

Figure 7.6 Normalized daily statistics for the Jungfraujoch installation on 23 May 1994. The hour-by-hour values for Y_F, L_S, L_C and PR are indicated. As for normalized monthly and annual statistics, the hourly values for Y_A and Y_R are readily apparent. All hourly Y and L values are expressed in kW/kWp and are in fact dimensionless. The daily values indicated comprise the sum total of the various hourly values

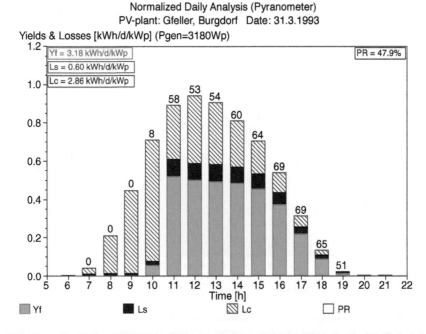

Figure 7.7 Normalized daily statistics at the Gfeller installation on 31 March 1993, when the installation's inverter failed to start automatically prior to 9 a.m. After being started manually, its solar tracker malfunctioned in that PR was too low and L_C was too high. The device failed a few days later

7.3.2   Normalized Daily Statistics Broken Down by Hours

The reference period τ for insolation and energy yield measurement amounting to less than a day can of course also be defined. If 'hour' is defined as the reference period τ, hourly Y values are obtained that indicate energy yield in kWh/kWp per hour, or the counterpart mean PV installation performance per hour relative to solar generator power output P_{Go}. Hourly L values can also be defined. If the hour unit (h) is expressed as a numerator and denominator, the metric for the Y and L values is kW/kWp, i.e. they are actually dimensionless, but relative to mean daily values expressed as kWh/kWp/d they are of course considerably lower and are for the most part less than 1 (although Y_R may slightly exceed 1 in exceptional cases). The relationship between daily and hourly values is simply that the daily value is the sum total of the hourly values for the relevant parameter during the reference day.

Using these newly defined hourly values for Y_F, Y_A, Y_R, L_C and L_S, the normalized daily statistics and the related hourly values can be visualized as is done for monthly or annual statistics, and the PR can be indicated in the bar chart in per cent via the hourly values for Y_F, L_S and L_C. The daily statistics bars in the examples shown here indicate the mean hourly value for each previous hour.

Figure 7.6 displays the normalized daily statistics and the related hourly values for the Jungfraujoch installation on 23 May 1994, at which time the installation's performance bounced back from 50% to normal following a five-day snowy period (see Figure 7.3). This bar chart clearly displays the exact course of events during the reference period. Up until 10 a.m., PR was 50% or less, because the snow covering the eastern solar generator had reduced its energy to near 0, as had been the case for the previous few days. But then PR rose abruptly between 10 and 11 a.m., reaching 85% between 11 a.m. and noon. Inasmuch as the solar generator is shaded during this period in the summer months, this phenomenon is attributable to human intervention – that is, the maintenance operator cleared the snow covering the solar generator on the east side after a heavy snowfall.

Figure 7.7 displays the normalized daily statistics for 31 March 1993 at the Burgdorf installation, which exhibited a number of operational anomalies in that month (see Figure 7.5). During the period leading up to 9 a.m., the installation's energy production remained at 0 because the inverter failed to start up automatically. L_C was correspondingly high during this period, i.e. nearly as high as Y_R as the inverter absorbed virtually none of the solar generator power output that was available to it. The inverter was started manually between 9 and 10 a.m., only to exhibit serious solar tracking problems at medium and high outputs, since the PR values were far too low. These anomalies were monitored into early April, until the inverter finally failed from a hardware malfunction.

Since most PV installations with reasonably efficient automatic data capture functions document the mean hourly values for the key parameters at a minimum, all such installations should be able to capture the normalized annual, monthly and daily statistics discussed thus far; this in turn allows for more precise assessments than those proposed in [7.1], using the same data.

7.4   Normalized PV Installation Power Output

In most PV installations with detailed measurement programmes, a series of values is documented per hour (at intervals of Δt) for each value of merit. For example, each value of merit is scanned at 1 s intervals at the 60 kWp installation in Burgdorf [4.5], [5.44]. Under normal circumstances (i.e. in the absence of any installation malfunction), mean 1 min values are derived from these data. Under normal circumstances, mean 5 min values are generated at the Jungfraujoch and Gfeller installations. In some cases, it may be desirable to reduce the reference period τ still further so that it equals the minimum time interval Δt for the stored readings – which for the Jungfraujoch and Gfeller installations would be 1 and 5 min respectively. This allows for a more precise assessment of the performance of such installations. To this end, a short-term mean value for each Y and L value can be set for the now very brief reference periods τ, and these values can then be regarded as normalized instantaneous y_F, y_A, y_R, l_C and l_S statistics. Like energy yield, these values are designated using the relevant lower-case letters.

In many installations, solar cell temperature T_Z is monitored in addition to ambient temperature. The capture losses L_C in such cases can be divided between the (unavoidable) thermal losses L_{CT} and the remaining miscellaneous losses L_{CM}, providing that on insertion of the above instantaneous values three additional instantaneous values, y_T, l_{CT} and l_{CM}, are defined. As a PV installation malfunction induces an instantaneous increase in L_{CM} and l_{CM}, these values of merit are ideal indicators of installation anomalies, particularly when a reference cell is monitoring global irradiance G_G at the solar generator surface. Temperature-corrected solar generator power output P_{GoT} is needed to define such values of merit.

Inasmuch as solar generator performance correlates with solar cell temperature, an otherwise ideal, loss-free solar generator with nominal power P_{Go} and solar cell temperature T_Z will exhibit the following performance on reaching its maximum power point at $G_o = 1\,\text{kW/m}^2$ irradiance (the symbols in this equation are highlighted below):

$$\text{Temperature-corrected nominal solar generator power output } P_{GoT} = P_{Go} \cdot [1 + c_T(T_Z - T_o)] \tag{7.7}$$

Hence the normalized instantaneous values for power, loss and instantaneous performance can be defined as follows:

$$\text{Normalized insolation (reference yield) } y_R = \frac{G_G}{G_o} = \frac{G_G}{1\,\text{kW/m}^2} \tag{7.8}$$

$$\text{Temperature-corrected insolation } y_T = y_R \cdot (P_{GoT}/PG_o) = y_R \cdot [1 + c_T(T_Z - T_o)] \tag{7.9}$$

$$\text{Normalized solar generator power output } y_A = P_A/PG_o \tag{7.10}$$

$$\text{Normalized output power } y_F = P_{use}/PG_o \tag{7.11}$$

$$\text{Normalized thermal capture losses } l_{CT} = y_R - y_T \tag{7.12}$$

$$\text{Normalized miscellaneous capture losses } l_{CM} = y_T - y_A \tag{7.13}$$

$$\text{Normalized system losses } l_S = y_A - y_F \tag{7.14}$$

$$\text{Instantaneous performance ratio } pr = y_F/y_R \tag{7.15}$$

where:

P_{GoT} = temperature-corrected nominal solar generator power output

P_{Go} = nominal solar generator capacity at STC power output

c_T = temperature coefficient at solar generator MPP (around -0.38% to -0.5% per K for crystalline solar cells)

T_Z = solar generator cell temperature (also referred to as T_C)

T_o = STC reference temperature ($25\,°\text{C}$) at which nominal solar generator power output P_{Go} is defined

G_G = overall irradiance on the solar generator surface in kW/m^2 (often referred to as G_I)

G_o = irradiance at STC power output ($1\,\text{kW/m}^2$)

P_A = overall solar generator power output

P_{use} = PV installation output power (for grid-connected systems, $P_{use} = P_{AC}$).

7.4.1 Normalized Daily Diagram with Instantaneous Values

Most of the normalized power outputs and losses defined here range from 0 to 1. In the case of transient cloud enhancements, PV installations in flat areas may occasionally exhibit transient y_R voltage spikes ranging up to 1.4, and in high mountain areas up to 2. This limitation of the fluctuation range of the instantaneous values thus defined eases the task of incorporating these functions into a single and optimally informative daily diagram containing normalized power output statistics. I shall refer to this type of diagram from here on as a normalized daily diagram. The main difficulty posed by the realization of such diagrams is ensuring that the curves for days with widely fluctuating insolation can be clearly differentiated from each other. If colours are supported, these curves should be colour coded, otherwise they should be differentiated using various symbols or various types of lines [7.4].

7.4.2 Derivation of Daily Energy Yield from Normalized Instantaneous Values

Daily Y_R, Y_T, Y_A, Y_F, L_{CT}, L_{CM} and L_S values can be determined by integrating the instantaneous y_R, y_T, y_A, y_F, l_{CT}, l_{CM} and l_S values over the relevant day:

$$Y_i = \int_0^T y_i \cdot dt = \sum_k y_{ik} \cdot \Delta t \quad \text{and} \quad L_i = \int_0^T l_i \cdot dt = \sum_k l_{ik} \cdot \Delta t \qquad (7.16)$$

This equation can be used to determine daily Y_T values (temperature-corrected reference yield), thermal capture losses L_{CT} and miscellaneous capture losses L_{CM}.

7.4.3 Definition of the Correction Factors k_G, k_T and of efficiency n_I

Using these normalized daily values, the following useful values of merit can be defined in addition to PR:

$$\text{Temperature correction factor } k_T = Y_T/Y_R \qquad (7.17)$$

$$\text{Solar generator correction factor } k_G = Y_A/Y_T \qquad (7.18)$$

The following equation applies to grid-connected systems:

$$\text{Inverter DC–AC conversion efficiency } n_I = Y_F/Y_A \qquad (7.19)$$

The correction factors k_G, k_T and n_I can of course also be defined for monthly and annual Y_R, Y_T, Y_A and Y_F values, from which monthly and annual PV installation energy yields can be readily derived (see Chapter 8).

7.4.4 Assessment Methods Using Normalized Daily Diagrams

Normalized daily diagrams containing the y_R, y_T, y_A, y_F, l_{CM} and pr values lend themselves particularly well to assessments of PV installation performance. Ideally, pr for a well-designed and properly functioning PV installation should be just below 1 throughout the day, whereby the following applies to normalized miscellaneous capture losses: $l_{CM} \ll 1$. If an installation's l_{CM} rises, it means that the installation is not absorbing all the available solar generator power output $y_T \cdot P_{Go}$.

Stand-alone installations do not of course exhibit this ideal scenario at all times as most of these installations are designed to provide energy in inclement weather via the power stored in their battery, which also needs to be fully charged from time to time so that it attains a satisfactory service life. But a fully charged battery does not fully use y_T, and thus l_{CM} rises along with the relevant daily value L_{CM}, whereas pr and the daily PR value decrease. This phenomenon is particularly pronounced in stand-alone installations that are operated all year round in temperate climates. Hence stand-alone installations always exhibit appreciably higher L_{CM} and appreciably lower PR than well-designed grid-connected systems.

On the other hand, in a properly functioning grid-connected system with a sufficiently large inverter (nominal DC-side inverter power output P_{DCn}, around P_{Go} or higher), l_{CM} exceeding 5–10% as a rule reliably indicates that the installation is experiencing problems.

Inasmuch as solar generator power output in flat areas hardly ever reaches P_{Go}, nominal solar generator power P_{Go} should be set somewhat higher (e.g. 10%) than nominal inverter power P_{DCn}, and for façade-mounted solar generators in such areas should be up to around 30% (also see Section 5.2.5.8). This of course presupposes that, in the presence of surplus power output, the inverter will not limit solar generator power and will allow it to reach its rated MPP and continue in operation, rather than simply shutting the solar generator down. This will of course induce a transient l_{CM} spike, which usually occurs around noon on sunny days in the presence of elevated y_T, whereby final yield y_F usually remains relatively constant. In such cases, L_{CM} and the solar generator correction factor k_G usually indicate consequent normalized daily yield losses. At other times of course (i.e. in the presence of lower y_T) such installations will normally exhibit $l_{CM} \ll 1$ and pr close to 1.

7.4.5 Specimen Normalized Daily Diagrams

The possibilities opened up by normalized daily diagrams will now be discussed in greater detail.

Figure 7.8 displays such a diagram for the Jungfraujoch installation on 22 November 1993. In this installation, P_{DCn} is around 1.8 kW, i.e. considerably higher than P_{Go}, owing to frequent insolation spikes of well over 1 kW/m² (see Section 5.2.5.8). According to the diagram, l_{CM} is very low throughout the day, and thus L_{CM} is also very low and k_G is close to 1. This installation exhibited trouble-free operation on the day in question.

Figure 7.9 displays an extended normalized daily diagram for the same installation on 3 May 1994, when, at around 9.40 a.m., l_{CM} rose abruptly and remained at a relatively high level until 12.15 p.m., whereby pr decreased accordingly. This phenomenon was attributable to a roof avalanche in front of the

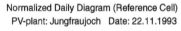

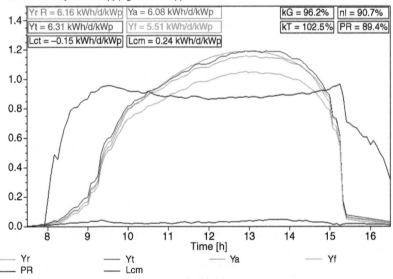

Figure 7.8 Normalized daily diagram for the Jungfraujoch installation on 22 November 1993, which was a sunny day with low l_{CM} and elevated k_G, i.e. the installation was functioning properly and was free of any power-reducing hindrance on the day in question. For a lengthy period in the early afternoon, $G \approx 1.2$ kW/m², $y_A \approx 1.16$ and $PDC \approx 1.31$ kW, i.e. they were considerably higher than PGo

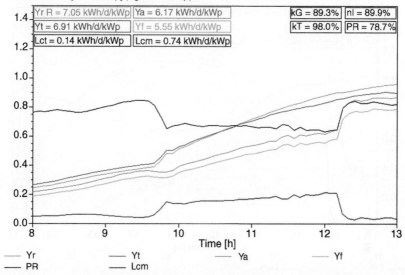

Figure 7.9 Extended normalized daily diagram for the Jungfraujoch installation for 3 May 1994 between the hours of 8 a.m. and 1 p.m., when I_{CM} rose appreciably (between 9.40 a.m. and 12.15 p.m.) owing to partial shading of the east-side solar generator resulting from a roof avalanche

east side of the solar generator. The resulting snow drift partially shaded this half of the solar generator. The snow slid off or was shovelled away shortly after noon, whereupon l_{CM} went back down and pr rose. Hence on the day in question, k_G fell far short of the value in Figure 7.8.

Figure 7.10 displays the normalized daily diagram for 4 January 1995 for the Birg installation (elevation 2670 m; $P_{Go} = 4.134$ kWp), which has a somewhat oversized solar generator whose inverter limits AC-side power input to around 3.5 kW or 0.85 for y_A in the normalized diagram. As y_T peaked at around 1.12 on this cold, sunny, winter day, l_{CM} rose sharply around noon, peaking at around 0.28; pr decreased appreciably around noon and did not recover until the afternoon. On the day in question, L_{CM} was 1.05 kWh/kWp/d instead of 0.35 kWh/kWp/d, which might have been the case with a non-oversized solar generator; in other words, the inverter power limitation induced capacity losses of around 0.7 kWh/kWp/d. k_G was accordingly lower (just under 85%). This installation was sized as for an installation in a flat area, but this is excessive for an Alpine installation.

Figure 7.11 displays the normalized daily diagram for the same installation on 15 January 1995 after a heavy snowfall. When the Sun came out, l_{CM} immediately increased to around 0.2, rising to around 0.37 at noon; pr was low throughout the day.

No power limitation occurred at noon despite the extremely sunny weather. The solar generator exhibited capacity losses on the day in question, being partially covered by snow. A similar normalized daily diagram would occur following the failure of a number of solar generator strings resulting from string diode or string fuse failure. As can be seen in the diagram, on the day in question L_{CM} was severely impaired (2.24 kWh/kWp/d), k_G was only just below 66%, PR was just below 57% and the inverter failed shortly before 10 a.m. owing to a voltage breakdown on the AC side.

Figure 7.12 displays the normalized daily diagram for the Aerni installation in Arisdorf on 23 May 1993, when the installation experienced a severe solar tracker problem and l_{CM} rose steadily under relatively low insolation conditions, reaching 0.4 at noon; pr was low accordingly. The daily L_{CM} of 2.24 kWh/kWp/d was very elevated, k_G reached only 60% and PR was only around 48%.

Figure 7.13 displays the normalized daily diagram for a triphase 20 kW inverter at the Bern University of Applied Sciences PV Lab in Burgdorf [4.5]. The solar modules of this shed-roof-mounted installation

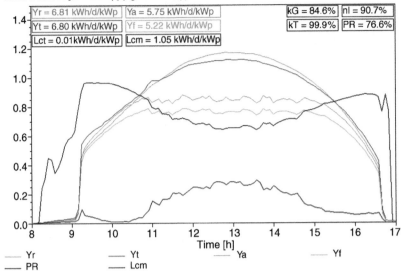

Figure 7.10   Extended normalized daily diagram for the Birg installation (elevation 2670 m) for 4 April 1995. As the solar generator of this façade-mounted installation is oversized (i.e. too powerful) for its inverter, I_{CM} increased at noon, resulting in appreciable capture losses. A k_G of around 85% and PR of around 77% were lower accordingly

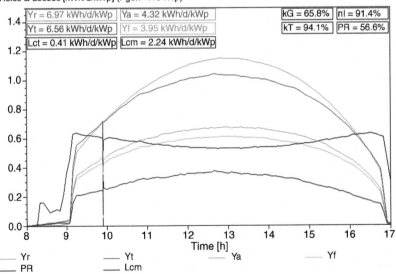

Figure 7.11   Normalized daily diagram for the Birg installation on 15 January 1995, when the installation's solar generator was partly shaded by heavy snowfall. I_{CM} increased immediately to around 0.2 when the Sun came out, reaching 0.37 by noon. Energy yield was severely impaired, k_G was only around 66% and PR was just under 57%

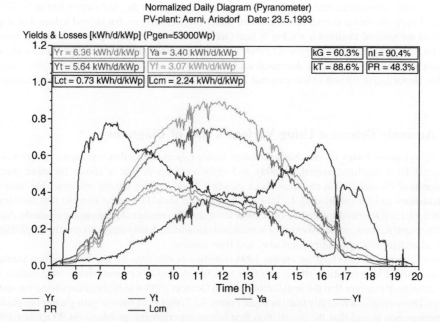

Figure 7.12 Normalized daily diagram for the Aerni installation in Arisdorf on 23 May 1993, when the installation's inverter experienced a severe solar tracking problem. As I_{CM} rose sharply over the course of the day, k_G and PR were both extremely low (around 60% and 48% respectively)

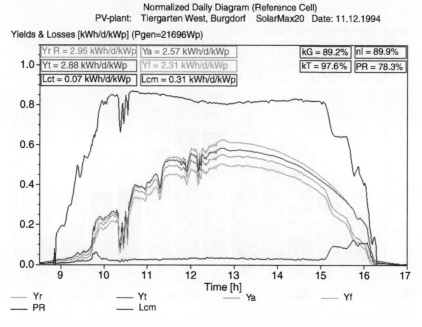

Figure 7.13 Normalized daily diagram for 11 December 1994 for a 20 kW installation on the west side of the newly built Electrical Engineering Department building at Bern University of Applied Sciences in Burgdorf [4.5]. As the figure shows, shading of the lower rows of modules induced a sharp rise in I_{CM}, which is thus a reliable indicator of partial solar generator shading

are horizontally series connected, such that late-afternoon shadows on the shed (which has no buildings in front of it) do not reduce overall solar generator power output in one go, but instead reduce it in stages as each of the rows of modules is shaded in turn (see Figure 1.11, which displays sunshine at around 5 p.m. on a day in February). As Figure 7.13 shows, l_{CM} abruptly increased to a new high after 3 p.m., and increased again after 3.30 p.m.; pr decreased accordingly owing to shading of the bottom two rows of modules. Hence l_{CM} registered a clear response to partial solar generator shading from nearby structures.

7.5 Anomaly Detection Using Various Types of Diagrams

Subdividing capture losses L_C into thermal capture losses L_{CT} and miscellaneous capture losses L_{CM} is also useful for normalized annual, monthly and daily statistics in that it allows for more precise assessments of PV installation problems. In cases where hourly values are the sole available basis for Y_T calculations and for subdividing L_C into L_{CT} and L_{CM}, it should be borne in mind that insolation-weighted solar cell temperature readings that allow for optimally precise calculations are available. As the following example shows, successive use of normalized statistics and daily diagrams greatly simplifies the task of identifying PV installation anomalies and their causes.

Figure 7.14 displays normalized annual 1994 statistics of this type for the Industriellen Betriebe Interlaken (IBI) PV installation, which uses four 1.8 kW inverters in a master–slave configuration. On closer scrutiny, it emerges that the installation's l_{CM} in October 1994 was higher than during the rest of the year. The normalized monthly statistics (see Figure 7.15) that were used to gain greater insight into this phenomenon reveal that the installation first began experiencing problems on 2 October 1994. According to the normalized daily statistics for this date (see Figure 7.16), the installation was down

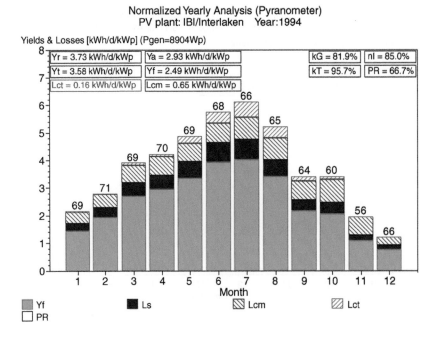

Figure 7.14 Normalized annual 1994 statistics for the IBI installation in Interlaken. The chart shows that L_{CM} loss in October was excessive relative to the other values, which means that the solar generator must have experienced a problem over the course of that month

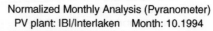

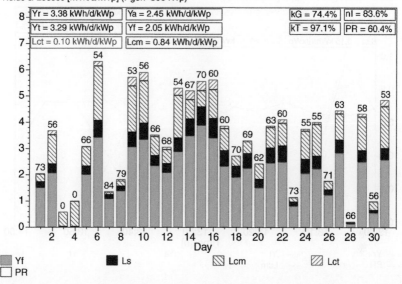

Figure 7.15 Normalized monthly statistics for the IBI Interlaken installation (October 1994), which clearly show that the installation experienced a problem on the 2nd of the month, was down on the 3rd and 4th, and resumed operation at reduced power on the 5th

Normalized Daily Analysis (Pyranometer)
PV plant: IBI/Interlaken   Date: 2.10.1994

Yields & Losses [kWh/d/kWp] (Pgen=8904Wp)

Yr = 3.67 kWh/d/kWp   Ya = 2.40 kWh/d/kWp   kG = 68.1%   nI = 85.3%
Yt = 3.52 kWh/d/kWp   Yf = 2.05 kWh/d/kWp   kT = 95.9%   PR = 55.7%
Lct = 0.15 kWh/d/kWp   Lcm = 1.12 kWh/d/kWp

Figure 7.16 Normalized daily statistics for the IBI Interlaken installation on 2 October 1994, on which date (as the chart shows) the installation was down from 1 to 2 p.m. As noted in Section 7.3.2, the bars in the daily statistics graphic indicate, for each hour, the mean hourly value of the previous hour, e.g. the 2 p.m. bar indicates the mean value for the 1–2 p.m. period

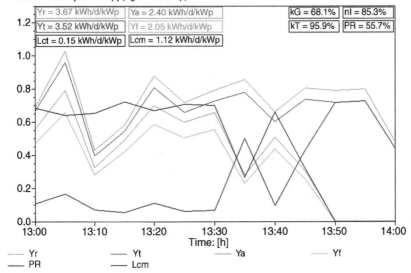

Figure 7.17 Detailed analytic representation of the 2 October 1994 failure, via an extended normalized daily diagram for the IBI Interlaken installation for the 1–2 p.m. period, with 5 min mean values. The failure was prefigured by the 1.35 p.m. readings resulting from a temporary solar tracker problem that induced an I_{CM} increase and a pr decrease. The master inverter definitively failed at 1.42 p.m., whereupon overall installation energy yield decreased to 0, resulting in y_F and pr of zero

between 1 and 2 p.m. owing to a master inverter failure, which was signalled by the abrupt and precipitous drop in hourly PR. At the same time, hourly L_{CM} rose to an hourly value of Y_T, which means that the inverter had ceased to absorb power from the solar generator.

This installation failure can be analysed in greater detail using the normalized daily diagram that has been extended to include the critical period (see Figure 7.17, which displays the 5 min readings). According to this diagram, the failure was prefigured by the 1.35 p.m. readings resulting from a temporary solar tracker problem (elevated I_{CM} increase in tandem with the corresponding pr decrease). The readings returned to normal at 1.40 p.m. The actual installation failure (which occurred at 1.42 p.m.) was initially signalled by the 1.45 p.m. readings, whereupon I_{CM} rose to y_T and pr decreased to y_F, i.e. the installation failed.

The installation was down on 3 and 4 October and was provisionally restarted on 5 October using two inverters, and on 14 October using three inverters (see Figure 7.15). As a result of this incident, the daily PR values continued to rise. Owing to the failure of a slave inverter on 16 October, the installation was operated with only two slave inverters for the rest of the month, which explains the low PR on sunny days.

Normalized diagrams also allow for the detection of high-risk solar generator incidents such as a smouldering fire in the generator circuit box of an 11-year-old solar generator (e.g. for the installation in Figure 7.18).

A normalized daily diagram for this installation for 16 April (Figure 7.19) reveals that an anomaly occurred shortly after 11 a.m. during a transient insolation spike (hefty and abrupt increase in I_{CM}). The installation was shut down manually on 20 April following detection of the anomaly.

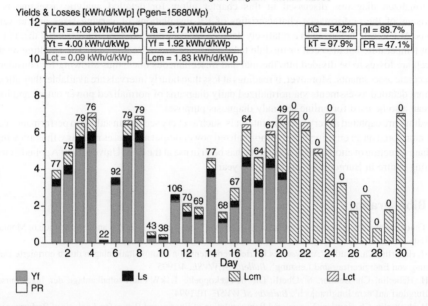

Figure 7.18 Normalized monthly statistics for April 2006 at the Localnet Power Company installation in Burgdorf. According to the chart, after the 16th, L_{CM} increased considerably and PR decreased. As this is an eight-string installation, these events signalled the failure of one string. A normalized daily diagram (see Figure 7.19) shed further light on this incident

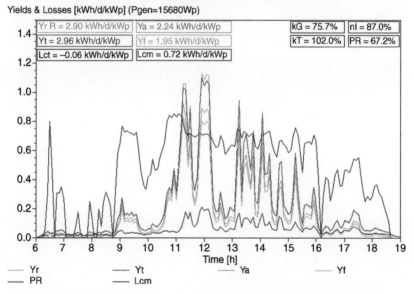

Figure 7.19 Detailed analysis on 16 April 2006 via a normalized daily diagram. The failure occurred at 11 a.m. during a cloud enhancement. This failure was presumably caused by a soldering joint defect that gradually worsened over the years and ultimately caused a smouldering fire in the junction box (see Figures 4.104–4.106)

7.6   Recap and Conclusions

The normalized diagrams discussed in this chapter allow for detailed analytic assessments and comparisons of the performance characteristics of PV installations of varying sizes and at various locations. At PV installations with a relatively detailed measurement system, it is essential that not only ambient temperature but also cell or module temperature be monitored as well. The resulting readings allow capture losses to be divided into thermal and miscellaneous ones, thus allowing for considerably more accurate assessments. Moreover, if readings at less than hourly intervals are available, they allow for even more detailed assessments via normalized daily diagrams of normalized power output and losses, which can also be used for online anomaly diagnosis purposes.

If readings are captured at very frequent intervals, such as every second, installation performance can be readily monitored on an ongoing basis via normalized power output and losses readings, thus allowing for immediate detection of anomalies. This method has been in use at the Bern University of Applied Sciences PV testing centre in Burgdorf [6.24] for many years.

7.7   Bibliography

[7.1] 'Guidelines for the Assessment of Photovoltaic Plants, Document B: Analysis and Presentation of Monitoring Data', Issue 4.1. JRC, Ispra, June 1993.

[7.2] H. Häberlin, Ch. Beutler:'Analyse des Betriebsverhaltens von Photovoltaikanlagen durch normierte Darstellung von Energieertrag und Leistung'. *Bulletin SEV/VSE*, 4/1995.

[7.3] H. Häberlin, Ch. Beutler, S. Oberli:'Die netzgekoppelte 1,1kW-Photovoltaikanlage der Ingenieurschule Burgdorf auf dem Jungfraujoch'. *Bulletin SEV/VSE*, 10/1994.

[7.4] H. Haeberlin, Ch. Beutler:'Normalized Representation of Energy and Power for Analysis of Performance and On-line Error Detection in PV-Systems'. 13th EU PV Conference on Photovoltaic Solar Energy Conversion, Nice, 1995.

Additional References

[Häb07].

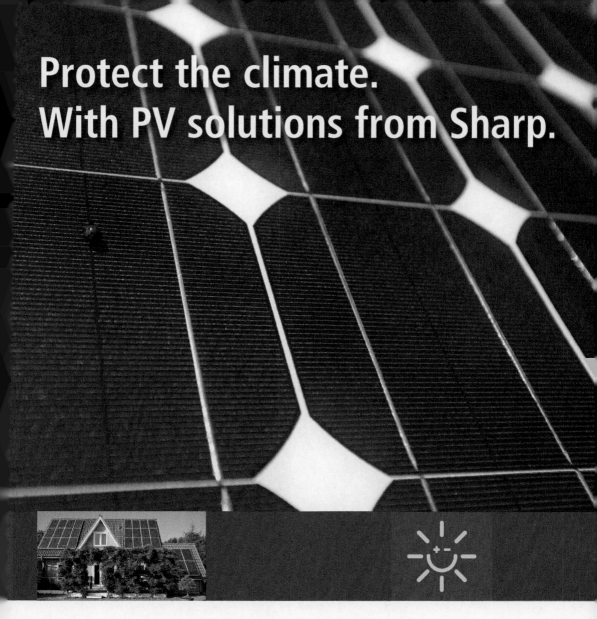

Protect the climate.
With PV solutions from Sharp.

Sharp has over 50 years of experience in photo-voltaics, its own PV factories as well as research and development facilities in Japan, America and Europe. Today's challenge is to establish photovoltaics worldwide as the most rational source of energy, both in ecological and eco-nomical terms. Sharp is ready for this chal-lenge: the company from the land of the rising sun offers the full range of the PV spectrum: monocrystalline, polycristalline and thin-film photovoltaic systems. Sharp is one of the global leading manufacturers of solar cells. Produc-tion capacity and quality standards are being increased continually. Consequently, Sharp has recently built the first 1 gigawatt thin-film solar cell plant in Sakai, near Osaka.

Welcome to Sharp: www.sharp.eu/solar

SHARP

LET'S FOCUS ON SAFETY FOR A MINUTE

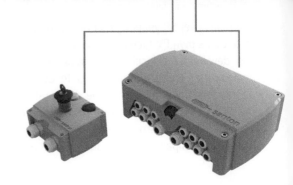

8

PV Installation Sizing

This chapter discusses the specific issues that arise in connection with PV installation sizing (see Chapter 5 for a discussion of PV installation configurations), and in so doing presupposes that the reader is conversant with general electrical engineering issues such as selecting the right wire gauges, installation fusing methods and calculating voltage drops. Of overarching importance for such matters is that the vendor-specified DC elements be used and that all insulation exhibits the requisite light and weathering resistance rating. Moreover, national PV installation regulations should be observed such as [4.10] ([4.13] in Switzerland).

8.1 Principal of and Baseline Values for Yield Calculations

The energy yield calculation method described here for all types of PV installations is based on the standardized energy yield diagrams and tables discussed in Chapter 7 and the relevant mean monthly values. Insolation yield $Y_R = H_G/(1\,\text{kW/m}^2)$ on the solar generator plane and the generation correction factor k_G are needed for all such calculations, and the temperature correction factor k_T is needed for grid-connected systems and stand-alone installations with charge controllers and solar trackers. For grid-connected systems it is also necessary to know the mean efficiency, $\eta_{WR} = n_I$, of the inverter being used (e.g. European efficiency η_{EU} as in Equation 5.28, or preferably overall efficiency $\eta_{tot\text{-}EU}$ as in Equation 5.33) as derived from the counterpart weighting factors. The mean efficiency $\eta_{LR\text{-}MPT}$ (η_{MPT}) calculated in an analogous manner should be known for charge controllers with solar trackers.

With sufficient practice, simulations for virtually any type of PV installation can be run using good simulation software (which costs a few hundred euros). However, such products are suboptimal for sizing actual PV installations as their array of options for comprehensive simulations renders them user unfriendly for novice or occasional users. Nor is such software suitable for teaching or learning purposes, as its functions lack transparency and prevent learners from gaining an all-round grasp of photovoltaics. Such tools are only suitable for installation optimization purposes for experienced users who possess a high level of expertise in photovoltaics. Unfortunately, the PV installation simulation software available today for the most part fails to take an accurate account of key parameters such as snow cover, soiling and capture losses.

This chapter discusses methods that allow for PV installation energy yield sizing and calculation with reasonable effort. However, it needs to be borne in mind here that, particularly in winter, insolation can fluctuate by more than a factor of 4 (see Figures 2.9 and 2.26) over the course of any given winter month, during which less than half of mean insolation may occur. Hence all PV installation energy yield

Photovoltaics: System Design and Practice, First Edition. Heinrich Häberlin.
© 2012 John Wiley & Sons, Ltd. Published 2012 by John Wiley & Sons, Ltd.

calculations are based on historical statistics that cannot possibly provide a wholly accurate energy yield forecast for any given month. Thus it is not always wise to spend extensive time and energy on determining very exact correction factors, since energy yield is bound to exhibit far greater fluctuations in any case on account of changing weather conditions.

8.1.1 Insolation Calculations

To size stand-alone installation solar generators, it is essential to have optimally precise insolation data for the envisaged installation site, so that irradiated energy H_G on the solar generator plane can be calculated (see Chapter 2).

As daily power balances are normally used for stand-alone installations, H_G should be expressed in kWh/m²/d for sizing purposes. Although monthly stand-alone installation H_G often needs to be determined, in many cases the determination of this value for the critical low-insolation months is sufficient.

The key guide values for grid-connected systems are usually annual energy yield and all monthly energy yield values. In cases where monthly insolation statistics are available for a grid-connected system, the monthly insolation values can be used without further ado; if not, the mean daily insolation values can simply be multiplied by n_d (number of days per month) for each month.

The data in Appendix A can be used to determine H_G at many European sites, although of course such calculations are much easier if done with a suitable software application [2.3], [2.4], [2.5]. Using Equation 7.3, $Y_R = H_G/(1\,\text{kW/m}^2)$ can readily be derived from H_G.

Using the simple methods described in Section 2.4, insolation on the solar generator plane can be determined with relatively little effort (see Table 2.4), although these methods take account of local differences to a limited degree only and cannot be used for sizing involving specific factors such as shading, the effect of adjacent buildings, and so on. If greater precision is required, the more labour-intensive methods involve the three-component model (see Section 2.5 and Table 2.8). Both of the tables here can be used to determine reference yield Y_R for a specific PV installation. The sections that follow that discuss sizing calculations for specific types of PV installations also contain tables showing the calculations for insolation and energy yield on one page.

8.1.2 Determination of the Temperature Correction Factor k_T

Extensive measurements at numerous installations realized by the Bern University of Applied Sciences PV Lab in various Swiss climate zones have shown that the solar generator correction factor k_T range is around $0.88 < k_T < 1.1$, which is also a representative range for PV installations elsewhere in Central Europe in view of the similarity between the climate in that region and in Switzerland. In Southern Europe, however, k_T tends to be around 0.8 during the summer.

Peak k_T in flat locations with low insolation is mostly observed in December or January, but is somewhat lower in mountainous areas during this period despite lower ambient temperatures, by virtue of the higher insolation-weighted temperatures that prevail owing to higher insolation. k_T tends to be the lowest in mountainous regions during February or March, and is lowest during the summer at flat locations with abundant sunshine. As summer insolation is lower in mountainous regions than in low-lying areas (due to storms) and the ambient temperatures are lower, k_T is far higher than in flat locations.

8.1.2.1 Estimation of Insolation-Weighted Solar Cell Temperature

Chapter 7 discusses the temperature correction factor $k_T = Y_T/Y_R$ for PV installations where irradiance $G_G = G_I$ on the solar module plane and solar cell temperature T_Z are measured using a monitoring programme. As is shown below, using the parameter known as insolation-weighted solar cell temperature T_{ZG}, $k_T = Y_T/Y_R$ can be determined for any PV installation based on the relevant weather data.

The following definition can be derived from Equation 7.9 (definition of y_T), Equation 7.16 (derivation of Y_i from instantaneous values y_i) and the designations and definitions in Chapter 7:

$$Y_T = \int_0^T y_R[1 + c_T(T_Z - T_o)] \cdot dt = Y_R - c_T \cdot T_o \cdot Y_R + {}^R c_T \int_0^T y_R \cdot T_Z \cdot dt \qquad (8.1)$$

$$\text{With insolation-weighted temperature inserted: } T_{ZG} = \frac{1}{Y_R} \int_0^T y_R \cdot T_Z \cdot dt \qquad (8.2)$$

The following can then be derived from these two equations:

$$k_T = \frac{Y_T}{Y_R} = 1 + c_T(T_{ZG} - T_o) \qquad (8.3)$$

Unfortunately, such insolation-weighted module temperatures are not widely available, and most weather data are limited to global insolation and ambient temperature T_U. Insolation-weighted module temperatures can only be determined via long-term monitoring at PV installations involving detailed measurement of insolation, module temperature and ambient temperature.

To do this, it is necessary to determine insolation on the solar generator plane based on mean aggregate daily Y_R of insolation on the solar generator plane and on mean cell temperature T_{ZG}.

A simple procedure for an estimate of this guide value will now be described.

The insolation curve on the horizontal plane over the course of a day is roughly the same as a sinusoidal half wave whose amplitude is G_{max}. On such a day that is t_{dh} long, around $H_G \approx (2/\pi) \cdot G_{max} \cdot t_{dh}$ of energy will be irradiated on this plane, where t_{dh} means the amount of time expressed in hours during which the Sun is more than 5° above the horizon.

Assuming (for reasons of simplicity) that irradiance G_G on the inclined solar generator plane will exhibit a roughly similar time curve, the following holds true for time $t_d \le t_{dh}$ while the Sun is a minimum of 5° above the horizon and with tilt angle β:

$$H_G = \frac{2}{\pi} G_{Gmax} \cdot t_d = Y_R \cdot G_o = Y_R \cdot 1 \, \text{kW/m}^2 \qquad (8.4)$$

A surface pointing due south with tilt angle β is parallel to the horizontal plane at the equivalent latitude $\varphi' = \varphi - \beta$ and the same longitude. Using Equation 2.2 and based on longitude φ and solar declination δ, the daily time t_d can be determined during which, at the equivalent latitude φ', the Sun is at least 5° above the horizon and a minimum of 5° above the horizon at the relevant latitude.

Based on this, the maximum insolation on a mean insolation day of any given month can be determined, if it is assumed that insolation will exhibit the characteristics of a sinusoidal half wave, i.e. that $G = G_{Gmax} \cdot \sin(\pi \cdot t/t_d)$:

$$G_{Gmax} = \frac{\pi \cdot Y_R \cdot G_o}{2 \cdot t_d} = \frac{\pi}{2 \cdot t_d} Y_R \cdot 1 \, \text{kW/m}^2 \qquad (8.5)$$

When $G_o = 1\,\text{kW/m}^2$, solar module temperature relative to ambient temperature T_U ranges from around $\Delta T_o \approx 23$ to $43\,°\text{C}$, depending on the solar generator configuration. Hence the following applies to the temperature increase ΔT_U relative to T_U in the presence of irradiance G:

$$\Delta T_U = T_Z - T_U = \Delta T_o \frac{G}{G_o} = \Delta T_o \cdot y_R \tag{8.6}$$

Hence the following applies using Equation 8.2:

$$T_{ZG} = \frac{1}{Y_R} \int_0^{t_d} y_R \cdot (T_U + \Delta T_o \cdot y_R) \cdot dt = T_U + \frac{1}{Y_R} \int_0^{t_d} y_R^2 \cdot \Delta T_o \cdot dt \tag{8.7}$$

Assuming that here too the insolation curve will equate to a sinusoidal half wave, the following applies:

$$\frac{1}{Y_R} \int_0^{t_d} y_R^2 \cdot \Delta T_o \cdot dt = \Delta T_o \cdot \frac{1}{2 \cdot Y_R} \cdot y_{R\max}^2 \cdot t_d \tag{8.8}$$

Hence the following applies for insolation-weighted temperature:

$$\Rightarrow T_{ZG} = T_U + \Delta T_o \frac{y_{R\max}^2 \cdot t_d}{2 \cdot Y_R} = T_U + \Delta T_o \frac{\pi^2}{8 \cdot t_d} Y_R = T_U + c_Y \cdot Y_R \tag{8.9}$$

Needless to say, the ambient temperatures T_U on which these equations are based vary over the course of a day, i.e. they are higher during the day than at night and higher in the afternoon than in the morning. But if it is nonetheless assumed that an average insolation curve will more or less equate to a sinusoidal half wave, the low morning temperature for the term on the left (T_U) of the integral in Equation 8.7 will be more or less cancelled out by the higher afternoon temperature, providing that Equations are applied in such a way that, in lieu of mean daily temperature T_{Um} over a 24-hour period (the value indicated by most weather stations), the mean temperature T_{Ud} during period t_d is applied (Sun at least 5° above the solar generator surface).

By adding a correction term T_K (mean temperature increase over the course of a day relative to the 24-hour mean daily temperature T_{Um}), T_{Ud} for mean ambient temperature T_{Um} can be obtained:

$$\Rightarrow T_{ZG} = T_{Um} + T_K + \Delta T_o \frac{\pi^2}{8 \cdot t_d} Y_R = T_{Um} + T_K + c_Y \cdot Y_R \tag{8.10}$$

Based on a series of measurements realized over a period of many years by the PV Lab at various Swiss sites, it has been found that a T_K of around 7 K is valid. The correction term T_K in Equation 8.10 takes into account the discrepancy between: (a) mean ambient temperature T_{Ud} during the day; (b) the 24-hour mean temperature indicated by the weather service; and (c) global warming since the mean values were determined.

Table 8.1 c_Y values expressed as K/(h/d) or °C/(h/d) for calculating monthly temperature increases based on Y_R for three different types of PV installations in Kloten ($\beta = 45°$)

c_Y for temperature calculation with Y_R		Jan	Feb	Mar	Apr	May	June	July	Aug	Sep	Oct	Nov	Dec
t_d in h		7.67	9.12	10.77	11.30	11.30	11.30	11.30	11.30	11.30	9.69	8.12	7.14
Free standing	$\Delta T = 23$K at 1kW/m²	3.72	3.13	2.65	2.52	2.52	2.52	2.52	2.52	2.52	2.94	3.51	4.00
On roof	$\Delta T = 33$K at 1kW/m²	5.34	4.49	3.80	3.62	3.62	3.62	3.62	3.62	3.62	4.22	5.04	5.73
Integrated	$\Delta T = 43$K at 1kW/m²	6.95	5.85	4.95	4.72	4.72	4.72	4.72	4.72	4.72	5.50	6.57	7.47

Before using Equation 8.10 to size an actual PV installation, the factor $c_Y = \Delta T_o \cdot (\pi^2/8t_d)$ should be determined for the site in question for an average insolation day in a month. Table 8.1 lists specimen monthly c_Y values for three different types of PV systems in Kloten with a 45° tilt angle, which more or less corresponds to the site's longitude.

$\Delta T_o \approx 23$ K for a ground-based PV installation; $\Delta T_o \approx 33$ K for a solar generator that exhibits reasonable rear ventilation and is mounted parallel to a nearby roof, façade or the like; and $\Delta T_o \approx 43$ K for a solar generator with no rear ventilation that is mounted on a roof or façade. ΔT_o can also be somewhat lower in the presence of frequent strong winds (e.g. in mountain locations).

The c_Y values correlate only very weakly with tilt angle. Although insolation decreases in the summer with a steeper tilt angle, the counterpart t_d of the solar generator surface is relatively low, which means that it has relatively little effect on c_Y. In as much as the purpose is merely to calculate a correction factor, T_{ZG} and thus k_T can be determined solely on the basis of a valid mean tilt angle for the envisaged site.

8.1.2.2 Determination of k_T and Y_T

Based on the insolation-weighted temperature T_{ZG} thus computed, using Equation 8.3 in conjunction with the temperature coefficient c_T for solar generator MPP, correction factor k_T can now be determined. In as much as ohmic loss increases at a higher temperature and additional mismatch losses can also occur owing to solar generator temperature differences, energy yield calculations for crystalline solar modules should be based on higher values, namely $c_T \approx -0.45$ to -0.5% per K, as follows:

$$k_T = 1 + c_T(T_{ZG} - T_o) \qquad (8.11)$$

where:

$c_T =$ temperature coefficient for (a) solar generator MPP and (b) wiring
    (around -0.0045 to -0.005 K^{-1} for crystalline solar cells)
$T_{ZG} =$ insolation-weighted solar generator cell temperature (see Section 8.1.2.1)
$T_o =$ STC reference temperature (25 °C) for which nominal solar generator power
    output P_{Go} is defined

Based on the above, using Equation 7.17 or 8.3 the temperature-corrected reference yield Y_T can be derived from Y_R as follows:

$$\text{Temperature corrected reference yield } Y_T = k_T \cdot Y_R \qquad (8.12)$$

Table 8.2 Temperature correction factors k_T (for low-, medium- and high-temperature effects) for the relevant reference stations for three types of PV installations. Temperature has only a minor effect on amorphous silicon modules

Temperature influence small

	Free field mounting of crystalline modules or plants with modules of amorphous Si												
	Jan	Feb	Mar	Apr	May	June	July	Aug	Sep	Oct	Nov	Dec	Year
Kloten	1.06	1.05	1.02	1.00	0.97	0.96	0.94	0.94	0.97	1.01	1.04	1.06	0.98
Davos	1.05	1.04	1.02	1.01	0.99	0.98	0.96	0.97	0.99	1.01	1.04	1.05	1.00
Locarno	1.03	1.03	1.01	1.00	0.98	0.95	0.93	0.93	0.96	0.99	1.02	1.03	0.98
Potsdam	1.07	1.06	1.03	1.00	0.97	0.95	0.94	0.95	0.98	1.01	1.04	1.06	0.98
Giessen	1.06	1.05	1.03	1.00	0.97	0.95	0.95	0.96	0.99	1.01	1.04	1.06	0.98
Munich	1.06	1.05	1.02	1.00	0.97	0.95	0.94	0.95	0.97	1.00	1.04	1.05	0.98
Marseilles	1.00	1.00	0.98	0.96	0.94	0.92	0.90	0.91	0.94	0.96	0.99	1.00	0.95
Seville	0.98	0.97	0.95	0.95	0.92	0.90	0.88	0.89	0.91	0.93	0.97	0.98	0.93
Cairo	0.96	0.94	0.93	0.91	0.89	0.88	0.88	0.88	0.89	0.90	0.93	0.95	0.91

Temperature influence medium

	Rooftop mounting with air space or façade integration of crystalline modules												
	Jan	Feb	Mar	Apr	May	June	July	Aug	Sep	Oct	Nov	Dec	Year
Kloten	1.05	1.03	1.01	0.98	0.95	0.93	0.91	0.92	0.95	0.99	1.04	1.05	0.96
Davos	1.03	1.02	1.00	0.98	0.97	0.96	0.94	0.95	0.97	0.98	1.02	1.03	0.98
Locarno	1.02	1.02	0.99	0.98	0.96	0.93	0.90	0.91	0.94	0.98	1.01	1.02	0.96
Potsdam	1.06	1.05	1.02	0.98	0.94	0.93	0.92	0.93	0.96	1.00	1.04	1.06	0.96
Giessen	1.06	1.04	1.01	0.98	0.94	0.93	0.92	0.93	0.97	1.00	1.04	1.06	0.96
Munich	1.05	1.03	1.01	0.98	0.94	0.92	0.92	0.93	0.95	0.99	1.02	1.04	0.96
Marseilles	0.98	0.98	0.95	0.93	0.91	0.89	0.87	0.88	0.91	0.93	0.97	0.98	0.92
Seville	0.95	0.95	0.93	0.92	0.89	0.87	0.85	0.85	0.88	0.91	0.95	0.96	0.90
Cairo	0.93	0.92	0.90	0.88	0.86	0.84	0.84	0.84	0.86	0.87	0.90	0.93	0.88

Temperature influence high

	Crystalline modules integrated without air space into roofs												
	Jan	Feb	Mar	Apr	May	June	July	Aug	Sep	Oct	Nov	Dec	Year
Kloten	1.04	1.02	0.99	0.96	0.93	0.91	0.89	0.90	0.93	0.98	1.03	1.04	0.94
Davos	1.00	0.99	0.97	0.96	0.94	0.93	0.91	0.92	0.94	0.96	1.00	1.01	0.96
Locarno	1.00	1.00	0.97	0.97	0.94	0.90	0.87	0.88	0.92	0.96	0.99	1.00	0.94
Potsdam	1.05	1.04	1.00	0.96	0.92	0.90	0.90	0.91	0.94	0.98	1.03	1.05	0.94
Giessen	1.05	1.03	1.00	0.96	0.92	0.91	0.90	0.91	0.96	0.99	1.03	1.05	0.95
Munich	1.03	1.01	0.99	0.95	0.92	0.90	0.89	0.90	0.93	0.97	1.01	1.03	0.94
Marseilles	0.96	0.96	0.93	0.91	0.88	0.86	0.84	0.85	0.89	0.91	0.95	0.96	0.90
Seville	0.93	0.92	0.90	0.89	0.86	0.84	0.81	0.82	0.86	0.88	0.93	0.94	0.87
Cairo	0.90	0.89	0.87	0.84	0.82	0.81	0.81	0.81	0.83	0.84	0.87	0.90	0.84

8.1.2.3   Use of k_T Values for Reference Stations

In as much as only one correction factor is determined based on insolation-weighted temperature T_{ZG}, a minor T_{ZG} calculation error will have little effect on the result obtained for total energy yield. Hence it often suffices to use the temperature correction factor k_T for the installation's reference station rather than calculating the exact T_{ZG} as in Section 8.1.2.1. Table 8.2 lists the k_T values determined using $c_T = -0.45\%$ per K for the nine reference stations for three different types of PV installations.

Note: For reasons of space, the southern hemisphere reference stations have been omitted here. The k_T values for reference stations at similar latitudes in the northern hemisphere are more or less applicable to the southern hemisphere if the six-month difference in the timing of the seasons is taken into account (for an example see the Alice Springs case, Problem 4, in Section 8.2.1.2).

8.1.3 Defining the Solar Generator Correction Factor k_G

8.1.3.1 Factors That Influence the Solar Generator Correction Factor k_G

As noted in Sections 7.2.2 and 7.4.3, the solar generator correction factor k_G characterizes the overall impact of various parameters on energy yield, which can be differentiated by expressing k_G as a product of various guide values – which of course should be as high as possible so as to obtain a high k_G value:

$$\text{Solar generator correction factor } k_G = k_{PM} \cdot k_{NG} \cdot k_{GR} \cdot k_{SP} \cdot k_{TB} \cdot k_{MM} \cdot k_R \cdot k_V \cdot k_S \cdot k_{MPP}$$

$$(8.13)$$

where:
k_{PM} = correction factor for module power loss (e.g. 0.9–1)
k_{NG} = correction factor for low insolation (e.g. 0.96–0.995)
k_{GR} = correction factor for glass reflection loss (e.g. 0.96–0.995)
k_{SP} = correction factor for spectral mismatch (e.g. 0.96–0.995)
k_{TB} = correction factor for partial module shading (e.g. 0.8–1)
k_{MM} = mismatch correction factor (e.g. 0.95–1)
k_R = correction factor for ohmic loss at resistances or string diodes (e.g. 0.96–0.998)
k_V = soiling correction factor (e.g. 0.8–1)
k_S = winter snow-covering correction factor (e.g. 0.5–1)
k_{MPP} = correction factor for non-MPP load in charge controllers, inverters or the like,
          if not otherwise taken into account by a parameter such as inverter η_{tot}.

8.1.3.1.1 Correction Factor k_{PM} for Module Power Loss

All solar module vendors define a specific manufacturing tolerance such as $\pm 3\%$, $\pm 5\%$ or $\pm 10\%$. In the past, it was common for the mean power output of most solar modules to fall far short of their nominal power output, which often was barely higher than the guaranteed minimum power output. But things have improved somewhat since the advent of standardized energy yield diagrams and tables, in that today's PV installations comprising modules from vendors whose products consistently exhibit low power output have lower PR values than installations from rival vendors; fortunately, word gets around after a while. As string power output is mainly determined by the lowest-output module, circumspect PV installation planners will base their k_{PM} calculations mainly on guaranteed minimum module power output. Thus, for example, $k_{PM} \approx 0.9$ for a module with $\pm 10\%$ tolerance and 0.95 for a module with $\pm 5\%$ tolerance. For modules that exhibit far better as-delivered quality than this, k_{PM} calculations can be based on a mean value that lies somewhere between the guaranteed minimum power output and the nominal power output of the module in question.

8.1.3.1.2 Correction Factor k_{NG} for Low Insolation

When insolation goes below $G_o = 1\,\text{kW/m}^2$, both open-circuit voltage and MPP voltage U_{MPP} decrease; as a result, solar cell efficiency η_{PV} decreases with diminishing insolation (see Figure 3.16). The resulting energy yield loss can be taken into account using the correction factor k_{NG}, which is lower at sites that exhibit a relatively high proportion of diffuse insolation (low-lying areas in Central Europe, particularly in winter) than at sites that exhibit a high proportion of direct beam radiation. Since in many thin-film modules η_{PV} barely decreases, even in the presence of low G, k_{NG} is nearly 1 at such sites.

8.1.3.1.3 Correction Factor k_{GR} for Glass Reflection Loss

Some of the radiation incident on a solar module that exhibits a highly unusual angle of incidence will be reflected (see Figure 8.1).

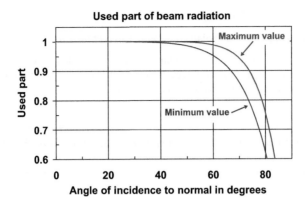

Figure 8.1 Absorbed direct beam radiation as a function of tilt angle relative to the norm. All such radiation is absorbed at small tilt angles, while some is lost with larger angles (owing to flatter incidence), depending on the type of glass used [8.1]

This phenomenon, which is particularly pronounced for direct beam radiation but also occurs with diffuse radiation, results in varying degrees of energy loss (depending on solar generator orientation and season) that is taken into account by k_{GR}. Particularly unfavourable conditions and thus low k_{GR} values are observed in southern-oriented façade-mounted installations during the summer, and with small solar generator tilt angles during the winter. Although glass reflection loss is usually quite low, it can amount to several per cent under such unfavourable conditions.

8.1.3.1.4 Correction Factor k_{SP} for Spectral Mismatch
As noted in Figures 2.46, 2.48 and 2.49, calibrated reference cells, which evaluate only a portion of the solar spectrum that solar cells have the power output to evaluate, normally register insolation that is several percentage points lower than the insolation registered by pyranometers on the same plane. A discrepancy of 2 to 3% in this regard persists even if the considerably lower accuracy of solar cells relative to pyranometers is taken into account. The fact that weather service insolation readings are pyranometer based gives rise to a spectral mismatch between natural sunlight and the light used for reference cell calibration and solar module power output measurements; the k_{SP} correction factor takes account of this mismatch.

8.1.3.1.5 Correction Factor k_{TB} for Partial Module Shading
Shading of an entire PV installation owing to a distant horizon can be taken into account using the shading correction factor k_B (see Section 2.5.4). As noted in Section 4.4.1, partial module shading by buildings, trees or the like results in a disproportionately elevated power loss in the string affected and thus to energy yield loss, which can be taken into account using the partial shading correction factor k_{TB}.

8.1.3.1.6 Mismatch Correction Factor k_{MM}
As noted in Section 4.4.2, string power output is mainly determined by the string module that produces the weakest current. Hence discrepancies in module power output or in the insolation at an individual string module result in disproportionate power loss in the string affected and thus energy yield loss, which can be taken into account using the mismatch correction factor k_{MM}. In as much as installations with module inverters exhibit no fine inter-module adjustments, $k_{MM} = 1$ for such installations.

8.1.3.1.7 Ohmic Loss Correction Factor k_R
Ohmic loss (see Section 4.5.4) in a well-sized PV installation ranges from 0.2 to 2%, but is lower in winter due to fact that solar generator power output is also lower during this period.

8.1.3.1.8 Correction Factor k_V for Module Soiling Loss

Module soiling can reduce PV installation energy yield, particularly in installations with framed modules and small tilt angles (see Section 4.5.8) [8.2], [8.3], [8.4]. Framed-module soiling is exacerbated by insufficient clearance (less than 1 cm) between module cells and frames; is mainly determined by local conditions in the vicinity of the installation; and can be relatively severe under unfavourable conditions – if, for example, the installation is near factory chimneys, railway lines or facilities with elevated air emissions – and result in energy yield loss ranging up to 10%, in isolated cases up to around 30%, but may be only around 2 to 4% at installations in the same city. Repeated heavy snowfalls exert a cleansing effect on tilted solar modules, thus counteracting soiling to some extent.

Soiling-induced power yield loss can be even higher in arid regions. For example, in a PV installation in Dakar, Senegal, during the dry season the energy yield decreased: (a) to only 18% of baseline within six months with no cleaning; (b) to 74% of baseline within a month when the modules were cleaned monthly; and (c) to 93% of baseline within a week when the modules were cleaned weekly [Wag06].

Hence in cases where a PV installation is not cleaned regularly, a soiling correction factor k_V of less than 1 should be set so as to allow for severe long-term power yield losses, which typically amount to a few per cent at a minimum.

8.1.3.1.9 Correction Factor k_S for Snow-Covering Loss

PV installation energy yield may decline or even cease altogether if the installation's solar modules are covered with snow for a number of days; hence, for PV installations at sites that are subject to winter snowfall, a snow correction factor k_S should be taken into account that can be above or below 1 depending on solar generator tilt angle, site location and site elevation. Snow does not slide off modules with small tilt angles as readily, and thus k_S is lower; for large tilt angles of around 45°, snow slides off the modules far more readily and thus k_S is higher. Except in high Alpine areas, snow covering can be disregarded for PV installations whose tilt angle exceeds 60° and that exhibit no hindrance to snow sliding off the solar modules.

For some time now, a working group at a German technical university has been researching the duration of snow covering on solar modules, based on weather satellite data. The group's findings have also been validated by data from terrestrial weather stations [8.5]. This same group also drew up and published snow-covering duration maps for Germany for November–April that are based on 2005–2007 NOAA/NESDIS weather satellite data [8.6]. At my behest, for the new edition of this book the group kindly created a series of snow-covering maps for Europe, the USA and East Asia based on 2005–2008 data (see Figures 8.2–8.7).

As Figures 8.2 and 8.4 show, low-lying areas near the Atlantic and North Sea coastlines seldom exhibit snow covering owing to the warming effects of the Gulf Stream, which extends to a relatively high latitude during the winter. On the other hand, lengthy snow-covering periods occur in the Alps, the Pyrenees, low mountain areas, and in regions with a continental climate by virtue of their being relatively far inland.

Figure 8.6 displays a snow-covering map for the USA, where relatively long snow-covering periods are prevalent in lower latitudes (in the middle of the country north of around the 35th parallel, and on the east coast at around the 40th parallel). This phenomenon is attributable to the absence of Gulf-Stream-type warming on the east coast, and to the fact that the mountains on the west coast protect the centre of the country against western winds.

Figure 8.7 displays a snow-covering map for the northern coasts of China, Korea and Japan and the southern Pacific coast of Russia. Here too, extended snow-covering periods are prevalent in low latitudes (from around the 35th to the 40th parallels). This part of the world also lacks a Gulf-Stream-type warming current on the various eastern coasts.

Figures 8.2–8.7 display snow-covering durations for the relevant terrains (which are often more or less horizontal) – although of course this does not mean that all solar generators in such areas are covered with snow in the winter, since many of them possess a tilt angle that enables the snow to slide off. And of course the steeper the angle, the more readily the snow slides off.

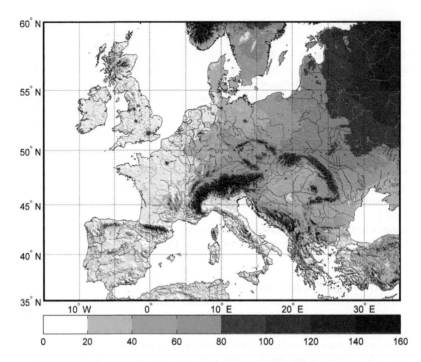

Figure 8.2 Snow-covering map for Europe based on NOAA and NESDIS weather satellite data for 2005–2008. The various tints indicate the respective mean number of snow-covering days (in classes of 20) registered from November to April. © G. Wirth, Work group of the laboratory for Solar Energy in the Division for Renewable Energies of the Munich University of Applied Sciences, http://www-lse.ee.hm.edu

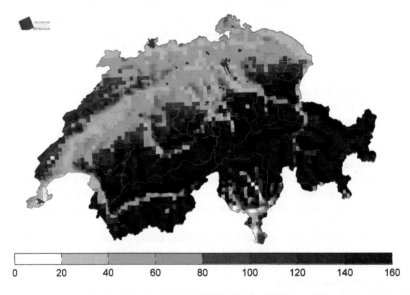

Figure 8.3 Snow-covering map for Switzerland based on NOAA and NESDIS weather satellite data for 2005–2008. The various tints indicate the respective mean number of snow-covering days (in classes of 20) registered from November to April. © G. Wirth, Work group of the laboratory for Solar Energy in the Division for Renewable Energies of the Munich University of Applied Sciences, http://www-lse.ee.hm.edu

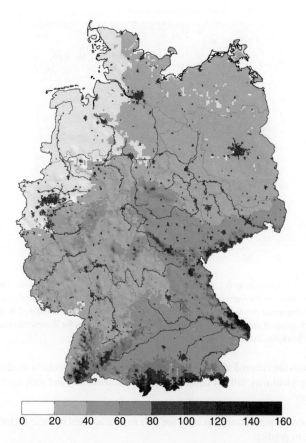

0   20   40   60   80   100   120   140   160

Figure 8.4   Snow-covering map for Germany based on NOAA and NESDIS weather satellite data for 2005–2008. The various tints indicate the respective mean number of snow-covering days (in classes of 20) registered from November to April. © G. Wirth, Work group of the laboratory for Solar Energy in the Division for Renewable Energies of the Munich University of Applied Sciences, http://www-lse.ee.hm.edu

8.1.3.1.10   Correction Factor k_{MPP} for Power Loss Attributable to MPP Tracking Malfunctions

Solar generators that lack an MPP tracker exhibit power and energy loss that are folded into the solar generator correction factor k_G (see Chapter 7) and thus must be taken into account in determining k_G. However, $\eta_{WR} = \eta_{tot}$ is applicable in cases where a grid-connected system inverter's overall efficiency is $\eta_{tot} = \eta \cdot \eta_{MPPT}$ for the installation's rated MPP voltage, and if this efficiency already takes account of MPP tracking characteristics. In such a case, $k_{MPP} = 1$ can be presupposed for energy yield calculation purposes, and only the k_G derived from the remaining parameters is used.

8.1.3.1.11   Key PV Installation Related Factors

The following factors are particularly important when it comes to sizing PV installations:

- k_{PM}: At one time, new solar modules tended to exhibit far lower power output than their nominal power output, which meant that their STC power only just equalled or was lower than the specified power output tolerances. This is rarely the case nowadays, however.
- k_V: Framed modules with insufficient spacing between module frames and solar cells and a flat tilt angle tend to exhibit power loss after a few years resulting from soiling, which is reversible for the most part via cleaning.

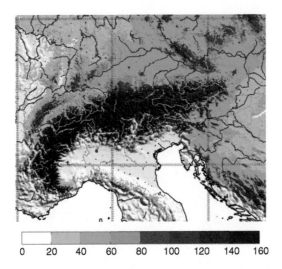

Figure 8.5   Detail of the map in Figure 8.2 showing snow covering in Alpine regions. National boundaries are shown in black, and the borders between the various Swiss cantons, German states or provinces are shown in grey. The various tints indicate the respective mean number of snow-covering days (in classes of 20) registered from November through April. © G. Wirth, Work group of the laboratory for Solar Energy in the Division for Renewable Energies of the Munich University of Applied Sciences, http://www-lse.ee.hm.edu

- k_S: If solar generators are covered with snow for extended periods, particularly modules with a small tilt angle, their energy yield may fall off considerably. Hence winter k_S and thus k_G values may exhibit considerable scatter.

Unfortunately, conventional simulation software fails to take account of these factors, whose impact is in any case difficult to predict.

Other factors also have a significant impact, particularly k_{SP}, k_{NG} and k_{GR}.

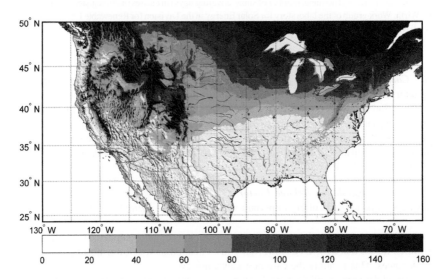

Figure 8.6   Snow-covering map for the USA based on NOAA and NESDIS weather satellite data for 2005–2008. The various tints indicate the respective mean number of snow-covering days (in classes of 20) registered from November to April. © G. Wirth, Work group of the laboratory for Solar Energy in the Division for Renewable Energies of the Munich University of Applied Sciences, http://www-lse.ee.hm.edu

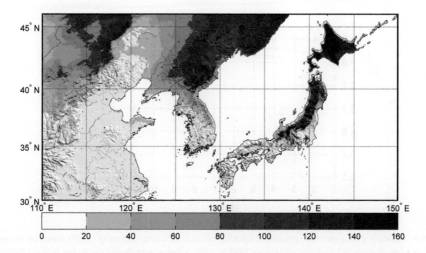

Figure 8.7 Snow-covering map for East Asia based on NOAA and NESDIS weather satellite data for 2005–2008. The various tints indicate the respective mean number of snow-covering days (in classes of 20) registered from November to April. © G. Wirth, Work group of the laboratory for Solar Energy in the Division for Renewable Energies of the Munich University of Applied Sciences, http://www.lse.ee.hm.edu

8.1.3.2 Guide Values for the Solar Generator Correction Factor k_G

Extensive measurements at numerous installations realized by the PV Lab since 1992 in various Swiss climate zones have shown that the monthly solar generator correction factor k_T ranges over $0.3 < k_G < 0.9$. Although these values are relatively high in summer, they fall off considerably during the snowy winter months.

Figures 8.8 and 8.9 display the monthly mean, maximum and minimum solar generator correction factor k_G readings for 1992–2006 (spring) for tilt angles ranging from 30° to 45°. These readings are based on monitoring numerous Swiss PV installations over a period of many years. In as much as from 1992 to 1995 most solar modules exhibited ±10% power tolerance and actual module power output tended to hover around the lower tolerance level, a curve is also shown that allows for the calculation of k_G values for new installations with modules whose STC power output matches the relevant nominal power output.

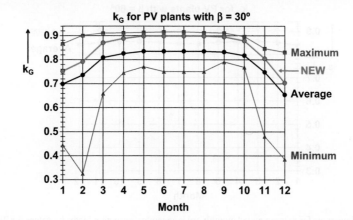

Figure 8.8 Monthly solar generator correction factor values for PV installations where $\beta = 30°$. These values are based on monitoring realized by the Bern University of Applied Sciences PV Lab in Switzerland from 1993 until the spring of 2006. The winter values exhibit extreme scatter

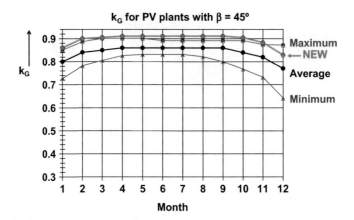

Figure 8.9 Monthly solar generator correction factor k_G values for PV installations ($\beta = 45°$), based on monitoring of two Swiss installations in low-lying areas from 1992 to 1996. Value scatter during the winter is considerably less severe than for $\beta = 30°$

In the interest of completeness, Figures 8.10 and 8.11 show the curves for $\beta = 60°$ and $90°$, based on experiential estimates.

Solar generators with a small tilt angle β exhibit low winter k_G in snowy areas, while solar generators that exhibit β ranging from $20°$ to $<35°$ and with no partial shading or soiling attain high k_G. Solar generators with small β exhibit k_G scatter during the winter. From April to October in low-lying Central European areas, k_G typically ranges from around 0.8 to 0.9 for PV installations whose β ranges from $20°$ to $<35°$, and is relatively high in winter in the presence of higher β values such as $>45°$. Summer k_G for façade-mounted installations tends to be lower on account of the elevated reflection loss on the glass module surface due to the fact that the Sun is higher in the sky during the summer ($k_{GR} < 1$). Building elements such as projecting roofs and stacked arrays of modules with $\beta < 90°$ (see Figure 2.39) can also induce partial shading.

These values can be applied throughout Central Europe without inducing major errors, as the climate there (particularly in Austria and Germany) is essentially the same as in Switzerland. Figures 8.2–8.7 can be used as rule-of-thumb values for other regions. For example, winter k_G for installations in Western and

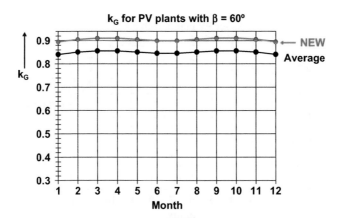

Figure 8.10 Calculations for recommended monthly solar generator correction factor k_G values for PV installations in low-lying areas ($\beta = 60°$), based on estimates derived from experience gained during a long-term monitoring project. Hence no minimum or maximum values were obtained

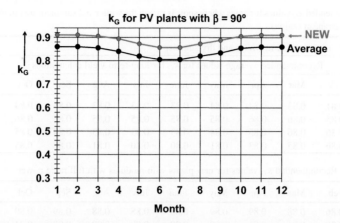

Figure 8.11   Calculations for recommended monthly solar generator correction factor k_G values for PV installations in low-lying areas ($\beta = 90°$), based on estimates derived from experience gained during a long-term monitoring project. Hence no minimum or maximum values were obtained. In this case, summer k_G was minimal

particularly Southern Europe is of course higher because there is little or no snow covering there during the winter.

Tables 8.3–8.5 contain recommended k_G values for power yield calculations. The tables are presented in pairs. Each upper table lists long-term energy yield projections that take account of solar generator soiling and degradation, as well as power output loss attributable to the excessively low power output exhibited by new modules. Each of the lower tables lists the k_G values that can be used to calculate energy yield for new PV installations, all of whose modules exhibit a power output that is at least as high as the nominal output.

Table 8.3 lists the recommended k_G values for Central European areas that are occasionally covered with snow in the winter. These values are essentially the same as those in Figures 8.8–8.11.

Table 8.4 indicates the recommended k_G values for locations where snow rarely falls and where summer precipitation is sparse (e.g. in the northern Mediterranean region).

Table 8.5 indicates the recommended k_G values for locations where snow never falls in the winter, rain rarely if ever falls during the summer, and precipitation occurs in the winter only (e.g. installations in Southern Europe and North Africa).

Table 8.3   Recommended k_G values for Central European areas that are occasionally covered with snow in the winter

Locations with occasional snowfall in winter

Recommended average values for k_G for long-term yield calculations

β	Jan	Feb	Mar	Apr	May	June	July	Aug	Sep	Oct	Nov	Dec
30°	0.69	0.73	0.81	0.83	0.84	0.84	0.84	0.84	0.84	0.82	0.75	0.66
45°	0.80	0.83	0.84	0.85	0.86	0.86	0.86	0.86	0.86	0.84	0.82	0.77
60°	0.84	0.85	0.86	0.86	0.85	0.85	0.85	0.85	0.86	0.86	0.85	0.84
90°	0.86	0.86	0.85	0.84	0.82	0.81	0.81	0.82	0.84	0.85	0.86	0.86

Recommended k_G values for new plants with modules with full rated power

β	Jan	Feb	Mar	Apr	May	June	July	Aug	Sep	Oct	Nov	Dec
30°	0.75	0.79	0.86	0.88	0.90	0.90	0.90	0.90	0.90	0.88	0.80	0.70
45°	0.85	0.88	0.89	0.90	0.91	0.91	0.91	0.91	0.91	0.89	0.87	0.82
60°	0.89	0.90	0.91	0.91	0.90	0.90	0.90	0.90	0.91	0.91	0.90	0.89
90°	0.91	0.91	0.91	0.89	0.87	0.86	0.86	0.87	0.89	0.91	0.91	0.91

Table 8.4 Recommended k_G values for areas where snow is rare in the winter and summers are relatively dry (e.g. in the northern Mediterranean region)

Locations with only rare snowfall in winter and dry summer

Recommended average values for k_G for long-term yield calculations

β	Jan	Feb	Mar	Apr	May	June	July	Aug	Sep	Oct	Nov	Dec
30°	0.80	0.81	0.83	0.84	0.84	0.83	0.83	0.83	0.84	0.84	0.81	0.80
45°	0.85	0.85	0.86	0.86	0.85	0.85	0.85	0.85	0.85	0.86	0.86	0.85
60°	0.86	0.86	0.86	0.85	0.84	0.83	0.83	0.83	0.84	0.85	0.86	0.86
90°	0.86	0.86	0.85	0.83	0.81	0.80	0.80	0.81	0.83	0.85	0.86	0.86

Recommended k_G values for new plants with modules with full rated power

β	Jan	Feb	Mar	Apr	May	June	July	Aug	Sep	Oct	Nov	Dec
30°	0.85	0.86	0.88	0.89	0.89	0.88	0.88	0.88	0.89	0.89	0.86	0.85
45°	0.90	0.90	0.91	0.91	0.90	0.90	0.90	0.90	0.90	0.91	0.91	0.90
60°	0.91	0.91	0.91	0.90	0.89	0.88	0.88	0.88	0.89	0.90	0.91	0.91
90°	0.91	0.91	0.90	0.88	0.86	0.85	0.85	0.86	0.88	0.90	0.91	0.91

For reasons of space, the southern hemisphere reference stations have been omitted here. The k_T values for reference stations at similar latitudes in the northern hemisphere are more or less applicable to the southern hemisphere if the six-month difference in the timing of the seasons is taken into account (for an example see the Alice Springs case, Problem 4, in Section 8.2.1.2).

Tables 8.4 and 8.5 presuppose that k_G will change slightly during the summer months owing to summer soiling attributable to the absence or near absence of precipitation. Depending on local conditions, this effect can of course be considerably more pronounced if the solar modules are not cleaned periodically, in which case k_G would be far lower [Wag06]. The farther south a façade-mounted installation ($β = 90°$), the lower the value of k_G during the summer and the greater the height of the Sun h_S at noon as a result of higher glass reflection loss (see Figure 8.1).

Based on the solar generator correction factor k_G and Y_T as in Equation 8.12, Equation 7.18 can be used to determine array yield Y_A as follows:

$$\text{Array yield } Y_A = k_G \cdot Y_T = k_G \cdot k_T \cdot Y_R \tag{8.14}$$

Table 8.5 Recommended k_G values for locations with no winter snowfall and very dry summers (e.g. Southern Europe and North Africa)

Locations without snowfall in winter and very dry summer

Recommended average values for k_G for long-term yield calculations

β	Jan	Feb	Mar	Apr	May	June	July	Aug	Sep	Oct	Nov	Dec
30°	0.84	0.84	0.84	0.84	0.83	0.82	0.82	0.82	0.83	0.84	0.84	0.84
45°	0.86	0.86	0.86	0.86	0.84	0.83	0.83	0.83	0.84	0.86	0.86	0.86
60°	0.86	0.86	0.86	0.85	0.83	0.82	0.82	0.82	0.83	0.85	0.86	0.86
90°	0.86	0.86	0.85	0.82	0.78	0.77	0.77	0.78	0.83	0.85	0.86	0.86

Recommended k_G values for new plants with modules with full rated power

β	Jan	Feb	Mar	Apr	May	June	July	Aug	Sep	Oct	Nov	Dec
30°	0.89	0.89	0.89	0.89	0.88	0.87	0.87	0.87	0.88	0.89	0.89	0.89
45°	0.91	0.91	0.91	0.91	0.89	0.88	0.88	0.88	0.89	0.91	0.91	0.91
60°	0.91	0.91	0.91	0.90	0.88	0.87	0.87	0.87	0.88	0.90	0.91	0.91
90°	0.91	0.91	0.90	0.87	0.83	0.82	0.82	0.83	0.88	0.90	0.91	0.92

The standardized energy yield and power output graph displays MPP tracking errors exhibited by inverters and by MPT charge controllers, expressed as k_G (see Chapter 7). However, as noted in Section 8.1.3.1.10, this calculation is best performed using k_G values containing solely the remaining effects that allow for maximum possible array yield when the system inverter or charge controller exhibits fault-free MPP tracking. Tables 8.3–8.5 list the values that impinge upon the other factors that affect k_G, but not those that affect MPP tracking. Its quality, which is a key characteristic of an inverter or MPT charge controller, along with conversion efficiency, should be determined solely on the basis of overall efficiency η_{tot} (see Sections 5.2.5 and 5.2.5.3, particularly the latter).

8.2   Energy Yield Calculation for Grid-Connected Systems

The monthly and annual energy yields of grid-connected systems are quite simple to determine using standardized diagrams and tables (see Chapter 7); Tables 8.6 and 8.7 or Tables A6.1 and A6.2 in Appendix A are best suited for this purpose. To do this, in a first step irradiation $HG = H_I$ is determined on the inclined plane (see Section 2.4 or 2.5). This value is then divided by $Go = 1\,kW/m^2$, which yields YR (full-load solar hours).

Y_R is then multiplied by the reference station k_T values (see Section 8.1.2), which yields YT, which is in turn multiplied by k_G (see Section 8.1.3), which yields Y_A. In a final step, Y_A is multiplied by mean European inverter efficiency η_{WR} (preferably by overall efficiency η_{tot}), which yields Y_F as follows:

$$\text{Final yield } Y_F = \eta_{WR} \cdot k_G \cdot k_T \cdot Y_R = \eta_{WR} \cdot k_G \cdot k_T \frac{H_G}{G_o} \qquad (8.15)$$

Monthly energy yield can be readily determined by multiplying Y_F by solar generator peak power P_{Go} and by n_d (number of days per month). Y_F is as follows:

$$\text{Monthly energy yield } E_{AC} = P_{Go} \cdot n_d \cdot Y_F \qquad (8.16)$$

Annual energy yield is obtained by tallying monthly yield.

Table 8.6   Energy yield calculation table for grid-connected systems using a simplified radiation calculation as in Section 2.4. The shaded fields are optional. A blank version of this table can also be found in Appendix A as Table A6.1

Energy yield calculation for grid-connected PV plants (simplified calculation of irradiation)

Location:				Reference station:								$\beta =$	
P_{Go} [kW]:				$P_{ACn} = k_{Gmax} \cdot P_{Go} \cdot \eta_{WR}$ [kW]:								$\gamma =$	

Month	Jan	Feb	Mar	April	May	June	July	Aug	Sept	Oct	Nov	Dec	Year	
H														kWh/m²
R(β,γ)														
$H_G = R(\beta,\gamma) \cdot H$														kWh/m²
$Y_R = H_G/1kWm^{-2}$														h/d
k_T														
$Y_T = k_T \cdot Y_R$														h/d
k_G														
$Y_A = k_G \cdot Y_T$														h/d
η_{WR} (η_{tot})														
$Y_F = \eta_{WR} \cdot Y_A$														h/d
n_d (days/month)	31	28	31	30	31	30	31	31	30	31	30	31	365	d
$E_{AC} = n_d \cdot P_{Go} \cdot Y_F$														kWh
$t_V = E_{AC}/P_{ACn}$														h
$PR = Y_F/Y_R$														

Table 8.7 Energy yield calculation table for grid-connected systems using the radiation calculation and three-component model from Section 2.5. The shaded fields are optional. A blank version of this table can also be found in Appendix A as Table A6.2

Energy yield calculation for grid-connected PV plants (with 3-component model)

Location:		Reference station:		$\beta =$
P_{Go} [kW]:		$P_{ACn} = k_{Gmax} \cdot P_{Go} \cdot \eta_{WR}$ [kW]:		$\gamma =$
	$R_D = \frac{1}{2}\cos \alpha_2 + \frac{1}{2}\cos (\alpha_1 + \beta) =$		$R_R = \frac{1}{2} - \frac{1}{2}\cos \beta =$	

Month	Jan	Feb	Mar	April	May	June	July	Aug	Sept	Oct	Nov	Dec	Year	
H														kWh/m^2
H_D														kWh/m^2
R_B														
k_B														
$H_{GB} = k_B \cdot R_B \cdot (H-H_D)$														kWh/m^2
$H_{GD} = R_D \cdot H_D$														kWh/m^2
ρ														
$H_{GR} = R_R \cdot \rho \cdot H$														kWh/m^2
$H_G = H_{GB} + H_{GD} + H_{GR}$														kWh/m^2
$Y_R = H_G/1kWm^{-2}$														h/d
k_T														
$Y_T = k_T \cdot Y_R$														h/d
k_G														
$Y_A = k_G \cdot Y_T$														h/d
$\eta_{WR} \; (\eta_{tot})$														
$Y_F = \eta_{WR} \cdot Y_A$														h/d
n_d (days/month)	31	28	31	30	31	30	31	31	30	31	30	31	365	d
$E_{AC} = n_d \cdot P_{Go} \cdot Y_F$														kWh
$t_V = E_{AC}/P_{ACn}$														h
$PR = Y_F / Y_R$														

The following can also be defined:

$$\text{Arithmetic nominal AC power output } P_{ACn} = k_{Gmax} \cdot P_{Go} \cdot \eta_{WR} \qquad (8.17)$$

$$\text{AC full-load hours } t_V = E_{AC}/P_{ACn} \qquad (8.18)$$

In Tables 8.6 and 8.7 the relevant equations are indicated in the left-hand column. Blank versions of both of these tables can be found in Appendix A.

Tables 8.6 and 8.7 can also be downloaded as Excel tables, with the equations already in them, from the download area for this book at www.electrosuisse.ch/photovoltaik or www.pvtest.ch. The specimen tables in this book were realized using these Excel tables and exact values. As a result of automatic rounding off to two decimal places for the tables, slight discrepancies may arise in connection with manual calculations that are based on these tables using numbers that have already been rounded off to two decimal places.

If monthly values H_G expressed in kWh/m^2 are available from weather data software, the Internet or other sources (see Section 8.4), these values can of course be entered directly in the H_G and Y_R rows and the aforementioned calculations can be dispensed with. In addition, if effective monthly tallies rather than mean daily tallies are given, multiplication by n_d can also be dispensed with.

In as much as inverter efficiency is unrelated to voltage (see Sections 5.2.5.1–5.2.5.3), the installation's optimal DC operating voltage should be determined prior to calculating the definitive yield (see Section 5.2.6.8.4).

8.2.1   Examples of Grid-Connected System Energy Yield

8.2.1.1   Exercises Involving the Simplified Insolation Calculation Procedure (see Section 2.4)

Problem 1 (Grid-connected system mounted on a building in Burgdorf)

Roof-mounted installation with the following characteristics: tilt angle $\beta = 30°$; solar generator azimuth $0°$ (S); solar generator peak power $P_{Go} = 3.18\,$kWp; mean inverter efficiency $\eta_{WR} = 93\%$. Capture losses attributable to mismatch, wiring and the like are taken into account by a solar generator correction factor k_G (see Table 8.3) for purposes of long-term energy yield projections. The reference station for $R(\beta,\gamma)$ and k_T calculation purposes is Kloten.

Determine the following:

(a) Monthly and annual global radiation on the solar generator plane (in kWh/m²/d).
(b) Monthly and annual energy yield in kWh.
(c) Percentage of total energy yield accounted for by energy yield from October to March.
(d) Annual energy yield Y_{Fa} for this installation in kWh/kWp/a.
(e) Specific full-load hours t_V for nominal inverter output P_{ACn} for the installation.
(f) The capacity factor $CF = t_V/(8760\,\text{h})$.

Solution:

(a) and (b)

Energy yield calculation for grid-connected PV plants (simplified calculation of irradiation)

Location:	Burgdorf			Reference station:		Kloten							$\beta =$	30°	
P_{Go} [kW]:		3.18		$P_{ACn} = k_{Gmax} \cdot P_{Go} \cdot \eta_{WR}$ [kW]:					2.484			$\gamma =$	0°	
Month	Jan	Feb	Mar	April	May	June	July	Aug	Sept	Oct	Nov	Dec	Year	
H	0.99	1.69	2.74	3.68	4.59	5.09	5.59	4.75	3.41	1.94	1.04	0.77	3.02	kWh/m²
$R(\beta,\gamma)$	1.34	1.28	1.16	1.06	1.01	0.98	1.00	1.05	1.13	1.21	1.26	1.34	1.087	
$H_G = R(\beta,\gamma)\cdot H$	1.33	2.16	3.18	3.90	4.64	4.99	5.59	4.99	3.85	2.35	1.31	1.03	3.283	kWh/m²
$Y_R = H_G/1\text{kWm}^{-2}$	1.33	2.16	3.18	3.90	4.64	4.99	5.59	4.99	3.85	2.35	1.31	1.03	3.283	h/d
k_T	1.05	1.03	1.01	0.98	0.95	0.93	0.91	0.92	0.95	0.99	1.04	1.05	0.961	
$Y_T = k_T \cdot Y_R$	1.39	2.23	3.21	3.82	4.40	4.64	5.09	4.59	3.66	2.32	1.36	1.08	3.155	h/d
k_G	0.69	0.73	0.81	0.83	0.84	0.84	0.84	0.84	0.84	0.82	0.75	0.66	0.815	
$Y_A = k_G \cdot Y_T$	0.96	1.63	2.60	3.17	3.70	3.90	4.27	3.85	3.07	1.91	1.02	0.72	2.572	h/d
η_{WR} (η_{tot})	0.93	0.93	0.93	0.93	0.93	0.93	0.93	0.93	0.93	0.93	0.93	0.93	0.930	
$Y_F = \eta_{WR}\cdot Y_A$	0.89	1.51	2.42	2.95	3.44	3.62	3.97	3.58	2.86	1.77	0.95	0.66	2.392	h/d
n_d (days/month)	31	28	31	30	31	30	31	31	30	31	30	31	365	d
$E_{AC} = n_d \cdot P_{Go} \cdot Y_F$	88	135	238	282	339	346	392	353	273	175	91	66	2776	kWh
$t_V = E_{AC}/P_{ACn}$	35	54	96	113	137	139	158	142	110	70	37	26	1118	h
$PR = Y_F/Y_R$	0.674	0.699	0.761	0.756	0.742	0.727	0.711	0.719	0.742	0.755	0.725	0.644	0.729	

(c) Percentage of total energy yield accounted for by energy yield from October to March: 793 kWh = 28.6%.
(d) Specific annual energy yield $Y_{Fa} = 2.392\,$kWh/kWp/d = 873 kWh/kWp/a.
(e) P_{ACn} full-load hours $t_V = 1118\,$h/a.
(f) Capacity factor $CF = 12.8\%$.

Note: Long-term monitoring by the Bern University of Applied Sciences PV Lab has shown that Kloten's H and $R(\beta,\gamma)$ values for Burgdorf, based on insolation measurements over past decades, are unduly pessimistic (see Section 10.1.1).

Problem 2 (Large grid-connected system on Mont Soleil)

Ground-based installation with the following characteristics: tilt angle $\beta = 45°$; solar generator azimuth 0° (S); solar generator peak power $P_{Go} = 560$ kWp; mean inverter efficiency $\eta_{WR} = 95\%$.

Capture losses attributable to mismatch, wiring and the like are taken into account by a solar generator correction factor k_G as in Table 8.3 for purposes of long-term energy yield projections. The reference station for $R(\beta,\gamma)$ and k_T calculations is Davos.

This installation is slightly idealized relative to the actual Mont Soleil installation in that the tilt angle is 45° rather than 50°, the solar generator azimuth is due south, and no shading is induced by the horizon or front module rows.

Determine the following:

(a) Monthly and annual global radiation on the solar generator plane (in kWh/m²/d).
(b) Monthly and annual energy yield in kWh.
(c) Percentage of total energy yield accounted for by energy yield from October to March.
(d) Annual energy yield for this installation in kWh/kWp/a.
(e) Specific full-load hours t_V for nominal inverter output P_{ACn} for the installation.
(f) The capacity factor $CF = t_V/(8760 \text{ h})$.

Solution:

(a) and (b)

Energy yield calculation for grid-connected PV plants (simplified calculation of irradiation)

Location:	Mt. Soleil			Reference station:			Davos				$\beta =$	45°	
P_{Go} [kW]:		560		$P_{ACn} = k_{Gmax} \cdot P_{Go} \cdot \eta_{WR}$ [kW]:					457.5		$\gamma =$	0°	
Month	**Jan**	**Feb**	**Mar**	**April**	**May**	**June**	**July**	**Aug**	**Sept**	**Oct**	**Nov**	**Dec**	**Year**
H	1.36	2.10	3.09	3.96	4.54	4.98	5.57	4.79	3.63	2.42	1.44	1.14	3.25	kWh/m²
$R(\beta,\gamma)$	1.90	1.59	1.33	1.10	0.95	0.91	0.92	1.00	1.15	1.39	1.69	1.98	1.164	
$H_G = R(\beta,\gamma) \cdot H$	2.58	3.34	4.11	4.36	4.31	4.53	5.12	4.79	4.17	3.36	2.43	2.26	3.784	kWh/m²
$Y_R = H_G/1kWm^{-2}$	2.58	3.34	4.11	4.36	4.31	4.53	5.12	4.79	4.17	3.36	2.43	2.26	3.784	h/d
k_T	1.05	1.04	1.02	1.01	0.99	0.98	0.96	0.97	0.99	1.01	1.04	1.05	1.002	
$Y_T = k_T \cdot Y_R$	2.71	3.47	4.19	4.40	4.27	4.44	4.92	4.65	4.13	3.40	2.53	2.37	3.792	h/d
k_G	0.80	0.83	0.84	0.85	0.86	0.86	0.86	0.86	0.86	0.84	0.82	0.77	0.843	
$Y_A = k_G \cdot Y_T$	2.17	2.88	3.52	3.74	3.67	3.82	4.23	4.00	3.55	2.85	2.08	1.82	3.196	h/d
η_{WR} (η_{tot})	0.95	0.95	0.95	0.95	0.95	0.95	0.95	0.95	0.95	0.95	0.95	0.95	0.950	
$Y_F = \eta_{WR} \cdot Y_A$	2.06	2.74	3.35	3.55	3.49	3.63	4.02	3.80	3.38	2.71	1.97	1.73	3.037	h/d
n_d (days/month)	31	28	31	30	31	30	31	31	30	31	30	31	365	d
$E_{AC} = n_d \cdot P_{Go} \cdot Y_F$	35.8	42.9	58.1	59.7	60.6	61.0	69.8	65.9	56.7	47.1	33.1	30.1	620.7	MWh
$t_V = E_{AC}/P_{ACn}$	78	94	127	130	132	133	153	144	124	103	72	66	1357	h
$PR = Y_F/Y_R$	0.798	0.820	0.814	0.816	0.809	0.801	0.784	0.792	0.809	0.806	0.810	0.768	0.802	

(c) Percentage of total energy yield accounted for by energy yield from October to March: 247.1 kWh = 39.8%.
(d) Specific annual energy yield $Y_{Fa} = 3.04$ kWh/kWp/d $= 1108$ kWh/kWp/a.
(e) P_{ACn} full-load hours $t_V = 1357$ h/a.
(f) Capacity factor $CF = 15.5\%$.

Problem 3 (Grid-connected system in Montana, Switzerland)

Ground-based installation with the following characteristics: tilt angle $\beta = 60°$; solar generator azimuth 0° (S); solar generator peak power $P_{Go} = 100$ kWp; mean inverter efficiency $\eta_{WR} = 96\%$.

Capture losses attributable to mismatch, wiring and the like are taken into account by a solar generator correction factor k_G as in Table 8.3 for new installations with modules whose output is on a par with STC nominal output. The reference station for $R(\beta,\gamma)$ and k_T calculations is Davos.

Determine the following:

(a) Monthly and annual global radiation on the solar generator plane (in kWh/m²/d).
(b) Monthly and annual energy yield in kWh.
(c) Percentage of total energy yield accounted for by energy yield from October to March.
(d) Specific annual energy yield for this installation in kWh/kWp/a.
(e) Full-load hours t_V for nominal inverter output P_{ACn} for the installation.
(f) The capacity factor CF $= t_V/(8760\,h)$.

Solution:

(a) and (b)

Energy yield calculation for grid-connected PV plants (simplified calculation of irradiation)

Location:	Montana			Reference station:		Davos						$\beta =$	60°	
P_{Go} [kW]:		100		$P_{ACn} = k_{Gmax} \cdot P_{Go} \cdot \eta_{WR}$ [kW]:					87.36			$\gamma =$	0°	
Month	Jan	Feb	Mar	April	May	June	July	Aug	Sept	Oct	Nov	Dec	Year	
H	1.61	2.53	3.88	4.77	5.61	5.74	6.08	5.28	4.13	2.80	1.68	1.35	**3.78**	kWh/m²
$R(\beta,\gamma)$	2.01	1.66	1.32	1.04	0.85	0.80	0.82	0.91	1.09	1.39	1.78	2.13	1.123	
$H_G = R(\beta,\gamma) \cdot H$	3.24	4.20	5.12	4.96	4.77	4.59	4.99	4.80	4.50	3.89	2.99	2.88	**4.244**	kWh/m²
$Y_R = H_G/1kWm^2$	3.24	4.20	5.12	4.96	4.77	4.59	4.99	4.80	4.50	3.89	2.99	2.88	**4.244**	h/d
k_T	1.05	1.04	1.02	1.01	0.99	0.98	0.96	0.97	0.99	1.01	1.04	1.05	1.005	
$Y_T = k_T \cdot Y_R$	3.40	4.37	5.22	5.01	4.72	4.50	4.79	4.66	4.46	3.93	3.11	3.02	**4.265**	h/d
k_G	0.89	0.90	0.91	0.91	0.90	0.90	0.90	0.90	0.91	0.91	0.90	0.89	0.902	
$Y_A = k_G \cdot Y_T$	3.02	3.93	4.75	4.56	4.25	4.05	4.31	4.19	4.06	3.58	2.80	2.69	**3.848**	h/d
η_{WR} (η_{tot})	0.96	0.96	0.96	0.96	0.96	0.96	0.96	0.96	0.96	0.96	0.96	0.96	0.960	
$Y_F = \eta_{WR} \cdot Y_A$	2.90	3.77	4.56	4.38	4.08	3.89	4.14	4.03	3.89	3.43	2.69	2.58	**3.694**	h/d
n_d (days/month)	31	28	31	30	31	30	31	31	30	31	30	31	365	d
$E_{AC} = n_d \cdot P_{Go} \cdot Y_F$	9.00	10.6	14.1	13.1	12.6	11.7	12.8	12.5	11.7	10.6	8.06	8.00	**134.8**	MWh
$t_V = E_{AC}/P_{ACn}$	103	121	162	150	145	134	147	143	134	122	92	92	**1543**	h
$PR = Y_F/Y_R$	0.897	0.899	0.891	0.882	0.855	0.847	0.829	0.838	0.865	0.882	0.899	0.897	0.870	

(c) Percentage of total energy yield accounted for by energy yield from October to March: 60.4 kWh $= 44.8\%$.
(d) Specific annual energy yield $Y_{Fa} = 3.694$ kWh/kWp/d $= 1348$ kWh/kWp/a.
(e) P_{ACn} full-load hours $t_V = 1543$ h/a.
(f) Capacity factor CF $= 17.6\%$.

Note: In as much as high-Alpine solar generators with a 60° tilt angle may be briefly covered with snow following a heavy snowfall, the above k_G values may be overly optimistic. In cases where the solar generator being covered with snow is completely inadmissible, the tilt angle should be 90°.

Problem 4 (Large PV installation near El Paso, Texas)

Large ground-based PV installation: tilt angle $\beta = 30°$; $\varphi = 31.8°$N; solar generator azimuth $0°$ (S); solar generator peak power $P_{Go} = 500$ MWp; mean inverter efficiency $\eta_{WR} = 97\%$, including losses in the medium-voltage transformer.

Capture losses attributable to mismatch, wiring and the like are taken into account by a solar generator correction factor k_G as in Table 8.5 for new installations with modules whose output is on a par with STC nominal output. The reference station for $R(\beta,\gamma)$ and k_T calculations is Cairo, Egypt, as it is on a similar latitude.

Determine the following:

(a) Monthly and annual global radiation on the solar generator plane (in kWh/m²/d).
(b) Monthly and annual energy yield in kWh.
(c) Percentage of total energy yield accounted for by energy yield from October to March.
(d) Specific annual energy yield for this installation in kWh/kWp/a.
(e) Full-load hours t_V for nominal inverter output P_{ACn} for the installation.
(f) The capacity factor CF $= t_V/(8760\,h)$.

Solution:

(a) and (b)

Energy yield calculation for grid-connected PV plants (simplified calculation of irradiation)

Location:	El Paso			Reference station:		Kairo							$\beta =$	30°	
P_{Go} [kW]:		500000		$P_{ACn} = k_{Gmax} \cdot P_{Go} \cdot \eta_{WR}$ [kW]:						431650			$\gamma =$	0°	
Month	Jan	Feb	Mar	April	May	June	July	Aug	Sept	Oct	Nov	Dec	Year	
H	3.69	4.90	6.34	7.63	8.25	8.43	7.71	7.18	6.09	5.26	3.98	3.41	6.07	kWh/m²
$R(\beta,\gamma)$	1.33	1.23	1.12	1.01	0.93	0.90	0.91	0.97	1.07	1.20	1.31	1.37	1.067	
$H_G = R(\beta,\gamma) \cdot H$	4.91	6.03	7.10	7.71	7.67	7.59	7.02	6.96	6.52	6.31	5.21	4.67	6.475	kWh/m²
$Y_R = H_G/1kWm^{-2}$	4.91	6.03	7.10	7.71	7.67	7.59	7.02	6.96	6.52	6.31	5.21	4.67	6.475	h/d
k_T	0.96	0.94	0.93	0.91	0.89	0.88	0.88	0.88	0.89	0.90	0.93	0.95	0.908	
$Y_T = k_T \cdot Y_R$	4.71	5.67	6.60	7.01	6.83	6.68	6.17	6.13	5.80	5.68	4.85	4.44	5.880	h/d
k_G	0.89	0.89	0.89	0.89	0.88	0.87	0.87	0.87	0.88	0.89	0.89	0.89	0.883	
$Y_A = k_G \cdot Y_T$	4.19	5.04	5.88	6.24	6.01	5.81	5.37	5.33	5.10	5.06	4.32	3.95	5.191	h/d
η_{WR} (η_{tot})	0.97	0.97	0.97	0.97	0.97	0.97	0.97	0.97	0.97	0.97	0.97	0.97	0.970	
$Y_F = \eta_{WR} \cdot Y_A$	4.07	4.89	5.70	6.05	5.83	5.63	5.21	5.17	4.95	4.90	4.19	3.83	5.035	h/d
n_d (days/month)	31	28	31	30	31	30	31	31	30	31	30	31	365	d
$E_{AC} = n_d \cdot P_{Go} \cdot Y_F$	63.0	68.5	88.4	90.8	90.3	84.5	80.8	80.2	74.3	76.0	62.8	59.4	918.9	GWh
$t_V = E_{AC}/P_{ACn}$	146	159	205	210	209	196	187	186	172	176	145	138	2129	h
$PR = Y_F/Y_R$	0.829	0.812	0.803	0.786	0.760	0.743	0.743	0.743	0.760	0.777	0.803	0.820	0.778	

(a) Percentage of total energy yield accounted for by energy yield from October to March: 418.1 kWh $= 45.5\%$.
(b) Specific annual energy yield $Y_{Fa} = 5.035$ kWh/kWp/d $= 1838$ kWh/kWp/a.
(c) P_{Acn} full-load hours $t_V = 2129$ h/a.
(d) Capacity factor CF $= 24.3\%$.

Note: A marginally higher energy yield is achievable in the summer using a 20° tilt angle. Around 40% more energy can be obtained using a biaxial solar tracker.

8.2.1.2 Examples Involving the Three-Component Model (see Section 2.5)

Problem 1 (New grid-connected system mounted on a building in Seville, Spain)

Roof-mounted installation with tilt angle $\beta = 35°$; $\varphi = 37.4°$N; solar generator azimuth 0° (S); $\alpha_1 = \alpha_2 = 0°$; year-round reflection factor $\rho = 0.35$; R_B values interpolated; solar generator peak power $P_{Go} = 60$ kWp; mean inverter efficiency $\eta_{WR} = 95\%$.

Capture losses attributable to mismatch, wiring, soiling and the like are taken into account by a solar generator correction factor k_G as in Table 8.5 (applies to a new installation whose power output equates to nominal output at a 30° tilt angle). The reference station is Seville, Spain, for purposes of k_T calculation.

Determine the following:

(a) Monthly and annual global radiation on the solar generator plane (in kWh/m²/d).
(b) Monthly and annual energy yield in kWh.
(c) Percentage of total energy yield accounted for by energy yield from October to March.
(d) Specific annual energy yield Y_{Fa} for this installation in kWh/kWp/a.
(e) Full-load hours t_V for P_{ACn}; also, verify that capacity factor CF $= t_V/(8760\,\mathrm{h})$.

Solution:

(a) and (b)

Energy yield calculation for grid-connected PV plants (with 3-component model)

Location:	Sevilla				Reference station:			Sevilla			β =	35°	
P_{Go} [kW]:		60		$P_{ACn} = k_{Gmax}·P_{Go}·\eta_{WR}$ [kW]:				50.73			γ =	0°	
$R_D = \frac{1}{2}\cos\alpha_2 + \frac{1}{2}\cos(\alpha_1+\beta) =$					0.910		$R_R = \frac{1}{2} - \frac{1}{2}\cos\beta =$				0.090		

Month	Jan	Feb	Mar	April	May	June	July	Aug	Sept	Oct	Nov	Dec	Year	
H	2.52	3.26	4.70	5.35	6.62	7.20	7.58	6.51	5.38	3.86	2.50	2.16	4.80	kWh/m²
H_D	1.08	1.41	1.75	2.22	2.37	2.40	2.15	2.11	1.82	1.51	1.19	0.99	1.75	kWh/m²
R_B	1.86	1.57	1.29	1.08	0.94	0.89	0.91	1.02	1.20	1.46	1.76	1.98		
k_B	1	1	1	1	1	1	1	1	1	1	1	1		
$H_{GB} = k_B·R_B·(H-H_D)$	2.68	2.90	3.81	3.38	4.01	4.25	4.95	4.47	4.27	3.44	2.31	2.32	3.57	kWh/m²
$H_{GD} = R_D·H_D$	0.98	1.28	1.59	2.02	2.16	2.18	1.96	1.92	1.66	1.37	1.08	0.90	1.59	kWh/m²
ρ	0.35	0.35	0.35	0.35	0.35	0.35	0.35	0.35	0.35	0.35	0.35	0.35		
$H_{GR} = R_R·\rho·H$	0.08	0.10	0.15	0.17	0.21	0.23	0.24	0.21	0.17	0.12	0.08	0.07	0.15	kWh/m²
$H_G = H_{GB}+H_{GD}+H_{GR}$	3.74	4.28	5.55	5.57	6.38	6.66	7.15	6.60	6.10	4.93	3.47	3.29	5.317	kWh/m²
$Y_R = H_G/1\mathrm{kWm^{-2}}$	3.74	4.28	5.55	5.57	6.38	6.66	7.15	6.60	6.10	4.93	3.47	3.29	5.317	h/d
k_T	0.95	0.95	0.93	0.92	0.89	0.87	0.85	0.85	0.88	0.91	0.95	0.96	0.900	
$Y_T = k_T·Y_R$	3.55	4.07	5.16	5.12	5.68	5.79	6.08	5.61	5.37	4.49	3.30	3.16	4.786	h/d
k_G	0.89	0.89	0.89	0.89	0.88	0.87	0.87	0.87	0.88	0.89	0.89	0.89	0.882	
$Y_A = k_G·Y_T$	3.16	3.62	4.59	4.56	5.00	5.04	5.29	4.88	4.72	3.99	2.93	2.81	4.221	h/d
η_{WR} (η_{tot})	0.95	0.95	0.95	0.95	0.95	0.95	0.95	0.95	0.95	0.95	0.95	0.95	0.950	
$Y_F = \eta_{WR}·Y_A$	3.00	3.44	4.36	4.33	4.75	4.79	5.02	4.64	4.49	3.79	2.79	2.67	4.010	h/d
n_d (days/month)	31	28	31	30	31	30	31	31	30	31	30	31	365	d
$E_{AC} = n_d·P_{Go}·Y_F$	5.59	5.78	8.12	7.80	8.83	8.62	9.34	8.62	8.08	7.06	5.02	4.97	87.81	MWh
$t_V = E_{AC}/P_{ACn}$	110	114	160	154	174	170	184	170	159	139	99	98	1731	h
$PR = Y_F/Y_R$	0.803	0.803	0.786	0.778	0.744	0.719	0.703	0.703	0.736	0.769	0.803	0.812	0.754	

(c) Percentage of total energy yield accounted for by energy yield from October to March: 36.52 kWh $= 41.6\%$.
(d) Specific annual energy yield $Y_{Fa} = 4.010$ kWh/kWp/d $= 1464$ kWh/kWp/a.
(e) P_{Acn} full-load hours $t_V = 1731$ h/a; capacity CF $= 19.8\%$.

Problem 2 (New grid-connected system in the desert near Aswan, Egypt)

Ground-based PV installation: tilt angle $\beta = 20°$; $\varphi = 24°N$; solar generator azimuth $0°$ (S); $\alpha_1 = \alpha_2 = 0°$; year-round reflection factor $\rho = 0.35$; solar generator peak power $P_{Go} = 10\,MWp$; mean inverter efficiency $\eta_{WR} = 97\%$.

Capture losses attributable to mismatch, wiring, soiling and the like are taken into account by a solar generator correction factor k_G as in Table 8.5 (applies to a new installation whose power output equates to nominal output at a 30° tilt angle). Although the reference station is Cairo, Egypt, for k_T calculation purposes, because of the ambient temperatures around 5 °C higher at the installation site, the k_T values should be 0.02 lower than those indicated for Cairo.

Determine the following:

(a) Monthly and annual global radiation on the solar generator plane (in kWh/m²/d).
(b) Monthly and annual energy yield in kWh.
(c) Percentage of total energy yield accounted for by energy yield from October to March.
(d) Specific annual energy yield Y_{Fa} for this installation in kWh/kWp/a.
(e) Full-load hours t_V for P_{ACn}; also, verify that capacity factor $CF = t_V/(8760\,h)$.

Solution:

(a) and (b)

Energy yield calculation for grid-connected PV plants (with 3-component model)

Location:	Aswan		Reference station:			Cairo			$\beta =$	20°	
P_{Go} [kW]:		10000	$P_{ACn} = k_{Gmax} \cdot P_{Go} \cdot \eta_{WR}$ [kW]:				8633		$\gamma =$	0°	
	$R_D = \frac{1}{2}\cos\alpha_2 + \frac{1}{2}\cos(\alpha_1+\beta) =$			0.970		$R_R = \frac{1}{2} - \frac{1}{2}\cos\beta =$			0.030		

Month	Jan	Feb	Mar	April	May	June	July	Aug	Sept	Oct	Nov	Dec	Year	
H	4.99	6.00	6.96	7.85	8.25	8.81	8.40	8.04	7.37	6.24	5.32	4.78	6.90	kWh/m²
H_D	1.14	1.23	1.46	1.56	1.67	1.44	1.61	1.56	1.44	1.34	1.14	1.05	1.39	kWh/m²
R_B	1.32	1.22	1.11	1.01	0.94	0.90	0.92	0.98	1.06	1.18	1.29	1.36		
k_B	1	1	1	1	1	1	1	1	1	1	1	1		
$H_{GB} = k_B \cdot R_B \cdot (H-H_D)$	5.10	5.82	6.09	6.34	6.16	6.66	6.23	6.32	6.31	5.77	5.41	5.08	5.939	kWh/m²
$H_{GD} = R_D \cdot H_D$	1.10	1.19	1.42	1.52	1.62	1.40	1.56	1.51	1.40	1.30	1.10	1.02	1.346	kWh/m²
ρ	0.35	0.35	0.35	0.35	0.35	0.35	0.35	0.35	0.35	0.35	0.35	0.35		
$H_{GR} = R_R \cdot \rho \cdot H$	0.05	0.06	0.07	0.08	0.09	0.09	0.09	0.08	0.08	0.07	0.06	0.05	0.073	kWh/m²
$H_G = H_{GB}+H_{GD}+H_{GR}$	6.25	7.07	7.58	7.94	7.87	8.15	7.88	7.91	7.79	7.14	6.57	6.15	7.358	kWh/m²
$Y_R = H_G/1kWm^{-2}$	6.25	7.07	7.58	7.94	7.87	8.15	7.88	7.91	7.79	7.14	6.57	6.15	7.358	h/d
k_T	0.94	0.92	0.91	0.89	0.87	0.86	0.86	0.86	0.87	0.88	0.91	0.93	0.889	
$Y_T = k_T \cdot Y_R$	5.88	6.50	6.90	7.07	6.85	7.01	6.78	6.80	6.78	6.28	5.98	5.72	6.543	h/d
k_G	0.89	0.89	0.89	0.89	0.88	0.87	0.87	0.87	0.88	0.89	0.89	0.89	0.883	
$Y_A = k_G \cdot Y_T$	5.23	5.79	6.14	6.29	6.03	6.10	5.90	5.92	5.96	5.59	5.32	5.09	5.778	h/d
η_{WR} (η_{tot})	0.97	0.97	0.97	0.97	0.97	0.97	0.97	0.97	0.97	0.97	0.97	0.97	0.970	
$Y_F = \eta_{WR} \cdot Y_A$	5.07	5.62	5.95	6.10	5.84	5.91	5.72	5.74	5.79	5.42	5.16	4.94	5.604	h/d
n_d (days/month)	31	28	31	30	31	30	31	31	30	31	30	31	365	d
$E_{AC} = n_d \cdot P_{Go} \cdot Y_F$	1.57	1.57	1.85	1.83	1.81	1.77	1.77	1.78	1.74	1.68	1.55	1.53	20.46	GWh
$t_V = E_{AC}/P_{ACn}$	182	182	214	212	210	206	205	206	201	195	179	177	2369	h
$PR = Y_F/Y_R$	0.812	0.794	0.786	0.768	0.743	0.726	0.726	0.726	0.743	0.760	0.786	0.803	0.762	

(c) Percentage of total energy yield accounted for by energy yield from October to March:
9.75 kWh = 47.7%.

(d) Specific annual energy yield $Y_{Fa} = 5.604\,kWh/kWp/d = 2046\,kWh/kWp/a$.

(e) P_{Acn} full-load hours $t_V = 2369\,h/a$; capacity $CF = 27.0\%$.

Problem 3 (Large grid-connected system near Munich, Germany)

New ground-based PV installation: tilt angle $\beta = 35°$; $\varphi = 48.2°N$; solar generator azimuth $\gamma = 30°$ (30° western deviation from due south); $\alpha_1 = \alpha_2 = 0°$; reflection factor ρ, April to October 0.3; March and November 0.35; February 0.45; December and January 0.5 (higher in the winter because of snow); R_B values interpolated; solar generator peak power $P_{Go} = 6\,MWp$; mean inverter efficiency $\eta_{WR} = 96\%$. Capture losses attributable to mismatch, wiring, soiling and the like are taken into account by a solar generator correction factor k_G as in Table 8.3 (applies to a new installation whose modules exhibit their nominal power at a 30° tilt angle). The reference station is Munich, Germany.

Determine the following:

(a) Monthly and annual global radiation on the solar generator plane (in $kWh/m^2/d$).
(b) Monthly and annual energy yield in kWh.
(c) Percentage of total energy yield accounted for by energy yield from October to March.
(d) Specific annual energy yield Y_{Fa} for this installation in kWh/kWp/a.
(e) Full-load hours t_V for P_{ACn}; also, verify that capacity factor $CF = t_V/(8760\,h)$.

Solution:

(a) and (b)

Energy yield calculation for grid-connected PV plants (with 3-component model)

Location:	Munich			Reference station:			Munich			$\beta =$	35°
P_{Go} [kW]:		6000		$P_{ACn} = k_{Gmax} \cdot P_{Go} \cdot \eta_{WR}$ [kW]:				5184		$\gamma =$	30°

| | $R_D = \frac{1}{2}\cos\alpha_2 + \frac{1}{2}\cos(\alpha_1+\beta) =$ | | | 0.910 | | | $R_R = \frac{1}{2} - \frac{1}{2}\cos\beta =$ | | | 0.090 | |

Month	Jan	Feb	Mar	April	May	June	July	Aug	Sept	Oct	Nov	Dec	Year	
H	1.03	1.80	2.88	4.01	5.04	5.43	5.40	4.61	3.53	2.13	1.13	0.79	3.14	kWh/m²
H_D	0.67	1.05	1.60	2.18	2.61	2.81	2.71	2.35	1.82	1.24	0.75	0.55	1.69	kWh/m²
R_B	2.29	1.80	1.42	1.18	1.03	0.97	1.00	1.11	1.31	1.65	2.11	2.52		
k_B	1	1	1	1	1	1	1	1	1	1	1	1		
$H_{GB} = k_B \cdot R_B \cdot (H-H_D)$	0.82	1.35	1.82	2.16	2.50	2.54	2.69	2.51	2.24	1.47	0.80	0.60	1.794	kWh/m²
$H_{GD} = R_D \cdot H_D$	0.61	0.96	1.46	1.98	2.38	2.56	2.47	2.14	1.66	1.13	0.68	0.50	1.547	kWh/m²
ρ	0.50	0.45	0.35	0.30	0.30	0.30	0.30	0.30	0.30	0.30	0.35	0.50		
$H_{GR} = R_R \cdot \rho \cdot H$	0.05	0.07	0.09	0.11	0.14	0.15	0.15	0.12	0.10	0.06	0.04	0.04	0.093	kWh/m²
$H_G = H_{GB}+H_{GD}+H_{GR}$	1.48	2.38	3.37	4.25	5.02	5.25	5.31	4.77	4.00	2.66	1.52	1.14	3.434	kWh/m²
$Y_R = H_G/1kWm^{-2}$	1.48	2.38	3.37	4.25	5.02	5.25	5.31	4.77	4.00	2.66	1.52	1.14	3.434	h/d
k_T	1.06	1.05	1.02	1.00	0.97	0.95	0.94	0.95	0.97	1.00	1.04	1.05	0.983	
$Y_T = k_T \cdot Y_R$	1.57	2.50	3.44	4.25	4.87	4.99	4.99	4.53	3.88	2.66	1.58	1.20	3.375	h/d
k_G	0.75	0.79	0.86	0.88	0.90	0.90	0.90	0.90	0.90	0.88	0.80	0.70	0.871	
$Y_A = k_G \cdot Y_T$	1.18	1.97	2.96	3.74	4.38	4.49	4.49	4.08	3.49	2.34	1.26	0.84	2.940	h/d
η_{WR} (η_{tot})	0.96	0.96	0.96	0.96	0.96	0.96	0.96	0.96	0.96	0.96	0.96	0.96	0.960	
$Y_F = \eta_{WR} \cdot Y_A$	1.13	1.90	2.84	3.59	4.21	4.31	4.31	3.92	3.35	2.25	1.21	0.80	2.822	h/d
n_d (days/month)	31	28	31	30	31	30	31	31	30	31	30	31	365	d
$E_{AC} = n_d \cdot P_{Go} \cdot Y_F$	210	318	528	646	783	776	802	728	603	418	219	150	6181	MWh
$t_V = E_{AC}/P_{ACn}$	41	61	102	125	151	150	155	140	116	81	42	29	1192	h
$PR = Y_F/Y_R$	0.763	0.796	0.842	0.845	0.838	0.821	0.812	0.821	0.838	0.845	0.799	0.706	0.822	

(c) Percentage of total energy yield accounted for by energy yield from October to March: 1843 kWh = 29.8%.

(d) Specific annual energy yield $Y_{Fa} = 2.822\,kWh/kWp/d = 1030\,kWh/kWp/a$.

(e) P_{ACn} full-load hours $t_V = 1192\,h/a$; capacity $CF = 13.6\%$.

Note: New PV installations in Germany can potentially exhibit annual energy yields in excess of 1000 kWh/kWp.

Problem 4 (Large grid-connected system near Alice Springs, Australia)

New ground-based PV installation: tilt angle $\beta = 20°$; $\varphi = 23.8°S$ ($\varphi = -23.8°$); solar generator azimuth $0°$ (S); $\alpha_1 = \alpha_2 = 0°$; year-round reflection factor $\rho = 0.35$; solar generator peak power $P_{Go} = 40\,MWp$; mean inverter efficiency $\eta_{WR} = 97\%$. Capture losses attributable to mismatch, wiring, soiling and the like are taken into account by a solar generator correction factor k_G as in Table 8.5 (applies to a new installation whose modules exhibit their nominal power at a 30° tilt angle). Reference station for k_T and k_G: Cairo. Note that the monthly values need to be shifted by six months as Cairo is located in the northern hemisphere and the Alice Springs installation is located in the southern hemisphere.

Determine the following:

(a) Monthly and annual global radiation on the solar generator plane (in $kWh/m^2/d$).
(b) Monthly and annual energy yield in kWh.
(c) Percentage of total energy yield accounted for by energy yield from October to March.
(d) Specific annual energy yield Y_{Fa} for this installation in kWh/kWp/a.
(e) Full-load hours t_V for P_{ACn}; also, verify that capacity factor $CF = t_V/(8760\,h)$.

Solution:

(a) and (b)

Energy yield calculation for grid-connected PV plants (with 3-component model)

Location: Alice Springs			Reference station:			Cairo (Δt= 6 months!)					$\beta =$	20°
P_{Go} [kW]:		40000	$P_{ACn} = k_{Gmax} \cdot P_{Go} \cdot \eta_{WR}$ [kW]:					34532			$\gamma =$	0°

| $R_D = \frac{1}{2}\cos\alpha_2 + \frac{1}{2}\cos(\alpha_1+\beta) =$ | | | | 0.970 | | $R_R = \frac{1}{2} - \frac{1}{2}\cos\beta =$ | | | | 0.030 | |

Month	Jan	Feb	Mar	April	May	June	July	Aug	Sept	Oct	Nov	Dec	Year	
H	7.46	7.15	6.36	5.56	4.46	4.30	4.66	5.52	6.62	7.15	7.37	7.39	6.16	kWh/m²
H_D	2.31	2.13	1.90	1.52	1.32	1.07	1.04	1.22	1.46	1.92	2.25	2.39	1.71	kWh/m²
R_B	0.92	0.99	1.08	1.19	1.30	1.37	1.34	1.24	1.12	1.02	0.94	0.91		
k_B	1	1	1	1	1	1	1	1	1	1	1	1		
$H_{GB} = k_B \cdot R_B \cdot (H-H_D)$	4.74	4.97	4.82	4.81	4.08	4.43	4.85	5.33	5.78	5.33	4.81	4.55	4.873	kWh/m²
$H_{GD} = R_D \cdot H_D$	2.24	2.07	1.84	1.47	1.28	1.04	1.01	1.18	1.42	1.86	2.18	2.32	1.657	kWh/m²
ρ	0.35	0.35	0.35	0.35	0.35	0.35	0.35	0.35	0.35	0.35	0.35	0.35		
$H_{GR} = R_R \cdot \rho \cdot H$	0.08	0.08	0.07	0.06	0.05	0.05	0.05	0.06	0.07	0.08	0.08	0.08	0.067	kWh/m²
$H_G = H_{GB}+H_{GD}+H_{GR}$	7.06	7.12	6.73	6.34	5.41	5.52	5.91	6.57	7.27	7.27	7.07	6.95	6.598	kWh/m²
$Y_R = H_G/1kWm^2$	7.06	7.12	6.73	6.34	5.41	5.52	5.91	6.57	7.27	7.27	7.07	6.95	6.598	h/d
k_T	0.88	0.88	0.89	0.90	0.93	0.95	0.96	0.94	0.93	0.91	0.89	0.88	0.910	
$Y_T = k_T \cdot Y_R$	6.21	6.27	5.99	5.71	5.03	5.24	5.67	6.18	6.76	6.62	6.29	6.12	6.005	h/d
k_G	0.87	0.87	0.88	0.89	0.89	0.89	0.89	0.89	0.89	0.89	0.88	0.87	0.883	
$Y_A = k_G \cdot Y_T$	5.41	5.45	5.27	5.08	4.48	4.67	5.05	5.50	6.02	5.89	5.54	5.32	5.304	h/d
η_{WR} (η_{tot})	0.97	0.97	0.97	0.97	0.97	0.97	0.97	0.97	0.97	0.97	0.97	0.97	0.970	
$Y_F = \eta_{WR} \cdot Y_A$	5.24	5.29	5.11	4.93	4.34	4.53	4.90	5.33	5.84	5.71	5.37	5.16	5.144	h/d
n_d (days/month)	31	28	31	30	31	30	31	31	30	31	30	31	365	d
$E_{AC} = n_d \cdot P_{Go} \cdot Y_F$	6.50	5.92	6.34	5.91	5.39	5.43	6.07	6.61	7.00	7.08	6.45	6.40	75.11	GWh
$t_V = E_{AC}/P_{ACn}$	188	171	184	171	156	157	176	191	203	205	187	185	2175	h
$PR = Y_F/Y_R$	0.743	0.743	0.760	0.777	0.803	0.820	0.829	0.812	0.803	0.786	0.760	0.743	0.780	

(c) Percentage of total energy yield accounted for by energy yield from October through March:
$36.42\,kWh = 48.5\%$.
(d) Specific annual energy yield $Y_{Fa} = 5.144\,kWh/kWp/d = 1878\,kWh/kWp/a$.
(e) P_{Acn} full-load hours $t_V = 2175\,h/a$; capacity $CF = 24.8\%$.

Note: The Cairo k_G and k_T values for January to June were used for July to December and vice versa.

8.3   Sizing PV Installations that Integrate a Battery Bank

In as much as the power output of a PV installation needs to equate to the average of the energy used by the installation's appliances plus the installation's capture losses, solar generator sizing is mainly based on the installation's energy balance.

In locations where insolation is far higher in summer than in winter (e.g. at middle and high latitudes, and particularly in low-lying foggy areas), a so-called hybrid installation that derives part of its energy from an auxiliary source such as a gas or diesel generator sometimes makes better economic sense than an installation composed solely of solar modules. Moreover, as the wind tends to be stronger during the winter, use of a wind turbine as an additional energy source at such locations is often advisable.

An installation that is to provide energy for a fixed amount of time in the absence of sunlight needs to integrate a battery bank that is rated for the requisite stand-alone period.

For stand-alone installations, only highly efficient appliances with minimally low power consumption should be used, since otherwise a larger and hence more expensive solar generator is necessary. DC appliances such as LED lamps, or energy-saving light bulbs with DC adaptors, already exhibit minimal power consumption. But as excellent and highly efficient sine wave inverters are available nowadays for stand-alone installations, standard 230 V AC appliances are often used whose power consumption is not optimized and which use additional energy by virtue of inverter conversion loss. Hence AC appliances with optimally low power consumption should also be used for stand-alone installations. The configuration of a 230 V stand-alone installation grid is in effect an invitation to use appliances that are not part of the installation's original energy balance. Although this poses no problem in buildings that are integrated into the power grid, since the only effect is a higher electricity bill, in stand-alone installations such extra appliances will sooner or later induce anomalies in and damage to the installation if additional energy is not fed into the system.

Once insolation on the solar generator plane has been determined (this is often done solely for the critical, low-insolation months), normally a stand-alone installation is sized by the following steps:

1. The mean daily power consumption of all appliances is determined.
2. The battery bank is sized.
3. The solar generator is sized, along with a possible auxiliary energy source.

8.3.1   Determination of Mean Daily Appliance Power Consumption

The calculation of mean daily appliance power consumption should be based on DC power consumption, although daily charge consumption for the various appliances can also be used instead.

In a PV installation, n different, intermittently activated appliances consume power P_i during time t_i, and each one draws a specific amount of energy E_i from the DC side of the installation. In determining AC appliance E_i for installations with inverters whose mean efficiency is $\eta_{WR} < 1$, the fact that these appliances draw power $P_{DCi} = P_{ACi}/\eta_{WR}$ on the DC side also needs to be taken into consideration (η_{WR} = mean efficiency).

The following equations apply to the DC energy needed by individual appliances (power consumption P_i):

$$\text{For DC appliances: } E_i = P_{DCi} \cdot t_i = P_i \cdot t_i \tag{8.19}$$

$$\text{For AC appliances: } E_i = P_{DCi} \cdot t_i = \frac{P_{ACi}}{\eta_{WR}} t_i = \frac{P_i}{\eta_{WR}} t_i \tag{8.20}$$

$$\text{For the DC energy used by all appliances } E_V = \sum_{i=1}^{n} E_i \tag{8.21}$$

Another factor here is the energy E_0 consumed by: (a) appliances that are always switched on (e.g. charge controller energy, inverter open-circuit loss, battery self-discharge); and (b) refrigerators and freezers that are always in operation.

In some installations, appliances that are switched on intermittently and exhibit energy consumption E_V are used at varying intervals h_B rather than every day. For example, $h_B = 1$ for a home that is permanently occupied, but $h_B = {}^2/_7$ for a weekend house that is occupied for only two days a week. Hence the following equation applies to mean daily energy consumption E_D:

> Mean general daily DC power consumption: $E_D = h_B \cdot E_V + E_0$ (8.22)

In this equation, h_B is the mean usage frequency ($0 \leq h_B \leq 1$).

For a permanently occupied house or an installation that is in continuous operation, $h_B = 1$ and thus mean daily energy consumption E_D is as follows:

> Mean daily DC power consumption: $E_D = E_V + E_0$ (8.23)

For a weekend house that is used for only two days a week, $h_B = {}^2/_7$, and thus its mean daily DC power consumption E_D is as follows:

> Mean daily DC power consumption: $E_D = {}^2/_7 E_V + E_0$ (8.24)

Mean daily charge consumption Q_D should be used for further PV installation sizing; this value can be readily derived from E_D using system voltage V_S (nominal battery voltage):

> Mean daily charge consumption $Q_D = \dfrac{E_D}{V_S}$ (8.25)

In addition to determining mean daily energy and charge consumption, maximum DC-side P_{DCmax} and AC-side P_{ACmax} power output need to be calculated so that the requisite wire gauges and fuses, as well as nominal inverter output (if applicable), can be calculated. It is also necessary in this regard to verify whether all connected loads need to be supplied with energy synchronously, or costs can be lowered through the use of lower maximum output.

Table 8.9 below or A6.3 (in Appendix A) should be used to determine the values discussed in this section and for battery sizing as described in the next section.

8.3.2 Requisite Battery Capacity K

As battery capacity is generally indicated in ampere-hours (Ah), PV installation battery sizing should be based on charge levels rather than energy.

An installation battery bank should have the capacity to supply all appliances autonomously for n_A days, and in so doing should be discharged solely to depth of discharge t_Z (see Section 5.1.1). The maximum available charge Q_E is referred to as usable capacity K_N:

> Usable battery capacity $K_N = K \cdot t_Z$ (8.26)

A battery must be able to meet all energy requirements during its autonomy period n_A, including when (as is often the case) the connected appliances need all available energy E_V on a daily basis. The following applies in accordance with the law of conservation of energy:

$$V_S \cdot K_N = n_A(E_V + E_0) \tag{8.27}$$

or using K_N,

$$K_N = \frac{n_A(E_V + E_0)}{V_S} \tag{8.28}$$

Hence the minimum required battery capacity is as follows, depending on whether K_{10}, K_{20} or K_{100} is used:

$$\text{Required battery capacity } K = \frac{n_A(E_V + E_0)}{V_S \cdot t_Z} \tag{8.29}$$

For installations with inverters, K also needs to exhibit the value as in Equation 5.19 at a minimum.

Once a battery capacity K has been selected and thus the battery's daily self-discharge rate is known (typically, for example, 3% of capacity per month and 0.1% of capacity per day), battery self-discharge can be taken into account by increasing E_0 by around $0.001 \cdot V_S \cdot K$. However, battery self-discharge is usually negligible and is normally taken into account by rounding off the relevant decimal numbers.

8.3.3 Solar Generator Sizing

Following a deep discharge event, a PV installation needs to be able to meet usable capacity K_N requirements within a specific system recovery time n_E (number of days until a full charge is reached), *plus* mean daily charge consumption Q_D. In addition, the amount of energy discharged by a battery is lower than the amount of its charge, the difference between these two values being ampere-hour efficiency η_{Ah} (see Section 5.1.1).

In order for a battery's load balance to maintain an equilibrium, it must receive the following mean daily charge Q_L:

$$\text{Mean daily charge consumption } Q_L = \frac{1}{\eta_{Ah}}\left(Q_D + \frac{K_N}{n_E}\right) \tag{8.30}$$

In cases where a system recovery time is not specified, $n_E = \infty$ and the minimum charge required to charge a battery is as follows:

$$Q_{Lmin} = \frac{Q_D}{\eta_{Ah}} \tag{8.31}$$

However, basing a PV-only installation on a solar generator that exhibits only Q_{Lmin} is a risky proposition as no power is left in reserve for contingencies such as two successive inclement weather periods.

On the other hand, the solar generator of a hybrid installation can be rated solely for Q_{Lmin} without any difficulty as any power deficit will be covered by the backup generator. Moreover, the fuel consumption that comes into play in such situations is very low as the backup generator will only be used in an emergency.

A PV installation's solar generator should output a mean daily charge Q_{PV} that equates to a Q_L load. In hybrid installations, a backup generator (e.g. a gas or diesel generator for large installation, or a thermoelectric or fuel cell device for small installations) can supply energy E_H during critical months (e.g. in winter).

The mean battery charging voltage exceeds system voltage V_S. This can roughly be taken into account by presupposing a mean charging voltage of $1.1 \cdot V_S$, which yields the following:

$$\text{Hybrid generator charge resulting from } E_H : \quad Q_H = \frac{E_H}{1.1 \cdot V_S} \qquad (8.32)$$

The following then applies to a hybrid system's backup generator:

$$Q_L = Q_{PV} + Q_H \text{ and } Q_{PV} = Q_L - Q_H \qquad (8.33)$$

8.3.3.1 Solar Generator Sizing for Standard Charge Controllers

Installations that use standard charge controllers usually have a solar generator comprising n_{SP} parallel-connected series strings whose requisite system voltage is generated by n_{MS} series-connected solar modules (see Figures 4.31 and 4.32). When a battery's charge limiting voltage V_G is reached (it usually ranges between $1.15 \cdot V_S$ and $1.2 \cdot V_S$) the voltage at each module's ohmic resistances, series controllers and diodes is as follows (excluding the voltage drop):

$$\text{Per-module voltage when the charge limiting voltage is reached}: V_{MG} = \frac{V_G}{n_{MS}} \qquad (8.34)$$

In most cases, 32–36 series-connected crystalline silicon solar cells per 12 V of system voltage are used for battery charging purposes, in which case the module voltage will be considerably lower than MPP voltage $V_{MPP\text{-}STC}$ at STC power output. In working PV installations, the module temperature usually far exceeds the STC power output temperature of 25 °C, and MPP voltage is always lower (see Figure 4.13) but still overshoots V_{MG}. Hence it can be assumed that module current I_{Mo} at $G_o = 1\,\text{kW/m}^2$ will always slightly overshoot module datasheet MPP current $I_{MPP\text{-}STC}$ at STC power output.

It can thus further be assumed that the following equation provides a close module current I_M estimate for use of 32–36 crystalline solar cells per 12 V of battery voltage:

$$I_M = I_{MPP\text{-}STC} \frac{G}{G_o} \qquad (8.35)$$

For other modules, the current I_{Mo} should be determined from the characteristic curves based on V_G and G_o, and the maximum anticipated module temperature.

Hence a module's charging capacity Q_M can be readily derived from reference yield Y_R as follows:

$$\text{Module charging capacity } Q_M = I_{Mo} \cdot Y_R \approx I_{MPP\text{-}STC} \cdot Y_R \qquad (8.36)$$

Most solar generators comprise a series of interconnected modules, which can exacerbate capture losses that are best taken into account using the generator correction factor k_G. Such losses are attributable to factors such as partial shading, soiling, snow cover, mismatch, spectral loss, glass reflection loss, module power loss and so on; as some modules exhibit a number of such effects, k_G should always be taken into account ($k_G < 1$).

Hence a solar generator string comprising n_{MS} identical series-connected modules will exhibit the following charging capacity Q_S:

$$\text{Per-string charging capacity } Q_S = k_G \cdot Q_M = k_G \cdot I_{Mo} \cdot Y_R \approx k_G \cdot I_{MPP\text{-}STC} \cdot Y_R \qquad (8.37)$$

In cases where a solar generator comprises n_{SP} parallel-connected series strings containing n_{MS} series-connected modules for generating the requisite system voltage (see Figures 4.31 and 4.32 in Section 4.3.3) and that exhibit mean daily per-string charging capacity Q_S, the following equation applies to the requisite number of parallel-connected strings:

$$\text{Requisite number of parallel-connected strings}: n_{SP} = \frac{Q_{PV}}{Q_S} = \frac{Q_L - Q_H}{Q_S} \qquad (8.38)$$

The following equation applies to a stand-alone installation with no backup generator:

$$n_{SP} = \frac{Q_L}{Q_S} \qquad (8.39)$$

8.3.3.2 Solar Generator Sizing for MPT Charge Controllers

A PV installation that contains a solar tracker charge controller is sized as for the grid-connected system Y_A. By multiplying various factors by η_{MPT} (overall MPT charge controller efficiency, which usually exceeds characteristic inverter η_{tot} by up to 99%), the maximum final yield capacity $Y_F{}'$ is obtained as with Equation 8.15:

$$\text{Final yield capacity } Y_F' = \eta_{MPT} \cdot k_G \cdot k_T \cdot Y_R = \eta_{MPT} \cdot k_G \cdot k_T \frac{H_G}{G_o} \qquad (8.40)$$

Hence $Y_F{}'$ is used here since, unlike a grid-connected system, a stand-alone installation does not always exhibit the full Y_F value. As in Equation 8.16, by multiplying final yield capacity $Y_F{}'$ by peak generator output P_{Go}, the mean daily energy yield capacity E_{DC} can be readily determined for each month. Multiplication by n_d can be dispensed with here as a daily energy yield is being determined:

$$\text{Mean daily energy yield of an MPT charge controller}: E_{DC} = P_{Go} \cdot Y_F' \qquad (8.41)$$

As the mean battery charging voltage is around $1.1 \cdot U_S$, the following applies:

$$\text{Solar generator charge resulting from } E_{DC}: Q_{PV} = \frac{E_{DC}}{1.1 \cdot U_S} \qquad (8.42)$$

An entire stand-alone installation can be dimensioned using Equations 8.19–8.42, which, though they are mainly based on the operational scenario for a house installation, also of course apply to other scenarios.

Stand-alone systems in temperate climates can often be sized by merely performing the relevant calculations for the conditions that obtain during the lowest-insolation period, which is November to February for installations operated all year round in the northern hemisphere.

The requisite sizing calculations can be realized using Tables 8.9, 8.10 and 8.11 (see examples).

Table 8.8 Recommend values for system autonomy n_A

n_A (days) Latitude	Winter (Months 1, 2, 11, 12)	Spring/autumn (Months 3, 4, 9, 10)	Summer (Months 5, 6, 7, 8)
30°N	3–6	2–5	2–4
40°N	4–8	3–6	2–4
50°N	7–15	5–8	3–5
60°N	10–25	7–10	3–6

For southern hemisphere at same latitude add/subtract 6 from month number indicated.

8.3.3.3 Guide Values for Selected Stand-alone Installation Parameters

Generator correction factor k_G: $0.5 < k_G < 0.9$, depending on month and solar generator orientation (see Section 8.1.3).

Ampere-hour efficiency η_{Ah}: 0.9 (0.8–0.98).

Depth of discharge t_Z: $0.3 < t_Z < 0.8$; t_Z normally ranges from 0.4 to 0.6. For lead–calcium batteries, $t_Z \leq 0.5$. For installations with inverters: $t_Z \leq 0.6$.

The values given here apply to a fully autonomous cycle extending over n_A days, whereby the daily depth of discharge is usually far lower than these values. Lower t_Z values increase battery service life.

System autonomy n_A See Table 8.8.

System recovery time n_E A multiple of n_A that is not inordinately high should be used for n_E.

$n_E \leq 5$ should be applied to a regularly used weekend house where $n_A = 2$ days. For higher n_A values, it is often better to use higher n_E values.

The more autonomy n_A a system exhibits and the shorter its recovery time n_E, the more reliable – but also expensive – its energy supply will be.

For a hybrid system (feasible only if a reliable backup generator is available) n_A and n_E can be considerably lower, which reduces system costs and allows for a higher ratio between available and actually used solar energy at middle and high latitudes, where PV-only installations will otherwise generate considerable surplus energy during the summer.

8.3.4 Stand-alone System Sizing Tables

As with grid-connected systems, stand-alone installations are best sized using suitable tables, which are provided in this section as well as in Appendix A. Table 8.9 (or Table A6.3) can be used to determine the requisite battery capacity K and mean daily charge Q_L.

First, the total daily DC energy E_{DC} required by all appliances is determined in the left-hand segments of the table. In order to calculate the DC energy needed by AC appliances, it is necessary to divide by stand-alone installation efficiency η_{WR}:

$$\text{DC energy needed by AC appliances: } E_{DC} = \frac{E_{AC}}{\eta_{WR}} \tag{8.43}$$

System voltage V_S, battery depth of discharge t_Z, Ah efficiency η_{Ah}, autonomy period n_A and system recovery time n_E are then entered in the right-hand segments of the table. The various interim and final values are then calculated successively using the equations indicated in the table. The outcome of this calculation procedure comprises the requisite battery capacity K, as well as the mean battery charge Q_L needed to obtain a good energy balance.

Tables 8.10 and 8.11 can be used to determine the requisite solar generator guide values. First, using the top segment of the table, the charge Q_H generated by a backup generator (if any) is subtracted from the Q_L

Table 8.9 Energy balance and battery capacity K calculation table (also available as Table A6.3 in Appendix A)

Energy balance sheet for PV stand-alone plants (layout of PV generator on separate sheet)

Consumers (switched)	AC/DC	P[W]	t [h]	E_{AC}	η_{WR}	E_{DC}		Location:		Month:	
1							Wh/d				
2							Wh/d	System voltage V_S =			V
3							Wh/d	Accumulator cycle depth t_z =			
4							Wh/d	Ah-efficiency η_{Ah} =			
5							Wh/d				
6							Wh/d	Autonomy days n_A =			d
7							Wh/d	System recovery time n_E =			d
8							Wh/d				
		$E_V = \Sigma\ E_{DC}$ switched cons. =					Wh/d	Frequency of use h_B =			

Permanent consumers	AC/DC	P[W]	t [h]	E_{AC}	η_{WR}	E_{DC}		Note:
1							Wh/d	This sheet with average daily energy
2							Wh/d	consumption is to be completed in most
3								cases only for the month with the highest
4							Wh/d	consumption. If energy consumption
		$E_0 = \Sigma\ E_{DC}$ perman. cons. =					Wh/d	varies considerably over the year, it may be neccessary to do use several sheets.

Usable accumulator capacity $K_N = n_A \cdot (E_V + E_0)/V_S$ =		Ah	Mean daily $E_D = h_B \cdot E_V + E_0$ =	Wh/d
Minimum accu capacity $K = K_N/t_z = n_A \cdot (E_V + E_0)/(V_S \cdot t_z)$ =		Ah	Mean daily $Q_D = E_D/V_S$ =	
Necessary daily charging $Q_L = (1/\eta_{Ah}) \cdot (Q_D + K_N/n_E)$ =		Ah/d		

Table 8.10 Sizing calculation table for a stand-alone installation solar generator using the simplified string calculation method of Section 2.4 (the shaded fields are optional; this table can also be found in Appendix A as Table A6.4)

Layout of PV generator for stand alone plants (simplified calculation of irradiation)

Location:		Reference station:		β =	γ =	°
I_{Mo} =	A P_{Mo} =	W System voltage V_S =	V	Modules/string n_{MS} =		

Month	Jan	Feb	Mar	April	May	June	July	Aug	Sept	Oct	Nov	Dec	
Req. daily charge Q_L													Ah/d
Auxillary energy E_H													Wh/d
$Q_H = E_H/(1.1 \cdot V_S)$													Ah/d
$Q_{PV} = Q_L - Q_H$ =													Ah/d

Month	Jan	Feb	Mar	April	May	June	July	Aug	Sept	Oct	Nov	Dec	
H													kWh/m²
$R(\beta,\gamma)$													
$H_G = R(\beta,\gamma) \cdot H$													kWh/m²
$Y_R = H_G/1kWm^{-2}$ =													h/d
Without MPT-charge-controller *(insert k_G-values in line below only in this case)*:													
k_G													
$Q_S = k_G \cdot I_{Mo} \cdot Y_R$ =													Ah/d
With MPT-charge-controller *(insert k_T- and k_G-values below only in this case)*:													
k_T													
$Y_T = k_T \cdot Y_R$													h/d
k_G													
$Y_A = k_G \cdot Y_T$													h/d
η_{MPT}													
$Y_F' = \eta_{MPT} \cdot Y_A$ =													h/d
$E_{DC-S} = n_{MS} \cdot P_{Mo} \cdot Y_F'$ =													Wh/d
$Q_S = E_{DC-S}/(1.1 \cdot V_S)$													Ah/d

$n_{SP}' = Q_{PV}/Q_S$													

Necessary number of parallel strings: Maximum(n_{SP}'), rounded up to the next integer :	n_{SP} =
Total necessary number of modules:	$n_M = n_{MS} \cdot n_{SP}$ =

Table 8.11 Sizing calculation table for a stand-alone installation solar generator using the three-component string calculation method (the shaded fields are optional; this table can also be found in Appendix A as Table A6.5)

Layout of PV generator for stand alone plants (with 3-component model)

Location:		Reference station:		$\beta =$		$\gamma =$		°

$R_D = \frac{1}{2}\cos\alpha_2 + \frac{1}{2}\cos(\alpha_1+\beta) =$

$R_R = \frac{1}{2} - \frac{1}{2}\cos\beta =$

$I_{Mo} =$	A	$P_{Mo} =$	W	System voltage $V_S =$	V	Modules / string $n_{MS} =$

Month	Jan	Feb	Mar	April	May	June	July	Aug	Sept	Oct	Nov	Dec	
Req. daily charge Q_L													Ah/d
Auxillary energy E_H													Wh/d
$Q_H = E_H/(1.1 \cdot V_S)$													Ah/d
$Q_{PV} = Q_L - Q_H =$													Ah/d

Month	Jan	Feb	Mar	April	May	June	July	Aug	Sept	Oct	Nov	Dec	
H													kWh/m²
H_D													kWh/m²
R_B													
k_B													
$H_{GB} = k_B \cdot R_B \cdot (H-H_D)$													kWh/m²
$H_{GD} = R_D \cdot H_D$													kWh/m²
ρ													
$H_{GR} = R_R \cdot \rho \cdot H$													kWh/m²
$H_G = H_{GB}+H_{GD}+H_{GR}$													kWh/m²
$Y_R = H_G/1kWm^{-2} =$													h/d

Without MPT-charge-controller *(insert k_G-values in line below only in this case):*

k_G													
$Q_S = k_G \cdot I_{Mo} \cdot Y_R =$													Ah/d

With MPT-charge-controller *(insert k_T- and k_G-values below only in this case):*

k_T													
$Y_T = k_T \cdot Y_R$													h/d
k_G													
$Y_A = k_G \cdot Y_T$													h/d
η_{MPT}													
$Y_F' = \eta_{MPT} \cdot Y_A =$													h/d
$E_{DC-S} = n_{MS} \cdot P_{Mo} \cdot Y_F' =$													Wh/d
$Q_S = E_{DC-S}/(1.1 \cdot V_S)$													Ah/d

$n_{SP}' = Q_{PV}/Q_S$													

Necessary number of parallel strings: Maximum(n_{SP}'), rounded up to the next integer :	$n_{SP} =$	
Total necessary number of modules:	$n_M = n_{MS} \cdot n_{SP} =$	

values (which may differ from one month to another) and the effective charge Q_{PV} generated by the PV installation is determined.

Then, using the next table segment, insolation is determined: for Table 8.10, using the simplified procedure as in Section 2.4; for Table 8.11, using the three-component model as in Section 2.5. Both of these tables can be used for calculations involving solar tracker or non-solar-tracker charge controllers, whereby figures only need be entered in the relevant row.

For non-solar-tracker charge controllers, only the k_G values for the relevant month need be entered (in the middle of the table). Per-string charging capacity Q_S can be derived from the relation $I_{Mo} = I_{MPP-STC}$ of the relevant module and the previously calculated reference yield Y_R.

As with grid-connected systems, monthly final yield capacity Y_F' can be determined for installations with MPT charge controllers by multiplying by k_T, k_G and η_{MPT} (in lieu of η_{WR} and η_{tot} as for grid-connected systems); from this, per-string DC energy capacity E_{DC-S} can be derived by multiplying by string module capacity $n_{MS} \cdot P_{Mo}$. In as much as the battery charging voltage averages around $1.1 \cdot U_S$, the string charging capacity Q_S can be derived from this value.

The requisite number n_{SP} of parallel-connected strings is obtained by: (a) dividing the requisite PV charge Q_{PV} by the string charge (determined using one method or the other) and rounding off to the nearest whole number; and (b) determining the maximum value for all months. In a final step, the requisite number of modules N_m is determined by multiplying the result of (a) by n_{MS}.

As with grid-connected systems, for installation MPT charge controllers the number n_{MS} of series-connected modules should be defined in such a way that the installation operates as much as possible within its optimal efficiency range, since, as with grid inverters, MPT charge controller efficiency is partly determined by the DC voltage used. MPT charge controllers are available from vendors such as Phocos and Outback Power Systems. To attain higher output, the solar generator can be subdivided if necessary into a series of arrays, each of which integrates an MPT charge controller.

8.3.5 Sizing Exercises for Stand-alone Installations

8.3.5.1 Sizing Exercises Involving the Simplified Insolation Calculation as in Section 2.4

Problem 1 (Weekend house near Biel that is used from March to October)

Non-maximum power tracker installation: reference station, Kloten; solar generator tilt angle $\beta = 45°$; solar generator azimuth $\gamma = 0°$ (S); system voltage 12 V; solar module used, Siemens M55; $P_{Mo} = 55$ W; $I_{Mo} = 3.15$ A; k_G as in Table 8.3 for long-term yield projections; autonomy period $n_A = 2$ days; system recovery time $n_E = 5$ days; maximum battery depth of discharge $t_Z = 50\%$; battery Ah efficiency $\eta_{Ah} = 90\%$; no inverter.

Connected appliances (all figures are per day):

#	Number	Mode	Designation	Energy need E	Power P	Time t
1	1	DC	Charge controller		0.5 W	24 h
2	3	DC	DC PLC lamps (11 W each)			2 h
3	1	DC	DC FL lamp (15 W each)			4 h
4	1	DC	DC TV set		36 W	2 h
5	1	DC	DC radio		4 W	3 h
6	1	DC	DC refrigerator	480 Wh		
7	Several	DC	Other small DC consumers	18 Wh		

Determine the following:

(a) Required battery capacity K.
(b) Required number of Siemens M55 and Isofoton I-55 modules.
(c) Number of solar modules needed if this installation were on Weissfluhjoch Mountain rather than near Biel.

Solution:

(a) K (and Q_L) are determined using Table 8.9 to establish the energy balance as follows:

Energy balance sheet for PV stand-alone plants (design of PV generator on separate sheet)

Consumers (switched)	AC/DC	P[W]	t [h]	E_{AC}	η_{WR}	E_{DC}	
1　3 PLC-lamps 11 W each	DC	33	2			66	Wh/d
2　1 FL-lamp 15 W	DC	15	4			60	Wh/d
3　TV set	DC	36	2			72	Wh/d
4　Radio	DC	4	3			12	Wh/d
5　Refrigerator	DC					480	Wh/d
6　Small consumers	DC					18	Wh/d
7							Wh/d
8							Wh/d
	$E_V = \Sigma E_{DC}$ switched cons. =					708	Wh/d

Location:	Biel	Month:	3. - 10.
System voltage V_S =			12	V
Accumulator cycle depth t_z =			0.5	
Ah-efficiency η_{Ah} =			0.9	
Autonomy days n_A =			2	d
System recovery time n_E =			5	d
Frequency of use h_B =			0.2857	

Permanent consumers	AC/DC	P[W]	t [h]	E_{AC}	η_{WR}	E_{DC}	
1　Charge controller	DC	0.5	24			12	Wh/d
2							Wh/d
3							
4							Wh/d
	$E_0 = \Sigma E_{DC}$ perman. cons. =					12	Wh/d

Note: This sheet with average daily energy consumption is to be completed in most cases only for the month with the highest consumption. If energy consumption varies considerably over the year, it may be neccessary to do use several sheets.

Usable accumulator capacity $K_N = n_A \cdot (E_V+E_0)/V_S$ =	120	Ah
Minimum accu capacity $K = K_N/t_z = n_A \cdot (E_V+E_0)/(V_S \cdot t_z)$ =	240	Ah
Necessary daily charging $Q_L = (1/\eta_{Ah}) \cdot (Q_0 + K_N/n_E)$ =	46.51	Ah/d

| Mean daily $E_D = h_B \cdot E_V + E_0$ = | 214.3 | Wh/d |
| Mean daily $Q_D = E_D/V_S$ = | 17.86 | Ah/d |

Hence the minimum required battery capacity K is 240 Ah for this installation, which is used for only eight months a year.

Note: The energy balance calculation procedure presupposes that, since the refrigerator is far and away the house's largest appliance, it will only be run on the weekends and will be shut off the rest of the time. If not, the refrigerator would have to be classified as a continuously operating appliance, which would make both K and Q_L considerably higher.

(b) Solar generator sizing using Table 8.10:

Layout of PV generator for stand alone plants (simplified calculation of irradiation)

Location: Biel		Reference station: Kloten			β =	45	γ =	0	°
I_{Mo} =	3.15 A	P_{Mo} =	55	W	System voltage V_S =		12	V	Modules/string n_{MS} =	1

Month	Jan	Feb	Mar	April	May	June	July	Aug	Sept	Oct	Nov	Dec	
Req. daily charge Q_L	0	0	46.51	46.51	46.51	46.51	46.51	46.51	46.51	46.51	0	0	Ah/d
Auxillary energy E_H	0	0	0	0	0	0	0	0	0	0	0	0	Wh/d
$Q_H = E_H /(1.1 \cdot V_S)$	0	0	0	0	0	0	0	0	0	0	0	0	Ah/d
$Q_{PV} = Q_L - Q_H$ =	0.0	0.0	46.5	46.5	46.5	46.5	46.5	46.5	46.5	46.5	0.0	0.0	Ah/d

Month	Jan	Feb	Mar	April	May	June	July	Aug	Sept	Oct	Nov	Dec	
H	0.91	1.65	2.67	3.73	4.65	5.21	5.68	4.78	3.42	2.03	0.99	0.73	kWh/m²
$R(\beta,\gamma)$	1.42	1.33	1.17	1.03	0.94	0.91	0.93	1.00	1.13	1.24	1.31	1.45	
$H_G = R(\beta,\gamma) \cdot H$	1.29	2.19	3.12	3.84	4.37	4.74	5.28	4.78	3.86	2.52	1.30	1.06	kWh/m²
$Y_R = H_G/1kWm^{-2}$ =	1.29	2.19	3.12	3.84	4.37	4.74	5.28	4.78	3.86	2.52	1.30	1.06	h/d
Without MPT-charge-controller (insert k_G-values in line below only in this case):													
k_G	0.80	0.83	0.84	0.85	0.86	0.86	0.86	0.86	0.86	0.84	0.82	0.77	
$Q_S = k_G \cdot I_{Mo} \cdot Y_R$ =	3.26	5.74	8.27	10.29	11.84	12.84	14.31	12.95	10.47	6.66	3.35	2.57	Ah/d
With MPT-charge charge-controller (insert k_T- and k_G-values below only in this case):													
k_T													
$Y_T = k_T \cdot Y_R$													h/d
k_G													
$Y_A = k_G \cdot Y_T$													h/d
η_{MPT}													
$Y_F' = \eta_{MPT} \cdot Y_A$ =													h/d
$E_{DC-S} = n_{MS} \cdot P_{Mo} \cdot Y_F'$ =													Wh/d
$Q_S = E_{DC-S}/(1.1 \cdot V_S)$													Ah/d

$n_{SP}' = Q_{PV}/Q_S$	0	0	5.63	4.52	3.93	3.62	3.25	3.59	4.44	6.98	0	0	
Necessary number of parallel strings: Maximum(n_{SP}'), rounded up to the next integer :											n_{SP} =	7	
Total necessary number of modules:											$n_M = n_{MS} \cdot n_{SP}$ =	7	

In this case, October is the critical month with the lowest insolation.

As $n_{MS} = 1$, then $n_M = n_{SP} = 7$ M55 modules are needed.

To determine the number of modules needed on the Weissfluhjoch as in task (c) above, the energy balance as in (a) can be used. However, an additional Table 8.10 needs to be completed for the insolation at this location.

(c) Same installation as on the Weissfluhjoch. The reference station for this installation would be Davos as the Weissfluhjoch is in the Alps.

Layout of PV generator for stand alone plants (simplified calculation of irradiation)

Location: Weissfluhjoch				Reference station: Davos					$\beta =$	45	$\gamma =$	0	°
$I_{Mo} =$	3.15 A	$P_{Mo} =$	55	W	System voltage $V_S =$			12	V	Modules/string $n_{MS} =$			1

Month	Jan	Feb	Mar	April	May	June	July	Aug	Sept	Oct	Nov	Dec	
Req. daily charge Q_L	0	0	46.5	46.5	46.5	46.5	46.5	46.5	46.5	46.5	0	0	Ah/d
Auxillary energy E_H	0	0	0	0	0	0	0	0	0	0	0	0	Wh/d
$Q_H = E_H/(1.1 \cdot V_S)$	0	0	0	0	0	0	0	0	0	0	0	0	Ah/d
$Q_{PV} = Q_L - Q_H =$	0	0	46.5	46.5	46.5	46.5	46.5	46.5	46.5	46.5	0	0	Ah/d

Month	Jan	Feb	Mar	April	May	June	July	Aug	Sept	Oct	Nov	Dec	
H	1.87	2.91	4.32	5.59	6.05	5.73	5.56	4.78	4.01	3.02	2.04	1.59	kWh/m²
$R(\beta,\gamma)$	1.90	1.59	1.33	1.10	0.95	0.91	0.92	1.00	1.15	1.39	1.69	1.98	
$H_G = R(\beta,\gamma) \cdot H$	3.55	4.63	5.75	6.15	5.75	5.21	5.12	4.78	4.61	4.20	3.45	3.15	kWh/m²
$Y_R = H_G/1kWm^2 =$	3.55	4.63	5.75	6.15	5.75	5.21	5.12	4.78	4.61	4.20	3.45	3.15	h/d
Without MPT-charge-controller *(insert k_G-values in line below only in this case)*:													
k_G		0.80	0.83	0.84	0.85	0.86	0.86	0.86	0.86	0.86	0.84	0.82	0.77
$Q_S = k_G \cdot I_{Mo} \cdot Y_R =$		8.95	12.1	15.2	16.5	15.6	14.1	13.9	12.9	12.5	11.1	8.91	7.64	Ah/d
With MPT-charge charge-controller *(insert k_T- and k_G-values below only in this case)*:													
k_T													
$Y_T = k_T \cdot Y_R$													h/d
k_G													
$Y_A = k_G \cdot Y_T$													h/d
η_{MPT}													
$Y_F' = \eta_{MPT} \cdot Y_A =$													h/d
$E_{DC-S} = n_{MS} \cdot P_{Mo} \cdot Y_F' =$													Wh/d
$Q_S = E_{DC-S}/(1.1 \cdot V_S)$													Ah/d

$n_{SP}' = Q_{PV}/Q_S$	0	0	3.06	2.82	2.99	3.29	3.36	3.59	3.72	4.19	0	0	

Necessary number of parallel strings: Maximum(n_{SP}'), rounded up to the next integer :	$n_{SP} =$	5
Total necessary number of modules:	$n_M = n_{MS} \cdot n_{SP} =$	5

In this case as well, the critical month with the lowest insolation is October, although insolation during this month on the Weissfluhjoch is considerably higher than in Biel. Hence only five (in lieu of seven) M55 modules are needed for this installation.

Note: The Excel tables in Problems 1 and 2 automatically compute all monthly values once the weather data have been entered. If these calculations were performed on a calculator, it would suffice to determine the relevant values for the critical months – namely, March and October for Problem 1 and November, December and January for Problem 2.

Problem 2 (Electricity supply for a zero-energy house in Burgdorf that is occupied all year round)

Roof installation with MPT charge controllers: $\eta_{MPT}=97\%$; optimal DC voltage around 130 V; reference station, Kloten; solar generator tilt angle $\beta=60°$; solar generator azimuth $\gamma=0°$ (S); system voltage 48 V; solar module characteristics: BP3160 (72 series-connected cells, $P_{Mo}=160\,W$, $I_{Mo}=4.55\,A$); k_T as in Table 8.2; k_G as in Table 8.3 for long-term yield projection; autonomy period $n_A=8$ days; system recovery time $n_E=30$ days; maximum battery depth of discharge $t_Z=80\%$; battery Ah efficiency $\eta_{Ah}=90\%$; inverter characteristics: open-circuit loss 12 W (DC, continuous); mean efficiency $\eta_{WR}=93\%$.

Connected appliances (all figures are per day):

#	Number	Mode	Designation	Energy need E	Power P	Time t
1	1	DC	Charge controller		1 W	24 h
2	6	AC	PLC lamps (11T/S I W each)			4 h
3	1	AC	TV set		60 W	2.5 h
4	1	AC	Cooking stove	1.9 kWh		
5	1	AC	Washing machine	0.9 kWh		
6	1	AC	Vacuum cleaner		800 W	0.2 h
7	1	AC	Refrigerator	450 Wh		
8	Several	AC	Other small consumers	250 Wh		

Determine the following:

(a) Required battery capacity K.
(b) Requisite number of modules n_M for use of BP3160 solar modules.
(c) Number of modules n_M needed if this installation were in St Moritz rather than Burgdorf.
(d) K and n_M if this installation were realized as a hybrid installation with the following characteristics: $n_A=4$ days; $n_E=\infty$ with an auxiliary gas or diesel generator; during the lowest-insolation period the backup generator produces 50% of the installation's energy.

Solution:

(a) K (and Q_L) are determined using Table 8.9 to establish the energy balance as follows:

Energy balance sheet for PV stand-alone plants (layout of PV generator on separate sheet)

Consumers (switched)	AC/DC	P[W]	t [h]	E_{AC}	η_{WR}	E_{DC}		Location: Burgdorf	Month:	
1 6 PLC-lamps 11 W each	AC	66	4	264	0.93	284	Wh/d			
2 TV set	AC	60	2.5	150	0.93	161	Wh/d	System voltage V_S =		48	V
3 Cooking stove	AC			1900	0.93	2043	Wh/d	Accumulator cycle depth t_Z =		0.8	
4 Washing machine	AC			900	0.93	968	Wh/d	Ah-efficiency η_{Ah} =		0.9	
5 Vacuum cleaner	AC	800	0.2	160	0.93	172	Wh/d			
6 Refrigerator	AC			450	0.93	484	Wh/d	Autonomy days n_A =		8	d
7 Small consumers	AC			250	0.93	269	Wh/d	System recovery time n_E =		30	d
8							Wh/d			
	$E_V = \Sigma\ E_{DC}$ switched cons. =					4381	Wh/d	Frequency of use h_B =		1	

Permanent consumers	AC/DC	P[W]	t [h]	E_{AC}	η_{WR}	E_{DC}		Note:
1 Charge controller	DC	1	24			24	Wh/d	This sheet with average daily energy
2 Inverter no-load losses	DC	12	24			288	Wh/d	consumption is to be completed in most
3								cases only for the month with the highest
4							Wh/d	consumption. If energy consumption
	$E_0 = \Sigma\ E_{DC}$ perman. cons. =					312	Wh/d	varies considerably over the year, it may

be neccessary to do use several sheets.

Usable accumulator capacity $K_N = n_A \cdot (E_V+E_0)/V_S$ =	782.1	Ah
Minimum accu capacity $K = K_N/t_Z = n_A \cdot (E_V+E_0)/(V_S \cdot t_Z)$ =	977.6	Ah
Necessary daily charging $Q_L = (1/\eta_{Ah}) \cdot (Q_D+K_N/n_E)$ =	137.6	Ah/d

Mean daily $E_D = h_B \cdot E_V + E_0$ =	4693	Wh/d
Mean daily $Q_D = E_D/V_S$ =	97.76	Ah/d

Hence the minimum required battery capacity K for operation of this year-round installation would be 978 Ah.

(b) Solar generator sizing using Table 8.10: As the optimal DC operating voltage of the envisaged MPT charge controller is around 130 V, each string will exhibit $n_{MS} = 4$ series-connected modules.

Layout of PV generator for stand alone plants (simplified calculation of irradiation)

Location: Burgdorf					Reference station: Kloten				$\beta =$ 60	$\gamma =$ 0	°	
$I_{Mo} =$ 4.55 A	$P_{Mo} =$	160 W		System voltage $V_S =$			48 V		Modules/string $n_{MS} =$		4	

Month	Jan	Feb	Mar	April	May	June	July	Aug	Sept	Oct	Nov	Dec	
Req. daily charge Q_L	137.6	137.6	137.6	137.6	137.6	137.6	137.6	137.6	137.6	137.6	137.6	137.6	Ah/d
Auxillary energy E_H	0	0	0	0	0	0	0	0	0	0	0	0	Wh/d
$Q_H = E_H/(1.1 \cdot V_S)$	0	0	0	0	0	0	0	0	0	0	0	0	Ah/d
$Q_{PV} = Q_L - Q_H =$	137.6	137.6	137.6	137.6	137.6	137.6	137.6	137.6	137.6	137.6	137.6	137.6	Ah/d

Month	Jan	Feb	Mar	April	May	June	July	Aug	Sept	Oct	Nov	Dec	
H	0.99	1.69	2.74	3.68	4.59	5.09	5.59	4.75	3.41	1.94	1.04	0.77	kWh/m²
$R(\beta,\gamma)$	1.47	1.33	1.12	0.95	0.84	0.80	0.82	0.91	1.06	1.21	1.31	1.48	
$H_G = R(\beta,\gamma) \cdot H$	1.46	2.25	3.07	3.50	3.86	4.07	4.58	4.32	3.61	2.35	1.36	1.14	kWh/m²
$Y_R = H_G/1kWm^{-2} =$	1.46	2.25	3.07	3.50	3.86	4.07	4.58	4.32	3.61	2.35	1.36	1.14	h/d

Without MPT-charge-controller *(insert k_G-values in line below only in this case)*:

	Jan	Feb	Mar	April	May	June	July	Aug	Sept	Oct	Nov	Dec	
k_G													
$Q_S = k_G \cdot I_{Mo} \cdot Y_R =$													Ah/d

With MPT-charge charge-controller *(insert k_T- and k_G-values below only in this case)*:

	Jan	Feb	Mar	April	May	June	July	Aug	Sept	Oct	Nov	Dec	
k_T	1.05	1.03	1.01	0.98	0.95	0.93	0.91	0.92	0.95	0.99	1.04	1.05	
$Y_T = k_T \cdot Y_R$	1.53	2.32	3.10	3.43	3.66	3.79	4.17	3.98	3.43	2.32	1.42	1.20	h/d
k_G	0.84	0.85	0.86	0.86	0.85	0.85	0.85	0.85	0.86	0.86	0.85	0.84	
$Y_A = k_G \cdot Y_T$	1.28	1.97	2.67	2.95	3.11	3.22	3.55	3.38	2.95	2.00	1.20	1.01	h/d
η_{MPT}	0.97	0.97	0.97	0.97	0.97	0.97	0.97	0.97	0.97	0.97	0.97	0.97	
$Y_F' = \eta_{MPT} \cdot Y_A =$	1.25	1.91	2.59	2.86	3.02	3.12	3.44	3.28	2.86	1.94	1.17	0.97	h/d
$E_{DC-S} = n_{MS} \cdot P_{Mo} \cdot Y_F' =$	796.8	1222	1655	1829	1933	1998	2201	2098	1833	1241	747.7	624	Wh/d
$Q_S = E_{DC-S}/(1.1 \cdot V_S)$	15.09	23.14	31.34	34.64	36.61	37.85	41.69	39.74	34.72	23.5	14.16	11.82	Ah/d

$n_{SP}' = Q_{PV}/Q_S$	9.12	5.95	4.39	3.97	3.76	3.64	3.30	3.46	3.96	5.86	9.72	11.6	

Necessary number of parallel strings: Maximum(n_{SP}'), rounded up to the next integer :	$n_{SP} =$	12
Total necessary number of modules:	$n_M = n_{MS} \cdot n_{SP} =$	48

Note: December is the critical (lowest-insolation) month for this installation, whose solar generator would be sized as in Problem 1 if the installation had no MPT charge controller. In this case, 32 strings and 64 BP3160 modules would be needed. An MPT charge controller would allow for cost savings on the solar generator side of the installation, but would at the same time render the installation more complex and would probably result in a high-frequency interference voltage on the installation's DC side.

The solar generator could be far smaller and still produce the same amount of energy if the installation were at a higher-insolation site (see task (c)).

(c) Same installation in St Moritz: The reference station for this installation would be Davos as St Moritz is in the Alps.

Layout of PV generator for stand alone plants (simplified calculation of irradiation)

Location: St. Moritz				Reference station: Davos				$\beta =$	60	$\gamma =$	0	°
$I_{Mo} =$	4.55 A	$P_{Mo} =$	160 W	System voltage $V_S =$			48 V	Modules / string $n_{MS} =$				4

Month	Jan	Feb	Mar	April	May	June	July	Aug	Sept	Oct	Nov	Dec	
Req. daily charge Q_L	137.6	137.6	137.6	137.6	137.6	137.6	137.6	137.6	137.6	137.6	137.6	137.6	Ah/d
Auxillary energy E_H	0	0	0	0	0	0	0	0	0	0	0	0	Wh/d
$Q_H = E_H/(1.1 \cdot V_S)$	0	0	0	0	0	0	0	0	0	0	0	0	Ah/d
$Q_{PV} = Q_L - Q_H =$	137.6	137.6	137.6	137.6	137.6	137.6	137.6	137.6	137.6	137.6	137.6	137.6	Ah/d

Month	Jan	Feb	Mar	April	May	June	July	Aug	Sept	Oct	Nov	Dec	
H	1.74	2.70	4.06	5.29	5.75	6.01	6.16	5.23	4.16	2.86	1.83	1.45	kWh/m²
$R(\beta,\gamma)$	2.01	1.66	1.32	1.04	0.85	0.80	0.82	0.91	1.09	1.39	1.78	2.13	
$H_G = R(\beta,\gamma) \cdot H$	3.50	4.48	5.36	5.50	4.89	4.81	5.05	4.76	4.53	3.98	3.26	3.09	kWh/m²
$Y_R = H_G/1kWm^{-2} =$	3.50	4.48	5.36	5.50	4.89	4.81	5.05	4.76	4.53	3.98	3.26	3.09	h/d

Without MPT-charge-controller *(insert k_G-values in line below only in this case)*:

k_G													
$Q_S = k_G \cdot I_{Mo} \cdot Y_R =$													Ah/d

With MPT-charge-controller *(insert k_T- and k_G-values below only in this case)*:

	Jan	Feb	Mar	April	May	June	July	Aug	Sept	Oct	Nov	Dec	
k_T	1.03	1.02	1.00	0.98	0.97	0.96	0.94	0.95	0.97	0.98	1.02	1.03	
$Y_T = k_T \cdot Y_R$	3.60	4.57	5.36	5.39	4.74	4.62	4.75	4.52	4.40	3.90	3.32	3.18	h/d
k_G	0.84	0.85	0.86	0.86	0.85	0.85	0.85	0.85	0.86	0.86	0.85	0.84	
$Y_A = k_G \cdot Y_T$	3.03	3.89	4.61	4.64	4.03	3.92	4.04	3.84	3.78	3.35	2.82	2.67	h/d
η_{MPT}	0.97	0.97	0.97	0.97	0.97	0.97	0.97	0.97	0.97	0.97	0.97	0.97	
$Y_F' = \eta_{MPT} \cdot Y_A =$	2.94	3.77	4.47	4.50	3.91	3.81	3.91	3.73	3.67	3.25	2.74	2.59	h/d
$E_{DC-S} = n_{MS} \cdot P_{Mo} \cdot Y_F' =$	1879	2412	2861	2878	2502	2436	2505	2386	2348	2080	1753	1659	Wh/d
$Q_S = E_{DC-S}/(1.1 \cdot V_S)$	35.58	45.69	54.19	54.52	47.38	46.13	47.45	45.19	44.47	39.39	33.21	31.42	Ah/d

$n_{SP}' = Q_{PV}/Q_S$	3.87	3.01	2.54	2.52	2.90	2.98	2.90	3.05	3.09	3.49	4.14	4.38	

Necessary number of parallel strings: Maximum(n_{SP}'), rounded up to the next integer :											$n_{SP} =$	5
Total necessary number of modules:											$n_M = n_{MS} \cdot n_{SP} =$	20

As the installation's appliance side is the same as for the Burgdorf installation, the latter's energy balance as in the table in (a) can be used.

However, 60% fewer solar modules are needed for the St Moritz installation than for the Burgdorf installation on account of the far higher insolation during the winter in St Moritz.

(d) Modified energy balance for the same installation, but with a hybrid solar generator: Realization of a hybrid installation can considerably cut costs in locations with widely fluctuating insolation. Although the actual energy consumption remains unchanged, a new energy balance needs to be drawn up for a hybrid installation since part of its wintertime energy is derived from a backup power source such as a gas generator; thus a proportionally smaller solar generator can be used for the remainder of the requisite energy (see next table). Using the solar generator values derived from the conditions that obtain during the lowest-insolation months, the power yield can be determined for the remaining months, as can the requisite amount of backup-generator output – which must of course be available at all times during the low-insolation winter months of November to January.

Energy balance sheet for PV stand-alone plants (layout of PV generator on separate sheet)

	Consumers (switched)	AC/DC	P[W]	t [h]	E_{AC}	η_{WR}	E_{DC}	
1	6 PLC-lamps 11 W each	AC	66	4	264	0.93	283.9	Wh/d
2	TV set	AC	60	2.5	150	0.93	161.3	Wh/d
3	Cooking stove	AC			1900	0.93	2043	Wh/d
4	Washing machine	AC			900	0.93	968	Wh/d
5	Vacuum cleaner	AC	800	0.2	160	0.93	172	Wh/d
6	Refrigerator	AC			450	0.93	484	Wh/d
7	Small consumers	AC			250	0.93	269	Wh/d
8								Wh/d
	$E_V = \Sigma E_{DC}$ switched cons. =						4381	Wh/d

Location: Burgdorf	Month:	
System voltage V_S =	48	V
Accumulator cycle depth t_z =	0.8	
Ah-efficiency η_{Ah} =	0.9	
Autonomy days n_A =	4	d
System recovery time n_E =	9999	d
Frequency of use h_B =	1	

	Permanent consumers	AC/DC	P[W]	t [h]	E_{AC}	η_{WR}	E_{DC}	
1	Charge controller	DC	1	24			24	Wh/d
2	Inverter no-load losses	DC	12	24			288	Wh/d
3								Wh/d
4								Wh/d
	$E_0 = \Sigma E_{DC}$ perman. cons. =						312	Wh/d

Note:
This sheet with average daily energy consumption is to be completed in most cases only for the month with the highest consumption. If energy consumption varies considerably over the year, it may be neccessary to do use several sheets.

Usable accumulator (accu) capacity $K_N = n_A \cdot (E_V + E_0)/V_S$ =	391.1	Ah
Minimum accu capacity $K = K_N/t_z = n_A \cdot (E_V + E_0)/(V_S \cdot t_z)$ =	488.8	Ah
Necessary daily charging $Q_L = (1/\eta_{Ah}) \cdot (Q_D + K_N/n_E)$ =	108.7	Ah/d

Mean daily $E_D = h_B \cdot E_V + E_0$ =	4693	Wh/d
Mean daily $Q_D = E_D/V_S$ =	97.76	Ah/d

Layout of PV generator for stand alone plants (simplified calculation of irradiation)

Location: Burgdorf			Reference station: Kloten				β =	60	γ =	0	°

I_{Mo} =	4.55	A	P_{Mo} =	160	W	System voltage V_S =		48	V	Modules / string n_{MS} =		4

Month	Jan	Feb	Mar	April	May	June	July	Aug	Sept	Oct	Nov	Dec	
Req. daily charge Q_L	108.7	108.7	108.7	108.7	108.7	108.7	108.7	108.7	108.7	108.7	108.7	108.7	Ah/d
Auxillary energy E_H	1760	0	0	0	0	0	0	0	0	0	2010	2870	Wh/d
$Q_H = E_H/(1.1 \cdot V_S)$	33.3	0	0	0	0	0	0	0	0	0	38.1	54.4	Ah/d
$Q_{PV} = Q_L - Q_H$ =	75.3	109	109	109	109	109	109	109	109	109	70.6	54.3	Ah/d

Month	Jan	Feb	Mar	April	May	June	July	Aug	Sept	Oct	Nov	Dec	
H	0.99	1.69	2.74	3.68	4.59	5.09	5.59	4.75	3.41	1.94	1.04	0.77	kWh/m²
$R(\beta,\gamma)$	1.47	1.33	1.12	0.95	0.84	0.80	0.82	0.91	1.06	1.21	1.31	1.48	
$H_G = R(\beta,\gamma) \cdot H$	1.46	2.25	3.07	3.50	3.86	4.07	4.58	4.32	3.61	2.35	1.36	1.14	kWh/m²
$Y_R = H_G/1kWm^{-2}$ =	1.46	2.25	3.07	3.50	3.86	4.07	4.58	4.32	3.61	2.35	1.36	1.14	h/d

Without MPT-charge-controller (insert k_G-values in line below only in this case):

	Jan	Feb	Mar	April	May	June	July	Aug	Sept	Oct	Nov	Dec	
k_G													
$Q_S = k_G \cdot I_{Mo} \cdot Y_R$ =													Ah/d

With MPT-charge-controller (insert k_T- and k_G-values below only in this case):

	Jan	Feb	Mar	April	May	June	July	Aug	Sept	Oct	Nov	Dec	
k_T	1.05	1.03	1.01	0.98	0.95	0.93	0.91	0.92	0.95	0.99	1.04	1.05	
$Y_T = k_T \cdot Y_R$	1.53	2.32	3.10	3.43	3.66	3.79	4.17	3.98	3.43	2.32	1.42	1.20	h/d
k_G	0.84	0.85	0.86	0.86	0.85	0.85	0.85	0.85	0.86	0.86	0.85	0.84	
$Y_A = k_G \cdot Y_T$	1.28	1.97	2.67	2.95	3.11	3.22	3.55	3.38	2.95	2.00	1.20	1.01	h/d
η_{MPT}	0.97	0.97	0.97	0.97	0.97	0.97	0.97	0.97	0.97	0.97	0.97	0.97	
$Y_F' = \eta_{MPT} \cdot Y_A$ =	1.25	1.91	2.59	2.86	3.02	3.12	3.44	3.28	2.86	1.94	1.17	0.97	h/d
$E_{DC-S} = n_{MS} \cdot P_{Mo} \cdot Y_F'$ =	797	1222	1655	1829	1933	1998	2201	2098	1833	1241	748	624	Wh/d
$Q_S = E_{DC-S}/(1.1 \cdot V_S)$	15.1	23.1	31.3	34.6	36.6	37.8	41.7	39.7	34.7	23.5	14.2	11.8	Ah/d

$n_{SP}' = Q_{PV}/Q_S$	4.99	4.70	3.47	3.14	2.97	2.87	2.61	2.73	3.13	4.63	4.99	4.60	

Necessary number of parallel strings: Maximum(n_{SP}'), rounded up to the next integer :	n_{SP} =	5
Total necessary number of modules:	$n_M = n_{MS} \cdot n_{SP}$ =	20

8.3.5.2   Sizing Exercise Using the Three-Component Model Insolation Calculation Procedure as in Section 2.5

Problem 1: Non-MPT charge controller PV installation in Tamanrasset, Algeria: $\varphi \approx 23°N$; reference station Cairo, Egypt; solar generator tilt angle $\beta = 30°$; solar generator azimuth $\gamma = 0°$ (S); $\alpha_1 = \alpha_2 = 0°$; values interpolated between 22°N and 24°N to make up for the lack of 23°N data in R_B (Table A6.4); year-round reflection factor $\rho = 0.35$; system voltage 24 V; solar module used, KC130GHT-2 (36 series-connected cells, $P_{Mo} = 130$ W, $I_{Mo} = 7.4$ A); k_G for long-term yield projections as in Table 8.5; autonomy period $n_A = 4$ days; system recovery time $n_E = 20$ days; maximum battery depth of discharge $t_Z = 80\%$; battery Ah efficiency $\eta_{Ah} = 90\%$; inverter characteristics: open-circuit loss 10 W (DC, continuous); mean efficiency $\eta_{WR} = 93\%$.

Connected appliances (all figures are per day):

#	Number	Mode	Designation	Energy need E	Power P	Time t
1	5	AC	PLC lamps (11 W each)			3 h
2	1	AC	FL lamp		36 W	4 h
3	1	AC	TV set		60 W	3.5 h
4	1	AC	Vacuum cleaner		800 W	0.2 h
5	1	AC	Radio		5 W	6 h
6	1	AC	Refrigerator	500 Wh		
7	Several	AC	Other small consumers	320 Wh		
8	1	DC	Charge controller		1 W	24 h

Determine the following:

(a) Required battery capacity K.
(b) Requisite number of modules n_M for use of KC130GHT-2 solar modules.

Solution:

(a) K (and Q_L) are determined using Table 8.9 to establish the energy balance as follows:

Energy balance sheet for PV stand-alone plants (layout of PV generator on separate sheet)

Consumers (switched)	AC/DC	P[W]	t [h]	E_{AC}	η_{WR}	E_{DC}				
1	5 PLC-lamps 11 W each	AC	55	3	165	0.93	177	Wh/d	Location: Tamanrasset	Month:
2	1 FL-lamp	AC	36	4	144	0.93	155	Wh/d		
3	TV set	AC	60	3.5	210	0.93	226	Wh/d	System voltage V_S =	24	V
4	Vacuum cleaner	AC	800	0.2	160	0.93	172	Wh/d	Accumulator cycle depth t_Z =	0.8	
5	Radio	AC	5	6	30	0.93	32	Wh/d	Ah-efficiency η_{Ah} =	0.9	
6	Refrigerator	AC			500	0.93	538	Wh/d		
7	Small consumers	AC			320	0.93	344	Wh/d	Autonomy days n_A =	4	d
8								Wh/d	System recovery time n_E =	20	d
		$E_V = \Sigma E_{DC}$ switched cons. =					1644	Wh/d	Frequency of use h_B =	1	

Permanent consumers	AC/DC	P[W]	t [h]	E_{AC}	η_{WR}	E_{DC}		
1	Charge controller	DC	1	24			24	Wh/d
2	Inverter no-load losses	DC	10	24			240	Wh/d
3								Wh/d
4								Wh/d
	$E_0 = \Sigma E_{DC}$ perman. cons. =						264	Wh/d

Note: This sheet with average daily energy consumption is to be completed in most cases only for the month with the highest consumption. If energy consumption varies considerably over the year, it may be neccessary to do use several sheets.

| Usable accumulator capacity $K_N = n_A \cdot (E_V + E_0)/V_S$ = | 318.0 | Ah | Mean daily $E_D = h_B \cdot E_V + E_0$ = | 1908 | Wh/d |
| Minimum accu capacity $K = K_N/t_Z = n_A \cdot (E_V + E_0)/(V_S \cdot t_Z)$ = | 397.5 | Ah | Mean daily $Q_D = E_D/V_S$ = | 79.5 | Ah/d |

| Necessary daily charging $Q_L = (1/\eta_{Ah}) \cdot (Q_D + K_N/n_E)$ = | | 106.0 | Ah/d |

Hence a battery with a minimum capacity of $K \approx 400$ Ah is needed.

(b) Solar generator sizing using Table 8.11:

Layout of PV generator for stand alone plants (with 3-component model)

| Location: Tamanrasset | | Reference station: Cairo | | | $\beta =$ | 30 | $\gamma =$ | 0 | ° |

| | | $R_D = \frac{1}{2}\cos\alpha_2 + \frac{1}{2}\cos(\alpha_1+\beta) =$ | 0.933 | $R_R = \frac{1}{2} - \frac{1}{2}\cos\beta =$ | 0.067 |

| $I_{Mo} =$ | 7.4 | A | $P_{Mo} =$ | 130 | W | System voltage $V_S =$ | 24 | V | Modules / string $n_{MS} =$ | 2 |

Month	Jan	Feb	Mar	April	May	June	July	Aug	Sept	Oct	Nov	Dec	
Req. daily charge Q_L	106	106	106	106	106	106	106	106	106	106	106	106	Ah/d
Auxillary energy E_H	0	0	0	0	0	0	0	0	0	0	0	0	Wh/d
$Q_H = E_H/(1.1 \cdot V_S)$	0	0	0	0	0	0	0	0	0	0	0	0	Ah/d
$Q_{PV} = Q_L - Q_H =$	106	106	106	106	106	106	106	106	106	106	106	106	Ah/d

Month	Jan	Feb	Mar	April	May	June	July	Aug	Sept	Oct	Nov	Dec	
H	5.30	6.34	6.98	7.44	7.28	7.80	7.61	7.08	6.44	6.02	4.92	4.59	kWh/m²
H_D	1.08	1.13	1.50	1.78	2.09	1.94	1.97	2.00	1.87	1.51	1.40	1.23	kWh/m²
R_B	1.41	1.27	1.10	0.96	0.85	0.81	0.83	0.91	1.04	1.21	1.37	1.47	
k_B	1	1	1	1	1	1	1	1	1	1	1	1	
$H_{GB} = k_B \cdot R_B \cdot (H-H_D)$	5.95	6.61	6.04	5.45	4.42	4.76	4.69	4.64	4.77	5.46	4.83	4.94	kWh/m²
$H_{GD} = R_D \cdot H_D$	1.01	1.05	1.40	1.66	1.95	1.81	1.84	1.87	1.74	1.41	1.31	1.15	kWh/m²
ρ	0.35	0.35	0.35	0.35	0.35	0.35	0.35	0.35	0.35	0.35	0.35	0.35	
$H_{GR} = R_R \cdot \rho \cdot H$	0.12	0.15	0.16	0.17	0.17	0.18	0.18	0.17	0.15	0.14	0.12	0.11	kWh/m²
$H_G = H_{GB}+H_{GD}+H_{GR}$	7.08	7.81	7.60	7.28	6.54	6.75	6.71	6.68	6.66	7.01	6.26	6.20	kWh/m²
$Y_R = H_G/1kWm^{-2} =$	7.08	7.81	7.60	7.28	6.54	6.75	6.71	6.68	6.66	7.01	6.26	6.20	h/d

Without MPT-charge-controller *(insert k_G-values in line below only in this case):*

	Jan	Feb	Mar	April	May	June	July	Aug	Sept	Oct	Nov	Dec	
k_G	0.84	0.84	0.84	0.84	0.83	0.82	0.82	0.82	0.83	0.84	0.84	0.84	
$Q_S = k_G \cdot I_{Mo} \cdot Y_R =$	44.0	48.5	47.2	45.3	40.2	41.0	40.7	40.5	40.9	43.6	38.9	38.54	Ah/d

With MPT-charge-controller *(insert k_T- and k_G-values below only in this case):*

	Jan	Feb	Mar	April	May	June	July	Aug	Sept	Oct	Nov	Dec	
k_T													
$Y_T = k_T \cdot Y_R$													h/d
k_G													
$Y_A = k_G \cdot Y_T$													h/d
η_{MPT}													
$Y_F' = \eta_{MPT} \cdot Y_A =$													h/d
$E_{DC-S} = n_{MS} \cdot P_{Mo} \cdot Y_F' =$													Wh/d
$Q_S = E_{DC-S}/(1.1 \cdot V_S)$													Ah/d

| $n_{SP}' = Q_{PV}/Q_S$ | 2.41 | 2.18 | 2.24 | 2.34 | 2.64 | 2.59 | 2.60 | 2.62 | 2.59 | 2.43 | 2.72 | 2.75 | |

| Necessary number of parallel strings: Maximum(n_{SP}'), rounded up to the next integer : | | | | $n_{SP} =$ | 3 |
| Total necessary number of modules: | | | | $n_M = n_{MS} \cdot n_{SP} =$ | 6 |

Only six KC130GHT-2 modules are needed at this high-insolation site, despite the elevated energy consumption that comes into play.

8.4 Insolation Calculation Freeware

Energy yield can be calculated using readily available freeware that provides insolation data for both the horizontal and inclined plane and which (after a bit of practice) allows for the generation of mean monthly and annual values in Wh/m²/d for many locations. By dividing these values by a factor of 1000, the counterpart figures in kWh/m²/d can then be obtained and plugged into the tables discussed in Sections 8.2 and 8.3. This saves a great deal of calculation time.

The figures thus obtained may differ slightly from those in the Appendix tables, since the baseline data used by the application (Meteonorm 4.0) that the latter figures are based on and the freeware available online differ and also employ different models. To be on the safe side, however, the data in this book should be used to size European PV installations, since these data tend to underestimate insolation, particularly for the winter months.

8.4.1   PVGIS Solar Irradiation Data

The European Commission Joint Research Centre (JRC) web site http://re.jrc.ec.europa.eu/pvgis allows for the interactive generation of the following insolation and related data for European, Middle East and African sites: insolation in the horizontal plane; insolation for southerly oriented solar generators in the northern hemisphere, and vice versa, with 15°, 25°, 40° and 90° tilt angles; and optimal tilt angles for annual yield. The PVGIS site also offers a tool for estimating (for Europe and environs) annual yield on solar generator surfaces with an optimal insolation orientation.

Maps containing annual global insolation tallies on the horizontal plane for all EU countries are available, as well as a few additional maps and links to other insolation databases.

8.4.2   The European Satel-Light Insolation Database

Satel-Light (http://www.satel-light.com) provides an interactive insolation database based on 1996–2000 satellite insolation data as measured at 30 min intervals by Meteosat. The groundwork for this database was laid under the aegis of an EU project from 1996 to 1998. This web site allows for the interactive generation of insolation data for locations in Europe, Morocco, Algeria and Tunisia for both the horizontal plane and a solar generator oriented in any direction. User-specified maps can also be realized.

To use Satel-Light it is necessary to register and provide an email address. The calculation results take a few minutes, following which the user is emailed a link showing where the requested data or map are available. Thus use of Satel-Light necessitates a bit more effort.

8.5   Simulation Software

Many applications of PV installation simulation software are commercially available, and a brief overview of many of them is available in [Mar03]. Most of these products cost anywhere from €400 to €1000, including the cost of obtaining a sufficient amount of weather data. With a little practice, these applications allow for detailed analytic assessments of various types of PV installation concepts during the planning phase and can take account of factors such as nearby and distant shading. However, some factors that have a substantial impact on PV installations are not taken into account, including substandard output of new solar modules, snow cover and gradual soiling accretions.

Unfortunately, the algorithms used in these programs are in the main disclosed allusively, if at all, and are thus virtually impossible for users to understand – which in effect means that users simply have to accept the simulation results on faith. Hence users of such applications would be well advised to acquire first a detailed understanding of the workings of PV installations, along the lines of the information contained in this book, otherwise the user will have no way of judging if the results of such simulations are valid or contain errors. It is also useful for users to compare their own software's interim or final results concerning factors such as insolation on the module plane with: (a) analogous findings from other sources, such as the insolation data mentioned in Section 8.4 or the results generated by other simulation programs; and/or (b) the results obtained using the calculation methods described in this book, as this is likely to allow for the detection of dubious simulation results.

The following is a partial listing of PV installation simulation applications that are available in Germany, Austria and Switzerland:

- PVSYST (www.pvsyst.com)
- PV*SOL (www.valentin.de)
- INSEL (www.inseldi.com).

Some of these programs integrate a certain amount of insolation data, while for others it is necessary to obtain such data elsewhere (e.g. from Meteonorm; see www.meteonorm.com) that provide insolation figures on the inclined plane at a sufficient number of locations to allow for PV installation simulations.

8.6 Bibliography

[8.1] A. Parretta, A. Sarno, R. Schioppo *et al.*: 'Analysis of Loss Mechanisms in Crystalline Silicon Modules in Outdoor Operation'. 14th EU PV Conference, Barcelona, 1997.

[8.2] H. Becker, W. Vaassen: 'Minderleistungen von Solargeneratoren aufgrund von Verschmutzungen'. 12. Symposium PV Solarenergie, Staffelstein, 1997.

[8.3] H. Häberlin, J. Graf: 'Gradual Reduction of PV Generator Yield due to Pollution'. Proceedings of the 2nd World Conference on Photovoltaic Energy Conversion, Vienna, 1998.

[8.4] H. Häberlin, Ch. Renken: 'Allmähliche Reduktion des Energieertrags von Photovoltaikanlagen durch permanente Verschmutzung'. 14. Symposium PV Solarenergie, Staffelstein, 1999.

[8.5] G. Wirth, M. Zehner, M. Homscheidt, G. Becker: 'Identifikation von Schneebedeckung auf Solaranlagen mit Hilfe von Satellitendaten'. 23. Symposium PV Solarenergie, Staffelstein, 2008.

[8.6] G. Wirth, M. Zehner, M. Homscheidt, G. Becker: 'Kartierung der Schneebedeckungsdauer zur Ertragsabschätzung und Auslegung von PV-Systemen'. 24. Symposium PV Solarenergie, Staffelstein, 2009.

Additional References

[Häb07], [Mar03], [Wag06].

S800-RSU. Remote switching unit for highest PV system availability.

The remote switching unit S800-RSU, winner of the Intersolar AWARD 2010 in photovoltaics, simplifies the control of PV systems by operating arrays or strings from remote. The combination of string protection switch S800PV-S and remote switching unit S800-RSU replaces the fuse and the switch disconnector, which eliminates a time- and cost-intensive replacement of the fuses. The remote switching units provide fastest switching performance at lowest power consumption. For more information please visit www.abb.com

ABB Switzerland Ltd
CMC Low Voltage Products
Fulachstrasse 150
CH-8201 Schaffhausen
Tel. +41 58 586 41 11
Fax +41 58 586 42 22
www.abb.ch

Power and productivity
for a better world™

9

The Economics of Solar Power

The solar modules used in PV installations entail major capital investment and fairly extensive use of energy resources. This chapter discusses the economics of such installations from both a financial and energy standpoint.

9.1 How Much Does Solar Energy Cost?

Inasmuch as PV installations are extremely capital intensive, the estimated cost of the electricity generated by them is largely determined by the interest rate and installation service suppositions on which such projections are based. Indeed, these projections can vary considerably for one and the same PV installation depending on which set of assumptions is applied. Moreover, in order to be valid such cost estimates need to specify the interest rate, depreciation period and other suppositions on which they are predicated.

It is safe to assume that the life span of solar modules and solar module wiring will range anywhere from 15 to 30 years, while the life span of inverters, charge controllers and other electronic components can be set at around 7 to 15 years. For stand-alone installations, battery replacement costs need to be figured in as the life span of these components is only around 2 to 10 years. Battery replacement engenders costs exceeding those of comparably sized grid-connected systems by anywhere from €0.33 to €0.67 per kilowatt hour; however, this applies solely to energy stored in the battery and not to non-stored energy that is used over the course of a day. The remaining upkeep costs mainly comprise servicing and repairs (particularly for inverters), insurance and in some cases periodic solar generator cleaning.

In order to obtain reasonably accurate projections of the cost of PV installation electricity, the total realization cost K_{TOT} for a complete PV installation should be broken down into the constituent costs K_G entailed by the following elements: solar generator; voltage surge protection and so on (15–30 years of depreciation); electronic components K_E such as inverters, charge controllers and the like (7–15 years of depreciation); and battery costs K_A (2–10 years of depreciation, depending on battery type and load). In addition, possible cost savings K_S entailed by the realization of a PV installation, such as roof tile costs for a roof-integrated installation, can be deducted from the realization costs K_G.

Hence the following equation applies to PV installation investment costs K_{TOT}:

$$\text{Total investment costs } K_{TOT} = (K_G - K_S) + K_E + K_A \qquad (9.1)$$

where:
K_G = costs attributable to the solar generator, site modification, wiring, voltage surge protection and so on

Photovoltaics: System Design and Practice. Heinrich Häberlin.
© 2012 John Wiley & Sons, Ltd. Published 2012 by John Wiley & Sons, Ltd.

Table 9.1 Requisite annual depreciation rates a, expressed as a percentage of invested capital, for depreciation in n years

	Amortization rate a versus interest rate and duration of amortization						
Duration n (years)	**Interest rate p**						
	2.5%	3%	4%	5%	6%	7%	8%
2	51.88%	52.26%	53.01%	53.78%	54.54%	55.31%	56.08%
3	35.01%	35.35%	36.03%	36.72%	37.41%	38.11%	38.80%
4	26.58%	26.90%	27.55%	28.20%	28.86%	29.52%	30.19%
5	21.52%	21.84%	22.46%	23.10%	23.74%	24.39%	25.05%
6	18.16%	18.46%	19.08%	19.70%	20.34%	20.98%	21.63%
7	15.75%	16.05%	16.66%	17.28%	17.91%	18.56%	19.21%
8	13.95%	14.25%	14.85%	15.47%	16.10%	16.75%	17.40%
10	11.43%	11.72%	12.33%	12.95%	13.59%	14.24%	14.90%
12	9.749%	10.05%	10.65%	11.28%	11.93%	12.59%	13.27%
15	8.077%	8.377%	8.994%	9.634%	10.30%	10.98%	11.68%
20	6.415%	6.722%	7.358%	8.024%	8.719%	9.439%	10.19%
25	5.428%	5.743%	6.401%	7.095%	7.823%	8.581%	9.368%
30	4.778%	5.102%	5.783%	6.505%	7.265%	8.059%	8.883%
35	4.321%	4.654%	5.358%	6.107%	6.897%	7.723%	8.580%
40	3.984%	4.326%	5.052%	5.828%	6.646%	7.501%	8.386%

K_S = cost savings attributable to roof tiles, façade cladding and so on; f or net solar generator costs $(K_G - K_S)$, 15–30 years of depreciation

K_E = costs arising from inverters, charge controllers and other electronic devices; mean depreciation period, 15–30 years

K_A = battery costs for stand-alone installations, 2–30 years of depreciation

Once the above figures have been set, annual costs k_J are derived from the depreciation rates a_i (as in Table 9.1) that apply to the various types of cost involved, and the annual installation operating and maintenance costs k_B are also determined:

$$\text{Annual costs } k_{J\ TOT} = (K_G - K_S) \cdot a_G + K_E \cdot a_E + K_A \cdot a_A + k_B \qquad (9.2)$$

where:

a_G = depreciation rate for net solar generator costs $(K_G - K_S)$ as in Table 9.1, 15–30 years of depreciation

a_E = electronic component depreciation rate as in Table 9.1, 7–15 years of depreciation

a_A = battery depreciation rate as in Table 9.1, 2–10 years of depreciation

k_B = annual operating, maintenance and repair costs (including insurance)

The effective energy price is obtained by dividing annual costs by the annual energy yield.

If arriving at an exact cost is too much effort, the guide values in Table 9.2 can be used instead, providing that great accuracy is not required.

Table 9.2 Guide values for PV electricity costs

Estimated prices for solar electricity as at 2009

For grid-connected systems, around €0.27 to €0.6 per kWh

Stored energy for stand-alone installations, around €0.85 to €1.55 per kWh

9.1.1   Examples of More Exact Energy Price Calculations

Example 1: 3.3 kWp Grid-Connected System at a Well-Situated Site in Mittelland, Switzerland (2009)

Total cost of the installation $K_{TOT}=$ €15 400 (equivalent). Depreciation interest rate $p=2.5\%$. The installation's depreciation period can be deemed to be 25 years, according to the KEV framework (compensatory feed-in remuneration for renewables; see Section 1.4.7), except for the inverters, which cost $K_E=$ €1700 and will depreciate in 10 years. The presumed annual operating and maintenance costs are €133.

At the well-situated site in this region, the installation will exhibit specific annual energy yield amounting to $Y_{Fa} \approx 1000$ kWh/kWp, based on solar generator peak power P_{Go} (see Section 1.4.4), which equates to an annual energy yield of 3300 kWh.

Annual costs k_J:

Examples 2010 (1 Euro = 1.5 CHF)

	Amortisation period	Investment cost K_i (Fr. / €)	Amortisation rate a_i	Annual cost K_J (CHF / €)
PV generator	25 years	20550 / 13700	5.43%	1115 / 744
Inverter	10 years	2550 / 1700	11.43%	291 / 194
Cost of operation				200 / 133
TOTAL		23100 / 15400		1606 / 1071

Energy price:
For annual yield of 1000 kWh/kWp: €0.32 per kWh.
For 1100 kWh/kWp (at a well-situated site):   €0.29 per kWh.
For 900 kWh/kWp (at a less well-situated site):   €0.36 per kWh.

Example 2: Installation Using Extremely Conservative Estimates for High-Interest-Rate Periods

This example concerns the same installation as for Example 1 (new 3.3 kWp grid-connected system in Mittelland, Switzerland). The presumed depreciation interest rate is $p=6\%$, and the very conservatively estimated life span is 15 and 7 years for the installation as a whole and its inverters respectively.

Annual costs k_J:

Examples 2010 (1 Euro = 1.5 CHF)

	Amortisation period	Investment cost K_i (Fr. / €)	Amortisation rate a_i	Annual cost K_J (CHF / €)
PV generator	15 years	20550 / 13700	10.30%	2117 / 1411
Inverter	7 years	2550 / 1700	17.91%	457 / 304
Cost of operation				200 / 133
TOTAL		23100 / 15400		2774 / 1848

Energy price:
For annual yield of 1000 kWh/kWp: €0.56 per kWh.
For 1100 kWh/kWp (at a well-situated site):   €0.51 per kWh.
For 900 kWh/kWp (at a less well-situated site):   €0.62 per kWh.

Example 3: 60 kWp Façade-Mounted Installation at an Average Site in Mittelland, Switzerland

Total cost of the installation K_{TOT} = €258 000 (equivalent). A portion of the façade cost savings will be used for aesthetically optimal integration of the solar generator into the building shell and for more expensive customized modules. The depreciation interest rate is $p = 3\%$. The installation's depreciation period can be deemed to be 25 years, except for the inverters, which cost K_E = €18 000 and will depreciate in 10 years. The presumed annual operating and maintenance costs are €666. Façade-mounted installations in this area exhibit specific annual energy yield amounting to $Y_{Fa} \approx 650$ kWh/kWp, based on solar generator peak power P_{Go} (see Section 1.4.4), which equates to an annual energy yield of 39 000 kWh.

Annual costs k_J:
Examples 2010 (1 Euro = 1.5 CHF)

	Amortisation period	Investment cost K_i (Fr. / €)	Amortisation rate a_i	Annual cost K_J (CHF / €)
PV generator (Netto: K_G-K_S)	25 years	360000 / 240000	5.74%	20675 / 13783
Inverter	10 years	27000 / 18000	11.72%	3164 / 2110
Cost of operation				1000 / 666
TOTAL		387000 / 258000		24839 / 16559

Energy price:
For presumed annual yield of 650 kWh/kWp: €0.42 per kWh.
For 1100 kWh/kWp (at a well-situated Alpine site):   €0.25 per kWh.
For 550 kWh/kWp (at a less well-situated site):   €0.50 per kWh.

Example 4: 1.2 kWp Stand-alone Installation at a Swiss Alpine Club (SAC) Lodge at 3000 m Elevation

The total installation costs are K_{TOT} = €20 400 (equivalent; relatively high construction and transport costs). The depreciation interest rate is $p = 3\%$. The presumed depreciation period is 30 years for the installation as a whole, except for: (a) the inverters and charge controllers, whose aggregate cost is K_E = €2200 and will depreciate in 12 years; and (b) the battery bank, which costs K_A = €7200 and will depreciate in 8 years, which is a relatively long period. The presumed operating and maintenance costs are €200. At a high Alpine site, the installation will exhibit specific annual energy yield amounting to $Y_{Fa} \approx 1250$ kWh/kWp, based on solar generator peak power P_{Go} (see Section 1.4.4), which equates to an annual energy yield of 1500 kWh.

Annual costs k_J :
Examples 2010 (1 Euro = 1.5 CHF)

	Amortisation period	Investment cost K_i (Fr. / €)	Amortisation rate a_i	Annual cost K_J (CHF / €)
PV generator wiring etc.	30 years	16500 / 11000	5.10%	842 / 561
Inverter Charge controller	12 years	3300 / 2200	10.05%	332 / 221
Accumulator	8 years	10800 / 7200	14.25%	1539 / 1026
Cost of operation				300 / 200
TOTAL		30600 / 20400		3013 / 2008

Energy price: **€1.34 per kWh.**

This estimate presupposes that any surplus energy generated by the installation owing to excess insolation will be used for heating or some other useful purpose. If this is not the case, the installation's energy costs for usable electricity will be higher. If the installation's elements are transported by truck rather than helicopter, the construction costs and thus energy costs will be appreciably lower. Electricity costs amounting to well under €1.33 are attainable for stand-alone installations in not overly remote areas such as this one.

9.1.2 Comparison of PV and Conventional Electricity Costs

Although stand-alone installations are considerably more cost intensive than grid-connected systems, the former are often the most economical solution in remote areas or for sites that are at a considerable distance from the public grid. PV electricity from stand-alone installations is far cheaper than power obtained from a battery bank and less expensive than using a diesel generator. Thus for sites in developing countries or remote mountainous areas, stand-alone installations are the best solution in terms not only of their cost, but also from an ecological and logistics standpoint, as they are emissions-free and require no grid energy. In many cases, a hybrid installation comprising solar modules and a backup power source such as a wind turbine or gas/diesel generator is far less expensive than a PV-only installation.

The PV energy prices indicated in the four examples above are the actual prices that would be incurred under current conditions, based on the stated suppositions. Although possible subsidies via tax breaks, power companies, or a pro rata surcharge on the electricity rates of all customers (according to consumption) will have no impact on the larger economic picture, they are likely to affect individual installation owners. However, a valid comparison of the price of conventional and PV electricity is not really possible since the prices of the former exclude certain external costs.

For example, there is simply no way of knowing whether the reserves (accruals) that are factored into electricity prices will be sufficient to cover the costs entailed by decommissioning and dismantling nuclear power plants and safe long-term storage of their spent fuel rods. Moreover, nuclear power plants are controversial and the known uranium deposits for conventional fission power plants will be depleted in a few decades. And although the risk of a serious accident occurring in a nuclear power plant that is built and maintained to western standards is minute, the depredations that could potentially be wrought by such an accident are so monumental that no insurance could possibly cover them – which means that the victims of such an accident would probably have to bear the lion's share of the resulting costs. In addition, international terrorism poses a major threat to the safe operation of nuclear power plants. The technology for plutonium breeder reactors, which use uranium more efficiently, is still under development, and such power plants would pose an even greater environmental threat than uranium power plants.

Although controlled nuclear fusion would generate far less radioactive waste, even after decades of extensive research this technology is far from being practicable.

The prices of fossil fuel electricity likewise exclude the costs arising from the health hazards and environmental damage attributable to air pollution or climate change; also, the price of oil and natural gas is rising steadily owing to a combination of political uncertainty and dwindling oil reserves. Moreover, coal cannot possibly be a viable long-term power generation solution on account of its high carbon emissions and the large amounts of ash it generates.

During the exceptionally sunny summer of July 2006 many Central European thermal power plants had to scale back their energy production substantially owing to excessively high water temperatures in their running water cooling systems, whereas installed PV capacity delivered peak yields during this period. On 27 July 2006, the Leipzig Power Exchange (LPX) spot price for peak load power reached a level (€0.54 per kilowatt hour according to DGS statistics) that for the first time ever exceeded the solar power export price, which, as mandated by Germany's Renewable Energy Act (EEG), in 2006 ranged from €0.406 to €0.518 per kilowatt hour in Germany, and up to €0.568 per kilowatt hour for façade-mounted installations. Comparable peak prices were registered in Switzerland as well. Owing to increased use of refrigerators and freezers, electricity demand peaks are becoming evermore prevalent on hot summer days in both Southern and Central Europe. PV electricity is ideally suited to meet demand during such peaks by virtue of its ready availability at such times.

Peak period electricity rates (aggregation of energy prices and grid use fees) in the low-voltage grid currently range from the equivalent of around €0.1 to €0.2 per kilowatt hour in Switzerland. At one time, some Swiss power companies required that only one electricity meter be used, which meant that customers who exported electricity to the grid received the same price for it as the price they paid for the energy they used – despite the fact that, owing to its uncertain availability, electricity exported to the grid is not worth as much to power companies as the peak-rate power provided by small hydro power plants. This policy constituted a welcome de facto subsidy for grid-connected systems on the part of the relevant power companies, for which this subsidy was a mere financial drop in the ocean. But unfortunately Swiss power companies have abandoned this practice since the advent of the KEV framework (compensatory feed-in remuneration for renewables). A more efficient way of promoting PV development would be through a cost-recovery feed-in tariff. Although such a programme has been in effect in Germany since 2000 and in Switzerland since 2009, it applies to only a handful of PV installations and entails a monumental amount of red tape (see Section 1.4.7).

Until early 2008 Switzerland lacked a national feed-in tariff statute for grid-connected systems, most of whose owners lost anywhere from the equivalent of €0.2 to €0.6 per kWh on the operation of these installations (depending on their location and the suppositions applied to the calculation). Break-even operation of such installations would be well within reach if PV system costs were subsidized locally and/or if tax incentives for such system were instituted, or if power companies offered high export tariffs.

Although cost trends are difficult to predict, one thing is certain: electricity prices are set to rise and solar cell prices will come down; this in turn means that the price gap between PV and conventional electricity will also narrow on account of higher prices for the former. Ever-growing production of solar cells spurred by national feed-in tariff statutes will also bring down solar module prices in an ever-growing number of countries. It is generally believed that under normal circumstances, doubling solar module production levels will bring down the price of modules by some 20%.

Numerous studies have accurately forecasted decreasing price trends, but have often overestimated the attendant timeline (see Figure 9.1). It was not until the price reductions occasioned by the financial crisis that such a timeline came close to matching reality, in that solar module prices in 2009 were close to those projected for 2007. In view of the fact that a technological breakthrough that would bring about a huge price reduction is not on the cards, price reductions going forward will be driven by vendor learning curves resulting from increased solar module production.

Subsidy programmes in a number of countries have enabled many solar module manufacturers to ramp up their production dramatically in recent years, an evolution that in 2009 occasioned heightened competition that drove prices down. For example, in the autumn of 2009, thin-film modules were selling

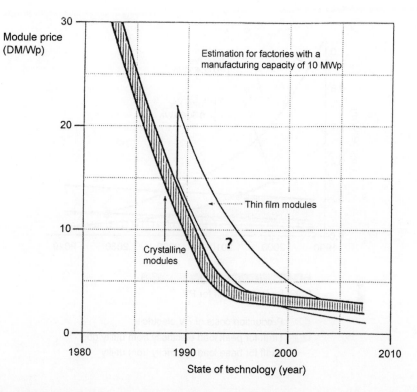

Figure 9.1   Solar module price projection from 1990 (1 DM is equivalent to about €0.5) [9.1]. This projection was far too optimistic, and the actual prices were far higher than those predicted. It was not until the major price reductions in 2009 that such a timeline came close to matching reality, in that solar module prices during that year were close to those forecast for 2007. However, this projection is accurate in that thin-film modules are now less expensive than crystalline modules (VDE Verlag)

for the equivalent of just $1 per Wp in Shanghai (www.solarplaza.com). These prices will of course reduce solar power costs, which, as noted in Section 9.1, are also strongly influenced by interest rates.

Figure 9.2 displays a 2008 European Photovoltaic Industry Association (EPIA) solar energy price forecast.

This graph has prompted the EPIA to predict that grid parity will be reached within a few years in Southern Europe and within a few years after that in Central Europe – grid parity being a state where consumer high-tariff electricity and PV power cost the same. This evolution will be spurred by higher annual insolation, first of course in Southern European countries such as Italy with higher electricity prices; then in places such as Spain, Greece and southern France, where electricity prices are lower; and finally in Central Europe (see Figure 9.3).

Figures 9.2 and 9.3 display a comparison between PV energy prices and the probable prices for grid electricity in the absence of PV installations. However, as noted in Section 5.2.7, a monumental expansion of European installed PV capacity would also affect electricity spot prices over the course of any given day. In the presence of abundant solar energy, consumer electricity prices will presumably fluctuate as well over the course of any given day in that they will decrease at noon on sunny days and rise after dark – which means that grid parity may be somewhat longer in coming relative to the timeline in Figure 9.3.

Increased PV exports to the grid will reduce power company operating revenue for the grid infrastructure, which is absolutely essential for grid-connected systems. Whereas increasing grid use fees for electricity used by PV installation owners would basically be out of the question, higher fees could

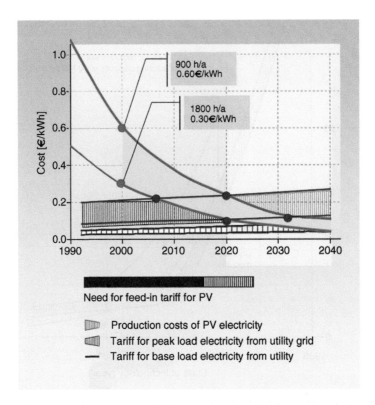

Figure 9.2 The 2008 EPIA solar energy price forecast. With the same investment costs, solar energy prices are strongly affected by specific annual energy yield Y_{Fa} (see Section 10.1.4.4). The upper $Y_{Fa} = 900$ h/a curve represents an average Central European site and the lower $Y_{Fa} = 1800$ h/a curve represents a high-insolation site. The red curve represents the current and future high-tariff consumer electricity price, including grid use (Diagram: Winfried Hoffmann (2000), SCHOTT Solar.)

be imposed on energy not used by battery-free installations and exported to the grid. Levying a reasonable grid hook-up fee would also be acceptable, since, as noted in Section 5.2.7, a massive PV installation build-out would necessitate expansion of the grid infrastructure and of energy storage installations, whose costs would also need to be covered.

9.1.3 PV Electricity Pump Storage System Costs

PV electricity costs will be driven up by energy storage costs if it becomes necessary to generate energy that exceeds grid energy absorption capacity (see Section 5.2.7).

The costs of the energy produced by a pump storage system can be determined as follows:

$$k_E = \frac{k_A \cdot (a + 2\%)}{t} + 1.3 \cdot k_P \qquad (9.3)$$

where:

k_E = cost per kilowatt hour of generated energy

k_A = specific pump storage system costs, which are currently around €788 per kW [9.2]

a = depreciation rate as in Table 9.1; the 2% supplement covers personnel costs and insurance

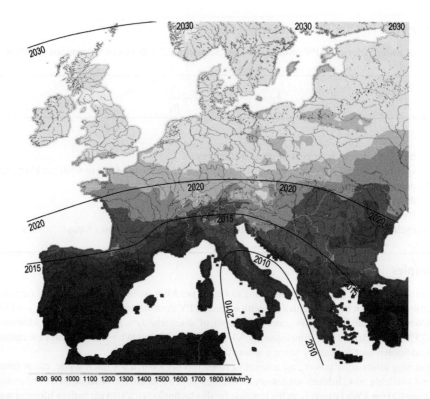

Figure 9.3 Possible spread of grid parity from Southern to Northern Europe. *Source*: NET (Nowak Energy +
Technologie AG) [9.9], based on Meteonorm solar insolation maps (www.meteonorm.com), which are derived from
European Photovoltaic Technology Platform data

t = operating time in generator mode

k_P = cost per kilowatt hour of pumping energy; the 1.3 factor applies because each kWh of
electricity uses around 1.3 kWh of pumping energy.

Example

Assuming a future solar power price of €0.166 per kWh, a 3% interest rate, a 40-year depreciation
period, and 2500 operating hours annually as in Equation 9.3, electricity will cost €0.233 per kWh,
which means that the additional cost attributable to pump storage will amount to €0.67 per kWh.

The pump storage system investment and operating costs for this example are by and large consistent
with the figures in [9.2], [9.3] and [9.4]. However, unlike solar module and inverter costs, pump storage
system costs are unlikely to decrease very much going forward.

Storing electricity in hydrogen has yet to be technologically optimized and is far more expensive.
Exact cost figures for this technology are hard to come by owing to the lack of commercial products
that exhibit sufficiently long life spans. Another problem with this technology is that the efficiency of
the hydrogen–water–hydrogen cycle is only around 35% and at best 50%, which means that the
additional costs attributable to storage loss are far higher than with pump storage systems.

9.1.4   PV Electricity Battery Storage Costs

The cost of using batteries in PV installations (see Section 5.2.7.5.2) can be determined as follows:

$$k_E = \frac{k_{A+E} \cdot a}{365 \cdot t_z} + k_{PV} \frac{1}{\eta_E} \tag{9.4}$$

where:

k_E = cost per kilowatt hour cost of generated energy

k_{A+E} = specific costs for a PV installation's battery bank (for Pb around €520 to €700 per kWh [5.92])

a = depreciation rate as in Table 9.1, for anticipated battery life span

t_z = daily depth of discharge

k_{PV} = PV electricity costs

η_E = cycling efficiency, which includes battery, inverter and rectifier loss ($<\eta_{Wh}$)

Examples

Assuming future solar power cost amounting to €0.166 per kWh, a 3% interest rate, use of currently available lithium-ion battery technology with a specific PV installation cost of around k_{A+E} = €520 per kWh, a 10-year battery life span with depth of discharge t_z = 30% and cycling efficiency η_E = 75%, the electricity cost from Equation 9.4 would be €0.78 per kWh – which in this case equates to an additional cost for battery storage amounting to €0.61 per kWh.

Assuming future solar power cost amounting to €0.166 per kWh, a 3% interest rate, use of future (i.e. not available now) lithium-ion battery technology with a specific PV installation cost of around k_{A+E} = €400 per kWh (which is on the low side for this technology), a 10-year battery life span with depth of discharge t_z = 89% and cycling efficiency η_E = 89%, the electricity cost from Equation 9.4 would be €0.347 per kWh – which in this case equates to an additional cost for battery storage amounting to only €0.18 per kWh.

9.2   Grey Energy; Energy Payback Time; Yield Factor

The solar cell and solar module production process is extremely energy intensive, particularly when it comes to electricity, as energy is needed to make a PV installation's inverters and other devices, as well as its wiring and mounting elements. A renewables-based energy generation system only makes ecological sense if manufacturing the system's components consumes far less energy than the energy the system produces during its lifetime.

Grey energy – the total energy that goes into making a product and its raw materials – is often something of an imponderable. In view of the importance of grey energy for PV installation assessments, various studies of grey energy use for solar modules that are available today and that will be available in the future have been conducted.

Although grey energy data are often lacking for many PV installation devices and components, grey energy can be roughly estimated via prices. Dividing Swiss energy consumption in 1989 by the country's GDP for the same year yields an electricity price equivalent to €0.63 per kWh, although figures ranging up to €0.85 or even €0.93 have been posited elsewhere on the grounds that a portion of GDP is not product related. The following guide value for grey energy can be posited, bearing in mind that this figure is basically an educated guess:

Grey energy for a given product GE ≈ €0.67 per kWh (or 1 SFr per kWh)

(unless other values are available). This guide value often provides reasonably useful results for factory-made products, which are subject to both labour and materials costs. For example, using this value, a freezer that costs €630 would consume around 800 kWh of grey energy. The value of €1.5 per kWh also provides useful results for solar modules and electronic devices.

However, scenarios such as energy resource costs engender major anomalies. Heating oil at €0.6 per litre provides around 10 kWh of thermal energy, which equates to grey energy of around 11 kWh per €0.79 of the product price. At the other extreme, however, the grey energy value for a painting worth millions of euros would be practically zero.

Calculations of grey energy for PV installation components should be based not on product price, but rather on peak power P_{Go} or on solar generator surface A_G:

$$\text{Peak-power-based grey energy}: e_P = GE/P_{Go} \quad (9.5)$$

$$\text{Solar generator surface-based grey energy}: e_A = GE/A_G \quad (9.6)$$

$$\text{Annual energy yield } E_a = P_{Go} \cdot t_{Vo} \quad (9.7)$$

$$\text{Total product life span } L \text{ energy yield } E_L = E_a \cdot L = P_{Go} \cdot t_{Vo} \cdot L \quad (9.8)$$

$$\text{Energy payback time ERZ} = GE/E_a \quad (9.9)$$

$$\text{Yield factor EF} = E_L/GE \quad (9.10)$$

where:
GE = grey energy in kWh
P_{Go} = solar generator peak power in Wp
A_G = solar generator surface in m^2
L = installation lifetimes (in years)

Product grey energy studies often indicate the primary energy needed to manufacture the product, whereby the electricity needed to make the product is converted to the relevant primary energy by dividing the former by the mean efficiency η_K of all the power stations that were involved in converting the primary energy to electricity. In Europe this figure is around $\eta_K = 35\%$.

The figures for primary energy per surface unit indicated in Table 9.3 [9.6] were converted to per watt peak power for grid-connected systems with monocrystalline Si modules (sc-Si, $\eta_M = 14\%$),

Table 9.3 Primary energy (grey energy) needed for the manufacture of monocrystalline modules (sc-Si), polycrystalline or multicrystalline (mc-Si) modules or amorphous modules (a-Si) for grid-connected systems [9.6]. Reproduced by permission of Elsevier

Embodied energy(*primary energy*)		Laminate	ALU frame	Inverter	Mounting	TOTAL
($\eta_{electrical} = 35\%$)	η_M	e_P[MJ/W]	e_P[MJ/W]	e_P[MJ/W]	e_P[MJ/W]	e_P[MJ/W]
Monocrystalline Si (sc-Si) (Roof/Façade-mounting)	14%	40.7	2.9	1.6	2.5	47.7
Monocrystalline Si (sc-Si) (Ground-mounting)	14%	40.7	2.9	1.6	13.2	58.4
Multicrystalline Si (mc-Si) (Roof/Façade-mounting)	13%	32.3	3.1	1.6	2.7	39.7
Multicrystalline Si (mc-Si) (Ground-mounting)	13%	32.3	3.1	1.6	14.2	51.2
Amorphous Si (a-Si) (Roof/Façade-mounting)	7%	17.1	5.7	1.6	5.0	29.5
Amorphous Si (a-Si) (Ground-mounting)	7%	17.1	5.7	1.6	26.4	50.9

polycrystalline or multicrystalline modules (mc-Si, $\eta_M = 13\%$) and amorphous Si modules (a-Si, $\eta_M = 7\%$). In performing this calculation, a distinction was made between grey energy for the manufacture of laminates, aluminium frames, inverters, wires and two different types of mounting structures. Inasmuch as grey energy for module frames and mounting structures is proportional to surface area, the peak energy consumption per watt for modules that exhibit lower efficiency η_M is higher.

Hence the electricity generated by a PV installation needs to be converted into primary energy in a similar manner; that is, this electricity needs to be quantified in order to obtain viable energy payback time and yield factor results.

However, it is simpler to determine energy payback time for various installations that exhibit differing numbers of full-load hours t_{Vo} and solar generator peak power P_{Go} if the installation's grey energy is converted to equivalent grey electrical energy E_e expressed as kWh$_e$, instead of basing the calculation on primary energy E_{Pr} expressed in MJ/W. Thus the following applies:

$$\text{Equivalent grey electrical energy } E_e\,[\text{kWh/W}] = \frac{\eta_K \cdot E_{Pr}}{3.6\,\text{MJ/W}} \qquad (9.11)$$

where η_K is the mean efficiency of all power plants that were involved in converting primary energy E_{Pr} into electrical current (in Europe, 35%).

This yields the grey energy values in Table 9.4, expressed in equivalent electrical energy.

Based on these values and using Equations , it is then a very simple matter directly to determine energy payback time and the yield factor EF for PV installations at various sites (see Figures 9.4–9.6).

The values in Tables 9.3 and 9.4, and in Figures 9.4–9.6, are based on data that probably date back to around 2002 [9.6], [9.7], since which time PV manufacturing technologies have improved. For example, crystalline solar cells are now made of far thinner wafers, thinner saw wires are used and production efficiency has improved. The energy needed to make thin-film solar cells, which was already relatively low, has been reduced still further, with the result that effective module production energy consumption has decreased in the interim. In addition, energy consumption can often be reduced relative to the values indicated in Figures 9.3 and 9.4, through the use of massive ground bolts in lieu of foundations.

Table 9.4 Grey energy (expressed as equivalent electrical energy E_e) for grid-connected systems with monocrystalline modules (sc-Si), polycrystalline or multicrystalline modules (mc-Si) or amorphous modules (a-Si). Data derivation method: values converted using Equation 9.11 [9.6]. Reproduced by permission of Elsevier

Embodied energy		Laminate	ALU frame	Inverter	Mounting	TOTAL
(equivalent electrical energy)	η_M	e_P[kWh$_e$/W]	e_P[kWh$_e$/W]	e_P[kWh$_e$/W]	e_P[kWh$_e$/W]	e_P[kWh$_e$/W]
Monocrystalline Si (sc-Si) (Roof/Façade-mounting)	14%	3.96	0.28	0.16	0.24	4.63
Monocrystalline Si (sc-Si) (Ground-mounting)	14%	3.96	0.28	0.16	1.28	5.68
Multicrystalline Si (mc-Si) (Roof/Façade-mounting)	13%	3.14	0.30	0.16	0.26	3.86
Multicrystalline Si (mc-Si) (Ground-mounting)	13%	3.14	0.30	0.16	1.38	4.98
Amorphous Si (a-Si) (Roof/Façade-mounting)	7%	1.67	0.56	0.16	0.49	2.86
Amorphous Si (a-Si) (Ground-mounting)	7%	1.67	0.56	0.16	2.57	4.95

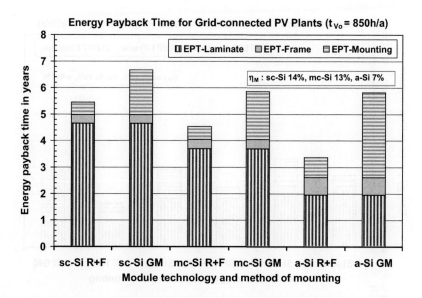

Figure 9.4 Energy payback times for grid-connected systems with monocrystalline modules (sc-Si), polycrystalline or multicrystalline modules (mc-Si) or amorphous modules (a-Si) for Central European sites, where $t_{Vo} = 850$ h/a. **R + F**, roof- or façade-mounted installation; **GM**, ground-based installation. *Note*: As these values date from 2002, the 2009 amounts will presumably be around 65% lower

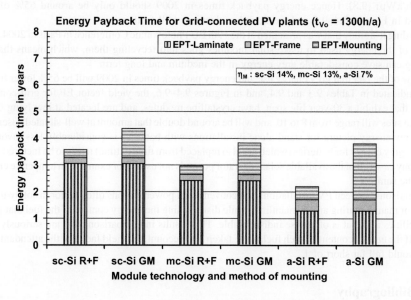

Figure 9.5 Energy payback times for grid-connected systems with monocrystalline modules (sc-Si), polycrystalline or multicrystalline modules (mc-Si) or amorphous modules (a-Si) for Southern European Alpine sites, where $t_{Vo} = 1300$ h/a. **R + F**, roof- or façade-mounted installation; **GM**, ground-based installation. *Note*: As these values date from 2002, the 2009 amounts will presumably be around 65% lower

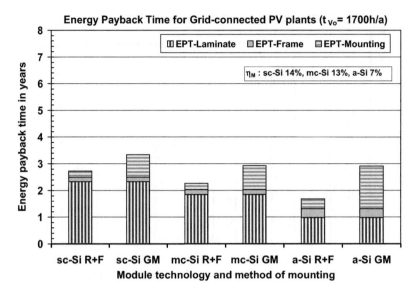

Figure 9.6   Energy payback times for grid-connected systems with monocrystalline modules (sc-Si), polycrystalline or multicrystalline modules (mc-Si) or amorphous modules (a-Si) for desert sites, where $t_{Vo} = 1700$ h/a. **R + F**, roof- or façade-mounted installation; **GM**, ground-based installation. *Note*: As these values date from 2002, the 2009 amounts will presumably be around 65% lower

For example, a recent publication reports that energy payback times of just under three years have been attained for crystalline module PV installations in low-lying areas in Switzerland (annual yield 920 kWh/kWp) [9.8]. Hence energy payback times in 2009 should only be around 65% of those indicated in Figures 9.4–9.6 and are bound to decrease still further.

According to a study that was presented at the 19th EU Photovoltaic Conference in Paris in 2004, only a fraction of the energy used to make solar modules goes into recycling them, which means that such recycling can save considerable grey energy in the medium and long term.

In view of the fact that grey energy use and energy payback times in 2009 will be 65% lower than the values indicated in Tables 9.3 and 9.4, and in Figures 9.4–9.6, the yield factor EF for grid-connected systems that exhibit a 30-year life span, have crystalline modules, and are located at low-lying Central European sites will range from 8 to 10, and will be around double that amount at well-situated desert sites.

The grey energy picture for stand-alone installations with batteries is considerably less favourable, since: (a) grey energy for batteries (which are also replaced from time to time) needs to be factored in; and (b) in many cases all of the available solar generator power is not used – for example, in temperate climates during the summer.

Comparisons between PV installations and conventional power stations often factor in solely the grey energy for manufacturing and dismantling, while disregarding the energy content of the fuel that powers such facilities and that is of course indispensable. This results in comparisons that are seriously out of whack. If the energy content of such fuels were taken into account, the yield factor of the attendant power plants would fall far short of 1.

9.3   Bibliography

[9.1] D. Strese:'Die Ludwig-Bölkow-Studie: Solarstrom wird rentabel'. In [Jäg90].

[9.2] 'Boom der Pumpspeicherwerke'. *Bulletin SEV/VSE*, 2/2006, p. 27.

[9.3] S. Grötzinger:'Das Potenzial der Wasserkraft–Szenarien im Spannungsfeld von Wirtschaft und Politik'. *Bulletin SEV/VSE*, 2/2006, pp. 22ff.

[9.4] M. Balmer, D. Möst, D. Spreng:'Schweizerische Wasserkraftwerke im Wettbewerb'. *Bulletin SEV/VSE*, 2/2006, pp.11ff.

[9.5] G. Hagedorn:'Kumulierter Energieverbrauch und Erntefaktoren von Photovoltaik-Systemen'. *Energiewirtschaftliche Tagesfragen*, 39. Jg (1989), Heft 11, pp. 712ff.

[9.6] E. Alsema:'Energy-Pay-Back Time and CO_2 Emissions of PV Systems'. In [Mar03].

[9.7] E. Alsema, M. de Wild-Scholten:'Environmental Impacts of Crystalline Silicon Photovoltaic Module Production'. 13th CIRP International Conference on Life Cycle Engineering, Leuven, 31 May–2 June 2006.

[9.8] G. Doka:'Energie- und Umweltbilanz der Solarenergie'. Herausgegeben von Swissolar, Zurich, 2008.

[9.9] S. Nowak, M. Gutschner:'Fotovoltaik in der Schweiz'. *Bulletin SEV/VSE*, 9/2009.

10

Performance Characteristics of Selected PV Installations

Section 10.1 discusses a series of PV installations for which performance characteristics extending over at least 12 months are available from monitoring their insolation on the solar generator plane and energy yield in each case, as well as all other key performance characteristics. For selected installations that were monitored by the Bern University of Applied Sciences PV Lab, all of the types of analyses discussed in Chapter 7 were realized, using extremely detailed data. This chapter also discusses selected PV installations from around the world for which the monthly performance ratio was determined, although less data were available for these installations. Only grid-connected systems are discussed as they are the only ones for which such yield data are available.

Section 10.2 presents a comparison of the energy yield of four Swiss PV installations of varying sizes and at various locations that have been in operation for more than 15 years and for which detailed long-term measurements are available for this period. Sections 10.4 and 10.5 centre around the long-term energy yield characteristics of all PV installations in Burgdorf, Switzerland, and the mean yields of German PV installations.

10.1 Energy Yield Data and Other Aspects of Selected PV Installations

In this section, each installation is first briefly described and its measurement readings are presented; this is followed by a discussion of the installation. It should be borne in mind here that a PV installation's performance ratio PR is partly determined by the nature of the insolation measurements that are realized, which for reference cells tend to be a few percentage points higher than for pyranometers.

10.1.1 Gfeller PV Installation in Burgdorf, Switzerland

The performance data of this roof installation on a house in Burgdorf have been monitored continuously since it went into operation in June 1992 (Table 10.1; for a photo of this installation, see Figure 1.10).

Recurrent breakdowns of the installation's first inverter, a Top Class 3000, resulted in substantial capture losses. This device was replaced in 1997 by a Top Class 4000/6 Grid III, necessitating the rewiring of the solar generator and modification of the DC voltage measurement technology. The installation has been trouble-free ever since and has an energy yield that is average for PV installations in Burgdorf.

The efficiency of the new inverter, which is free of start-up problems, is around 2% higher than its predecessor, and DC-side wiring loss is lower owing to the higher voltage exhibited by the installation, which now comprises six series-connected modules rather than the original four. These measures improved the installation's solar generator correction factor k_G and performance ratio PR by several percentage points (see Table 10.2).

Photovoltaics: System Design and Practice. Heinrich Häberlin.
© 2012 John Wiley & Sons, Ltd. Published 2012 by John Wiley & Sons, Ltd.

Table 10.1   Main technical characteristics of the Gfeller installation

Location:	CH-3400 Burgdorf, 540 m, $\varphi = 47.0°$N	**Commissioning: 24.9.1992**
PV generator:	$P_{Go} = 3.18$ kWp, $A_G = 25.6$ m^2, 60 sc-Si modules M55,	rooftop installation
	inclination $\beta = 28°$, azimuth $\gamma = -10°$	*(modules with frames)*
	(E deviation from S)	
Inverter:	ASP Top Class 3000 until 14.04.1997	
	ASP Top Class 4000 Grid III since 14.04.1997	
Measured values:	• Irradiance in module plane (with pyranometer)	
	• Ambient and module temperature	
	• DC current and DC voltage	
	• produced AC active power	

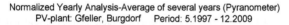

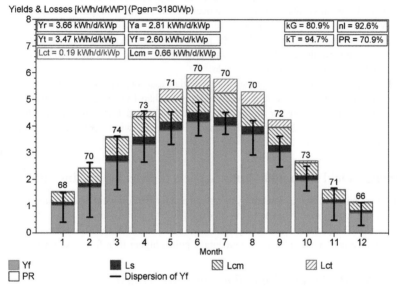

Figure 10.1   Normalized multi-year statistics (with scatter shown) since the new ASP Top Class 4000/6 Grid III inverter was installed in the Gfeller installation, whose yield curve is typical for a PV installation in Mittelland. Some 29.3% of the installation's annual energy was produced in the winter, and from November to March energy yield exhibited extreme scatter. This is typical for Mittelland installations, which are subject to occasional snow shading

Table 10.2   Specific annual energy yield and performance ratio for the Gfeller PV installation from 1993 to 2009. The PR figures are derived from pyranometer readings

Year	1993	1994	1995	1996	1997	1998	1999	2000	2001	2002	2003	2004	2005	2006	2007	2008	2009
Y_{Fa} [kWh/ kWp]	825	854	898	792	1004	982	897	994	915	931	1048	919	944	932	967	908	921
PR in %	67.1	70.5	67.0	61.7	72.3	72.4	72.3	73.9	72.6	73.2	70.7	70.4	68.7	70.8	69.2	68.9	66.9

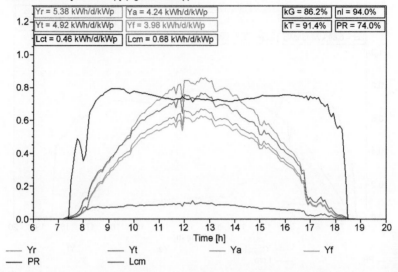

Figure 10.2 Performance curve for the Gfeller installation on a sunny September day in 2000. The curve exhibits no abnormalities, and no solar generator shading is observed. The solar generator correction factor is $k_G = 86.2\%$

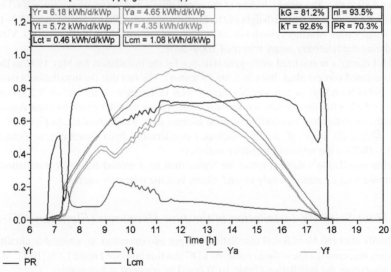

Figure 10.3 A similar day two years later. The solar generator was shaded between 9 and 11 a.m. by the tree near the installation that had grown during the intervening two years, resulting in: (a) far higher capture losses L_{CM} than had been the case in 2002; and (b) a decrease in the solar generator correction factor to $k_G = 81.2\%$

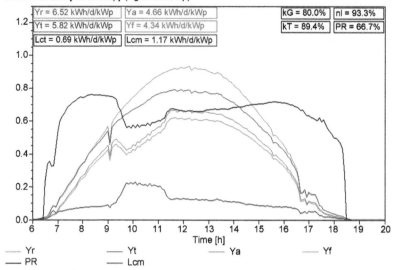

Figure 10.4 Similar day in 2006, when, already in mid September, the solar generator was shaded between 9 and 11 a.m. by the aforementioned tree, whose height had of course increased during the intervening four years. In addition, on account of solar module soiling, afternoon L_{cm} is higher than in Figures 10.2 and 10.3, causing a decrease in the solar generator correction factor to $k_G = 80\%$

As the installation is located near the city limits at a site whose environs are devoid of factories or railway lines, it exhibits permanent soiling at a far slower rate than at less well-situated sites in Burgdorf (see Figures 4.89–4.98). The steady decline of the installation's performance ratio since 2003 is mainly attributable to the gradual increase in height of a tree in front of the installation (see Figures 10.2–10.4) and a series of lengthy snow-shading episodes in 2003, 2004 and particularly in 2005 and 2009. Virtually no snow fell during the relatively sunny winter of 2006–2007.

Figure 10.1 displays normalized multi-year statistics for the installation for May 1997 to December 2009 (the bracketed vertical black lines indicate Y_F scatter). The fact that the installation's mean multi-year yield (949 kWh/kWp) is considerably higher than in Example 1 in Section 8.2.1.1 is mainly attributable to the fact that the effective registered insolation Y_R on the solar generator plane is around 10% higher than the insolation that was derived from the exercise data, which are older. As Figure 10.1 clearly shows, Y_F with a tilt angle of $\beta = 28°$ fluctuates considerably from November to February (see Figures 10.5–10.7 for the relevant December statistics).

The Gfeller installation's thermal losses are higher than for a ground-based PV installation, and its modules, whose roof clearance is only around 10 cm, become relatively warm.

10.1.2   Mont Soleil PV Installation in the Jura Mountains (Elevation 1270 m)

In 1992, shortly after the Mont Soleil installation was put into operation, an unsuitable circuit breaker induced arcing that caused a fire in the installation's DC junction box (see Figure 1.13) [10.1]. As a result, it was months before the installation (Table 10.3) could be operated at full power.

Owing to the fact that from 1992 to 1999 the installation was monitored by an institution other than the Bern University of Applied Sciences, using a sophisticated but unfortunately unreliable measuring instrument, some of the electricity meter readings from this period are missing, all of the remaining

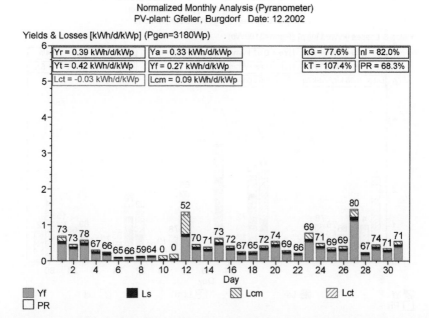

Figure 10.5 Normalized monthly statistics for December 2002, during which there was virtually no snow shading except on the 10th and 12th. On account of frequent heavy fog, insolation on the solar generator plane amounting to only $Y_R = 0.39$ h/d is extremely low, whereas k_G of 77.6% is relatively high for December (attributable to the virtual absence of snow shading). $Y_F = 0.27$ h/d is likewise very low on account of the minimal Y_R

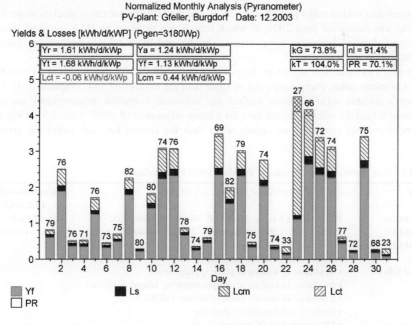

Figure 10.6 Normalized monthly statistics for December 2003, during which there was virtually no snow shading except on the 16th, 23rd and 24th; thus the 78% for k_G is relatively high for December, as is $Y_R = 1.61$ h/d. Hence Y_F with $Y_F = 1.13$ h/d is also elevated – a factor of more than 4 times higher than in Figure 10.5

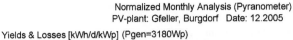

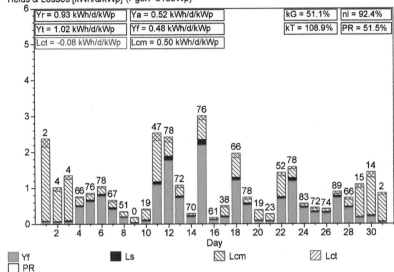

Figure 10.7 Normalized monthly statistics for December 2005, when insolation on the generator plane averaged about $Y_R = 0.93$ h/d. However, the solar generator was covered with snow on 1–3, 9–11, 18, 22 and 29–31 December, which explains the relatively low k_G of only 51.1%. The performance ratio of PR = 51.5% and final yield $Y_F = 0.48$ h/d were also quite low

measurement data exhibit major gaps in some cases, and the performance ratio calculations are inexact. Monitoring was suspended until 2001, at which time it was resumed by HTA Burgdorf (now Bern University of Applied Sciences) using a new, simpler and more reliable measuring system (see Figure 10.8).

The system's measurement signals are logged by a Campbell CR10X data logger at 2 s intervals and are saved as mean values every 5 min. All of these data are transferred and computed daily. Despite the system's sizeable solar generator surface, the reference insolation measurements are realized immediately behind the solar generator field for a mean orientation ($\beta = 50°$, $\gamma = -6.5°$). Only overall solar generator direct current is measured, rather than the current for each individual array field.

Table 10.3 Main technical characteristics of the Mont Soleil PV installation

Location:	CH-2610 Mont Soleil, 1270 m, $\varphi = 47.2°$N	**Commissioning:** 19.2.1992
PV-Generator:	$P_{Go} = 554.6$ kWp, $A_G = 4465$ m^2, 10 464 modules M55,	free field installation
Array field 1:	inclination $\beta = 50°$, azimuth $\gamma = -20°$ (E deviation from S)	*(frameless modules)*
Array field 1:	inclination $\beta = 50°$, azimuth $\gamma = -35°$ (E deviation from S)	
Inverter:	ABB (special development), $P_{ACn} = 500$ kW	
Measured values:	• Irradiance in horizontal plane (pyranometer)	
	• Irradiance in module plane (pyranometer, heated + vented)	
	• Irradiance in module plane (reference cell M1R)	
	• Ambient and module temperature	
	• DC current and DC voltage	
	• produced AC active power	
	• line voltage	

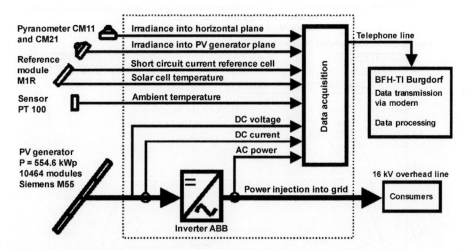

Figure 10.8 Configuration of the Mont Soleil PV installation monitoring system. Of the system's 110 module 'tables', each of which comprises 96 modules, 109 are currently connected to the PV installation's large ABB inverter; the remaining modules are operated using small inverters and are not monitored

This monitoring system, which has been in operation since June 2001 (Figure 10.9), provides the desired information concerning the Mont Soleil PV installation's performance at a relatively low cost.

The reliability and hence energy yield of the Mont Soleil installation have greatly improved since the inception of HTA Burgdorf monitoring in mid 2001. Reliable monitoring allows for timely detection and resolution of incidents.

Figure 10.9 Weather station for the Mont Soleil installation measuring system, which has been in operation since 2001. Insolation is monitored on the solar generator plane using a heated CM-21 pyranometer and an M1R reference cell. Horizontal insolation is measured using a CM-11 pyranometer, while module temperature is monitored via reference cell temperature readings. The ambient temperature is also monitored

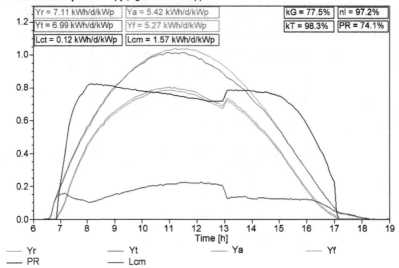

Figure 10.10 Normalized daily diagram for the Mont Soleil installation for 2 October 2001 showing the failure and manual restart of one of the installation's 11 sub-arrays. One sub-array Q_{DC} switch was tripped until 1 p.m., resulting in array yield loss L_{cm} during this period. When the sub-array was restarted, the performance ratio in the daily diagram increased by around 7% and the sub-array loss decreased accordingly

However, until an installation overhaul in mid 2005, the presence of undersized circuit breakers caused the 11 sub-arrays to shut down repeatedly every so often (particularly during cloud enhancements), resulting in yield losses (see Figure 10.10).

The system's inverter also shut down occasionally, necessitating a manual restart on each occasion. The sub-array failures ceased to occur in 2006, but the inverter began exhibiting occasional failures that were quickly detected by the monitoring system, but that caused an energy loss on each occasion until the inverter was restarted. This loss amounts to several percentage points per year.

Despite these problems, the Mont Soleil installation has performed extremely well for the past decade or so (Table 10.4). From June 2001, when the installation was commissioned, until March 2005, when the installation was overhauled and a new monitoring programme was begun, the installation's mean energy yield was $Y_F = 2.86$ kWh/d/kWp or 1044 kWh/kWp (see Figure 10.11). This exceeds the current Swiss mean of 833 kWh/kWp by 25.3% and is far higher than the long-term mean of 958 kWh/kWp registered by the installation between 1993 and 2001. The installation's winter output between June 2001 and March 2005 accounted for 38.2% of its annual yield and is thus

Table 10.4 Specific annual energy yield and performance ratio for the Mont Soleil PV installation from 1993 to 2009. The pre-2002 performance ratio is inexact owing to sporadic measurement reading gaps during the previous monitoring period, and no performance ratio is available for 1999–2001 because no insolation readings are available for that period

Year	1993	1994	1995	1996	1997	1998	1999	2000	2001	2002	2003	2004	2005	2006	2007	2008	2009
Y_{Fa} [kWh/ kWp]	956	929	938	1063	1135	1060	872	732	933	965	1165	988	1000	979	1039	959	951
PR in %	77.8	76.6	75.7	78.9	78.0	78.5	—	—	—	75.3	73.6	74.3	69.0	69.6	73.8	72.9	68.7

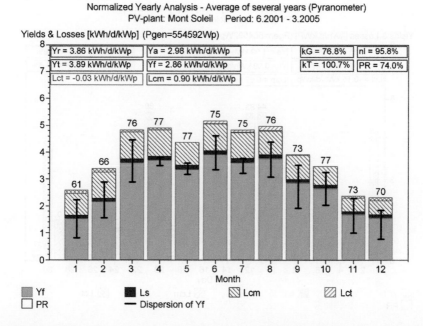

Normalized Yearly Analysis - Average of several years (Pyranometer)
PV-plant: Mont Soleil Period: 6.2001 - 3.2005

Figure 10.11 Normalized multi-year statistics from the date of commissioning of the new Bern University of Applied Sciences PV Lab monitoring programme (2001) until the 2005 renovation. The installation's winter output during this period accounted for 38.2% of its annual yield. November–January energy yield exhibited extreme scatter, but the mean yield was considerably higher than for installations in low-lying areas (see Figure 10.1)

considerably higher than the average of around 25 to 30% registered by Mittelland installations, but lower than the 45 to 58% at high Alpine installations.

Owing to lengthy snow-shading periods in January, February, March and December, the installation's 2009 annual yield was extremely low, as was the proportion of annual energy produced during the winter (30.7%). The installation's monthly performance ratio since monitoring began has ranged from 28.5 to 79.7%. The lower of these ratios occurred in the very snowy month of February 2009, and was only 38.9% in June 2006 on account of an inverter failure. The peak monthly performance ratio (79.7%) registered by the Mont Soleil installation is for the most part topped only by Alpine installations such as the Jungfraujoch and Birg systems (see Sections 10.13 and 10.14) or by installations whose effective STC module output is equal to or greater than nominal output.

The Mont Soleil installation's 500 kW ABB inverter has thus far exhibited outstanding energy conversion efficiency n_I amounting to 84–96.9% a month. Owing to the relatively low ambient temperatures at the installation site (1270 m above sea level), the module temperature correction factor k_T ranging from 94.8 to 108.6% is on the high side. However, the solar generator's 50° tilt angle is not sufficiently steep at this elevation to prevent occasional winter snow shading, and thus the monthly solar generator correction factor k_G from 2001 to 2009 has ranged from 30.6% in February 2009 due to extensive snow shading (see Figure 4.67) to 82.6% in April 2007. Figure 10.12 displays the normalized statistics for December 2005, when k_G was 55.4%.

One highly positive aspect of the Mont Soleil installation's performance is that its modules have thus far exhibited next to no soiling and only very little delamination. Moreover, the laminates (which are glued to a mounting structure) have by and large worked out quite well, although they are sometimes damaged by the considerable snow pressure exerted on them in winter (see Section 4.5.2.1).

As with many Central European PV installations, the Mont Soleil installation registered in 2003 far and away its highest yield to date (see Figure 10.13), although its performance was marred somewhat by yield

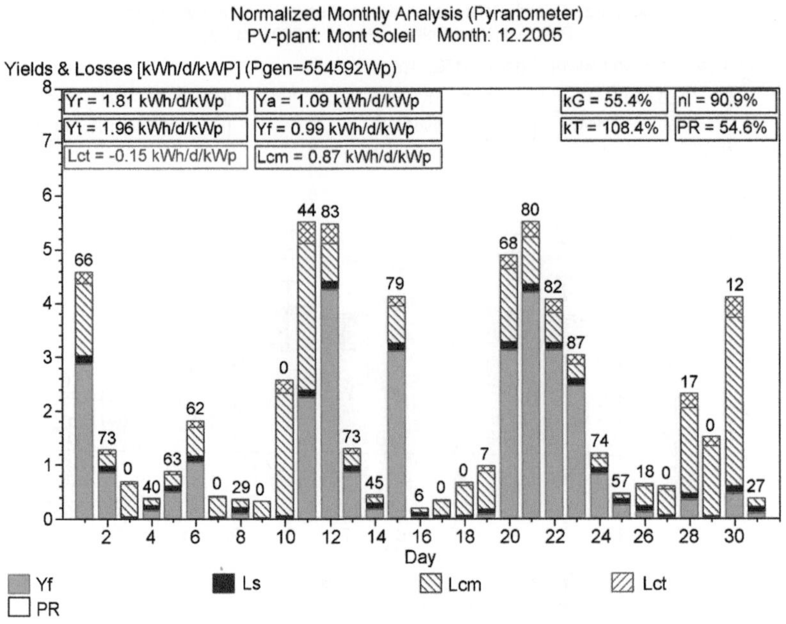

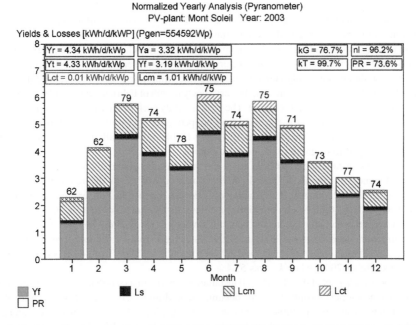

Figure 10.12 Normalized monthly statistics for December 2005, when insolation on the generator plane was about average ($Y_R = 1.81$ h/d). However, the solar generator was partly or wholly covered with snow on December 1–3, 9–11, 17, 21 and 25–31, which explains the relatively low k_G of only 55.4%. The performance ratio of PR = 54.6% and final yield $Y_F = 0.99$ h/d were also quite low, although this is considerably higher than for low-lying areas (see Figure 10.7)

Figure 10.13 Normalized annual statistics for 2003, during which the installation attained its highest annual yield to date ($Y_F = 3.19$ h/d, 1165 kWh/kWp/a). This figure would have been 1234 kWh/kWp had it not been for the energy loss resulting from snow shading, scheduled installation downtime, adventitious sub-array shutdown, and trouble in the 16 kV line

Figure 10.14   View of half of the Jungfraujoch installation's 1.13 kWp solar generator, which is mounted on the façade of the high-Alpine Jungfraujoch Research Station (HFSJG, 3454 m, ≈46.5°N). Two insolation sensors (a reference cell and a heated pyranometer) are visible on the right

loss attributable to the following events (the numbers below indicate the percentage of loss attributable to the event in question):

- Inverter failure: 0
- Snow cover: 2.11
- Scheduled maintenance downtime for the 16 kV medium-voltage cable: 0.91
- Adventitious tripping of the DC string switch and trouble in the 16 kV line: 2.76

10.1.3   Jungfraujoch PV Installation (Elevation: 3454 m)

The Jungfraujoch installation (Figure 10.14), which at the time of its realization was the world's highest PV system (3454 m), was designed and realized by the Bern University of Applied Sciences PV Lab in 1993. The installation has exhibited both trouble-free performance during its operation of more than 16 years and 100% availability of its energy production and measurement readings. Inasmuch as all PV installation components are subjected to extreme stress at such heights, any components that perform well under such extreme ambient conditions will in all likelihood also be highly reliable under normal operating conditions (also see [10.2]–[10.4], [10.7]–[10.9]).

Following a number of improvements, the installation's energy yield has been optimized relative to the early years of its operation. Its energy yield decreased between 1999 and 2001 (see Table 10.5) owing to the replacement of research station windows, and its yield was relatively low in the spring of 2001 on account of a lengthy snow-shading event that affected the eastern half of the solar generator. In 2005, despite a power outage on 23 August resulting from valley flooding, the installation attained a record specific annual energy yield of 1537 kWh/kWp owing to its effective peak output of 1.13 kWp (the previous high was 1504 kWh/kWp in 1997); its winter energy yield accounted for 48.5% of annual yield. The 2005 yield would have been 1540 kWh/kWp had it not been for the power outage. The installation's annual energy yield from 1994 to 2009 was 1408 kWh/kWp, and on average it produced 46.2% of its annual energy during the winter (range 43.2–50.7%).

In the 16 years of its operation, the installation has on average produced around 70% more energy than comparable installations in the Mittelland region, where annual energy yield is around 833 kWh/kWp.

Table 10.5 Specific annual energy yield (in terms of effective STC power output) and performance ratio for the Jungfraujoch PV installation from 1994 to 2009 (including mean values)

Year	1994	1995	1996	1997	1998	1999	2000	2001	2002	2003	2004	2005	2006	2007	2008	2009	Average
Y_{Fa} [kWh/ kWp]	1272	1404	1454	1504	1452	1330	1372	1325	1400	1467	1376	1537	1449	1453	1375	1358	1408
PR in	81.8	84.1	84.7	85.3	87.0	84.8	84.6	78.6	85.2	84.9	86.2	86.9	85.5	85.9	86.4	87.0	84.9

Table 10.6 Main technical characteristics of the Jungfraujoch PV installation

Location:	CH-3801 Jungfraujoch, 3454 m m.ü.M, $\varphi = 46.5°$N	**Commissioning: 29.10.1993**
PV generator:	$P_{Go} = 1.152$ kWp, $A_G = 9.65$ m^2, 24 modules M75 (48Wp)	façade installation
Subarray 1:	inclination $\beta = 90°$, azimuth $\gamma = 12°$ (W deviation from S)	*(framed modules)*
Subarray 2:	inclination $\beta = 90°$, azimuth $\gamma = 27°$ (W deviation from S)	
Inverter:	ASP Top Class 1800 from 29.10.1993 until 16.7.1996	
	ASP Top Class 2500/4 Grid III since 16.7.1996	
Measured values:	• Irradiance in module plane 1 (pyranometer, heated + vented)	
	• Irradiance in module plane 1 (reference cell M1R)	
	• Irradiance in module plane 2 (pyranometer, heated + vented)	
	• Irradiance in module plane 2 (reference cell M1R)	
	• Ambient and module temperature of both subarrays	
	• DC voltage and current of each sub-array 1 and 2	
	• produced AC active power	
	• line voltage	

Such yields are outstanding for a Central European PV installation and would even be excellent for a system in Southern Europe.

The installation's solar generator (see Table 10.6) comprises 24 Siemens M75 modules (48 Wp; nominal output 1152 Wp) that are mounted vertically on the façade of the Jungfraujoch Research Station (Figure 10.15). As STC conditions occasionally occur at this elevation, effective solar generator power output can be determined directly from the measurement readings, by increasing the inverter DC output readings under STC by the arithmetic amount of wiring and string diode loss, which yields an effective STC power output of 1130 Wp. After 32 months of excellent performance, it proved possible to increase

Block Diagram PV-Plant Jungfraujoch

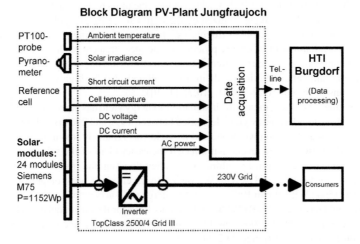

Figure 10.15 Block diagram for the Jungfraujoch PV installation (1.152 kWp nominal, 1.13 kWp effective), which was realized by the Bern University of Applied Sciences PV Lab on the Jungfraujoch at a height of 3454 m

the installation's energy yield by eliminating the string diodes and replacing the existing inverter with an optimized model (Top Class 2500/4 Grid III).

All of the above readings were logged at 2 s intervals, were then cached in a Campbell CR10 data logger, and under normal conditions were calculated and saved at 5 min intervals. All original data are available in an error file in the event of a malfunction, thus allowing for detailed analysis of the error. The data are automatically transferred (via a phone modem) to the PV Lab in Burgdorf early each morning for storage and further analysis.

In order for such an installation to exhibit optimal reliability, it is essential that its working components and electrical elements be sized accurately. Also indispensable for such an installation, in view of the high winds to which the site is subject, are efficient lightning and voltage surge protection devices.

10.1.3.1 Performance and Reliability

Since inception, the Jungfraujoch installation has withstood the following high-Alpine conditions without incurring any damage whatsoever:

- Severe storms with winds up to 250 km/h, which constitute a rigorous test for the installation's working components, structure and design.
- Lightning accompanied by severe lightning strikes that induced voltage surge damage in inadequately protected installations.
- Cloud enhancements ranging up to 1720 W/m^2 resulting from diffuse reflection of radiation off the glacier in front of the solar generator. Such voltage peaks put the system's inverters to the test in that insolation and DC power output are directly proportional to each other.
- Major temperature fluctuations. The installation's solar generator temperature readings range from -29 to $+66\,°C$, and at sunset on a cold winter day the solar generator temperature can drop $40\,°C$ in 30 min.
- Snow and ice covering of the solar generator. Snowfall ranging to upwards of 3 m is not unheard of in the spring. The snow depth in such cases is determined not by snowfall but rather by wind velocity and direction during and after the snowfall. However, the installation's energy yield sometimes also declines owing to extremely heavy frost, as well as overshadowing by huge icicles that form in front of the array.

Window-replacement scaffolding that was in place from August to October 1999, 2000 and 2001 partly shaded the solar generator during these periods, and one module incurred mechanical damage in 2001 as a result of the work being performed. When this module was replaced, incipient delamination on the bottom edge of another module in the western half of the solar generator was observed that had not been detected during a visual inspection two years previously. Hence this delamination appeared to be progressing at a relatively rapid rate. It was probably attributable to moisture penetrating the module edge, combined with incipient electrolytic breakdown of the adjacent solar cells. Although no measurable solar generator yield loss was observed, the module was replaced in 2001 as a precautionary measure.

In more than 16 years of operation under these extreme conditions, only one of the solar generator's 24 modules has exhibited signs of wear and tear from natural factors; no power output decrease was registered prior to replacement of this module. The only problem with the installation has been the major snowfalls that sometimes occur in spring, resulting in the eastern half of the solar generator being covered with snow (see Figure 10.16); this in turn induces an energy loss in the affected element lasting anywhere from a few days to a few weeks.

The lengthiest and most severe shading of the solar generator to date occurred from 28 March to 27 June 2001 on account of accumulations of snow in front of the installation's building that reduced output in the eastern half of the solar generator to nearly nil for a period of many days. This also resulted in a performance ratio decrease to around 22% in April, around 23% in May (Figure 10.17) and around 16% in June, which equates to a yield loss amounting to around 120 kWh. The installation's solar generator correction factor ($k_G = 84.5\%$) and thus its performance ratio (PR $= 78.6\%$) attained a record low in 2001

Figure 10.16   East generator at the Jungfraujoch PV installation on 7 May 2001. All of the solar modules are mounted horizontally to the strings in groups of four. With the rear module shaded, the individual strings barely produced any power

(see Table 10.5). The clearance between the solar generator and ground is an extremely important parameter for the design of Alpine PV installations.

10.1.3.1.1   Mean Normalized Annual Energy Yield and Performance Ratio from 1994 to 2009

Figure 10.18 displays the normalized annual statistics, based on effective solar generator power output, for the average year between 1994 and 2009, together with monthly Y_F and Y_A, insolation yield Y_R and temperature-corrected insolation yield Y_T.

Figures 10.19, 10.20 and 10.21 display the normalized annual statistics for the year with the lowest and highest annual yield, as well as for 2003, which was the third-highest energy yield year. All of the readings used for these diagrams were obtained using a reference cell.

Springtime solar generator snow shading resulted in higher L_{CM} values and a lower performance ratio, particularly in May and June. Yield was also somewhat off from August to October on account of shading that occurred between 1999 and 2001.

Peak energy yield and insolation in the 1994 to 2009 period occurred in 2005. In 2003, record summer temperatures and insolation were registered in the Mittelland region. However, 2003 was only the third-best year for energy yield and insolation during this period; on the other hand, insolation-weighted module temperature was the highest and thus k_T the lowest during this same year.

Note: The performance ratios shown in the diagrams and tables here are based on reference cell Y_R readings, which in most cases were a few percentage points lower than the counterpart pyranometer readings (see Section 2.8.3). Hence these performance ratios are several percentage points higher than those derived from pyranometer measurements. This means, for example, that the performance ratios in Figure 10.18 would have been only 81% had they been based on pyranometer readings.

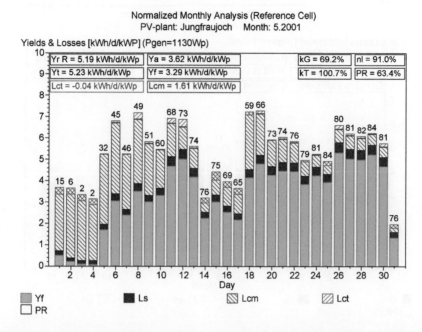

Figure 10.17 The shading of the installation's solar generator is readily apparent in this normalized monthly statistics diagram for May 2001. On the date for which Figure 10.16 was realized (7 May), the installation's performance ratio was only 46%. The diagram also clearly shows the slow melting of the snow over the course of the month in that field losses L_{cm} decreased and the performance ratio recovered

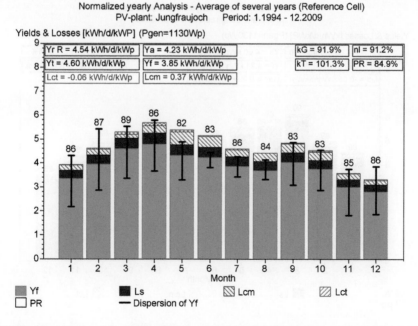

Figure 10.18 Normalized statistics (2004–2009) for the Jungfraujoch PV installation, with monthly scatter. The monthly performance ratio ranged from 82 to 89%, and the mean annual performance ratio was 84.9%; 46.2% of total annual energy was produced during the winter. *Note*: Had the figures been based on a solar generator nominal output of 1.152 kWp, Y_F and PR would have been around 2% lower

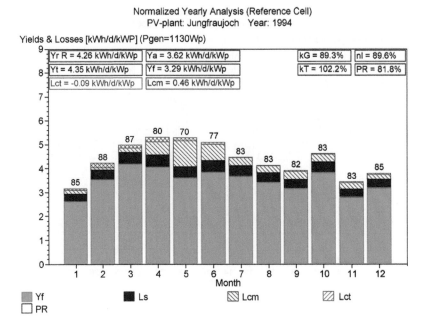

Figure 10.19 Normalized annual statistics for 1994, when annual yield was the lowest in 16 years, and from April to June was reduced by snow. As a result, the mean k_G and PR were on the low side. The installation produced 48% of its energy during the winter, mainly due to relatively low summer yield resulting from snow shading from April to June

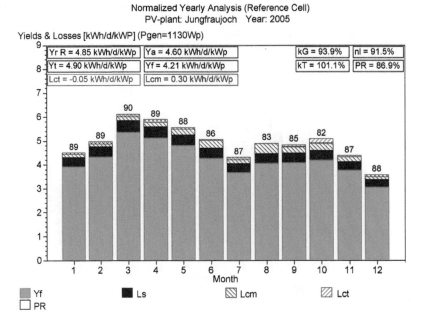

Figure 10.20 Normalized annual statistics for 2005, when annual yield was the highest in 16 years. Insolation yield Y_R reached a record high during this year and there was virtually no snow-induced shading. Hence the mean annual k_G and PR were on the high side; 48.5% of the installation's energy was produced during the winter

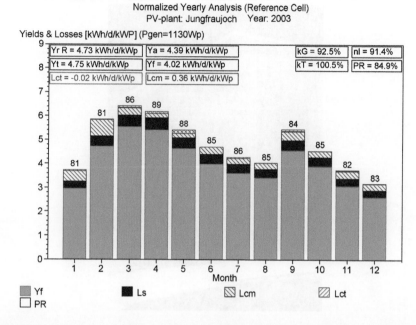

Figure 10.21 Normalized annual statistics for 2003, during which many Central European PV installations attained record annual yields. However, in 2003 the Jungfraujoch installation only attained its third-highest yield due to the following factors: winter snow shading; lower Y_R than in 2005; and a record low k_T resulting from high temperatures

10.1.4 Birg PV Installation (Elevation: 2670 m)

The Birg PV installation (see Figures 10.22 and 10.23; for the key technical data see Table 10.7) was designed and realized over the course of 1992 and was put into operation on 21 December of that year. The installation is only a few kilometres from the Jungfraujoch installation, but is farther north of the main Alpine ridge.

After a number of problems were resolved in 1993 and 1994 (inverter failure; disturbance of the Schilthornbahn cable car's radio signals by the installation's inverter; lengthy inverter downtime resulting from substandard grid conditions), for more than eight years the installation exhibited 100% reliability and better than average energy yield (see Figure 10.24).

When one of the installation's inverters failed in January 2003, the installation operator decided to replace it in view of its advanced age and the fact that authorized vendor servicing was not readily available. Hence the existing Solcon 3400HE device was replaced by a new ASP Top Class 4000/6 Grid III inverter, which has exhibited no problems since installation.

Table 10.8 displays the installation's specific annual energy yield Y_{Fa} and performance ratio PR since January 1993, which were lower: (a) during the first two years of operation on account of frequent inverter failures in 2003 and 2004; and (b) in 2003 owing to a lengthy downtime episode in that year.

Apart from the 2003 inverter failure (see Figures 5.140 and 5.141), in the more than 17 years of the installation's operation it has exhibited only one instance of yield-reducing ageing. The installation has exhibited consistent solar generator correction factor k_G and performance ratio in all years except for 1999 and 2005, when the performance ratio was somewhat lower on account of a lengthy snowfall-induced shading episode in the spring (compare Figures 10.25 and 10.26). In addition, no energy was produced during a two-day power outage in August 2005 resulting from valley flooding, further reducing the performance ratio for the year.

The Birg installation's solar generator power output P_{Go} sizing, which was realized in accordance with the standard recommendations at the time for installations in low-lying areas, failed to take account of the

Figure 10.22 Birg installation solar generator, whose upper portion is shaded by the building's roof in the early morning and late afternoon. The pyranometer and ambient temperature sensor are visible on the right

Figure 10.23 The lower module and field junction box are prone to snow shading in the snowy winters that prevail here, owing to the relatively small clearance between the solar generator and the roof above it

Table 10.7 Main technical characteristics of the Birg PV installation

Location:	Station Birg of Schilthorn cableway, 2670 m, $\varphi = 46.5°N$	**Commissioning:** **21.12.1992**
PV-Generator:	$P_{Go} = 4.134\,kWp$, $A_G = 33.3\,m^2$, 78 modules M55, inclination $\beta = 90°$, azimuth $\gamma = 5°$ (W deviation from S)	façade installation *(framed modules)*
Inverter:	Solcon 3400HE from 21.12.1992 until 25.1.2003 ASP Top Class 4000 Grid III since 21.2.2003	
Measured values:	• Irradiance in module plane (pyranometer) • Ambient and module temperature • DC current and DC voltage • produced AC active power • line voltage	

fact that such high-Alpine installations may be subject to insolation spikes in excess of 1.3 kW/m^2 (also see Figures 5.144 and 5.145). However, as the system's inverter can only handle a DC power input that equates to nominal DC output P_{DCn} (4 kW), it limits the input to this amount, and this in turn induces an energy loss (see Figure 10.27).

By virtue of (a) the installation being façade mounted ($\beta = 90°$) and (b) the high summer Sun, summer insolation is relatively low and reflection losses (see Figure 8.1) are on the high side. Moreover, the absence of a frontal glacier as at the Jungfraujoch installation means that the Birg installation does not benefit from snow reflections. Nonetheless, the measurement readings revealed that the snow reflections resulting from sporadic summer snow can temporarily increase energy yield. Hence installation sites whose solar generators exhibit frontal snow all year round (as is the case at the Jungfraujoch site) are extremely beneficial for energy yield.

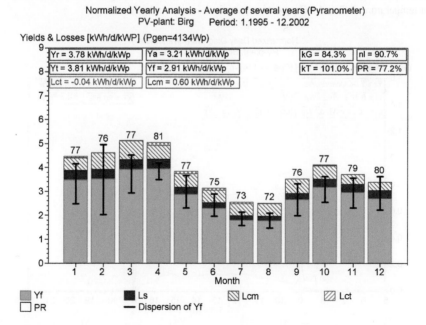

Normalized Yearly Analysis - Average of several years (Pyranometer)
PV-plant: Birg   Period: 1.1995 - 12.2002

Yields & Losses [kWh/d/kWP] (Pgen=4134Wp)

Yr = 3.78 kWh/d/kWp	Ya = 3.21 kWh/d/kWp	kG = 84.3%	nI = 90.7%
Yt = 3.81 kWh/d/kWp	Yf = 2.91 kWh/d/kWp	kT = 101.0%	PR = 77.2%
Lct = -0.04 kWh/d/kWp	Lcm = 0.60 kWh/d/kWp		

■ Yf   ■ Ls   ▧ Lcm   ▨ Lct
□ PR   — Dispersion of Yf

Figure 10.24 Normalized statistics (mean figures for 1995–2002) for the Birg PV installation, with monthly scatter. The monthly performance ratio ranged from 72 to 81%, and the mean annual performance ratio was 77.2%. As this installation, unlike the Jungfraujoch installation, has no frontal reflecting glacier and is instead in a snow-free environment in summer, a high-Alpine façade-mounted installation of this type exhibits extremely low summer energy yield. The proportion of annual energy yield accounted for by the installation's winter output averages 56.3%

Table 10.8 Specific annual energy yield and performance ratios for the Birg PV installation (1993–2009)

Year	1993	1994	1995	1996	1997	1998	1999	2000	2001	2002	2003	2004	2005	2006	2007	2008	2009
Y_{Fa} [kWh/kWp]	575	885	1090	1079	1111	1103	991	1056	1074	1010	998	1072	1098	1089	1052	1071	1011
PR in %	44.5	72.3	78.2	79.3	76.7	77.0	74.6	76.7	77.7	77.0	67.6	76.8	74.9	76.1	77.0	77.8	76.0

From 1995 to 2002, during which the installation was free of inverter failure, its winter energy output accounted for an average of 56.3% of annual yield, which is exceptionally high for a Central European PV installation. For the monitoring period as a whole (1993–2009), winter energy output relative to total annual yield ranged from 53.4 to 59.7%. Annual energy yield during the installation's inverter failure-free periods (1995–2002 and 2004–2009) averaged 1065 kWh/kWp.

For further details, including comparisons between the Birg and Jungfraujoch installations, see Section 10.2 and [10.5], [10.10] and [10.12].

10.1.5 Stade de Suisse PV Installation in Bern

The power company BKW's Stade de Suisse PV installation (STC nominal power output 855 kWp) at the new Wankdorf football stadium in Bern ($\varphi = 46.9°$N), which went into operation on 18 March 2005, comprises 5122 Kyocera KC167GH-2 modules (167 Wp, mean effective power output around 170 Wp). The solar generator exhibits a 6.8° tilt angle and three orientations ($\gamma = -63°$, 27° and 117°; see Figure 10.28), and is subdivided into seven arrays each of which is connected to a Sputnik Solarmax 125 inverter.

The Bern University of Applied Sciences PV Lab has been monitoring and analysing the following installation operating data since 1 April 2005: insolation on the three solar generator planes; module and ambient temperature; DC and AC power output of all arrays; and grid voltage.

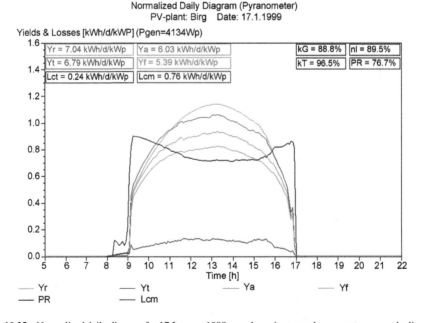

Figure 10.25 Normalized daily diagram for 17 January 1999, one day prior to a solar generator snow-shading event. No power limitation occurred, owing to the installation's moderate G_{max} of around 1.12 kW/m². L_{CM} did not rise excessively during the course of the day, and the k_G of 88.8% and PR of 76.7% are relatively high

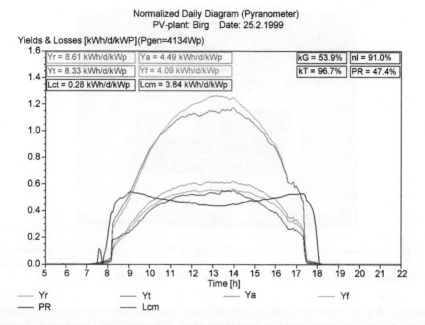

Figure 10.26   Normalized daily diagram for 25 February 1999, one day before the lower portion of the solar generator was covered with snow (also see Figure 10.23). L_{CM} increased abruptly early in the morning, remaining high throughout the day and only slightly below Y_A. Hence k_G (53.9%) and PR (47.4%) are extremely low, and the installation exhibited a substantial energy loss

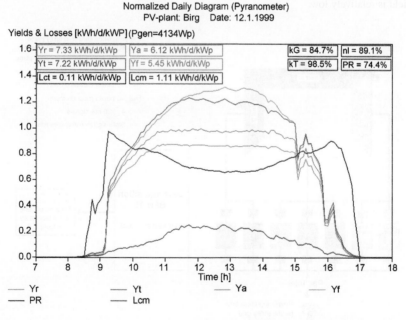

Figure 10.27   Normalized daily diagram for 12 December 1999, one day prior to a solar generator snow-shading event. Insolation was extremely high, reaching a lengthy peak of around 1.3 kW/m² around noon. No power limitation occurred between 10.45 a.m. and 12.30 p.m., during which time L_{CM} rose considerably relative to Figure 10.25, k_G was only 84.7% and PR was only 74.4%

Figure 10.28   Aerial view (from a hot-air balloon) of the Stade de Suisse PV installation, whose relatively small tilt angle $\beta \approx 7°$ causes winter snow to remain on the solar modules for very lengthy periods. The modules are cleaned periodically using a special cleaning vehicle (Photo: www.bergfoto.ch/Simon Oberli)

Figure 10.29 displays a block diagram for the installation, which has been working perfectly since June 2005 (a few start-up kinks were ironed out prior to this time). The installation's energy yield is relatively high in the summer and on snow-shading-free overcast winter days owing to its flat tilt angle. However, the downside to this arrangement is that it takes a relatively long time for module snow to melt and for the installation to resume energy production; further, the installation derives virtually no benefit from insolation reflected off the stadium's snow-shaded environs. Hence winter energy output relative to total annual yield is relatively low.

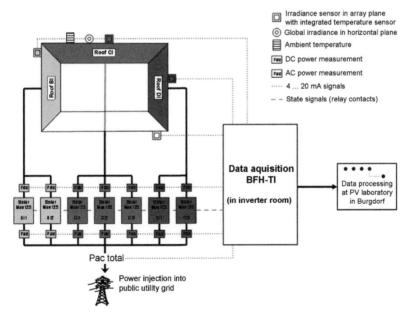

Figure 10.29   Block diagram of the 855 kWp Stade de Suisse PV installation, which is currently the world's largest football stadium PV installation. The system was upgraded to 1342 MWp in 2007 by the addition of four inverters

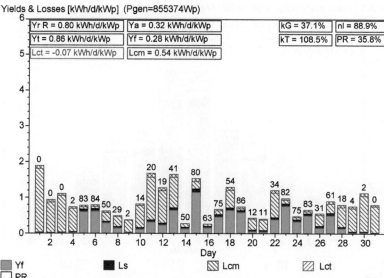

Figure 10.30 Normalized monthly statistics for the Stade de Suisse PV installation for December 2005, during which time insolation on the solar generator plane was only $Y_R = 0.80$ h/d on account of snow shading on 1–4, 7–13, 18, 20, 22 and 26–31 December. Consequently, k_G was only 37.1%, PR only 35.3% and the final yield only $Y_F = 0.28$ h/d – and thus far lower than for the installation as in Figure 10.7

Figure 10.30 displays the installation's normalized monthly statistics for December 2005. Comparing these statistics with those of a nearby installation ($\beta = 28°$; see Figure 10.7) clearly shows that the Stade de Suisse installation is far more sensitive to snow shading: k_G during this month was only 37.1% versus 51.1% for the other installation.

Hence the Stade de Suisse installation produces the lion's share of its energy in the summer. Figure 10.31 displays the normalized monthly statistics for 2006. In contrast to the winter of 2005–2006 with its lengthy snow-shading episodes in November–December 2005 and March–April 2006, the installation's winter 2006–2007 energy yield was relatively high owing to the virtual absence of snow shading.

The mean monthly readings and the annual solar generator correction factor k_G for the months without snow shading were extremely high for the following reasons:

- Mean module STC power output was relatively close to nominal power (167 Wp). The post-production mean, which according to the vendor's post-manufacturing data ranges from around 161 to 177 Wp, averages 1.6% higher than nominal output. In contrast, the readings in this regard obtained by an independent measurement organization (TUEV Rhineland) using a large sample of the system modules revealed a mean output that was around 1.6% lower than nominal output; this means that, allowing for measurement accuracy, the results obtained by the vendor and this organization differed only slightly and were close to nominal power.
- As noted in Section 2.8, monthly and annual reference cell insolation readings are usually a few percentage points higher than those obtained with a pyranometer; accordingly, k_G and PR will be several percentage points higher for the same installation.
- Moreover, the solar module surfaces and unheated reference cells are sometimes shaded by snow in winter, thus reducing insolation readings and upping the solar generator correction factor during this period. This installation exhibits an extremely high performance ratio for the same reasons.

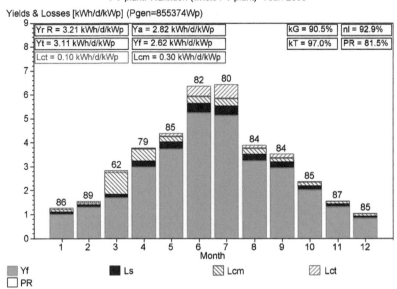

Normalized Yearly Analysis (Reference Cell)
PV-plant: Wankdorf (whole PV plant) Year: 2006

Yields & Losses [kWh/d/kWp] (Pgen=855374Wp)

Yr R = 3.21 kWh/d/kWp	Ya = 2.82 kWh/d/kWp	kG = 90.5%	nI = 92.9%
Yt = 3.11 kWh/d/kWp	Yf = 2.62 kWh/d/kWp	kT = 97.0%	PR = 81.5%
Lct = 0.10 kWh/d/kWp	Lcm = 0.30 kWh/d/kWp		

Figure 10.31 Normalized annual statistics for the Stade de Suisse PV installation for 2006, which was a year free of inverter problems. Hence the miscellaneous capture losses L_{CM} were associated with lengthy snow-shading episodes in March and April and the consequent yield loss. The installation's winter energy output relative to total annual yield was only 25.8% in 2006, and because of severe weather was only 24.2% for the winter of 2005–2006

The second solar generator upgrading phase (in 2007, to 1342 kWp) was realized at the outer southeast and southwest sides of the building (see Figure 10.28). As these modules exhibit a steeper tilt angle ($\beta = 20°$), it takes less time for winter snow to slide off them. Figure 10.32 displays the normalized annual statistics for the upgraded Stade de Suisse PV installation for 2008, during which the annual yield was 966 kWh/kWp.

10.1.6 Newtech PV Installation with Thin-Film Solar Cell Modules

This installation – a collaborative effort by ADEV Burgdorf and the Bern University of Applied Sciences PV Lab – is a pilot system featuring three new thin-film solar cell technologies that was mounted on the roof of the Ypsomed Company building in Burgdorf. The modules are oriented due south, are virtually never shaded and were mounted just prior to commissioning. The installation, which went into operation on 17 December 2001, has been monitored ever since, thus allowing revealing comparisons with monocrystalline and polycrystalline solar cell systems.

10.1.6.1 Technical Characteristics

The Newtech PV installation ($\varphi = 47.1°$N; 540 m above sea level) comprises three grid-connected 1 kWp solar generators, each of which exhibits a different, new, thin-film solar cell technology. The total power output of the installation is 2844 Wp; module tilt angle $\beta = 30°$; orientation is due south ($\gamma = 0°$). The solar generators feed their energy into the building grid via ASP Top Class Spark inverters, each of which integrates a transformer. Use of such transformer inverters in thin-film solar cells helps to avoid long-term degradation of thin-film solar modules [10.15]:

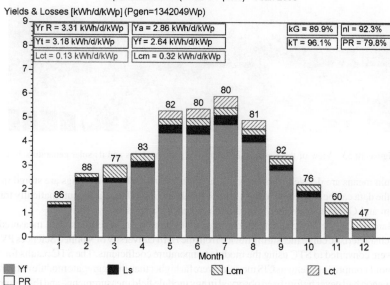

Figure 10.32 Normalized 2008 annual statistics for the Stade de Suisse PV installation. The miscellaneous capture losses L_{CM} were associated with brief snow-shading periods in March and April and longer such periods in late October and December and the attendant yield losses. Nonetheless, winter energy output relative to total annual yield was 27.2% owing to higher yield in the spring

- **Newtech installation 1: Copper indium diselenide (CuInSe$_2$) cells (CIS cells):** 24 framed 40 Wp Siemens ST40 modules; three strings, each comprising eight series-connected modules; at STC nominal power output $P_{STC\text{-}Nominal} = 960$ Wp; $T_K \approx -0.4\%$ per K. Reading: $P_{STC} \approx 1015$ Wp.
- **Newtech installation 2: Tandem amorphous silicon cells:** 20 framed 43 Wp Solarex MST 43-LV modules; two strings, each comprising 10 series-connected modules; STC nominal power output $P_{STC\text{-}Nominal} = 860$ Wp; $T_K \approx -0.22\%$ per K. Reading: $P_{STC} \approx 810$ Wp.
- **Newtech installation 3: Triple amorphous silicon cells:** 16 framed 64 Wp Uni-Solar US 64 modules; two strings, each comprising eight series-connected modules; STC nominal power output $P_{STC\text{-}Nominal} = 1024$ Wp; $T_K \approx -0.21\%$ per K. Reading: $P_{STC} \approx 1000$ Wp.

Figure 10.33 displays a picture of the installation, while Figure 10.34 displays a block diagram of the installation's monitoring system.

The monitoring system logs the following readings at 2 s intervals:

- Insolation on the module plane using a heated CM11 pyranometer.
- Solar cell temperature of the three solar arrays using PT100 sensors.
- Ambient temperature using a PT100 device.

The following readings are logged for the three solar generators:

- Direct current and DC voltage, which are used to determine DC output.
- Effective output fed into the building grid.
- Grid voltage at the feed-in point for phase 1.

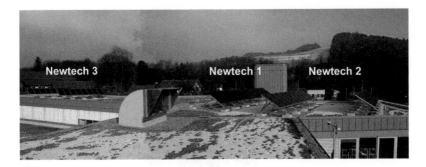

Figure 10.33   View of the Newtech installation's solar modules and its solar generators 1–3

Then, 5 min means are derived from these readings and stored. The 2 s readings are stored in an error file, so that the data can be recovered in the event of a malfunction. The data are automatically transferred via a modem and GSM, and are then stored and readied for analysis.

In late March 2002, the *I–V* characteristic curves of all three solar generators were measured using a proprietary characteristic curve measuring device from the Bern University of Applied Sciences PV Lab and the results were converted to STC using the module temperature coefficients. The STC results for Newtech solar generator 1 comprising Siemens CIS modules were far higher than the aggregate module nominal power. This phenomenon had never before been observed in any module field measurements, and thus represented a major departure on Siemens' part from standard industry practice at the time, which was to provide modules whose baseline output marginally exceeded the guaranteed minimum level but was considerably lower than nominal output. On the other hand, the amorphous modules exhibited a power output which, according to the industry standard, was a few percentage points lower than the aggregate STC nominal power.

10.1.6.2   Energy Yield of the Three Thin-Film Solar Cell Installations

The three solar generators and their monitoring systems have exhibited virtually trouble-free operation to date. As the solar generators were commissioned in winter because of construction delays, owing to lengthy

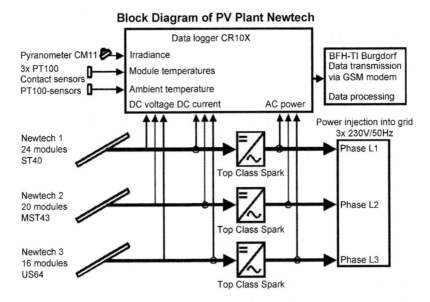

Figure 10.34   Block diagram of the Newtech PV installation and its monitoring system

Table 10.9 Normalized annual energy yield and performance ratio for the Newtech 1 (CIS), Newtech 2 (a-Si tandem) and Newtech 3 (triple a-Si) solar generators from 2002 to 2009

	Newtech 1 (CIS)								(CIS)
Year	2002	2003	2004	2005	2006	2007	2008	2009	
Y_{Fa} (kWh/kWp)	1092	1258	1102	1149	1086	1107	1030	1052	
PR (%)	82.6	81.9	81.2	78.7	78.9	76.2	76.0	74.2	

	Newtech 2								(a-Si Tandem)
Year	2002	2003	2004	2005	2006	2007	2008	2009	
Y_{Fa} (kWh/kWp)	964	1037	883	930	882	926	855	904	
PR (%)	73.0	67.6	65.0	63.7	64.1	63.7	63.1	63.8	

	Newtech 3								(a-Si Triple)
Year	2002	2003	2004	2005	2006	2007	2008	2009	
Y_{Fa} (kWh/kWp)	1033	1103	957	1015	977	1014	935	983	
PR (%)	78.2	71.9	70.5	69.4	71.0	69.7	69.0	69.4	

snow-shading episodes in December 2001, January 2002 and March 2002 that impaired measurement accuracy it was not possible to document accurately the exact initial degradation curves during the first few months of operation despite the fact that the devices were monitored from the day of commissioning.

The performance ratio of all three installations (see Table 10.9) has gradually decreased but cannot be attributable to soiling as the arrays are cleaned every autumn and spring.

The performance ratio decline in the Newtech 1 CIS installation was relatively minor, but thereafter was relatively constant.

The Newtech 2 amorphous silicon tandem-cell array and the Newtech 3 amorphous triple-cell array each exhibited most of their overall performance ratio reduction during the first year of operation (in keeping with the Staebler–Wronski effect; see Section 3.5.3.1). Their respective performance ratio decreases have diminished greatly since 2004, and both exhibit the seasonal efficiency variation that is characteristic of amorphous Si cells.

As the Newtech 2 modules are no longer made, for reasons of space they are discussed here in less detail than the other array modules.

Figures 10.35, 10.36, 10.37 and 10.38 display the normalized annual statistics for the Newtech 1 and 3 installations for 2002 and 2009, which respectively exhibited: (a) very little snow; (b) very low November and December insolation; (c) snow-shading periods in January, February and December; and (d) a very brief snow-shading period in March. It should also be noted that the solar generator correction factor k_G and performance ratio PR have decreased over time.

Minor cell structure changes were observed in the Newtech 1 and 3 installations, whereas the Newtech 2 installation exhibited pronounced delamination.

Figure 10.39 displays monthly DC efficiency and Figure 10.40 displays the solar generator correction factor $k_G = Y_A/Y_T$ for the three Newtech installations and for two crystalline silicon cell installations. As noted, $k_G = 1$ in an ideal installation.

Additional normalized monthly and annual statistics for all three Newtech installations and for the other installations discussed in Sections 10.1.1–10.16 are available under *PV-Messdaten* at www.pvtest.ch.

The normalized energy efficiency of the Newtech 1 CIS installation from 2002 to 2007 exceeded that of the best monocrystalline cell installation, mainly because the modules exhibited far greater effective STC nominal power than the nominal power indicated on the vendor datasheet (also see Figure 10.65). Hence this installation exhibited far and away the best k_G. Another positive aspect of the Newtech 1 installation is that it exhibits elongated cells and high module edges, which ensure that snow and dirt affect all cells

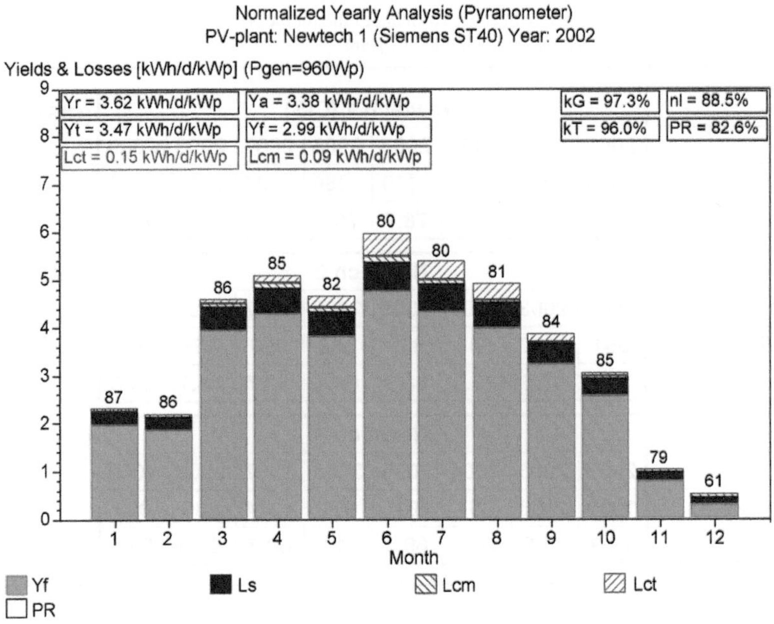

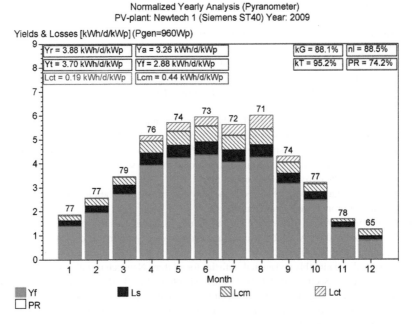

Figure 10.35 Normalized annual 2002 statistics for the Newtech 1 installation comprising Siemens ST40 CIS modules. Miscellaneous capture losses L_{CM} were very minor as baseline module power output was considerably higher than the specified nominal output, and thus k_G and PR were very high (also due to a lack of snow shading over the course of the year). Winter energy output relative to total annual yield was 31.8%

Figure 10.36 Normalized annual 2009 statistics for the Newtech 1 installation comprising Siemens ST40 modules. Miscellaneous capture losses L_{CM} decreased appreciably during the summer owing to degradation of the modules, which are cleaned at six-month intervals. Consequently, k_G and PR were also lower than in Figure 10.35. Winter energy output relative to total annual yield was 30.5%

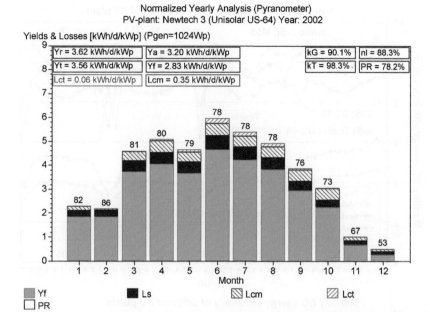

Figure 10.37   Normalized annual 2002 statistics for the Newtech 3 installation (a-Si triple system with Uni-Solar US-64 modules). This system also exhibited relatively high k_G owing to elevated baseline output, high thin-film cell efficiency under lower insolation conditions, and the absence of snow shading during the year. Winter energy output relative to total annual yield was 31.1%

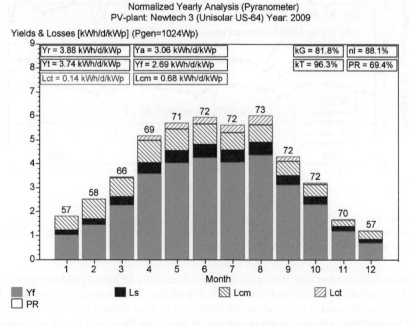

Figure 10.38   Normalized annual 2009 statistics for the Newtech 3 installation (a-Si triple system with Uni-Solar US-64 modules). Module degradation was almost completely halted from this point on by thermal insulation that was installed on the back of the solar generator in the autumn of 2003 (also see Figures 10.39 and 10.40). This installation is particularly sensitive to snow as its module surfaces are rippled (see Figure 10.36, months 1, 3 and 12 in 2009). Winter energy output relative to total annual yield was 27.5%

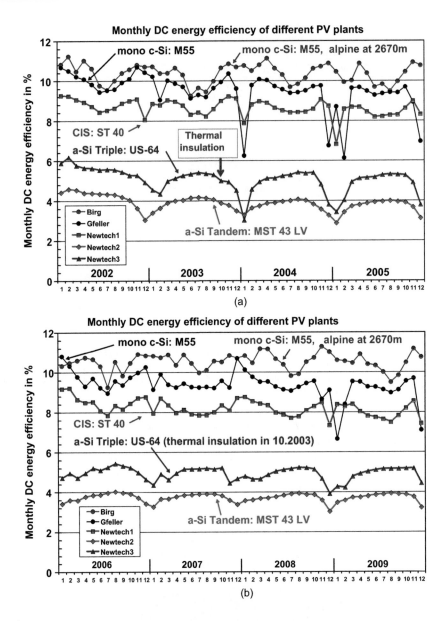

Figure 10.39 Monthly DC efficiency from 2002 to 2009 for the three Newtech installations relative to two monocrystalline c-SI installations, one in Burgdorf 540 m above sea level and the other in the Alps at 2670 m above sea level. Tilt angles: Newtech installations, $\beta = 30°$; Gfeller, $\beta = 28°$; Birg, $\beta = 90°$

equally and to a minor extent only. Winter energy output relative to total annual yield ranged from 30.5 to 35.4%. This installation has exhibited around 1.5% degradation a year since 2003 (see Figure 10.40).

The yield of the Newtech 2 a-Si tandem-cell installation is about average for a monocrystalline cell system. The installation's thermal losses in summer are lower than for crystalline cell installations. But under weak insolation conditions, module output voltage drops off sharply, causing the inverter to operate outside of the MPP, which is not good. This installation exhibited clearly discernible degradation from 2002 to 2004 that induced a DC efficiency loss, as well as a solar generator correction factor k_G and

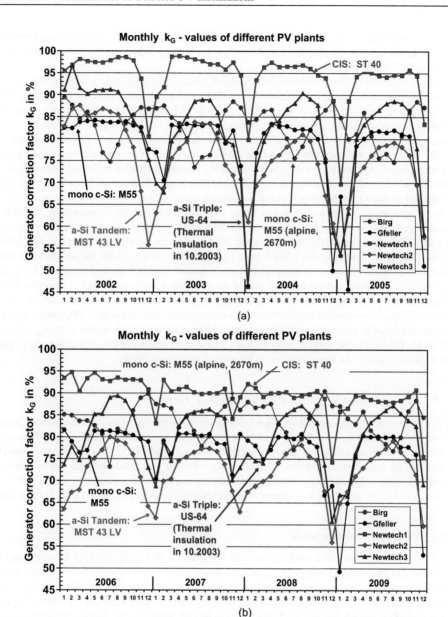

Figure 10.40 Monthly solar generator correction factor k_G from 2002 to 2009 for the three Newtech installations relative to two monocrystalline c-SI installations, one in Burgdorf at 540 m above sea level and the other in the Alps at 2670 m above sea level. Tilt angles: Newtech installations, $\beta = 30°$; Gfeller, $\beta = 28°$; Birg, $\beta = 90°$. $k_G = 100\%$ in an ideal installation

performance ratio PR (around 11%). This power loss has slowed considerably since then, however. Winter energy output relative to total annual yield ranged from 27.8 to 32.7% during the period.

The yield of the Newtech 3 a-Si Uni-Solar triple-cell installation is on a par with that of good monocrystalline installations. This system also benefits from far lower temperature coefficients in the summer and is notable for its good performance ratio on low-insolation days. The slightly rippled module surfaces which prevent snow from sliding off and the high module edges can induce complete snow shading of the bottom row of cells. The installation's winter energy output relative to total annual yield ranges from 27.3 to 32%.

As Figures 10.39 and 10.40 show, although the Newtech 3 installation exhibited considerable degradation until the autumn of 2003 (around 3% from July 2002 to 2003 alone), thermal insulation of the rear side of the installation (using foam around 2 cm thick) has by and large halted the degradation. As a result, the installation's mean operating temperature has greatly increased, reaching a peak of around 75 °C in summer, up from 60 °C previously. These higher temperatures promote thermal annealing in summer, which by virtue of the Wronski effect allows for almost full reversal of the degradation induced by low winter temperatures. As a result, ever since September 2003, this installation has recovered far more quickly than the Newtech 2 installation and its peak summer yield is far higher. Thus it would appear that Uni-Solar a-Si triple-cell modules lend themselves ideally to being mounted on insulation and for the manufacture of the relevant roof and façade joining products. Some products of this type are commercially available.

The solar generator correction factor k_G of the Newtech 1 CIS installation has decreased by an average of around 1.5% a year since 2003, whereas, since 2004, this factor has decreased hardly at all for the amorphous Newtech 2 and 3 modules (see Figure 10.40), both of which exhibit the seasonal efficiency variation that is typical of such systems (peak efficiency around August, lowest efficiency in the spring, depending on snow conditions) and that lags behind the temperature curve by around 1–2 months. For all three installations, following module cleaning each spring and autumn the module characteristic curves are measured and are harmonized with their peak power output, as converted to peak STC nominal power. The curves exhibited by the solar generator peak outputs thus harmonized are similar to the k_G curves in Figure 10.40. The use of a modern, higher-efficiency transformer inverter has improved the energy yield and performance ratio of all three installation by several percentage points.

10.1.7 Neue Messe PV Installation in Munich, Germany

In 1997 a large-scale grid-connected system was installed on the roof of Munich's Neue Messe, a trade show and conference centre ($\varphi = 48.3°$N, 1 MWp). Test operation and optimization began in November 1997 and the installation, which was officially accepted in August 1998 [10.11], is now jointly owned by the solar energy organization Solarenergieförderverein Bayern and the municipal utility company Stadtwerke München. The facility's measurement data are recorded and evaluated at regular intervals. An installation similar to this one was mounted on neighbouring roofs in 2002 by Phoenix Sonnenstrom.

10.1.7.1 Technical Characteristics

The installation comprises 7812 laminates of 130 Wp ($V_{MPP} = 20.4$ V, $I_{MPP} = 6.35$ A) containing 84 Siemens Solar monocrystalline solar cells each and a module surface of $A_M = 1012$ m^2. The installation (including the requisite inter-module clearance) occupies 38 100 m^2 of the roof's 66 000 m^2 surface area, comprises a total of 372 22-module series-connected strings, and is subdivided into 12 arrays. The total power output of the installation is 1016 MWp. The module tilt angle $\beta = 28°$ and the orientation is due south ($\gamma = 0°$). Figure 10.41 displays a partial view of one of the arrays. The installation's direct current is converted into alternating current by three 330 kVA Siemens inverters in a master–slave configuration (with a rotary master) and is exported to the 20 kV medium-voltage grid via a transformer (see Figure 10.42).

10.1.7.2 Energy Yield

The installation was formally accepted in August 1998 following a 10-month testing and optimization period, since which time the installation's specific annual energy yield Y_{Fa} and performance ratio PR have been impressive, despite sporadic technical problems (see Table 10.10).

Figure 10.41   Partial view of an array on the roof of Munich's Neue Messe, a trade show and conference centre. Additional arrays on the roofs of other trade show and conference centres are visible in the background (Photo: Solardach München-Riem GmbH)

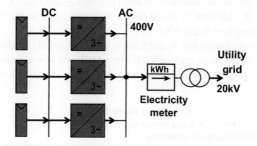

Figure 10.42   Block diagram of the Neue Messe PV installation. The power produced by three master–slave inverters is fed into the grid via a transformer

The installation attained exceptionally high yields from 2002 to 2004 owing to relatively high insolation and for the most part trouble-free operation during this period (see Figures 10.43 and 10.44).

The installation's co-owner Solarenergieförderverein Bayern kindly provided mean multi-year monthly insolation, DC and AC energy yield figures for purposes of realizing a more in-depth analysis of the system's performance. Inasmuch as only mean ambient temperature readings, but not insolation-weighted module temperatures, were available, it was not possible to break down capture losses any

Table 10.10   Specific annual energy yield and performance ratio of the Neue Messe installation from 1999 to 2005. These figures show that robust annual energy yield amounting to more than 1000 kWh/kWp is obtainable in low-lying areas of Germany

Year	1999	2000	2001	2002	2003	2004	2005
Y_{Fa} in kWh/kWp	973	990	932	1026	1113	1026	952
PR in %	79.5	77.0	77.2	78.9	78.5	80.6	73.6

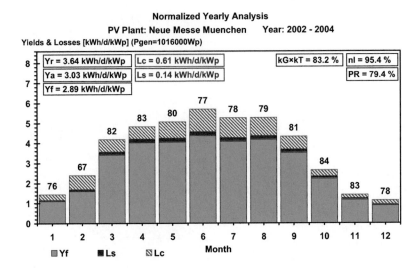

Figure 10.43 Normalized 2002–2004 figures for the Neue Messe PV installation, whose operation was for the most part trouble-free during this period. Energy yield was lower during part of February owing to snow shading. Winter energy output relative to total annual yield was 30% (*Data source*: Solarenergieförderverein Bayern)

further for the normalized annual statistics; thus only the Y_R, Y_A and Y_F figures are shown in the diagrams (see Chapter 7). Since L_C could not be broken down into L_{CM} and L_{CT}, only $k_G \cdot k_T$ but not k_G and k_T could be determined. However, if necessary it would be possible to determine the approximate insolation weighted module temperature (see Section 8.1.2.1).

As with the Mont Soleil installation, the Neue Messe system's originally installed DC switches caused problems at first that necessitated their replacement. The installation also experienced sporadic inverter problems, some of the modules were damaged during a severe storm, and a terminal fracture and defective

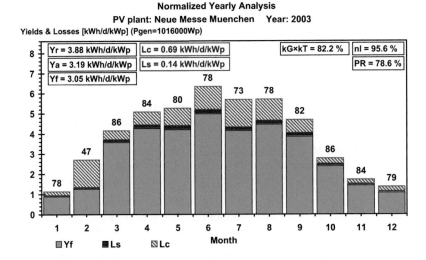

Figure 10.44 Normalized 2003 statistics for the Neue Messe PV installation, whose energy yield reached a record 1113 kWh/kWp during the period. Energy yield was considerably lower during part of February owing to snow shading. Winter energy output relative to total annual yield was 29% (*Data source*: Solarenergieförderverein Bayern)

insulation were observed in one string [10.11]. As with many other PV installations, the Neue Messe system has exhibited capture losses resulting from snow shading. From 1999 to 2004, capture losses amounting to around 0.3 to 2.7% of annual energy yield were registered [10.14].

10.1.8 Leipziger Land PV Installation

This 5 MWp ground-based PV system was installed in 2004 by Geosol Gesellschaft für Solarenergie, Shell Solar and WestFonds Immobilien-Anlage-gesellschaft, on the site of a former coal dust dump 30 km southwest of Leipzig, Germany. The installation occupies 21.6 hectares and comprises four 1.25 MWp arrays, each of whose DC outputs is exported to the medium-voltage grid by two 2 MVA transformers after being converted to alternating current by three Sinvert 400 master–slave inverters. For illustrations of the installation see the following: block diagram (Figure 10.45); partial view (Figure 1.15); aerial view of the entire installation (Figure 10.46); and two detail views (Figures 10.47 and 10.48). The installation's main technical data are given in Table 10.11.

10.1.8.1 Energy Yield

Despite the installation's northern location (51.2°N), the available data indicate that it has performed extremely well. Geosol was good enough to provide detailed monitoring data for 2005 for the purposes of an in-depth analysis of the installation's performance. From the excellent quality of these data and the fact that Geosol also included module temperature readings, all of the evaluations discussed in Chapter 7 could be realized for the Leipziger Land installation as well (as was done for the installations discussed in Sections 10.1.1–10.1.6). Figure 10.49 displays the normalized figures for 2005.

As the Leipziger Land system is a ground-based installation, the modules exhibit less calefaction and thus the temperature correction factor k_T (annual mean 96.7%) is relatively high. Energy yield was

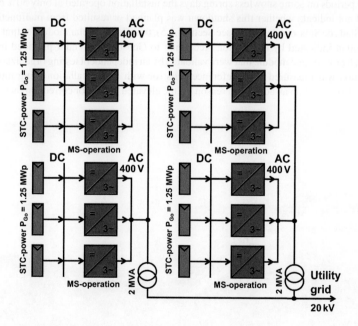

Figure 10.45 Block diagram of the ground-based Leipziger Land PV installation, each of whose two arrays feeds three 400 kVA master–slave inverters (the master inverter is a rotary type). The energy generated by the arrays is exported to the 20 kV medium-voltage grid via a 2 MVA transformer (one per array). The transformers are shut down at night on the medium-voltage side to avoid unnecessary power loss

Figure 10.46   Aerial view of the 5 MWp Leipziger Land PV installation (Photo: Geosol GmbH, www.geosol.de)

somewhat lower in January and February on account of snow, resulting in an exceptionally high solar generator correction factor k_G (annual mean 89.2%) and thus an exceptionally high performance ratio (annual mean 82.4%). Inverter efficiency attained a similarly high level (95–96%) as was the case with the Mont Soleil and Neue Messe installations (see Sections 10.1.2 and 10.1.7 respectively). The data revealed that for certain periods on some snowless spring days the installation operated at only 50% of its normal output, but did not indicate whether this shutdown was planned or resulted from malfunctions.

As with the Stade de Suisse installation (see Section .1.5), the effective solar module output was close to the nominal output indicated by the vendor. According to Geosol, the vendor provided measurement readings for each provisioned module that were validated by an independent testing organization. The fact that the insolation was measured using reference cells (or with this installation using mini reference modules) may also constitute an additional reason for the elevated generator correction factor k_G and the

Figure 10.47   Detail view of an array, whose laminates are installed on a wood mounting frame. Figure 10.48 displays a detail view of the installation's lightning protection system (Photo: Geosol)

Figure 10.48   Rear view of an array, whose modules are installed on insulated mounting racks within the protection zone of the lightning rods installed around 50 cm above the modules. A down-conductor is visible in the lower left-hand corner of the photo (Photo: Geosol)

attendant performance ratio PR. Insolation monitored using reference cells tends to be lower than with pyranometers (see Section 2.8.3), and after a snowfall both the solar generator and reference cells are shaded; thus Y_R is lower, rendering k_G and PR higher. In addition, reference cell measurement accuracy is usually far poorer than that of pyranometers. Nonetheless, the performance ratio (annual mean 82.4%) exhibited by the Leipziger Land installation is outstanding for a system in a low-lying area.

The 2008 data kindly provided by Solar Asset Management GmbH (Figure 10.50) show that at that time the installation was still performing extremely well and the generator correction factor k_G and performance ratio PR were even higher than in 2005, presumably because of less snow shading. However, annual energy yield amounting to 978 kWh/kWp was lower than in 2005 owing to lower insolation.

Since it was necessary to use different software to analyse the 2008 data, the presentation of thermal capture losses L_{CT} from November to March (which were below zero during this period, i.e. Y_T was higher than Y_R) differs somewhat from the presentation in Figure 10.49.

Energy yield in November and December 2008 was lower than for this period in 2005, whereas the reverse held true for January and February in 2008. The Leipziger Land installation is of particular interest as it is only a few kilometres away from the biaxial solar tracked Borna PV installation which integrates

Table 10.11   Key technical data for the ground-based 5 MWp Leipziger Land PV installation

Location:	Espenhain/Sachsen, 164 m, $\varphi = 51.2°$N, $\lambda = 12.5°$E	**Commissioning: Aug. 2004**
PV-generator:	$P_{Go} = 4998$ kWp, $A_G = 43\,918$ m^2, 33 264 modules SQ150,	free field installation
	inclination $\beta = 30°$, azimuth $\gamma = 0°$ (S), consisting of	*(frameless modules)*
	4 sub-plants with 462 parallel strings with	(installation on wooden
	18 monocrystalline modules SQ150 (72 cells in series)	support structures)
Inverter:	Per sub-plant 3 Siemens Sinvert 400 kVA (Master–Slave)	
Measured values:	• Irradiance in horizontal plane (pyranometer)	
(among others)	• Irradiance in module plane (reference cell)	
	• Ambient and module temperature	
	• DC current and DC voltage	
	• produced AC active power	

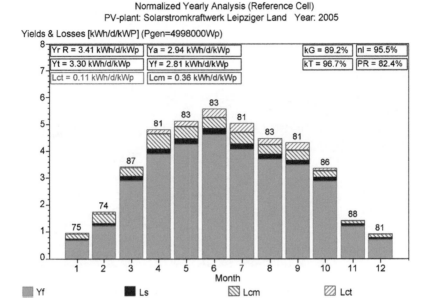

Figure 10.49 Normalized 2005 figures for the Leipziger Land PV installation, whose specific annual energy yield was an impressive 1023 kWh/kWp during the period. Owing to high yields in March and October, winter energy output relative to total annual yield amounted to an impressive 28.5%, despite the installation's northerly location (*Data source*: Geosol)

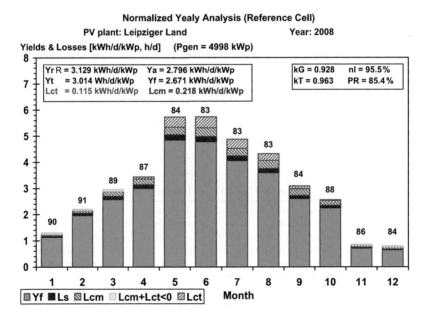

Figure 10.50 Normalized 2008 figures for the Leipziger Land PV installation, whose specific annual energy yield was an impressive 978 kWh/kWp. Winter energy output relative to total annual yield was 28.7% for the period (*Data source*: Solar Asset Management GmbH)

Solon Movers (see Section 10.1.9) and exhibits essentially the same configuration as the likewise biaxial solar tracked Erlasee Solar Park (see Section 10.1.10). Thus a comparison of the energy yields for 2008 allows for the determination of the additional power afforded by biaxial solar tracking.

10.1.9 Borna PV Installation with Biaxial Solar Trackers

As noted, the Borna installation is located in close proximity to the fixed solar panel Leipziger Land PV installation (see Section 10.1.8), which has been in operation for a number of years. The Borna system, which comprises more than 400 Solon Mover biaxial solar trackers, each of which integrates inverters, exhibits an installed capacity of 3447 MWp. Inasmuch as no DC output or temperature data were available for this installation, only the L_C and L_S tally (i.e. total capture and system losses) are shown in Figure 10.51; nor was it possible to break down the figures into L_{CT}, L_{CM} and L_S or compute k_T, k_G or n_I.

During the year in question, the installation with solar trackers produced nearly 30% more electricity than the fixed solar panel system (Figure 10.52), which is not only an impressive figure for a northerly site, but even higher than the mean of 28% indicated in [4.16] for installations of this type in Central Europe.

10.1.10 Erlasee Solar Park with Biaxial Solar Trackers

As noted in Sections 1.4.4 and 4.5.1.7, biaxial solar trackers can increase a PV installation's energy yield and its full-load hour tally. The 12 MWp Erlasee Solar Park, which was built by Solon, comprises around 1500 Solon Mover biaxial solar trackers (total 12 MWp) with integrated inverters. Figure 10.53 displays an aerial view of the installation, Figure 4.61 displays a close-up of a Solon Mover and Figure 4.62 displays a partial view of the installation. As can be seen in Figure 10.53, the gaps between the solar modules must be sufficiently large to avoid reciprocal shading, which means that such an installation takes

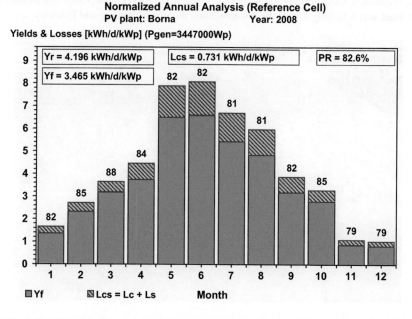

Figure 10.51 Normalized 2008 figures for the Borna PV installation, which has Solon Mover biaxial solar trackers with integrated inverters. The installation's specific annual energy yield for the period was 1268 kWh/kWp, with winter energy output relative to total annual yield amounting to 27.2% (*Data source*: Solar Asset Management GmbH)

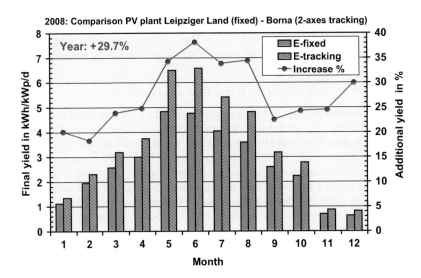

Figure 10.52 Comparison of normalized 2008 energy yield for (a) the biaxial solar tracked Borna PV installation and (b) the fixed solar panel Leipziger Land PV installation, which are in close proximity to each other. Relative to the Leipziger Land (ground-based) installation, the Borna (solar tracked) installation produced between 18% and 38% more electricity, and its annual yield was 29.7% higher

up considerably more space than a fixed solar panel system. In addition, PV installations with solar trackers are subjected to greater mechanical stress during storms.

Figure 10.54 displays the normalized 2007 energy yield for the Erlasee Solar Park compared with that of a fixed solar panel installation at the same site. Owing to exceptionally high direct beam radiation in April, the installation's additional power output ascribable to solar tracking was also very elevated, but on the other hand was relatively low in the low-insolation months of January and February.

Figure 10.53 Aerial view of the 12 MWp Erlasee Solar Park comprising 1500 Solon Mover biaxial solar trackers with integrated inverters. The Solon Mover allows for higher energy yield and a greater number of full-load hours relative to fixed solar panel installations (Photo: Solon AG, © www.paul-langrock.de)

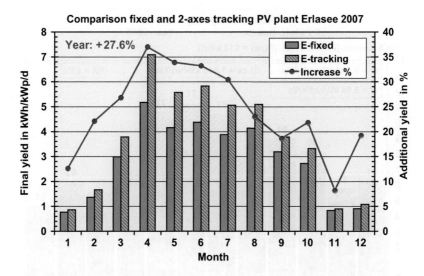

Figure 10.54 Normalized 2007 energy yield for the solar tracker Erlasee Solar Park installation compared with the energy yield of a fixed solar panel installation at the same site. The installation's additional monthly power output ascribable to solar tracking ranged from 8 to 37%, while annual energy yield was 27.6% higher (1342 kWh/kWp)

10.1.11 Guadix PV Installation in Southern Spain, with Biaxial Solar Trackers

Biaxial solar trackers are particularly useful in Southern European or desert locations with elevated direct beam radiation. However, at the time this book went to press no extensive data concerning a fixed solar panel installation in Southern Europe were available, although data concerning the small-scale Guadix biaxial solar tracker installation in southern Spain (Figure 10.55) were available (an approximately 612 kWp system with Solon Mover biaxial solar trackers, six 100 kWp arrays and located at around 37.3°N, approximately 80 km northwest of Almeria).

The installation's solar trackers provide a considerable amount of additional solar generator energy amounting to a whopping 2817 kWh/m². With the higher temperatures in Southern Europe (also see Section 10.1.13) and probably the dust shading in the summer, the Guadix installation's performance ratio is lower than that of the Borna installation, which is located at 51°N. However, it was not possible to compare the figures directly here as no data are available for a fixed solar panel installation at this location.

Annual energy yields of up to 2200 kWh/kWp can be attained using biaxial solar trackers at Southern European sites with high direct beam radiation. Biaxial solar trackers deliver even higher energy at desert sites, and more than $t_{Vm} = 3000$ full-load hours for AC peak output P_{ACmax} (see Section 5.2.7.3).

10.1.12 Biaxial Solar Tracker ENEA PV Installation near Naples, Italy

Located at the ENEA Research Centre in Portici (near Naples; latitude 40.8°N), this 9.6 kWp biaxial solar tracker installation has been in operation since mid 2006 and comprises two 4.8 kWp arrays with 24 Sanyo HIP-200NHE1 modules; each array is mounted on two biaxial solar trackers (12 modules per tracker) [10.18] and integrates a transformerless Sunny Boy SB4200TL inverter.

The measurement readings shown here, which were registered from July 2006 to June 2008, demonstrate that high specific energy yields are obtainable with biaxial solar tracker PV installations in Southern Europe (see Figure 10.56).

The annual energy yield was only slightly lower than for the Guadix installation, but the performance ratio was considerably higher. This may be attributable to the tolerances exhibited by the reference cells at both locations, as well as the following factors:

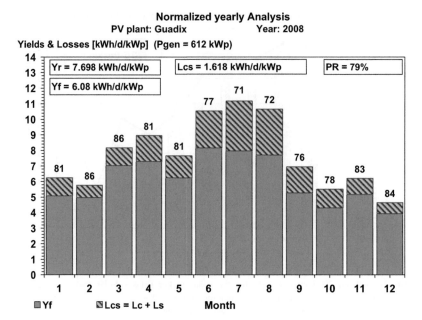

Figure 10.55 Normalized 2008 figures for the Guadix PV installation in southern Spain (around 37°N) with Solon Mover biaxial solar trackers with integrated inverters. The installation's specific annual energy yield was 2225 kWh/kWp, while winter energy output relative to total annual yield was 41.6% (*Data source*: [10.17])

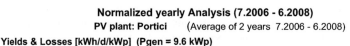

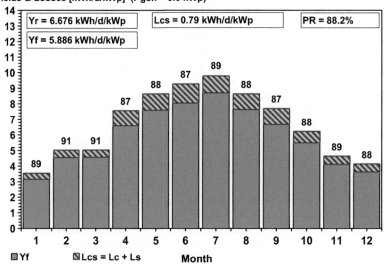

Figure 10.56 Normalized annual figures for July 2006 to June 2008 for the ENEA Research Centre experimental PV installation in southern Italy (around 41°N) with biaxial Deger solar trackers, Sanyo HIP-200NHE1 solar modules and Sunny Boy SB4200TL inverters. Mean specific annual energy yield was 2148 kWh/kWp, while winter energy output relative to total annual yield was 35.8%, which is considerably lower than for the Guadix installation (These figures were obtained from a diagram in [10.18].)

- Greater inverter efficiency: The transformerless inverters used exhibit higher efficiency than those in the Solon Mover biaxial solar trackers.
- Lower thermal capture losses: The temperature coefficient for the installation's solar modules (the front and rear of whose solar cells possess additional silicon film coatings) indicated by the vendor is −0.3% per K, which is considerably lower than the −0.45 to 0.5% per K figure for standard crystalline silicon solar cells. In addition, the installation is located near the coast, where ambient temperatures may be lower than those to which the (inland) Guadix installation is exposed.

10.1.13　PV Installation in Mudgee, Australia

This installation is located in a small inland town around 200 km northwest of Sydney (Figure 10.57) on the other side of the coastal mountain range at 32.6°S ($\varphi = -32.6°$).

The installation comprises 10 BP3160 modules (two parallel-connected strings comprising five series-connected modules each; $P_{Go} = 1.6$ kWp; tilt angle $\beta = 25°$, $\gamma = 0°$; the installation faces *north* despite its southern hemisphere location) and integrates an SMA SB1700 inverter. The monitoring data were recorded by SMA using a commercially available measurement instrument. Peak solar generator power is 1.6 kWp [10.16]. The University of New South Wales kindly provided the installation's 2005 and 2006 monitoring data (Figure 10.58).

Owing to the installation's southern hemisphere location at a latitude corresponding to a Southern European or North African site, the minimum and maximum values are shifted by six months, i.e. the lowest yields and thermal capture losses occur in June and July, while the maximum yields and thermal capture losses occur in December and January.

The installation's insolation and yield are higher than in more southerly coastal Sydney, probably because of the installation's inland location on the leeward side of the coastal mountain range.

As no DC output readings were available for the installation, DC power was calculated using the vendor's AC power efficiency curve. Using the available module temperature data, capture losses L_C

Figure 10.57　Photo of a 1.6 kWp PV installation on a single family home in Mudgee, Australia, whose latitude is around 32.6°S and which is located around 200 km northwest of Sydney (Photo: Muriel Watt, UNSW)

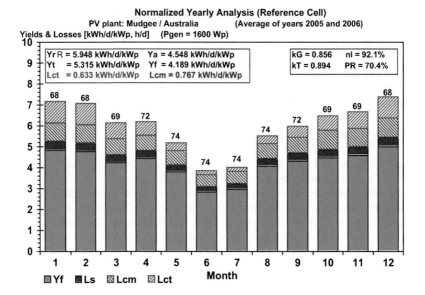

Figure 10.58 Normalized annual figures for 2005 and 2006 (mean monthly figures) for the 1.6 kWp installation in Mudgee. Mean annual yield amounted to 1529 kWh/kWp, while winter energy output relative to total annual yield was 44.7% (*Data source*: UNSW, Sydney, Australia)

could be broken down into their L_{CT} and L_{CM} constituents, thus also allowing for estimation of the yield loss induced by temperature increases.

The monthly and annual k_T readings match up relatively well with the Southern European figures in Table 8.2, if the fact that the seasons experienced by the northern and southern hemispheres differ by six months is taken into account.

Figures 10.59 and 10.60 show the normalized annual bar charts for 2005 and 2006. The values for these two years differ very little, with the highest fluctuations occurring in the lowest-insolation month (in this case June).

10.1.14 PV Installation in Springerville, Arizona

This installation is sited in the Arizona Desert (longitude $\varphi = 34°$N, latitude $\lambda = 109°$W, 2000 m above sea level). The installation was built in various stages, beginning in 2001, by the Tucson Electric Power Company (TEP), which is also the installation operator (see Figure 1.16). The installation's 2006 output was 4.59 MWp, at which time it comprised 34 arrays of 130–135 kWp, each of which is connected to a Xantrex PV-150 triphase inverter. All of these inverters are mounted outdoors; 26 of the arrays integrate solar generators comprising 300 Wp RWE Schott Solar crystalline solar generators; four of the arrays integrate 45–50 Wp First Solar CdTe modules; and the remaining four have 43 Wp MST-43 BP Solarex amorphous silicon tandem-cell modules (like the Newtech 2 installation). All of the modules are oriented due south and exhibit a tilt angle of $\beta = 34°$ ($\beta = \varphi$ is ideal for sites with direct beam radiation). The installation's electricity is exported to the 34.5 kV medium-voltage grid. The installation can be upgraded at the same site to a nominal AC power of 8 MW [10.13].

TEP kindly provided, for the purposes of analysis, multi-year monitoring data concerning insolation and exported electricity. As only insolation and AC power yield data were available, it was only possible to determine Y_R, Y_F and PR; capture losses could not be broken down into their L_{CT}, L_{CM} and L_S components.

Figure 10.61 displays the normalized figures for 2005, throughout which the installation was operated under the same output conditions. As with the Jungfraujoch installation, energy production was relatively

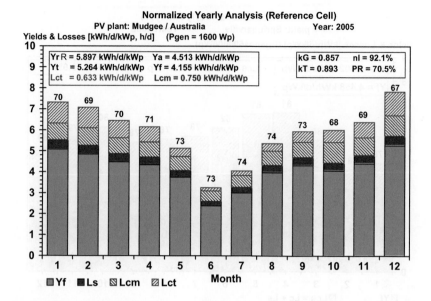

Figure 10.59 Normalized 2005 figures for the 1.6 kWp PV installation in Mudgee. Mean annual yield amounted to 1517 kWh/kWp, while winter energy output relative to total annual yield was 43.7% (*Data source*: UNSW, Sydney, Australia)

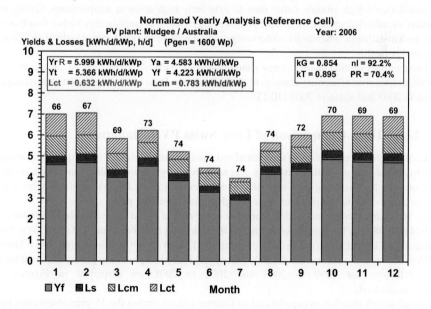

Figure 10.60 Normalized 2006 figures for the 1.6 kWp PV installation in Mudgee. Mean annual yield amounted to 1541 kWh/kWp, while winter energy output relative to total annual yield was 45.6% (*Data source*: UNSW, Sydney, Australia)

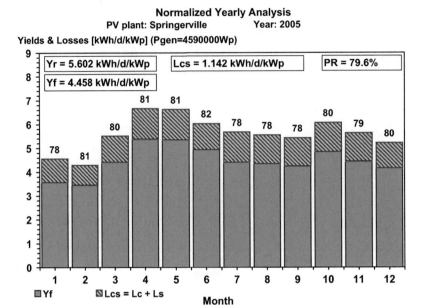

Figure 10.61 Normalized 2005 figures for the Springerville, Arizona, PV installation (latitude $\varphi = 34°$N; 2000 m above sea level). The installation's specific annual energy yield amounted to 1628 kWh/kWp for the period; the performance ratio was 79.6%; the winter energy output relative to total annual yield was 46.4% (*Data source*: Tucson Electric Power Company)

homogeneous over the course of the year, and the nearly 80% performance ratio was partially attributable to the installation's high altitude rather than to extremely high ambient temperatures. Owing to the installation's southerly location, its specific annual energy yield was considerably higher than that of the Jungfraujoch installation, whereas its winter energy output relative to a total annual yield of 46.4% was on a par with this facility.

The installation's specific annual energy yield reached 1720 kWh/kWp in 2004. Mean specific annual energy yield since commissioning amounts to 1673 kWh/kWp. Parts of the installation were damaged by lightning in 2003 and again in 2004 [10.13].

10.2 Long-Term Comparison of Four Swiss PV Installations

This section presents a comparison of the normalized monthly energy yield of four Swiss PV installations for which detailed monitoring data extending over more than 15 years are available and which are sited as follows: one in a low-lying region, a second at a high elevation in the Jura Mountains, a third at a high Alpine location, and a fourth on an extremely high Alpine mountain peak.

Figure 10.62 displays the normalized monthly energy yield based on peak solar generator output from 1994 to 1998 for a PV installation in each of the following locations: Burgdorf (3.18 kWp, 540 m); Mont Soleil (555 kWp, 1270 m); Birg cable car station (4134 kWp, 2670 m); and Jungfraujoch Mountain (1.15 kWp, 3454 m). The normalized monthly energy yield based on peak solar generator output for these same installations for 1999 to 2003 and 2004 to 2008 are displayed in Figures 10.63 and 10.64 respectively.

The Jungfraujoch installation experienced no inverter failures during the 15-year observation period, whereas the Burgdorf, Birg and Mont Soleil installations experienced sporadic breakdowns (in some cases involving inverter failures) primarily during the first two years of operation. The Birg and Mont Soleil installations each experienced a relatively lengthy inverter failure thereafter – Birg in the spring of

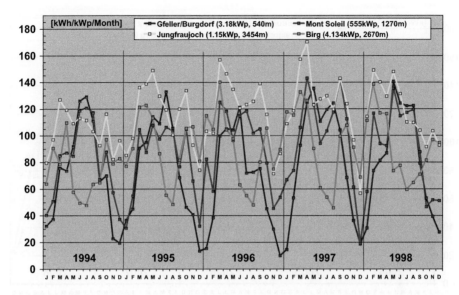

Figure 10.62 Normalized monthly energy yield (derived from peak nominal solar generator output) from 1994 to 1998 for the following PV installations: Jungfraujoch (1152 kWp), Birg (4134 kWp), Mont Soleil (555 kWp) and Burgdorf (3.18 kWp)

2003 and Mont Soleil in summer of 2006. A mid 1996 inverter failure (which went unnoticed at first because the owner was on holiday at the time) cut into the Burgdorf installation's energy production for that year. Energy production of PV installations in Switzerland's Mittelland region, which often experiences fog or low stratus clouds, fluctuates greatly over the course of any given year between a

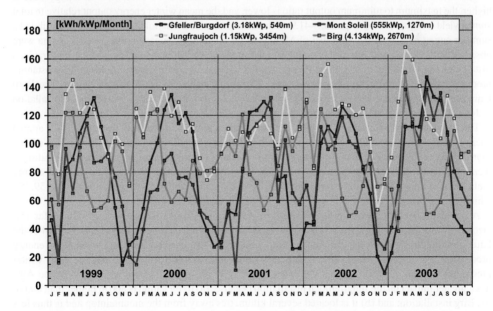

Figure 10.63 Normalized monthly energy yield (derived from peak nominal solar generator output) from 1999 to 2003 for the following PV installations: Jungfraujoch (1152 kWp), Birg (4134 kWp), Mont Soleil (555 kWp) and Burgdorf (3.18 kWp)

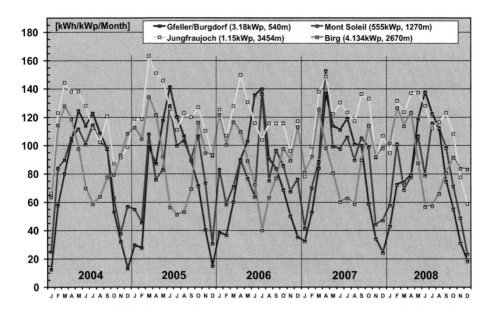

Figure 10.64   Normalized monthly energy yield (derived from peak nominal solar generator output) from 2004 to 2008 for the following PV installations: Jungfraujoch (1152 kWp), Birg (4134 kWp), Mont Soleil (555 kWp) and Burgdorf (3.18 kWp)

pronounced high in summer and the reverse in winter. The mean winter energy output relative to total annual yield for the Burgdorf installation is 29%, and the ratio of the maximum to minimum output ranges from about 5:1 to 15:1.

The situation is similar for the higher-elevation Mont Soleil installation, except that its energy yield is higher, the maximum to minimum output ratio is lower, and the mean winter energy output relative to total annual yield is higher (38%; range 30.7–41.9%). In some years, the Mont Soleil installation has exhibited a summer output peak on a par with that of Mittelland installations, whereas in other years maxima were observed in the spring and autumn, as with the Jungfraujoch installation. Monitoring of the Jungfraujoch installation from the autumn of 1999 to the spring of 2001 was only rudimentary, resulting in a substantial production decrease owing to string breakdowns that went undetected for a lengthy period. The Bern University of Applied Sciences PV Lab implemented a new measuring system at the Jungfraujoch installation in June 2001, since which time production has risen appreciably with the continuous monitoring.

The two façade installations in the Alps exhibit a far superior performance, however. Their annual energy production is considerably higher than that of the other installations, and their performance is far more consistent by virtue of their lower monthly yield scatter. The minimum to maximum yield ratio for these installations normally ranges from around 2 to 3, with two production peaks annually and low energy yield in the summer. The winter energy output relative to total annual yield ranges from 43.2 to 59.7%, which is on a par with that of desert installations (see Section 10.1.14).

A comparison of the two Alpine installations reveals that: (a) the Birg system produces far less electricity during the summer than the Jungfraujoch facility owing to the reflecting glacier in front of the latter installation; and (b) the Birg installation sometimes registers higher specific monthly energy production than the Jungfraujoch installation. This is attributable to the fact that the high winter precipitation often observed at sites with southwesterly exposure on the southern slopes of the Alps has a greater impact on the Jungfraujoch installation because: (a) its elevation is 800 m higher than that of the Birg installation; and (b) it is located several kilometres away from the mountaintop and is thus less affected by this phenomenon. However, owing to the Jungfraujoch's reflecting glacier its low yield period in the summer is far less pronounced than that of the Birg installation, and thus its annual energy yield is far higher than that of the latter facility.

The only real problem faced by both of these high Alpine installations is the transient snow shading that often occurs in spring, although the 90° tilt angle exhibited by both facilities mitigates this problem for the most part. With greater clearance between the ground and solar generator (e.g. 5 to 7 m in lieu of 3 m or 2–3 m rather than less than 1 m for the Jungfraujoch and Birg installations respectively), this snow shading would be a rarity.

Alpine building façades with good Sun exposure are ideal for grid-connected PV systems, whose energy is far more compatible with the Swiss energy supply load profile than is the case with Mittelland region PV installations and dovetails extremely well with run-of-river station energy output. Between the months of November and February, such Alpine PV installations exhibit per installed kWp solar generator power that is several magnitudes higher than that of the equivalent building roof or façade installations in Mittelland. Solar generators in Mittelland need to be oversized, whereas those installed at Alpine sites do not. Inasmuch as most façade-mounted PV installations are well protected against lightning strikes, the frequent atmospheric voltage surges in Alpine locations can be readily handled by implementing suitable protective measures. Hence such installations should be mounted on as many Alpine buildings as possible, particularly those used for tourism purposes, providing that the grid conditions allow for an inverter hook-up. Various installations of this nature have been realized in recent years, which is a welcome development.

The performance monitoring that has been conducted at the Jungfraujoch and Birg installations for more than 16 years now shows that: (a) once the initial kinks have been ironed out and if quality inverters are used, the projected energy yields amounting from 1000 to 1500 kWh/kWp can in fact be attained; (b) grid-connected systems can deliver reliable performance and high to extremely high energy yields with high winter energy output relative to total annual yield under the extreme climatic conditions of the Alps; and (c) under optimal conditions, high Alpine PV installations can attain annual energy yields that are on a par with those of Southern European installations.

10.3 Long-Term Energy Yield of the Burgdorf Installation

Following the introduction of the Burgdorf model back in 1991, a relatively large number of PV installations were installed in this small town between 1991 and 1996 (see Figure 1.17), for which, with Bern University of Applied Sciences PV Lab monitoring, detailed data extending over a 17-year period are available.

Expressing energy yield in kilowatt hours per kilowatt peak (kWh/kWp) greatly eases the task of comparing the performance of various PV installations, and has the further advantage of rendering the size of the installation irrelevant since the energy yield of each installation is expressed relative to 1 kWp (1 kW peak output under STC) (see Chapter 7).

Figure 10.65 displays the annual energy yields normalized in this fashion for all Burgdorf PV installations.

The figure contains the following information concerning all of the relevant installations: mean energy yield; maximum energy yield (initially based on regularly cleaned installations with transformerless inverters; from 2002 onwards based on the Newtech 1 CIS installation, see Section 10.1.6); and minimum energy yield (based on a suboptimal façade installation that was partially shaded during the summer by building elements and trees; tilt angle $\beta = 60°$; $\gamma = 20°$; the installation also experienced inverter problems in 2008). The tilt angle of most of the other installation ranges from 30° to 35°, and virtually all them have framed modules. Mean energy yield was determined by weighting energy yield of each of the installations with their nominal installation output. In this process, a 62.5 kWp PV installation with solar roof tiles and transformer inverters that is owned by a bankrupt company and thus has not been serviced in several years and exhibits a very low and atypical energy yield was excluded. Since (a) a c-Si PV installation realized in 2007 and a second in 2008 exhibited modules that likewise provided the specified nominal output, and (b) the Newtech 1 CIS installation had undergone continuous degradation (see Section 10.1.6), this installation was displaced from its leading position (annual energy yield of the best installation in 2008 and 2009: 1060 and 1096 kWh/kWp respectively).

In order to analyse the long-term performance and possible degradation exhibited by solar generators, it is necessary to monitor the relevant installations for a sufficiently lengthy period. Hence my further

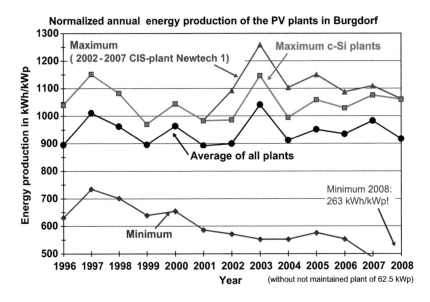

Figure 10.65 Normalized annual energy yield from 1996 to 2008 for all inverter PV installations in Burgdorf (mean weighted with the installation output under STC). From 2002 to 2007, the Newtech 1 CIS installation exhibited the highest output as its effective baseline output was higher than the specified nominal output (see Section 10.1.6). Owing to insolation fluctuations from one year to the next, the mean and maximum production figures for the PV installations monitored are for the most part keyed to the annual insolation registered in Burgdorf (see Figure 11.1)

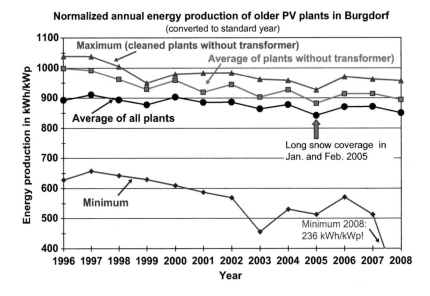

Figure 10.66 Normalized annual energy yield from 1996 to 2008 for all inverter PV installations in Burgdorf that had been in operation for at least 10 years at the time of the analysis (mean value weighted using output under STC, converted to a standard insolation year using $H = 1163\,\text{kWh/m}^2$). The maximum and minimum energy yield fluctuations are considerably lower than in Figure 10.65, although the minimum yield for a neglected installation whose inverter malfunctions went undetected for a lengthy period is extremely low

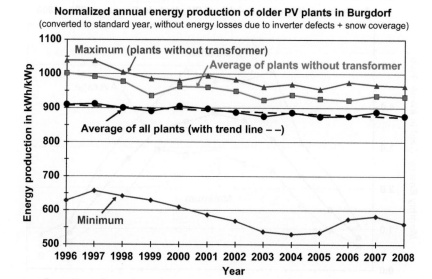

Figure 10.67   Normalized annual energy yield from 1996 to 2008 for all inverter PV installations in Burgdorf that had been in operation for at least 10 years at the time of the analysis (mean value weighted via output under STC), converted to a standard insolation year using $H = 1163$ kWh/m². The mean energy yield for this period was around 0.3% lower, whereby the energy losses occasioned here by inverter failure and snow shadowing were eliminated via an interim calculation

analyses (see Figures 10.66 and 10.67) only included installations that had been in service for at least 10 years, i.e. a total of 32 installation with 39 inverters and aggregate output of 171 kWp.

In the interest of eliminating the effect of inter-annual insolation fluctuations, the energy yields of the various installations were first converted to a standard insolation year $H = 1163$ kWh/m². This now unduly low figure (see Figure 11.1) was defined in the late 1990s as the mean between the data in [2.1] and [2.2] and for measurements realized since late 1991, and has been retained here for reasons of comparability. This in turn constitutes the method that was used to derive the normalized annual energy yield curve in Figure 10.66.

In order to gain some insight into the ageing and degradation of the relevant solar modules, the energy losses for the installations that exhibited inverter failures were calculated, whereby the installation energy loss occasioned by snow shadowing was also determined. Figure 10.67 displays the resulting corrected normalized annual energy curves.

As the curve shows, the mean value decreased by around 3.7% over a 12-year period, which equates to an annual decline of around 0.3%. This evolution is likely to be mainly attributable to installation degradation and ageing, as many of the PV installations in Burgdorf are cleaned periodically. But some have probably never been cleaned and the decline is partially attributable to tree shading in some cases (see Figures 10.2–10.4).

10.4   Mean PV Installation Energy Yield in Germany

The web site of the German solar energy organization Solarenergie-Förderverein Deutschland (www. pv-ertraege.de) contains an energy yield statistics database where owners of (primarily) small-scale PV installations can enter their monthly energy yield figures. The number of installations whose statistics are included in this database has steadily increased since October 1991, and at the time this book went to press the database offered data from more than 5000 PV installations with an aggregate output of more than

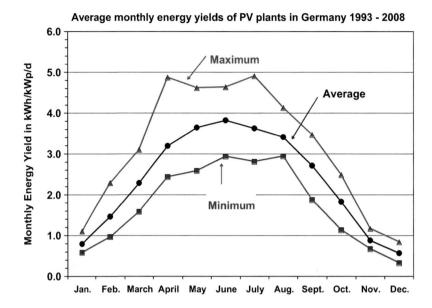

Figure 10.68 Normalized monthly energy yield in kWh/kWp/d from 1993 to 2008 for German PV installations according to the data available at www.pv-ertraege.de. The fact that the maximum exceptionally occurred in April in both Germany and Switzerland is attributable to the extremely sunny conditions during this month in low-lying areas (see Figure 10.54)

50 MWp. The mean aggregate output of the installations represented in this database increased from around 2 kWp in 1992 to more than 10 kWp in 2008.

Figure 10.68 displays the normalized mean monthly yield in kWh/kWp/d (mean, minimum and maximum values) derived from this database. Most of these installations are probably located in low-lying areas. As noted in Section 10.1, the energy yield of Swiss PV installations exhibits particularly elevated scatter during the winter. However, as these are nationwide figures, the scatter between the observed minimum and maximum figures is less pronounced than for the individual installations, whose maximum amounts to a factor of close to 5 as opposed to 2.6 (see Figure 10.1).

10.5 Bibliography

[10.1] R. Minder: 'Das Solarkraftwerk Phalk Mont-Soleil: Betriebserfahrungen und erste Bilanz'. *Bulletin SEV/VSE*, 10/1993.

[10.2] H. Häberlin, Ch. Beutler, S. Oberli: 'Die netzgekoppelte 1.1 kW-Photovoltaikanlage der Ingenieurschule Burgdorf auf dem Jungfraujoch'. *Bulletin SEV*, 10/1994.

[10.3] H. Häberlin, Ch. Beutler: 'Die netzgekoppelte Photovoltaikanlage der ISB auf dem Jungfraujoch'. 10. Symposium PV Sonnenenergie, Staffelstein, March 1995.

[10.4] H. Häberlin, Ch. Beutler: 'Highest Grid Connected PV Plant in the World at Jungfraujoch (3454m): Excellent Performance in the First Two Years of Operation'. Proceedings of the 13th EU PV Conference, Nice, 1995.

[10.5] H. Häberlin, Ch. Beutler: 'Energieertrag hochalpiner netzgekoppelter Photovoltaikanlagen'. *Elektrotechnik*, 9/1996.

[10.6] H. Häberlin, Ch. Renken: 'Hoher Energieertrag auf Jungfraujoch: Die ersten fünf Betriebsjahre der netzge-koppelten 1.1 kWp-Photovoltaikanlage der HTA Burgdorf'. 14. Symposium Photovoltaische Solarenergie, Staffelstein, March 1999.

[10.7] H. Häberlin: 'Grid-connected PV plant on Jungfraujoch in the Swiss Alps', in Michael Ross and Jimmy Royer (eds), *Photovoltaics in Cold Climates*. James & James, London, 1999.

[10.8] H. Häberlin:'Netzgekoppelte Photovoltaikanlage Jungfraujoch: 10 Jahre störungsfreier Betrieb mit Rekord-Energieerträgen'. *Bulletin SEV/VSE*, 10/2004.

[10.9] H. Häberlin:'Grid Connected PV Plant Jungfraujoch (3454m) in the Swiss Alps: 10 Years of Trouble-free Operation with Record Energy Yields'. Proceedings of the 19th EU PV Conference, Paris, 2004.

[10.10] H. Häberlin:'Hochalpine Photovoltaikanlagen–Langzeiterfahrungen mit Fassadenanlagen'. *Elektrotechnik*, 6–7/2004.

[10.11] G. Becker, B. Gisler, G. Kummerle *et al*.: 'More than 6 Years of Operation Experience with a 1 MW Photovoltaic Power Plant – Highlights and Weak Points'. 19th EU PV Conference, Paris, 2004.

[10.12] H. Häberlin:'Langzeiterfahrungen mit zwei hochalpinen Photovoltaikanlagen'. 20. Symposium Photovoltaische Solarenergie, Staffelstein, 2005.

[10.13] T. Hansen, L. Moore, T. Mysak, H. Post:'Photovoltaic Power Plant Experience at Tucson Electric Power'. November 2005 (available from www.greenwatts.com).

[10.14] G. Becker, B. Schniebelsberger, W. Weber *et al*.: 'An Approach to the Impact of Snow on the Yield of Grid-Connected PV Systems'. 21st EU PV Conference, Dresden, 2006.

[10.15] H. Schmidt, B. Burger, K. Kiefer:'Welcher Wechselrichter für welche Modultechnologie'. 21. Symposium Photovoltaische Solarenergie, Staffelstein, 2006.

[10.16] M. Watt, R. Morgan, R. Passey:'Experiences with Residential Grid-Connected Photovoltaic Systems in Australia', Proceedings of Solar 2006, the 44th Annual Conference of the Australian and New Zealand Solar Energy Society, Canberra, 13–15 September 2006.

[10.17] U. Rindelhardt, A. Dietrich, Ch. Rösner:'Tracked Megawatt PV Plants: Operation Results 2008 in Germany and Spain'. 24th EU PV Conference, Hamburg, 2009.

[10.18] G. Graditi, C. Cancro, A. Merola, C. Privato:'One or Two Axes Tracking? The Case Study of a 10 kWp Tracking PV System Installed in Southern Italy'. 24th EU PV Conference, Hamburg, 2009.

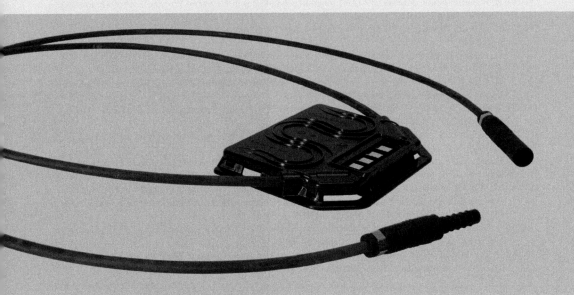

«Renewable energy – an excellent idea deserves excellent connections»

Solar Energy Solutions

Connect with RADOX® SOLAR connectivity solutions using our quality to secure reliable and maintenance free connections throughout the entire lifetime of your solar module.

RADOX® SOLAR cables are extremely robust and resist high mechanical load and abrasion.

RADOX® SOLAR connectors contribute significantly to high performance and long service life.

RADOX® SOLAR junction boxes guarantee excellent heat dissipation by their patented design.

HUBER+SUHNER AG
8330 Pfäffikon Switzerland
info@hubersuhner.com

hubersuhner.com

11

In Conclusion. . .

Despite major price decreases since 1990, PV installations are still relatively high-ticket items, and manufacturing them is an energy-intensive process. Nonetheless, there have long been numerous applications from an economic, social and ecological standpoint where PV installations are the most suitable electricity generation solution.

PV installations are used in many places around the world to provide small to medium amounts of electricity for appliances in domains such as the following: telecommunications, navigation, weather and environmental monitoring (e.g. for hurricane forecasting), holiday homes, and mountain and Alpine cabins. In developing countries, PV installations provide electricity for water pumps in arid regions and rural hospitals, as well as basic power for households in remote rural areas. As little as 0.1 to 0.2 kWh per day often suffices to provide a household with electricity for lighting, a radio, CD player, mobile pone and TV. If a small, well-insulated refrigerator is added to this, basic household electricity needs can be met with 0.5 to 1 kWh a day. Owing to the modularity of PV installations, these amounts of electricity can be provided for individual households or entire villages. And as most developing countries enjoy high daily insolation that is subject to considerably less seasonal fluctuation than in moderate climate zones, PV electricity generation is considerably less expensive than in regions such as Central Europe. The high investment costs entailed by PV installations still pose a problem, but one that can be overcome through measures such as development aid projects. These measures allow for the establishment of a modicum of modern comforts in the rural areas of developing countries, while at the same helping to stem the tide of migration from rural areas to large cities, as well as illegal immigration into industrial countries. And as the households potentially in need of this type of electricity supply number in the hundreds of millions, the potential market for PV installations is monumental.

Although stand-alone installations (see Section 5.1) could help to slow spiralling energy consumption or help meet electricity demand in developing countries, the electricity produced by such installations is a mere drop in the ocean compared with the amounts of electricity that are produced today. Hence larger-scale installations and grid-connected systems are needed, which, with the declining costs of batteries, allow for far lower electricity prices and have a more positive impact on the energy balance (yield factor greater than 1; see Section 9.2) in terms of cost than stand-alone installations whose energy depreciation periods are longer on account of the need to replace the batteries of these installations periodically.

Nonetheless, because of output fluctuations over the course of a day or because of weather conditions, as well as the related output peak at around noon on sunny days, the amount of PV energy that can be exported to the public grid is limited for reasons having to do with the laws of physics. In the absence of specific measures in this regard, in Switzerland only about 10 to 15% of total electricity demand can be exported to the public grid (see Section 5.2.7). However, today's information technologies allow for the implementation of additional tariff-related and technical measures such as load balancing during periods of excess supply, which should enable these restrictions to be rolled back appreciably and relatively easily.

Photovoltaics: System Design and Practice. Heinrich Häberlin.
© 2012 John Wiley & Sons, Ltd. Published 2012 by John Wiley & Sons, Ltd.

Another possibility in this regard is greater use of the existing European grid system for load-balancing purposes in cases where electricity production varies from one region to another.

If greater amounts of electricity are to be generated over the long term within the context of a grid system, some of this electricity will need to be stored in order to maintain grid stability, using technologies such as pump storage systems, compressed-air energy accumulators (CAESs) and chemical storage in batteries (short term) or hydrogen (longer term). Pump storage systems are extremely useful for storing large amounts of fluctuating solar and wind power as they help to reduce output peaks and make up for PV electricity production shortfalls; this in turn enhances grid stability (see Section 5.2.7). Certain environmental groups that are fundamentally opposed to pump storage systems would do well to rethink their position in this regard.

Though stand-alone installations entail higher electricity prices ranging from around €0.85 to €1.55 per kWh, they constitute a wholly viable economic solution for their application domain. However, the same cannot be said of grid-connected systems, which, based on realistic life span and interest-rate assumptions relative to current electricity prices (peak rate ranging from around €0.1 to €0.2 per kWh), will never be an economically viable solution unless higher feed-in tariffs are adopted into law, as has been in done in Germany, Spain and Switzerland.

Considerably less energy is needed to manufacture PV installation components than was once the case, and the energy payback times for Central European PV systems currently installed in flat areas amount to around three years (see Section 9.2). Hence the contention that manufacturing grid-connected system components makes no sense from an energy standpoint is simply untrue, since such installations exhibit a far longer life span than stand-alone installations.

The favourable PV installation energy export conditions that prevail in Japan, Germany and Spain have resulted in a decade-long period of steadily increasing demand for solar modules; this in turn has prompted the industry to expand considerably its production capacities and has increased tremendously the economic importance of these modules. Aggregate PV industry sales probably amount to around €30 billion a year by now. And on the learning curve, each time the manufacturing volume of an industrial product doubles, an approximately 20% price reduction for that product normally ensues. Such price reductions have in fact occurred for inverters, but not to the same extent for solar modules. Inasmuch as production capacities for pure silicon and for solar modules have been greatly expanded recently, while at the same time the financial crisis has slowed the growth in demand for these products, in 2009 solar module prices finally came down considerably; as a result feed-in tariffs (see Tables 1.2 and 1.3) are now fairly high relative to current solar module prices. Improved efficiency as well as reduced energy and materials use will promote a steady decrease in the prices of solar modules going forward, as will the annual reductions of feed-in tariffs under Germany's Renewable Energy Act (see Tables 1.2 and 1.3). The way the cost trends appear now, it looks as though, as from 2010, prices may decrease further than was originally predicted.

Steadily growing demand and sales revenue will also encourage module vendors to invest more resources in R&D, with the aim of optimizing their products and the efficiency of them, and/or reducing the amount of silicon per watt needed to make them. Such optimizations also reduce electricity generation costs (see Section 9.1) and the per-watt amount of energy needed to make solar modules (see Section 9.2). The silicon shortage that has afflicted the PV industry in recent years has resulted in an increased interest in thin-film solar cells that use less silicon, or even none at all, while still providing good efficiency. Such solutions include amorphous triple solar cells, modules with micromorph tandem cells, and CIS or CdTe modules.

From an environmental protection standpoint, legislating a steady increase in the price of environmentally harmful fossil fuels would be ideal, in that it would promote the implementation of energy-saving measures and would render wind and solar power technologies more competitive via simple and highly efficient market mechanisms. But such legislation would be extremely difficult to push through. Hence the second-best solution would be to expedite this evolution by subsidizing PV electricity production via statutory financial incentives that are subject to time limits.

However, subsidies based on per kilowatt of installed capacity are a less desirable approach, as this solely promotes the installation of PV systems as opposed to operating them with a view to obtaining the highest possible energy yield. The automatic annual degression framework called for by Germany's Renewable Energy Act has been highly successful and since the law's enactment in 2000 has been instituted in one form or another by many European countries. Experience has shown that when such subsidy frameworks (which of course temporarily entail a substantial financial outlay) remain in force for long enough, the relevant markets, industries and technologies evolve in such a way that the economic viability of the relevant technologies steadily improves, thus allowing for their implementation with steadily declining subsidies, and eventually with none at all. This is exactly what occurred in Germany's wind turbine industry. In the early and middle 1990s, Switzerland was in the vanguard of PV installation development and realization. However, Switzerland's feed-in tariffs (KEV) law enacted in 2008, which sets a number of low limits, will not be enough to spur a major expansion of PV electricity and should solely be regarded as a first step. As Figure 9.2 shows, the need for market subsidy programmes will increase exponentially once grid parity is reached.

Production capacity expansion, whether it involves expanding existing facilities or building new ones, cannot be realized over night. In other words, we cannot shut down our nuclear power plants and transition to an all solar and/or wind power energy system at the drop of a hat. How lengthy a process this might be is clearly illustrated by Figure 1.23, which shows world solar cell production from 1997 to 2008 and contains an estimate of possible production levels for 2009.

In view of the fact that (a) PV installations in moderate climate zones are likely to use monocrystalline and polycrystalline solar modules for the most part in the coming years, and (b) today's thin-film triple-solar-cell modules are considerably more efficient than the amorphous modules of the past, but are still relatively inefficient, the cause of fostering the development of photovoltaics will best be served if price reductions for monocrystalline and polycrystalline solar cells are passed along to end customers as technical advances are made and if the amount of energy expended to manufacture solar modules is further reduced. Another desirable advance would be for the efficiency of high-production-volume products to move several percentage points closer to the efficiency now achieved with laboratory PV installations, providing that this evolution does not result in a significant price increase.

On the other hand, today's amorphous silicon solar cell technologies are ideal for small and very low-power portable devices such as pocket calculators that use solar energy when sufficient light is available and switch to a battery when it is not. The spectral sensitivity of amorphous silicon solar cells lends itself extremely well to such applications. Primary batteries are a suboptimal solution from an energy and ecological standpoint as the amount of energy needed to manufacture them exceeds the energy they provide by up to a factor of 50, and disposing of them also poses a problem.

Product researchers are working feverishly on the development of improved, low-priced thin-film solar cells that require little in the way of manufacturing energy, and many researchers are attempting to develop tandem solar cells that make more efficient use of the available insolation and thus exhibit improved efficiency. It is likely that one or more breakthroughs in these fields will ultimately allow for the mass production of solar cells whose efficiency is at least on a par with that of today's monocrystalline and polycrystalline solar cells but that are less expensive and consume less manufacturing energy. The materials used in such products should also be environmentally friendly, which means that the use of cadmium telluride (CdTe) for mass-produced solar cells is a suboptimal solution. It is essential that the recycling concept for damaged and decommissioned CdTe solar modules rules out ecological damage of any kind. As of October 2009, these were by far the least expensive modules available (manufacturing cost around €0.65 per Wp) and are thus coming into increasing use for large-scale ground-based PV installations. CIS modules are also a promising solution, as their efficiency is now close to that of crystalline silicon modules (see Figure 10.39), but shortages of indium would pose a problem if these modules were produced in far greater numbers than is currently the case.

The fact that PV products now represent a market worth more than €30 billion has prompted the publication of a number of studies concerning the future of this market. According to one such study, by Photon Consulting (*Photon*, April 2007), the manufacturing costs of PV installation components will decrease by 2010 to a point where in some countries solar power costs will be on a par with utility company rates (see Figure 9.2).

In view of the fact that global warming has sparked a greater interest in renewables, the question arises as to the impact of climate change on the various energy resources that come into play here.

In terms of hydro power, it is generally assumed that winter watercourse flows will rise as the result of an anticipated trend towards greater precipitation, but will substantially decline in the summer, particularly if Alpine glaciers continue to melt. Hence hydro power output will tend to be higher in the winter, but lower in the summer. A slight increase in winter hydro power production would be beneficial for the electricity industry, since electricity demand is higher in winter than in summer. However, global warming will probably result in a net decrease in hydro power output over the long term.

Overall electricity demand is set to rise, particularly during the summer owing to the increased use of air-conditioners, refrigerators and freezers in hot weather. And although the power consumption of electrically heated homes will decline considerably on account of global warming, the replacement of fossil fuels with technologies such as heat pumps (in lieu of oil and gas heating) and ventilation systems and rotary pumps (in low-energy buildings) will probably preclude a substantial decrease in overall winter energy demand. And of course a population increase will translate into increased electricity demand as well.

Climate change is likely to be good for the PV sector. As noted in Chapter 10 and elsewhere, since 1992 the Bern University of Applied Sciences PV Lab has been conducting long-term monitoring of numerous PV installations. The measurement readings for some of these installations, which extend over uninterrupted periods ranging from 16 to 18 years, all indicate that insolation on the inclined plane of solar generators, as well as energy yield, have been rising by an average of around 0.1 to 0.6% per year, depending on the installation (see Figures 11.1–11.3). The two curves in Figure 11.1 substantially exceed the mean values that were obtained using Meteonorm 5 solar insolation software, whose underlying database is mainly derived from readings that date back to the 1980s and 1990s [2.5]. Inasmuch as the values in Tables A2.1–A2.10 and A3.1–A3.3 are derived from relatively old solar generator insolation readings (i.e. using the Meteonorm 3 and 4 applications), as a rule they can be regarded as a conservative basis for the energy yield calculations described in Section 2.6 and Chapter 8. In other words, the actual yields will in most cases exceed the calculated yields by several percentage points.

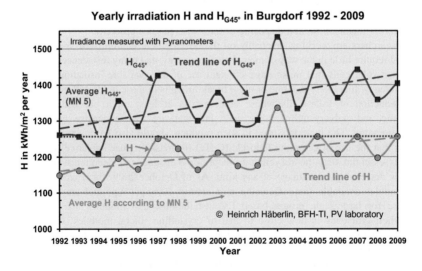

Figure 11.1 Annual irradiation H values for the horizontal plane and irradiation $H_{G45°}$ for a 45° plane pointed due south, at the main Bern University of Applied Sciences weather measurement station from 1992 to 2009. Also shown are the mean values of H and $H_{G45°}$ as computed by the Meteonorm 5 application [2.5], which supports insolation computations for surfaces pointed in any direction

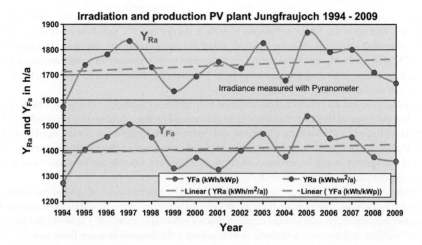

Figure 11.2   Annual energy yield Y_{Ra} on the solar generator plane and final yield Y_{Fa} for the Jungfraujoch PV installation (see Section 10.1.3), from 1994 to 2009

As global warming will induce increased ambient temperatures, the increase in overall PV installation output will probably not be on a par with the level of increase in insolation. Summer soiling of solar modules is also likely to rise, while winter snow-shading periods will grow shorter and PV installation operation under medium- and higher-capacity conditions – where overall inverter efficiency is often higher – will become more prevalent. Hence the overall increase in PV installation energy yield will probably keep pace with the increase in insolation. An analogous steady rise in energy yield and insolation was observed in certain PV installations (Birg, Burgdorf and Jungfraujoch; see Chapter 10) whose performance was monitored continuously for a minimum of 16 years (see Figures 11.2 and 11.3).

The figures derived from this monitoring show that PV installation and hydro power energy production dovetail extremely well with each other. For example, during the 1950s and 1960s when Switzerland did not yet have nuclear power plants, electricity was sometimes in short supply during springtime

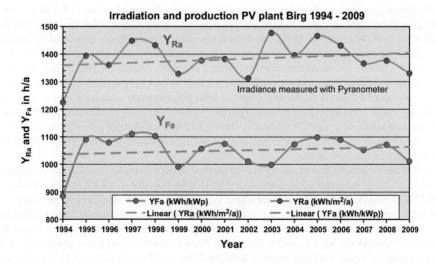

Figure 11.3   Annual energy yield Y_{Ra} on the solar generator plane and final yield Y_{Fa} for the Birg PV installation (see Section 10.1.4), from 1994 to 2009

fair-weather periods because storage lakes were almost completely depleted to the point where the government called upon the Swiss population to reduce energy use. PV installations could prove to be highly useful in such situations, or in the event of power shortages during dry summers. The advantage of installing PV systems at Alpine or Southern European sites is that, relative to lower-insolation regions, their energy yield is several magnitudes higher from November to February, and of course their overall energy yield is higher as well.

PV energy is set to make a substantial contribution to worldwide electricity output – of that there can be no doubt. But the scope of this contribution will be determined by the extent to which: (a) solar cell mass production is optimized and its costs are lowered; and (b) the crucial problem of storing surplus electricity is solved. If greater amounts of electricity are not generated by renewable (but fluctuating) resources such as solar and wind power, it will be necessary to develop an efficient, reliable and low-cost solution for the storage of surplus energy from these sources.

As noted in Section 5.2.7.5.6, this problem could be solved by instituting a supergrid whose north–south and east–west links would compensate for seasonal and daily energy output imbalances, respectively, by providing the requisite electricity storage capacity. Such a solution is technically feasible, but for reasons of political stability will become a realistic option at best only decades or more from now.

Small to medium-size PV installations are set to become more prevalent worldwide, except in polar regions. Around the equator (from around 38°N to 38°S), not only PV installations but also solar thermal power plants make for excellent sources of electricity, as long as such facilities are large enough to be cost effective.

Indeed, in some cases, parabolic trough solar thermal power plants are already the most economical solution for 10–1000 MW grid-connected systems in desert regions with a high proportion of direct beam radiation, providing that water is available for system cooling purposes. The drawback of solar thermal power plants, however, is that they use only direct beam radiation and cannot use diffuse radiation. Such systems should be composed of blocks comprising around 30 or 40 MW each and should be neither too small, because of the capital investment entailed by the basic infrastructure (collector field, steam generators, steam turbines, electrical generators and transformers), nor too large, on account of the thermal loss that occurs in the extensive cable runs between the collector field and steam turbines. If equipped with a suitable short-term thermal accumulator, solar thermal power plants can also produce electricity at night after a sunny day. In other words, a relatively simple solution for the storage problem is available. And on less sunny days, electricity could be produced by an auxiliary natural gas or oil-fired installation.

A solar thermal power plant of this type has been in operation at Kramer Junction in the Mojave Desert in California since 1991. In this 354 MW facility, which comprises nine solar arrays (SEGS I–IX, 1 · 14 MW, 6 · 30 MW and 2 · 80 MW), a turbine is operated by steam that is generated by heating to 400 °C a metal pipe through which oil flows and that is arranged at the focal point of a parabolic trough. The heated oil can also be stored temporarily. During low solar insolation or peak electricity demand periods, the oil is heated using natural gas (this is referred to as a hybrid system) – which, however, produces only around 25% of the facility's total output. Solar thermal electricity is already a nearly economically viable solution in California owing to the extensive use of air-conditioners and the resulting peak demand on hot summer days. If oil prices continue to rise, solar thermal power plants are bound to become economically viable in the near future, and in combination with geothermal energy would exhibit an optimal daily production curve.

The first of three parabolic trough collector, solar thermal power plants (Andasol 1, 2 and 3) has been in operation near Guadix in southern Spain since late 2008 (see Section 10.1.11). Each of these facilities will have 50 MW of output. Andasol 2 and 3 are slated to go on line in 2009 and 2011 respectively.

By using thermal accumulators containing hot liquid salt (a 28 500 ton mixture of $NaNO_3$ and KNO_3), the Andasol power plants can produce their nominal output for 7.5 hours after dusk on a sunny day.

Figure 11.4 displays the basic overall layout of the Andasol solar thermal power plants; Figure 11.5 displays a parabolic trough collector used for these facilities; and Figures 11.6 and 11.7 display aerial views of the station's central element with its two salt accumulators, the usual technical installations and the three cooling towers.

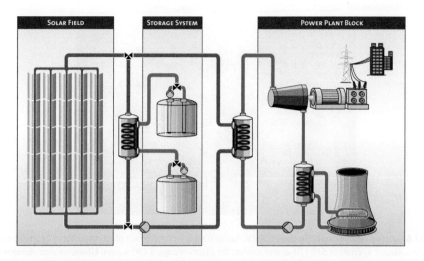

Figure 11.4 Layout of an Andasol solar thermal power plant. The solar heated thermal oil in the parabolic trough collector field channels its heat directly into the steam circuit via a heat exchanger (right). The steam generated in the heat exchanger drives the turbines, which in turn drive the plant's generator. On sunny days, the left heat exchanger heats a salt accumulator using a portion of the oil, whereby the salt solution is pumped back and forth between the accumulators and heat exchangers (upper accumulator temperature around 390 °C, lower accumulator temperature around 290 °C). Energy can be produced at night by heating the thermal oil using the heat stored in the salt accumulator; this oil then generates steam for the turbines via the heat exchangers (Photo: Solar Millennium AG)

Inasmuch as parabolic trough collectors reach a peak temperature of only around 300 to 400 °C, in keeping with the second law of thermodynamics, the Carnot efficiency (which represents an upper and virtually unattainable limit) is also relatively low.

Hence solar thermal power plants are not overly efficient (peak 28%, annual mean 15%) and a substantial amount of the solar energy absorbed is dissipated into the atmosphere as heat. This in turn means that solar thermal power plants need to have sufficient cooling water available to allow for the realization of evaporative cooling. Air cooling is also an option, but is less efficient than evaporative cooling.

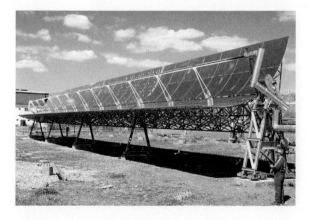

Figure 11.5 Parabolic trough collector used for commercial solar thermal power plants such as the Andasol plants. The collectors are arranged in a north–south direction. They need to be spaced sufficiently far apart to avoid shading (Photo: Forschungsverbund Erneuerbare Energien, www.fvee.de)

Figure 11.6 Aerial view of two of the three 50 MW Andasol solar thermal power plants in southern Spain. The surface area of the collector aperture is 510 120 square metres; the plants occupy a 1.95 square kilometre area; mean nominal direct beam radiation is 2136 kWh/m^2; and the plants operate for 3500 full-load hours annually (*Data source*: www.solarmillennium.de; photo: Solar Millennium AG/Paul Langrock)

Inasmuch as a substantial portion of the insolation in Central Europe comprises diffuse radiation and snow falls only sporadically, PV installations are better suited for this region than solar thermal power plants, and normally provide robust energy yields in low-fog sunny locations – albeit at a higher price per kWh. Although PV installations take up a fair amount of space, this does not pose a major obstacle to their realization. Owing to their modularity, PV installations can be mounted anywhere, across the whole range of power outputs.

In my view, it is counterproductive to couch the debate on the relative merits of PV and solar thermal power in ideological terms, as the Desertec project has done (see Section 5.2.7.5.6; Desertec argues that PV installations are a better solution by virtue of their decentralized nature, thus making solar thermal power plants the 'bad guys' owing to their large size). There are pros and cons to both technologies; in any case the major challenge of transitioning to a wholly renewable electricity supply within the next few decades can only be met through intelligent intermeshing of all available renewable resources.

In order for a substantial portion of electricity demand to be met within this time, we need not only to develop better solar cells in the testing lab (as is being done today), but also to mass-produce solar cells and

Figure 11.7 View of the central element of the Andasol 1 solar thermal power plant, with its two salt accumulators (left), various other structures and installations (centre), and its three cooling towers (upper right). For further information visit www.solarmillennium.de (Photo: Solar Millennium AG/Paul Langrock)

PV installations and incrementally roll back the production costs for these elements. Moreover, in order for the relevant vendors to effect the requisite multi-billion-dollar capital investments, stable economic conditions need to be established and these companies need to have relatively good assurance of a reasonable ROI. All new energy technologies are less profitable out of the starting gate than their conventional counterparts, and it always takes a few decades for such technologies genuinely to pay off. In other words, we need to use, promote, subsidize and optimize PV technologies, familiarize ourselves with the problems entailed by these technologies and endeavour to resolve them. An integrated grid comprising PV installations, desert solar thermal power plants, wind turbines, hydro power, biomass power stations and geothermal facilities would meet all of our energy needs in the long run. Fossil fuel and uranium reserves are finite, and despite extensive research extending over decades, it is by no means certain that nuclear fusion will come into its own as a viable energy source anytime soon.

Appendix A

Calculation Tables and Insolation Data

A1 Insolation Calculation Tables (see Chapter 2)

A1.1 Basic Insolation Calculation (see Section 2.4)

According to Equation 2.7, irradiation energy H_G on the solar generator plane is as follows:

$$H_G = R(\beta,\gamma) \cdot H$$

where H = global radiation incident on the horizontal plane and $R(\beta,\gamma)$ = global radiation factor for the relevant reference station.

The data for H are listed in Section A2 (bolded figures); the $R(\beta,\gamma)$ data are in Section A3.

A1.2 Insolation Calculation Using the Three-Component Model (see Section 2.5)

According to Equation 2.7, irradiation energy H_G on the solar generator plane is as follows:

$$H_G = H_{GB} + H_{GD} + H_{GR} = k_B \cdot R_B \cdot (H - H_D) + R_D \cdot H_D + R_{R \cdot \rho} \cdot H$$

where:

H = global radiation incident on the horizontal plane
H_D = diffuse radiation incident on the horizontal plane
k_B = shading correction factor as in Section 2.5.4 (unshaded $k_B = 1$)
R_B = direct beam radiation factor as in Section A4
R_D = corrected diffuse radiation factor $R_D = \frac{1}{2} \cos \alpha_2 + \frac{1}{2} \cos(\alpha_1 + \beta)$
R_R = effective portion of reflective radiation $R_R = \frac{1}{2} - \frac{1}{2} \cos \beta$
α_1 = horizon elevation in the γ direction
α_2 = façade/roof edge elevation relative to the solar generator plane
β = angle of incidence of the inclined surface relative to the horizontal plan
ρ = reflection factor (albedo) of the ground in front of the solar generator

The data for H (bolded figures) and H_D are listed in Section A2; see Section A5 for the k_B calculation.

Photovoltaics: System Design and Practice. Heinrich Häberlin.
© 2012 John Wiley & Sons, Ltd. Published 2012 by John Wiley & Sons, Ltd.

Table A1.1   Blank table for H_G calculation using the basic method described in Section 2.4

	Jan	Feb	Mar	April	May	June	July	Aug	Sept	Oct	Nov	Dec	Year	Unit
H														kWh/m^2
$R(\beta,\gamma)$														
H_G														kWh/m^2

Table A1.2   Calculation table for H_G using the three-component method

$R_D = \frac{1}{2}\cos\alpha_2 + \frac{1}{2}\cos(\alpha_1+\beta) =$						$R_R = \frac{1}{2} - \frac{1}{2}\cos\beta =$							

	Jan	Feb	Mar	Apr	May	June	July	Aug	Sep	Oct	Nov	Dec	Year	Unit
H														kWh/m^2
H_D														kWh/m^2
R_B														
k_B														
$H_{GB}= k_B \cdot R_B \cdot (H\text{-}H_D)$														kWh/m^2
$H_{GD}= R_D \cdot H_D$														kWh/m^2
ρ														
$H_{GR}= R_R \cdot \rho \cdot H$														kWh/m^2
$H_G=H_{GB}+H_{GD}+H_{GR}$														kWh/m^2

A2   Aggregate Monthly Horizontal Global Insolation

The global irradiation H figures are bolded. *Source*: Meteonorm [2.4]; also (in part) [2.2], [2.3].

A3   Global Insolation for Various Reference Locations

As noted in Section 2.4, for reasons of space it was not possible to provide reference station data for tropical regions and the southern hemisphere, for which the three-component model should be used (see Section 2.5).

In cases where reference stations are used as a calculation aid for PV installation sites in the southern hemisphere, an approximate figure can be arrived at by: (a) using a northern hemisphere reference station that is at about the same geographic latitude; and (b) shifting the relevant monthly figures by six months to compensate for the fact that the seasons experienced by the northern and southern hemisphere always differ by six months. Hence the figures for December in the northern hemisphere apply to figures for June in the southern hemisphere, while the figures for January in the northern hemisphere apply to July in the southern hemisphere and so on. This method can also be used for the correction factors k_T and k_G (see Sections 8.1.2 and 8.1.3). Lower correction factors k_G than those displayed in Table 8.5 (see Section 4.5.8) apply in arid regions with PV installations that are not cleaned regularly, owing to the layer of dust that accumulates on the solar modules over time.

Approximate year-round k_T figures can be obtained for low-lying equatorial regions using the summer figures for southern Europe.

Table A2.1 Horizontal global insolation in kWh/m²/d at various places in Switzerland (part 1)

Location		Jan	Feb	Mar	Apr	May	June	July	Aug	Sep	Oct	Nov	Dec	Year
Adelboden	H	1.46	2.33	3.52	4.46	5.27	5.36	5.85	5.07	3.89	2.60	2.49	1.26	3.55
46.5°N, 7.6°E, 1350m	H_D	0.68	1.03	1.52	2.00	2.37	2.49	2.50	2.18	1.70	1.18	0.75	0.58	1.58
Airolo	H	1.56	2.44	3.80	4.58	5.20	5.75	5.96	5.04	3.84	2.41	1.60	1.25	3.62
46.5°N, 8.6°E, 1175m	H_D	0.70	1.04	1.54	2.01	2.36	2.55	2.51	2.18	1.69	1.15	0.76	0.58	1.59
Basel (-Binningen)	H	0.96	1.64	2.54	3.57	4.61	5.16	5.59	4.73	3.38	2.02	1.13	0.82	3.00
47.6°N, 7.6°E, 316m	H_D	0.67	1.05	1.59	2.16	2.62	2.81	2.70	2.36	1.84	1.25	0.77	0.56	1.69
Beatenberg	H	1.31	2.10	3.29	4.14	4.84	4.97	5.47	4.79	3.59	2.39	1.41	1.09	3.28
46.7°N, 7.8°E, 1150m	H_D	0.81	1.17	1.68	2.29	2.73	2.94	2.74	2.38	1.92	1.36	0.91	0.69	1.80
Bern (-Liebefeld)	H	1.06	1.76	2.79	3.72	4.68	5.20	5.69	4.82	3.56	2.06	1.13	0.84	3.12
46.9°N, 7.4°E, 565m	H_D	0.71	1.09	1.63	2.19	2.63	2.81	2.69	2.36	1.86	1.27	0.78	0.58	1.72
Biasca	H	1.45	2.11	3.15	3.81	4.53	5.49	5.87	5.04	3.69	2.12	1.44	1.14	3.32
46.3°N, 9.0°E, 300m	H_D	0.69	1.00	1.47	1.88	2.23	2.51	2.50	2.18	1.68	1.08	0.73	0.56	1.54
Biel/Bienne	H	0.91	1.65	2.67	3.73	4.65	5.21	5.68	4.78	3.42	2.03	0.99	0.73	3.03
47.1°N, 7.2°E, 435m	H_D	0.65	1.07	1.61	2.19	2.63	2.81	2.69	2.36	1.86	1.26	0.73	0.54	1.70
Brig	H	1.48	2.23	3.54	4.66	5.65	6.10	6.32	5.39	4.02	2.62	1.55	1.21	3.72
46.3°N, 8.0°E, 680m	H_D	0.80	1.16	1.63	2.19	2.56	2.71	2.52	2.24	1.83	1.34	0.90	0.68	1.71
Brugg	H	0.82	1.57	2.60	3.67	4.76	5.22	5.60	4.70	3.34	1.85	0.87	0.65	2.97
47.5°N, 8.2°E, 350m	H_D	0.62	1.04	1.59	2.17	2.63	2.81	2.70	2.36	1.84	1.22	0.67	0.50	1.68
Buchs (SG)	H	1.06	1.86	3.16	3.90	4.88	5.10	5.48	4.58	3.42	2.13	1.15	0.81	3.12
47.1°N, 9.5°E, 450m	H_D	0.71	1.10	1.64	2.19	2.63	2.81	2.71	2.37	1.86	1.28	0.78	0.57	1.72
Burgdorf	H	0.99	1.69	2.74	3.68	4.59	5.09	5.59	4.75	3.41	1.94	1.04	0.77	3.02
47.1°N, 7.6°E, 540m	H_D	0.69	1.08	1.62	2.18	2.62	2.81	2.70	2.37	1.86	1.25	0.75	0.56	1.70
Chur (-Ems)	H	1.43	2.20	3.29	4.36	5.13	5.35	5.69	4.78	3.69	2.51	1.46	1.11	3.41
46.9°N, 9.5°E, 560m	H_D	0.79	1.15	1.67	2.24	2.68	2.89	2.68	2.38	1.89	1.34	0.90	0.68	1.53
Davos	H	1.68	2.66	4.02	5.01	5.58	5.70	5.88	5.01	3.95	2.78	1.72	1.34	3.77
46.8°N, 9.8°E, 1560m	H_D	0.78	1.12	1.62	2.23	2.68	2.87	2.78	2.43	1.91	1.34	0.88	0.66	1.77
Disentis	H	1.52	2.43	3.71	4.54	5.17	5.55	5.73	4.85	3.75	2.48	1.56	1.21	3.54
46.7°N, 8.8°E, 1150m	H_D	0.77	1.10	1.57	2.19	2.64	2.82	2.68	2.37	1.87	1.33	0.88	0.66	1.74
Fribourg/Freiburg	H	1.08	1.75	2.83	3.78	4.77	5.26	5.76	4.88	3.55	2.02	1.10	0.86	3.13
46.8°N, 7.2°E, 620m	H_D	0.72	1.09	1.64	2.19	2.63	2.81	2.68	2.36	1.87	1.27	0.78	0.59	1.72
Genève	H	0.95	1.75	2.91	4.08	5.05	5.67	6.11	5.16	3.89	2.28	1.08	0.82	3.31
46.2°N, 6.2°E, 375m	H_D	0.69	1.11	1.67	2.22	2.63	2.80	2.63	2.33	1.86	1.33	0.78	0.59	1.72
Grächen	H	1.68	2.66	4.09	4.97	5.68	6.09	6.26	5.47	4.29	2.91	1.76	1.41	3.94
46.2°N, 7.9°E, 1610m	H_D	0.80	1.10	1.50	2.18	2.65	2.82	2.63	2.30	1.82	1.34	0.93	0.69	1.73
Interlaken	H	1.13	1.88	3.03	3.81	4.82	5.12	5.66	4.75	3.49	2.20	1.26	0.90	3.17
46.7°N, 7.9°E, 565m	H_D	0.73	1.12	1.66	2.19	2.63	2.81	2.70	2.37	1.87	1.30	0.82	0.61	1.73
Kloten (Airport)	H	0.91	1.66	2.69	3.77	4.78	5.20	5.59	4.73	3.38	1.92	0.94	0.70	3.02
47.5°N, 8.5°E, 436m	H_D	0.65	1.06	1.60	2.18	2.63	2.81	2.70	2.36	1.85	1.24	0.70	0.52	1.69
La Chaux-de-Fonds	H	1.32	2.06	3.05	3.86	4.39	4.99	5.62	4.75	3.56	2.28	1.39	1.08	3.19
47.1°N, 6.8°E, 1018m	H_D	0.79	1.16	1.70	2.31	2.76	2.94	2.70	2.39	1.91	1.36	0.89	0.67	1.79
Lausanne	H	1.03	1.73	2.89	3.94	4.89	5.44	5.94	5.02	3.69	2.13	1.13	0.86	3.22
46.5°N, 6.6°E, 450m	H_D	0.71	1.10	1.66	2.21	2.63	2.81	2.65	2.35	1.87	1.30	0.79	0.60	1.72
Locarno	H	1.42	2.00	3.06	3.57	4.39	5.50	5.95	5.17	3.72	2.17	1.44	1.13	3.29
46.2°N, 8.8°E, 200m	H_D	0.82	1.21	1.74	2.33	2.77	2.86	2.59	2.28	1.92	1.40	0.93	0.71	1.79
Lugano	H	1.36	1.98	2.97	3.49	4.35	5.39	5.82	5.15	3.71	2.17	1.44	1.14	3.25
46.0°N, 9.0°E, 273m	H_D	0.84	1.22	1.76	2.32	2.77	2.89	2.64	2.29	1.92	1.41	0.94	0.72	1.81
Luzern	H	0.91	1.56	2.62	3.56	4.42	4.63	5.13	4.49	3.21	1.92	0.98	0.72	2.85
47.0°N, 8.3°E, 450m	H_D	0.66	1.05	1.61	2.17	2.60	2.77	2.72	2.37	1.85	1.25	0.73	0.54	1.69
Martigny	H	1.39	2.17	3.33	4.61	5.50	5.94	6.41	5.42	4.01	2.61	1.52	1.17	3.67
46.1°N, 7.1°E, 475m	H_D	0.81	1.17	1.68	2.19	2.60	2.76	2.48	2.24	1.84	1.34	0.91	0.69	1.72

Table A2.2   Horizontal global insolation in kWh/m^2/d at various places in Switzerland (part 2)

Location		Jan	Feb	Mar	Apr	May	June	July	Aug	Sep	Oct	Nov	Dec	Year
Montana	H	1.61	2.53	3.88	4.77	5.61	5.74	6.08	5.28	4.13	2.80	1.68	1.35	3.78
46.3°N, 7.5°E, 1506m	H$_D$	0.73	1.04	1.46	2.03	2.42	2.61	2.46	2.15	1.70	1.23	0.83	0.63	1.61
Olten	H	0.85	1.60	2.63	3.61	4.62	5.15	5.60	4.68	3.34	1.87	0.92	0.67	2.96
47.4°N, 7.9°E, 400m	H$_D$	0.63	1.05	1.60	2.17	2.62	2.81	2.70	2.37	1.85	1.23	0.69	0.51	1.68
Payerne	H	0.99	1.71	2.88	3.94	4.92	5.52	5.98	5.06	3.65	2.04	1.03	0.77	3.21
46.8°N, 7.0°E, 490m	H$_D$	0.69	1.09	1.65	2.20	2.63	2.81	2.64	2.33	1.86	1.27	0.75	0.56	1.71
Schaffhausen	H	0.85	1.60	2.62	3.74	4.85	5.31	5.65	4.71	3.36	1.89	0.89	0.67	3.01
47.7°N, 8.6°E, 400m	H$_D$	0.63	1.04	1.59	2.18	2.63	2.81	2.69	2.36	1.84	1.22	0.68	0.51	1.68
Sion/Sitten	H	1.38	2.18	3.35	4.57	5.55	5.88	6.31	5.34	4.00	2.67	1.50	1.13	3.65
46.2°N, 7.3°E, 500m	H$_D$	0.80	1.17	1.68	2.19	2.59	2.77	2.52	2.26	1.84	1.34	0.90	0.68	1.73
St. Gallen	H	0.99	1.75	3.10	3.82	4.77	5.03	5.47	4.63	3.34	2.01	1.07	0.80	3.06
47.4°N, 9.4°E, 680m	H$_D$	0.68	1.08	1.63	2.19	2.63	2.81	2.71	2.37	1.85	1.25	0.75	0.56	1.71
St. Moritz	H	1.74	2.70	4.06	5.29	5.75	6.01	6.16	5.23	4.16	2.86	1.83	1.45	3.93
46.5°N, 9.8°E, 1835m	H$_D$	0.77	1.06	1.50	2.05	2.63	2.84	2.66	2.37	1.86	1.33	0.89	0.65	1.71
Winterthur	H	0.86	1.62	2.66	3.82	4.83	5.23	5.57	4.70	3.34	1.89	0.90	0.65	3.00
47.5°N, 8.7°E, 440m	H$_D$	0.64	1.05	1.60	2.19	2.63	2.81	2.70	2.36	1.84	1.23	0.69	0.50	1.68
Zermatt	H	1.67	2.74	4.15	5.16	5.84	6.19	6.39	5.57	4.35	2.95	1.79	1.42	4.01
46.0°N, 7.8°E, 1640m	H$_D$	0.82	1.08	1.49	2.12	2.61	2.80	2.57	2.26	1.81	1.33	0.93	0.70	1.71
Zürich	H	0.84	1.59	2.66	3.79	4.77	5.17	5.53	4.65	3.31	1.88	0.89	0.64	2.97
47.4°N, 8.5°E, 415m	H$_D$	0.63	1.04	1.60	2.19	2.63	2.81	2.71	2.37	1.85	1.23	0.69	0.49	1.68
Montain locations:		Jan	Feb	Mar	Apr	May	June	July	Aug	Sep	Oct	Nov	Dec	Year
Chasseral	H	1.59	2.33	3.33	4.27	4.59	4.94	5.59	4.85	3.74	2.66	1.68	1.37	3.40
47.1°N, 7.1°E, 1599m	H$_D$	0.68	1.01	1.49	1.96	2.24	2.41	2.46	2.15	1.68	1.17	0.76	0.58	1.55
Cimetta (ob Locarno)	H	1.70	2.54	3.72	4.27	4.73	5.69	6.22	5.42	3.96	2.54	1.80	1.44	3.69
46.2°N, 8.8°E, 1672m	H$_D$	0.73	1.06	1.54	1.97	2.28	2.54	2.53	2.21	1.72	1.18	0.80	0.62	1.60
Corvatsch	H	2.02	3.05	4.59	5.85	6.45	6.31	6.16	5.28	4.44	3.29	2.19	1.76	4.27
46.4°N, 9.8°E, 3315m	H$_D$	0.55	0.75	1.02	1.49	1.99	2.31	2.24	1.99	1.48	0.99	0.65	0.45	1.32
Evolène-Villa (VS)	H	1.68	2.75	4.19	5.19	5.88	6.06	6.29	5.42	4.29	2.95	1.82	1.42	3.99
46.1°N, 7.5°E, 1825m	H$_D$	0.81	1.07	1.45	2.10	2.60	2.83	2.62	2.32	1.82	1.33	0.92	0.70	1.71
Grand-St-Bernard	H	1.74	2.83	4.32	5.37	6.09	6.19	6.18	5.32	4.15	2.90	1.87	1.51	4.04
45.9°N, 7.2°E, 2472m	H$_D$	0.81	1.06	1.41	2.04	2.52	2.80	2.65	2.36	1.89	1.35	0.93	0.69	1.71
Grimsel-Hospiz	H	1.68	2.66	4.09	4.91	5.64	5.72	5.92	5.07	4.01	2.67	1.73	1.42	3.79
46.6°N, 8.3°E, 1980m	H$_D$	0.79	1.14	1.62	2.24	2.68	2.87	2.78	2.43	1.91	1.34	0.89	0.67	1.78
Gütsch (ob Andermatt)	H	1.80	2.86	4.35	5.56	5.91	5.91	6.05	5.16	4.15	2.80	1.92	1.53	4.00
46.7°N, 8.6°E, 2287m	H$_D$	0.73	0.97	1.33	1.90	2.58	2.87	2.70	2.39	1.85	1.34	0.85	0.61	1.68
Jungfraujoch	H	1.65	2.65	3.97	5.53	6.15	6.42	6.31	5.47	4.46	3.19	2.16	1.57	4.12
46.6°N, 8.0°E, 3580m	H$_D$	0.66	0.91	1.30	1.63	2.10	2.29	2.20	1.94	1.47	1.01	0.65	0.52	1.39
La Dôle	H	1.44	2.11	3.12	3.86	4.29	4.68	5.33	4.61	3.62	2.45	1.56	1.27	3.19
46.4°N, 6.1°E, 1670m	H$_D$	0.68	1.00	1.46	1.89	2.18	2.35	2.44	2.12	1.67	1.16	0.75	0.59	1.52
Moléson	H	1.51	2.38	3.39	4.42	4.78	4.78	5.40	4.80	3.74	2.66	1.68	1.34	3.40
46.6°N, 7.0°E, 1972m	H$_D$	0.69	1.04	1.51	1.99	2.28	2.37	2.45	2.15	1.69	1.19	0.78	0.60	1.56
Mont Soleil	H	1.36	2.10	3.09	3.96	4.54	4.98	5.57	4.79	3.63	2.42	1.44	1.14	3.25
47.2°N, 7.0°E, 1270m	H$_D$	0.65	0.98	1.45	1.91	2.23	2.42	2.46	2.14	1.66	1.14	0.71	0.55	1.53
Napf	H	1.32	1.94	2.83	3.45	4.08	4.46	5.13	4.42	3.31	2.28	1.39	1.10	2.97
47.0°N, 7.9°E, 1406m	H$_D$	0.84	1.25	1.85	2.44	2.90	3.12	3.03	2.64	2.08	1.46	0.95	0.72	1.94
Pilatus	H	1.49	2.30	3.32	4.20	4.52	4.18	4.66	4.35	3.53	2.72	1.63	1.27	3.19
47.0°N, 8.2°E, 2106m	H$_D$	0.68	1.02	1.49	1.95	2.23	2.21	2.30	2.07	1.65	1.18	0.76	0.57	1.51
San Bernardino	H	1.68	2.71	4.05	5.03	5.54	5.85	6.03	5.13	3.97	2.61	1.76	1.39	3.81
46.5°N, 9.2°E, 1639m	H$_D$	0.79	1.06	1.51	2.15	2.69	2.89	2.71	2.40	1.91	1.38	0.91	0.67	1.75
Säntis	H	1.60	2.54	3.65	4.92	5.42	5.16	5.45	4.78	3.89	2.95	1.85	1.42	3.62
47.3°N, 9.4°E, 2490m	H$_D$	0.76	1.12	1.66	2.22	2.68	2.83	2.78	2.42	1.89	1.30	0.85	0.64	1.76
Weissfluhjoch	H	1.87	2.91	4.32	5.59	6.05	5.73	5.56	4.78	4.01	3.02	2.04	1.59	3.96
46.8°N, 9.8°E, 2690m	H$_D$	0.75	1.07	1.56	2.12	2.64	2.87	2.79	2.43	1.90	1.31	0.85	0.64	1.74

Table A2.3 Horizontal global insolation in kWh/m²/d at various locations in Germany

Location		Jan	Feb	Mar	Apr	May	June	July	Aug	Sep	Oct	Nov	Dec	Year
Berlin	H	0.60	1.20	2.28	3.57	4.85	5.25	5.04	4.32	2.95	1.63	0.72	0.43	2.74
52.5°N, 13.3°E, 33m	H_D	0.45	0.82	1.42	2.06	2.57	2.80	2.69	2.28	1.69	1.05	0.54	0.34	1.56
Bocholt	H	0.63	1.39	2.19	3.80	4.85	4.82	4.87	4.13	2.76	1.73	0.86	0.48	2.71
51.8°N, 6.6°E, 24m	H_D	0.47	0.89	1.42	2.09	2.58	2.78	2.69	2.28	1.69	1.08	0.60	0.37	1.58
Bonn	H	0.72	1.47	2.23	3.56	4.44	4.63	4.85	4.08	2.73	1.76	0.96	0.56	2.66
50.7°N, 7.1°E, 65m	H_D	0.52	0.93	1.45	2.09	2.57	2.76	2.70	2.30	1.70	1.12	0.65	0.41	1.60
Braunschweig	H	0.65	1.30	2.26	3.53	4.80	5.11	4.75	4.15	2.83	1.63	0.77	0.46	2.68
52.3°N, 10.5°E, 81m	H_D	0.47	0.85	1.42	2.06	2.57	2.80	2.68	2.28	1.68	1.05	0.56	0.35	1.56
Bremen	H	0.60	1.30	2.09	3.57	4.78	4.54	4.61	3.96	2.67	1.56	0.77	0.41	2.57
53.1°N, 8.8°E, 24m	H_D	0.44	0.84	1.36	2.05	2.56	2.74	2.67	2.26	1.64	1.01	0.55	0.32	1.54
Coburg	H	0.70	1.51	2.30	3.77	4.89	4.92	5.33	4.35	2.97	1.77	0.84	0.50	2.83
50.3°N, 11.0°E, 357m	H_D	0.52	0.95	1.48	2.12	2.60	2.79	2.70	2.32	1.75	1.13	0.62	0.39	1.61
Dresden	H	0.77	1.47	2.35	3.65	4.85	4.82	4.87	4.20	2.79	1.87	0.91	0.57	2.76
51.1°N, 13.7°E, 271m	H_D	0.54	0.92	1.47	2.10	2.58	2.79	2.70	2.30	1.70	1.13	0.63	0.42	1.60
Fichtelberg	H	0.77	1.51	2.35	3.56	4.52	4.32	4.52	4.06	2.76	1.90	1.03	0.57	2.64
50.5°N, 13.0°E, 1214m	H_D	0.64	1.08	1.68	2.37	2.90	3.09	3.02	2.59	1.94	1.30	0.77	0.50	1.82
Frankfurt (Main)	H	0.73	1.52	2.38	3.81	4.83	5.06	5.22	4.34	3.10	1.77	0.94	0.57	2.85
50.1°N, 8.8°E, 92m	H_D	0.55	0.97	1.50	2.13	2.60	2.80	2.71	2.33	1.77	1.14	0.66	0.43	1.63
Freiburg	H	0.89	1.59	2.62	3.72	4.73	5.40	5.52	4.80	3.41	2.04	1.15	0.63	3.04
48.0°N, 7.9°E, 308m	H_D	0.64	1.03	1.59	2.17	2.62	2.81	2.70	2.34	1.83	1.24	0.77	0.51	1.68
Giessen	H	0.65	1.42	2.23	3.80	4.70	4.90	5.09	4.13	2.85	1.63	0.81	0.48	2.71
50.6°N, 8.7°E, 201m	H_D	0.49	0.93	1.45	2.12	2.59	2.79	2.70	2.31	1.72	1.09	0.60	0.38	1.60
Hamburg	H	0.53	1.15	2.09	3.53	4.73	5.11	4.68	4.08	2.73	1.51	0.69	0.41	2.61
53.6°N, 10.0°E, 10m	H_D	0.40	0.78	1.35	2.04	2.55	2.79	2.67	2.26	1.63	0.99	0.51	0.31	1.52
Hohenpeissenberg	H	1.34	2.08	3.17	4.10	4.87	5.31	5.40	4.68	3.72	2.52	1.42	1.10	3.31
47.8°N, 11.0°E, 990m	H_D	0.75	1.11	1.65	2.26	2.71	2.89	2.75	2.39	1.84	1.28	0.85	0.64	1.76
Kassel	H	0.65	1.42	2.26	3.65	4.75	4.75	4.85	4.15	2.79	1.59	0.81	0.48	2.68
51.3°N, 9.5°E, 233m	H_D	0.48	0.91	1.44	2.09	2.58	2.77	2.69	2.29	1.69	1.07	0.59	0.38	1.58
Konstanz	H	0.89	1.59	2.79	4.01	4.89	5.43	5.52	4.70	3.31	1.94	1.01	0.70	3.07
47.7°N, 9.2°E, 450m	H_D	0.64	1.04	1.61	2.19	2.63	2.81	2.71	2.36	1.84	1.24	0.72	0.52	1.69
Köln	H	0.72	1.46	2.23	3.56	4.44	4.63	4.85	4.08	2.74	1.75	0.95	0.55	2.66
50.9°N, 7.0°E, 39m	H_D	0.52	0.93	1.45	2.09	2.57	2.76	2.70	2.30	1.70	1.11	0.65	0.41	1.60
Mannheim	H	0.74	1.54	2.35	3.80	4.85	5.04	5.26	4.35	3.12	1.77	0.98	0.57	2.85
49.5°N, 8.5°E, 106m	H_D	0.56	0.98	1.51	2.14	2.61	2.80	2.71	2.33	1.78	1.16	0.69	0.44	1.64
München (Flugh.)	H	1.03	1.80	2.88	4.01	5.04	5.43	5.40	4.61	3.53	2.13	1.13	0.79	3.14
48.2°N, 11.7°E, 447m	H_D	0.67	1.05	1.60	2.18	2.61	2.81	2.71	2.35	1.82	1.24	0.75	0.55	1.69
Nürnberg	H	0.77	1.61	2.33	3.77	4.70	4.90	5.23	4.36	3.17	1.87	1.01	0.57	2.85
49.5°N, 11.1°E, 312m	H_D	0.56	0.99	1.50	2.14	2.60	2.80	2.71	2.33	1.79	1.17	0.69	0.44	1.64
Osnabrück	H	0.60	1.34	2.11	3.65	4.82	4.75	4.75	4.10	2.69	1.63	0.81	0.43	2.64
52.3°N, 8.0°E, 104m	H_D	0.45	0.87	1.39	2.07	2.57	2.77	2.68	2.28	1.66	1.05	0.57	0.34	1.56
Passau	H	0.86	1.73	2.64	4.03	5.13	5.16	5.47	4.59	3.24	2.09	1.01	0.65	3.04
48.6°N, 13.5°E, 412m	H_D	0.62	1.04	1.58	2.17	2.61	2.81	2.70	2.35	1.81	1.23	0.71	0.48	1.67
Potsdam	H	0.60	1.20	2.28	3.57	4.85	5.26	5.04	4.32	2.95	1.63	0.72	0.43	2.73
52.4°N, 13.0°E, 81m	H_D	0.45	0.82	1.42	2.06	2.57	2.80	2.69	2.28	1.69	1.05	0.54	0.34	1.56
Schleswig	H	0.50	1.18	1.97	3.65	4.92	4.85	4.95	4.22	2.67	1.44	0.69	0.39	2.61
54.5°N, 9.6°E, 59m	H_D	0.38	0.77	1.30	2.02	2.54	2.78	2.67	2.24	1.60	0.95	0.49	0.29	1.50
Stuttgart	H	0.91	1.66	2.57	3.77	4.73	5.14	5.30	4.49	3.27	1.94	1.18	0.72	2.97
48.8°N, 9.2°E, 318m	H_D	0.63	1.02	1.56	2.16	2.61	2.80	2.71	2.35	1.81	1.20	0.75	0.51	1.67
Trier	H	0.70	1.49	2.45	3.72	4.78	5.14	5.13	4.35	3.15	1.83	0.86	0.57	2.85
49.8°N, 6.7°E, 278m	H_D	0.53	0.96	1.51	2.13	2.60	2.80	2.71	2.33	1.78	1.16	0.64	0.44	1.63
Weihenstephan	H	1.00	1.80	2.83	3.96	4.96	5.35	5.35	4.56	3.48	2.11	1.10	0.79	3.11
48.4°N, 11.7°E, 476m	H_D	0.66	1.05	1.59	2.17	2.62	2.81	2.71	2.35	1.81	1.24	0.74	0.54	1.69
Würzburg	H	0.82	1.59	2.62	3.91	4.96	5.38	5.28	4.46	3.31	1.92	0.94	0.65	2.97
49.8°N, 10.0°E, 275m	H_D	0.57	0.98	1.54	2.14	2.60	2.80	2.71	2.33	1.78	1.17	0.66	0.47	1.64

Table A2.4 Horizontal global insolation in kWh/m²/d at various locations in Austria, Northern Europe and Eastern Europe

Locations in Austria		Jan	Feb	Mar	Apr	May	June	July	Aug	Sep	Oct	Nov	Dec	Year
Feuerkogel	H	1.23	1.99	3.07	3.86	4.49	4.27	4.39	3.92	3.12	2.50	1.42	1.03	2.92
47.8°N, 13.7°E, 1598m	H$_D$	0.81	1.21	1.80	2.44	2.93	3.09	3.02	2.63	2.06	1.40	0.92	0.68	1.91
Graz	H	1.23	1.80	2.95	3.89	4.70	5.09	5.38	4.66	3.36	2.23	1.30	0.96	3.12
47.1°N, 15.5°E, 342m	H$_D$	0.84	1.25	1.85	2.46	2.93	3.14	2.99	2.63	2.08	1.45	0.94	0.71	1.94
Innsbruck	H	1.23	1.99	3.17	4.03	4.82	4.92	4.92	4.32	3.31	2.37	1.39	0.96	3.12
47.3°N, 11.4°E, 582m	H$_D$	0.83	1.24	1.81	2.45	2.93	3.14	3.04	2.64	2.07	1.44	0.94	0.70	1.93
Klagenfurt	H	1.34	2.16	3.24	4.20	5.11	5.47	5.71	4.95	3.65	2.20	1.25	0.96	3.36
46.7°N, 14.3°E, 452m	H$_D$	0.81	1.17	1.69	2.28	2.68	2.87	2.66	2.35	1.91	1.38	0.89	0.68	1.78
Salzburg	H	1.03	1.73	2.72	3.65	4.52	4.73	4.82	4.15	3.19	2.11	1.10	0.79	2.88
47.8°N, 13.0°E, 435m	H$_D$	0.78	1.21	1.81	2.44	2.93	3.14	3.04	2.64	2.06	1.42	0.87	0.64	1.91
Sonnblick	H	1.87	2.59	4.15	5.20	5.64	5.50	5.38	4.63	3.84	3.10	1.94	1.49	3.76
47.1°N, 13.0°E, 3106m	H$_D$	0.73	1.12	1.59	2.19	2.67	2.86	2.78	2.42	1.90	1.29	0.85	0.65	1.75
Wien	H	0.86	1.54	2.71	3.81	5.12	5.38	5.45	4.67	3.22	2.09	0.96	0.65	3.03
48.2°N, 16.4°E, 170m	H$_D$	0.62	1.01	1.59	2.17	2.61	2.81	2.71	2.35	1.81	1.24	0.69	0.48	1.67
Locations in N Europe		Jan	Feb	Mar	Apr	May	June	July	Aug	Sep	Oct	Nov	Dec	Year
Göteborg / S	H	0.39	0.90	1.69	3.56	4.85	5.33	5.41	3.94	2.55	1.25	0.58	0.27	2.56
57.5°N, 12.0°E, 5m	H$_D$	0.30	0.63	1.16	1.94	2.51	2.77	2.63	2.18	1.52	0.84	0.40	0.21	1.42
Helsinki /SF	H	0.25	0.88	2.08	3.58	5.28	6.06	5.45	4.03	2.31	1.05	0.31	0.14	2.62
60.2°N, 24.9°E, 5m	H$_D$	0.20	0.57	1.17	1.87	2.43	2.69	2.60	2.12	1.41	0.73	0.25	0.12	1.34
Kobenhavn / DK	H	0.52	1.13	2.02	3.84	5.04	5.14	5.34	4.25	2.67	1.43	0.66	0.36	2.69
55.7°N, 12.5°E, 5m	H$_D$	0.37	0.73	1.29	1.99	2.53	2.78	2.65	2.21	1.57	0.92	0.45	0.26	1.48
Olso / N	H	0.34	0.97	2.09	3.50	5.47	5.49	5.38	4.13	2.55	1.16	0.49	0.23	2.65
59.9°N, 10.7°E, 5m	H$_D$	0.27	0.66	1.32	2.09	2.58	3.04	2.86	2.35	1.61	0.86	0.37	0.18	1.51
Stockholm / S	H	0.34	0.96	2.19	3.63	5.30	5.91	5.22	4.18	2.57	1.23	0.46	0.24	2.68
59.2°N, 18.0°E, 5m	H$_D$	0.24	0.61	1.20	1.90	2.44	2.71	2.63	2.13	1.46	0.79	0.32	0.17	1.38
Locations in E Europe		Jan	Feb	Mar	Apr	May	June	July	Aug	Sep	Oct	Nov	Dec	Year
Beograd	H	1.25	2.11	3.28	4.41	5.56	5.99	6.21	5.58	4.12	2.71	1.51	1.13	3.65
44.7°N, 20.5°E, 76m	H$_D$	0.88	1.27	1.77	2.29	2.59	2.72	2.49	2.14	1.85	1.42	0.99	0.77	1.76
Bratislava	H	0.89	1.66	2.73	4.22	5.47	5.91	6.16	4.99	3.65	2.35	1.08	0.67	3.31
48.1°N, 17.1°E, 289m	H$_D$	0.56	0.93	1.44	2.01	2.43	2.60	2.49	2.19	1.69	1.16	0.66	0.44	1.55
Bucuresti	H	1.32	1.96	2.86	4.44	5.40	6.40	6.31	5.66	4.08	2.72	1.40	0.91	3.62
44.5°N, 26.2°E, 88m	H$_D$	0.74	1.07	1.54	2.08	2.46	2.58	2.50	2.20	1.79	1.28	0.81	0.57	1.63
Budapest	H	0.96	1.66	2.79	4.25	5.23	5.74	5.91	4.97	3.65	2.28	1.18	0.74	3.28
47.5°N, 19.1°E, 105m	H$_D$	0.59	0.95	1.47	2.03	2.43	2.60	2.52	2.20	1.71	1.16	0.69	0.47	1.57
Istanbul	H	1.60	2.39	3.44	5.02	6.18	6.80	6.77	6.12	4.61	3.07	1.93	1.39	4.11
41.0°N, 29°E, 5m	H$_D$	1.01	1.38	1.87	2.25	2.46	2.47	2.36	2.10	1.90	1.51	1.12	0.89	1.77
Ljubliana	H	1.10	1.76	2.86	3.77	4.78	5.14	5.45	4.63	3.24	1.97	1.10	0.77	3.04
46.1°N, 14.5°E, 300m	H$_D$	0.74	1.12	1.67	2.20	2.63	2.81	2.72	2.39	1.88	1.27	0.79	0.57	1.73
Moskva	H	0.48	1.27	2.47	3.35	5.27	5.21	5.11	4.11	2.37	1.24	0.54	0.33	2.64
55.8°N, 37.5°E, 150m	H$_D$	0.40	0.84	1.48	2.21	2.72	3.09	2.96	2.46	1.73	0.99	0.47	0.29	1.64
Praha	H	0.68	1.33	2.31	3.80	4.79	4.86	4.65	4.39	2.91	1.83	0.84	0.48	2.74
50.1°N, 14.5°E, 187m	H$_D$	0.52	0.92	1.49	2.13	2.60	2.79	2.69	2.33	1.75	1.16	0.63	0.39	1.61
Pristina	H	1.63	2.50	3.39	4.61	5.56	5.97	6.38	5.91	4.44	2.95	1.68	1.25	3.86
42.7°N, 21.2°E, 575m	H$_D$	0.98	1.33	1.85	2.31	2.61	2.74	2.42	2.04	1.83	1.47	1.08	0.86	1.79
Sarajevo	H	1.44	2.16	3.33	4.39	5.33	5.52	5.91	5.02	3.69	2.66	1.54	1.20	3.52
43.9°N, 18.5°E, 630m	H$_D$	0.92	1.31	1.80	2.31	2.66	2.86	2.62	2.40	2.01	1.47	1.03	0.81	1.85
Sofia	H	1.16	1.99	2.86	3.96	4.73	5.31	5.59	4.99	3.65	2.40	1.32	0.96	3.24
42.8°N, 23.4°E, 550m	H$_D$	0.82	1.25	1.75	2.28	2.65	2.81	2.72	2.41	1.98	1.43	0.93	0.71	1.81
Warszawa / PL	H	0.64	1.34	2.33	3.66	5.15	5.06	5.22	4.43	2.79	1.75	0.78	0.45	2.79
52.2°N, 21.0°E, 80m	H$_D$	0.47	0.87	1.44	2.08	2.56	2.80	2.69	2.28	1.69	1.08	0.56	0.35	1.57
Zagreb	H	1.07	1.92	2.81	4.12	5.18	5.52	5.85	5.19	3.90	2.27	1.31	0.77	3.32
45.8°N, 16.0°E, 148m	H$_D$	0.74	1.16	1.68	2.23	2.63	2.81	2.67	2.35	1.89	1.34	0.86	0.58	1.74

Table A2.5 Horizontal global insolation in kWh/m²/d at various locations in France and northwest Europe

Locations in France		Jan	Feb	Mar	Apr	May	June	July	Aug	Sep	Oct	Nov	Dec	Year
Ajaccio (Korsika)	H	1.73	2.50	3.69	5.02	6.00	7.08	7.15	6.22	4.80	3.22	2.06	1.53	4.24
41.9°N, 8.8°E, 5m	H_D	0.85	1.16	1.58	1.96	2.22	2.14	1.99	1.84	1.59	1.27	0.93	0.75	1.52
Bordeaux	H	1.25	1.99	3.10	4.32	4.99	5.64	5.91	5.13	4.13	2.57	1.58	1.10	3.48
44.8°N, 0.7°W, 46m	H_D	0.87	1.27	1.79	2.31	2.72	2.82	2.61	2.33	1.85	1.43	0.98	0.76	1.81
Clermont-Ferrand	H	1.20	1.94	3.05	4.20	4.63	5.56	5.98	4.96	3.86	2.33	1.44	1.03	3.36
45.8°N, 3.1°E, 329m	H_D	0.83	1.23	1.76	2.30	2.76	2.84	2.57	2.37	1.89	1.42	0.94	0.72	1.80
Dijon	H	0.91	1.71	2.76	4.13	4.87	5.67	5.95	4.89	3.65	2.11	1.30	0.74	3.21
47.3°N, 5.1°E, 222m	H_D	0.65	1.08	1.62	2.19	2.63	2.79	2.64	2.35	1.85	1.27	0.81	0.55	1.70
Lyon	H	0.99	1.69	2.93	4.18	4.86	5.82	6.14	5.09	3.80	2.11	1.19	0.80	3.29
45.8°N, 4.9°E, 163m	H_D	0.71	1.10	1.68	2.22	2.63	2.78	2.62	2.35	1.88	1.31	0.82	0.58	1.72
Marseille	H	1.80	2.45	3.89	5.14	6.19	6.96	7.05	6.09	4.63	3.00	1.92	1.49	4.21
43.3°N, 5.4°E, 5m	H_D	0.79	1.11	1.49	1.90	2.16	2.18	2.02	1.85	1.58	1.24	0.87	0.70	1.49
Millau	H	1.42	2.06	3.39	4.46	5.06	6.17	6.62	5.45	4.32	2.54	1.56	1.23	3.69
44.1°N, 3.1°E, 715m	H_D	0.91	1.30	1.78	2.30	2.72	2.66	2.28	2.23	1.81	1.47	1.02	0.80	1.77
Nancy	H	0.79	1.59	2.57	3.86	4.78	5.28	5.59	4.66	3.29	1.87	1.03	0.65	3.00
48.7°N, 6.2°E, 212m	H_D	0.58	1.01	1.56	2.16	2.62	2.81	2.69	2.34	1.81	1.19	0.71	0.48	1.66
Nice	H	1.76	2.40	3.65	4.70	5.91	6.56	6.68	5.91	4.51	2.90	1.85	1.49	4.03
43.7°N, 7.2°E, 5m	H_D	0.90	1.27	1.73	2.25	2.53	2.58	2.39	2.12	1.81	1.42	1.00	0.79	1.73
Paris	H	0.77	1.49	2.37	3.72	4.39	5.02	5.11	4.42	3.15	1.90	1.08	0.63	2.83
48.80°N, 2.3°E, 75m	H_D	0.57	0.99	1.52	2.15	2.58	2.81	2.72	2.35	1.80	1.19	0.73	0.47	1.65
Perpignan	H	1.76	2.50	3.82	4.73	5.33	6.19	6.43	5.64	4.46	3.00	1.99	1.59	3.96
42.7°N, 2.9°E, 42m	H_D	0.82	1.15	1.64	2.08	2.39	2.59	2.54	2.27	1.83	1.33	0.92	0.73	1.69
Rennes	H	0.93	1.71	2.76	4.18	4.87	5.50	5.49	4.68	3.53	2.11	1.22	0.77	3.14
48.1°N, 1.7°W, 35m	H_D	0.65	1.05	1.60	2.18	2.62	2.81	2.70	2.35	1.83	1.25	0.78	0.54	1.69
Strasbourg	H	0.79	1.61	2.57	3.81	4.82	5.26	5.52	4.66	3.29	1.80	1.06	0.65	2.97
48.5°N, 7.8°E, 150m	H_D	0.59	1.02	1.57	2.17	2.62	2.81	2.70	2.34	1.81	1.18	0.72	0.48	1.66
Locations in UK + EIR		Jan	Feb	Mar	Apr	May	June	July	Aug	Sep	Oct	Nov	Dec	Year
Birmingham / UK	H	0.63	1.17	2.01	3.47	4.35	4.54	4.43	3.87	2.68	1.49	0.82	0.47	2.49
52.5°N, 2.2°W, 100m	H_D	0.46	0.81	1.35	2.06	2.54	2.75	2.65	2.26	1.65	1.01	0.57	0.35	1.54
Dublin / EIR	H	0.65	1.17	2.26	3.61	4.64	4.78	4.78	3.68	2.77	1.59	0.78	0.47	2.59
53.1°N, 6.1°W, 55m	H_D	0.53	0.92	1.59	2.31	2.86	3.12	3.00	2.52	1.87	1.16	0.63	0.39	1.74
Efford /UK	H	0.82	1.64	2.50	4.22	5.11	5.31	5.23	4.52	3.17	1.90	1.08	0.65	3.00
50.8°N, 1.6°W, 16m	H_D	0.64	1.08	1.69	2.32	2.85	3.11	2.98	2.55	1.95	1.29	0.77	0.52	1.81
Glasgow / UK	H	0.45	1.04	1.94	3.41	4.49	4.71	4.35	3.49	2.33	1.25	0.61	0.32	2.36
55.7°N, 4.5°W, 10m	H_D	0.39	0.81	1.45	2.23	2.83	3.11	2.97	2.46	1.73	1.00	0.51	0.29	1.65
London / UK	H	0.65	1.20	2.25	3.42	4.44	4.87	4.59	4.00	2.94	1.69	0.87	0.50	2.61
51.6°N, 0.0°W, 5m	H_D	0.48	0.84	1.43	2.06	2.56	2.79	2.68	2.28	1.70	1.08	0.61	0.38	1.57
Locations in Holland		Jan	Feb	Mar	Apr	May	June	July	Aug	Sep	Oct	Nov	Dec	Year
Amsterdam /NL	H	0.64	1.47	2.39	4.00	5.22	5.28	5.10	4.37	2.94	1.66	0.81	0.47	2.86
52.6°N, 4.8°E, 5m	H_D	0.47	0.88	1.43	2.07	2.55	2.80	2.69	2.28	1.69	1.05	0.56	0.35	1.56
Vlissingen / NL	H	0.72	1.54	2.45	4.01	4.99	5.28	5.26	4.42	3.00	1.77	0.94	0.53	2.90
51.5°N, 3.6°E, 8m	H_D	0.51	0.92	1.47	2.10	2.58	2.80	2.70	2.29	1.71	1.10	0.63	0.39	1.60
Location in Belgium		Jan	Feb	Mar	Apr	May	June	July	Aug	Sep	Oct	Nov	Dec	Year
Uccle / Bruxelles	H	0.65	1.32	2.13	3.41	4.52	4.70	4.59	3.99	2.81	1.73	0.81	0.50	2.59
50.8°N, 4.4°E, 105m	H_D	0.49	0.90	1.42	2.07	2.58	2.77	2.68	2.29	1.71	1.11	0.60	0.39	1.58

Table A2.6 Horizontal global insolation in kWh/m²/d at various locations in Southern Europe

Locations in Italy		Jan	Feb	Mar	Apr	May	June	July	Aug	Sep	Oct	Nov	Dec	Year
Ancona	H	1.27	2.01	3.41	4.70	5.78	6.29	6.53	5.56	4.34	2.66	1.58	1.17	3.76
43.6°N, 13.5°E, 105m	H$_D$	0.91	1.32	1.80	2.25	2.52	2.60	2.34	2.19	1.82	1.48	1.04	0.81	1.75
Bolzano (Bozen)	H	1.27	2.08	3.19	4.42	4.52	5.88	5.86	5.11	3.81	2.47	1.49	1.03	3.43
46.5°N, 11.3°E, 241m	H$_D$	0.82	1.19	1.71	2.24	2.76	2.75	2.63	2.29	1.88	1.36	0.92	0.69	1.77
Brindisi	H	1.73	2.30	3.55	4.94	6.09	6.80	6.82	5.98	4.75	3.24	2.02	1.60	4.15
40.7°N, 18.0°E, 10m	H$_D$	1.03	1.39	1.87	2.28	2.49	2.48	2.34	2.18	1.86	1.51	1.13	0.92	1.79
Cagliari (Sardinia)	H	2.11	2.88	4.20	5.35	6.19	7.20	6.91	6.15	4.94	3.39	2.28	1.87	4.46
39.2°N, 9.1°E, 18m	H$_D$	0.94	1.24	1.59	1.95	2.20	2.09	2.08	1.92	1.66	1.35	1.03	0.84	1.57
Gela (Sicily)	H	2.54	3.48	4.63	5.93	6.75	7.30	7.28	6.41	5.40	3.99	2.81	2.19	4.89
37.1°N, 14.2°E, 33m	H$_D$	1.09	1.39	1.77	2.10	2.35	2.37	2.28	2.15	1.83	1.49	1.17	0.99	1.75
Genova	H	1.39	2.13	3.17	3.96	4.75	5.62	6.12	5.23	3.68	2.47	1.56	1.17	3.43
44.4°N, 8.9°E, 3m	H$_D$	0.91	1.28	1.80	2.36	2.76	2.84	2.54	2.30	1.99	1.46	1.01	0.78	1.83
Messina (Sicily)	H	1.92	2.71	4.15	5.56	6.75	7.25	7.39	6.53	5.23	3.62	2.19	1.70	4.58
38.2°N, 15.6°E, 59m	H$_D$	0.99	1.29	1.64	1.92	2.04	2.06	1.89	1.80	1.60	1.36	1.06	0.88	1.54
Milano	H	1.06	1.80	3.10	4.32	5.23	5.97	6.08	5.28	3.89	2.35	1.20	0.89	3.43
45.5°N, 9.3°E, 103m	H$_D$	0.65	1.02	1.56	2.06	2.45	2.61	2.52	2.21	1.77	1.23	0.74	0.56	1.61
Monte Terminillo	H	1.85	2.45	3.46	3.89	4.80	5.11	6.12	5.11	4.03	2.88	1.99	1.63	3.60
42.5°N, 13.0°E, 1875m	H$_D$	0.84	1.15	1.59	1.94	2.30	2.44	2.53	2.22	1.80	1.31	0.92	0.73	1.65
Napoli	H	1.66	2.45	3.69	5.04	6.00	6.86	6.86	6.08	4.63	3.41	1.94	1.59	4.17
40.8°N, 14.3°E, 72m	H$_D$	0.88	1.20	1.61	1.98	2.23	2.19	2.09	1.92	1.68	1.28	0.97	0.78	1.57
Pisa	H	1.44	2.21	3.33	4.75	5.81	6.62	7.22	6.02	4.44	2.79	1.70	1.20	3.96
43.7°N, 10.4°E, 5m	H$_D$	0.91	1.28	1.77	2.24	2.55	2.56	2.12	2.07	1.83	1.43	1.01	0.78	1.71
Roma	H	1.87	2.62	3.96	5.14	6.19	6.74	6.96	6.12	4.75	3.26	2.11	1.60	4.27
41.8°N, 12.6°E, 131m	H$_D$	0.85	1.16	1.54	1.94	2.18	2.24	2.05	1.87	1.61	1.26	0.94	0.75	1.53
Torino	H	1.59	2.11	3.32	4.37	5.09	5.79	5.88	5.11	3.89	2.57	1.58	1.30	3.55
45.2°N, 7.7°E, 282m	H$_D$	0.85	1.25	1.75	2.29	2.71	2.80	2.63	2.33	1.91	1.42	0.97	0.75	1.80
Trapani (Sicily)	H	2.02	2.93	4.25	5.59	6.51	7.39	7.05	6.15	4.99	3.58	2.38	1.87	4.55
38.0°N, 12.5°E, 14m	H$_D$	0.99	1.29	1.63	1.93	2.12	2.00	2.03	1.94	1.69	1.38	1.07	0.90	1.58
Locations in Spain		Jan	Feb	Mar	Apr	May	June	July	Aug	Sep	Oct	Nov	Dec	Year
Almeria	H	2.50	3.45	4.68	5.83	6.51	6.84	6.94	6.26	5.11	3.96	2.81	2.16	4.75
36.8°N, 2.5°W, 7m	H$_D$	1.19	1.53	1.94	2.35	2.65	2.75	2.64	2.43	2.10	1.65	1.29	1.08	1.97
Barcelona	H	1.72	2.45	3.75	4.75	5.53	6.31	6.49	5.65	4.44	3.00	1.94	1.53	3.96
41.4°N, 2.2°E, 5m	H$_D$	0.99	1.36	1.81	2.30	2.63	2.66	2.47	2.28	1.92	1.51	1.09	0.88	1.82
Madrid	H	2.13	2.76	4.56	5.09	6.57	7.44	7.41	6.48	5.02	3.39	2.14	1.59	4.54
40.5°N, 3.6°W, 668m	H$_D$	0.88	1.20	1.43	1.97	2.08	1.99	1.87	1.77	1.58	1.29	0.98	0.79	1.49
Palma de Mallorca	H	2.09	2.81	4.08	5.31	6.34	6.94	6.91	6.08	4.66	3.41	2.20	1.85	4.39
39.5°N, 2.6°E, 8m	H$_D$	0.93	1.23	1.60	1.95	2.16	2.18	2.08	1.94	1.71	1.34	1.01	0.83	1.58
Santander	H	1.44	2.08	3.33	4.15	5.21	5.59	5.47	4.75	3.98	2.62	1.66	1.23	3.45
43.5°N, 3.8°W, 65m	H$_D$	0.95	1.32	1.82	2.37	2.70	2.84	2.76	2.47	1.96	1.49	1.05	0.82	1.88
Sevilla	H	2.52	3.26	4.70	5.35	6.62	7.20	7.58	6.51	5.38	3.86	2.50	2.16	4.80
37.4°N, 6.0°W, 30m	H$_D$	1.08	1.41	1.75	2.22	2.37	2.40	2.15	2.11	1.82	1.51	1.19	0.99	1.75
Locations in Portugal		Jan	Feb	Mar	Apr	May	June	July	Aug	Sep	Oct	Nov	Dec	Year
Lisboa	H	2.13	2.81	4.66	5.28	6.43	7.20	7.25	6.68	5.19	3.65	2.16	1.87	4.60
38.7°N, 9.2°W, 77m	H$_D$	0.96	1.27	1.50	1.98	2.13	2.08	1.95	1.73	1.59	1.34	1.05	0.87	1.54
Porto	H	1.92	2.54	4.08	5.04	6.22	6.94	6.65	6.19	4.85	3.26	2.02	1.53	4.27
41.1°N, 8.6°W, 100m	H$_D$	0.88	1.19	1.54	1.98	2.18	2.19	2.17	1.86	1.61	1.30	0.96	0.78	1.55
Location in Greece		Jan	Feb	Mar	Apr	May	June	July	Aug	Sep	Oct	Nov	Dec	Year
Athinai (Athens)	H	2.13	2.67	3.36	4.87	5.86	6.68	6.86	6.45	5.19	3.43	2.20	1.70	4.27
38.0°N, 23.7°E, 107m	H$_D$	0.99	1.30	1.69	2.06	2.28	2.26	2.10	1.83	1.63	1.40	1.07	0.89	1.62
Location in Malta		Jan	Feb	Mar	Apr	May	June	July	Aug	Sep	Oct	Nov	Dec	Year
Luga/Qrendi	H	2.64	3.60	5.06	5.95	7.39	7.78	7.92	6.96	5.62	4.10	3.09	2.30	5.20
35.8°N, 14.5°E, 135m	H$_D$	1.15	1.43	1.72	2.12	2.13	2.19	2.01	1.97	1.81	1.52	1.19	1.05	1.69

Table A2.7 Horizontal global insolation in kWh/m²/d at various locations in Africa

Location		Jan	Feb	Mar	Apr	May	June	July	Aug	Sep	Oct	Nov	Dec	Year
Abidjan	H	4.37	5.01	5.44	5.33	5.19	4.37	4.14	4.01	4.61	5.25	5.47	4.51	4.80
5.3°N, 3.9°W, 10m	H$_D$	2.16	2.25	2.30	2.32	2.25	2.25	2.26	2.30	2.37	2.26	2.03	2.11	2.24
Addis Abeba	H	5.52	6.07	6.16	5.71	5.38	4.68	3.84	4.06	4.73	5.86	6.19	5.49	5.30
9.1°N, 38.7°E, 2408m	H$_D$	2.11	2.23	2.42	2.56	2.55	2.51	2.38	2.46	2.54	2.36	1.98	2.03	2.34
Alger	H	2.09	3.03	4.10	5.35	6.34	6.80	7.32	6.22	5.08	3.46	2.38	2.11	4.52
36.5°N, 3.0°E, 5m	H$_D$	1.05	1.33	1.70	2.01	2.18	2.21	1.93	1.94	1.70	1.46	1.12	0.94	1.63
Aswan (Assuan)	H	4.99	6.00	6.96	7.85	8.25	8.81	8.40	8.04	7.37	6.24	5.32	4.78	6.90
24.1°N, 32.8°E, 192m	H$_D$	1.14	1.23	1.46	1.56	1.67	1.44	1.61	1.56	1.44	1.34	1.14	1.05	1.39
Asyut	H	4.15	5.28	6.34	7.42	8.11	8.50	8.28	7.82	7.08	5.76	4.61	3.89	6.42
27.1°N, 31.2°E, 69m	H$_D$	1.25	1.36	1.61	1.72	1.76	1.70	1.74	1.66	1.50	1.40	1.23	1.17	1.51
Bamako	H	5.65	6.45	6.49	6.40	6.22	5.99	5.35	5.19	5.72	5.80	5.79	5.34	5.86
12.7°N, 7.8°W, 480m	H$_D$	1.89	1.95	2.29	2.47	2.54	2.55	2.60	2.60	2.49	2.26	1.93	1.86	2.28
Cairo	H	3.42	4.41	5.56	6.59	7.46	7.96	7.81	7.23	6.28	5.06	3.78	3.10	5.72
30.1°N, 31.2°E, 16m	H$_D$	1.26	1.47	1.76	1.99	2.05	2.01	1.99	1.89	1.73	1.50	1.30	1.18	1.68
Casablanca	H	2.79	3.57	4.85	5.88	6.62	6.89	6.92	6.41	5.38	4.08	2.95	2.47	4.89
33.5°N, 7.5°W, 5m	H$_D$	1.24	1.54	1.86	2.20	2.38	2.45	2.37	2.21	1.95	1.64	1.31	1.14	1.86
Dakar	H	5.19	6.24	6.68	7.06	6.94	6.29	5.59	5.47	5.47	5.83	5.19	4.70	5.87
14.4°N, 17.3°W, 20m	H$_D$	1.42	1.28	1.50	1.57	1.68	1.96	2.18	2.18	2.07	1.64	1.50	1.47	1.70
Dar Es Salam	H	5.09	5.23	4.70	3.94	4.39	4.37	4.44	4.80	4.92	5.28	5.76	5.30	4.84
6.9°S, 39.2°E, 55m	H$_D$	2.44	2.44	2.39	2.21	2.03	1.89	1.94	2.08	2.30	2.40	2.34	2.39	2.23
Durban	H	5.83	5.78	5.09	4.10	3.46	3.07	3.24	3.89	4.25	4.78	5.19	6.05	4.55
29.9°S, 31°E, 8m	H$_D$	2.95	2.69	2.30	1.83	1.42	1.24	1.31	1.61	2.11	2.54	2.84	3.01	2.15
Kampala	H	4.87	4.99	5.16	4.80	4.61	4.51	4.42	4.68	5.00	4.93	4.82	4.69	4.78
0.1°N, 32.6°E, 1140m	H$_D$	2.20	2.26	2.30	2.20	2.10	2.04	2.03	2.14	2.25	2.24	2.19	2.14	2.17
Kapstadt	H	8.02	7.27	6.02	4.32	3.07	2.50	2.73	3.50	4.85	6.15	7.66	8.23	5.35
34.0°S, 18.5°E, 44m	H$_D$	2.21	1.97	1.68	1.41	1.12	0.98	1.04	1.32	1.68	2.06	2.17	2.26	1.66
Khartoum	H	5.86	6.68	7.25	7.56	7.12	6.84	6.82	6.68	6.50	6.09	5.95	5.73	6.59
15.6°N, 32.5°E, 380m	H$_D$	1.34	1.43	1.63	1.77	2.01	2.09	2.09	2.10	1.99	1.78	1.44	1.25	1.74
Kinshasa	H	4.92	5.02	5.45	5.26	4.42	3.68	3.53	3.58	4.42	4.78	5.14	4.87	4.58
4.3°S, 15.7°E, 340m	H$_D$	2.62	2.65	2.47	2.31	2.28	2.24	2.28	2.43	2.59	2.64	2.56	2.58	2.47
Lagos	H	4.44	4.78	4.90	4.90	4.86	4.33	3.73	3.62	4.06	4.38	4.84	4.45	4.43
6.5°N, 3.6°E, 5m	H$_D$	2.51	2.66	2.79	2.81	2.72	2.70	2.68	2.71	2.79	2.72	2.45	2.41	2.66
Luanda	H	5.16	5.30	5.44	5.23	4.93	4.41	4.32	4.45	5.04	5.11	5.32	5.17	4.98
9.1°S, 13.2°E, 5m	H$_D$	2.48	2.46	2.32	2.10	1.85	1.80	1.86	2.08	2.27	2.43	2.46	2.47	2.21
Lusaka	H	4.85	5.48	5.66	5.31	5.28	5.14	5.55	6.26	6.60	6.43	5.95	5.66	5.68
15.4°S, 28.3°E, 1283m	H$_D$	2.76	2.73	2.51	2.19	1.74	1.51	1.45	1.62	2.09	2.53	2.76	2.80	2.22
Nairobi	H	5.98	6.07	5.95	5.14	4.61	4.06	3.79	4.08	5.04	5.49	5.23	5.49	5.08
1.3°S, 36.9°E, 1624m	H$_D$	2.19	2.28	2.33	2.40	2.30	2.25	2.28	2.41	2.48	2.45	2.44	2.28	2.34
Ndjamena	H	5.93	6.51	6.75	6.80	6.82	6.48	5.91	5.38	6.27	6.55	6.50	6.12	6.33
12.1°N, 15°E, 297m	H$_D$	1.54	1.67	1.92	2.06	2.04	2.10	2.23	2.30	2.10	1.76	1.38	1.32	1.87
Maputo	H	6.82	6.51	5.76	4.75	4.08	3.74	3.86	4.46	4.99	5.55	5.91	6.72	5.25
25.9°S, 32.6°E, 70m	H$_D$	2.24	2.03	1.78	1.47	1.11	0.94	1.00	1.27	1.74	2.13	2.41	2.35	1.70
Pretoria	H	6.79	6.51	5.83	4.78	4.36	3.98	4.29	5.04	6.07	6.38	6.45	6.89	5.61
25.7°S, 28.2°E, 1369m	H$_D$	2.50	2.31	2.00	1.67	1.22	1.06	1.06	1.30	1.64	2.14	2.46	2.54	1.82
Tamanrasset	H	5.30	6.34	6.98	7.44	7.28	7.80	7.61	7.08	6.44	6.02	4.92	4.59	6.47
22.8°N, 5.5°E, 1380m	H$_D$	1.08	1.13	1.50	1.78	2.09	1.94	1.97	2.00	1.87	1.51	1.40	1.23	1.62
Tripoli	H	2.44	3.33	4.67	5.71	6.88	7.47	7.48	6.58	5.38	3.94	2.86	2.20	4.91
32.7°N, 13.3°E, 5m	H$_D$	1.18	1.46	1.73	1.99	2.02	1.94	1.84	1.86	1.75	1.52	1.25	1.09	1.63
Tunis	H	2.52	3.19	4.51	5.47	6.72	7.51	7.63	6.72	5.52	4.12	3.00	2.42	4.94
36.8°N, 10.3°E, 5m	H$_D$	1.10	1.44	1.81	2.21	2.36	2.29	2.13	2.05	1.81	1.47	1.16	0.99	1.73
Windhoek	H	7.28	7.05	6.36	6.03	5.40	4.97	5.19	5.95	6.86	7.46	7.46	7.78	6.47
22.5°S, 17.1°E, 1728m	H$_D$	2.34	2.17	1.94	1.39	0.97	0.82	0.84	1.05	1.41	1.81	2.21	2.25	1.60

Table A2.8 Horizontal global insolation in kWh/m²/d at various locations in Asia and the Middle East

Location		Jan	Feb	Mar	Apr	May	June	July	Aug	Sep	Oct	Nov	Dec	Year
Bangalore	H	5.39	6.24	6.67	6.63	6.19	5.18	5.01	5.21	5.35	5.02	4.49	4.66	5.50
12.9°N, 77.8°E, 905m	H$_D$	1.96	2.03	2.22	2.42	2.54	2.60	2.60	2.60	2.53	2.37	2.16	2.00	2.33
Bagdad	H	2.46	3.41	4.66	5.80	7.14	7.70	7.63	7.02	5.77	4.39	3.16	2.39	5.12
33.2°N, 44.5°E, 37m	H$_D$	1.16	1.43	1.72	1.96	1.91	1.83	1.77	1.66	1.59	1.39	1.18	1.07	1.55
Bangkok	H	4.56	4.92	5.66	5.62	5.23	4.80	4.80	4.61	4.22	4.13	4.44	4.75	4.79
13.7°N, 100.6°E, 20m	H$_D$	1.85	2.03	2.13	2.31	2.40	2.42	2.42	2.41	2.33	2.18	1.93	1.67	2.17
Bejing (Peking)	H	2.09	2.88	3.72	5.02	5.45	5.47	4.20	4.22	3.91	3.17	2.20	1.80	3.67
39.8°N, 116.4°E,40m	H$_D$	0.99	1.31	1.76	2.15	2.50	2.64	2.52	2.33	1.94	1.46	1.08	0.88	1.79
Calcutta	H	4.18	4.90	5.40	6.41	6.25	4.82	4.35	4.46	4.19	4.72	4.40	3.83	4.82
22.5°N, 88.4°E, 10m	H$_D$	1.73	1.96	2.31	2.46	2.67	2.71	2.62	2.58	2.39	2.13	1.79	1.66	2.25
Colombo	H	5.59	6.15	6.57	6.16	5.56	4.56	5.17	5.48	5.66	5.40	4.79	5.16	5.52
7.0°N, 80.1°E, 5m	H$_D$	2.19	2.27	2.37	2.50	2.51	2.46	2.48	2.54	2.54	2.46	2.33	2.20	2.40
Damaskus	H	2.84	3.86	5.22	6.88	7.68	8.54	8.41	7.63	6.51	5.04	3.78	2.73	5.75
33.5°N, 36.5°E, 700m	H$_D$	1.17	1.43	1.71	1.82	1.96	1.76	1.73	1.69	1.50	1.32	1.09	1.06	1.52
Dehli	H	3.82	4.90	6.08	6.89	7.18	6.56	5.38	5.16	5.69	5.33	4.30	3.72	5.41
28.5°N, 77.3°E, 215m	H$_D$	1.26	1.43	1.64	1.93	2.15	2.40	2.47	2.34	1.95	1.48	1.24	1.13	1.78
Djakarta	H	4.44	5.14	4.82	4.58	4.29	4.37	4.44	4.56	4.46	4.42	3.94	3.99	4.45
6.2°S, 106.7°E, 5m	H$_D$	2.20	2.34	2.25	2.12	1.96	1.90	1.94	2.06	2.14	2.18	2.06	2.07	2.10
Guangzou (Canton)	H	2.73	2.18	2.11	2.61	3.55	4.14	4.89	4.44	4.15	3.95	3.44	3.08	3.44
23.2°N, 114.0°E, 10m	H$_D$	1.68	1.65	1.69	2.03	2.45	2.62	2.69	2.57	2.37	2.08	1.78	1.61	2.10
Hanoi	H	3.60	3.67	3.92	4.35	4.63	4.74	4.98	4.46	4.32	3.96	3.86	3.71	4.18
20.9°N, 105.8°E, 16m	H$_D$	1.61	1.77	1.96	2.16	2.27	2.30	2.33	2.20	2.08	1.87	1.69	1.57	1.98
Irkutsk	H	0.99	1.96	3.36	4.46	5.30	5.23	4.68	4.08	3.15	1.97	1.13	0.72	3.08
52.5°N, 104.3°E, 468m	H$_D$	0.54	0.86	1.34	2.01	2.54	2.83	2.73	2.32	1.72	1.11	0.64	0.43	1.59
Islamabad	H	2.97	4.03	4.95	5.85	6.84	7.13	6.26	5.96	5.68	4.56	3.42	2.94	5.04
34.1°N, 73.0°E, 515m	H$_D$	1.04	1.20	1.57	1.88	1.95	2.00	2.24	2.01	1.53	1.22	1.02	0.88	1.54
Jerusalem	H	3.10	3.60	4.92	6.21	7.46	8.56	8.47	7.61	6.74	5.38	3.89	2.95	5.74
31.8°N, 35.2°E, 790m	H$_D$	1.24	1.53	1.85	2.07	2.05	1.73	1.69	1.73	1.46	1.29	1.18	1.11	1.58
Lhasa	H	4.18	5.06	5.11	6.21	6.21	5.06	4.25	4.06	4.25	3.68	4.27	4.49	4.08	4.55
29.6°N, 91.2°E, 3700m	H$_D$	1.28	1.53	2.15	2.42	2.73	2.65	2.58	2.53	2.22	1.94	1.31	1.11	2.04
Manila	H	4.15	4.82	5.81	6.27	5.62	5.52	4.82	4.20	4.70	4.27	4.27	4.08	4.87
14.7°N, 121.1°E, 5m	H$_D$	1.86	2.02	2.10	2.15	2.36	2.35	2.42	2.39	2.33	2.15	1.92	1.78	2.15
Mumbai (Bombay)	H	4.60	5.42	6.21	6.62	6.52	4.91	3.86	3.92	4.72	5.08	4.65	4.27	5.06
19.3°N, 73.1°E, 5m	H$_D$	1.80	1.98	2.22	2.43	2.58	2.69	2.48	2.46	2.46	2.18	1.89	1.75	2.24
Riyad	H	4.03	4.35	5.71	5.91	6.89	6.50	5.86	6.29	5.88	5.35	4.44	3.75	5.41
24.7°N, 46.7°E, 585m	H$_D$	1.25	1.59	1.65	2.02	1.88	2.12	2.28	1.95	1.75	1.39	1.21	1.20	1.69
Seoul	H	2.00	2.91	3.70	4.58	4.61	4.58	3.63	3.56	3.58	3.00	2.11	1.68	3.33
37.5°N, 126.9°E, 86m	H$_D$	1.28	1.65	2.16	2.65	3.00	3.13	2.89	2.71	2.36	1.84	1.39	1.16	2.18
Shanghai	H	1.92	2.42	2.88	3.94	4.18	4.44	4.92	4.08	3.53	3.50	2.71	2.28	3.40
31.1°N, 121.6°E, 5m	H$_D$	1.50	1.85	2.26	2.74	2.97	3.09	3.06	2.83	2.49	2.06	1.66	1.46	2.33
Singapore	H	4.44	5.14	4.82	4.58	4.29	4.37	4.44	4.56	4.46	4.42	3.94	3.99	4.45
1.2°N, 104.2°E, 5m	H$_D$	2.51	2.50	2.62	2.56	2.45	2.33	2.37	2.48	2.60	2.62	2.55	2.48	2.50
Taipei	H	2.56	2.54	3.32	4.10	4.54	5.42	5.97	5.34	4.83	3.88	2.81	2.38	3.97
25.0°N, 121.5°E, 10m	H$_D$	1.43	1.56	1.93	2.24	2.40	2.53	2.48	2.39	2.19	1.88	1.53	1.34	1.99
Tashkent	H	1.76	2.71	2.68	5.50	7.01	7.87	7.68	6.89	5.50	3.58	2.09	1.51	4.63
41.3°N, 69.2°E, 460m	H$_D$	0.88	1.18	1.60	1.87	1.91	1.80	1.75	1.56	1.36	1.22	0.95	0.77	1.40
Teheran	H	2.45	3.31	4.08	5.43	6.43	7.66	7.49	6.94	5.73	4.25	2.93	2.26	4.91
35.5°N, 52.0°E, 1132m	H$_D$	1.08	1.34	1.74	2.01	2.16	1.88	1.85	1.67	1.52	1.32	1.10	0.98	1.55
Tokyo	H	2.56	3.09	3.71	4.27	4.85	4.21	4.13	4.54	3.24	2.81	2.31	2.21	3.49
35.7°N, 139.8°E, 5m	H$_D$	1.33	1.72	2.23	2.70	3.01	3.07	3.01	2.82	2.36	1.91	1.46	1.22	2.23
Ulan Bator	H	1.59	2.62	4.15	5.09	5.76	5.79	5.12	4.42	3.86	2.95	1.82	1.27	3.70
47.8°N, 106.8°E, 1330m	H$_D$	0.73	1.07	1.54	2.19	2.65	2.87	2.76	2.39	1.87	1.26	0.82	0.62	1.73
Vladivostok	H	2.19	3.12	4.18	4.78	5.45	4.51	4.10	3.96	3.96	3.22	2.20	1.77	3.62
43.2°N, 132.0°E, 5m	H$_D$	0.91	1.25	1.77	2.33	2.72	2.74	2.60	2.38	2.01	1.46	1.03	0.81	1.83

Table A2.9 Horizontal global insolation in kWh/m^2/d at various locations in North America

Location		Jan	Feb	Mar	Apr	May	June	July	Aug	Sep	Oct	Nov	Dec	Year
Anchorage	H	0.34	0.99	2.38	3.80	4.77	5.06	4.69	3.55	2.31	1.16	0.45	0.18	2.47
61.3°N, 149.6°W, 5m	H$_D$	0.22	0.61	1.19	1.98	2.72	3.08	2.93	2.34	1.56	0.82	0.32	0.13	1.49
Atlanta	H	2.62	3.44	4.51	5.65	6.25	6.39	6.11	5.70	4.87	4.07	2.96	2.37	4.57
33.9°N, 83.9°W, 320m	H$_D$	1.21	1.49	1.85	2.14	2.36	2.46	2.46	2.28	1.97	1.56	1.26	1.10	1.84
Boston	H	1.87	2.69	3.72	4.71	5.62	6.12	6.04	5.35	4.25	3.00	1.90	1.51	3.89
42.4°N, 70.9°W, 5m	H$_D$	0.94	1.29	1.77	2.28	2.61	2.71	2.63	2.35	1.94	1.46	1.06	0.84	1.82
Chicago	H	1.85	2.64	3.52	4.57	5.72	6.33	6.12	5.41	4.23	3.04	1.82	1.46	3.89
42.0°N, 87.7°W, 177m	H$_D$	0.97	1.32	1.83	2.31	2.59	2.65	2.60	2.35	1.95	1.48	1.07	0.86	1.83
Dallas	H	2.88	3.65	4.70	5.59	6.22	6.91	7.01	6.34	5.20	4.20	3.12	2.62	4.86
32.9°N, 96.9°W, 146m	H$_D$	1.12	1.39	1.68	2.00	2.18	2.06	1.94	1.89	1.75	1.43	1.19	1.04	1.64
Denver	H	2.38	3.27	4.40	5.55	6.19	6.84	6.70	5.99	4.97	3.80	2.62	2.14	4.57
39.8°N, 104.9°W, 1585m	H$_D$	0.86	1.12	1.50	1.84	2.16	2.14	2.09	1.91	1.58	1.20	0.93	0.75	1.51
Edmonton	H	1.08	1.96	3.48	4.97	5.88	6.41	6.39	5.33	3.56	2.23	1.13	0.77	3.59
53.5°N, 113.4°W, 668m	H$_D$	0.48	0.82	1.25	1.85	2.45	2.68	2.48	2.11	1.68	1.06	0.62	0.40	1.49
El Paso	H	3.69	4.90	6.34	7.63	8.25	8.43	7.71	7.18	6.09	5.26	3.98	3.41	6.07
31.8°N,106.4°W, 1205m	H$_D$	1.09	1.22	1.38	1.51	1.72	1.82	2.04	1.90	1.75	1.33	1.14	1.02	1.49
Honolulu	H	3.92	4.70	5.40	5.91	6.34	6.53	6.59	6.45	5.93	5.02	4.13	3.72	5.38
21.4°N, 157.8°W, 5m	H$_D$	1.50	1.64	1.86	2.06	2.09	2.06	1.99	1.91	1.80	1.68	1.53	1.42	1.79
Houston	H	2.64	3.41	4.23	5.02	5.62	6.03	5.93	5.62	4.85	4.20	3.05	2.50	4.42
29.9°N, 95.5°W, 10m	H$_D$	1.37	1.67	2.01	2.32	2.49	2.51	2.48	2.33	2.08	1.70	1.44	1.27	1.97
Los Angeles	H	2.83	3.65	4.78	6.06	6.44	6.65	7.28	6.71	5.36	4.18	3.14	2.62	4.97
33.9°N, 118.2°W, 10m	H$_D$	1.22	1.50	1.87	2.14	2.44	2.53	2.27	2.10	1.95	1.60	1.28	1.11	1.83
Miami	H	3.49	4.21	5.12	5.99	5.96	5.61	5.83	5.59	4.89	4.33	3.63	3.27	4.82
25.9°N, 80.1°W, 5m	H$_D$	1.46	1.71	1.96	2.17	2.42	2.54	2.47	2.36	2.17	1.83	1.53	1.37	2.00
Minneapolis	H	1.73	2.66	3.75	4.70	5.66	6.20	6.32	5.37	4.03	2.75	1.66	1.32	3.84
45.3°N, 93.7°W, 214m	H$_D$	0.82	1.12	1.63	2.20	2.56	2.69	2.52	2.28	1.87	1.36	0.94	0.72	1.72
Montreal	H	1.59	2.60	3.85	4.58	5.68	6.21	6.02	4.89	4.03	2.54	1.24	1.18	3.70
45.7°N, 73.6°W, 41m	H$_D$	0.81	1.12	1.58	2.21	2.55	2.69	2.60	2.38	1.85	1.37	0.87	0.71	1.73
New Orleans	H	2.72	3.55	4.44	5.52	6.05	6.12	5.73	5.45	4.90	4.32	3.12	2.59	4.54
29.9°N, 90.1°W, 10m	H$_D$	1.37	1.65	2.00	2.23	2.41	2.51	2.52	2.35	2.08	1.68	1.43	1.27	1.96
New York	H	1.87	2.72	3.73	4.74	5.66	5.99	5.85	5.40	4.32	3.19	1.86	1.48	3.90
40.8°N, 74.0°W, 5m	H$_D$	1.03	1.37	1.85	2.31	2.62	2.74	2.68	2.37	1.98	1.51	1.12	0.91	1.87
Phoenix	H	3.29	4.37	5.62	7.22	8.01	8.16	7.38	6.88	5.97	4.82	3.57	2.96	5.68
33.7°N, 112.2°W, 340m	H$_D$	1.08	1.30	1.59	1.68	1.82	1.94	2.16	2.00	1.71	1.38	1.15	1.01	1.57
Salt Lake City	H	2.11	3.26	4.73	5.85	7.15	7.78	8.04	6.94	5.59	3.96	2.32	1.80	4.96
40.8°N, 111.9°W, 1285m	H$_D$	0.87	1.09	1.36	1.77	1.85	1.84	1.56	1.55	1.33	1.12	0.95	0.79	1.34
San Diego	H	3.06	3.91	4.96	6.18	6.35	6.57	6.98	6.57	5.45	4.40	3.38	2.82	5.05
33.1°N, 117.0°W, 5m	H$_D$	1.23	1.49	1.86	2.12	2.46	2.54	2.37	2.16	1.94	1.59	1.27	1.12	1.84
San Francisco	H	2.16	2.98	4.25	5.67	6.71	7.17	7.36	6.51	5.38	3.89	2.49	1.96	4.71
37.8°N, 122.3°W, 5m	H$_D$	0.99	1.29	1.64	1.91	2.06	2.09	1.90	1.81	1.56	1.31	1.07	0.90	1.54
Seattle	H	0.86	1.70	2.91	4.04	5.34	6.07	6.25	5.27	3.80	2.39	1.09	0.71	3.37
47.9°N, 122.0°W, 5m	H$_D$	0.55	0.95	1.48	2.01	2.43	2.60	2.49	2.18	1.70	1.16	0.67	0.46	1.55
St. Louis	H	2.18	2.93	3.89	5.02	5.86	6.42	6.36	5.69	4.61	3.48	2.28	1.82	4.20
38.7°N, 90.1°W, 142m	H$_D$	1.03	1.35	1.78	2.17	2.44	2.49	2.41	2.20	1.88	1.46	1.12	0.92	1.77
Toronto	H	1.57	2.54	3.57	4.63	5.77	6.30	6.29	5.45	4.05	2.67	1.36	1.15	3.77
43.8°N, 79.3°W, 75m	H$_D$	0.91	1.24	1.75	2.26	2.55	2.67	2.54	2.28	1.93	1.44	0.95	0.76	1.77
Tucson	H	3.43	4.41	5.62	7.06	7.89	8.14	7.05	6.69	6.00	4.99	3.80	3.12	5.68
32.1°N, 111.0°W, 779m	H$_D$	1.14	1.38	1.65	1.79	1.88	1.94	2.25	2.08	1.76	1.41	1.19	1.07	1.62
Vancouver	H	0.85	1.74	2.98	4.25	6.05	6.50	6.50	5.43	3.81	2.07	1.03	0.64	3.48
49.5°N, 123.0°W, 5m	H$_D$	0.52	0.92	1.44	1.99	2.37	2.56	2.45	2.13	1.66	1.08	0.62	0.41	1.51
Washington D.C.	H	2.07	2.91	3.92	4.94	5.67	6.20	6.04	5.36	4.44	3.33	2.24	1.79	4.07
39.1°N, 76.8°W, 5m	H$_D$	1.01	1.33	1.76	2.18	2.47	2.54	2.48	2.27	1.90	1.46	1.10	0.91	1.78
Winnipeg	H	1.51	2.59	4.15	4.97	5.76	6.48	6.41	5.52	3.74	2.33	1.27	1.10	3.81
49.9°N, 97.2°W, 240m	H$_D$	0.58	0.82	1.17	2.03	2.57	2.66	2.52	2.16	1.81	1.25	0.78	0.53	1.57

Table A2.10   Horizontal global insolation in kWh/m²/d at various locations in the rest of the world

Global und diffuse irradiation at some locations in Central and South America in kWh/m²/d

Location		Jan	Feb	Mar	Apr	May	June	July	Aug	Sep	Oct	Nov	Dec	Year
Asuncion	H	5.70	5.71	5.60	3.30	2.27	2.71	3.49	4.09	3.76	4.69	6.56	5.46	4.44
25.1°S, 57.4°W, 98m	H$_D$	2.86	2.60	2.14	1.99	1.57	1.46	1.39	1.66	2.22	2.59	2.55	2.94	2.16
Bahia Blanca	H	6.37	6.28	4.83	3.31	2.24	2.10	1.96	2.97	3.87	5.80	6.56	6.95	4.43
38.8°S, 61.8°W, 5m	H$_D$	2.87	2.28	1.96	1.54	1.16	0.87	1.03	1.30	1.86	2.10	2.61	2.83	1.86
Bogota	H	4.82	5.02	4.97	4.81	4.88	4.70	4.95	5.14	5.18	4.78	4.44	4.52	4.85
4.4°N, 73.8°W, 2650m	H$_D$	2.12	2.22	2.27	2.23	2.20	2.13	2.17	2.25	2.28	2.20	2.07	2.04	2.18
Brasilia	H	4.68	5.50	4.53	4.99	4.72	4.75	4.96	5.49	5.25	4.68	4.75	4.72	4.91
15.7°S, 47.5°W, 912m	H$_D$	2.73	2.74	2.51	2.22	1.87	1.64	1.67	1.91	2.39	2.62	2.71	2.74	2.31
Buenos Aires	H	7.10	6.47	5.15	3.79	2.76	2.16	2.28	3.19	4.30	5.30	6.76	6.98	4.68
34.8°S, 58.4°W, 5m	H$_D$	2.57	2.29	1.97	1.54	1.18	1.02	1.09	1.38	1.81	2.28	2.51	2.72	1.86
Caracas	H	4.74	5.20	5.42	5.18	4.94	4.97	5.11	5.16	5.19	4.88	4.62	4.51	4.99
10.6°N, 66.8°W, 920m	H$_D$	1.94	2.07	2.24	2.37	2.38	2.34	2.34	2.36	2.31	2.20	2.01	1.89	2.20
Guatemala City	H	5.25	5.60	5.87	5.79	5.40	5.18	5.74	5.41	4.81	4.92	4.94	4.90	5.31
14.7°N, 90.7°W, 1488m	H$_D$	1.89	2.14	2.41	2.59	2.64	2.62	2.60	2.61	2.53	2.33	2.04	1.85	2.35
La Paz	H	7.39	6.41	6.48	6.53	5.91	5.50	5.79	6.51	7.34	8.01	8.04	7.51	6.78
16.4°S, 67.6°W, 3630m	H$_D$	2.20	2.32	2.02	1.47	1.17	1.06	1.02	1.17	1.39	1.59	1.89	2.20	1.62
Lima	H	5.74	5.98	6.13	6.08	5.43	5.24	4.79	5.56	6.29	6.51	6.22	6.15	5.84
12.2°S, 76.8°W, 10m	H$_D$	2.30	2.18	1.93	1.51	1.33	1.14	1.45	1.51	1.66	1.91	2.15	2.20	1.77
Manaus	H	4.66	5.14	4.92	4.90	4.73	4.42	4.66	5.13	5.55	5.23	4.56	4.63	4.87
3.1°S, 59.9°W, 26m	H$_D$	2.38	2.41	2.40	2.28	2.11	2.03	2.03	2.13	2.24	2.37	2.38	2.35	2.26
Mexico City	H	4.56	5.40	5.93	6.07	5.56	5.44	5.13	4.92	4.42	4.56	4.34	4.32	5.05
19.3°N, 99.1°W, 2277m	H$_D$	1.81	1.98	2.28	2.54	2.70	2.71	2.69	2.63	2.44	2.22	1.94	1.74	2.31
Porto Allegre	H	6.20	5.47	4.65	3.42	2.50	2.06	2.20	2.90	3.74	4.96	5.79	6.41	4.18
30.1°S, 51.1°W, 5m	H$_D$	2.77	2.63	2.25	1.88	1.50	1.31	1.38	1.70	2.15	2.51	2.78	2.80	2.14
Quito	H	5.71	5.97	6.12	6.07	5.42	5.23	4.78	5.55	6.27	6.48	6.19	6.12	5.83
0.2°S, 78.3°W, 2818m	H$_D$	1.95	1.97	1.97	1.87	1.88	1.81	1.95	1.93	1.85	1.79	1.81	1.76	1.88
Recife	H	5.72	5.72	5.17	4.63	4.71	4.44	4.91	5.57	5.65	6.09	6.17	6.06	5.40
8.1°S, 34.9°W, 5m	H$_D$	2.63	2.62	2.62	2.46	2.12	2.01	1.92	1.96	2.34	2.42	2.45	2.48	2.33
Rio de Janeiro	H	5.85	5.59	5.22	4.06	3.84	3.21	3.79	4.06	4.22	4.87	5.28	5.61	4.63
22.8°S, 43.5°W, 5m	H$_D$	2.57	2.44	2.21	1.82	1.56	1.39	1.47	1.70	1.97	2.27	2.46	2.56	2.03
Santiago de Chile	H	6.43	5.60	4.32	3.07	1.88	1.32	1.52	2.27	2.69	4.57	5.79	6.48	3.82
33.4°S, 70.7°W, 520m	H$_D$	2.80	2.57	2.20	1.71	1.24	0.97	1.08	1.46	1.83	2.46	2.77	2.88	1.99

Global und diffuse irradiation at some locations in Australia and New Zealand in kWh/m²/d

Location		Jan	Feb	Mar	Apr	May	June	July	Aug	Sep	Oct	Nov	Dec	Year
Adelaide	H	7.69	6.45	5.30	3.81	2.47	2.08	2.42	3.14	4.66	5.65	6.78	7.40	4.81
34.9°S, 139.1°E, 5m	H$_D$	2.04	2.00	1.70	1.38	1.09	0.94	0.99	1.25	1.54	1.98	2.20	2.28	1.61
Alice Springs	H	7.46	7.15	6.36	5.56	4.46	4.30	4.66	5.52	6.62	7.15	7.37	7.39	6.16
23.8°S, 133.9°E, 545m	H$_D$	2.31	2.13	1.90	1.52	1.32	1.07	1.04	1.22	1.46	1.92	2.25	2.39	1.71
Auckland (Neuseeland)	H	6.59	5.61	4.60	3.40	2.29	1.82	2.18	2.81	3.87	5.08	5.92	6.42	4.21
37.3°S, 175.1°E, 5m	H$_D$	2.56	2.35	1.91	1.44	1.08	0.92	0.97	1.28	1.73	2.19	2.57	2.72	1.81
Brisbaine	H	6.35	5.70	4.80	3.69	2.90	2.44	2.91	3.61	4.92	5.44	6.34	6.34	4.61
27.9°S, 153.4°E, 5m	H$_D$	2.60	2.43	2.15	1.77	1.42	1.26	1.31	1.57	1.85	2.30	2.50	2.67	1.98
Darwin	H	5.04	5.21	5.21	5.67	5.40	5.40	5.56	6.15	6.56	6.48	6.17	5.38	5.68
12.4°S, 130.9°E, 31m	H$_D$	2.74	2.72	2.58	2.21	1.86	1.61	1.64	1.84	2.18	2.51	2.69	2.76	2.27
Melbourne	H	7.13	6.55	4.93	3.21	2.14	1.93	2.00	2.71	3.86	5.27	6.11	6.69	4.37
37.9°S, 145.0°E, 5m	H$_D$	2.40	2.06	1.81	1.45	1.08	0.88	0.96	1.27	1.71	2.13	2.53	2.67	1.74
Perth	H	7.75	6.92	5.73	4.18	3.14	2.74	2.86	3.74	5.03	6.31	7.50	7.92	5.31
32.1°S, 116.1°E, 5m	H$_D$	2.35	2.20	1.90	1.58	1.25	1.07	1.15	1.41	1.77	2.11	2.30	2.43	1.79
Sydney	H	6.04	5.56	4.22	3.07	2.63	2.33	2.55	3.56	4.62	5.86	6.50	6.13	4.42
33.8°S, 151.4°E, 5m	H$_D$	2.68	2.40	2.07	1.60	1.21	1.03	1.08	1.30	1.71	2.09	2.45	2.76	1.86
Wellington (Neuseeland)	H	6.77	5.64	4.17	2.94	1.85	1.33	1.57	2.25	3.46	4.61	5.61	6.23	3.87
41,3°S, 175,1°E, 5m	H$_D$	2.53	2.34	1.97	1.44	1.05	0.86	0.93	1.28	1.78	2.33	2.77	2.94	1.85

Table A3.1 Global insolation factors $R(\beta,\gamma)$ for locations in Switzerland, Austria and southwest Germany

Location	β	γ	Jan	Feb	Mar	Apr	May	Jun	Jul	Aug	Sep	Oct	Nov	Dec	Year

Location	β	γ	Jan	Feb	Mar	Apr	May	Jun	Jul	Aug	Sep	Oct	Nov	Dec	Year
Kloten (airport)	20°	0°	1.24	1.20	1.13	1.06	1.03	1.00	1.02	1.06	1.11	1.16	1.18	1.24	1.07
(φ = 47,2°)	20°	±30°	1.21	1.17	1.11	1.05	1.02	1.00	1.01	1.04	1.09	1.14	1.15	1.21	1.06
(for northern	20°	±45°	1.16	1.13	1.08	1.04	1.01	1.00	1.00	1.03	1.07	1.10	1.13	1.17	1.04
and central	30°	0°	1.34	1.28	1.16	1.06	1.01	0.98	1.00	1.05	1.13	1.21	1.26	1.34	1.08
Switzerland,	30°	±30°	1.29	1.23	1.13	1.04	0.99	0.97	0.99	1.04	1.11	1.18	1.21	1.28	1.06
SW-Germany	30°	±45°	1.21	1.17	1.09	1.03	0.98	0.97	0.98	1.02	1.08	1.13	1.15	1.24	1.04
and locations	45°	0°	1.42	1.33	1.17	1.03	0.94	0.91	0.93	1.00	1.13	1.24	1.31	1.45	1.05
in lower parts	45°	±30°	1.37	1.26	1.13	1.01	0.93	0.90	0.92	0.98	1.09	1.19	1.23	1.34	1.02
of Austria)	45°	±45°	1.26	1.19	1.07	0.98	0.92	0.90	0.91	0.96	1.04	1.13	1.18	1.28	0.99
	60°	0°	1.47	1.33	1.12	0.95	0.84	0.80	0.82	0.91	1.06	1.21	1.31	1.48	0.98
	60°	±30°	1.37	1.25	1.07	0.92	0.84	0.80	0.82	0.89	1.02	1.14	1.23	1.38	0.95
	60°	±45°	1.26	1.16	1.02	0.90	0.83	0.80	0.81	0.87	0.98	1.08	1.13	1.28	0.92
	90°	0°	1.34	1.16	0.89	0.68	0.56	0.52	0.53	0.63	0.81	1.00	1.13	1.34	0.71
	90°	±30°	1.24	1.07	0.84	0.68	0.58	0.54	0.55	0.63	0.77	0.93	1.05	1.24	0.71
	90°	±45°	1.11	0.97	0.79	0.66	0.58	0.55	0.56	0.62	0.74	0.85	0.95	1.10	0.68
Davos	20°	0°	1.47	1.33	1.20	1.10	1.03	1.00	1.02	1.05	1.13	1.24	1.39	1.52	1.13
(φ = 46,5°)	20°	±30°	1.40	1.28	1.17	1.08	1.02	1.00	1.01	1.04	1.11	1.20	1.33	1.45	1.11
(for alpine	20°	±45°	1.33	1.23	1.13	1.06	1.01	1.00	1.00	1.03	1.08	1.16	1.26	1.36	1.08
regions)	30°	0°	1.67	1.46	1.27	1.11	1.01	0.98	0.99	1.05	1.16	1.32	1.53	1.73	1.16
	30°	±30°	1.57	1.39	1.23	1.09	1.00	0.97	0.99	1.03	1.13	1.27	1.46	1.63	1.13
	30°	±45°	1.46	1.31	1.17	1.06	0.99	0.97	0.98	1.01	1.09	1.21	1.36	1.50	1.09
	45°	0°	1.90	1.59	1.33	1.10	0.95	0.91	0.92	1.00	1.15	1.39	1.69	1.98	1.17
	45°	±30°	1.76	1.50	1.26	1.07	0.94	0.90	0.92	0.98	1.11	1.31	1.58	1.82	1.13
	45°	±45°	1.59	1.38	1.20	1.04	0.93	0.89	0.91	0.96	1.07	1.22	1.44	1.64	1.08
	60°	0°	2.01	1.66	1.32	1.04	0.85	0.80	0.82	0.91	1.09	1.39	1.78	2.13	1.12
	60°	±30°	1.84	1.53	1.25	1.01	0.85	0.80	0.81	0.89	1.05	1.29	1.64	1.95	1.08
	60°	±45°	1.64	1.40	1.17	0.98	0.84	0.79	0.81	0.87	0.99	1.19	1.47	1.73	1.03
	90°	0°	1.96	1.53	1.14	0.80	0.57	0.51	0.52	0.62	0.83	1.17	1.65	2.09	0.89
	90°	±30°	1.76	1.39	1.07	0.78	0.58	0.53	0.54	0.62	0.79	1.07	1.50	1.88	0.85
	90°	±45°	1.53	1.24	0.99	0.77	0.59	0.55	0.56	0.62	0.75	0.97	1.31	1.63	0.80
Locarno	20°	0°	1.34	1.20	1.13	1.04	1.01	1.00	1.02	1.06	1.12	1.16	1.25	1.34	1.09
(φ = 46,1°)	20°	±30°	1.29	1.17	1.10	1.03	1.01	1.00	1.01	1.05	1.09	1.13	1.20	1.30	1.07
(for region	20°	±45°	1.24	1.13	1.08	1.02	1.00	1.00	1.00	1.03	1.07	1.10	1.17	1.23	1.05
lake Geneva,	30°	0°	1.47	1.27	1.16	1.03	0.99	0.98	0.99	1.06	1.14	1.20	1.33	1.49	1.10
southern side	30°	±30°	1.41	1.23	1.13	1.02	0.98	0.97	0.98	1.04	1.11	1.16	1.28	1.40	1.08
of the Alps,	30°	±45°	1.32	1.17	1.09	1.01	0.97	0.97	0.98	1.02	1.08	1.12	1.22	1.32	1.05
valleys in the	45°	0°	1.61	1.33	1.16	0.99	0.93	0.90	0.92	1.00	1.13	1.22	1.42	1.62	1.08
Alps)	45°	±30°	1.51	1.26	1.12	0.97	0.92	0.90	0.92	0.99	1.09	1.17	1.33	1.51	1.05
	45°	±45°	1.39	1.19	1.07	0.95	0.91	0.90	0.90	0.96	1.05	1.11	1.25	1.38	1.01
	60°	0°	1.68	1.32	1.11	0.91	0.83	0.79	0.81	0.91	1.06	1.19	1.43	1.68	1.01
	60°	±30°	1.54	1.24	1.06	0.89	0.83	0.79	0.81	0.90	1.02	1.12	1.33	1.55	0.98
	60°	±45°	1.39	1.15	1.01	0.87	0.82	0.79	0.80	0.87	0.97	1.06	1.22	1.40	0.93
	90°	0°	1.54	1.13	0.87	0.66	0.56	0.51	0.51	0.61	0.80	0.97	1.25	1.55	0.75
	90°	±30°	1.39	1.04	0.82	0.65	0.57	0.53	0.53	0.62	0.77	0.90	1.13	1.40	0.73
	90°	±45°	1.22	0.94	0.77	0.64	0.58	0.55	0.55	0.62	0.73	0.83	1.02	1.23	0.70

Global irradiation factors $R(\beta,\gamma)$ for tilted planes

Table A3.2   Global insolation factors $R(\beta,\gamma)$ for locations in Germany and its neighbours to the east and west

Global irradiation factors $R(\beta,\gamma)$ for tilted planes															
Location	β	γ	Jan	Feb	Mar	Apr	May	Jun	Jul	Aug	Sep	Oct	Nov	Dec	Year
Potsdam	20°	0°	1.28	1.22	1.15	1.08	1.04	1.02	1.02	1.07	1.13	1.21	1.27	1.33	1.09
($\varphi = 52,4°$)	20°	±30°	1.24	1.20	1.13	1.07	1.03	1.01	1.02	1.06	1.11	1.18	1.23	1.28	1.07
(for North	20°	±45°	1.20	1.16	1.09	1.05	1.02	1.00	1.01	1.04	1.08	1.13	1.17	1.22	1.05
and East)	30°	0°	1.40	1.30	1.19	1.09	1.03	1.00	1.01	1.07	1.16	1.26	1.33	1.44	1.10
	30°	±30°	1.36	1.26	1.16	1.07	1.01	0.99	1.00	1.05	1.13	1.22	1.30	1.39	1.08
	30°	±45°	1.28	1.20	1.12	1.05	1.00	0.98	0.99	1.03	1.10	1.18	1.23	1.28	1.05
	45°	0°	1.56	1.38	1.21	1.07	0.98	0.94	0.95	1.03	1.16	1.32	1.43	1.56	1.08
	45°	±30°	1.44	1.32	1.17	1.04	0.97	0.93	0.94	1.01	1.12	1.26	1.37	1.50	1.05
	45°	±45°	1.36	1.24	1.12	1.01	0.95	0.92	0.93	0.98	1.07	1.19	1.27	1.39	1.02
	60°	0°	1.60	1.40	1.18	0.99	0.89	0.84	0.85	0.95	1.11	1.31	1.47	1.61	1.01
	60°	±30°	1.52	1.32	1.13	0.97	0.88	0.83	0.85	0.93	1.07	1.24	1.37	1.50	0.98
	60°	±45°	1.36	1.22	1.06	0.93	0.86	0.83	0.84	0.91	1.02	1.15	1.27	1.39	0.95
	90°	0°	1.56	1.28	0.97	0.74	0.61	0.56	0.58	0.68	0.87	1.12	1.33	1.56	0.76
	90°	±30°	1.40	1.18	0.92	0.72	0.61	0.57	0.59	0.67	0.83	1.03	1.23	1.44	0.75
	90°	±45°	1.28	1.08	0.85	0.70	0.61	0.58	0.59	0.66	0.78	0.94	1.10	1.28	0.72
Giessen	20°	0°	1.26	1.22	1.13	1.08	1.03	1.01	1.02	1.05	1.11	1.18	1.24	1.25	1.08
($\varphi = 50,3°$)	20°	±30°	1.22	1.19	1.11	1.06	1.03	1.00	1.01	1.04	1.09	1.15	1.21	1.20	1.07
(for central part	20°	±45°	1.19	1.15	1.08	1.04	1.02	1.00	1.00	1.03	1.07	1.12	1.15	1.15	1.05
and western)	30°	0°	1.33	1.31	1.16	1.09	1.02	0.99	1.00	1.05	1.13	1.22	1.32	1.35	1.09
Germany)	30°	±30°	1.30	1.25	1.13	1.07	1.01	0.98	1.00	1.03	1.10	1.19	1.26	1.30	1.07
	30°	±45°	1.22	1.20	1.10	1.04	0.99	0.98	0.98	1.02	1.08	1.13	1.21	1.25	1.04
	45°	0°	1.44	1.39	1.16	1.06	0.96	0.92	0.94	1.01	1.13	1.26	1.38	1.45	1.06
	45°	±30°	1.37	1.31	1.12	1.03	0.95	0.92	0.93	0.99	1.09	1.21	1.32	1.35	1.04
	45°	±45°	1.26	1.24	1.08	1.01	0.93	0.91	0.92	0.97	1.05	1.15	1.24	1.30	1.01
	60°	0°	1.48	1.41	1.12	0.99	0.87	0.82	0.84	0.92	1.07	1.24	1.41	1.50	0.99
	60°	±30°	1.37	1.31	1.08	0.96	0.86	0.82	0.83	0.91	1.03	1.18	1.32	1.40	0.96
	60°	±45°	1.26	1.22	1.02	0.92	0.85	0.81	0.83	0.88	0.98	1.09	1.21	1.30	0.93
	90°	0°	1.37	1.25	0.90	0.72	0.59	0.54	0.56	0.66	0.82	1.03	1.26	1.40	0.74
	90°	±30°	1.26	1.15	0.85	0.71	0.60	0.56	0.58	0.65	0.79	0.96	1.15	1.30	0.73
	90°	±45°	1.15	1.03	0.81	0.68	0.60	0.57	0.58	0.65	0.75	0.88	1.03	1.15	0.70
Munich	20°	0°	1.33	1.25	1.15	1.07	1.03	1.01	1.02	1.06	1.12	1.20	1.28	1.33	1.09
(airport)	20°	±30°	1.28	1.21	1.13	1.06	1.02	1.00	1.01	1.04	1.10	1.17	1.23	1.30	1.08
($\varphi = 48,2°$)	20°	±45°	1.23	1.17	1.10	1.04	1.01	1.00	1.00	1.03	1.08	1.13	1.19	1.24	1.06
(for southern	30°	0°	1.47	1.35	1.19	1.08	1.01	0.98	1.00	1.05	1.16	1.27	1.38	1.48	1.11
Bavaria, valleys	30°	±30°	1.40	1.29	1.16	1.06	1.00	0.98	0.99	1.04	1.12	1.22	1.32	1.42	1.08
in the Alps in	30°	±45°	1.30	1.23	1.12	1.04	0.99	0.97	0.98	1.02	1.09	1.17	1.23	1.33	1.06
Austria)	45°	0°	1.60	1.44	1.21	1.05	0.95	0.92	0.93	1.01	1.15	1.31	1.47	1.64	1.09
	45°	±30°	1.51	1.36	1.17	1.02	0.94	0.91	0.92	0.99	1.11	1.25	1.38	1.55	1.06
	45°	±45°	1.40	1.27	1.11	0.99	0.93	0.90	0.91	0.96	1.06	1.18	1.30	1.42	1.02
	60°	0°	1.70	1.47	1.18	0.97	0.85	0.81	0.83	0.92	1.10	1.30	1.51	1.73	1.02
	60°	±30°	1.58	1.37	1.12	0.95	0.85	0.81	0.82	0.90	1.05	1.22	1.40	1.61	0.99
	60°	±45°	1.42	1.27	1.06	0.92	0.84	0.81	0.82	0.88	0.99	1.13	1.28	1.45	0.95
	90°	0°	1.63	1.33	0.95	0.70	0.57	0.52	0.54	0.64	0.84	1.09	1.36	1.64	0.78
	90°	±30°	1.49	1.23	0.89	0.69	0.59	0.54	0.56	0.64	0.80	1.00	1.23	1.52	0.76
	90°	±45°	1.33	1.11	0.83	0.67	0.59	0.56	0.56	0.63	0.76	0.91	1.11	1.33	0.73

Table A3.3 Global insolation factors for $R(\beta, \gamma)$ for locations in Southern Europe, North Africa and the Middle East

Global irradiation factors R(β,γ) for tilted planes															
Location	β	γ	Jan	Feb	Mar	Apr	May	Jun	Jul	Aug	Sep	Oct	Nov	Dec	Year
Marseille	20°	0°	1.41	1.26	1.17	1.08	1.03	1.00	1.02	1.06	1.15	1.23	1.35	1.44	1.11
(φ = 43,3°)	20°	±30°	1.35	1.23	1.15	1.07	1.02	1.00	1.01	1.05	1.12	1.20	1.30	1.37	1.10
(Southern	20°	±45°	1.28	1.18	1.11	1.05	1.01	0.99	1.00	1.04	1.09	1.15	1.24	1.31	1.07
France, northern	30°	0°	1.57	1.35	1.22	1.09	1.01	0.97	0.99	1.06	1.18	1.31	1.49	1.61	1.14
Mediterranean	30°	±30°	1.48	1.29	1.18	1.07	1.00	0.97	0.98	1.04	1.14	1.26	1.40	1.52	1.11
coast, central	30°	±45°	1.37	1.23	1.14	1.04	0.98	0.96	0.97	1.02	1.10	1.20	1.31	1.40	1.07
Italy,Region of	45°	0°	1.73	1.43	1.24	1.06	0.94	0.89	0.91	1.01	1.18	1.37	1.61	1.79	1.12
Adriatic sea	45°	±30°	1.61	1.35	1.19	1.03	0.93	0.89	0.90	0.99	1.13	1.30	1.50	1.66	1.08
etc.)	45°	±45°	1.47	1.25	1.12	1.00	0.91	0.88	0.89	0.96	1.08	1.21	1.38	1.50	1.04
	60°	0°	1.80	1.43	1.20	0.97	0.82	0.76	0.79	0.91	1.11	1.35	1.65	1.89	1.05
	60°	±30°	1.65	1.33	1.13	0.94	0.82	0.77	0.79	0.89	1.06	1.26	1.53	1.73	1.01
	60°	±45°	1.47	1.22	1.06	0.91	0.81	0.77	0.79	0.87	1.00	1.17	1.38	1.53	0.96
	90°	0°	1.64	1.22	0.93	0.66	0.50	0.43	0.45	0.58	0.82	1.12	1.48	1.74	0.76
	90°	±30°	1.47	1.10	0.86	0.65	0.53	0.47	0.49	0.59	0.78	1.02	1.33	1.55	0.73
	90°	±45°	1.27	0.98	0.80	0.64	0.54	0.50	0.51	0.59	0.74	0.92	1.15	1.34	0.70
Sevilla	20°	0°	1.31	1.21	1.14	1.05	1.00	0.98	0.99	1.03	1.11	1.19	1.26	1.33	1.09
(φ = 37,3°)	20°	±30°	1.27	1.18	1.11	1.04	1.00	0.98	0.99	1.03	1.09	1.16	1.22	1.29	1.08
(for southern	20°	±45°	1.21	1.14	1.09	1.03	0.99	0.98	0.98	1.01	1.06	1.12	1.17	1.22	1.06
Europe,	30°	0°	1.43	1.28	1.17	1.04	0.97	0.94	0.95	1.01	1.13	1.25	1.35	1.46	1.10
southern	30°	±30°	1.36	1.23	1.13	1.03	0.97	0.94	0.95	1.00	1.09	1.20	1.29	1.39	1.08
Turkey and	30°	±45°	1.28	1.18	1.09	1.01	0.96	0.94	0.95	0.99	1.06	1.15	1.22	1.30	1.05
nothern coast	45°	0°	1.53	1.32	1.16	0.99	0.89	0.84	0.86	0.95	1.10	1.28	1.42	1.58	1.07
of Morocco,	45°	±30°	1.44	1.26	1.12	0.97	0.88	0.85	0.86	0.93	1.06	1.21	1.34	1.48	1.04
Algeria and	45°	±45°	1.32	1.18	1.06	0.95	0.88	0.85	0.86	0.92	1.02	1.14	1.24	1.36	1.00
Tunisia)	60°	0°	1.56	1.30	1.10	0.89	0.76	0.71	0.72	0.83	1.02	1.24	1.42	1.61	0.98
	60°	±30°	1.44	1.22	1.04	0.87	0.77	0.72	0.73	0.83	0.98	1.16	1.32	1.49	0.95
	60°	±45°	1.30	1.13	0.98	0.85	0.77	0.73	0.74	0.81	0.93	1.08	1.21	1.34	0.91
	90°	0°	1.35	1.06	0.81	0.58	0.44	0.38	0.38	0.50	0.71	0.98	1.21	1.42	0.68
	90°	±30°	1.22	0.96	0.76	0.59	0.48	0.43	0.43	0.53	0.69	0.89	1.10	1.28	0.67
	90°	±45°	1.07	0.88	0.72	0.59	0.50	0.46	0.47	0.54	0.66	0.82	0.97	1.11	0.65
Cairo	20°	0°	1.25	1.18	1.10	1.03	0.98	0.95	0.96	1.01	1.07	1.16	1.24	1.27	1.06
(φ =30,1°)	20°	±30°	1.21	1.15	1.08	1.02	0.97	0.95	0.96	1.00	1.06	1.13	1.20	1.23	1.05
(for northern	20°	±45°	1.17	1.12	1.06	1.01	0.97	0.96	0.96	0.99	1.04	1.10	1.15	1.18	1.03
Africa and Near	30°	0°	1.33	1.23	1.12	1.01	0.93	0.90	0.91	0.97	1.07	1.20	1.31	1.37	1.06
and Middle	30°	±30°	1.27	1.19	1.09	1.00	0.93	0.90	0.91	0.97	1.05	1.16	1.26	1.30	1.04
East)	30°	±45°	1.21	1.14	1.06	0.98	0.93	0.90	0.91	0.96	1.03	1.12	1.19	1.23	1.02
	45°	0°	1.40	1.26	1.08	0.94	0.83	0.78	0.80	0.88	1.03	1.20	1.36	1.44	1.01
	45°	±30°	1.32	1.19	1.05	0.92	0.83	0.79	0.81	0.88	1.00	1.15	1.28	1.36	0.99
	45°	±45°	1.22	1.12	1.01	0.91	0.83	0.80	0.81	0.87	0.97	1.09	1.20	1.25	0.96
	60°	0°	1.39	1.21	1.00	0.82	0.68	0.62	0.65	0.75	0.93	1.15	1.34	1.44	0.91
	60°	±30°	1.29	1.13	0.96	0.81	0.70	0.65	0.67	0.76	0.90	1.08	1.25	1.34	0.89
	60°	±45°	1.17	1.06	0.92	0.80	0.71	0.67	0.69	0.76	0.87	1.01	1.14	1.21	0.86
	90°	0°	1.14	0.93	0.69	0.48	0.34	0.28	0.30	0.40	0.60	0.85	1.09	1.22	0.60
	90°	±30°	1.04	0.85	0.66	0.51	0.40	0.35	0.37	0.45	0.59	0.79	0.99	1.10	0.60
	90°	±45°	0.91	0.78	0.64	0.52	0.44	0.40	0.42	0.48	0.59	0.73	0.88	0.96	0.59

A4 R_B Factors for Insolation Calculations Using the Three-Component Model

The R_B factors are used to calculate direct beam radiation on the solar generator plane, as derived from direct beam radiation $H_B = (H - H_D)$ on the horizontal plane (see Section 2.5). These factors are determined solely by the solar generator's angle of incidence β and orientation γ, as well as by the geographic latitude of the site for which insolation on the solar generator plane is being determined. The figures indicated below were generated using Meteonorm 95 software [2.2].

The latitude φ for each location can be found in the tables in Section A2, which also contains the monthly H figures (bolded) and H_D figures. The latitude for southern hemisphere locations (indicated as °S in Section A2) is $\varphi < 0°$. The angles of incidence β listed in the tables apply to a southern orientation in the northern hemisphere and vice versa.

The tables that follow indicate the R_B factor for each of 54 solar generator orientations (angle of incidence $\beta = 20°$, $30°$, $35°$, $45°$, $60°$ and $90°$; in the tropics $\beta = 10°$, $20°$, $30°$, $35°$, $45°$, $60°$; $\gamma = -60°$, $-45°$, $-30°$, $0°$, $+30°$, $+45°$ and $+60°$ for each of these values) for locations at the following latitudes: $60°$, $58°$, $56°$, $54°$, $53°$, $52°$, $51°$, $50°$, $49°$, $48°$, $47°$, $46°$, $45°$, $44°$, $43°$, $42°$, $41°$, $40°$, $39°$, $38°$, $37°$, $36°$, $35°$, $34°$, $32°$, $30°$, $28°$, $26°$, $24°$, $22°$, $20°$, $15°$, $10°$, $5°$, $0°$, $-5°$, $-10°$, $-15°$, $-20°$, $-22°$, $-24°$, $-26°$, $-28°$, $-30°$, $-32°$, $-34°$, $-36°$, $-38°$ and $-40°$.

Regions located between 34°N and 54°N are broken down more precisely so as to allow for optimal accuracy in such cases. Accuracy can be increased further for the β and γ indicated, via an interpolation.

Inasmuch as Section A3 does not list reference stations for the tropics and the southern hemisphere, only the three-component model will provide exact insolation figures on an inclined plane for such regions using the tables in this book.

R_B - factors for latitude 60 degrees

β	γ	Jan	Feb	Mar	Apr	May	Jun	Jul	Aug	Sep	Oct	Nov	Dec
20°	0°	3.35	2.15	1.59	1.30	1.15	1.10	1.12	1.23	1.45	1.90	2.84	4.29
20°	±30°	3.03	1.99	1.50	1.26	1.13	1.08	1.10	1.19	1.38	1.78	2.59	3.84
20°	±45°	2.64	1.80	1.40	1.20	1.10	1.06	1.08	1.15	1.31	1.63	2.28	3.31
20°	±60°	2.15	1.56	1.28	1.13	1.06	1.03	1.04	1.10	1.21	1.43	1.90	2.62
30°	0°	4.39	2.64	1.81	1.39	1.18	1.10	1.14	1.29	1.61	2.28	3.65	5.76
30°	±30°	3.92	2.40	1.69	1.33	1.15	1.08	1.11	1.24	1.51	2.09	3.27	5.11
30°	±45°	3.36	2.13	1.55	1.26	1.11	1.05	1.08	1.19	1.41	1.87	2.83	4.33
30°	±60°	2.64	1.78	1.37	1.16	1.05	1.01	1.03	1.11	1.27	1.60	2.27	3.32
35°	0°	4.86	2.85	1.90	1.42	1.18	1.09	1.13	1.31	1.67	2.44	4.01	6.44
35°	±30°	4.32	2.58	1.76	1.36	1.15	1.07	1.10	1.25	1.56	2.22	3.58	5.68
35°	±45°	3.68	2.26	1.61	1.27	1.10	1.04	1.07	1.19	1.44	1.97	3.08	4.79
35°	±60°	2.85	1.87	1.41	1.17	1.04	1.00	1.02	1.11	1.29	1.67	2.44	3.63
45°	0°	5.69	3.22	2.05	1.45	1.16	1.05	1.10	1.31	1.76	2.70	4.64	7.63
45°	±30°	5.02	2.88	1.88	1.38	1.12	1.03	1.07	1.25	1.63	2.44	4.11	6.70
45°	±45°	4.23	2.49	1.69	1.28	1.07	1.00	1.03	1.18	1.49	2.14	3.49	5.60
45°	±60°	3.22	2.02	1.45	1.16	1.01	0.95	0.98	1.09	1.31	1.77	2.71	4.18
60°	0°	6.60	3.57	2.14	1.42	1.06	0.94	0.99	1.24	1.78	2.94	5.32	8.98
60°	±30°	5.78	3.16	1.93	1.33	1.03	0.92	0.97	1.18	1.64	2.62	4.67	7.85
60°	±45°	4.81	2.69	1.71	1.23	0.98	0.89	0.93	1.11	1.48	2.26	3.91	6.50
60°	±60°	3.58	2.13	1.45	1.10	0.92	0.85	0.88	1.01	1.28	1.83	2.96	4.75
90°	0°	7.04	3.55	1.89	1.06	0.67	0.54	0.60	0.87	1.48	2.82	5.56	9.79
90°	±30°	6.10	3.07	1.67	1.00	0.67	0.55	0.60	0.83	1.34	2.45	4.82	8.48
90°	±45°	4.98	2.54	1.44	0.91	0.65	0.56	0.60	0.79	1.18	2.05	3.93	6.93
90°	±60°	3.58	1.93	1.18	0.81	0.62	0.55	0.58	0.72	1.00	1.60	2.87	4.92

R$_B$ - factors for latitude 58 degrees

β	γ	Jan	Feb	Mar	Apr	May	Jun	Jul	Aug	Sep	Oct	Nov	Dec
20°	0°	2.91	2.02	1.54	1.28	1.14	1.09	1.11	1.21	1.41	1.81	2.55	3.52
20°	±30°	2.64	1.87	1.46	1.23	1.12	1.07	1.09	1.18	1.35	1.69	2.33	3.17
20°	±45°	2.33	1.70	1.37	1.18	1.09	1.05	1.07	1.14	1.28	1.56	2.08	2.76
20°	±60°	1.93	1.49	1.25	1.12	1.05	1.02	1.03	1.09	1.19	1.39	1.75	2.23
30°	0°	3.74	2.44	1.74	1.36	1.16	1.09	1.12	1.26	1.55	2.14	3.22	4.63
30°	±30°	3.36	2.23	1.62	1.30	1.13	1.07	1.09	1.22	1.47	1.97	2.90	4.13
30°	±45°	2.90	1.98	1.49	1.23	1.09	1.04	1.06	1.16	1.37	1.77	2.53	3.53
30°	±60°	2.32	1.68	1.33	1.14	1.04	1.00	1.02	1.09	1.24	1.53	2.06	2.76
35°	0°	4.12	2.62	1.82	1.38	1.16	1.08	1.11	1.27	1.61	2.28	3.52	5.14
35°	±30°	3.68	2.38	1.69	1.32	1.13	1.05	1.09	1.23	1.51	2.08	3.15	4.56
35°	±45°	3.15	2.10	1.54	1.24	1.09	1.03	1.05	1.17	1.40	1.86	2.73	3.87
35°	±60°	2.49	1.75	1.36	1.14	1.03	0.98	1.00	1.09	1.26	1.58	2.19	2.99
45°	0°	4.78	2.93	1.94	1.41	1.13	1.03	1.07	1.27	1.68	2.51	4.03	6.03
45°	±30°	4.23	2.63	1.78	1.33	1.10	1.01	1.05	1.22	1.56	2.27	3.59	5.32
45°	±45°	3.58	2.29	1.61	1.24	1.05	0.98	1.01	1.15	1.43	2.00	3.06	4.47
45°	±60°	2.77	1.87	1.39	1.13	0.99	0.93	0.96	1.06	1.27	1.67	2.40	3.38
60°	0°	5.48	3.22	2.01	1.36	1.02	0.91	0.96	1.19	1.69	2.70	4.57	7.02
60°	±30°	4.82	2.86	1.82	1.28	0.99	0.89	0.94	1.14	1.56	2.41	4.03	6.15
60°	±45°	4.02	2.44	1.62	1.18	0.95	0.87	0.90	1.07	1.41	2.09	3.38	5.11
60°	±60°	3.03	1.95	1.37	1.06	0.89	0.83	0.86	0.98	1.22	1.70	2.59	3.78
90°	0°	5.75	3.14	1.74	0.99	0.63	0.50	0.56	0.81	1.38	2.54	4.70	7.53
90°	±30°	4.98	2.72	1.54	0.93	0.63	0.52	0.57	0.78	1.24	2.21	4.07	6.52
90°	±45°	4.07	2.26	1.33	0.86	0.62	0.53	0.57	0.74	1.10	1.86	3.33	5.32
90°	±60°	2.95	1.73	1.10	0.76	0.59	0.52	0.55	0.68	0.94	1.46	2.45	3.80

R$_B$ - factors for latitude 56 degrees

β	γ	Jan	Feb	Mar	Apr	May	Jun	Jul	Aug	Sep	Oct	Nov	Dec
20°	0°	2.60	1.90	1.49	1.25	1.13	1.08	1.10	1.19	1.38	1.73	2.33	3.02
20°	±30°	2.38	1.77	1.42	1.22	1.10	1.06	1.08	1.16	1.32	1.62	2.14	2.74
20°	±45°	2.11	1.62	1.33	1.17	1.08	1.04	1.06	1.12	1.25	1.50	1.92	2.41
20°	±60°	1.77	1.43	1.22	1.11	1.04	1.02	1.03	1.08	1.17	1.34	1.64	1.99
30°	0°	3.29	2.27	1.67	1.33	1.14	1.07	1.10	1.24	1.51	2.02	2.90	3.91
30°	±30°	2.97	2.09	1.57	1.27	1.11	1.05	1.08	1.20	1.43	1.87	2.62	3.50
30°	±45°	2.58	1.87	1.45	1.21	1.08	1.03	1.05	1.15	1.33	1.69	2.30	3.02
30°	±60°	2.09	1.59	1.29	1.12	1.03	0.99	1.01	1.08	1.21	1.47	1.90	2.40
35°	0°	3.60	2.43	1.74	1.35	1.14	1.06	1.09	1.25	1.55	2.14	3.15	4.31
35°	±30°	3.23	2.22	1.62	1.29	1.11	1.04	1.07	1.20	1.46	1.97	2.84	3.85
35°	±45°	2.79	1.97	1.49	1.22	1.07	1.01	1.04	1.14	1.36	1.77	2.47	3.29
35°	±60°	2.23	1.66	1.32	1.12	1.01	0.97	0.99	1.07	1.22	1.51	2.01	2.58
45°	0°	4.13	2.70	1.85	1.36	1.10	1.01	1.05	1.23	1.61	2.34	3.58	5.02
45°	±30°	3.68	2.43	1.70	1.29	1.07	0.99	1.02	1.18	1.50	2.12	3.19	4.44
45°	±45°	3.13	2.13	1.54	1.21	1.03	0.96	0.99	1.12	1.38	1.88	2.74	3.75
45°	±60°	2.45	1.75	1.34	1.10	0.97	0.92	0.94	1.04	1.23	1.58	2.18	2.88
60°	0°	4.70	2.94	1.89	1.30	0.99	0.88	0.93	1.15	1.61	2.50	4.02	5.78
60°	±30°	4.14	2.61	1.72	1.23	0.96	0.86	0.91	1.10	1.48	2.24	3.54	5.07
60°	±45°	3.47	2.24	1.53	1.14	0.92	0.84	0.88	1.03	1.34	1.94	2.99	4.23
60°	±60°	2.64	1.80	1.31	1.02	0.87	0.81	0.83	0.95	1.17	1.59	2.32	3.17
90°	0°	4.85	2.82	1.61	0.93	0.59	0.47	0.52	0.76	1.28	2.31	4.06	6.09
90°	±30°	4.20	2.44	1.42	0.88	0.59	0.49	0.54	0.74	1.16	2.01	3.52	5.28
90°	±45°	3.43	2.03	1.24	0.81	0.59	0.50	0.54	0.70	1.03	1.69	2.88	4.31
90°	±60°	2.51	1.57	1.03	0.72	0.56	0.50	0.52	0.64	0.88	1.34	2.14	3.10

R_B - factors for latitude 54 degrees

β	γ	Jan	Feb	Mar	Apr	May	Jun	Jul	Aug	Sep	Oct	Nov	Dec
20°	0°	2.37	1.81	1.45	1.23	1.11	1.07	1.09	1.18	1.35	1.66	2.16	2.68
20°	±30°	2.18	1.69	1.38	1.20	1.09	1.05	1.07	1.15	1.29	1.56	1.99	2.45
20°	±45°	1.95	1.56	1.30	1.15	1.07	1.04	1.05	1.11	1.23	1.45	1.80	2.17
20°	±60°	1.66	1.38	1.20	1.10	1.03	1.01	1.02	1.07	1.15	1.31	1.55	1.82
30°	0°	2.95	2.14	1.61	1.30	1.12	1.06	1.09	1.21	1.46	1.92	2.65	3.42
30°	±30°	2.67	1.97	1.51	1.25	1.10	1.04	1.06	1.17	1.39	1.78	2.41	3.07
30°	±45°	2.34	1.77	1.40	1.19	1.06	1.02	1.04	1.13	1.30	1.62	2.13	2.67
30°	±60°	1.92	1.52	1.26	1.11	1.02	0.98	1.00	1.06	1.19	1.41	1.77	2.15
35°	0°	3.21	2.28	1.67	1.31	1.11	1.04	1.07	1.22	1.50	2.03	2.86	3.74
35°	±30°	2.89	2.08	1.56	1.26	1.09	1.02	1.05	1.17	1.42	1.87	2.59	3.35
35°	±45°	2.51	1.86	1.44	1.19	1.05	1.00	1.02	1.12	1.32	1.68	2.27	2.89
35°	±60°	2.03	1.58	1.28	1.10	1.00	0.96	0.98	1.05	1.20	1.45	1.86	2.29
45°	0°	3.66	2.51	1.76	1.32	1.07	0.98	1.02	1.20	1.55	2.20	3.23	4.31
45°	±30°	3.26	2.26	1.63	1.25	1.04	0.97	1.00	1.15	1.45	2.00	2.89	3.83
45°	±45°	2.80	1.99	1.48	1.18	1.01	0.94	0.97	1.09	1.33	1.78	2.49	3.26
45°	±60°	2.21	1.65	1.29	1.08	0.95	0.90	0.93	1.02	1.19	1.50	2.00	2.53
60°	0°	4.12	2.70	1.79	1.25	0.95	0.85	0.89	1.11	1.53	2.33	3.58	4.92
60°	±30°	3.63	2.41	1.63	1.18	0.93	0.84	0.88	1.06	1.42	2.09	3.17	4.32
60°	±45°	3.06	2.08	1.46	1.10	0.90	0.82	0.85	1.00	1.29	1.82	2.69	3.62
60°	±60°	2.36	1.68	1.25	0.99	0.84	0.79	0.81	0.92	1.13	1.50	2.10	2.74
90°	0°	4.18	2.54	1.49	0.87	0.55	0.43	0.48	0.71	1.19	2.11	3.56	5.10
90°	±30°	3.62	2.20	1.32	0.82	0.56	0.46	0.50	0.69	1.08	1.83	3.08	4.42
90°	±45°	2.96	1.84	1.15	0.76	0.55	0.48	0.51	0.66	0.97	1.55	2.53	3.61
90°	±60°	2.18	1.43	0.96	0.68	0.53	0.48	0.50	0.61	0.83	1.23	1.89	2.61

R_B - factors for latitude 53 degrees

β	γ	Jan	Feb	Mar	Apr	May	Jun	Jul	Aug	Sep	Oct	Nov	Dec
20°	0°	2.27	1.77	1.43	1.22	1.11	1.06	1.08	1.17	1.33	1.63	2.09	2.55
20°	±30°	2.10	1.66	1.37	1.19	1.09	1.05	1.07	1.14	1.28	1.54	1.93	2.33
20°	±45°	1.88	1.53	1.29	1.15	1.06	1.03	1.05	1.11	1.22	1.43	1.75	2.08
20°	±60°	1.61	1.36	1.19	1.09	1.03	1.01	1.02	1.06	1.15	1.29	1.52	1.75
30°	0°	2.82	2.08	1.58	1.28	1.11	1.05	1.08	1.20	1.44	1.88	2.54	3.22
30°	±30°	2.56	1.92	1.49	1.23	1.09	1.03	1.06	1.16	1.37	1.74	2.32	2.90
30°	±45°	2.25	1.73	1.38	1.17	1.06	1.01	1.03	1.12	1.29	1.59	2.05	2.53
30°	±60°	1.85	1.49	1.25	1.10	1.01	0.98	0.99	1.06	1.18	1.39	1.72	2.05
35°	0°	3.06	2.21	1.64	1.30	1.10	1.03	1.06	1.20	1.48	1.98	2.74	3.52
35°	±30°	2.76	2.02	1.54	1.24	1.08	1.01	1.04	1.16	1.40	1.82	2.48	3.16
35°	±45°	2.40	1.81	1.42	1.18	1.04	0.99	1.01	1.11	1.30	1.65	2.18	2.73
35°	±60°	1.96	1.54	1.26	1.09	0.99	0.96	0.97	1.05	1.18	1.43	1.80	2.18
45°	0°	3.47	2.42	1.72	1.30	1.06	0.97	1.01	1.18	1.52	2.14	3.08	4.04
45°	±30°	3.10	2.19	1.59	1.23	1.03	0.96	0.99	1.14	1.42	1.95	2.76	3.59
45°	±45°	2.66	1.93	1.45	1.16	1.00	0.93	0.96	1.08	1.31	1.73	2.39	3.06
45°	±60°	2.12	1.61	1.27	1.06	0.94	0.90	0.92	1.01	1.17	1.47	1.93	2.39
60°	0°	3.88	2.60	1.74	1.22	0.94	0.83	0.88	1.08	1.50	2.25	3.40	4.58
60°	±30°	3.43	2.32	1.59	1.16	0.92	0.82	0.86	1.04	1.38	2.02	3.01	4.03
60°	±45°	2.89	2.00	1.42	1.08	0.88	0.81	0.84	0.98	1.26	1.76	2.56	3.38
60°	±60°	2.24	1.63	1.22	0.97	0.83	0.78	0.80	0.91	1.11	1.46	2.01	2.58
90°	0°	3.90	2.42	1.44	0.84	0.53	0.42	0.46	0.68	1.15	2.02	3.35	4.71
90°	±30°	3.38	2.10	1.27	0.80	0.54	0.45	0.49	0.67	1.04	1.76	2.90	4.08
90°	±45°	2.76	1.75	1.11	0.74	0.54	0.46	0.50	0.64	0.94	1.49	2.38	3.33
90°	±60°	2.04	1.37	0.93	0.67	0.52	0.46	0.49	0.60	0.81	1.19	1.79	2.42

R_B - factors for latitude 52 degrees

β	γ	Jan	Feb	Mar	Apr	May	Jun	Jul	Aug	Sep	Oct	Nov	Dec
20°	0°	2.19	1.73	1.41	1.21	1.10	1.06	1.08	1.16	1.32	1.60	2.02	2.43
20°	±30°	2.02	1.62	1.35	1.18	1.08	1.04	1.06	1.13	1.27	1.51	1.88	2.23
20°	±45°	1.82	1.50	1.28	1.14	1.06	1.03	1.04	1.10	1.21	1.41	1.71	2.00
20°	±60°	1.57	1.34	1.18	1.09	1.03	1.01	1.02	1.06	1.14	1.28	1.49	1.69
30°	0°	2.69	2.02	1.56	1.27	1.10	1.04	1.07	1.19	1.42	1.83	2.45	3.05
30°	±30°	2.45	1.87	1.47	1.22	1.08	1.02	1.05	1.15	1.35	1.71	2.24	2.76
30°	±45°	2.16	1.69	1.37	1.16	1.05	1.00	1.02	1.11	1.27	1.56	1.99	2.41
30°	±60°	1.79	1.46	1.23	1.09	1.01	0.97	0.99	1.05	1.17	1.37	1.67	1.97
35°	0°	2.92	2.15	1.61	1.28	1.09	1.02	1.05	1.19	1.46	1.93	2.63	3.33
35°	±30°	2.64	1.97	1.51	1.23	1.07	1.00	1.03	1.15	1.38	1.78	2.39	2.99
35°	±45°	2.30	1.76	1.39	1.17	1.03	0.98	1.01	1.10	1.29	1.61	2.10	2.59
35°	±60°	1.89	1.51	1.25	1.08	0.99	0.95	0.97	1.04	1.17	1.40	1.75	2.09
45°	0°	3.29	2.34	1.69	1.28	1.05	0.96	1.00	1.17	1.49	2.08	2.94	3.80
45°	±30°	2.95	2.12	1.56	1.22	1.02	0.95	0.98	1.12	1.40	1.89	2.64	3.38
45°	±45°	2.54	1.87	1.42	1.14	0.99	0.92	0.95	1.07	1.29	1.69	2.29	2.89
45°	±60°	2.03	1.57	1.25	1.05	0.94	0.89	0.91	1.00	1.16	1.44	1.86	2.27
60°	0°	3.67	2.50	1.70	1.20	0.92	0.82	0.86	1.06	1.46	2.18	3.24	4.29
60°	±30°	3.24	2.24	1.55	1.13	0.90	0.81	0.85	1.02	1.35	1.96	2.87	3.78
60°	±45°	2.74	1.94	1.39	1.06	0.87	0.80	0.83	0.97	1.23	1.71	2.45	3.18
60°	±60°	2.13	1.58	1.20	0.96	0.82	0.77	0.79	0.89	1.09	1.42	1.93	2.43
90°	0°	3.66	2.31	1.38	0.81	0.51	0.40	0.45	0.66	1.11	1.94	3.16	4.37
90°	±30°	3.17	2.00	1.23	0.77	0.52	0.43	0.47	0.65	1.01	1.69	2.74	3.79
90°	±45°	2.59	1.68	1.08	0.72	0.52	0.45	0.48	0.63	0.91	1.43	2.25	3.09
90°	±60°	1.92	1.31	0.90	0.65	0.51	0.45	0.48	0.58	0.78	1.14	1.70	2.26

R_B - factors for latitude 51 degrees

β	γ	Jan	Feb	Mar	Apr	May	Jun	Jul	Aug	Sep	Oct	Nov	Dec
20°	0°	2.12	1.70	1.40	1.21	1.09	1.05	1.07	1.15	1.31	1.57	1.96	2.33
20°	±30°	1.96	1.59	1.34	1.17	1.08	1.04	1.06	1.13	1.26	1.49	1.83	2.15
20°	±45°	1.77	1.48	1.27	1.13	1.05	1.02	1.04	1.10	1.20	1.39	1.66	1.93
20°	±60°	1.53	1.32	1.18	1.08	1.02	1.00	1.01	1.05	1.13	1.26	1.46	1.64
30°	0°	2.59	1.97	1.53	1.25	1.09	1.03	1.06	1.18	1.40	1.79	2.36	2.90
30°	±30°	2.36	1.82	1.45	1.21	1.07	1.02	1.04	1.14	1.34	1.67	2.16	2.63
30°	±45°	2.08	1.65	1.35	1.15	1.04	1.00	1.02	1.10	1.26	1.53	1.93	2.31
30°	±60°	1.74	1.44	1.22	1.08	1.00	0.97	0.98	1.04	1.16	1.35	1.63	1.89
35°	0°	2.79	2.09	1.58	1.26	1.08	1.01	1.04	1.18	1.44	1.88	2.54	3.16
35°	±30°	2.53	1.92	1.49	1.21	1.06	1.00	1.02	1.14	1.36	1.74	2.31	2.84
35°	±45°	2.22	1.72	1.37	1.15	1.03	0.98	1.00	1.09	1.27	1.58	2.04	2.47
35°	±60°	1.82	1.48	1.23	1.07	0.98	0.94	0.96	1.03	1.16	1.38	1.70	2.00
45°	0°	3.14	2.27	1.65	1.26	1.03	0.95	0.99	1.15	1.47	2.02	2.82	3.59
45°	±30°	2.81	2.06	1.53	1.20	1.01	0.93	0.97	1.11	1.38	1.85	2.54	3.20
45°	±45°	2.43	1.82	1.40	1.13	0.97	0.91	0.94	1.06	1.27	1.65	2.21	2.75
45°	±60°	1.95	1.53	1.23	1.04	0.93	0.88	0.90	0.99	1.14	1.41	1.80	2.17
60°	0°	3.48	2.41	1.66	1.17	0.90	0.80	0.85	1.04	1.43	2.11	3.09	4.03
60°	±30°	3.08	2.16	1.51	1.11	0.89	0.80	0.84	1.00	1.33	1.90	2.75	3.56
60°	±45°	2.61	1.87	1.36	1.04	0.86	0.79	0.82	0.95	1.21	1.66	2.34	3.00
60°	±60°	2.04	1.53	1.17	0.94	0.81	0.76	0.78	0.88	1.07	1.39	1.85	2.30
90°	0°	3.44	2.21	1.33	0.78	0.49	0.38	0.43	0.64	1.07	1.86	2.99	4.07
90°	±30°	2.98	1.92	1.19	0.75	0.51	0.42	0.46	0.63	0.98	1.62	2.59	3.53
90°	±45°	2.44	1.60	1.04	0.70	0.51	0.44	0.47	0.61	0.88	1.37	2.13	2.88
90°	±60	1.82	1.26	0.87	0.63	0.50	0.44	0.47	0.57	0.76	1.10	1.61	2.11

R$_B$ - factors for latitude 50 degrees

β	γ	Jan	Feb	Mar	Apr	May	Jun	Jul	Aug	Sep	Oct	Nov	Dec
20°	0°	2.05	1.66	1.38	1.20	1.09	1.05	1.07	1.14	1.29	1.55	1.91	2.24
20°	±30°	1.90	1.57	1.32	1.16	1.07	1.03	1.05	1.12	1.25	1.47	1.78	2.07
20°	±45°	1.72	1.45	1.25	1.13	1.05	1.02	1.03	1.09	1.19	1.37	1.63	1.86
20°	±60°	1.50	1.31	1.17	1.08	1.02	1.00	1.01	1.05	1.12	1.25	1.43	1.60
30°	0°	2.49	1.92	1.51	1.24	1.08	1.02	1.05	1.17	1.38	1.76	2.29	2.77
30°	±30°	2.27	1.78	1.43	1.20	1.06	1.01	1.03	1.13	1.32	1.64	2.10	2.52
30°	±45°	2.01	1.62	1.33	1.14	1.03	0.99	1.01	1.09	1.24	1.50	1.87	2.21
30°	±60°	1.69	1.41	1.21	1.07	1.00	0.96	0.98	1.04	1.15	1.33	1.59	1.83
35°	0°	2.68	2.03	1.56	1.25	1.07	1.00	1.03	1.16	1.41	1.84	2.45	3.01
35°	±30°	2.43	1.87	1.46	1.20	1.05	0.99	1.01	1.13	1.34	1.71	2.23	2.71
35°	±45°	2.14	1.68	1.35	1.14	1.02	0.97	0.99	1.08	1.26	1.55	1.97	2.37
35°	±60°	1.77	1.45	1.22	1.06	0.97	0.94	0.96	1.02	1.15	1.36	1.65	1.93
45°	0°	3.00	2.20	1.62	1.24	1.02	0.94	0.97	1.13	1.44	1.97	2.71	3.40
45°	±30°	2.69	2.00	1.50	1.18	1.00	0.92	0.96	1.09	1.35	1.80	2.45	3.04
45°	±45°	2.33	1.77	1.37	1.12	0.96	0.91	0.93	1.04	1.25	1.61	2.13	2.61
45°	±60°	1.88	1.50	1.21	1.03	0.92	0.88	0.89	0.98	1.13	1.38	1.74	2.08
60°	0°	3.31	2.33	1.61	1.15	0.89	0.79	0.83	1.02	1.40	2.04	2.96	3.80
60°	±30°	2.93	2.09	1.48	1.09	0.87	0.79	0.82	0.99	1.30	1.84	2.63	3.36
60°	±45°	2.49	1.81	1.33	1.02	0.84	0.77	0.81	0.94	1.19	1.62	2.25	2.84
60°	±60°	1.95	1.49	1.15	0.93	0.80	0.75	0.77	0.87	1.05	1.35	1.79	2.19
90°	0°	3.24	2.11	1.29	0.76	0.47	0.37	0.41	0.61	1.04	1.78	2.84	3.81
90°	±30°	2.81	1.83	1.14	0.72	0.49	0.40	0.44	0.61	0.94	1.55	2.46	3.30
90°	±45°	2.30	1.54	1.01	0.68	0.50	0.43	0.46	0.59	0.85	1.32	2.03	2.70
90°	±60°	1.72	1.21	0.85	0.62	0.49	0.43	0.46	0.55	0.74	1.06	1.54	1.98

R$_B$ - factors for latitude 49 degrees

β	γ	Jan	Feb	Mar	Apr	May	Jun	Jul	Aug	Sep	Oct	Nov	Dec
20°	0°	1.99	1.63	1.36	1.19	1.08	1.04	1.06	1.14	1.28	1.53	1.86	2.16
20°	±30°	1.85	1.54	1.31	1.16	1.07	1.03	1.05	1.11	1.24	1.45	1.74	2.00
20°	±45°	1.68	1.43	1.24	1.12	1.05	1.02	1.03	1.08	1.19	1.36	1.59	1.81
20°	±60°	1.47	1.29	1.16	1.07	1.02	1.00	1.01	1.05	1.12	1.24	1.41	1.56
30°	0°	2.40	1.88	1.49	1.23	1.08	1.02	1.04	1.16	1.37	1.72	2.21	2.66
30°	±30°	2.19	1.74	1.41	1.19	1.05	1.00	1.03	1.12	1.30	1.61	2.03	2.42
30°	±45°	1.95	1.59	1.31	1.13	1.03	0.99	1.00	1.08	1.23	1.48	1.82	2.13
30°	±60°	1.64	1.39	1.20	1.07	0.99	0.96	0.97	1.03	1.14	1.31	1.55	1.77
35°	0°	2.58	1.98	1.53	1.23	1.06	0.99	1.02	1.15	1.39	1.80	2.37	2.87
35°	±30°	2.34	1.83	1.44	1.19	1.04	0.98	1.01	1.12	1.32	1.67	2.16	2.60
35°	±45°	2.07	1.65	1.34	1.13	1.01	0.96	0.98	1.07	1.24	1.52	1.92	2.27
35°	±60°	1.72	1.42	1.20	1.06	0.97	0.93	0.95	1.01	1.14	1.34	1.61	1.86
45°	0°	2.88	2.14	1.59	1.22	1.01	0.92	0.96	1.12	1.42	1.92	2.61	3.24
45°	±30°	2.59	1.95	1.47	1.17	0.98	0.91	0.95	1.08	1.33	1.76	2.36	2.90
45°	±45°	2.24	1.73	1.35	1.10	0.95	0.90	0.92	1.03	1.23	1.58	2.06	2.50
45°	±60°	1.82	1.46	1.20	1.02	0.91	0.87	0.89	0.97	1.11	1.36	1.69	2.00
60°	0°	3.16	2.25	1.58	1.13	0.87	0.77	0.82	1.01	1.37	1.98	2.84	3.60
60°	±30°	2.80	2.02	1.44	1.07	0.86	0.77	0.81	0.97	1.27	1.79	2.52	3.19
60°	±45°	2.38	1.76	1.30	1.00	0.83	0.76	0.79	0.92	1.16	1.58	2.16	2.70
60°	±60°	1.88	1.45	1.13	0.92	0.79	0.74	0.76	0.86	1.03	1.32	1.72	2.09
90°	0°	3.07	2.02	1.24	0.73	0.45	0.35	0.39	0.59	1.00	1.71	2.70	3.58
90°	±30°	2.66	1.76	1.11	0.70	0.47	0.39	0.43	0.59	0.91	1.49	2.34	3.10
90°	±45°	2.18	1.48	0.97	0.66	0.48	0.41	0.44	0.57	0.83	1.27	1.93	2.54
90°	±60°	1.63	1.16	0.82	0.60	0.47	0.42	0.45	0.54	0.72	1.03	1.47	1.87

R$_B$ - factors for latitude 48 degrees

β	γ	Jan	Feb	Mar	Apr	May	Jun	Jul	Aug	Sep	Oct	Nov	Dec
20°	0°	1.93	1.60	1.35	1.18	1.08	1.04	1.05	1.13	1.27	1.50	1.82	2.09
20°	±30°	1.80	1.51	1.30	1.15	1.06	1.03	1.04	1.11	1.23	1.43	1.70	1.94
20°	±45°	1.64	1.41	1.23	1.11	1.04	1.01	1.02	1.08	1.18	1.34	1.56	1.76
20°	±60°	1.44	1.28	1.15	1.07	1.01	0.99	1.00	1.04	1.11	1.23	1.38	1.52
30°	0°	2.32	1.84	1.47	1.22	1.07	1.01	1.03	1.14	1.35	1.69	2.15	2.55
30°	±30°	2.13	1.71	1.39	1.17	1.05	1.00	1.02	1.11	1.29	1.58	1.98	2.33
30°	±45°	1.90	1.56	1.30	1.12	1.02	0.98	1.00	1.08	1.22	1.45	1.78	2.06
30°	±60°	1.60	1.37	1.18	1.06	0.99	0.96	0.97	1.02	1.13	1.29	1.52	1.72
35°	0°	2.49	1.93	1.51	1.22	1.05	0.98	1.01	1.14	1.37	1.76	2.29	2.75
35°	±30°	2.26	1.78	1.42	1.17	1.03	0.97	1.00	1.11	1.31	1.64	2.09	2.50
35°	±45°	2.00	1.61	1.32	1.12	1.00	0.95	0.98	1.06	1.23	1.50	1.86	2.19
35°	±60°	1.67	1.40	1.19	1.05	0.96	0.93	0.94	1.01	1.12	1.32	1.57	1.80
45°	0°	2.76	2.08	1.55	1.20	0.99	0.91	0.95	1.10	1.39	1.87	2.52	3.09
45°	±30°	2.49	1.90	1.45	1.15	0.97	0.90	0.93	1.07	1.31	1.72	2.28	2.77
45°	±45°	2.16	1.69	1.33	1.09	0.94	0.89	0.91	1.02	1.22	1.55	2.00	2.40
45°	±60°	1.76	1.43	1.18	1.01	0.90	0.86	0.88	0.96	1.10	1.33	1.65	1.92
60°	0°	3.02	2.18	1.54	1.11	0.86	0.76	0.80	0.99	1.34	1.93	2.72	3.42
60°	±30°	2.68	1.96	1.41	1.05	0.84	0.76	0.80	0.95	1.24	1.74	2.43	3.03
60°	±45°	2.29	1.71	1.27	0.99	0.82	0.75	0.78	0.91	1.14	1.54	2.08	2.57
60°	±60°	1.81	1.41	1.11	0.90	0.78	0.73	0.75	0.85	1.01	1.29	1.67	2.00
90°	0°	2.91	1.94	1.20	0.70	0.43	0.33	0.37	0.57	0.97	1.65	2.57	3.37
90°	±30°	2.52	1.69	1.07	0.68	0.46	0.38	0.41	0.57	0.88	1.44	2.22	2.92
90°	±45°	2.07	1.42	0.94	0.64	0.47	0.40	0.43	0.56	0.80	1.23	1.84	2.39
90°	±60°	1.56	1.12	0.80	0.59	0.46	0.42	0.44	0.53	0.70	0.99	1.40	1.77

R$_B$ - factors for latitude 47 degrees

β	γ	Jan	Feb	Mar	Apr	May	Jun	Jul	Aug	Sep	Oct	Nov	Dec
20°	0°	1.88	1.58	1.34	1.17	1.07	1.03	1.05	1.12	1.26	1.48	1.78	2.03
20°	±30°	1.76	1.49	1.28	1.14	1.05	1.02	1.04	1.10	1.22	1.41	1.66	1.88
20°	±45°	1.61	1.39	1.22	1.11	1.04	1.01	1.02	1.07	1.17	1.33	1.53	1.71
20°	±60°	1.42	1.26	1.14	1.06	1.01	0.99	1.00	1.04	1.11	1.22	1.36	1.49
30°	0°	2.25	1.80	1.44	1.20	1.06	1.00	1.03	1.13	1.33	1.66	2.09	2.46
30°	±30°	2.06	1.67	1.37	1.16	1.04	0.99	1.01	1.10	1.27	1.55	1.93	2.25
30°	±45°	1.84	1.53	1.28	1.12	1.01	0.97	0.99	1.07	1.21	1.43	1.73	1.99
30°	±60°	1.57	1.35	1.17	1.05	0.98	0.95	0.96	1.02	1.12	1.28	1.49	1.67
35°	0°	2.40	1.89	1.48	1.21	1.04	0.97	1.00	1.13	1.35	1.73	2.22	2.65
35°	±30°	2.19	1.75	1.40	1.16	1.02	0.96	0.99	1.09	1.29	1.61	2.04	2.40
35°	±45°	1.94	1.58	1.30	1.11	0.99	0.95	0.97	1.05	1.21	1.47	1.82	2.11
35°	±60°	1.63	1.37	1.18	1.04	0.96	0.92	0.94	1.00	1.11	1.30	1.54	1.75
45°	0°	2.66	2.02	1.53	1.18	0.98	0.90	0.94	1.09	1.37	1.83	2.44	2.96
45°	±30°	2.40	1.85	1.42	1.13	0.96	0.89	0.92	1.05	1.29	1.68	2.21	2.66
45°	±45°	2.09	1.65	1.31	1.07	0.93	0.88	0.90	1.01	1.20	1.51	1.94	2.30
45°	±60°	1.71	1.40	1.16	1.00	0.90	0.86	0.87	0.95	1.09	1.31	1.60	1.86
60°	0°	2.89	2.11	1.50	1.08	0.84	0.75	0.79	0.97	1.31	1.87	2.62	3.26
60°	±30°	2.57	1.90	1.38	1.03	0.83	0.75	0.78	0.94	1.22	1.69	2.34	2.89
60°	±45°	2.20	1.66	1.25	0.97	0.81	0.74	0.77	0.89	1.12	1.50	2.01	2.45
60°	±60°	1.74	1.38	1.09	0.89	0.77	0.72	0.75	0.83	1.00	1.26	1.62	1.92
90°	0°	2.76	1.86	1.16	0.68	0.41	0.32	0.36	0.55	0.93	1.59	2.45	3.19
90°	±30°	2.39	1.62	1.03	0.66	0.44	0.36	0.40	0.55	0.86	1.38	2.12	2.76
90°	±45°	1.97	1.36	0.91	0.62	0.46	0.39	0.42	0.54	0.78	1.18	1.75	2.26
90°	±60°	1.48	1.08	0.78	0.57	0.45	0.41	0.43	0.52	0.68	0.96	1.34	1.68

R_B - factors for latitude 46 degrees

β	γ	Jan	Feb	Mar	Apr	May	Jun	Jul	Aug	Sep	Oct	Nov	Dec
20°	0°	1.84	1.55	1.32	1.16	1.06	1.02	1.04	1.12	1.25	1.46	1.74	1.97
20°	±30°	1.72	1.47	1.27	1.13	1.05	1.02	1.03	1.09	1.21	1.39	1.63	1.83
20°	±45°	1.58	1.37	1.21	1.10	1.03	1.00	1.02	1.07	1.16	1.31	1.51	1.67
20°	±60°	1.39	1.25	1.14	1.06	1.01	0.99	1.00	1.03	1.10	1.21	1.34	1.46
30°	0°	2.18	1.76	1.42	1.19	1.05	0.99	1.02	1.12	1.32	1.63	2.03	2.37
30°	±30°	2.00	1.64	1.35	1.15	1.03	0.98	1.00	1.09	1.26	1.53	1.88	2.17
30°	±45°	1.80	1.50	1.27	1.11	1.01	0.97	0.99	1.06	1.19	1.41	1.69	1.93
30°	±60°	1.53	1.33	1.16	1.05	0.98	0.95	0.96	1.01	1.11	1.26	1.46	1.63
35°	0°	2.33	1.85	1.46	1.19	1.03	0.97	0.99	1.11	1.34	1.70	2.16	2.55
35°	±30°	2.12	1.71	1.38	1.15	1.01	0.96	0.98	1.08	1.27	1.58	1.98	2.32
35°	±45°	1.89	1.55	1.28	1.10	0.99	0.94	0.96	1.05	1.20	1.45	1.77	2.04
35°	±60°	1.59	1.35	1.16	1.03	0.95	0.92	0.93	0.99	1.10	1.28	1.51	1.70
45°	0°	2.57	1.97	1.50	1.17	0.97	0.89	0.92	1.07	1.35	1.79	2.36	2.84
45°	±30°	2.32	1.80	1.40	1.12	0.95	0.88	0.91	1.04	1.27	1.64	2.14	2.55
45°	±45°	2.02	1.61	1.29	1.06	0.92	0.87	0.89	1.00	1.18	1.48	1.88	2.22
45°	±60°	1.66	1.38	1.15	0.99	0.89	0.85	0.87	0.94	1.07	1.29	1.56	1.80
60°	0°	2.78	2.05	1.47	1.06	0.82	0.73	0.77	0.95	1.28	1.82	2.52	3.11
60°	±30°	2.47	1.84	1.35	1.01	0.81	0.74	0.77	0.92	1.19	1.65	2.25	2.76
60°	±45°	2.12	1.62	1.22	0.96	0.79	0.73	0.76	0.88	1.10	1.46	1.94	2.35
60°	±60°	1.69	1.34	1.07	0.88	0.76	0.72	0.74	0.82	0.98	1.24	1.57	1.85
90°	0°	2.63	1.79	1.12	0.65	0.39	0.30	0.34	0.53	0.90	1.53	2.34	3.02
90°	±30°	2.28	1.55	1.00	0.63	0.43	0.35	0.38	0.54	0.83	1.33	2.02	2.61
90°	±45°	1.87	1.31	0.89	0.61	0.44	0.38	0.41	0.53	0.75	1.14	1.68	2.14
90°	±60°	1.42	1.04	0.75	0.56	0.44	0.40	0.42	0.50	0.66	0.93	1.29	1.60

R_B - factors for latitude 45 degrees

β	γ	Jan	Feb	Mar	Apr	May	Jun	Jul	Aug	Sep	Oct	Nov	Dec
20°	0°	1.80	1.53	1.31	1.15	1.06	1.02	1.04	1.11	1.24	1.44	1.70	1.92
20°	±30°	1.68	1.45	1.26	1.13	1.04	1.01	1.03	1.09	1.20	1.38	1.60	1.79
20°	±45°	1.55	1.36	1.20	1.09	1.03	1.00	1.01	1.06	1.15	1.30	1.48	1.63
20°	±60°	1.37	1.24	1.13	1.05	1.00	0.99	0.99	1.03	1.09	1.20	1.33	1.43
30°	0°	2.12	1.73	1.41	1.18	1.04	0.98	1.01	1.11	1.30	1.60	1.98	2.30
30°	±30°	1.95	1.61	1.34	1.14	1.02	0.97	1.00	1.09	1.25	1.50	1.83	2.11
30°	±45°	1.75	1.48	1.25	1.10	1.00	0.96	0.98	1.05	1.18	1.39	1.66	1.88
30°	±60°	1.50	1.31	1.15	1.04	0.97	0.94	0.95	1.01	1.10	1.25	1.44	1.59
35°	0°	2.26	1.81	1.44	1.18	1.02	0.96	0.98	1.10	1.32	1.66	2.10	2.46
35°	±30°	2.06	1.67	1.36	1.14	1.00	0.95	0.97	1.07	1.26	1.55	1.93	2.24
35°	±45°	1.84	1.52	1.27	1.09	0.98	0.93	0.95	1.04	1.19	1.42	1.73	1.98
35°	±60°	1.55	1.33	1.15	1.02	0.95	0.91	0.93	0.99	1.09	1.26	1.48	1.65
45°	0°	2.48	1.92	1.47	1.15	0.95	0.88	0.91	1.06	1.32	1.75	2.29	2.73
45°	±30°	2.24	1.76	1.37	1.10	0.94	0.87	0.90	1.03	1.25	1.61	2.08	2.46
45°	±45°	1.96	1.58	1.27	1.05	0.92	0.86	0.89	0.99	1.17	1.46	1.83	2.14
45°	±60°	1.62	1.35	1.13	0.98	0.88	0.84	0.86	0.93	1.06	1.27	1.53	1.74
60°	0°	2.67	1.99	1.43	1.04	0.81	0.72	0.76	0.93	1.25	1.77	2.44	2.98
60°	±30°	2.38	1.79	1.32	1.00	0.80	0.72	0.76	0.90	1.17	1.61	2.18	2.65
60°	±45°	2.04	1.57	1.20	0.94	0.78	0.72	0.75	0.87	1.08	1.43	1.88	2.26
60°	±60°	1.63	1.31	1.05	0.87	0.75	0.71	0.73	0.81	0.97	1.21	1.52	1.78
90°	0°	2.50	1.72	1.08	0.63	0.37	0.28	0.32	0.51	0.87	1.47	2.24	2.86
90°	±30°	2.17	1.50	0.96	0.61	0.41	0.34	0.37	0.52	0.80	1.29	1.94	2.48
90°	±45°	1.79	1.26	0.86	0.59	0.43	0.37	0.40	0.51	0.73	1.10	1.60	2.03
90°	±60°	1.36	1.01	0.73	0.55	0.43	0.39	0.41	0.49	0.65	0.90	1.24	1.52

R$_B$ - factors for latitude 44 degrees

β	γ	Jan	Feb	Mar	Apr	May	Jun	Jul	Aug	Sep	Oct	Nov	Dec
20°	0°	1.76	1.51	1.30	1.15	1.05	1.01	1.03	1.10	1.23	1.42	1.67	1.87
20°	±30°	1.65	1.43	1.25	1.12	1.04	1.01	1.02	1.08	1.19	1.36	1.57	1.75
20°	±45°	1.52	1.34	1.19	1.09	1.02	1.00	1.01	1.06	1.15	1.28	1.46	1.60
20°	±60°	1.35	1.23	1.12	1.05	1.00	0.98	0.99	1.03	1.09	1.19	1.31	1.41
30°	0°	2.06	1.69	1.39	1.17	1.03	0.98	1.00	1.10	1.29	1.58	1.94	2.23
30°	±30°	1.90	1.58	1.32	1.13	1.01	0.97	0.99	1.08	1.23	1.48	1.79	2.04
30°	±45°	1.71	1.46	1.24	1.09	0.99	0.96	0.97	1.04	1.17	1.37	1.62	1.83
30°	±60°	1.47	1.29	1.14	1.03	0.97	0.94	0.95	1.00	1.09	1.24	1.41	1.56
35°	0°	2.19	1.77	1.42	1.16	1.01	0.95	0.97	1.09	1.30	1.63	2.05	2.38
35°	±30°	2.01	1.64	1.34	1.12	0.99	0.94	0.96	1.06	1.24	1.53	1.88	2.17
35°	±45°	1.79	1.50	1.25	1.08	0.97	0.93	0.95	1.03	1.17	1.40	1.69	1.92
35°	±60°	1.52	1.31	1.14	1.02	0.94	0.91	0.92	0.98	1.08	1.25	1.45	1.61
45°	0°	2.40	1.88	1.44	1.13	0.94	0.87	0.90	1.04	1.30	1.71	2.22	2.63
45°	±30°	2.17	1.72	1.35	1.09	0.93	0.86	0.89	1.01	1.23	1.58	2.02	2.37
45°	±45°	1.91	1.55	1.25	1.04	0.91	0.85	0.88	0.98	1.15	1.43	1.78	2.07
45°	±60°	1.58	1.33	1.12	0.97	0.87	0.84	0.85	0.92	1.05	1.25	1.49	1.69
60°	0°	2.57	1.93	1.40	1.02	0.79	0.70	0.74	0.91	1.23	1.73	2.35	2.86
60°	±30°	2.29	1.74	1.29	0.98	0.79	0.71	0.74	0.89	1.15	1.57	2.10	2.54
60°	±45°	1.97	1.53	1.18	0.93	0.77	0.71	0.74	0.85	1.06	1.39	1.82	2.17
60°	±60°	1.58	1.28	1.03	0.86	0.74	0.70	0.72	0.80	0.95	1.19	1.48	1.72
90°	0°	2.39	1.66	1.04	0.61	0.36	0.27	0.31	0.49	0.84	1.42	2.14	2.72
90°	±30°	2.07	1.44	0.93	0.59	0.40	0.32	0.36	0.50	0.78	1.24	1.85	2.36
90°	±45°	1.71	1.22	0.83	0.57	0.42	0.36	0.39	0.50	0.71	1.07	1.54	1.93
90°	±60°	1.30	0.98	0.71	0.53	0.42	0.38	0.40	0.48	0.63	0.87	1.19	1.45

R$_B$ - factors for latitude 43 degrees

β	γ	Jan	Feb	Mar	Apr	May	Jun	Jul	Aug	Sep	Oct	Nov	Dec
20°	0°	1.72	1.48	1.28	1.14	1.05	1.01	1.02	1.09	1.22	1.41	1.64	1.83
20°	±30°	1.62	1.41	1.24	1.11	1.03	1.00	1.02	1.08	1.18	1.35	1.55	1.71
20°	±45°	1.49	1.33	1.19	1.08	1.02	0.99	1.00	1.05	1.14	1.27	1.44	1.57
20°	±60°	1.33	1.22	1.12	1.04	1.00	0.98	0.99	1.02	1.08	1.18	1.29	1.39
30°	0°	2.01	1.66	1.37	1.16	1.02	0.97	0.99	1.09	1.27	1.55	1.89	2.16
30°	±30°	1.86	1.56	1.30	1.12	1.01	0.96	0.98	1.07	1.22	1.46	1.75	1.99
30°	±45°	1.68	1.43	1.23	1.08	0.99	0.95	0.97	1.04	1.16	1.35	1.59	1.78
30°	±60°	1.45	1.28	1.13	1.03	0.96	0.93	0.95	1.00	1.08	1.22	1.39	1.52
35°	0°	2.13	1.73	1.40	1.15	1.00	0.94	0.96	1.08	1.28	1.60	2.00	2.31
35°	±30°	1.95	1.61	1.32	1.11	0.98	0.93	0.95	1.05	1.23	1.50	1.84	2.11
35°	±45°	1.75	1.47	1.24	1.07	0.96	0.92	0.94	1.02	1.16	1.38	1.65	1.87
35°	±60°	1.49	1.30	1.13	1.01	0.93	0.90	0.92	0.97	1.08	1.23	1.42	1.58
45°	0°	2.32	1.83	1.42	1.12	0.93	0.86	0.89	1.03	1.28	1.67	2.16	2.54
45°	±30°	2.11	1.68	1.33	1.07	0.92	0.85	0.88	1.00	1.21	1.55	1.96	2.29
45°	±45°	1.85	1.51	1.23	1.02	0.90	0.85	0.87	0.96	1.14	1.40	1.74	2.01
45°	±60°	1.54	1.30	1.10	0.96	0.87	0.83	0.85	0.92	1.04	1.23	1.46	1.65
60°	0°	2.48	1.88	1.37	1.00	0.78	0.69	0.73	0.90	1.20	1.68	2.28	2.74
60°	±30°	2.21	1.70	1.27	0.96	0.77	0.70	0.73	0.87	1.13	1.53	2.04	2.44
60°	±45°	1.91	1.50	1.15	0.91	0.76	0.70	0.73	0.84	1.04	1.36	1.77	2.09
60°	±60°	1.54	1.25	1.02	0.84	0.74	0.69	0.71	0.79	0.94	1.16	1.44	1.66
90°	0°	2.29	1.59	1.01	0.58	0.34	0.25	0.29	0.47	0.81	1.37	2.05	2.59
90°	±30°	1.98	1.39	0.90	0.57	0.38	0.31	0.34	0.48	0.75	1.20	1.78	2.24
90°	±45°	1.63	1.18	0.81	0.56	0.41	0.35	0.37	0.48	0.69	1.03	1.48	1.84
90°	±60°	1.25	0.95	0.70	0.52	0.41	0.37	0.39	0.47	0.62	0.85	1.15	1.39

R_B - factors for latitude 42 degrees

β	γ	Jan	Feb	Mar	Apr	May	Jun	Jul	Aug	Sep	Oct	Nov	Dec
20°	0°	1.69	1.46	1.27	1.13	1.04	1.00	1.02	1.09	1.21	1.39	1.61	1.79
20°	±30°	1.59	1.39	1.23	1.11	1.03	1.00	1.01	1.07	1.17	1.33	1.52	1.67
20°	±45°	1.47	1.31	1.18	1.08	1.01	0.99	1.00	1.05	1.13	1.26	1.42	1.54
20°	±60°	1.32	1.21	1.11	1.04	0.99	0.98	0.98	1.02	1.08	1.17	1.28	1.37
30°	0°	1.96	1.63	1.35	1.14	1.01	0.96	0.98	1.08	1.26	1.53	1.85	2.10
30°	±30°	1.81	1.53	1.29	1.11	1.00	0.95	0.97	1.06	1.21	1.44	1.72	1.94
30°	±45°	1.64	1.41	1.22	1.07	0.98	0.94	0.96	1.03	1.15	1.34	1.56	1.74
30°	±60°	1.42	1.26	1.12	1.02	0.96	0.93	0.94	0.99	1.08	1.21	1.37	1.49
35°	0°	2.07	1.70	1.38	1.14	0.99	0.93	0.95	1.07	1.27	1.58	1.95	2.24
35°	±30°	1.91	1.58	1.30	1.10	0.97	0.92	0.95	1.04	1.21	1.48	1.80	2.05
35°	±45°	1.71	1.45	1.22	1.06	0.96	0.91	0.93	1.01	1.15	1.36	1.62	1.82
35°	±60°	1.46	1.28	1.12	1.00	0.93	0.90	0.91	0.97	1.07	1.22	1.40	1.54
45°	0°	2.25	1.79	1.39	1.10	0.92	0.84	0.88	1.01	1.26	1.64	2.10	2.46
45°	±30°	2.05	1.65	1.31	1.06	0.91	0.84	0.87	0.99	1.20	1.52	1.91	2.22
45°	±45°	1.80	1.48	1.21	1.01	0.89	0.84	0.86	0.95	1.12	1.38	1.70	1.95
45°	±60°	1.51	1.28	1.09	0.95	0.86	0.82	0.84	0.91	1.03	1.21	1.43	1.60
60°	0°	2.39	1.83	1.34	0.98	0.76	0.67	0.71	0.88	1.18	1.64	2.20	2.64
60°	±30°	2.14	1.65	1.24	0.94	0.76	0.69	0.72	0.86	1.10	1.49	1.98	2.35
60°	±45°	1.85	1.46	1.13	0.90	0.75	0.69	0.72	0.83	1.02	1.33	1.72	2.02
60°	±60°	1.49	1.23	1.00	0.83	0.73	0.68	0.70	0.78	0.92	1.14	1.40	1.61
90°	0°	2.19	1.54	0.97	0.56	0.32	0.24	0.27	0.45	0.78	1.32	1.97	2.47
90°	±30°	1.89	1.34	0.87	0.56	0.37	0.30	0.33	0.47	0.73	1.16	1.70	2.14
90°	±45°	1.57	1.14	0.78	0.54	0.39	0.34	0.36	0.47	0.67	1.00	1.42	1.76
90°	±60°	1.20	0.92	0.68	0.51	0.41	0.36	0.38	0.46	0.60	0.82	1.10	1.33

R_B - factors for latitude 41 degrees

β	γ	Jan	Feb	Mar	Apr	May	Jun	Jul	Aug	Sep	Oct	Nov	Dec
20°	0°	1.66	1.45	1.26	1.12	1.03	1.00	1.01	1.08	1.20	1.38	1.59	1.75
20°	±30°	1.56	1.38	1.22	1.10	1.02	0.99	1.01	1.06	1.16	1.32	1.50	1.64
20°	±45°	1.45	1.30	1.17	1.07	1.01	0.98	1.00	1.04	1.13	1.25	1.40	1.51
20°	±60°	1.30	1.20	1.11	1.04	0.99	0.97	0.98	1.02	1.07	1.16	1.27	1.35
30°	0°	1.91	1.61	1.34	1.13	1.00	0.95	0.98	1.07	1.24	1.50	1.81	2.05
30°	±30°	1.77	1.51	1.27	1.10	0.99	0.95	0.97	1.05	1.20	1.42	1.68	1.89
30°	±45°	1.61	1.39	1.20	1.06	0.97	0.94	0.95	1.02	1.14	1.32	1.54	1.70
30°	±60°	1.40	1.25	1.11	1.02	0.95	0.92	0.94	0.99	1.07	1.20	1.35	1.47
35°	0°	2.02	1.67	1.36	1.13	0.98	0.92	0.95	1.06	1.25	1.55	1.90	2.17
35°	±30°	1.86	1.55	1.29	1.09	0.97	0.91	0.94	1.03	1.20	1.45	1.76	1.99
35°	±45°	1.67	1.42	1.21	1.05	0.95	0.91	0.93	1.00	1.14	1.34	1.59	1.78
35°	±60°	1.43	1.26	1.11	1.00	0.92	0.89	0.91	0.96	1.06	1.20	1.38	1.51
45°	0°	2.19	1.75	1.37	1.08	0.90	0.83	0.86	1.00	1.24	1.61	2.04	2.38
45°	±30°	1.99	1.61	1.29	1.05	0.89	0.83	0.86	0.98	1.18	1.49	1.86	2.15
45°	±45°	1.76	1.46	1.19	1.00	0.88	0.83	0.85	0.94	1.11	1.36	1.66	1.89
45°	±60°	1.47	1.26	1.08	0.94	0.85	0.82	0.83	0.90	1.02	1.19	1.40	1.56
60°	0°	2.32	1.78	1.31	0.96	0.75	0.66	0.70	0.86	1.15	1.60	2.14	2.55
60°	±30°	2.07	1.61	1.21	0.93	0.75	0.67	0.71	0.84	1.08	1.46	1.92	2.27
60°	±45°	1.79	1.43	1.11	0.88	0.74	0.68	0.71	0.81	1.01	1.31	1.67	1.95
60°	±60°	1.45	1.20	0.98	0.82	0.72	0.68	0.70	0.77	0.91	1.12	1.37	1.56
90°	0°	2.10	1.48	0.94	0.54	0.30	0.22	0.26	0.43	0.76	1.27	1.89	2.36
90°	±30°	1.82	1.29	0.84	0.54	0.36	0.29	0.32	0.45	0.70	1.12	1.64	2.05
90°	±45°	1.50	1.10	0.76	0.52	0.38	0.33	0.35	0.46	0.65	0.97	1.36	1.68
90°	±60°	1.16	0.89	0.66	0.50	0.40	0.36	0.37	0.45	0.59	0.80	1.07	1.28

R_B - factors for latitude 40 degrees

β	γ	Jan	Feb	Mar	Apr	May	Jun	Jul	Aug	Sep	Oct	Nov	Dec
20°	0°	1.63	1.43	1.25	1.11	1.03	0.99	1.01	1.07	1.19	1.36	1.56	1.71
20°	±30°	1.54	1.36	1.21	1.09	1.02	0.99	1.00	1.06	1.16	1.30	1.48	1.61
20°	±45°	1.43	1.29	1.16	1.07	1.01	0.98	0.99	1.04	1.12	1.24	1.38	1.49
20°	±60°	1.29	1.19	1.10	1.03	0.99	0.97	0.98	1.01	1.07	1.15	1.25	1.33
30°	0°	1.87	1.58	1.32	1.12	1.00	0.94	0.97	1.06	1.23	1.48	1.77	2.00
30°	±30°	1.74	1.48	1.26	1.09	0.98	0.94	0.96	1.04	1.18	1.40	1.65	1.84
30°	±45°	1.58	1.37	1.19	1.06	0.97	0.93	0.95	1.02	1.13	1.31	1.51	1.67
30°	±60°	1.38	1.23	1.11	1.01	0.95	0.92	0.93	0.98	1.06	1.19	1.33	1.44
35°	0°	1.97	1.64	1.34	1.11	0.97	0.91	0.94	1.05	1.24	1.52	1.86	2.12
35°	±30°	1.82	1.53	1.27	1.08	0.96	0.91	0.93	1.02	1.19	1.43	1.72	1.94
35°	±45°	1.64	1.40	1.20	1.04	0.94	0.90	0.92	0.99	1.13	1.32	1.56	1.74
35°	±60°	1.41	1.25	1.10	0.99	0.92	0.89	0.90	0.96	1.05	1.19	1.35	1.48
45°	0°	2.13	1.72	1.35	1.07	0.89	0.82	0.85	0.99	1.22	1.58	1.99	2.30
45°	±30°	1.94	1.58	1.27	1.03	0.88	0.82	0.85	0.96	1.16	1.46	1.82	2.09
45°	±45°	1.72	1.43	1.18	0.99	0.87	0.82	0.84	0.93	1.09	1.33	1.62	1.84
45°	±60°	1.44	1.24	1.07	0.93	0.85	0.81	0.83	0.89	1.01	1.18	1.38	1.53
60°	0°	2.24	1.74	1.28	0.94	0.73	0.65	0.68	0.84	1.13	1.56	2.07	2.46
60°	±30°	2.01	1.57	1.19	0.91	0.73	0.66	0.69	0.83	1.06	1.43	1.86	2.19
60°	±45°	1.74	1.39	1.09	0.87	0.73	0.67	0.69	0.80	0.99	1.28	1.62	1.89
60°	±60°	1.42	1.18	0.97	0.81	0.71	0.67	0.69	0.76	0.90	1.10	1.34	1.52
90°	0°	2.01	1.43	0.91	0.52	0.29	0.20	0.24	0.41	0.73	1.23	1.82	2.26
90°	±30°	1.74	1.24	0.82	0.52	0.34	0.27	0.30	0.43	0.68	1.08	1.57	1.96
90°	±45°	1.44	1.06	0.74	0.51	0.37	0.32	0.34	0.44	0.63	0.94	1.31	1.61
90°	±60°	1.11	0.86	0.64	0.49	0.39	0.35	0.37	0.44	0.57	0.78	1.03	1.23

R_B - factors for latitude 39 degrees

β	γ	Jan	Feb	Mar	Apr	May	Jun	Jul	Aug	Sep	Oct	Nov	Dec
20°	0°	1.60	1.41	1.24	1.11	1.02	0.99	1.00	1.07	1.18	1.35	1.54	1.68
20°	±30°	1.51	1.35	1.20	1.09	1.01	0.98	1.00	1.05	1.15	1.29	1.46	1.58
20°	±45°	1.41	1.27	1.15	1.06	1.00	0.98	0.99	1.03	1.11	1.23	1.36	1.46
20°	±60°	1.27	1.18	1.09	1.03	0.99	0.97	0.98	1.01	1.06	1.15	1.24	1.31
30°	0°	1.83	1.55	1.30	1.11	0.99	0.94	0.96	1.05	1.22	1.46	1.74	1.95
30°	±30°	1.70	1.46	1.25	1.08	0.98	0.93	0.95	1.03	1.17	1.38	1.62	1.80
30°	±45°	1.55	1.36	1.18	1.05	0.96	0.93	0.94	1.01	1.12	1.29	1.49	1.63
30°	±60°	1.36	1.22	1.10	1.00	0.94	0.92	0.93	0.97	1.06	1.18	1.31	1.42
35°	0°	1.93	1.61	1.32	1.10	0.96	0.90	0.93	1.03	1.22	1.50	1.82	2.06
35°	±30°	1.78	1.50	1.26	1.07	0.95	0.90	0.92	1.01	1.17	1.41	1.69	1.89
35°	±45°	1.60	1.38	1.18	1.03	0.93	0.89	0.91	0.99	1.12	1.31	1.53	1.70
35°	±60°	1.39	1.23	1.09	0.98	0.91	0.88	0.90	0.95	1.04	1.18	1.33	1.45
45°	0°	2.07	1.68	1.33	1.05	0.88	0.81	0.84	0.97	1.20	1.55	1.94	2.24
45°	±30°	1.89	1.55	1.25	1.02	0.87	0.81	0.84	0.95	1.15	1.44	1.78	2.03
45°	±45°	1.68	1.41	1.16	0.98	0.86	0.81	0.83	0.92	1.08	1.31	1.59	1.79
45°	±60°	1.41	1.22	1.05	0.92	0.84	0.80	0.82	0.88	1.00	1.16	1.35	1.49
60°	0°	2.17	1.69	1.26	0.93	0.72	0.63	0.67	0.83	1.11	1.53	2.01	2.37
60°	±30°	1.95	1.53	1.17	0.89	0.72	0.65	0.68	0.81	1.04	1.40	1.81	2.12
60°	±45°	1.69	1.36	1.07	0.85	0.72	0.66	0.68	0.79	0.97	1.25	1.58	1.83
60°	±60°	1.38	1.16	0.95	0.80	0.70	0.66	0.68	0.75	0.89	1.08	1.31	1.48
90°	0°	1.93	1.38	0.87	0.50	0.27	0.19	0.22	0.39	0.70	1.19	1.75	2.16
90°	±30°	1.67	1.20	0.79	0.50	0.33	0.26	0.29	0.42	0.66	1.04	1.51	1.87
90°	±45°	1.39	1.03	0.72	0.50	0.36	0.31	0.33	0.43	0.62	0.91	1.27	1.54
90°	±60°	1.08	0.84	0.63	0.48	0.38	0.34	0.36	0.43	0.56	0.76	0.99	1.18

R_B - factors for latitude 38 degrees

β	γ	Jan	Feb	Mar	Apr	May	Jun	Jul	Aug	Sep	Oct	Nov	Dec
20°	0°	1.57	1.39	1.23	1.10	1.02	0.98	1.00	1.06	1.17	1.33	1.51	1.65
20°	±30°	1.49	1.33	1.19	1.08	1.01	0.98	0.99	1.05	1.14	1.28	1.44	1.55
20°	±45°	1.39	1.26	1.15	1.06	1.00	0.97	0.98	1.03	1.11	1.22	1.35	1.44
20°	±60°	1.26	1.17	1.09	1.02	0.98	0.97	0.97	1.01	1.06	1.14	1.23	1.30
30°	0°	1.79	1.53	1.29	1.10	0.98	0.93	0.95	1.04	1.20	1.44	1.71	1.90
30°	±30°	1.67	1.44	1.23	1.07	0.97	0.93	0.94	1.02	1.16	1.36	1.59	1.76
30°	±45°	1.52	1.34	1.17	1.04	0.96	0.92	0.94	1.00	1.11	1.28	1.46	1.60
30°	±60°	1.34	1.21	1.09	1.00	0.94	0.91	0.92	0.97	1.05	1.16	1.30	1.39
35°	0°	1.88	1.58	1.30	1.09	0.95	0.89	0.92	1.02	1.21	1.48	1.78	2.01
35°	±30°	1.74	1.48	1.24	1.06	0.94	0.89	0.91	1.00	1.16	1.39	1.65	1.85
35°	±45°	1.57	1.36	1.17	1.02	0.93	0.89	0.90	0.98	1.10	1.29	1.50	1.66
35°	±60°	1.36	1.22	1.08	0.98	0.91	0.88	0.89	0.94	1.03	1.17	1.32	1.43
45°	0°	2.02	1.65	1.30	1.04	0.87	0.80	0.83	0.96	1.19	1.52	1.90	2.17
45°	±30°	1.84	1.52	1.23	1.01	0.86	0.80	0.83	0.94	1.13	1.41	1.74	1.98
45°	±45°	1.64	1.38	1.15	0.97	0.85	0.80	0.82	0.91	1.07	1.29	1.55	1.75
45°	±60°	1.39	1.21	1.04	0.92	0.83	0.80	0.81	0.88	0.99	1.15	1.33	1.46
60°	0°	2.11	1.65	1.23	0.91	0.70	0.62	0.66	0.81	1.09	1.49	1.96	2.30
60°	±30°	1.89	1.50	1.14	0.88	0.71	0.64	0.67	0.80	1.02	1.36	1.76	2.06
60°	±45°	1.64	1.33	1.05	0.84	0.70	0.65	0.67	0.78	0.96	1.23	1.54	1.78
60°	±60°	1.35	1.13	0.94	0.79	0.69	0.65	0.67	0.75	0.87	1.06	1.28	1.44
90°	0°	1.85	1.33	0.84	0.47	0.25	0.17	0.21	0.37	0.68	1.15	1.68	2.08
90°	±30°	1.61	1.16	0.77	0.48	0.32	0.25	0.28	0.40	0.64	1.01	1.46	1.80
90°	±45°	1.34	0.99	0.70	0.48	0.35	0.30	0.32	0.42	0.60	0.88	1.22	1.48
90°	±60°	1.04	0.81	0.61	0.46	0.37	0.33	0.35	0.42	0.55	0.74	0.96	1.13

R_B - factors for latitude 37 degrees

β	γ	Jan	Feb	Mar	Apr	May	Jun	Jul	Aug	Sep	Oct	Nov	Dec
20°	0°	1.55	138	1.22	1.09	1.01	0.98	0.99	1.05	1.16	1.32	1.49	1.62
20°	±30°	1.47	1.32	1.18	1.07	1.00	0.97	0.99	1.04	1.13	1.27	1.42	1.53
20°	±45°	1.37	1.25	1.14	1.05	0.99	0.97	0.98	1.02	1.10	1.21	1.33	1.42
20°	±60°	1.25	1.16	1.08	1.02	0.98	0.96	0.97	1.00	1.06	1.13	1.22	1.28
30°	0°	1.76	1.51	1.27	1.09	0.97	0.92	0.94	1.03	1.19	1.42	1.68	1.86
30°	±30°	1.64	1.42	1.22	1.06	0.96	0.92	0.94	1.02	1.15	1.35	1.57	1.73
30°	±45°	1.50	1.32	1.16	1.03	0.95	0.92	0.93	0.99	1.10	1.26	1.44	1.57
30°	±60°	1.32	1.20	1.08	0.99	0.93	0.91	0.92	0.96	1.04	1.15	1.28	1.37
35°	0°	1.84	1.55	1.29	1.08	0.94	0.88	0.91	1.01	1.19	1.45	1.75	1.96
35°	±30°	1.71	1.46	1.23	1.05	0.93	0.88	0.90	0.99	1.15	1.37	1.62	1.81
35°	±45°	1.55	1.34	1.16	1.01	0.92	0.88	0.90	0.97	1.09	1.27	1.48	1.63
35°	±60°	1.34	1.20	1.07	0.97	0.90	0.87	0.89	0.94	1.03	1.15	1.30	1.40
45°	0°	1.97	1.61	1.28	1.02	0.86	0.79	0.82	0.95	1.17	1.49	1.85	2.12
45°	±30°	1.80	1.49	1.21	0.99	0.85	0.79	0.82	0.93	1.11	1.39	1.70	1.93
45°	±45°	1.60	1.36	1.13	0.96	0.84	0.80	0.82	0.90	1.05	1.27	1.52	1.71
45°	±60°	1.36	1.19	1.03	0.91	0.83	0.79	0.81	0.87	0.98	1.13	1.30	1.43
60°	0°	2.04	1.61	1.21	0.89	0.69	0.60	0.64	0.79	1.06	1.46	1.90	2.22
60°	±30°	1.84	1.46	1.12	0.86	0.69	0.63	0.66	0.78	1.00	1.34	1.71	1.99
60°	±45°	1.60	1.30	1.03	0.83	0.69	0.64	0.66	0.77	0.94	1.20	1.50	1.73
60°	±60°	1.32	1.11	0.93	0.78	0.69	0.65	0.66	0.74	0.86	1.04	1.25	1.40
90°	0°	1.78	1.28	0.81	0.45	0.24	0.16	0.19	0.35	0.65	1.11	1.62	1.99
90°	±30°	1.55	1.12	0.74	0.47	0.30	0.24	0.27	0.39	0.62	0.97	1.40	1.72
90°	±45°	1.29	0.96	0.68	0.47	0.34	0.29	0.31	0.41	0.58	0.85	1.18	1.42
90°	±60°	1.00	0.79	0.60	0.45	0.36	0.33	0.34	0.41	0.53	0.72	0.93	1.09

RB - factors for latitude 36 degrees

β	γ	Jan	Feb	Mar	Apr	May	Jun	Jul	Aug	Sep	Oct	Nov	Dec
20°	0°	1.53	1.36	1.21	1.09	1.00	0.97	0.99	1.05	1.15	1.31	1.47	1.59
20°	±30°	1.45	1.31	1.17	1.07	1.00	0.97	0.98	1.04	1.13	1.26	1.40	1.51
20°	±45°	1.36	1.24	1.13	1.05	0.99	0.96	0.98	1.02	1.09	1.20	1.32	1.40
20°	±60°	1.24	1.16	1.08	1.02	0.98	0.96	0.97	1.00	1.05	1.13	1.21	1.27
30°	0°	1.72	1.49	1.26	1.08	0.96	0.91	0.93	1.02	1.18	1.40	1.65	1.82
30°	±30°	1.61	1.40	1.21	1.06	0.95	0.91	0.93	1.01	1.14	1.33	1.54	1.69
30°	±45°	1.47	1.31	1.15	1.03	0.94	0.91	0.92	0.99	1.09	1.25	1.42	1.54
30°	±60°	1.30	1.19	1.08	0.99	0.93	0.90	0.91	0.96	1.04	1.14	1.27	1.35
35°	0°	1.80	1.53	1.27	1.06	0.93	0.87	0.90	1.00	1.18	1.43	1.71	1.92
35°	±30°	1.67	1.43	1.21	1.04	0.92	0.87	0.90	0.98	1.13	1.35	1.59	1.77
35°	±45°	1.52	1.33	1.15	1.01	0.91	0.87	0.89	0.96	1.08	1.26	1.46	1.60
35°	±60°	1.32	1.19	1.06	0.96	0.90	0.87	0.88	0.93	1.02	1.14	1.28	1.38
45°	0°	1.92	1.58	1.26	1.01	0.84	0.78	0.81	0.93	1.15	1.46	1.81	2.06
45°	±30°	1.76	1.47	1.19	0.98	0.84	0.78	0.81	0.92	1.10	1.36	1.66	1.88
45°	±45°	1.57	1.34	1.12	0.95	0.83	0.79	0.81	0.89	1.04	1.25	1.49	1.67
45°	±60°	1.34	1.17	1.02	0.90	0.82	0.79	0.80	0.86	0.97	1.12	1.28	1.40
60°	0°	1.99	1.57	1.18	0.87	0.67	0.59	0.63	0.78	1.04	1.43	1.85	2.16
60°	±30°	1.79	1.43	1.10	0.85	0.68	0.61	0.64	0.77	0.99	1.31	1.67	1.93
60°	±45°	1.56	1.28	1.02	0.82	0.68	0.63	0.65	0.75	0.93	1.18	1.47	1.68
60°	±60°	1.29	1.09	0.91	0.77	0.68	0.64	0.66	0.73	0.85	1.03	1.22	1.37
90°	0°	1.72	1.24	0.79	0.43	0.22	0.15	0.18	0.33	0.63	1.07	1.56	1.91
90°	±30°	1.49	1.08	0.72	0.45	0.29	0.23	0.26	0.37	0.60	0.94	1.35	1.66
90°	±45°	1.24	0.93	0.66	0.46	0.33	0.28	0.30	0.40	0.56	0.83	1.14	1.37
90°	±60°	0.97	0.77	0.58	0.44	0.36	0.32	0.34	0.40	0.52	0.70	0.90	1.06

RB - factors for latitude 35 degrees

β	γ	Jan	Feb	Mar	Apr	May	Jun	Jul	Aug	Sep	Oct	Nov	Dec
20°	0°	1.51	1.35	1.20	1.08	1.00	0.97	0.98	1.04	1.15	1.29	1.45	1.57
20°	±30°	1.43	1.29	1.16	1.06	0.99	0.96	0.98	1.03	1.12	1.25	1.39	1.48
20°	±45°	1.34	1.23	1.13	1.04	0.98	0.96	0.97	1.02	1.09	1.19	1.30	1.38
20°	±60°	1.23	1.15	1.07	1.01	0.97	0.96	0.96	1.00	1.05	1.12	1.20	1.26
30°	0°	1.69	1.46	1.25	1.07	0.95	0.90	0.93	1.02	1.17	1.38	1.62	1.79
30°	±30°	1.58	1.38	1.20	1.05	0.95	0.90	0.92	1.00	1.13	1.31	1.52	1.66
30°	±45°	1.45	1.29	1.14	1.02	0.94	0.90	0.92	0.98	1.09	1.24	1.40	1.52
30°	±60°	1.29	1.18	1.07	0.98	0.92	0.90	0.91	0.96	1.03	1.14	1.25	1.33
35°	0°	1.77	1.51	1.25	1.05	0.92	0.86	0.89	0.99	1.16	1.41	1.68	1.87
35°	±30°	1.64	1.41	1.20	1.03	0.91	0.87	0.89	0.97	1.12	1.33	1.57	1.73
35°	±45°	1.49	1.31	1.14	1.00	0.91	0.87	0.88	0.95	1.07	1.24	1.43	1.57
35°	±60°	1.31	1.18	1.06	0.96	0.89	0.86	0.88	0.93	1.01	1.13	1.26	1.36
45°	0°	1.88	1.55	1.24	1.00	0.83	0.76	0.79	0.92	1.13	1.44	1.77	2.01
45°	±30°	1.72	1.44	1.18	0.97	0.83	0.77	0.80	0.90	1.08	1.34	1.63	1.83
45°	±45°	1.54	1.31	1.10	0.94	0.82	0.78	0.80	0.88	1.03	1.24	1.47	1.63
45°	±60°	1.31	1.16	1.01	0.89	0.81	0.78	0.79	0.86	0.96	1.10	1.26	1.38
60°	0°	1.93	1.54	1.16	0.85	0.66	0.58	0.61	0.76	1.02	1.40	1.80	2.09
60°	±30°	1.74	1.40	1.08	0.83	0.67	0.60	0.63	0.76	0.97	1.28	1.63	1.88
60°	±45°	1.52	1.25	1.00	0.80	0.67	0.62	0.64	0.74	0.91	1.16	1.43	1.63
60°	±60°	1.26	1.07	0.90	0.76	0.67	0.63	0.65	0.72	0.84	1.01	1.20	1.34
90°	0°	1.65	1.20	0.76	0.41	0.21	0.13	0.16	0.31	0.60	1.03	1.50	1.84
90°	±30°	1.43	1.04	0.69	0.43	0.28	0.22	0.24	0.36	0.58	0.91	1.30	1.59
90°	±45°	1.20	0.90	0.64	0.44	0.32	0.27	0.29	0.38	0.55	0.80	1.10	1.32
90°	±60°	0.94	0.75	0.57	0.44	0.35	0.31	0.33	0.39	0.51	0.68	0.88	1.02

R_B - factors for latitude 34 degrees

β	γ	Jan	Feb	Mar	Apr	May	Jun	Jul	Aug	Sep	Oct	Nov	Dec
20°	0°	1.48	1.33	1.19	1.07	0.99	0.96	0.97	1.04	1.14	1.28	1.44	1.55
20°	±30°	1.41	1.28	1.16	1.06	0.99	0.96	0.97	1.02	1.11	1.24	1.37	1.46
20°	±45°	1.33	1.22	1.12	1.04	0.98	0.96	0.97	1.01	1.08	1.18	1.29	1.37
20°	±60°	1.21	1.14	1.07	1.01	0.97	0.95	0.96	0.99	1.04	1.11	1.19	1.24
30°	0°	1.66	1.44	1.23	1.06	0.94	0.90	0.92	1.01	1.16	1.37	1.59	1.75
30°	±30°	1.56	1.37	1.18	1.04	0.94	0.90	0.92	0.99	1.12	1.30	1.49	1.63
30°	±45°	1.43	1.28	1.13	1.01	0.93	0.90	0.91	0.97	1.08	1.22	1.38	1.49
30°	±60°	1.27	1.16	1.06	0.98	0.92	0.90	0.91	0.95	1.02	1.13	1.24	1.32
35°	0°	1.73	1.48	1.24	1.04	0.91	0.85	0.88	0.98	1.15	1.39	1.65	1.83
35°	±30°	1.61	1.39	1.19	1.02	0.91	0.86	0.88	0.97	1.11	1.32	1.54	1.70
35°	±45°	1.47	1.29	1.12	0.99	0.90	0.86	0.88	0.95	1.06	1.23	1.41	1.54
35°	±60°	1.29	1.17	1.05	0.95	0.89	0.86	0.87	0.92	1.00	1.12	1.25	1.34
45°	0°	1.83	1.52	1.22	0.98	0.82	0.75	0.78	0.91	1.12	1.41	1.73	1.96
45°	±30°	1.68	1.42	1.16	0.96	0.82	0.76	0.79	0.89	1.07	1.32	1.60	1.79
45°	±45°	1.51	1.29	1.09	0.93	0.82	0.77	0.79	0.88	1.02	1.22	1.44	1.60
45°	±60°	1.29	1.14	1.00	0.88	0.81	0.77	0.79	0.85	0.95	1.09	1.24	1.35
60°	0°	1.88	1.50	1.13	0.84	0.64	0.56	0.60	0.75	1.00	1.36	1.76	2.03
60°	±30°	1.70	1.37	1.06	0.82	0.66	0.59	0.62	0.74	0.95	1.25	1.59	1.83
60°	±45°	1.49	1.23	0.98	0.79	0.66	0.61	0.63	0.73	0.90	1.14	1.40	1.59
60°	±60°	1.23	1.06	0.89	0.75	0.66	0.62	0.64	0.71	0.83	0.99	1.18	1.31
90°	0°	1.59	1.16	0.73	0.39	0.19	0.12	0.15	0.29	0.58	1.00	1.45	1.77
90°	±30°	1.38	1.01	0.67	0.42	0.27	0.21	0.23	0.35	0.56	0.88	1.26	1.53
90°	±45°	1.15	0.87	0.62	0.43	0.31	0.26	0.28	0.37	0.53	0.78	1.06	1.27
90°	±60°	0.91	0.73	0.56	0.43	0.34	0.31	0.32	0.39	0.50	0.66	0.85	0.99

R_B - factors for latitude 32 degrees

β	γ	Jan	Feb	Mar	Apr	May	Jun	Jul	Aug	Sep	Oct	Nov	Dec
20°	0°	1.45	1.31	1.17	1.06	0.98	0.95	0.96	1.02	1.12	1.26	1.40	1.50
20°	±30°	1.38	1.26	1.14	1.04	0.98	0.95	0.96	1.01	1.10	1.22	1.34	1.43
20°	±45°	1.30	1.20	1.11	1.03	0.97	0.95	0.96	1.00	1.07	1.17	1.27	1.34
20°	±60°	1.20	1.13	1.06	1.00	0.97	0.95	0.96	0.99	1.03	1.10	1.17	1.22
30°	0°	1.61	1.41	1.21	1.04	0.93	0.88	0.90	0.99	1.13	1.33	1.54	1.69
30°	±30°	1.51	1.33	1.16	1.02	0.92	0.88	0.90	0.98	1.10	1.27	1.45	1.58
30°	±45°	1.39	1.25	1.11	1.00	0.92	0.89	0.90	0.96	1.06	1.20	1.35	1.45
30°	±60°	1.24	1.15	1.05	0.97	0.91	0.89	0.90	0.94	1.01	1.11	1.21	1.28
35°	0°	1.67	1.44	1.21	1.02	0.89	0.84	0.86	0.96	1.13	1.35	1.59	1.76
35°	±30°	1.56	1.36	1.16	1.00	0.89	0.84	0.86	0.95	1.09	1.28	1.49	1.63
35°	±45°	1.42	1.26	1.10	0.97	0.88	0.85	0.86	0.93	1.05	1.20	1.37	1.49
35°	±60°	1.26	1.14	1.03	0.94	0.88	0.85	0.86	0.91	0.99	1.10	1.22	1.30
45°	0°	1.76	1.47	1.19	0.95	0.80	0.73	0.76	0.88	1.08	1.37	1.66	1.87
45°	±30°	1.61	1.37	1.13	0.93	0.80	0.74	0.77	0.87	1.04	1.28	1.54	1.71
45°	±45°	1.45	1.26	1.06	0.91	0.80	0.75	0.77	0.86	0.99	1.18	1.39	1.53
45°	±60°	1.25	1.12	0.98	0.87	0.79	0.76	0.78	0.84	0.93	1.07	1.21	1.31
60°	0°	1.78	1.43	1.09	0.80	0.61	0.54	0.57	0.71	0.96	1.31	1.67	1.92
60°	±30°	1.61	1.31	1.02	0.79	0.63	0.57	0.60	0.71	0.92	1.20	1.51	1.73
60°	±45°	1.42	1.18	0.95	0.77	0.64	0.59	0.61	0.71	0.87	1.09	1.34	1.51
60°	±60°	1.18	1.02	0.86	0.73	0.65	0.61	0.63	0.69	0.81	0.96	1.13	1.25
90°	0°	1.48	1.08	0.68	0.35	0.16	0.09	0.12	0.26	0.53	0.93	1.35	1.64
90°	±30°	1.28	0.94	0.63	0.39	0.24	0.19	0.21	0.32	0.52	0.82	1.17	1.42
90°	±45°	1.08	0.82	0.59	0.41	0.29	0.25	0.27	0.35	0.50	0.73	0.99	1.18
90°	±60°	0.86	0.69	0.53	0.41	0.33	0.29	0.31	0.37	0.48	0.63	0.80	0.92

R_B - factors for latitude 30 degrees

β	γ	Jan	Feb	Mar	Apr	May	Jun	Jul	Aug	Sep	Oct	Nov	Dec
20°	0°	1.41	1.28	1.15	1.05	0.97	0.94	0.95	1.01	1.11	1.24	1.37	1.46
20°	±30°	1.35	1.24	1.13	1.03	0.97	0.94	0.95	1.00	1.09	1.20	1.31	1.39
20°	±45°	1.27	1.18	1.09	1.02	0.96	0.94	0.95	0.99	1.06	1.15	1.25	1.31
20°	±60°	1.18	1.11	1.05	1.00	0.96	0.94	0.95	0.98	1.03	1.09	1.16	1.20
30°	0°	1.56	1.37	1.18	1.02	0.91	0.86	0.89	0.97	1.11	1.30	1.50	1.63
30°	±30°	1.46	1.30	1.14	1.00	0.91	0.87	0.89	0.96	1.08	1.24	1.41	1.53
30°	±45°	1.36	1.23	1.09	0.98	0.91	0.87	0.89	0.95	1.05	1.18	1.31	1.41
30°	±60°	1.22	1.13	1.03	0.96	0.90	0.88	0.89	0.93	1.00	1.09	1.19	1.25
35°	0°	1.61	1.40	1.18	1.00	0.87	0.82	0.84	0.94	1.10	1.32	1.54	1.70
35°	±30°	1.51	1.32	1.13	0.98	0.87	0.83	0.85	0.93	1.07	1.25	1.45	1.58
35°	±45°	1.38	1.23	1.08	0.96	0.87	0.83	0.85	0.92	1.03	1.18	1.33	1.44
35°	±60°	1.23	1.12	1.02	0.93	0.87	0.84	0.85	0.90	0.98	1.08	1.19	1.27
45°	0°	1.68	1.42	1.15	0.93	0.77	0.71	0.74	0.86	1.05	1.32	1.60	1.79
45°	±30°	1.55	1.33	1.10	0.91	0.78	0.72	0.75	0.85	1.01	1.24	1.48	1.64
45°	±45°	1.40	1.22	1.04	0.89	0.78	0.74	0.76	0.84	0.97	1.15	1.34	1.47
45°	±60°	1.22	1.09	0.96	0.86	0.78	0.75	0.76	0.82	0.92	1.04	1.18	1.27
60°	0°	1.70	1.37	1.04	0.77	0.58	0.51	0.54	0.68	0.92	1.25	1.59	1.82
60°	±30°	1.54	1.26	0.98	0.76	0.61	0.54	0.57	0.69	0.88	1.16	1.45	1.65
60°	±45°	1.36	1.14	0.92	0.74	0.62	0.57	0.59	0.69	0.84	1.06	1.29	1.44
60°	±60°	1.14	0.99	0.84	0.72	0.63	0.60	0.61	0.68	0.79	0.94	1.09	1.20
90°	0°	1.38	1.01	0.63	0.32	0.13	0.06	0.09	0.22	0.49	0.87	1.26	1.53
90°	±30°	1.20	0.88	0.59	0.36	0.22	0.17	0.19	0.29	0.48	0.77	1.10	1.32
90°	±45°	1.01	0.77	0.55	0.38	0.27	0.23	0.25	0.33	0.48	0.69	0.93	1.10
90°	±60°	0.81	0.65	0.51	0.39	0.31	0.28	0.30	0.35	0.46	0.60	0.76	0.87

R_B - factors for latitude 28 degrees

β	γ	Jan	Feb	Mar	Apr	May	Jun	Jul	Aug	Sep	Oct	Nov	Dec
20°	0°	1.38	1.26	1.14	1.03	0.96	0.93	0.94	1.00	1.09	1.22	1.34	1.43
20°	±30°	1.32	1.22	1.11	1.02	0.96	0.93	0.94	0.99	1.07	1.18	1.29	1.36
20°	±45°	1.25	1.17	1.08	1.01	0.96	0.93	0.94	0.98	1.05	1.14	1.23	1.28
20°	±60°	1.16	1.10	1.04	0.99	0.95	0.94	0.94	0.97	1.02	1.08	1.14	1.19
30°	0°	1.51	1.34	1.16	1.00	0.89	0.85	0.87	0.95	1.09	1.27	1.46	1.58
30°	±30°	1.42	1.27	1.12	0.99	0.90	0.86	0.87	0.95	1.06	1.22	1.38	1.48
30°	±45°	1.32	1.20	1.08	0.97	0.90	0.86	0.88	0.94	1.03	1.16	1.28	1.37
30°	±60°	1.20	1.11	1.02	0.95	0.89	0.87	0.88	0.92	0.99	1.08	1.17	1.23
35°	0°	1.56	1.36	1.15	0.98	0.85	0.80	0.82	0.92	1.08	1.28	1.50	1.64
35°	±30°	1.46	1.29	1.11	0.96	0.86	0.81	0.83	0.91	1.04	1.22	1.40	1.53
35°	±45°	1.34	1.21	1.06	0.94	0.86	0.82	0.84	0.90	1.01	1.15	1.30	1.40
35°.	±60°	1.20	1.10	1.00	0.92	0.86	0.83	0.84	0.89	0.96	1.07	1.17	1.24
45°	0°	1.62	1.37	1.12	0.90	0.75	0.69	0.71	0.83	1.02	1.28	1.54	1.71
45°	±30°	1.50	1.28	1.07	0.89	0.76	0.71	0.73	0.83	0.99	1.21	1.43	1.58
45°	±45°	1.36	1.19	1.01	0.87	0.77	0.72	0.74	0.82	0.95	1.12	1.30	1.42
45°	±60°	1.18	1.07	0.94	0.84	0.77	0.74	0.75	0.81	0.90	1.02	1.15	1.23
60°	0°	1.62	1.31	1.00	0.74	0.56	0.48	0.51	0.65	0.89	1.20	1.52	1.73
60°	±30°	1.47	1.21	0.95	0.73	0.58	0.52	0.55	0.66	0.85	1.11	1.38	1.57
60°	±45°	1.30	1.10	0.89	0.72	0.60	0.55	0.58	0.67	0.82	1.02	1.23	1.38
60°	±60°	1.10	0.96	0.82	0.70	0.62	0.58	0.60	0.66	0.77	0.91	1.06	1.15
90°	0°	1.29	0.94	0.58	0.28	0.10	0.04	0.06	0.19	0.45	0.81	1.18	1.42
90°	±30°	1.12	0.83	0.55	0.33	0.20	0.15	0.17	0.27	0.45	0.72	1.02	1.23
90°	±45°	0.95	0.73	0.52	0.36	0.26	0.22	0.23	0.31	0.45	0.65	0.88	1.03
90°	±60°	0.76	0.62	0.49	0.37	0.30	0.27	0.28	0.34	0.44	0.57	0.72	0.82

RB - factors for latitude 26 degrees

β	γ	Jan	Feb	Mar	Apr	May	Jun	Jul	Aug	Sep	Oct	Nov	Dec
20°	0°	1.35	1.24	1.12	1.02	0.95	0.92	0.93	0.99	1.08	1.20	1.32	1.39
20°	±30°	1.30	1.20	1.10	1.01	0.95	0.92	0.93	0.98	1.06	1.16	1.27	1.33
20°	±45°	1.23	1.15	1.07	1.00	0.95	0.93	0.94	0.98	1.04	1.12	1.21	1.26
20°	±60°	1.15	1.09	1.03	0.98	0.95	0.93	0.94	0.97	1.01	1.07	1.13	1.17
30°	0°	1.47	1.31	1.13	0.98	0.88	0.83	0.85	0.94	1.07	1.24	1.42	1.53
30°	±30°	1.39	1.25	1.10	0.97	0.88	0.84	0.86	0.93	1.04	1.19	1.34	1.44
30°	±45°	1.29	1.18	1.06	0.96	0.88	0.85	0.87	0.92	1.02	1.14	1.26	1.34
30°	±60°	1.17	1.09	1.01	0.94	0.89	0.86	0.87	0.91	0.98	1.06	1.15	1.20
35°	0°	1.51	1.32	1.13	0.96	0.83	0.78	0.80	0.90	1.05	1.25	1.45	1.58
35°	±30°	1.42	1.26	1.09	0.94	0.84	0.79	0.81	0.90	1.02	1.20	1.37	1.48
35°	±45°	1.31	1.18	1.04	0.93	0.84	0.81	0.82	0.89	0.99	1.13	1.27	1.36
35°	±60°	1.17	1.08	0.99	0.91	0.85	0.82	0.83	0.88	0.95	1.05	1.15	1.21
45°	0°	1.56	1.33	1.09	0.88	0.73	0.66	0.69	0.81	0.99	1.24	1.49	1.65
45°	±30°	1.44	1.25	1.04	0.86	0.74	0.69	0.71	0.81	0.96	1.17	1.38	1.52
45°	±45°	1.31	1.16	0.99	0.85	0.75	0.71	0.73	0.80	0.93	1.10	1.26	1.37
45°	±60°	1.15	1.04	0.93	0.83	0.76	0.73	0.74	0.80	0.88	1.00	1.12	1.19
60°	0°	1.54	1.26	0.96	0.71	0.53	0.45	0.49	0.62	0.85	1.15	1.45	1.65
60°	±30°	1.40	1.16	0.91	0.70	0.56	0.50	0.53	0.64	0.82	1.07	1.33	1.50
60°	±45°	1.25	1.06	0.86	0.70	0.58	0.54	0.56	0.65	0.79	0.99	1.19	1.32
60°	±60°	1.06	0.93	0.80	0.69	0.60	0.57	0.59	0.65	0.75	0.88	1.02	1.11
90°	0°	1.20	0.88	0.54	0.24	0.07	0.02	0.04	0.16	0.41	0.75	1.10	1.33
90°	±30°	1.04	0.77	0.51	0.31	0.18	0.13	0.15	0.25	0.42	0.68	0.96	1.15
90°	±45°	0.89	0.69	0.49	0.34	0.24	0.20	0.22	0.29	0.43	0.62	0.82	0.97
90°	±60°	0.72	0.59	0.47	0.36	0.29	0.26	0.27	0.33	0.42	0.55	0.68	0.77

R_B - factors for latitude 24 degrees

β	γ	Jan	Feb	Mar	Apr	May	Jun	Jul	Aug	Sep	Oct	Nov	Dec
20°	0°	1.32	122	1.11	1.01	0.94	0.90	0.92	0.98	1.06	1.18	1.29	1.36
20°	±30°	1.27	1.18	1.09	1.00	0.94	0.91	0.92	0.97	1.05	1.15	1.24	1.31
20°	±45°	1.21	1.14	1.06	0.99	0.94	0.92	0.93	0.97	1.03	1.11	1.19	1.24
20°	±60°	1.13	1.08	1.03	0.98	0.94	0.93	0.93	0.96	1.01	1.06	1.12	1.15
30°	0°	1.43	1.28	1.11	0.97	0.86	0.82	0.84	0.92	1.05	1.22	1.38	1.49
30°	±30°	1.35	1.22	1.08	0.96	0.87	0.83	0.85	0.92	1.03	1.17	1.31	1.40
30°	±45°	1.27	1.16	1.04	0.94	0.87	0.84	0.85	0.91	1.00	1.12	1.23	1.31
30°	±60°	1.15	1.08	1.00	0.93	0.88	0.85	0.86	0.90	0.97	1.05	1.13	1.18
35°	0°	1.46	1.29	1.10	0.93	0.81	0.76	0.79	0.88	1.03	1.22	1.41	1.53
35°	±30°	1.38	1.23	1.07	0.92	0.82	0.78	0.80	0.88	1.00	1.17	1.33	1.44
35°	±45°	1.28	1.16	1.03	0.91	0.83	0.79	0.81	0.87	0.98	1.11	1.24	1.32
35°	±60°	1.15	1.07	0.98	0.90	0.84	0.81	0.82	0.87	0.94	1.03	1.12	1.18
45°	0°	1.50	1.29	1.05	0.85	0.70	0.64	0.67	0.78	0.97	1.20	1.44	1.58
45°	±30°	1.40	1.21	1.01	0.84	0.72	0.67	0.69	0.79	0.94	1.14	1.34	1.47
45°	±45°	1.27	1.13	0.97	0.83	0.73	0.69	0.71	0.79	0.91	1.07	1.23	1.33
45°	±60°	1.12	1.02	0.91	0.82	0.75	0.72	0.73	0.78	0.87	0.98	1.09	1.16
60°	0°	1.47	1.21	0.93	0.68	0.50	0.42	0.46	0.59	0.82	1.11	1.39	1.57
60°	±30°	1.34	1.12	0.88	0.68	0.54	0.47	0.50	0.61	0.79	1.03	1.27	1.43
60°	±45°	1.20	1.02	0.84	0.68	0.56	0.52	0.54	0.63	0.77	0.95	1.14	1.27
60°	±60°	1.03	0.91	0.78	0.67	0.59	0.56	0.57	0.63	0.73	0.86	0.99	1.07
90°	0°	1.13	0.82	0.49	0.21	0.05	0.00	0.02	0.13	0.37	0.70	1.03	1.24
90°	±30°	0.98	0.72	0.48	0.28	0.16	0.11	0.13	0.22	0.39	0.63	0.90	1.08
90°	±45°	0.83	0.65	0.47	0.32	0.23	0.19	0.20	0.28	0.40	0.58	0.78	0.91
90°	±60°	0.69	0.57	0.45	0.35	0.28	0.25	0.26	0.31	0.40	0.52	0.65	0.73

R_B - factors for latitude 22 degrees

β	γ	Jan	Feb	Mar	Apr	May	Jun	Jul	Aug	Sep	Oct	Nov	Dec
10°	0°	1.17	1.12	1.06	1.01	0.98	0.96	0.97	1.00	1.04	1.10	1.15	1.19
10°	±30°	1.14	1.10	1.05	1.01	0.98	0.96	0.97	1.00	1.03	1.08	1.13	1.16
10°	±45°	1.11	1.08	1.04	1.01	0.98	0.97	0.97	0.99	1.03	1.06	1.10	1.13
10°	±60°	1.08	1.05	1.02	1.00	0.98	0.97	0.98	0.99	1.01	1.04	1.07	1.09
20°	0°	1.30	1.20	1.09	1.00	0.92	0.89	0.91	0.96	1.05	1.16	1.27	1.34
20°	±30°	1.25	1.17	1.07	0.99	0.93	0.90	0.91	0.96	1.04	1.13	1.22	1.28
20°	±45°	1.19	1.13	1.05	0.98	0.93	0.91	0.92	0.96	1.02	1.10	1.17	1.22
20°	±60°	1.12	1.07	1.02	0.97	0.94	0.92	0.93	0.96	1.00	1.05	1.11	1.14
30°	0°	1.39	1.25	1.09	0.95	0.84	0.80	0.82	0.90	1.03	1.19	1.35	1.45
30°	±30°	1.32	1.20	1.06	0.94	0.85	0.81	0.83	0.90	1.01	1.15	1.28	1.37
30°	±45°	1.24	1.14	1.03	0.93	0.86	0.83	0.84	0.90	0.99	1.10	1.21	1.28
30°	±60°	1.14	1.06	0.99	0.92	0.87	0.85	0.86	0.90	0.96	1.04	1.11	1.16
35°	0°	1.42	1.26	1.08	0.91	0.80	0.74	0.77	0.86	1.01	1.19	1.37	1.48
35°	±30°	1.34	1.20	1.04	0.91	0.81	0.76	0.78	0.86	0.99	1.14	1.30	1.40
35°	±45°	1.25	1.13	1.01	0.90	0.82	0.78	0.80	0.86	0.96	1.09	1.21	1.29
35°	±60°	1.13	1.05	0.96	0.89	0.83	0.80	0.81	0.86	0.93	1.02	1.11	1.16
45°	0°	1.45	1.25	1.02	0.82	0.68	0.62	0.64	0.76	0.94	1.17	1.39	1.53
45°	±30°	1.35	1.18	0.99	0.82	0.70	0.65	0.67	0.76	0.91	1.11	1.30	1.42
45°	±45°	1.24	1.10	0.95	0.81	0.72	0.67	0.69	0.77	0.89	1.04	1.19	1.29
45°	±60°	1.10	1.00	0.90	0.80	0.73	0.70	0.72	0.77	0.86	0.96	1.07	1.13
60°	0°	1.41	1.16	0.89	0.64	0.47	0.40	0.43	0.56	0.78	1.06	1.33	1.50
60°	±30°	1.29	1.08	0.85	0.65	0.51	0.45	0.48	0.59	0.76	0.99	1.22	1.37
60°	±45°	1.16	0.99	0.81	0.66	0.55	0.50	0.52	0.61	0.74	0.92	1.10	1.22
60°	±60°	1.00	0.88	0.76	0.65	0.58	0.54	0.56	0.62	0.72	0.84	0.96	1.04

R_B - factors for latitude 20 degrees

β	γ	Jan	Feb	Mar	Apr	May	Jun	Jul	Aug	Sep	Oct	Nov	Dec
10°	0°	1.16	1.11	1.06	1.01	0.97	0.96	0.96	0.99	1.03	1.09	1.14	1.17
10°	±30°	1.13	1.09	1.05	1.00	0.97	0.96	0.97	0.99	1.03	1.08	1.12	1.15
10°	±45°	1.11	1.07	1.04	1.00	0.98	0.96	0.97	0.99	1.02	1.06	1.10	1.12
10°	±60°	1.07	1.05	1.02	1.00	0.98	0.96	0.97	0.99	1.01	1.04	1.06	1.08
20°	0°	1.28	1.18	1.08	0.98	0.91	0.88	0.90	0.95	1.04	1.15	1.25	1.31
20°	±30°	1.23	1.15	1.06	0.98	0.92	0.89	0.90	0.95	1.03	1.12	1.21	1.26
20°	±45°	1.18	1.11	1.04	0.97	0.92	0.90	0.91	0.95	1.01	1.09	1.16	1.20
20°	±60°	1.11	1.06	1.01	0.97	0.93	0.92	0.92	0.95	0.99	1.05	1.10	1.13
30°	0°	1.36	1.22	1.07	0.93	0.83	0.78	0.80	0.88	1.01	1.17	1.32	1.41
30°	±30°	1.29	1.17	1.04	0.93	0.84	0.80	0.82	0.89	0.99	1.13	1.26	1.34
30°	±45°	1.22	1.12	1.02	0.92	0.85	0.82	0.83	0.89	0.97	1.08	1.19	1.25
30°	±60°	1.12	1.05	0.98	0.91	0.86	0.84	0.85	0.89	0.95	1.03	1.10	1.14
35°	0°	1.38	1.23	1.05	0.89	0.78	0.72	0.75	0.84	0.98	1.17	1.34	1.44
35°	±30°	1.31	1.17	1.02	0.89	0.79	0.75	0.77	0.84	0.97	1.12	1.27	1.36
35°	±45°	1.22	1.11	0.99	0.88	0.80	0.77	0.78	0.85	0.95	1.07	1.19	1.26
35°	±60°	1.11	1.03	0.95	0.87	0.82	0.79	0.80	0.85	0.92	1.00	1.09	1.14
45°	0°	1.40	1.21	1.00	0.80	0.66	0.59	0.62	0.73	0.91	1.13	1.34	1.47
45°	±30°	1.31	1.14	0.96	0.80	0.68	0.63	0.65	0.74	0.89	1.08	1.26	1.37
45°	±45°	1.20	1.07	0.93	0.80	0.70	0.66	0.68	0.75	0.87	1.02	1.16	1.25
45°	±60°	1.07	0.98	0.88	0.79	0.72	0.69	0.71	0.76	0.84	0.95	1.05	1.11
60°	0°	1.35	1.12	0.85	0.61	0.44	0.37	0.40	0.53	0.75	1.02	1.28	1.44
60°	±30°	1.24	1.04	0.82	0.63	0.49	0.43	0.46	0.56	0.73	0.96	1.18	1.31
60°	±45°	1.11	0.96	0.79	0.64	0.53	0.48	0.50	0.59	0.72	0.89	1.06	1.17
60°	±60°	0.97	0.86	0.74	0.64	0.56	0.53	0.55	0.61	0.70	0.82	0.93	1.01

R_B - factors for latitude 15 degrees

β	γ	Jan	Feb	Mar	Apr	May	Jun	Jul	Aug	Sep	Oct	Nov	Dec
10°	0°	1.13	1.09	1.04	0.99	0.96	0.94	0.95	0.98	1.02	1.07	1.12	1.14
10°	±30°	1.11	1.07	1.03	0.99	0.96	0.95	0.95	0.98	1.01	1.06	1.10	1.12
10°	±45°	1.09	1.06	1.02	0.99	0.97	0.95	0.96	0.98	1.01	1.05	1.08	1.10
10°	±60°	1.06	1.04	1.01	0.99	0.97	0.95	0.97	0.98	1.00	1.03	1.05	1.06
20°	0°	1.22	1.14	1.05	0.95	0.88	0.85	0.87	0.92	1.01	1.11	1.20	1.25
20°	±30°	1.19	1.11	1.03	0.95	0.89	0.87	0.88	0.93	1.00	1.09	1.16	1.21
20°	±45°	1.14	1.08	1.02	0.95	0.90	0.88	0.89	0.93	0.99	1.06	1.12	1.16
20°	±60°	1.08	1.04	1.00	0.95	0.92	0.90	0.91	0.94	0.98	1.03	1.07	1.10
30°	0°	1.28	1.16	1.02	0.89	0.79	0.74	0.76	0.84	0.96	1.11	1.25	1.33
30°	±30°	1.23	1.12	1.00	0.89	0.80	0.76	0.78	0.85	0.95	1.08	1.19	1.26
30°	±45°	1.16	1.08	0.98	0.89	0.82	0.79	0.80	0.86	0.94	1.04	1.14	1.19
30°	±60°	1.08	1.02	0.95	0.89	0.84	0.82	0.83	0.87	0.93	1.00	1.06	1.10
35°	0°	1.30	1.16	1.00	0.84	0.73	0.68	0.70	0.79	0.93	1.10	1.25	1.35
35°	±30°	1.23	1.11	0.98	0.85	0.75	0.70	0.72	0.80	0.92	1.06	1.20	1.28
35°	±45°	1.16	1.06	0.95	0.85	0.77	0.73	0.75	0.81	0.91	1.02	1.13	1.19
35°	±60°	1.07	1.00	0.92	0.85	0.79	0.77	0.78	0.82	0.89	0.97	1.05	1.09
45°	0°	1.30	1.12	0.93	0.74	0.60	0.53	0.56	0.67	0.85	1.05	1.24	1.36
45°	±30°	1.22	1.07	0.90	0.75	0.63	0.58	0.60	0.69	0.84	1.01	1.17	1.27
45°	±45°	1.13	1.01	0.88	0.75	0.66	0.62	0.64	0.71	0.82	0.96	1.09	1.17
45°	±60°	1.02	0.94	0.85	0.76	0.69	0.67	0.68	0.73	0.81	0.91	0.99	1.05
60°	0°	1.22	1.01	0.77	0.54	0.37	0.30	0.33	0.46	0.67	0.93	1.16	1.30
60°	±30°	1.13	0.95	0.75	0.56	0.43	0.37	0.40	0.50	0.67	0.87	1.07	1.19
60°	±45°	1.02	0.88	0.73	0.59	0.48	0.44	0.46	0.54	0.67	0.83	0.98	1.07
60°	±60°	0.90	0.81	0.70	0.60	0.53	0.50	0.51	0.57	0.66	0.77	0.87	0.94

R_B - factors for latitude 10 degrees

β	γ	Jan	Feb	Mar	Apr	May	Jun	Jul	Aug	Sep	Oct	Nov	Dec
10°	0°	1.11	1.07	1.02	0.98	0.94	0.92	0.93	0.96	1.00	1.05	1.09	1.12
10°	±30°	1.09	1.06	1.02	0.98	0.95	0.93	0.94	0.96	1.00	1.04	1.08	1.10
10°	±45°	1.07	1.04	1.01	0.98	0.95	0.94	0.95	0.97	1.00	1.03	1.06	1.08
10°	±60°	1.05	1.03	1.00	0.98	0.96	0.94	0.96	0.97	0.99	1.02	1.04	1.05
20°	0°	1.18	1.10	1.01	0.92	0.85	0.82	0.84	0.89	0.98	1.07	1.16	1.21
20°	±30°	1.15	1.08	1.00	0.93	0.87	0.84	0.85	0.90	0.97	1.05	1.13	1.17
20°	±45°	1.11	1.06	0.99	0.93	0.88	0.86	0.87	0.91	0.97	1.03	1.09	1.13
20°	±60°	1.06	1.02	0.98	0.94	0.90	0.89	0.89	0.92	0.96	1.01	1.05	1.07
30°	0°	1.22	1.11	0.97	0.84	0.74	0.70	0.72	0.80	0.92	1.06	1.18	1.25
30°	±30°	1.17	1.07	0.96	0.85	0.76	0.72	0.74	0.81	0.91	1.03	1.14	1.20
30°	±45°	1.12	1.04	0.95	0.86	0.79	0.76	0.77	0.83	0.91	1.01	1.09	1.14
30°	±60°	1.05	0.99	0.93	0.87	0.82	0.80	0.81	0.85	0.90	0.97	1.03	1.07
35°	0°	1.22	1.09	0.94	0.79	0.68	0.62	0.65	0.74	0.88	1.04	1.18	1.27
35°	±30°	1.17	1.06	0.93	0.80	0.71	0.66	0.68	0.76	0.88	1.01	1.13	1.21
35°	±45°	1.11	1.02	0.91	0.81	0.74	0.70	0.72	0.78	0.87	0.98	1.08	1.14
35°	±60°	1.03	0.97	0.90	0.83	0.77	0.74	0.76	0.80	0.87	0.94	1.01	1.05
45°	0°	1.20	1.05	0.86	0.68	0.53	0.47	0.50	0.61	0.78	0.98	1.15	1.26
45°	±30°	1.14	1.00	0.85	0.69	0.58	0.52	0.55	0.64	0.78	0.95	1.10	1.18
45°	±45°	1.06	0.96	0.83	0.71	0.62	0.58	0.60	0.67	0.78	0.91	1.03	1.10
45°	±60°	0.97	0.90	0.81	0.73	0.67	0.64	0.65	0.70	0.78	0.87	0.95	1.00
60°	0°	1.11	0.91	0.69	0.46	0.30	0.22	0.26	0.39	0.59	0.84	1.05	1.17
60°	±30°	1.03	0.86	0.68	0.50	0.37	0.32	0.34	0.44	0.60	0.80	0.98	1.08
60°	±45°	0.94	0.82	0.67	0.54	0.44	0.39	0.41	0.49	0.62	0.77	0.90	0.99
60°	±60°	0.84	0.76	0.66	0.57	0.50	0.47	0.48	0.54	0.62	0.73	0.82	0.87

R$_B$ - factors for latitude 5 degrees

β	γ	Jan	Feb	Mar	Apr	May	Jun	Jul	Aug	Sep	Oct	Nov	Dec
10°	0°	1.09	1.05	1.01	0.96	0.93	0.91	0.92	0.95	0.99	1.04	1.08	1.10
10°	±30°	1.07	1.04	1.00	0.97	0.93	0.92	0.93	0.95	0.99	1.03	1.06	1.08
10°	±45°	1.06	1.03	1.00	0.97	0.94	0.93	0.94	0.96	0.99	1.02	1.05	1.07
10°	±60°	1.04	1.02	1.00	0.97	0.96	0.93	0.95	0.97	0.99	1.01	1.03	1.04
20°	0°	1.14	1.07	0.98	0.90	0.82	0.79	0.81	0.86	0.95	1.04	1.12	1.16
20°	±30°	1.11	1.05	0.98	0.90	0.84	0.81	0.83	0.88	0.95	1.03	1.09	1.13
20°	±45°	1.08	1.03	0.97	0.91	0.86	0.84	0.85	0.89	0.95	1.01	1.07	1.10
20°	±60°	1.04	1.01	0.96	0.92	0.89	0.87	0.88	0.91	0.95	0.99	1.03	1.05
30°	0°	1.16	1.05	0.93	0.80	0.70	0.65	0.67	0.75	0.88	1.01	1.13	1.19
30°	±30°	1.12	1.03	0.92	0.81	0.73	0.68	0.70	0.77	0.88	0.99	1.09	1.15
30°	±45°	1.07	1.00	0.92	0.83	0.76	0.72	0.74	0.80	0.88	0.97	1.05	1.10
30°	±60°	1.02	0.97	0.91	0.85	0.80	0.77	0.78	0.82	0.88	0.95	1.00	1.03
35°	0°	1.15	1.04	0.89	0.74	0.63	0.57	0.60	0.69	0.83	0.99	1.12	1.19
35°	±30°	1.11	1.01	0.88	0.76	0.66	0.62	0.64	0.72	0.83	0.97	1.08	1.14
35°	±45°	1.06	0.98	0.88	0.78	0.70	0.66	0.68	0.74	0.84	0.94	1.03	1.09
35°	±60°	0.99	0.94	0.87	0.80	0.75	0.72	0.73	0.78	0.84	0.91	0.98	1.01
45°	0°	1.12	0.97	0.80	0.62	0.47	0.40	0.43	0.55	0.72	0.91	1.08	1.17
45°	±30°	1.06	0.94	0.79	0.64	0.52	0.47	0.50	0.59	0.73	0.89	1.03	1.11
45°	±45°	1.00	0.91	0.79	0.67	0.58	0.54	0.55	0.63	0.74	0.86	0.97	1.04
45°	±60°	0.93	0.86	0.78	0.70	0.64	0.61	0.62	0.67	0.75	0.84	0.91	0.95
60°	0°	1.00	0.83	0.61	0.39	0.22	0.15	0.18	0.31	0.52	0.75	0.95	1.07
60°	±30°	0.94	0.79	0.61	0.44	0.32	0.26	0.29	0.39	0.54	0.73	0.89	0.99
60°	±45°	0.87	0.76	0.62	0.49	0.39	0.35	0.37	0.45	0.57	0.71	0.84	0.91
60°	±60°	0.79	0.72	0.63	0.54	0.47	0.44	0.45	0.51	0.59	0.69	0.77	0.82

R$_B$ - factors for latitude 0 degrees (β towards S)

β	γ	Jan	Feb	Mar	Apr	May	Jun	Jul	Aug	Sep	Oct	Nov	Dec
10°	0°	1.07	1.03	0.99	0.95	0.91	0.89	0.90	0.93	0.97	1.02	1.06	1.08
10°	±30°	1.06	1.03	0.99	0.95	0.92	0.90	0.91	0.94	0.97	1.01	1.05	1.07
10°	±45°	1.04	1.02	0.99	0.96	0.93	0.92	0.92	0.95	0.98	1.01	1.04	1.05
10°	±60°	1.03	1.01	0.99	0.97	0.95	0.92	0.94	0.96	0.98	1.00	1.02	1.03
20°	0°	1.10	1.04	0.95	0.86	0.79	0.75	0.77	0.83	0.92	1.01	1.08	1.12
20°	±30°	1.08	1.02	0.95	0.88	0.81	0.78	0.80	0.85	0.92	1.00	1.06	1.10
20°	±45°	1.06	1.01	0.95	0.89	0.84	0.81	0.82	0.87	0.93	0.99	1.04	1.07
20°	±60°	1.02	0.99	0.95	0.91	0.87	0.85	0.86	0.89	0.93	0.98	1.01	1.03
30°	0°	1.10	1.01	0.89	0.76	0.65	0.60	0.62	0.71	0.83	0.97	1.07	1.14
30°	±30°	1.07	0.99	0.88	0.78	0.68	0.64	0.66	0.74	0.84	0.95	1.05	1.10
30°	±45°	1.04	0.97	0.88	0.80	0.72	0.69	0.71	0.77	0.85	0.94	1.02	1.06
30°	±60°	0.99	0.94	0.89	0.82	0.77	0.75	0.76	0.80	0.86	0.92	0.98	1.01
35°	0°	1.09	0.98	0.84	0.69	0.57	0.51	0.54	0.64	0.78	0.93	1.06	1.13
35°	±30°	1.06	0.96	0.84	0.72	0.62	0.57	0.59	0.67	0.79	0.92	1.03	1.09
35°	±45°	1.01	0.94	0.84	0.74	0.66	0.62	0.64	0.71	0.80	0.91	0.99	1.04
35°	±60°	0.96	0.91	0.85	0.78	0.72	0.69	0.71	0.75	0.82	0.89	0.95	0.98
45°	0°	1.04	0.91	0.73	0.55	0.40	0.33	0.37	0.49	0.66	0.85	1.00	1.09
45°	±30°	1.00	0.88	0.74	0.59	0.47	0.42	0.44	0.54	0.68	0.83	0.96	1.04
45°	±45°	0.95	0.86	0.75	0.63	0.54	0.49	0.51	0.59	0.70	0.82	0.92	0.98
45°	±60°	0.89	0.83	0.75	0.67	0.61	0.58	0.59	0.65	0.72	0.80	0.87	0.91
60°	0°	0.91	0.74	0.53	0.31	0.14	0.08	0.11	0.24	0.44	0.67	0.86	0.97
60°	±30°	0.86	0.72	0.55	0.39	0.26	0.20	0.23	0.33	0.48	0.66	0.82	0.91
60°	±45°	0.81	0.70	0.57	0.45	0.35	0.31	0.33	0.40	0.52	0.66	0.77	0.84
60°	±60°	0.75	0.68	0.60	0.51	0.44	0.41	0.42	0.48	0.56	0.65	0.73	0.77

R_B - factors for latitude 0 degrees (β towards N)

β	γ	Jan	Feb	Mar	Apr	May	Jun	Jul	Aug	Sep	Oct	Nov	Dec
10°	0°	0.90	0.94	0.98	1.02	1.06	1.08	1.07	1.04	1.00	0.95	0.91	0.89
10°	±30°	0.91	0.94	0.98	1.02	1.05	1.07	1.06	1.03	1.00	0.95	0.92	0.90
10°	±45°	0.93	0.95	0.98	1.01	1.04	1.05	1.05	1.02	0.99	0.96	0.93	0.92
10°	±60°	0.94	0.96	0.98	1.00	1.02	1.05	1.03	1.01	0.99	0.97	0.95	0.94
20°	0°	0.78	0.84	0.93	1.01	1.09	1.13	1.11	1.05	0.96	0.87	0.80	0.76
20°	±30°	0.80	0.86	0.93	1.00	1.07	1.10	1.09	1.03	0.96	0.88	0.82	0.78
20°	±45°	0.83	0.87	0.93	0.99	1.05	1.07	1.06	1.02	0.96	0.89	0.84	0.81
20°	±60°	0.86	0.90	0.94	0.98	1.02	1.03	1.03	1.00	0.95	0.91	0.87	0.85
30°	0°	0.63	0.73	0.85	0.98	1.08	1.14	1.11	1.02	0.90	0.77	0.66	0.60
30°	±30°	0.67	0.75	0.85	0.96	1.06	1.10	1.08	1.00	0.90	0.78	0.69	0.64
30°	±45°	0.71	0.78	0.86	0.95	1.02	1.06	1.04	0.98	0.89	0.80	0.73	0.69
30°	±60°	0.76	0.81	0.87	0.93	0.98	1.01	0.99	0.95	0.89	0.83	0.78	0.75
35°	0°	0.55	0.66	0.80	0.94	1.07	1.13	1.10	1.00	0.86	0.70	0.58	0.51
35°	±30°	0.60	0.69	0.80	0.93	1.04	1.09	1.07	0.98	0.86	0.73	0.62	0.57
35°	±45°	0.65	0.72	0.81	0.91	1.00	1.04	1.02	0.95	0.85	0.75	0.67	0.63
35°	±60°	0.71	0.76	0.82	0.89	0.95	0.98	0.97	0.92	0.85	0.78	0.73	0.69
45°	0°	0.38	0.51	0.68	0.86	1.02	1.09	1.06	0.93	0.76	0.57	0.42	0.34
45°	±30°	0.45	0.56	0.69	0.84	0.98	1.04	1.01	0.90	0.75	0.60	0.48	0.42
45°	±45°	0.52	0.60	0.71	0.83	0.93	0.98	0.96	0.87	0.76	0.64	0.54	0.49
45°	±60°	0.60	0.66	0.73	0.81	0.88	0.91	0.90	0.84	0.76	0.68	0.61	0.58
60°	0°	0.12	0.26	0.47	0.69	0.88	0.97	0.93	0.77	0.56	0.33	0.16	0.08
60°	±30°	0.24	0.35	0.50	0.67	0.83	0.91	0.87	0.74	0.57	0.40	0.27	0.21
60°	±45°	0.33	0.42	0.53	0.67	0.79	0.84	0.82	0.72	0.59	0.46	0.36	0.31
60°	±60°	0.43	0.49	0.57	0.66	0.74	0.77	0.76	0.69	0.61	0.52	0.45	0.41

RB - factors for latitude −5 degrees

β	γ	Jan	Feb	Mar	Apr	May	Jun	Jul	Aug	Sep	Oct	Nov	Dec
10°	0°	0.92	0.95	0.99	1.04	1.08	1.10	1.09	1.06	1.01	0.97	0.93	0.91
10°	±30°	0.93	0.96	0.99	1.03	1.07	1.08	1.08	1.05	1.01	0.97	0.94	0.92
10°	±45°	0.94	0.96	0.99	1.02	1.05	1.07	1.06	1.04	1.00	0.97	0.95	0.93
10°	±60°	0.95	0.97	0.99	1.01	1.03	1.07	1.04	1.02	1.00	0.98	0.96	0.95
20°	0°	0.81	0.87	0.96	1.05	1.13	1.16	1.15	1.08	0.99	0.90	0.83	0.79
20°	±30°	0.83	0.88	0.95	1.03	1.10	1.13	1.12	1.06	0.99	0.91	0.85	0.81
20°	±45°	0.85	0.90	0.95	1.02	1.07	1.10	1.09	1.04	0.98	0.92	0.87	0.84
20°	±60°	0.88	0.91	0.95	1.00	1.03	1.05	1.04	1.01	0.97	0.92	0.89	0.87
30°	0°	0.68	0.77	0.89	1.02	1.14	1.19	1.17	1.07	0.94	0.81	0.71	0.65
30°	±30°	0.71	0.79	0.89	1.00	1.10	1.15	1.13	1.04	0.94	0.82	0.73	0.69
30°	±45°	0.75	0.81	0.89	0.98	1.06	1.10	1.08	1.01	0.93	0.83	0.76	0.73
30°	±60°	0.79	0.83	0.89	0.95	1.01	1.03	1.02	0.98	0.91	0.85	0.80	0.77
35°	0°	0.61	0.71	0.85	1.00	1.13	1.20	1.17	1.06	0.91	0.76	0.64	0.57
35°	±30°	0.65	0.73	0.85	0.98	1.09	1.14	1.12	1.02	0.90	0.77	0.67	0.62
35°	±45°	0.69	0.76	0.85	0.95	1.04	1.09	1.07	0.99	0.89	0.79	0.71	0.66
35°	±60°	0.74	0.79	0.85	0.92	0.98	1.01	1.00	0.95	0.88	0.81	0.75	0.72
45°	0°	0.45	0.57	0.74	0.93	1.09	1.17	1.13	1.00	0.82	0.63	0.48	0.41
45°	±30°	0.51	0.61	0.75	0.90	1.04	1.11	1.08	0.96	0.81	0.65	0.54	0.47
45°	±45°	0.56	0.64	0.75	0.87	0.98	1.04	1.01	0.92	0.80	0.68	0.59	0.54
45°	±60°	0.63	0.68	0.76	0.84	0.92	0.95	0.94	0.87	0.79	0.71	0.64	0.61
60°	0°	0.20	0.34	0.54	0.77	0.97	1.07	1.02	0.86	0.64	0.41	0.24	0.15
60°	±30°	0.30	0.41	0.56	0.74	0.91	0.99	0.95	0.81	0.63	0.46	0.33	0.26
60°	±45°	0.38	0.46	0.58	0.72	0.85	0.91	0.88	0.77	0.64	0.50	0.40	0.35
60°	±60°	0.46	0.52	0.60	0.69	0.78	0.82	0.80	0.73	0.64	0.55	0.48	0.44

RB - factors for latitude −10 degrees

β	γ	Jan	Feb	Mar	Apr	May	Jun	Jul	Aug	Sep	Oct	Nov	Dec
10°	0°	0.94	0.97	1.01	1.06	1.10	1.12	1.11	1.07	1.03	0.98	0.94	0.93
10°	±30°	0.94	0.97	1.01	1.05	1.08	1.10	1.09	1.06	1.02	0.98	0.95	0.93
10°	±45°	0.95	0.97	1.00	1.04	1.07	1.08	1.07	1.05	1.02	0.98	0.96	0.94
10°	±60°	0.96	0.98	1.00	1.02	1.04	1.08	1.05	1.03	1.01	0.98	0.97	0.96
20°	0°	0.84	0.90	0.99	1.08	1.16	1.21	1.19	1.12	1.02	0.93	0.86	0.82
20°	±30°	0.86	0.91	0.98	1.06	1.13	1.17	1.15	1.09	1.01	0.93	0.87	0.84
20°	±45°	0.87	0.92	0.97	1.04	1.10	1.13	1.12	1.07	1.00	0.94	0.89	0.86
20°	±60°	0.90	0.93	0.97	1.01	1.05	1.07	1.06	1.03	0.98	0.94	0.90	0.89
30°	0°	0.73	0.81	0.93	1.07	1.19	1.26	1.23	1.12	0.99	0.85	0.75	0.70
30°	±30°	0.75	0.83	0.93	1.04	1.15	1.20	1.18	1.09	0.97	0.86	0.77	0.73
30°	±45°	0.78	0.84	0.92	1.01	1.10	1.14	1.12	1.05	0.96	0.87	0.79	0.76
30°	±60°	0.81	0.85	0.91	0.98	1.04	1.07	1.05	1.00	0.94	0.87	0.82	0.80
35°	0°	0.66	0.76	0.90	1.05	1.20	1.27	1.23	1.11	0.96	0.81	0.69	0.63
35°	±30°	0.69	0.77	0.89	1.02	1.15	1.21	1.18	1.08	0.94	0.81	0.71	0.66
35°	±45°	0.72	0.79	0.88	0.99	1.09	1.14	1.11	1.03	0.93	0.82	0.74	0.70
35°	±60°	0.76	0.81	0.87	0.95	1.02	1.05	1.03	0.98	0.90	0.83	0.78	0.75
45°	0°	0.51	0.63	0.80	1.00	1.17	1.26	1.22	1.07	0.88	0.69	0.55	0.47
45°	±30°	0.56	0.66	0.80	0.96	1.11	1.19	1.15	1.02	0.86	0.71	0.59	0.53
45°	±45°	0.60	0.68	0.79	0.92	1.04	1.10	1.07	0.97	0.85	0.72	0.63	0.58
45°	±60°	0.66	0.71	0.79	0.88	0.96	1.00	0.98	0.91	0.82	0.74	0.67	0.64
60°	0°	0.27	0.41	0.62	0.85	1.07	1.18	1.13	0.95	0.71	0.48	0.31	0.23
60°	±30°	0.35	0.46	0.62	0.81	0.99	1.09	1.04	0.89	0.70	0.52	0.38	0.32
60°	±45°	0.42	0.51	0.63	0.78	0.92	0.99	0.95	0.84	0.69	0.55	0.45	0.39
60°	±60°	0.49	0.55	0.64	0.73	0.83	0.88	0.85	0.77	0.67	0.58	0.51	0.47

R_B - factors for latitude −15 degrees

β	γ	Jan	Feb	Mar	Apr	May	Jun	Jul	Aug	Sep	Oct	Nov	Dec
10°	0°	0.95	0.98	1.02	1.07	1.12	1.15	1.13	1.09	1.04	1.00	0.96	0.94
10°	±30°	0.96	0.98	1.02	1.06	1.10	1.12	1.11	1.08	1.04	0.99	0.96	0.95
10°	±45°	0.96	0.98	1.01	1.05	1.08	1.10	1.09	1.06	1.03	0.99	0.97	0.95
10°	±60°	0.97	0.98	1.00	1.03	1.05	1.10	1.06	1.04	1.01	0.99	0.97	0.96
20°	0°	0.87	0.93	1.02	1.12	1.21	1.26	1.23	1.15	1.06	0.96	0.89	0.85
20°	±30°	0.88	0.94	1.01	1.09	1.17	1.21	1.19	1.13	1.04	0.96	0.90	0.87
20°	±45°	0.90	0.94	1.00	1.07	1.13	1.16	1.15	1.09	1.02	0.96	0.91	0.88
20°	±60°	0.91	0.94	0.98	1.03	1.08	1.10	1.09	1.05	1.00	0.95	0.92	0.90
30°	0°	0.77	0.86	0.98	1.12	1.26	1.33	1.30	1.18	1.04	0.90	0.79	0.74
30°	±30°	0.79	0.86	0.97	1.09	1.21	1.27	1.24	1.14	1.02	0.90	0.81	0.76
30°	±45°	0.81	0.87	0.95	1.05	1.14	1.19	1.17	1.09	0.99	0.90	0.83	0.79
30°	±60°	0.83	0.87	0.93	1.00	1.07	1.10	1.09	1.03	0.96	0.89	0.84	0.82
35°	0°	0.71	0.81	0.95	1.11	1.27	1.35	1.31	1.18	1.02	0.86	0.74	0.68
35°	±30°	0.73	0.82	0.94	1.08	1.21	1.28	1.25	1.13	0.99	0.86	0.76	0.71
35°	±45°	0.76	0.83	0.92	1.03	1.14	1.20	1.17	1.08	0.96	0.86	0.78	0.74
35°	±60°	0.79	0.83	0.90	0.98	1.05	1.09	1.07	1.01	0.93	0.86	0.80	0.77
45°	0°	0.57	0.70	0.87	1.07	1.26	1.36	1.31	1.15	0.95	0.75	0.61	0.53
45°	±30°	0.61	0.71	0.85	1.02	1.19	1.27	1.23	1.09	0.92	0.76	0.64	0.58
45°	±45°	0.65	0.73	0.84	0.97	1.10	1.17	1.14	1.03	0.89	0.76	0.67	0.62
45°	±60°	0.68	0.74	0.82	0.91	1.00	1.05	1.03	0.95	0.86	0.77	0.70	0.67
60°	0°	0.34	0.49	0.70	0.95	1.18	1.30	1.24	1.04	0.80	0.56	0.38	0.30
60°	±30°	0.41	0.52	0.69	0.89	1.09	1.19	1.14	0.98	0.77	0.58	0.44	0.37
60°	±45°	0.47	0.55	0.68	0.84	0.99	1.07	1.04	0.90	0.74	0.60	0.49	0.44
60°	±60°	0.52	0.58	0.67	0.78	0.88	0.94	0.91	0.82	0.71	0.61	0.54	0.50

R$_B$ - factors for latitude −20 degrees

β	γ	Jan	Feb	Mar	Apr	May	Jun	Jul	Aug	Sep	Oct	Nov	Dec
10°	0°	0.97	1.00	1.04	1.09	1.15	1.17	1.16	1.12	1.06	1.01	0.97	0.96
10°	±30°	0.97	1.00	1.03	1.08	1.12	1.15	1.14	1.10	1.05	1.01	0.98	0.96
10°	±45°	0.97	0.99	1.02	1.06	1.10	1.12	1.11	1.08	1.04	1.00	0.98	0.96
10°	±60°	0.98	0.99	1.01	1.04	1.07	1.12	1.07	1.05	1.02	1.00	0.98	0.97
20°	0°	0.90	0.96	1.05	1.15	1.26	1.31	1.29	1.20	1.09	0.99	0.92	0.88
20°	±30°	0.91	0.96	1.04	1.13	1.21	1.26	1.24	1.16	1.07	0.99	0.92	0.89
20°	±45°	0.92	0.96	1.02	1.09	1.16	1.20	1.19	1.12	1.05	0.98	0.93	0.90
20°	±60°	0.92	0.96	1.00	1.05	1.10	1.13	1.12	1.07	1.02	0.97	0.93	0.92
30°	0°	0.81	0.90	1.03	1.18	1.33	1.41	1.37	1.24	1.09	0.94	0.84	0.78
30°	±30°	0.82	0.90	1.01	1.14	1.27	1.34	1.31	1.19	1.06	0.93	0.85	0.80
30°	±45°	0.84	0.90	0.99	1.09	1.20	1.25	1.23	1.14	1.03	0.93	0.85	0.82
30°	±60°	0.85	0.90	0.96	1.03	1.10	1.14	1.13	1.06	0.99	0.92	0.86	0.84
35°	0°	0.76	0.86	1.00	1.18	1.35	1.44	1.40	1.25	1.07	0.91	0.79	0.73
35°	±30°	0.77	0.86	0.98	1.13	1.28	1.36	1.32	1.19	1.04	0.90	0.80	0.75
35°	±45°	0.79	0.86	0.96	1.08	1.20	1.26	1.23	1.13	1.01	0.89	0.81	0.77
35°	±60°	0.81	0.86	0.93	1.01	1.10	1.14	1.12	1.05	0.96	0.88	0.82	0.79
45°	0°	0.63	0.76	0.93	1.15	1.36	1.48	1.42	1.24	1.02	0.81	0.67	0.59
45°	±30°	0.66	0.76	0.91	1.09	1.28	1.37	1.33	1.17	0.98	0.81	0.69	0.63
45°	±45°	0.69	0.77	0.89	1.03	1.18	1.26	1.22	1.09	0.94	0.81	0.71	0.66
45°	±60°	0.71	0.77	0.85	0.95	1.05	1.11	1.08	1.00	0.89	0.80	0.73	0.69
60°	0°	0.41	0.56	0.78	1.04	1.30	1.44	1.38	1.15	0.88	0.63	0.46	0.37
60°	±30°	0.47	0.58	0.76	0.98	1.20	1.32	1.26	1.07	0.84	0.64	0.50	0.43
60°	±45°	0.51	0.60	0.74	0.91	1.08	1.18	1.13	0.98	0.80	0.65	0.54	0.48
60°	±60°	0.55	0.62	0.71	0.83	0.95	1.01	0.98	0.88	0.76	0.65	0.57	0.53

R$_B$ - factors for latitude −22 degrees

β	γ	Jan	Feb	Mar	Apr	May	Jun	Jul	Aug	Sep	Oct	Nov	Dec
10°	0°	0.97	1.00	1.05	1.10	1.16	1.19	1.17	1.13	1.07	1.02	0.98	0.96
10°	±30°	0.97	1.00	1.04	1.09	1.13	1.16	1.15	1.11	1.06	1.01	0.98	0.96
10°	±45°	0.98	1.00	1.03	1.07	1.11	1.13	1.12	1.08	1.04	1.01	0.98	0.97
10°	±60°	0.98	0.99	1.02	1.04	1.07	1.13	1.08	1.06	1.03	1.00	0.98	0.97
20°	0°	0.91	0.97	1.06	1.17	1.28	1.34	1.31	1.22	1.11	1.00	0.93	0.89
20°	±30°	0.92	0.97	1.05	1.14	1.23	1.28	1.26	1.18	1.08	1.00	0.93	0.90
20°	±45°	0.92	0.97	1.03	1.10	1.18	1.22	1.20	1.14	1.06	0.99	0.94	0.91
20°	±60°	0.93	0.96	1.00	1.06	1.11	1.14	1.13	1.08	1.03	0.98	0.94	0.92
30°	0°	0.83	0.92	1.05	1.20	1.36	1.45	1.41	1.27	1.11	0.96	0.85	0.80
30°	±30°	0.84	0.91	1.02	1.16	1.30	1.37	1.34	1.22	1.08	0.95	0.86	0.81
30°	±45°	0.85	0.91	1.00	1.11	1.22	1.28	1.25	1.16	1.04	0.94	0.87	0.83
30°	±60°	0.86	0.90	0.97	1.04	1.12	1.16	1.14	1.08	1.00	0.92	0.87	0.85
35°	0°	0.78	0.88	1.03	1.21	1.39	1.49	1.44	1.28	1.10	0.93	0.81	0.74
35°	±30°	0.79	0.88	1.00	1.16	1.31	1.40	1.36	1.22	1.06	0.92	0.81	0.76
35°	±45°	0.80	0.87	0.97	1.10	1.22	1.29	1.26	1.15	1.02	0.91	0.82	0.78
35°	±60°	0.82	0.87	0.94	1.03	1.11	1.16	1.14	1.06	0.97	0.89	0.83	0.80
45°	0°	0.66	0.78	0.96	1.19	1.41	1.53	1.47	1.28	1.05	0.84	0.69	0.62
45°	±30°	0.68	0.78	0.93	1.12	1.32	1.42	1.37	1.20	1.01	0.83	0.71	0.65
45°	±45°	0.70	0.78	0.91	1.06	1.21	1.29	1.25	1.12	0.96	0.82	0.73	0.68
45°	±60°	0.72	0.78	0.87	0.97	1.08	1.14	1.11	1.02	0.91	0.81	0.74	0.71
60°	0°	0.44	0.59	0.81	1.09	1.36	1.51	1.44	1.20	0.92	0.66	0.48	0.40
60°	±30°	0.49	0.61	0.79	1.01	1.25	1.37	1.31	1.11	0.87	0.67	0.52	0.45
60°	±45°	0.53	0.62	0.76	0.94	1.12	1.22	1.17	1.01	0.83	0.67	0.56	0.50
60°	±60°	0.57	0.63	0.73	0.85	0.97	1.04	1.01	0.90	0.78	0.66	0.58	0.54

R$_B$ - factors for latitude −24 degrees

β	γ	Jan	Feb	Mar	Apr	May	Jun	Jul	Aug	Sep	Oct	Nov	Dec
20°	0°	0.92	0.99	1.08	1.19	1.30	1.37	1.34	1.24	1.12	1.02	0.94	0.91
20°	±30°	0.93	0.98	1.06	1.16	1.25	1.31	1.28	1.20	1.10	1.01	0.94	0.91
20°	±45°	0.93	0.98	1.04	1.12	1.20	1.24	1.22	1.15	1.07	1.00	0.94	0.92
20°	±60°	0.94	0.97	1.01	1.07	1.12	1.15	1.14	1.09	1.03	0.98	0.95	0.93
30°	0°	0.84	0.93	1.07	1.23	1.40	1.49	1.45	1.30	1.13	0.98	0.87	0.82
30°	±30°	0.85	0.93	1.04	1.18	1.33	1.41	1.37	1.24	1.10	0.97	0.87	0.83
30°	±45°	0.86	0.92	1.01	1.13	1.24	1.31	1.28	1.18	1.06	0.95	0.88	0.84
30°	±60°	0.87	0.91	0.98	1.06	1.14	1.18	1.16	1.09	1.01	0.93	0.88	0.86
35°	0°	0.80	0.90	1.05	1.24	1.43	1.53	1.48	1.32	1.12	0.95	0.82	0.76
35°	±30°	0.81	0.89	1.02	1.18	1.35	1.44	1.40	1.25	1.08	0.93	0.83	0.78
35°	±45°	0.82	0.89	0.99	1.12	1.25	1.33	1.29	1.18	1.04	0.92	0.84	0.80
35°	±60°	0.83	0.88	0.95	1.04	1.13	1.19	1.16	1.08	0.99	0.90	0.84	0.81
45°	0°	0.68	0.80	0.99	1.22	1.46	1.59	1.53	1.32	1.08	0.86	0.71	0.64
45°	±30°	0.70	0.80	0.96	1.16	1.36	1.47	1.42	1.24	1.04	0.85	0.73	0.67
45°	±45°	0.72	0.80	0.93	1.08	1.24	1.33	1.29	1.15	0.99	0.84	0.74	0.69
45°	±60°	0.74	0.79	0.88	0.99	1.10	1.16	1.14	1.04	0.92	0.82	0.75	0.72
60°	0°	0.47	0.62	0.85	1.13	1.42	1.58	1.50	1.25	0.96	0.69	0.51	0.43
60°	±30°	0.51	0.63	0.82	1.05	1.30	1.43	1.37	1.15	0.91	0.69	0.55	0.48
60°	±45°	0.55	0.64	0.79	0.97	1.16	1.27	1.22	1.05	0.86	0.69	0.57	0.52
60°	±60°	0.58	0.65	0.75	0.87	1.00	1.08	1.04	0.93	0.79	0.68	0.60	0.56
90°	0°	0.03	0.15	0.40	0.73	1.06	1.25	1.16	0.87	0.53	0.23	0.06	0.00
90°	±30°	0.14	0.24	0.41	0.65	0.92	1.08	1.01	0.76	0.50	0.29	0.17	0.11
90°	±45°	0.21	0.29	0.42	0.60	0.80	0.91	0.86	0.68	0.49	0.33	0.23	0.19
90°	±60°	0.27	0.32	0.41	0.53	0.66	0.73	0.70	0.59	0.46	0.35	0.28	0.25

R$_B$ - factors for latitude −26 degrees

β	γ	Jan	Feb	Mar	Apr	May	Jun	Jul	Aug	Sep	Oct	Nov	Dec
20°	0°	0.94	1.00	1.09	1.21	1.33	1.40	1.36	1.26	1.14	1.03	0.95	0.92
20°	±30°	0.94	0.99	1.07	1.17	1.28	1.33	1.31	1.22	1.11	1.02	0.95	0.92
20°	±45°	0.94	0.98	1.05	1.13	1.22	1.26	1.24	1.17	1.08	1.00	0.95	0.93
20°	±60°	0.94	0.97	1.02	1.08	1.14	1.17	1.15	1.10	1.04	0.99	0.95	0.93
30°	0°	0.86	0.95	1.09	1.26	1.43	1.53	1.49	1.33	1.15	1.00	0.89	0.83
30°	±30°	0.87	0.94	1.06	1.21	1.36	1.44	1.40	1.27	1.12	0.98	0.89	0.84
30°	±45°	0.87	0.94	1.03	1.15	1.27	1.34	1.31	1.20	1.07	0.96	0.89	0.85
30°	±60°	0.88	0.92	0.99	1.07	1.16	1.21	1.18	1.11	1.02	0.94	0.89	0.86
35°	0°	0.81	0.92	1.07	1.27	1.47	1.58	1.53	1.35	1.15	0.97	0.84	0.78
35°	±30°	0.82	0.91	1.04	1.21	1.38	1.48	1.44	1.28	1.11	0.95	0.85	0.80
35°	±45°	0.83	0.90	1.01	1.14	1.28	1.36	1.32	1.20	1.06	0.94	0.85	0.81
35°	±60°	0.84	0.89	0.96	1.06	1.16	1.21	1.18	1.10	1.00	0.91	0.85	0.82
45°	0°	0.70	0.83	1.02	1.26	1.51	1.65	1.58	1.36	1.11	0.89	0.74	0.66
45°	±30°	0.72	0.83	0.98	1.19	1.40	1.52	1.47	1.28	1.06	0.88	0.75	0.69
45°	±45°	0.73	0.82	0.95	1.11	1.28	1.38	1.33	1.18	1.01	0.86	0.76	0.71
45°	±60°	0.75	0.81	0.90	1.01	1.13	1.20	1.16	1.06	0.94	0.84	0.76	0.73
60°	0°	0.50	0.65	0.88	1.18	1.48	1.65	1.57	1.31	1.00	0.72	0.54	0.45
60°	±30°	0.54	0.66	0.85	1.09	1.35	1.50	1.43	1.20	0.94	0.72	0.57	0.50
60°	±45°	0.57	0.66	0.81	1.00	1.21	1.32	1.27	1.09	0.88	0.71	0.59	0.54
60°	±60°	0.59	0.66	0.76	0.90	1.04	1.11	1.08	0.95	0.81	0.69	0.61	0.57
90°	0°	0.05	0.19	0.44	0.78	1.14	1.33	1.24	0.93	0.57	0.26	0.08	0.02
90°	±30°	0.16	0.27	0.44	0.70	0.99	1.15	1.08	0.82	0.54	0.32	0.19	0.13
90°	±45°	0.23	0.31	0.44	0.63	0.85	0.97	0.91	0.72	0.51	0.35	0.25	0.20
90°	±60°	0.28	0.34	0.43	0.56	0.70	0.78	0.74	0.61	0.48	0.37	0.29	0.26

R_B - factors for latitude −28 degrees

β	γ	Jan	Feb	Mar	Apr	May	Jun	Jul	Aug	Sep	Oct	Nov	Dec
20°	0°	0.95	1.01	1.11	1.23	1.36	1.43	1.39	1.28	1.15	1.04	0.97	0.93
20°	±30°	0.95	1.00	1.08	1.19	1.30	1.36	1.33	1.23	1.12	1.03	0.96	0.93
20°	±45°	0.95	0.99	1.06	1.14	1.23	1.29	1.26	1.18	1.09	1.01	0.96	0.93
20°	±60°	0.95	0.98	1.03	1.09	1.15	1.19	1.17	1.11	1.05	0.99	0.96	0.94
30°	0°	0.88	0.97	1.11	1.29	1.47	1.58	1.53	1.36	1.18	1.01	0.90	0.85
30°	±30°	0.88	0.96	1.08	1.23	1.39	1.48	1.44	1.30	1.14	1.00	0.90	0.86
30°	±45°	0.88	0.95	1.04	1.17	1.30	1.37	1.34	1.22	1.09	0.98	0.90	0.86
30°	±60°	0.89	0.93	1.00	1.09	1.18	1.23	1.20	1.12	1.03	0.95	0.90	0.87
35°	0°	0.83	0.94	1.10	1.30	1.52	1.64	1.58	1.39	1.18	0.99	0.86	0.80
35°	±30°	0.84	0.93	1.06	1.24	1.42	1.53	1.48	1.31	1.13	0.97	0.86	0.81
35°	±45°	0.84	0.92	1.02	1.17	1.32	1.40	1.36	1.23	1.08	0.95	0.86	0.82
35°	±60°	0.85	0.90	0.98	1.07	1.18	1.24	1.21	1.12	1.01	0.92	0.86	0.83
45°	0°	0.73	0.85	1.05	1.30	1.57	1.72	1.65	1.41	1.15	0.92	0.76	0.69
45°	±30°	0.74	0.85	1.01	1.22	1.45	1.58	1.52	1.32	1.09	0.90	0.77	0.71
45°	±45°	0.75	0.84	0.97	1.14	1.32	1.42	1.38	1.21	1.03	0.88	0.77	0.72
45°	±60°	0.76	0.82	0.91	1.03	1.16	1.23	1.20	1.08	0.96	0.85	0.78	0.74
60°	0°	0.53	0.68	0.92	1.23	1.55	1.74	1.65	1.36	1.04	0.76	0.57	0.48
60°	±30°	0.56	0.69	0.88	1.14	1.41	1.57	1.50	1.25	0.98	0.75	0.60	0.52
60°	±45°	0.58	0.68	0.84	1.04	1.26	1.38	1.32	1.13	0.91	0.73	0.61	0.55
60°	±60°	0.61	0.68	0.78	0.92	1.07	1.16	1.12	0.98	0.84	0.71	0.63	0.58
90°	0°	0.08	0.22	0.48	0.84	1.22	1.43	1.33	1.00	0.62	0.30	0.11	0.04
90°	±30°	0.18	0.29	0.48	0.75	1.06	1.24	1.15	0.87	0.58	0.35	0.21	0.15
90°	±45°	0.24	0.33	0.47	0.67	0.90	1.04	0.97	0.76	0.54	0.37	0.27	0.22
90°	±60°	0.29	0.35	0.45	0.58	0.73	0.82	0.78	0.65	0.50	0.38	0.31	0.27

R_B - factors for latitude −30 degrees

β	γ	Jan	Feb	Mar	Apr	May	Jun	Jul	Aug	Sep	Oct	Nov	Dec
20°	0°	0.96	1.02	1.12	1.25	1.39	1.46	1.43	1.30	1.17	1.05	0.98	0.94
20°	±30°	0.96	1.01	1.10	1.21	1.33	1.39	1.36	1.26	1.14	1.04	0.97	0.94
20°	±45°	0.96	1.00	1.07	1.16	1.26	1.31	1.28	1.20	1.10	1.02	0.97	0.94
20°	±60°	0.95	0.99	1.03	1.10	1.17	1.20	1.19	1.12	1.06	1.00	0.96	0.94
30°	0°	0.89	0.99	1.13	1.32	1.52	1.63	1.58	1.40	1.20	1.03	0.92	0.87
30°	±30°	0.90	0.98	1.10	1.26	1.43	1.53	1.48	1.33	1.16	1.01	0.92	0.87
30°	±45°	0.90	0.96	1.06	1.19	1.33	1.41	1.37	1.25	1.11	0.99	0.91	0.88
30°	±60°	0.89	0.94	1.01	1.10	1.20	1.26	1.23	1.14	1.04	0.96	0.91	0.88
35°	0°	0.85	0.96	1.12	1.34	1.57	1.70	1.64	1.43	1.20	1.01	0.88	0.82
35°	±30°	0.86	0.95	1.08	1.27	1.47	1.58	1.53	1.35	1.15	0.99	0.88	0.83
35°	±45°	0.86	0.93	1.04	1.19	1.35	1.44	1.40	1.26	1.10	0.97	0.88	0.83
35°	±60°	0.86	0.91	0.99	1.09	1.20	1.27	1.24	1.14	1.03	0.93	0.87	0.84
45°	0°	0.75	0.88	1.08	1.34	1.63	1.79	1.71	1.46	1.18	0.94	0.79	0.71
45°	±30°	0.76	0.87	1.04	1.26	1.51	1.65	1.58	1.36	1.12	0.92	0.79	0.73
45°	±45°	0.77	0.85	0.99	1.17	1.36	1.48	1.42	1.25	1.06	0.90	0.79	0.74
45°	±60°	0.77	0.83	0.93	1.05	1.19	1.27	1.23	1.11	0.98	0.86	0.79	0.75
60°	0°	0.56	0.71	0.96	1.28	1.63	1.83	1.73	1.42	1.08	0.79	0.60	0.51
60°	±30°	0.58	0.71	0.91	1.18	1.48	1.65	1.57	1.30	1.01	0.77	0.62	0.55
60°	±45°	0.60	0.71	0.86	1.08	1.31	1.45	1.38	1.17	0.94	0.75	0.63	0.57
60°	±60°	0.62	0.69	0.80	0.95	1.11	1.20	1.16	1.01	0.86	0.73	0.64	0.60
90°	0°	0.10	0.26	0.53	0.90	1.30	1.53	1.42	1.07	0.67	0.34	0.14	0.07
90°	±30°	0.20	0.32	0.51	0.80	1.13	1.33	1.23	0.93	0.62	0.37	0.23	0.17
90°	±45°	0.26	0.35	0.50	0.71	0.96	1.11	1.04	0.81	0.58	0.39	0.28	0.23
90°	±60°	0.30	0.37	0.47	0.61	0.78	0.87	0.83	0.68	0.52	0.40	0.32	0.28

R$_B$ - factors for latitude −32 degrees

β	γ	Jan	Feb	Mar	Apr	May	Jun	Jul	Aug	Sep	Oct	Nov	Dec
20°	0°	0.97	1.04	1.14	1.27	1.42	1.50	1.46	1.33	1.19	1.07	0.99	0.95
20°	±30°	0.97	1.02	1.11	1.23	1.35	1.43	1.39	1.28	1.15	1.05	0.98	0.95
20°	±45°	0.96	1.01	1.08	1.17	1.28	1.34	1.31	1.22	1.12	1.03	0.98	0.95
20°	±60°	0.96	0.99	1.04	1.11	1.18	1.22	1.20	1.14	1.07	1.01	0.97	0.95
30°	0°	0.91	1.01	1.15	1.35	1.56	1.69	1.63	1.44	1.23	1.05	0.94	0.88
30°	±30°	0.91	0.99	1.12	1.28	1.47	1.58	1.53	1.36	1.18	1.03	0.93	0.88
30°	±45°	0.91	0.97	1.07	1.21	1.36	1.45	1.41	1.27	1.13	1.01	0.93	0.89
30°	±60°	0.90	0.95	1.02	1.12	1.22	1.29	1.26	1.16	1.06	0.97	0.92	0.89
35°	0°	0.87	0.98	1.15	1.37	1.62	1.76	1.70	1.47	1.23	1.03	0.90	0.84
35°	±30°	0.87	0.96	1.11	1.30	1.51	1.64	1.58	1.39	1.18	1.01	0.90	0.84
35°	±45°	0.87	0.95	1.06	1.22	1.39	1.49	1.44	1.29	1.12	0.98	0.89	0.85
35°	±60°	0.87	0.92	1.00	1.11	1.23	1.30	1.27	1.16	1.04	0.95	0.88	0.85
45°	0°	0.77	0.91	1.11	1.39	1.69	1.87	1.79	1.51	1.22	0.97	0.81	0.73
45°	±30°	0.78	0.89	1.06	1.30	1.56	1.72	1.64	1.41	1.15	0.95	0.81	0.75
45°	±45°	0.78	0.87	1.01	1.20	1.41	1.53	1.48	1.29	1.08	0.92	0.81	0.76
45°	±60°	0.78	0.85	0.95	1.08	1.22	1.31	1.27	1.14	1.00	0.88	0.80	0.76
60°	0°	0.58	0.74	1.00	1.34	1.71	1.93	1.82	1.49	1.13	0.82	0.63	0.54
60°	±30°	0.61	0.74	0.94	1.23	1.55	1.74	1.65	1.36	1.05	0.80	0.64	0.57
60°	±45°	0.62	0.73	0.89	1.11	1.37	1.52	1.44	1.22	0.98	0.78	0.65	0.59
60°	±60°	0.63	0.71	0.82	0.98	1.15	1.25	1.20	1.05	0.88	0.74	0.65	0.61
90°	0°	0.13	0.29	0.57	0.97	1.40	1.65	1.53	1.14	0.72	0.38	0.17	0.09
90°	±30°	0.22	0.34	0.55	0.85	1.21	1.43	1.32	1.00	0.66	0.40	0.25	0.19
90°	±45°	0.28	0.37	0.53	0.75	1.02	1.19	1.11	0.86	0.61	0.42	0.30	0.25
90°	±60°	0.31	0.38	0.49	0.64	0.82	0.93	0.88	0.71	0.55	0.42	0.33	0.29

R$_B$ - factors for latitude −34 degrees

β	γ	Jan	Feb	Mar	Apr	May	Jun	Jul	Aug	Sep	Oct	Nov	Dec
20°	0°	0.98	1.05	1.15	1.29	1.45	1.55	1.50	1.36	1.21	1.08	1.00	0.96
20°	±30°	0.98	1.03	1.12	1.25	1.38	1.47	1.43	1.30	1.17	1.06	0.99	0.96
20°	±45°	0.97	1.02	1.09	1.19	1.30	1.37	1.34	1.24	1.13	1.04	0.99	0.96
20°	±60°	0.96	1.00	1.05	1.12	1.20	1.25	1.22	1.15	1.08	1.01	0.97	0.95
30°	0°	0.93	1.02	1.18	1.38	1.62	1.75	1.69	1.48	1.25	1.07	0.95	0.90
30°	±30°	0.92	1.01	1.14	1.31	1.52	1.64	1.58	1.40	1.20	1.05	0.95	0.90
30°	±45°	0.92	0.99	1.09	1.24	1.40	1.50	1.45	1.30	1.15	1.02	0.94	0.90
30°	±60°	0.91	0.96	1.03	1.14	1.25	1.32	1.28	1.18	1.07	0.98	0.92	0.90
35°	0°	0.89	1.00	1.17	1.41	1.68	1.84	1.76	1.52	1.26	1.06	0.92	0.86
35°	±30°	0.89	0.98	1.13	1.33	1.56	1.70	1.64	1.43	1.21	1.03	0.91	0.86
35°	±45°	0.88	0.96	1.08	1.24	1.43	1.54	1.49	1.32	1.14	1.00	0.91	0.86
35°	±60°	0.88	0.93	1.02	1.13	1.26	1.34	1.30	1.18	1.06	0.96	0.89	0.86
45°	0°	0.80	0.93	1.15	1.44	1.77	1.96	1.87	1.57	1.26	1.00	0.83	0.75
45°	±30°	0.80	0.91	1.09	1.34	1.63	1.80	1.71	1.46	1.19	0.97	0.83	0.77
45°	±45°	0.80	0.89	1.04	1.24	1.46	1.60	1.53	1.33	1.11	0.94	0.83	0.77
45°	±60°	0.79	0.86	0.96	1.10	1.26	1.36	1.31	1.17	1.02	0.89	0.81	0.77
60°	0°	0.61	0.78	1.04	1.40	1.80	2.04	1.92	1.56	1.17	0.86	0.66	0.57
60°	±30°	0.63	0.77	0.98	1.28	1.63	1.83	1.73	1.42	1.09	0.83	0.67	0.59
60°	±45°	0.64	0.75	0.92	1.16	1.43	1.59	1.52	1.27	1.01	0.80	0.67	0.61
60°	±60°	0.65	0.72	0.84	1.01	1.20	1.31	1.25	1.08	0.91	0.76	0.67	0.63
90°	0°	0.16	0.33	0.62	1.03	1.50	1.78	1.64	1.22	0.78	0.42	0.20	0.12
90°	±30°	0.24	0.37	0.59	0.91	1.30	1.54	1.42	1.07	0.71	0.44	0.28	0.21
90°	±45°	0.29	0.39	0.56	0.80	1.09	1.28	1.19	0.92	0.65	0.44	0.32	0.26
90°	±60°	0.33	0.40	0.51	0.68	0.87	0.99	0.93	0.75	0.57	0.43	0.35	0.31

R$_B$ - factors for latitude −36 degrees

β	γ	Jan	Feb	Mar	Apr	May	Jun	Jul	Aug	Sep	Oct	Nov	Dec
20°	0°	0.99	1.06	1.17	1.32	1.49	1.60	1.55	1.39	1.23	1.09	1.01	0.97
20°	±30°	0.99	1.05	1.14	1.27	1.42	1.51	1.47	1.33	1.19	1.07	1.00	0.97
20°	±45°	0.98	1.03	1.10	1.21	1.33	1.40	1.37	1.26	1.14	1.05	0.99	0.97
20°	±60°	0.97	1.00	1.06	1.13	1.22	1.27	1.25	1.17	1.09	1.02	0.98	0.96
30°	0°	0.94	1.04	1.20	1.42	1.67	1.83	1.75	1.52	1.28	1.09	0.97	0.91
30°	±30°	0.94	1.02	1.16	1.35	1.56	1.70	1.63	1.43	1.23	1.07	0.96	0.91
30°	±45°	0.93	1.00	1.11	1.26	1.44	1.55	1.49	1.33	1.17	1.03	0.95	0.91
30°	±60°	0.92	0.97	1.05	1.15	1.28	1.35	1.32	1.20	1.09	0.99	0.93	0.90
35°	0°	0.91	1.02	1.20	1.45	1.74	1.92	1.84	1.57	1.30	1.08	0.94	0.87
35°	±30°	0.90	1.00	1.16	1.37	1.62	1.77	1.70	1.47	1.24	1.05	0.93	0.88
35°	±45°	0.90	0.98	1.10	1.27	1.48	1.60	1.54	1.36	1.17	1.02	0.92	0.87
35°	±60°	0.89	0.94	1.03	1.15	1.29	1.38	1.34	1.21	1.08	0.97	0.90	0.87
45°	0°	0.82	0.96	1.18	1.49	1.85	2.07	1.96	1.63	1.30	1.03	0.86	0.78
45°	±30°	0.82	0.94	1.12	1.39	1.70	1.88	1.79	1.51	1.22	1.00	0.85	0.78
45°	±45°	0.82	0.91	1.06	1.27	1.52	1.67	1.60	1.37	1.14	0.96	0.84	0.79
45°	±60°	0.81	0.88	0.98	1.13	1.30	1.41	1.36	1.20	1.04	0.91	0.83	0.79
60°	0°	0.64	0.81	1.08	1.46	1.90	2.16	2.04	1.64	1.22	0.89	0.69	0.59
60°	±30°	0.66	0.79	1.02	1.34	1.71	1.94	1.83	1.49	1.14	0.86	0.69	0.62
60°	±45°	0.66	0.77	0.95	1.20	1.50	1.68	1.59	1.32	1.04	0.83	0.69	0.63
60°	±60°	0.66	0.74	0.87	1.04	1.25	1.37	1.31	1.12	0.93	0.78	0.69	0.64
90°	0°	0.19	0.36	0.67	1.11	1.61	1.92	1.77	1.31	0.83	0.46	0.24	0.15
90°	±30°	0.27	0.40	0.63	0.97	1.40	1.66	1.54	1.14	0.75	0.47	0.30	0.23
90°	±45°	0.31	0.42	0.59	0.85	1.17	1.38	1.28	0.98	0.68	0.47	0.34	0.28
90°	±60°	0.34	0.42	0.54	0.71	0.93	1.06	1.00	0.80	0.60	0.45	0.36	0.32

R$_B$ - factors for latitude −38 degrees

β	γ	Jan	Feb	Mar	Apr	May	Jun	Jul	Aug	Sep	Oct	Nov	Dec
20°	0°	1.00	1.07	1.19	1.35	1.53	1.65	1.60	1.42	1.25	1.11	1.02	0.98
20°	±30°	1.00	1.06	1.15	1.29	1.46	1.56	1.51	1.36	1.21	1.09	1.01	0.98
20°	±45°	0.99	1.04	1.12	1.23	1.36	1.44	1.40	1.28	1.16	1.06	1.00	0.97
20°	±60°	0.98	1.01	1.07	1.15	1.24	1.30	1.27	1.18	1.10	1.03	0.99	0.97
30°	0°	0.96	1.06	1.23	1.46	1.74	1.91	1.83	1.57	1.31	1.11	0.99	0.93
30°	±30°	0.95	1.04	1.18	1.38	1.62	1.77	1.70	1.48	1.26	1.08	0.98	0.93
30°	±45°	0.94	1.01	1.13	1.29	1.48	1.60	1.55	1.37	1.19	1.05	0.96	0.92
30°	±60°	0.93	0.98	1.06	1.17	1.31	1.40	1.35	1.23	1.10	1.00	0.94	0.91
35°	0°	0.93	1.04	1.23	1.50	1.82	2.01	1.92	1.63	1.33	1.10	0.96	0.89
35°	±30°	0.92	1.02	1.18	1.41	1.68	1.85	1.77	1.52	1.27	1.07	0.95	0.89
35°	±45°	0.91	0.99	1.12	1.31	1.53	1.67	1.60	1.39	1.19	1.03	0.93	0.89
35°	±60°	0.90	0.95	1.05	1.18	1.33	1.43	1.38	1.24	1.10	0.98	0.91	0.88
45°	0°	0.84	0.99	1.22	1.55	1.94	2.18	2.06	1.70	1.34	1.06	0.88	0.80
45°	±30°	0.84	0.96	1.16	1.44	1.77	1.98	1.88	1.57	1.26	1.02	0.87	0.80
45°	±45°	0.83	0.93	1.09	1.31	1.58	1.75	1.67	1.42	1.17	0.98	0.86	0.80
45°	±60°	0.82	0.89	1.00	1.16	1.35	1.46	1.41	1.23	1.06	0.92	0.84	0.80
60°	0°	0.67	0.84	1.12	1.53	2.01	2.31	2.16	1.72	1.28	0.93	0.72	0.62
60°	±30°	0.68	0.82	1.06	1.40	1.81	2.06	1.94	1.56	1.18	0.89	0.72	0.64
60°	±45°	0.68	0.80	0.98	1.25	1.58	1.78	1.68	1.38	1.08	0.85	0.72	0.65
60°	±60°	0.68	0.76	0.89	1.08	1.30	1.44	1.37	1.17	0.96	0.80	0.70	0.65
90°	0°	0.22	0.40	0.72	1.19	1.74	2.08	1.92	1.41	0.90	0.50	0.27	0.18
90°	±30°	0.29	0.43	0.67	1.04	1.51	1.81	1.66	1.22	0.81	0.50	0.33	0.25
90°	±45°	0.33	0.44	0.62	0.90	1.26	1.49	1.38	1.04	0.73	0.50	0.36	0.30
90°	±60°	0.36	0.44	0.56	0.75	0.99	1.14	1.07	0.85	0.63	0.47	0.38	0.33

RB - factors for latitude −40 degrees

β	γ	Jan	Feb	Mar	Apr	May	Jun	Jul	Aug	Sep	Oct	Nov	Dec
20°	0°	1.01	1.09	1.20	1.38	1.58	1.72	1.65	1.46	1.27	1.12	1.03	0.99
20°	±30°	1.01	1.07	1.17	1.32	1.50	1.61	1.56	1.39	1.22	1.10	1.02	0.99
20°	±45°	1.00	1.05	1.13	1.25	1.40	1.49	1.44	1.31	1.17	1.07	1.01	0.98
20°	±60°	0.98	1.02	1.08	1.16	1.26	1.33	1.30	1.20	1.11	1.04	0.99	0.97
30°	0°	0.98	1.08	1.25	1.50	1.81	2.00	1.91	1.62	1.35	1.13	1.01	0.95
30°	±30°	0.97	1.06	1.20	1.42	1.68	1.85	1.77	1.52	1.28	1.10	0.99	0.94
30°	±45°	0.96	1.03	1.15	1.32	1.53	1.67	1.60	1.40	1.21	1.07	0.98	0.93
30°	±60°	0.94	0.99	1.07	1.20	1.35	1.44	1.40	1.26	1.12	1.02	0.95	0.92
35°	0°	0.95	1.07	1.26	1.55	1.90	2.12	2.01	1.69	1.37	1.13	0.98	0.91
35°	±30°	0.94	1.04	1.21	1.45	1.75	1.95	1.85	1.57	1.30	1.09	0.97	0.91
35°	±45°	0.93	1.01	1.14	1.34	1.59	1.74	1.67	1.44	1.22	1.05	0.95	0.90
35°	±60°	0.91	0.97	1.06	1.20	1.37	1.48	1.43	1.27	1.11	1.00	0.92	0.89
45°	0°	0.87	1.01	1.26	1.61	2.04	2.31	2.18	1.78	1.39	1.09	0.91	0.82
45°	±30°	0.86	0.99	1.19	1.49	1.86	2.10	1.98	1.63	1.30	1.05	0.90	0.82
45°	±45°	0.85	0.95	1.12	1.36	1.65	1.84	1.75	1.47	1.20	1.00	0.88	0.82
45°	±60°	0.83	0.91	1.02	1.19	1.40	1.53	1.47	1.27	1.08	0.94	0.85	0.81
60°	0°	0.70	0.88	1.17	1.60	2.13	2.47	2.30	1.81	1.33	0.97	0.75	0.65
60°	±30°	0.71	0.85	1.10	1.46	1.91	2.20	2.06	1.64	1.23	0.93	0.75	0.66
60°	±45°	0.71	0.82	1.02	1.30	1.66	1.90	1.78	1.44	1.12	0.88	0.74	0.67
60°	±60°	0.69	0.78	0.92	1.12	1.36	1.52	1.45	1.21	0.99	0.82	0.72	0.67
90°	0°	0.26	0.44	0.78	1.27	1.88	2.27	2.08	1.51	0.96	0.54	0.31	0.21
90°	±30°	0.32	0.46	0.72	1.12	1.63	1.97	1.80	1.32	0.86	0.54	0.36	0.27
90°	±45°	0.35	0.47	0.66	0.96	1.36	1.62	1.49	1.12	0.77	0.52	0.38	0.32
90°	±60°	0.37	0.46	0.59	0.80	1.06	1.23	1.15	0.90	0.67	0.50	0.40	0.35

A5 Shading Diagrams for Various Latitudes

The approximate effect of shading (see Section 2.5.4) can be determined using the three-component model.

To calculate the shading correction factor k_B, use Equation 2.23 as follows:

$$\text{Shading correction factor } k_B = 1 - \Sigma GPB/\Sigma GPS$$

where:

ΣGPB = aggregate weighting of the shaded points along the Sun's path for insolation at noon for each such point and for the relevant month

ΣGPS = total weighting for all points along the Sun's path for the month of interest

For unshaded PV installations, $k_B = 1$; for fully shaded installations, $k_B = 0$.

Assessment diagram for shading diagrams:

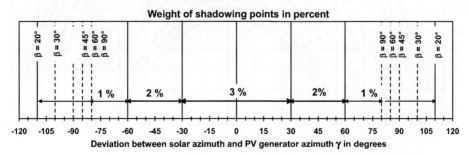

Weight of shadowing points in percent

Deviation between solar azimuth and PV generator azimuth γ in degrees

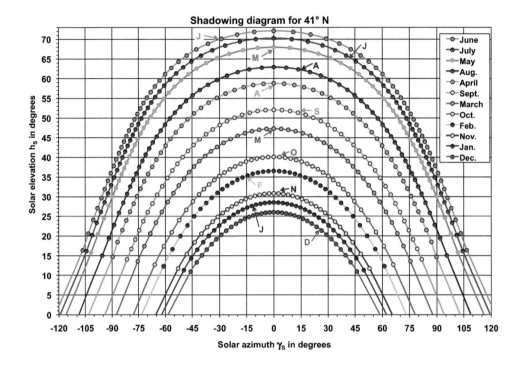

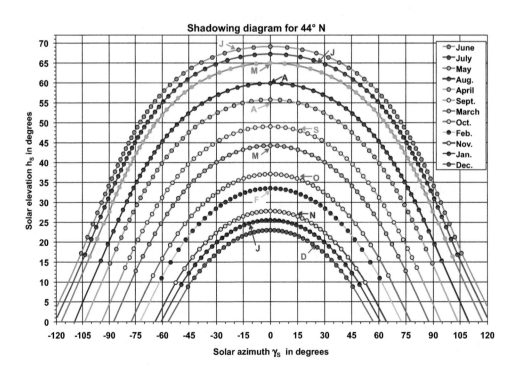

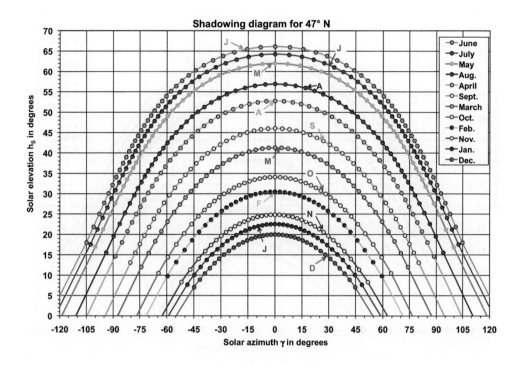

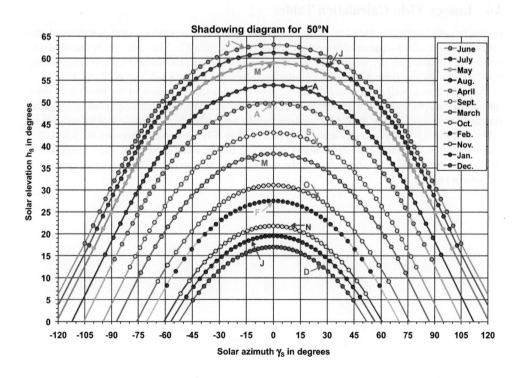

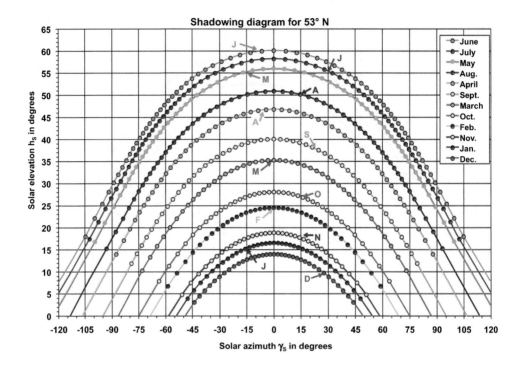

A6   Energy Yield Calculation Tables

For reasons of convenience, the tables in this section also contain the insolation calculation element, thus obviating the need for a separate table for this purpose (as in Section A1). The tables can be enlarged prior to use with a photocopier if need be. The insolation data for these calculations can be found in Section A2. For the k_T values see Table 8.2 or A7.1; for the k_G values see Tables 8.3–8.5 or A7.2–A7.4.

The tables here as well as other useful tables can be downloaded (by readers of this book) from www. electrosuisse.ch/photovoltaik or www.pvtest.ch. These tables contain the correct equations (pre-realized via the relevant links), while the fields containing results are read-only. A software tool for direct beam radiation R_B using the three-component model (see Section A4) for locations between 66°S and 66°N (including sites with different orientations) is also available from these web sites. However, no instructions are provided for this software; the user needs to be familiar with the relevant material in the present book.

A6.1 Energy Yield Calculation Tables for Grid-Connected Systems (see Section 8.2)

Table A6.1 Energy yield calculation table for grid-connected systems using a simplified radiation calculation as in Section 2.4. The global insolation factor $R(\beta,\gamma)$ data are listed in Section A3. The shaded fields are optional

Energy yield calculation for grid-connected PV plants (simplified calculation of irradiation)

Location:				Reference station:								$\beta =$	
P_{Go} [kW]:				$P_{ACn} = k_{Gmax} \cdot P_{Go} \cdot \eta_{WR}$ [kW]:								$\gamma =$	

Month	Jan	Feb	Mar	April	May	June	July	Aug	Sept	Oct	Nov	Dec	Year	
H														kWh/m²
$R(\beta,\gamma)$														
$H_G = R(\beta,\gamma) \cdot H$														kWh/m²
$Y_R = H_G/1\,kWm^{-2}$														h/d
k_T														
$Y_T = k_T \cdot Y_R$														h
k_G														
$Y_A = k_G \cdot Y_T$														h
η_{WR} (η_{tot})														
$Y_F = \eta_{WR} \cdot Y_A$														h
n_d (days/month)	31	28	31	30	31	30	31	31	30	31	30	31	365	d
$E_{AC} = n_d \cdot P_{Go} \cdot Y_F$														kWh
$t_V = E_{AC}/P_{ACn}$														h
$PR = Y_F/Y_R$														

Table A6.2 Energy yield calculation table for grid-connected systems with the insolation calculation as in Section 2.5. The direct beam radiation factors R_B for 60°N to 40°S are listed in Section A4. The values for the ground reflection factor ρ are listed in Table 2.6. The shaded fields are optional

Energy yield calculation for grid-connected PV plants (with 3-component model)

Location: | | Reference station: | | $\beta =$ |
P_{Go} [kW]: | | $P_{ACn} = k_{Gmax} \cdot P_{Go} \cdot \eta_{WR}$ [kW]: | | $\gamma =$ |

$R_D = \frac{1}{2}\cos \alpha_2 + \frac{1}{2}\cos (\alpha_1 + \beta) =$ | $R_R = \frac{1}{2} - \frac{1}{2}\cos\beta =$

Month	Jan	Feb	Mar	April	May	June	July	Aug	Sept	Oct	Nov	Dec	Year	
H														kWh/m^2
H_D														kWh/m^2
R_B														
k_B														
$H_{GB} = k_B \cdot R_B \cdot (H-H_D)$														kWh/m^2
$H_{GD} = R_D \cdot H_D$														kWh/m^2
ρ														
$H_{GR} = R_R \cdot \rho \cdot H$														kWh/m^2
$H_G = H_{GB} + H_{GD} + H_{GR}$														kWh/m^2
$Y_R = H_G/1\text{kWm}^{-2}$														h/d
k_T														
$Y_T = k_T \cdot Y_R$														h/d
k_G														
$Y_A = k_G \cdot Y_T$														h/d
η_{WR} (η_{tot})														
$Y_F = \eta_{WR} \cdot Y_A$														h/d
n_d (days/month)	31	28	31	30	31	30	31	31	30	31	30	31	365	d
$E_{AC} = n_d \cdot P_{Go} \cdot Y_F$														kWh
$t_V = E_{AC}/P_{ACn}$														h
$PR = Y_F/Y_R$														

Table A6.3 Energy balance and battery capacity calculation table

Energy balance sheet for PV stand-alone plants (layout of PV generator on separate sheet)

Consumers (switched)	AC/DC	P[W]	t [h]	E_{AC}	η_{WR}	E_{DC}		Location:	Month:
1							Wh/d		
2							Wh/d	System voltage $V_S =$	V
3							Wh/d	Accumulator cycle depth $t_z =$	
4							Wh/d	Ah-efficiency $\eta_{Ah} =$	
5							Wh/d		
6							Wh/d	Autonomy days $n_A =$	d
7							Wh/d	System recovery time $n_E =$	d
8							Wh/d		
		$E_V = \Sigma E_{DC}$ switched cons. =					Wh/d	Frequency of use $h_B =$	

Permanent consumers	AC/DC	P[W]	t [h]	E_{AC}	η_{WR}	E_{DC}		Note:
1							Wh/d	This sheet with average daily energy consumption is to be completed in most
2							Wh/d	cases only for the month with the highest
3								consumption. If energy consumption
4							Wh/d	varies considerably over the year, it may
		$E_0 = \Sigma E_{DC}$ perman. cons. =					Wh/d	be neccessary to do use several sheets.

Usable accumulator capacity $K_N = n_A \cdot (E_V+E_0)/V_S =$		Ah	Mean daily $E_D = h_B \cdot E_V + E_0 =$		Wh/d
Minimum accu capacity $K = K_N/t_z = n_A \cdot (E_V+E_0)/(V_S \cdot t_z) =$		Ah	Mean daily $Q_D = E_D/V_S =$		Ah/d
Necessary daily charging $Q_L = (1/\eta_{Ah}) \cdot (Q_D+K_N/n_E) =$		Ah/d			

A6.2   Stand-alone Installation Sizing Tables (see Section 8.3)

Table A6.4   Sizing table for a stand-alone installation solar generator for use with the basic insolation calculation as in Section 2.4. The global insolation factor $R(\beta,\gamma)$ data are listed in Section A3. The shaded fields are optional

---

Layout of PV generator for stand alone plants (simplified calculation of irradiation)

| Location: | | Reference station: | | | $\beta =$ | | $\gamma =$ | | ° |

| $I_{Mo} =$ | A | $P_{Mo} =$ | | W | System voltage $V_S =$ | | V | Modules / string $n_{MS} =$ | |

Month	Jan	Feb	Mar	April	May	June	July	Aug	Sept	Oct	Nov	Dec	
Req. daily charge Q_L													Ah/d
Auxillary energy E_H													Wh/d
$Q_H = E_H /(1.1 \cdot V_S)$													Ah/d
$Q_{PV} = Q_L - Q_H =$													Ah/d

Month	Jan	Feb	Mar	April	May	June	July	Aug	Sept	Oct	Nov	Dec	
H													kWh/m²
$R(\beta,\gamma)$													
$H_G = R(\beta,\gamma) \cdot H$													kWh/m²
$Y_R = H_G/1kWm^{-2} =$													h/d

Without MPT-charge-controller *(insert k_G-values in line below only in this case)* :

k_G													
$Q_S = k_G \cdot I_{Mo} \cdot Y_R =$													Ah/d

With MPT-charge-controller *(insert k_T- and k_G-values below only in this case)* :

k_T													
$Y_T = k_T \cdot Y_R$													h/d
k_G													
$Y_A = k_G \cdot Y_T$													h/d
η_{MPT}													
$Y_F' = \eta_{MPT} \cdot Y_A =$													h/d
$E_{DC-S} = n_{MS} \cdot P_{Mo} \cdot Y_F' =$													Wh/d
$Q_S = E_{DC-S}/(1.1 \cdot V_S)$													Ah/d

$n_{SP}' = Q_{PV}/ Q_S$													

Necessary number of parallel strings: Maximum(n_{SP}'), rounded up to the next integer :	$n_{SP} =$	
Total necessary number of modules:	$n_M = n_{MS} \cdot n_{SP} =$	

Table A6.5   Sizing table for a stand-alone installation solar generator for use with the three-component (insolation calculation) model as in Section 2.5. The direct beam radiation factors R_B for 60°N to 40°S are listed in Section A4. The values for the ground reflection factor ρ are listed in Table 2.6. The shaded fields are optional.

Layout of PV generator for stand alone plants (with 3-component model

Location:				Reference station:					$\beta =$			$\gamma =$		°

$R_D = \frac{1}{2}\cos\alpha_2 + \frac{1}{2}\cos(\alpha_1+\beta) =$		$R_R = \frac{1}{2} - \frac{1}{2}\cos\beta =$	

$I_{Mo} =$		A	$P_{Mo} =$		W	System voltage $V_S =$		V	Modules / string $n_{MS} =$		

Month	Jan	Feb	Mar	April	May	June	July	Aug	Sept	Oct	Nov	Dec	
Req. daily charge Q_L													Ah/d
Auxillary energy E_H													Wh/d
$Q_H = E_H/(1.1 \cdot V_S)$													Ah/d
$Q_{PV} = Q_L - Q_H =$													Ah/d

Month	Jan	Feb	Mar	April	May	June	July	Aug	Sept	Oct	Nov	Dec	
H													kWh/m²
H_D													kWh/m²
R_B													
k_B													
$H_{GB} = k_B \cdot R_B \cdot (H-H_D)$													kWh/m²
$H_{GD} = R_D \cdot H_D$													kWh/m²
ρ													
$H_{GR} = R_R \cdot \rho \cdot H$													kWh/m²
$H_G = H_{GB}+H_{GD}+H_{GR}$													kWh/m²
$Y_R = H_G/1kWm^{-2} =$													h/d

Without MPT-charge-controller *(insert k_G-values in line below only in this case):*

k_G													
$Q_S = k_G \cdot I_{Mo} \cdot Y_R =$													Ah/d

With MPT-charge-controller *(insert k_T- and k_G-values below only in this case):*

k_T													
$Y_T = k_T \cdot Y_R$													h/d
k_G													
$Y_A = k_G \cdot Y_T$													h/d
η_{MPT}													
$Y_F' = \eta_{MPT} \cdot Y_A =$													h/d
$E_{DC-S} = n_{MS} \cdot P_{Mo} \cdot Y_F' =$													Wh/d
$Q_S = E_{DC-S}/(1.1 \cdot V_S)$													Ah/d

$n_{SP}' = Q_{PV}/Q_S$													

Necessary number of parallel strings: Maximum(n_{SP}'), rounded up to the next integer:	$n_{SP} =$	
Total necessary number of modules:	$n_M = n_{MS} \cdot n_{SP} =$	

A7 k_T and k_G Figures for Energy Yield Calculations

Table A7.1 Temperature correction factors k_T (for low-, medium- and high-temperature effects) for the relevant reference stations for three types of PV installations (see Section 8.1.2.3). The k_T figures were determined for temperature coefficients amounting to $c_T = -0.45\%$ per K, which is typical for standard crystalline silicon solar cells. To obtain more accurate temperature correction factors for PV installations at higher elevations than the reference stations, increase the figures by approximately 0.025 per 1000 m of elevation. When modules with other temperature coefficients are used, the difference relative to 1 can be adjusted in accordance with the temperature coefficient ratios. The upper table can be used to obtain approximate figures for amorphous silicon modules.

Temperature influence small:
Free field mounting of crystalline modules or plants with modules of amorphous Si

	Jan	Feb	Mar	Apr	May	June	July	Aug	Sep	Oct	Nov	Dec	Year
Kloten	1.06	1.05	1.02	1.00	0.97	0.96	0.94	0.94	0.97	1.01	1.04	1.06	0.98
Davos	1.05	1.04	1.02	1.01	0.99	0.98	0.96	0.97	0.99	1.01	1.04	1.05	1.00
Locarno	1.03	1.03	1.01	1.00	0.98	0.95	0.93	0.93	0.96	0.99	1.02	1.03	0.98
Potsdam	1.07	1.06	1.03	1.00	0.97	0.95	0.94	0.95	0.98	1.01	1.04	1.06	0.98
Giessen	1.06	1.05	1.03	1.00	0.97	0.95	0.95	0.96	0.99	1.01	1.04	1.06	0.98
München	1.06	1.05	1.02	1.00	0.97	0.95	0.94	0.95	0.97	1.00	1.04	1.05	0.98
Marseille	1.00	1.00	0.98	0.96	0.94	0.92	0.90	0.91	0.94	0.96	0.99	1.00	0.95
Sevilla	0.98	0.97	0.95	0.95	0.92	0.90	0.88	0.89	0.91	0.93	0.97	0.98	0.93
Kairo	0.96	0.94	0.93	0.91	0.89	0.88	0.88	0.88	0.89	0.90	0.93	0.95	0.91

Temperature influence medium:
Rooftop mounting with air space or façade integration of crystalline modules

	Jan	Feb	Mar	Apr	May	June	July	Aug	Sep	Oct	Nov	Dec	Year
Kloten	1.05	1.03	1.01	0.98	0.95	0.93	0.91	0.92	0.95	0.99	1.04	1.05	0.96
Davos	1.03	1.02	1.00	0.98	0.97	0.96	0.94	0.95	0.97	0.98	1.02	1.03	0.98
Locarno	1.02	1.02	0.99	0.98	0.96	0.93	0.90	0.91	0.94	0.98	1.01	1.02	0.96
Potsdam	1.06	1.05	1.02	0.98	0.94	0.93	0.92	0.93	0.96	1.00	1.04	1.06	0.96
Giessen	1.06	1.04	1.01	0.98	0.94	0.93	0.92	0.93	0.97	1.00	1.04	1.06	0.96
München	1.05	1.03	1.01	0.98	0.94	0.92	0.92	0.93	0.95	0.99	1.02	1.04	0.96
Marseille	0.98	0.98	0.95	0.93	0.91	0.89	0.87	0.88	0.91	0.93	0.97	0.98	0.92
Sevilla	0.95	0.95	0.93	0.92	0.89	0.87	0.85	0.85	0.88	0.91	0.95	0.96	0.90
Kairo	0.93	0.92	0.90	0.88	0.86	0.84	0.84	0.84	0.86	0.87	0.90	0.93	0.88

Temperature influence high:
Crystalline modules integrated without air space into roofs

	Jan	Feb	Mar	Apr	May	June	July	Aug	Sep	Oct	Nov	Dec	Year
Kloten	1.04	1.02	0.99	0.96	0.93	0.91	0.89	0.90	0.93	0.98	1.03	1.04	0.94
Davos	1.00	0.99	0.97	0.96	0.94	0.93	0.91	0.92	0.94	0.96	1.00	1.01	0.96
Locarno	1.00	1.00	0.97	0.97	0.94	0.90	0.87	0.88	0.92	0.96	0.99	1.00	0.94
Potsdam	1.05	1.04	1.00	0.96	0.92	0.90	0.90	0.91	0.94	0.98	1.03	1.05	0.94
Giessen	1.05	1.03	1.00	0.96	0.92	0.91	0.90	0.91	0.96	0.99	1.03	1.05	0.95
München	1.03	1.01	0.99	0.95	0.92	0.90	0.89	0.90	0.93	0.97	1.01	1.03	0.94
Marseille	0.96	0.96	0.93	0.91	0.88	0.86	0.84	0.85	0.89	0.91	0.95	0.96	0.90
Sevilla	0.93	0.92	0.90	0.89	0.86	0.84	0.81	0.82	0.86	0.88	0.93	0.94	0.87
Kairo	0.90	0.89	0.87	0.84	0.82	0.81	0.81	0.81	0.83	0.84	0.87	0.90	0.84

A7.1 k_T Figures for Various Reference Stations

Table A7.2 Recommended k_G figures for Central European areas that are occasionally covered with snow in the winter (see Section 8.1.3.2)

Locations with occasional snowfall in winter
Recommended average values for k_G for long-term yield calculations

β	Jan	Feb	Mar	Apr	May	June	July	Aug	Sep	Oct	Nov	Dec
30°	0.69	0.73	0.81	0.83	0.84	0.84	0.84	0.84	0.84	0.82	0.75	0.66
45°	0.80	0.83	0.84	0.85	0.86	0.86	0.86	0.86	0.86	0.84	0.82	0.77
60°	0.84	0.85	0.86	0.86	0.85	0.85	0.85	0.85	0.86	0.86	0.85	0.84
90°	0.86	0.86	0.85	0.84	0.82	0.81	0.81	0.82	0.84	0.85	0.86	0.86

Recommended k_G-values for new plants with modules with full rated power

β	Jan	Feb	Mar	Apr	May	June	July	Aug	Sep	Oct	Nov	Dec
30°	0.75	0.79	0.86	0.88	0.90	0.90	0.90	0.90	0.90	0.88	0.80	0.70
45°	0.85	0.88	0.89	0.90	0.91	0.91	0.91	0.91	0.91	0.89	0.87	0.82
60°	0.89	0.90	0.91	0.91	0.90	0.90	0.90	0.90	0.91	0.91	0.90	0.89
90°	0.91	0.91	0.91	0.89	0.87	0.86	0.86	0.87	0.89	0.91	0.91	0.91

A7.2 k_G Figures for Various Reference Stations

In determining which figures to apply to a specific site (including those not referenced in the tables) the latitude and local climate should always be taken into consideration. The snow-covering maps in Section 8.1.3.1 can also be used for this purpose.

Tables A7.3 and A7.4 presuppose that k_G will change slightly during the summer months owing to summer soiling from the lack of precipitation. Depending on local conditions, this effect can of course be considerably more pronounced if the solar modules are not cleaned periodically, in which case k_G would be far lower. The farther south a façade-mounted installation ($\beta = 90°$) is, the lower kG will go during the summer and the greater the height of the Sun h_s will be at noon.

For reasons of space, the southern hemisphere reference stations have been omitted here. The k_T values for reference stations at similar latitudes in the northern hemisphere are more or less applicable to the southern hemisphere if the six-month difference in the timing of the seasons is taken into account (for an example see the Alice Springs case in Section 8.2.1.2).

Table A7.3 Recommended k_G values for areas where snow is rare in the winter and summers are relatively dry (e.g. in the northern Mediterranean region; see Section 8.1.3.2)

Locations with only rare snowfall in winter and dry summer
Recommended average values for k_G for long-term yield calculations

β	Jan	Feb	Mar	Apr	May	June	July	Aug	Sep	Oct	Nov	Dec
30°	0.80	0.81	0.83	0.84	0.84	0.83	0.83	0.83	0.84	0.84	0.81	0.80
45°	0.85	0.85	0.86	0.86	0.85	0.85	0.85	0.85	0.85	0.86	0.86	0.85
60°	0.86	0.86	0.86	0.85	0.84	0.83	0.83	0.83	0.84	0.85	0.86	0.86
90°	0.86	0.86	0.85	0.83	0.81	0.80	0.80	0.81	0.83	0.85	0.86	0.86

Recommended k_G-values for new plants with modules with full rated power

β	Jan	Feb	Mar	Apr	May	June	July	Aug	Sep	Oct	Nov	Dec
30°	0.85	0.86	0.88	0.89	0.89	0.88	0.88	0.88	0.89	0.89	0.86	0.85
45°	0.90	0.90	0.91	0.91	0.90	0.90	0.90	0.90	0.90	0.91	0.91	0.90
60°	0.91	0.91	0.91	0.90	0.89	0.88	0.88	0.88	0.89	0.90	0.91	0.91
90°	0.91	0.91	0.90	0.88	0.86	0.85	0.85	0.86	0.88	0.90	0.91	0.91

Table A7.4 Recommended k_G values for locations with no winter snowfall and very dry summers (e.g. Southern Europe and North Africa; see Section 8.1.3.2)

Locations without snowfall in winter and very dry summer
Recommended average values for k_G for long-term yield calculations

β	Jan	Feb	Mar	Apr	May	June	July	Aug	Sep	Oct	Nov	Dec
30°	0.84	0.84	0.84	0.84	0.83	0.82	0.82	0.82	0.83	0.84	0.84	0.84
45°	0.86	0.86	0.86	0.86	0.84	0.83	0.83	0.83	0.84	0.86	0.86	0.86
60°	0.86	0.86	0.86	0.85	0.83	0.82	0.82	0.82	0.83	0.85	0.86	0.86
90°	0.86	0.86	0.85	0.82	0.78	0.77	0.77	0.78	0.83	0.85	0.86	0.86

Recommended k_G-values for new plants with modules with full rated power

β	Jan	Feb	Mar	Apr	May	June	July	Aug	Sep	Oct	Nov	Dec
30°	0.89	0.89	0.89	0.89	0.88	0.87	0.87	0.87	0.88	0.89	0.89	0.89
45°	0.91	0.91	0.91	0.91	0.89	0.88	0.88	0.88	0.89	0.91	0.91	0.91
60°	0.91	0.91	0.91	0.90	0.88	0.87	0.87	0.87	0.88	0.90	0.91	0.91
90°	0.91	0.91	0.90	0.87	0.83	0.82	0.82	0.83	0.88	0.90	0.91	0.92

A8   Insolation and Energy Yield Calculation Maps

A8.1   Specimen Polar Shading Diagram (Figure A8.1)

A8.2   Insolation Maps

The following pages contain a series of maps indicating aggregate annual global irradiation H on a horizontal surface for the following: worldwide; Alpine countries; Germany; Africa; Asia; the Oceanic

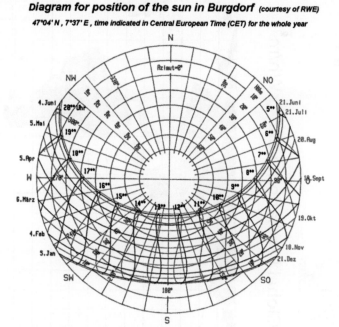

Diagram for position of the sun in Burgdorf *(courtesy of RWE)*

47°04′ N, 7°37′ E, time indicated in Central European Time (CET) for the whole year

Figure A8.1 Polar shading diagram for Burgdorf, Switzerland (47.1°N, 7.6°E). *Source*: RWE. RWE-Sonnenstands-diagramm Burgdorf. Diagram showing position of the Sun in Burgdorf Latitude 47°04′N, longitude 7°37′E, time indicated in CET. (Azimuth = azimuth, Höhe = elevation)

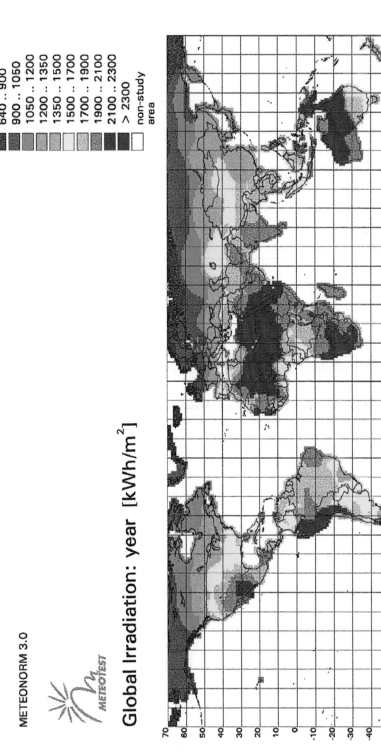

Figure A8.2 Aggregate global irradiation H on a horizontal surface for the entire world, in kWh/m^2 per year. This map is from Meteotest and was generated using Meteonorm 3.0 [2.3].

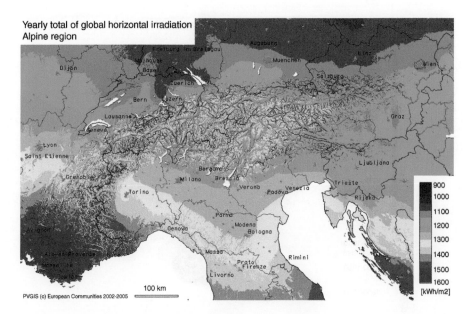

Figure A8.3   Aggregate global irradiation H on a horizontal surface for Alpine countries, in kWh/m^2 per year. Map from PV-GIS, © European Communities [2.7].

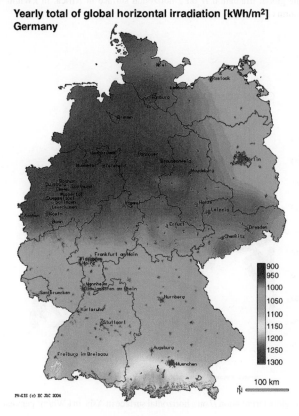

Figure A8.4   Aggregate global irradiation H on a horizontal surface in Germany, in kWh/m^2 per year. Map from PV-GIS, © European Communities, 2002–2005 [2.7].

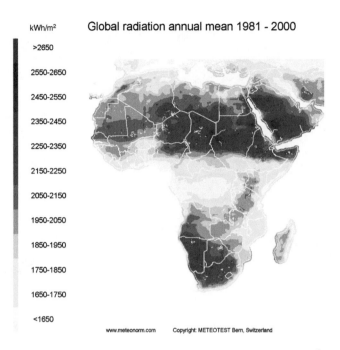

Figure A8.5   Aggregate global irradiation H on a horizontal surface in Africa, in kWh/m^2 per year. Map from Meteotest (www.meteonorm.com).

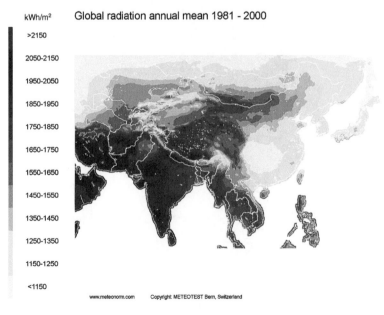

Figure A8.6   Aggregate global irradiation H on a horizontal surface in Asia, in kWh/m^2 per year. Map from Meteotest (www.meteonorm.com).

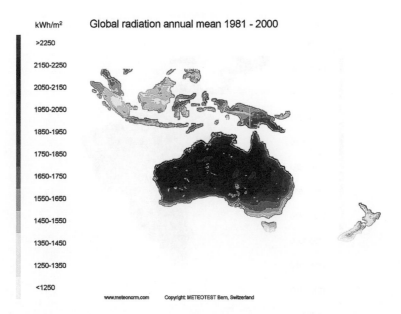

Figure A8.7 Aggregate global irradiation H on a horizontal surface in Oceania, in kWh/m^2 per year. Map from Meteotest (www.meteonorm.com).

region; and North and South America. When using these maps (Figures A8.2–A8.9), bear in mind that the exact same colours are not always used for a specific insolation. The scale for each map is indicated on the map itself.

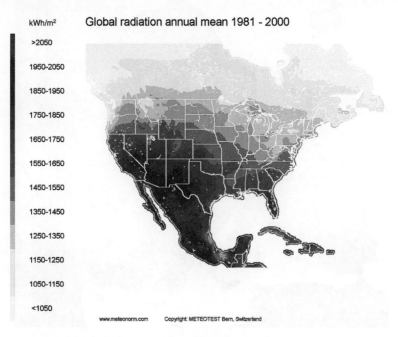

Figure A8.8 Aggregate global irradiation H on a horizontal surface in North America, in kWh/m^2 per year. Map from Meteotest (www.meteonorm.com).

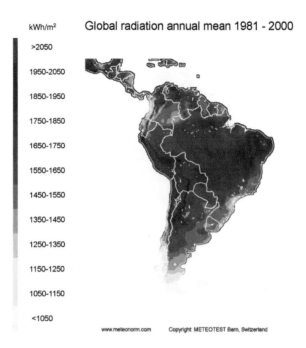

Figure A8.9   Aggregate global irradiation H on a horizontal surface in South America, in kWh/m^2 per year. Map from Meteotest (www.meteonorm.com).

Figure A8.10   Annual energy yield in kWh/kWp for an optimally oriented PV installation in Europe and environs. Map from PV-GIS, © European Communities, 2002–2005 [2.7].

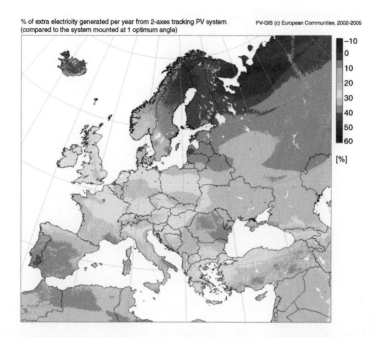

Figure A8.11 Annual additional energy generated (in per cent) using a biaxial solar tracker in a PV installation, relative to an optimally oriented, fixed solar module system (see Figure A.10). Map from PVGIS, © European Communities, 2002–2005 [2.11].

A8.3   Maps for Estimates of Annual PV Energy Yield in Europe and Environs

The map in Figure A8.10 allows for rough calculations of energy yield in Europe and environs for an installed PV installation with an optimal angle of incidence β and solar generator orientation γ. The optimal angle of incidence β ranges from 25° to 45° depending on location, while the optimal orientation γ is usually around 0° (southern orientation).

The map in Figure A8.11 allows for estimation of the additional annual energy yield that can potentially be obtained using a biaxial solar tracker in a PV installation in Europe or environs.

Additional maps showing insolation and possible PV installation energy yield can be downloaded from http://re.jrc.ec.europa.eu/pvgis/. These maps, which were created by the European Joint Research Centre (JRC) in Ispra, Italy, can be reproduced, providing that the source is acknowledged.

Appendix B

Links; Books; Acronyms; etc.

B1   Links to PV Web Sites

B1.1   Organizations

http://www.aee.at	Arbeitsgemeinschaft Erneuerbare Energie in Austria
http://www.dgs.de	Deutsche Gesellschaft für Sonnenenergie
http://www.energytech.at	Contains information concerning renewable energy and energy efficiency in Austria
http://www.epia.org	European Photovoltaic Industry Association (EPIA)
http://www.eurosolar.at	Web site of the Austrian section of Eurosolar
http://www.eurosolar.org	Eurosolar (European solar energy lobbying group)
http://www.iea-pvps.org	PV programme of the International Energy Agency (IEA)
http://www.ises.org	International Solar Energy Society (ISES)
http://www.photovoltaik.ch	Swiss PV research web site containing many downloads, as well as links to Swiss PV organizations
http://www.pv-ertraege.de	PV installation energy yield database for Germany
http://www.pvgap.org	PV Global Approval Program, which strives to institute PV quality standards worldwide
http://www.pvresources.com	Contains a wealth of PV information, particularly concerning large PV installations
http://www.satel-light.com	Irradiance data for Europe; allows users to generate individualized maps
http://www.seia.org	Solar Energy Industries Association (SEIA); mainly pertains to the USA
http://www.sfv.de	Web site for a solar energy advocacy organization containing a wealth of information for PV installation owners
http://www.soda-is.com	Irradiance data
http://www.solarbuzz.com	Lists of and links to PV vendors, mainly in the USA
http://www.solarenergy.org	Solar Energy International (USA)
http://www.solarinfo.de	Information concerning PV vendors in Germany
http://www.solarplaza.com	Lists of and links to PV vendors
http://www.solarserver.de	A wealth of PV information, including books, lists of vendors and so on
http://www.solarwirtschaft.de	Bundesverband Solarwirtschaft (German PV industry association)

Photovoltaics: System Design and Practice. Heinrich Häberlin.
© 2012 John Wiley & Sons, Ltd. Published 2012 by John Wiley & Sons, Ltd.

http://www.sses.ch   Schweizerische Vereinigung für Sonnenenergie (Swiss
                                            PV organization)
http://www.swissolar.ch   Schweizerischer Fachverband für Sonnenenergie (Swiss
                                            association of PV professionals)

B1.2   Government Organizations

http://www.eere.energy.gov   Energy Efficiency and Renewable Energy (US Department
                                            of Energy)
http://www.energie-schweiz.ch   Swiss Ministry of Energy
http://www.bfe.admin.ch
http://www.erneuerbare-energien.de   Renewable energy web site of Germany's Ministry of the
                                            Environment
http://www.nrel.gov/ncpv   PV web site of the National Renewable Energy
                                            Laboratory (USA)
http://www.pv.bnl.gov   Web site concerning the health and environmental effects of
                                            PV installations
http://www.swissgrid.ch   Swissgrid web site; contains information concerning the
                                            KEV law

B1.3   Research Organizations

http://re.jrc.ec.europa.eu/esti/   European Commission Joint Research Centre in Ispra, Italy
   index_en.htm
http://www.arsenal.ac.at   Austrian Institute of Technology
   http://www.ait.at
http://www.fvee.de   Forschungsverbund Erneuerbare Energie in Deutschland
                                            (German renewable energy research association)
http://www.hmi.de   Helmholtz Zentrum in Germany
http://www.isaac.supsi.ch   Swiss energy, ecology and economy research organization,
                                            which also has a PV lab
http://www.ise.fraunhofer.de   Fraunhofer Institut für Solare Energiesysteme (Fraunhofer
                                            Solar Energy System Institute)
http://www.iwes.fraunhofer.de   Fraunhofer Institut für Windenergie und Energie-
                                            systemtechnik IWES (Fraunhofer Institute for Wind Power
                                            Energy System Technologies)
http://www.isfh.de   Institut für Solarenergieforschung Hameln (solar energy
                                            research organization)
http://www.pv.unsw.edu.au   Centre for PV engineering at UNSW in Sydney, Australia
http://www.pvtest.ch   Bern University of Applied Sciences PV Lab
http://www.zsw-bw.de   Zentrum für Sonnenenergie- und Wasserstoffforschung
                                            (Solar energy and water research organization)

B1.4   Specialized Journals

http://www.bva-bielefeld.de   *Sonne Wind & Wärme* magazine
http://www.photon.de   PV magazine (in German)
http://www.photon-magazine.com   *Photon* magazine (in English)
http://www.sonnenenergie.de   *Sonnenenergie* magazine, published by DGS

http://www.sses.ch	*Zeitschrift Erneuerbare Energien* (renewable energy magazine published by the Swiss organization SSES)
http://www.sunwindenergy.com	*Sun & Wind Energy* magazine

Note: The list above is intended solely to enable readers to contact organizations that I felt were of interest at the time this book went to press, and represents merely a selection of the worthwhile information that is available online. I of course have no control over the content or availability of any of the aforementioned web sites and cannot therefore assume any responsibility or liability for such availability or content.

B2 Books on Photovoltaics and Related Areas

[Bas87]   U. Bastiansen: *Wasserstoff – Der Energieträger der Zukunft*. Eichborn-Verlag, Frankfurt, 1987, ISBN 3-8218-1111-0.

[Bur83]   M. Buresch: *Photovoltaic Energy Systems*. McGraw-Hill, New York, 1983, ISBN 0-07-008952-3.

[Deh05]   Dehn + Söhne: *Blitzplaner*. Dehn + Söhne, Neumarkt, 2005, ISBN 3-00-015976-2.

[DGS05]   R. Haselhuhn, C. Hemmerle: *Leitfaden Photovoltaische Anlagen*. Deutsche Gesellschaft für Sonnenenergie (DGS), Berlin, 2005, ISBN 3-9805738-3-4.

[Eic01]   U. Eicker: *Solare Technologien für Gebäude*. B.G. Teubner, Stuttgart, 2001, ISBN 3-519-05057-9.

[FIZ83]   *Informationspaket, Photovoltaik*. Fachinformationszentrum Energie, Physik, Mathematik, Karlsruhe, 1983.

[Gre86]   *Silicon Solar Cells – Operating Principles, Technology and System, Applications"*. Centre for PV Devices and Systems, University of New South Wales, Sydney, 1986, ISBN 0-85823-580-3.

[Gre95]   *Silicon Solar Cells – Advanced Principles &, Practice*. Centre for PV Devices and Systems, University of New South Wales, Sydney, 1995, ISBN 0-7334-0994-6.

[Häb91]   H. Häberlin: *Photovoltaik – Strom aus Sonnenlicht für Inselanlagen und Verbundnetz*. AT-Verlag, Aarau, 1991, ISBN 3-85502-434-0.

[Häb07]   H. Häberlin: *Photovoltaik – Strom aus Sonnenlicht für Verbundnetz und Inselanlagen*. AZ-Verlag, Aarau and VDE-Verlag, Berlin, 2007, ISBN 978-3-905214-53-6 and 978-3-8007-3003-2.

[Hag02]   I.B. Hagemann: *Gebäudeintegrierte Photovoltaik – Architektonische Integration der Photovoltaik in die Gebäudehülle*. Rudolf Müller-Verlag, Cologne, 2002, ISBN 3-481-01776-6.

[Han95]   B. Hanus: *Das grosse Anwenderbuch der Solartechnik*. Franzis-Verlag, 1995, ISBN 3-7723-7791-2.

[Has89]   P. Hasse, J. Wiesinger: *Handbuch für Blitzschutz und Erdung*. Pflaum-Verlag, Munich, 1989, ISBN 3-7905-0559-5.

[Has05]   R. Haselhuhn: *Photovoltaik – Gebäude liefern Strom*. TUEV-Verlag, Cologne, 2005, ISBN 3-8249-0854-9.

[Her92]   L. Herzog, U. Muntwyler: *Photovoltaik – Planungsunterlagen für autonome und netzgekoppelte Anlagen*. PACER/Bundesamt für Konjunkturfragen, Bern, 1992, ISBN 3-905232-12-X.

[Hof96]   V.U. Hofmann: *Photovoltaik – Strom aus Licht*. vdf-Verlag, Zurich, 1996, ISBN 3-7281-2211-4.

[Hu83]   C. Hu, R.W. White: *Solar Cells: From Basics to Advanced Systems*. McGraw-Hill, New York, 1983, ISBN 0-07-030745-8.

[Hum93]   O. Humm, P. Toggweiler: *Photovoltaik und Architektur*. Birkhäuser-Verlag, Basle, 1993, ISBN 3-7643-2891-6.

[Ima92]   M.S. Imamura, P. Helm, W. Palz: *Photovoltaic System Technology – A European Handbook*. Commission of the European Communities, H S Stephens, Felmersham, 1992, ISBN 0-9510271-9-0.

[Jac89]   P. Jacobs: *Strom aus Sonnenlicht*. Wagner, Solartechnik GmbH, Marburg, 1989, ISBN 3-923129-03-3.

[Jäg90]   F. Jäger, A. Räuber (eds): *Photovoltaik – Strom aus der Sonne*. C.F. Müller-Verlag, Karlsruhe, 1990, ISBN 3-7880-7337-3.

[Joh80]   W.D. Johnston, Jr: *Solar Voltaic Cells*. Marcel Dekker, New York, 1980, ISBN 0-8247-6992-9.

[Kar88]   S. Karamanolis: *Alles über Solarzellen*. Elektra-Verlag, Neubiberg, 1988, ISBN 3-922238-78-5.

[Köt94]   H.K. Köthe: *Stromversorgung mit Solarzellen*. Franzis-Verlag, Munich, 1994, ISBN 3-7723-9434-5.

[Kri92]   B. Krieg: *Strom aus der Sonne*. Elektor-Verlag, Aachen, 1992, ISBN 3-928051-17-2.

[Kur03]   P. Kurzweil: *Brennstoffzellentechnik*. Vieweg-Verlag, Wiesbaden, 2003, ISBN 3-528-03965-5.

[Lad86]   H. Ladener: *Solare Stromversorgung für Geräte, Fahrzeuge und Häuser*. Oekobuch-Verlag, Freiburg (Breisgau), 1986, ISBN 3-922964-28-1.

[Lew95]   H.J. Lewerenz, H. Jungblut: *Photovoltaik – Grundlagen und Anwendungen*. Springer-Verlag, Berlin, 1995, ISBN 3-540-58539-7.

[Luq03]   A. Luque, S. Hegedus (eds): *Handbook of Photovoltaic Science and Engineering*. John Wiley & Sons, Ltd, Chichester, 2003, ISBN 0-471-49196-9.

[Mar94]   T. Markvart: *Solar Electricity*. John Wiley & Sons, Ltd, Chichester, 1994, ISBN 0-471-94161-1.

[Mar00]   T. Markvart: *Solar Electricity*, 2nd edition. John Wiley & Sons, Ltd, Chichester, 2000, ISBN 0-471-98852-9.

[Mar03]   T. Markvart, L. Castaner (eds): *Photovoltaics – Fundamentals and Applications*. Elsevier, Oxford, 2003, ISBN 1856173909.

[Mun90]   U. Muntwyler: *Praxis mit Solarzellen*. Franzis-Verlag, Munich, 1990, ISBN 3-7723-2043-0.

[Nor00]   T. Nordmann, Ch. Schmidt: *Im Prinzip Sonne*. Kontrast-Verlag, Zurich, 2000, ISBN 3-9521287-6-7.

[Pan86]   P. Panzer: *Praxis des Überspannungs- und Störspannungsschutzes elektronischer Geräte und Anlagen*. Vogel-Verlag, Würzburg, 1986, ISBN 3-8023-0887-5.

[Par95]   L.D. Partain (ed.): *Solar Cells and their Applications*. John Wiley & Sons, Inc., New York, 1995, ISBN 0-471-57420-1.

[Qua03]   V. Quaschning: *Regenerative Energiesysteme – Technologie, Berechnung, Simulation*. Carl Hanser-Verlag, Munich, 2003, ISBN 3-446-21983-8.

[Räu86]   A. Räuber, F. Jäger: *Photovoltaische Solarenergienutzung. Vergleichende Studie der Entwicklungstendenzen in der Bundesrepublik Deutschland, in Europa, den USA und Japan*, 0340-7608, BMFT-FB-T86-048. Fachinformationszentrum Energie, Physik, Mathematik, Karlsruhe, 1986.

[Rin01]   U. Rindelhardt: *Photovoltaische Stromversorgung*. B.G. Teubner, Stuttgart, 2001, ISBN 3-519-00411-9.

[Rod89]   A. Rodewald: *Elektromagnetische Verträglichkeit – Grundlagen, Experimente, Praxis*. Vieweg-Verlag, Wiesbaden, 1989, ISBN 3-528-04924-3.

[Ros99]   M. Ross, J. Royer (eds): *Photovoltaics in Cold Climates*. James & James, London, 1999, ISBN 1-873936-89-3.

[Rot97]   W. Roth, H. Schmidt (eds): *Photovoltaikanlagen*, Begleitbuch zum OTTI-Seminar Photovoltaikanlagen. Fraunhofer Institut fur Solare Energiesysteme, Freiburg, 1997.

[Sch87]   A. Schwarz, K. Schnuer: *Stromquelle Tageslicht*. Orac-Verlag, Vienna, 1987, ISBN 3-7015-0091-6.

[Sch88]   J. Schmid: *Photovoltaik. Direktumwandlung von Sonnenlicht in Strom*. TUEV Rheinland, Cologne, 1988, ISBN 3-88585-396-5.

[Sch93]   S. Schodel: *Photovoltaik – Grundlagen und Komponenten für Projektierung und Installation*. Pflaum-Verlag, Munich, 1993, ISBN 3-7905-0621-4.

[Sch99]   J. Schmid (ed.): *Photovoltaik – Strom aus der Sonne*. C.F. Müller, Heidelberg, 1999, ISBN 3-7880-7589-9.

[Sche87]   H. Scheer (ed.): *Die gespeicherte Sonne*. Piper-Verlag, Munich, 1987, ISBN 3-492-10828-8.

[See93]   T. Seemann, R. Wiechmann: *Solare Hausstromversorgung mit Netzverbund*. VDE-Verlag, Berlin, 1993, ISBN 3-8007-1849-9.

[Sel00]   T. Seltmann: *Fotovoltaik: Strom ohne Ende*. Solarpraxis, Berlin, 2000, ISBN 3-934595-02-2.

[Sic96]   F. Sick, Th.Erge (eds): *Photovoltaics in Buildings*. James & James, London, 1996, ISBN 1 873936 59 1.

[SNV88]   Sonnenergie: *Begriffe und Definitionen*, SN 165000. Schweizerischen Normen-Vereinigung, Zurich, 1988.

[SOL]   Sonnenenergie Fachverband, Schweiz: *Empfehlungen zur Nutzung der Sonnenenergie*. Available from SOLAR, Hopfenweg 21, 3007 Bern, Switzerland (www.solarpro.ch).

[Sta87]   M.R. Starr, W. Halcrow and Partners and W. Palz: *Photovoltaischer Strom für Europa*. TUEV Rheinland, Cologne, 1987, ISBN 3-88585-223-3.

[Wag06]   A. Wagner: *Photovoltaik-Engineering*. Springer-Verlag, Berlin, 2006, ISBN 3-540-30732-X.

[Web91]   R. Weber: *Der sauberste Brennstoff – Der Weg zur Wasserstoff-Wirtschaft*. Olynthus-Verlag, Oberbözberg, 1991, ISBN 3-907175-13-1.

[Wen95]   S.R. Wenham, M.A. Green, M.E. Watt: *Applied Photovoltaics*. Centre for PV Devices and Systems, University of New South Wales, Sydney, 1995, ISBN 0-86758-909-4.

[Wil94]   H. Wilk: *Solarstrom – Handbuch zur Planung und Ausführung von Photovoltaikanlagen*. Arbeitsgemeinschaft Erneuerbare Energien, Gleisdorf, 1994, ISBN 3-901425-01-2.

[Win89]   C.J. Winter, J. Nitsch (eds): *Wasserstoff als Energieträger*. Springer-Verlag, Berlin, 1989, ISBN 3-540-50221-1.

[Win91]   C.J. Winter, R.L. Sizmann, L.L. Vant-Hull: *Solar Power Plants*. Springer-Verlag, Berlin, 1991, ISBN 3-540-18897-5.

[Wür95]   P. Würfel: *Physik der Solarzellen*. Spektrum-Verlag, Heidelberg, 1995, ISBN 3-86025-717-X.

[Wür05]   P. Würfel: *Physics of Solar Cells*. Wiley-VCH Verlag GmbH, Weinheim, 2005, ISBN 3-527-40428-7.

[Zah04]   R. Zahoransky: *Energietechnik*. Vieweg, Wiesbaden, 2004, ISBN 3-528-13925-0.

B3   Acronyms

AC	Alternating current
AM	Relative air mass number
a-Si	Amorphous silicon
BFE	Bundesamt für Energie (Swiss Deparment of Energy)
BFH	Bern Technical University
BKW	Bernische Kraftwerke AG (Swiss utility)
BMU	Germany's Ministry of the Environment
CdTe	Cadmium telluride
CIGS	Copper indium gallium diselenide
CIS	Copper indium diselenide
c-si	Crystalline silicon
Cz	Czochralski method (for Si monocrystal production)
DC	Direct current
DGS	Deutsche Gesellschaft für Sonnenenergie (German Solar Energy Society)
DOE	US Department of Energy
EEG	Germany's Renewable Energy Act (Erneuerbare-Energien-Gesetz; see www.erneuerbare-energien.de)
EMC	Electromagnetic Compatibility
ENS	German acronym for *two, independent, parallel grid monitoring devices to each of which is assigned a switching element in series.* (In most cases this simply refers to installations that allow for continuous grid impedance monitoring to detect undesirable stand-alone operation of grid-connected inverters.)
ETH	Eidgenössische Technische Hochschule (a Swiss technical university)
EVU	German acronym for an electrical utility company
EW	German acronym for electrical utility company (mostly used in Switzerland)
GaAs	Gallium arsenide
GAK	German acronym for generator circuit box
GDP	Gross domestic product
Ge	Germanium
HF	High frequency
KEV	Swiss law on feed-in fees (see www.swissgrid.ch)
LF	Low frequency
MPP	Maximum power point
MPPT	Maximum power point tracker
MPT	Maximum power tracker (short for maximum power point tracker)
NOCT	Nominal cell operating temperature at $G = 800 \, \text{W/m}^2$ for an outdoor module under open-circuit conditions, with ambient temperature of 20 °C and wind velocity of 1 m/s.
NREL	National Renewable Energy Laboratory (USA)
PAL	German acronym for equipotential conductor
PAS	German acronym for equipotential bonding bus bar
PV	Photovoltaic(s) (adj.)
PWM	German acronym for pulse width modulation
SEV	Schweizerischer Elektrotechnischer Verein (now known as Electrosuisse)
Si	Silicon
SLF	Schweizerisches Lawinenforschungsinstitut (Swiss Avalanche Research Institute)
SPD	Surge protection device
STC	Standard test conditions ($G = G_o = 1 \, \text{kW/m}^2$, AM1.5 spectrum, cell temperature 25 °C)
TGAK	German acronym for array circuit box
THD	Total harmonic distortion

B4 Prefixes for Decimal Fractions and Metric Multiples

Prefix	Symbol	Meaning	Example
exa	E	10^{18}	$1\,EJ = 10^{18}\,J$
peta	P	10^{15}	$1\,PJ = 10^{15}\,J$
tera	T	10^{12}	$1\,TWh = 10^{12}\,Wh$
giga	G	10^{9}	$1\,GW = 10^{9}\,W$
mega	M	10^{6}	$1\,MV = 10^{6}\,V$
kilo	k	10^{3}	$1\,kHz = 10^{3}\,Hz$
centi	c	10^{-2}	$1\,cm = 10^{-2}\,m$
milli	m	10^{-3}	$1\,m\Omega = 10^{-3}\,\Omega$
micro	μ	10^{-6}	$1\,\mu H = 10^{-6}\,H$
nano	n	10^{-9}	$1\,ns = 10^{-9}\,s$
pico	p	10^{-12}	$1\,pF = 10^{-12}\,F$
femto	f	10^{-15}	$1\,fA = 10^{-15}\,A$
atto	a	10^{-18}	$1\,aA = 10^{-18}\,A$

B5 Conversion Factors

In this book, the SI system of units has been used wherever possible. However, certain additional metrics are better suited for certain domains, for didactic reasons. Specifically, the atomic processes entailed by the electron volt (eV) energy metric are highly useful for solar cell physics calculations, while kilowatt hours (kWh) are the most practical energy metric for irradiance and electrical process calculations. Hence these metrics have been used in the relevant sections.

Conversion factors:
$$1\,eV = 1.602 \cdot 10^{-19}\,J$$
$$1\,J = 1\,W\,s = 1\,N\,m$$
$$1\,Wh = 3.6\,kJ = 3.6 \cdot 10^{3}\,W\,s$$
$$1\,kWh = 3.6\,MJ = 3.6 \cdot 10^{6}\,W\,s$$
$$1\,MWh = 3.6\,GJ = 1000\,kWh$$
$$1\,GWh = 3.6\,TJ = 10^{6}\,kWh = 1\,million\,kWh$$
$$1\,TWh = 3.6\,PJ = 10^{9}\,kWh = 1\,billion\,kWh$$
$$1\,PJ = 277.8\,GWh = 277.8\,million\,kWh$$
$$1\,EJ = 277.8\,TWh = 277.8\,billion\,kWh$$
$$1\,Ah = 3600\,A\,s$$

B6 Key Physical Constants

e Electron charge: $e = 1.602 \cdot 10^{-19}\,A\,s$

c Speed of light: $c = 299\,800\,km/s$

h Planck's constant: $h = 6.626 \cdot 10^{-34}\,J\,s$

k Boltzmann's constant: $k = 1.38 \cdot 10^{-23}\,J/K$

ε_0 Electrical constant: $\varepsilon_0 = 8.854\,pF/m$

μ_0 Magnetic constant: $\mu_0 = 0.4\pi\,\mu H/m$

Index

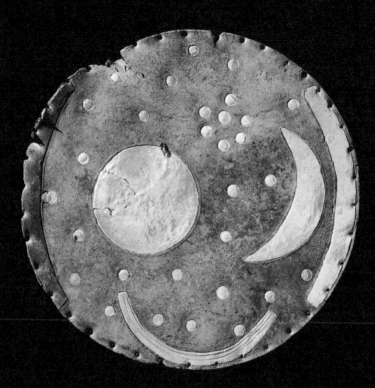

Printed and bound by CPI Group (UK) Ltd, Croydon, CR0 4YY

16/04/2025

14658382-0004